NMR of Paramagnetic Molecules

Applications to Metallobiomolecules and Models

NMR of Paramagnetic Molecules

Applications to Metallobiomolecules and Models

SECOND EDITION

Ivano Bertini†

Claudio Luchinat
Department of Chemistry "Ugo Schiff" and
Magnetic Resonance Center (CERM)
University of Florence, and Interuniversity
Consortium for Magnetic Resonance on
Metallo Proteins (CIRMMP), Italy

Giacomo Parigi
Department of Chemistry "Ugo Schiff" and
Magnetic Resonance Center (CERM)
University of Florence, and Interuniversity
Consortium for Magnetic Resonance on
Metallo Proteins (CIRMMP), Italy

Enrico Ravera
Magnetic Resonance Center (CERM), and
Interuniversity Consortium for Magnetic Resonance on
Metallo Proteins (CIRMMP), Italy

†Deceased

ELSEVIER

AMSTERDAM • BOSTON • HEIDELBERG • LONDON • NEW YORK • OXFORD • PARIS
SAN DIEGO • SAN FRANCISCO • SINGAPORE • SYDNEY • TOKYO

Elsevier
Radarweg 29, PO Box 211, 1000 AE Amsterdam, Netherlands
The Boulevard, Langford Lane, Kidlington, Oxford OX5 1GB, United Kingdom
50 Hampshire Street, 5th Floor, Cambridge, MA 02139, United States

Library of Congress Cataloging-in-Publication Data
A catalog record for this book is available from the Library of Congress

British Library Cataloguing-in-Publication Data
A catalogue record for this book is available from the British Library

ISBN: 978-0-444-63436-8

Publisher: Cathleen Sether
Acquisition Editor: Kathryn Morrissey
Editorial Project Manager: Jill Cetel
Production Project Manager: Paul Prasad Chandramohan
Designer: Maria Ines Cruz

Typeset by Thomson Digital

Contents

Preface

Applications of NMR to paramagnetic molecules and biomolecules, both in solution and in the solid state, keep growing both in number and sophistication. They represent a respectable share of all NMR activity.

For NMR experiments on paramagnetic systems, one should understand the general theory of NMR and, on top of this, the theory of the electron-nucleus interaction and its consequences for the NMR parameters. For these reasons, the field of NMR of paramagnetic molecules has its own niche in the entire scientific panorama.

With this book, the authors aim to provide an up-to-date tool for everyone wishing to understand, predict, and exploit-or minimize-the effects of paramagnetic centers on the NMR spectra of molecules and biomolecules. The scientific activity of the authors focuses on structural and dynamic studies of paramagnetic metalloproteins, in the Magnetic Resonance Center (CERM) at the University of Florence. The laboratory is an NMR Research Infrastructure which is one of the main nodes of the European Infrastructure for integrated biology (INSTRUCT). We are thus exposed to the needs of the scientific community and have responded to them in several ways from the development of new instruments or parts of them to the description of new phenomena and development of new software. Since 1984, with the help of colleagues from the universities of Pisa and Siena, we have organized fourteen "Chianti Workshops," a series of conferences well known to the scientific community in the field that have always had a special focus on electron and nuclear relaxation.

The recent refocusing in the interest for the NMR of paramagnetic molecules is witnessed by several European funded activities, among which three COST Actions – on hyperpolarized systems (Cost Action TD1103), on iron-sulfur proteins (Cost Action CA15133), and on relaxometry (Cost Action CA15209)-and the Marie Curie ITN program "pNMR," aimed at "Pushing the Envelope of Nuclear Magnetic Resonance Spectroscopy for Paramagnetic Systems" (grant agreement no. 317127), to which the authors' Institutions are partners. In this frame, the authors acknowledge the constructive collaboration with the coordinator of the pNMR project, Guido Pintacuda, and with the other partners, in particular Martin Kaupp, Juha Vaara, and Marcellus Ubbink.

This book largely capitalizes on our previous books, of which we here maintain (and try to improve) the pictorial way of presenting theoretical aspects:

I. Bertini, C. Luchinat (1986), *NMR of Paramagnetic Molecules in Biological Systems.* Benjamin/Cummings, Menlo Park, CA.

L. Banci, I. Bertini, C. Luchinat (1991), *Nuclear and Electron Relaxation. The Magnetic Nucleus-Unpaired Electron Coupling in Solution.* VCH, Weinheim.

I. Bertini, C. Luchinat (1996), *NMR of Paramagnetic Substances,* Coord. Chem. Rev. 150, Elsevier, Amsterdam.

I. Bertini, C. Luchinat and G. Parigi (2001), *NMR of Paramagnetic Molecules, 1 edn.* Elsevier, Amsterdam.

With respect to the previous books, the field of paramagnetic molecules is now more extensively projected into the domains of solid-state NMR, partially oriented systems, cross correlations, protein structure calculation, contrast agents for magnetic resonance imaging, and dynamic nuclear polarization.

Chapter 1 describes the interactions between a spin, electronic or nuclear, and a magnetic field: just some basic physics which cannot be avoided. Chapter 2 deals with contact and dipolar shifts. The aim here is to be clear and rigorous. Chapter 3 describes the effects determined by molecular partial orientation (self-orientation) in high magnetic fields as induced by magnetic anisotropy, where paramagnetic residual dipolar couplings feature prominently. Chapter 4 deals with relaxation: a complex subject that we have tried to make simple and pictorial, but also exhaustive and rigorous. In this chapter the nuclear Overhauser effect is also presented, also in the light of its implications for the developing field of dynamic nuclear polarization. In Chapter 5 the basics of solid-state NMR in paramagnetic systems are provided, and advanced and updated experimental approaches are presented to show how anisotropy of the magnetic interactions affects the spectra and how it can be averaged. Dynamic nuclear polarization, which stems from the understanding of electron and nuclear relaxation in solids, is also discussed. Chapter 6 covers chemical exchange, the effect of diffusion on relaxation, and the effect of bulk magnetic susceptibility on chemical shifts. Theory and experiments of Nuclear Magnetic Relaxation Dispersions are described.

In Chapters 7 and 8 the electron relaxation properties of various transition metal ions and lanthanoid ions are introduced, and the consequent nuclear relaxation properties discussed in the context of their suitability for NMR. Chapters 9 and 10 present some applications. Chapter 9 deals with the information provided by paramagnetic data in connection with molecular structure and dynamic features, and presents guidelines for their use. Chapter 10 deals with the relaxation properties of molecules of interest as contrast agents for magnetic resonance imaging. In Chapter 11 the effects of electron-electron exchange coupling on the shifts and relaxation are presented theoretically, and examples are given. Finally, in Chapter 12, experiments and strategies necessary to achieve the highest level of performance in NMR of paramagnetic systems are presented, strategies to minimize adverse paramagnetic effects, as well as ways to exploit such effects to extract structural and dynamic properties are discussed.

We thank our colleagues, doctors, and students who have collaborated with us at CERM for reading and discussing various parts of the books. Interactions with Lucia Banci, Simone Ciofi-Baffoni, Isabella C. Felli, Marco Fragai, Mario Piccioli, Roberta Pierattelli, Antonio Rosato, and Paola Turano are acknowledged in particular. Interactions with other colleagues at the University of Florence, and in particular with Gabriele Spina and Maurizio Romanelli, are greatly acknowledged.

We want to pay special tribute here to Ivano Bertini, the senior author of all the preceding books, who sadly left us four years ago while we were already planning this new edition. And we want to acknowledge the discussions that we had with him on the new topics and chapters that feature in the present edition. We hope that his inspiring contribution will still be clearly detectable!

On a historical perspective, we take the opportunity to pay tribute to the late Luigi Sacconi, the founder of the inorganic chemistry school in Florence, to William DeW. Horrocks Jr., Bruce R. McGarvey, and Seymour H. Koenig for their influence on the early scientific careers of Ivano Bertini and Claudio Luchinat. A special role in our understanding of the intricacies of paramagnetic relaxation has been played over the years by Jozef Kowalewski, a friend and a brilliant scientist with whom we collaborated at several projects. More recently we have enjoyed discussing the perspectives of the field with David A. Case, Robert G. Griffin, Shimon Vega. Fruitful collaborations were also established over the years with Silvio Aime, Marina Bennati, Carlos F.G.C. Geraldes, Daniella Goldfarb, Thomas J. Meade, Thomas F. Prisner, Luca Sgheri. Last but not least, the authors want to acknowledge the constructive interactions with Bruker Biospin and Stelar s.r.l.

CHAPTER

INTRODUCTION

1

This chapter is intended to recall the principles of magnetism, the definition of magnetic induction and of magnetic induction in a vacuum, which is referred to as magnetic field. Readers may not recollect that the molar magnetic susceptibility is expressed in cubic meters per mol. Some properties of electron and nuclear spins are reviewed and finally some basic concepts of the magnetic resonance experiments are refreshed. In summary, this chapter should introduce the readers into the language used by the authors.

1.1 MAGNETIC MOMENTS AND MAGNETIC FIELDS

This book deals with NMR experiments on systems which contain unpaired electrons. Unpaired electrons disturb the experiment to such an extent that quite different conditions are needed. However, since we have to live with molecules bearing unpaired electrons, we do our best to take advantage from these properly designed NMR experiments in order to learn as much as possible regarding the properties of the unpaired electrons and the structure and dynamics or the substance. To be more precise, we are going to exploit NMR in order to learn how the unpaired electron(s) interacts with the resonating nucleus and how these perturbed nuclei provide information typical of NMR experiments.

The nucleus under investigation must have a magnetic moment in order to make the NMR experiment possible. An unpaired electron also has a magnetic moment. A magnetic moment μ (J T^{-1}) can be visualized as a magnetic dipole (Fig. 1.1). Such magnetic moment causes a magnetic dipolar field. In electromagnetism, this vector is provided by a continuous current in a coil. If a second magnetic moment μ_2 (which we take to be of smaller vector intensity without loss of generality) is within the dipolar field created by the former magnetic moment μ_1 anchored at a distance r, it will orient accordingly, as represented in Fig. 1.2. We can refer to the electronic magnetic moment as the large magnetic moment μ_1 and to the nuclear magnetic moment as the small magnetic moment μ_2. The absolute value of the magnetic moment associated with the electron is 658 times (Section 1.2) that for a proton, which has the largest magnetic moment among the magnetic nuclei (except tritium). The orientation of the small magnetic moment along the dipolar field of the large magnetic moment shown in Fig. 1.2 represents the minimum energy situation. In general, the energy of the interaction between the two magnetic bars depends on the relative orientation of the two vectors, if at equilibrium or fixed by external forces, according to Eq. (1.1):

$$E^{\text{dip}} = -\frac{\mu_0}{4\pi}\left[\frac{3(\mu_1 \cdot \boldsymbol{r})(\mu_2 \cdot \boldsymbol{r})}{r^5} - \frac{\mu_1 \cdot \mu_2}{r^3}\right] \tag{1.1}$$

where μ_0 is the magnetic permeability of a vacuum (J^{-1} T^2 m^3), $\boldsymbol{r}$ is the vector connecting the two point dipoles, and r is its magnitude. The energy can be negative (stabilization) or positive (destabilization) according to the relative magnitude of the two terms in parentheses.

NMR of Paramagnetic Molecules. http://dx.doi.org/10.1016/B978-0-444-63436-8.00001-6

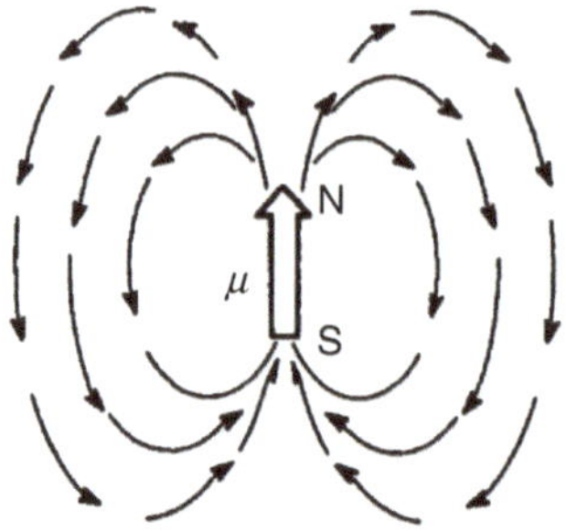

FIGURE 1.1 A magnetic moment can be seen as a magnetic dipole $\boldsymbol{\mu}$ characterized by north (N) and south (S) polarities. It gives rise to a magnetic field, which is indicated by force lines. The dipolar nature provides the vectorial nature of this moment, whose intensity is indicated by μ.

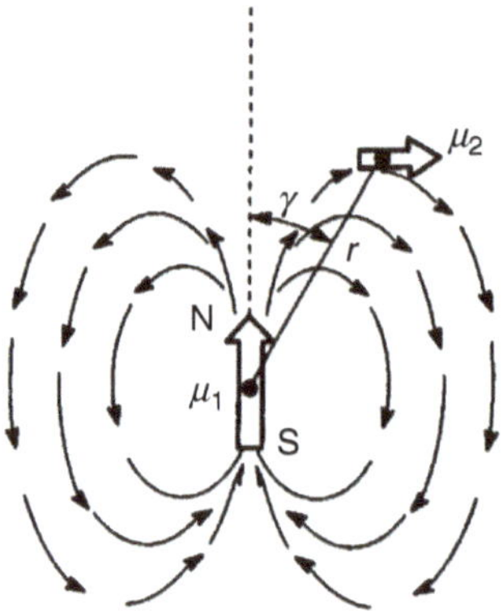

FIGURE 1.2 Orientation of a small magnetic moment $\boldsymbol{\mu}_2$ (e.g., that of the nucleus) within a magnetic field generated by a larger magnetic moment $\boldsymbol{\mu}_1$ (e.g., that of the electron) at distance r. Here γ is the angle between $\boldsymbol{\mu}_1$ and $\boldsymbol{r}$.

In all our experiments the two magnetic bars are immersed in an external magnetic field. The intensity of the magnetic field is proportional to the density of force lines. Later, we will be interested in the effective field in a given region of space, which is referred to as magnetic induction $\boldsymbol{B}$ (expressed in tesla):

$$\boldsymbol{B} = \mu_0(\boldsymbol{H} + \boldsymbol{M}) = \boldsymbol{B}_0 + \mu_0\boldsymbol{M} \tag{1.2}$$

where $\boldsymbol{H}$ and $\boldsymbol{M}$ (J T^{-1} m^{-3}) are the magnetic field strength and the magnetization of the medium referred to unit volume respectively, and μ_0 is already defined [Eq. (1.1)]. The magnetic induction is thus given by the magnetic induction in a vacuum ($\mu_0\boldsymbol{H} = \boldsymbol{B}_0$) plus a contribution ($\mu_0\boldsymbol{M}$) depending on the kind of substance constituting the medium. In this book, the magnetic induction in a vacuum $\boldsymbol{B}_0$ will always be referred to as the *external magnetic field*.

The energy E of a magnetic moment $\boldsymbol{\mu}$ immersed in a magnetic field $\boldsymbol{B}_0$ is given by

$$E = -\boldsymbol{\mu} \cdot \boldsymbol{B}_0. \tag{1.3}$$

Eq. (1.3) shows that the energy is at a minimum when $\boldsymbol{\mu}$ is aligned along $\boldsymbol{B}_0$.

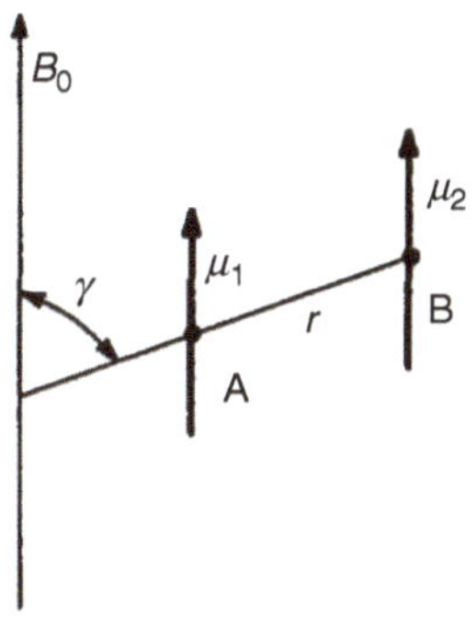

FIGURE 1.3 Two magnetic bars, μ_1 and μ_2, anchored at generic points A and B at distance r in a magnetic field B_0. γ is the angle between the magnetic field and the AB vector.

In the absence of further limit conditions which may hold in the case of the electron (see later), we can now think of an electron spin and a nuclear spin anchored at points A and B, both aligned along the external magnetic field $\boldsymbol{B}_0$, as shown in Fig. 1.3. Since the two magnetic moments are forced to be parallel by the strong external field, the energy of the interaction between them, given by Eq. (1.1), simplifies to

$$E^{\text{dip}} = -\frac{\mu_0}{4\pi}\frac{\mu_1\mu_2}{r^3}\left(3\cos^2\gamma - 1\right) \tag{1.4}$$

where γ is the angle between the direction of $\boldsymbol{B}_0$ and that of the AB vector (Appendix II).

1.2 ABOUT THE SPIN MOMENTS

Electron and nuclear magnetic moments can be regarded as arising from a property of the particles, i.e., they possess an intrinsic angular momentum as if they were spinning. This is not to be interpreted literally: the intrinsic angular momentum can acquire a semi-integer spin, which may not arise from rotation; electrons simply do have spin just as they do have charge. For the nuclei the spin results from the combination of the spins of the nucleons.

The intrinsic spin angular momenta are given by

$$\boldsymbol{J}_S = \hbar\boldsymbol{S} \quad \boldsymbol{J}_I = \hbar\boldsymbol{I} \tag{1.5}$$

for the electron and nucleus respectively, where $\boldsymbol{S}$ and $\boldsymbol{I}$ are dimensionless *spin* angular momentum vectors and $\hbar = h/2\pi$ is the Planck constant (J s rad^{-1}). The moduli of the vectors are given by

$$J_S = \hbar\sqrt{S(S+1)} \quad J_I = \hbar\sqrt{I(I+1)} \tag{1.6}$$

where S and I are quantum numbers associated with the spinning particle. For a single electron or for a single nucleon (proton or neutron), $S = I = ½$. S or I identify sets of spin wave functions for the above particles. Note that the values of the angular momenta are not related to the nature of the particles.

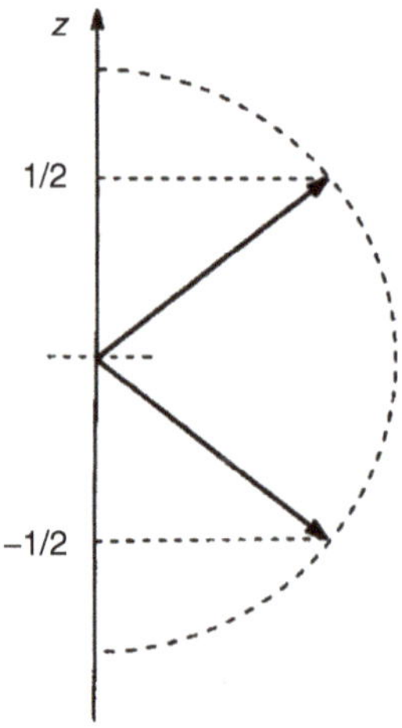

FIGURE 1.4 Allowed orientations of an $I = ½$ angular momentum relative to the z direction defined by the external magnetic field. The vector has modulus $\sqrt{3}/2$, and its projections on the z-axis are ½ and −½.

The projection of $\boldsymbol{S}$ and $\boldsymbol{I}$ along a z direction (defined by an external magnetic field or otherwise) are +½ or −½ (Fig. 1.4). Thus we have two wave functions, one with S (or I) = ½ and with a component along z = ½ and another with S (or I) = ½ and with a component along z = −½.

The component is indicated in quantum mechanics as M_S or M_I. The notation to indicate the wave function is thus

$$|S, M_S\rangle \quad \text{or} \quad |I, M_I\rangle$$

where $|\,\rangle$ is the "ket" notation for wave functions.

These wave functions are eigenfunctions of the operators S^2 (I^2) and S_z (I_z):

$$\begin{aligned} S^2|S, M_S\rangle &= S(S+1)|S, M_S\rangle \\ I^2|I, M_I\rangle &= I(I+1)|I, M_I\rangle \\ S_z|S, M_S\rangle &= M_S|S, M_S\rangle \\ I_z|I, M_I\rangle &= M_I|I, M_I\rangle. \end{aligned} \tag{1.7}$$

Physically, it means that it is possible to know simultaneously the square of the intensity of the spin angular momentum and its component along z. Since the spin wave functions are not eigenfunctions of the operators S or I, it is impossible to know intensity and orientation of the angular momentum vector simultaneously. We will learn how to live with it!

Since the electron and the proton are charged particles, there is a magnetic moment associated with the angular momenta. The latter is related to a motion, and a motion of a charged particle produces a magnetic moment. The neutron is not charged as a result of balancing of quarks with charges of different sign; but its quarks provide the neutron with a non-zero magnetic moment.

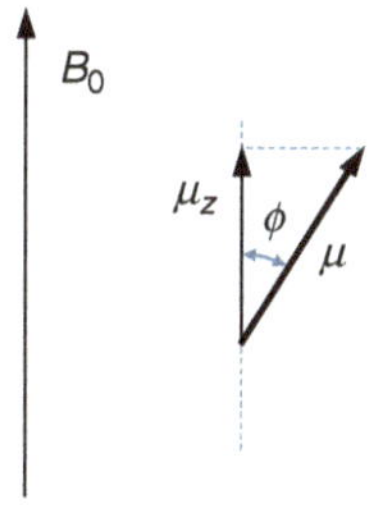

FIGURE 1.5 A magnetic moment μ in a magnetic field forming an angle ϕ with the magnetic field direction.

The intrinsic angular momentum $\boldsymbol{S}$ is related to the intrinsic magnetic moment $\boldsymbol{\mu}_S$ through the relation

$$\mu_S = -g_e \mu_B \boldsymbol{S}$$

and therefore the moduli of $\boldsymbol{\mu}_S$ and $\boldsymbol{\mu}_I$ are given by

$$|\mu_S| = |g_e| \frac{|e|\hbar}{2m_e} \sqrt{S(S+1)} = |g_e| \mu_B \sqrt{S(S+1)} \tag{1.8}$$

$$|\mu_I| = |g_I| \frac{|e|\hbar}{2m_p} \sqrt{I(I+1)} = |g_I| \mu_N \sqrt{I(I+1)} \tag{1.9}$$

where e is the elementary charge of the electron, g_e is the so-called free electron g value, which is 2.0023.[1] μ_B and μ_N are the electron Bohr magneton and nuclear magneton, m_e and m_p are the electron and proton masses, and g_I depends on the nucleus under consideration (see later). The ratio of $|\mu_S|$ and $|\mu_I|$ for the proton is 658.2107 [1]. Analogously to the angular momenta, only the projection along z of the magnetic moment and its modulus are known, but not its direction.

Sometimes the magnetogyric ratio γ is used to indicate the ratio between magnetic moments and angular momenta

$$\gamma_S = -\frac{g_e \mu_B}{\hbar} \quad \gamma_I = \frac{g_I \mu_N}{\hbar}, \tag{1.10}$$

where γ_S and γ_I for the proton have opposite signs, their ratio thus being –658.2107.

If reference is made to Fig. 1.5, it appears that the angle ϕ is known because the modulus of vector $\boldsymbol{\mu}$ is known, as well as its projection along the z axis, but the orientation of $\boldsymbol{\mu}$ cannot be known

$$\mu_z = \mu \cos\phi. \tag{1.11}$$

This nicely reconciles the quanto-mechanical picture with classical physics, which shows that a magnetic moment which must form an angle ϕ with the direction of an external magnetic field *precesses* about it with an angular frequency

$$\omega = -\gamma B_0. \tag{1.12}$$

[1] In this book g_e is taken positive, and the equations containing g_e are explicitly written in such a way as to contain a positive g_e.

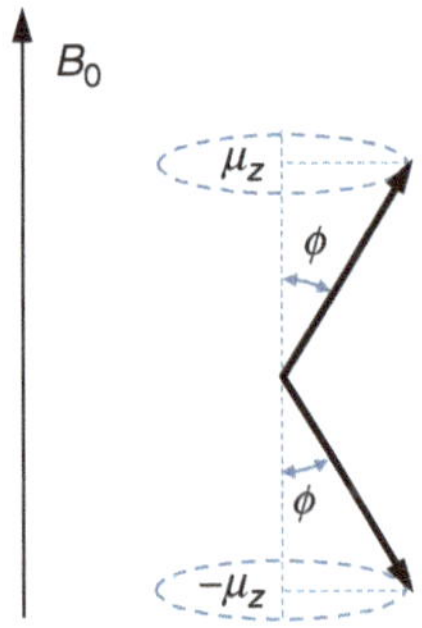

FIGURE 1.6 The allowed precessions of a spin $I = ½$ with negative γ (positive ω) in a magnetic field.

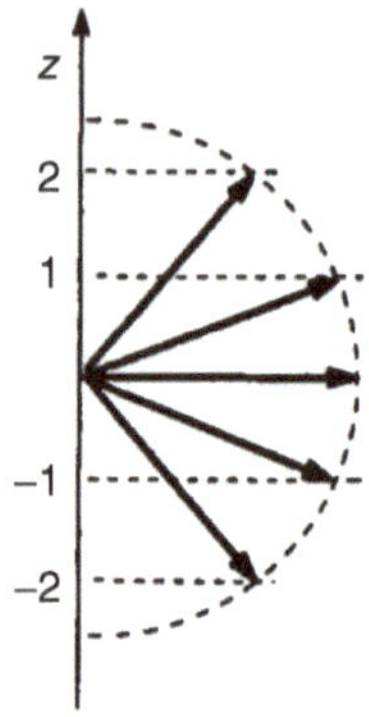

FIGURE 1.7 Allowed orientations and z projections of a spin $S = 2$ (or $I = 2$) in a magnetic field.

The resulting picture of a spin moment (magnetic or angular) in a magnetic field $\boldsymbol{B}_0$ is that it precesses about the $\boldsymbol{B}_0$ direction with an angular frequency proportional to the intensity of $\boldsymbol{B}_0$ and to its own magnetic moment (Fig. 1.6). The sign of ω, which is related to the sign of g [Eqs. (1.10) and (1.12)], gives the direction of precession.

In the case of more than one unpaired electron the total spin value S is ½ the number of unpaired electrons. Commonly, we will deal with one to seven unpaired electrons and S can thus take values from ½ to 7⁄2.

In the case of odd numbers of protons and/or neutrons, a total spin value I varying from ½ to 7 occurs. Owing to the complex intranuclear forces, the g_I values also vary from 5.96 for ^{3}H to 0.097 for ^{191}Ir. The g_I values for magnetically active nuclei are summarized in Appendix I.

The number of allowed values of M_S (or M_I) is $2S + 1$ (or $2I + 1$), and the values range from S to $-S$ (or from I to $-I$), differing by one unit. Fig. 1.7 shows the allowed orientations for a spin $S = 2$ (or $I = 2$). It is just an extension of the $I = S = ½$ case.

1.3 SOMETHING MORE ABOUT THE NUCLEAR SPIN

Nuclear spin vectors are localized on the nucleus, at least for the purposes discussed here. Therefore they can be treated as point dipoles. We have already shown that they are described by $2I + 1$ wave functions, each characterized by the value I and a value of M_I.

We have already seen that in a magnetic field there are different allowed spin orientations. We want now to point out that the different spin orientations in a magnetic field correspond to different energies. This is quite intuitive by looking at Fig. 1.7. If the orientation of the magnetic spin dipole is different in a magnetic field from case to case, then the interaction energies will be different. Actually, according to Eq. (1.3), the energy will be given by the product of the projection along z of the spin magnetic moment and the external magnetic field

$$E = -g_I \mu_N M_I B_0 \tag{1.13}$$

where $g_I \mu_N M_I$ is the projection of the spin magnetic moment along $\boldsymbol{B}_0$.

In quantum mechanical terms the energy is given by the Hamiltonian operator, which in this case is called the nuclear Zeeman Hamiltonian

$$H = -g_I \mu_N \boldsymbol{I} \cdot \boldsymbol{B}_0 \tag{1.14}$$

where $\boldsymbol{I}$ is the spin operator. Now, if we define the z axis along $\boldsymbol{B}_0$, the nuclear spin vectors have a nonzero projection along z, whereas they have generally zero time average in the xy plane. Therefore, we can write

$$H = -g_I \mu_N B_0 I_z. \tag{1.15}$$

Since the application of I_z on a wave function $|I, M_I\rangle$ gives M_I [Eq. (1.7)], the energies of interaction between the spin and the magnetic field given in Eq. (1.13) are obtained. Such energies are dependent on the magnitude of the external magnetic field (Fig. 1.8) and the energy separation ΔE between two adjacent levels is

$$\Delta E = g_I \mu_N B_0 (M_I - (M_I - 1)) = g_I \mu_N B_0. \tag{1.16}$$

In the absence of an external magnetic field the Zeeman Hamiltonian provides zero energy and all the $|I, M_I\rangle$ levels (termed as I manifold) have the same energy. However, this may not be true for nuclei with $I > ½$. In this case, the nonspherical distribution of the charge causes the presence of a quadrupole moment. Whereas a dipole can be described by a vector with two polarities, a quadrupole can be visualized by two dipoles as in Fig. 1.9.

The presence of a quadrupole moment can make the $|I, M_I\rangle$ levels inequivalent even in the absence of an external magnetic field, provided there is an electric field gradient. Only the wave functions with the same absolute value of M_I are pairwise degenerate in axial symmetry, i.e., $M_I = \pm 1, \pm 2$, etc. An example is reported in Fig. 1.10 for $I = 3/2$.

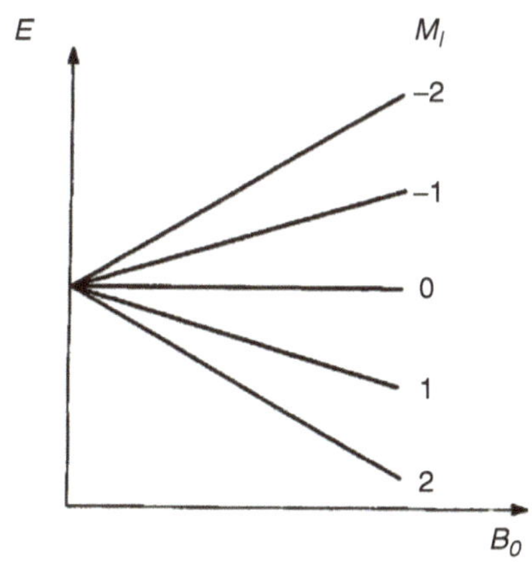

FIGURE 1.8 The Zeeman energies of a nuclear spin ($I = 2$) as a function of the external magnetic field B_0.

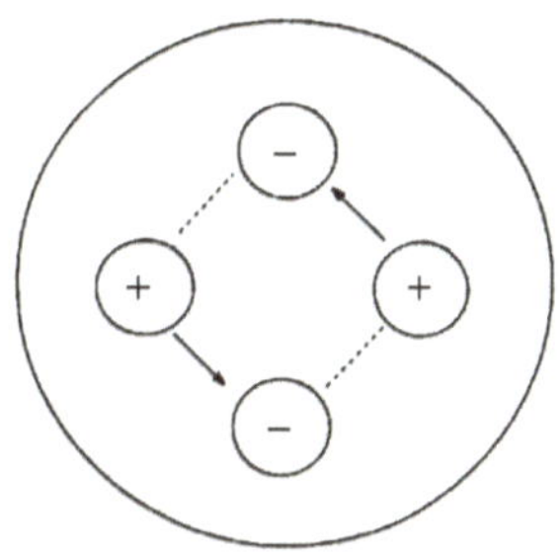

FIGURE 1.9 Schematic drawing of a quadrupole moment.

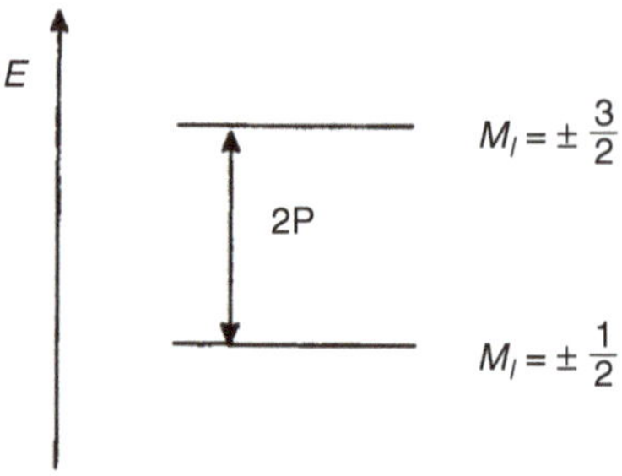

FIGURE 1.10 The energy levels of a spin $I = \frac{3}{2}$ at zero magnetic field in axial symmetry, with P being the product of the quadrupole moment with the electric field gradient.

1.4 A LOT MORE ABOUT THE ELECTRON SPIN

At variance with the nucleus, the electron is associated with an orbital, i.e., a wave function which is related to the distribution in space of the electron cloud, and which displays an angular momentum $\boldsymbol{L}$ and a magnetic moment $\boldsymbol{\mu}_L$. In analogy with the spin operators [Eq. (1.7)], the following relations hold

$$L^2|l,m_l\rangle = l(l+1)|l,m_l\rangle$$
$$L_z|l,m_l\rangle = m_l|l,m_l\rangle$$

where n, l, m_l are the quantum numbers describing the electron orbital, with $l = 0,\ldots, n$ and $m_l = -l,\ldots, l$.

In a naive and incorrect way, we can say that the electron with $S = ½$ senses the orbital magnetic moment. Actually, a charged particle cannot sense the orbital magnetic moment due to its own movement. However, the electron moves in the electric potential of the charged nucleus. If we change the system of reference, the movement of the electron around the nucleus can be seen as a movement of the nucleus around the electron (Fig. 1.11). The "motion" of the charged nucleus then generates a magnetic field which is sensed by the electron. A convenient way to describe the relative movement of the nucleus with respect to the electron is that of using the same n, l, and m_l quantum numbers describing

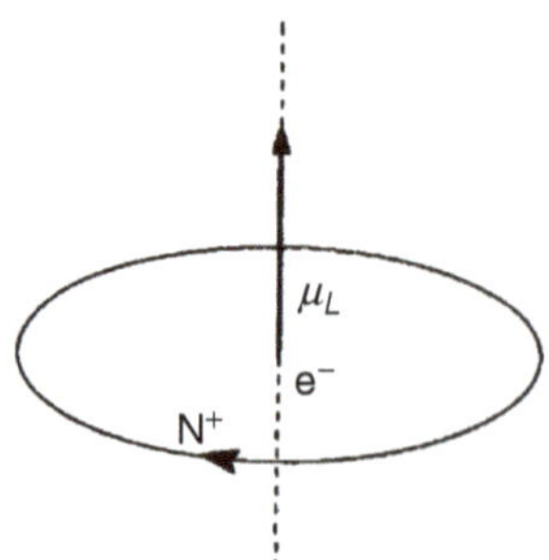

FIGURE 1.11 The electron "senses" the orbital magnetic moment.

the electron. The resulting angular and magnetic properties will depend on the values of m_s for the spin, m_l for the orbital, and on their interaction. The latter phenomenon is called *spin–orbit coupling* and is of paramount importance in understanding the electronic properties. The spin–orbit coupling increases for increasing number of electrons, i.e., it increases from left to right in a period and from top to bottom in a group within the periodic table. Traditionally, two different formalisms are used for transition metal ions of the first series on one side and lanthanoids on the other. In the latter case, spin–orbit coupling is strong, and l and m_l are not good quantum numbers. This case will be treated later (Chapter 8). In the former case, spin–orbit coupling is small enough to be considered a perturbation. For more than one unpaired electron, total L and M_L can be defined. In a molecule, the ligand field defines internal direction(s) along which the orbital angular momentum is preferentially aligned (quantized). Other orientations have higher energies.

We now let the molecule interact with an external magnetic field $\boldsymbol{B}_0$. The interaction energy, as far as the orbital is concerned, is given by the orbital Zeeman operator

$$H = -\mu_L \cdot \boldsymbol{B}_0 = \mu_B \boldsymbol{L} \cdot \boldsymbol{B}_0. \tag{1.17}$$

This interaction will tend to disalign $\boldsymbol{L}$ from its internal axes (Fig. 1.12A). As a result, when the molecule rotates with respect to $\boldsymbol{B}_0$, the interaction energy of Eq. (1.17) is orientation dependent.

In coordination chemistry, reference is often made to the limiting case in which the orbital contribution tends to zero. In this case, the treatment is equal to the nuclear case and the same Hamiltonian is used [the opposite sign with respect to Eq. (1.14) is justified by the positive g_e]:

$$H = g_e \mu_B \boldsymbol{S} \cdot \boldsymbol{B}_0 \tag{1.18}$$

and

$$E = g_e \mu_B M_S B_0. \tag{1.19}$$

In such a system, the external magnetic field defines the molecular z-axis. If we rotate the molecule with respect to $\boldsymbol{B}_0$, the spin and its magnetic moment are not affected (Fig. 1.12B). However, in the molecule of Fig. 1.12A, a molecular z-axis can be defined. When rotating the molecule, the orbital contribution to the overall magnetic moment changes, whereas the spin contribution is constant. The total Zeeman Hamiltonian is

$$H = \mu_B (\boldsymbol{L} + g_e \boldsymbol{S}) \cdot \boldsymbol{B}_0. \tag{1.20}$$

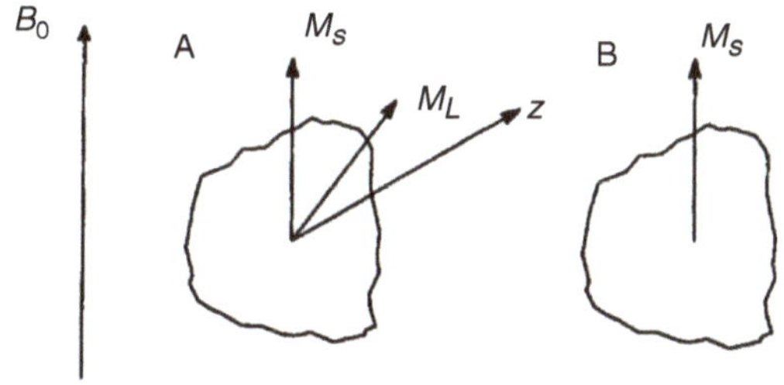

FIGURE 1.12 (A) Orientation of M_S and M_L in the presence of internal molecular axes. (B) A case in which the external magnetic field determines the quantization axes.

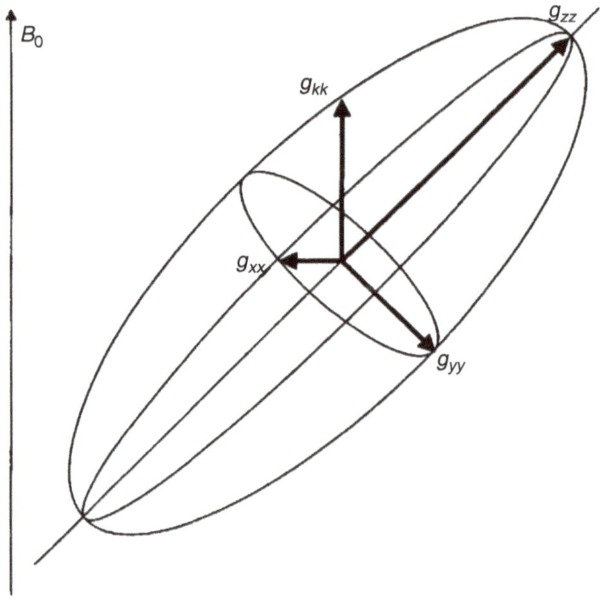

FIGURE 1.13 The ellipsoid representing the components of the g tensor in every direction. The molecule to which the tensor is associated has a generic orientation in the magnetic field B_0.

A convenient way to handle Eq. (1.20) is that of defining a tensor $\boldsymbol{g}$ which couples the magnetic moment $\boldsymbol{S}$ with the external magnetic field. Such a tensor defines the coupling between $\boldsymbol{S}$ and $\boldsymbol{B_0}$ for all molecular directions. We can represent the tensor as a solid ellipsoid (Fig. 1.13) with three principal directions defining the axes of the ellipsoid and of the molecule. In any kk direction we have a value of g_{kk} such that

$$g_{kk}^2 = g_{xx}^2 \cos^2 \alpha + g_{yy}^2 \cos^2 \beta + g_{zz}^2 \cos^2 \gamma \tag{1.21}$$

where cos α, cos β, and cos γ are the direction cosines of the kk vector. The projections of the total electron magnetic moment along any kk direction defined by $\boldsymbol{B_0}$ are given by $\mu_B g_{kk} M_S$. The energy of the $|S, M_S\rangle$ function, when the magnetic field is along the kk direction is

$$E = \mu_B g_{kk} M_S B_0. \tag{1.22}$$

As we can see, the expression of the energy does not contain L. The Hamiltonian has the form

$$H = \mu_B \boldsymbol{S} \cdot \boldsymbol{g} \cdot \boldsymbol{B_0} \tag{1.23}$$

which is the scalar product of the $\boldsymbol{S}$ vector (defined in Section 1.2), the $\boldsymbol{g}$ tensor, and the $\boldsymbol{B_0}$ vector.

This new formalism, known as spin-Hamiltonian formalism, does not contain the $\boldsymbol{L}$ operator, which would require more laborious calculations. It's effects are parametrically included in the $\boldsymbol{g}$ tensor, which would revert from ellipsoidal to spherical in the absence of orbital angular momentum.

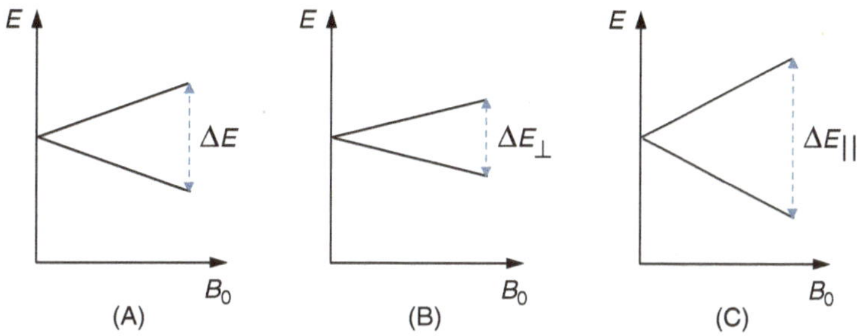

FIGURE 1.14 The splitting of the $S = ½$ manifold in a magnetic field B_0 when (A) g is isotropic and there are only two energy values independent of the orientation of the molecule in the magnetic field and (B,C) the energies depend on the orientation of the molecule in the magnetic field ($\Delta E_{\parallel} > \Delta E_{\perp}$).

When the molecule under investigation rotates fast with respect to the g anisotropy (i.e., the reorientation rate τ_r^{-1} (Section 4.2) is larger than the spreading of the different orientation-dependent energies of the spin ($\tau_r^{-1} > \Delta E/\hbar$)), we measure an average g value $\bar{g}$, which is also different from g_e. The two limit situations of isotropic and anisotropic g are illustrated in Fig. 1.14. When $g = 2.0023$, and therefore the orbital contribution is zero, the splitting of any S manifold is as in Fig. 1.14A and independent of the orientation of the molecule with respect to the external magnetic field; when there is an orbital contribution, a different splitting of the S manifold in any direction occurs (Fig. 1.14B–C), and upon rapid rotation there is an average splitting of the levels.

Besides providing a different effective magnetic moment for each orientation, spin–orbit coupling is also able to cause a splitting of an S manifold with $S > ½$ at zero magnetic field. When S is, let us say, $^3/_2$, spin–orbit coupling and low symmetry effects split the quartet in a way similar to that depicted in Fig. 1.10. When S is half integer, at least twofold degeneracies remain (so-called Kramers doublets, $M_S = \pm n/2$, n integer), whereas when S is integer the splitting can remove any degeneracy (Fig. 1.15).

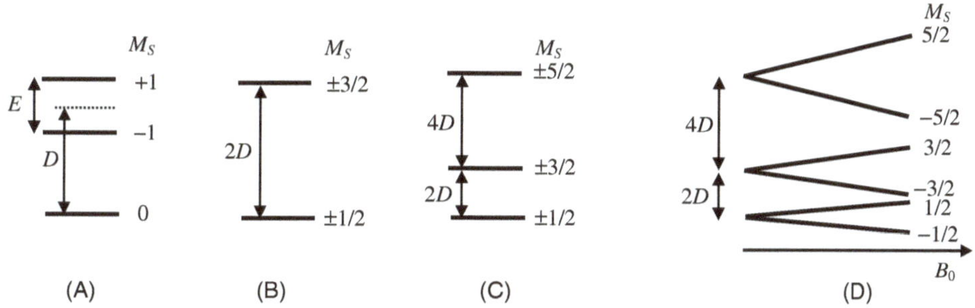

FIGURE 1.15 The splitting of an $S = 1$ (A), $S = {}^3/_2$ (B) and $S = 5/2$ (C) manifold in the presence of spin orbit coupling and low symmetry components. D is the axial and E the rhombic ZFS parameter (the latter only shown in case A). The wave functions are labeled as high field eigenfunctions. (D) Electron energy levels in the presence of axial ZFS for B_0 along z, in the $S = 5/2$ case.

Such splitting is called zero field splitting and indicated as ZFS. It adds up to the Zeeman energy. In the spin-Hamiltonian formalism, i.e, when the effects of the orbital angular momentum are parameterized, it is indicated as

$$H = \boldsymbol{S} \cdot \boldsymbol{D} \cdot \boldsymbol{S} \tag{1.24}$$

where $\boldsymbol{D}$ is the ZFS tensor. It is traceless, in the sense that its effect upon rapid rotation is zero (rapid means that the rotation rate (s^{-1}) is larger than the maximum energy splitting ($\Delta E/\hbar$ (s^{-1}))). However, its appearance is of paramount importance in electron relaxation and in determining the magnetic properties of metal complexes. The comparison of Fig. 1.10 with Fig. 1.15B shows that the nuclear quadrupole splitting and the ZFS are formally similar. In general, the ZFS is defined by two parameters, D (axial anisotropy) and E (rhombic anisotropy), that characterize the $\boldsymbol{D}$ tensor (Fig. 1.15).

Hamiltonian Eq. (1.24) is formally equivalent to that describing the dipolar interaction between two spins s_1 and s_2, whose sum is S. Actually, in organic radicals where spin–orbit interactions are negligibly small, it is the dipolar interaction between the two electron spins in an $S = 1$ system that causes ZFS.

1.5 ABOUT THE ENERGIES

Up to now we have seen that S or I manifolds split in an external magnetic field according to their M_S or M_I values. The latter are the allowed components of the S or I vectors along the external magnetic field. When we said that the spins orient in a magnetic field as in Fig. 1.3, we actually referred to the projection along z relative to the low energy orientation, which is the only populated at $T = 0$ K. The excited levels are separated by the Zeeman energy [Eqs. (1.15) and (1.16)]. Such energies are about 0.3 cm^{-1} at 0.3 T for the electron, and 658 times smaller for the proton. The thermal energy kT is about 200 cm^{-1} at 300 K and about 0.7 cm^{-1} at 1 K. So, the population of the two levels is almost the same at every temperature above a few Kelvin. The Boltzmann population P_i of each M_I level is

$$P_i = \frac{e^{-E_i/(kT)}}{\sum_i e^{-E_i/(kT)}} \tag{1.25}$$

where E_i is the energy of the ith level with respect to the ground level and the sum is extended to all levels. When $kT \gg E_i$, as happens at room temperature, $E_i/(kT)$ tends to zero, the exponential tends to unity and each level is almost equally populated. The magnetic resonance experiments are based on the small population differences. The energy of the system (for instance, an ensemble of N_A spins) is given by the sum of the energies for each level weighted by the population of the level.

1.6 MAGNETIZATION AND MAGNETIC SUSCEPTIBILITY

The effect of the external magnetic field is that of splitting the energies of the S or I manifolds (eg, Fig. 1.8) and, therefore, of making different the populations of the levels. The difference in population according to the Boltzmann law [Eq. (1.25)] tells us that the magnetic field has indeed changed the energy of the system. By making reference for simplicity to Fig. 1.4 (two orientations), the spins with the lower energy orientation are more than those with the higher energy orientation. As a consequence, an

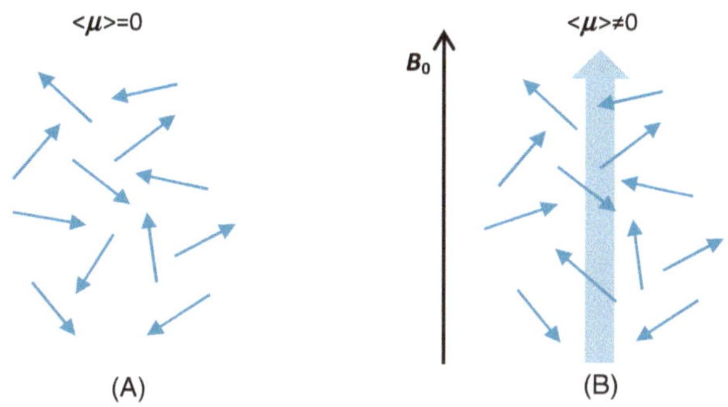

FIGURE 1.16 An ensemble of magnetic moments (A) orient themselves along the applied magnetic field B_0 (B). The partial orientation determines a resultant non-zero magnetic moment.

induced magnetic moment μ_{ind} is established. The net interaction energy of the whole system with the magnetic field is the product of the induced magnetic moment and the magnetic field. The magnetization per unit volume M [Eq. (1.2)] corresponds to the induced magnetic moment per unit volume and, for many substances, is found to be proportional to the applied magnetic field B_0:

$$M = \frac{\mu_{\text{ind}}}{V} = \chi_V H = \frac{1}{\mu_0}\chi_V B_0 \tag{1.26}$$

where χ_V, the dimensionless proportionality constant between M and H, is the magnetic susceptibility per unit volume. This is easily obtained from Eq. (1.2) since $\chi_V \ll 1$ (except for ferromagnetic systems, see Section 10.8).

Classically, this effect can be seen as an ensemble of magnetic moments randomly oriented in the absence of a magnetic field with resultant equal to zero. When an external magnetic field is applied, it tends to orient the magnetic moments and to provide a resultant different from zero (Fig. 1.16). The larger the magnetic field, the larger is the resultant induced magnetic moment. From Eq. (1.26), $\chi_V = \mu_0 M/B_0 = \mu_0\mu_{\text{ind}}/(B_0 V)$: for N_A particles, $\mu_{\text{ind}} = N_A\langle\mu\rangle V/V_M$, where $\langle\mu\rangle$ is the average induced magnetic moment per particle (see later), and

$$\chi_M = V_M\chi_V = V_M\frac{\mu_0 M}{B_0} = \frac{\mu_0 N_A\langle\mu\rangle}{B_0} \tag{1.27}$$

where χ_M ($\text{m}^3\,\text{mol}^{-1}$) is the magnetic susceptibility per mole, and V_M is the molar volume; χ_M is magnetic field independent, just like χ_V.

Let us now take an S manifold, unsplit at zero magnetic field, with no orbital angular momentum. The sum of the energies in a magnetic field would be zero [Eq. (1.19)] if the levels were equally populated:

$$E = g_e\mu_B B_0\sum_{M_S=-S}^{S} M_S = 0. \tag{1.28}$$

However, if the populations are considered, in an ensemble of N_A particles, and by recalling that $M_S = \langle S, M_S|S_z|S, M_S\rangle$ (Eq. (1.7)), the energy is

$$E = N_A g_e \mu_B B_0 \frac{\sum \langle S, M_S|S_z|S, M_S\rangle e^{-g_e\mu_B B_0 M_S/(kT)}}{\sum e^{-g_e\mu_B B_0 M_S/(kT)}}. \tag{1.29}$$

If we consider that $g_e\mu_B B_0 M_S \ll kT$, the exponential is well approximated by $(1 - g_e\mu_B B_0 M_S/kT)$. It follows from Eq. (1.28) that the denominator is just $2S + 1$, and the energy is given by (Appendix VI)

$$E = -N_A g_e^2 \mu_B^2 \frac{S(S+1)B_0^2}{3kT} = N_A g_e \mu_B B_0 \langle S_z \rangle. \tag{1.30}$$

The quantity

$$\langle S_z \rangle = \frac{\sum \langle S, M_S|S_z|S, M_S\rangle e^{-g_e\mu_B B_0 M_S/(kT)}}{\sum e^{-g_e\mu_B B_0 M_S/(kT)}} = -\frac{g_e\mu_B S(S+1)B_0}{3kT} \tag{1.31}$$

is called the expectation value of S_z. By operating with S_z on each $|S, M_S\rangle$ level, considering the population, and summing up over all the levels, we obtain an expectation value different from zero.

From the classical treatment, the energy of a system composed of N_A particles is also given by the product of the induced magnetic moment per mole along the field, μ_{ind} ($\mu_{\text{ind}} = N_A\langle\mu\rangle$) and the external magnetic field B_0 [cf. Eq. (1.3)]:

$$E = -\mu_{\text{ind}} B_0. \tag{1.32}$$

Therefore, by combining Eqs. (1.30) and (1.32):

$$\langle \mu \rangle = \frac{\mu_{\text{ind}}}{N_A} = \mu_B^2 g_e^2 \frac{S(S+1)B_0}{3kT} = -\mu_B g_e \langle S_z \rangle. \tag{1.33}$$

In other words, the induced magnetic moment per particle is just proportional to the expectation value $\langle S_z\rangle$. Note that the value of $\langle S_z\rangle$ is referred to a single spin S and to its fractional occupancy of the energy level ladder. In the case of a spin ensemble, the value of $\langle S_z\rangle$ provides the average value of S_z of the ensemble. $\langle S_z\rangle$ is a dimensionless number. From Eqs. (1.27) and (1.33), the magnetic susceptibility per mole is given by the Curie's law:

$$\chi_M = \mu_0 N_A \mu_B^2 g_e^2 \frac{S(S+1)}{3kT}. \tag{1.34}$$

By using Eq. (1.8) we obtain

$$\chi_M = \mu_0 N_A \frac{\mu_S^2}{3kT}. \tag{1.35}$$

This result of the quantum mechanical treatment coincides with the derivation of magnetic susceptibility in terms of classical magnetic moments if their moduli are taken equal to those associated to the individual spins.

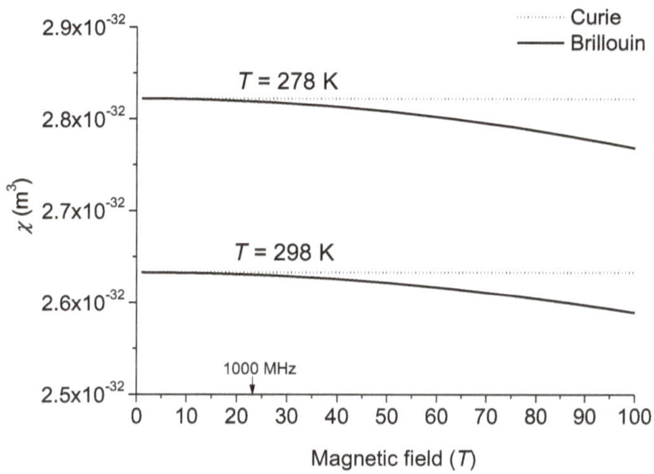

FIGURE 1.17 χ as a function of the magnetic field, as described by Eq. (1.37) (Brillouin function). The dotted lines report the value of χ calculated with Eq. (1.34) (Curie's law). $S = 1/2$, $T = 278$ and 298 K. A proton Larmor frequency of 1000 MHz corresponds to a magnetic field of 23.5 T.

A magnetic susceptibility per molecule, χ, can be defined as the molar magnetic susceptibility χ_M divided by Avogadro's constant:

$$\chi = \chi_M / N_A. \tag{1.36}$$

This definition implies that the average performed on a large number of molecules is the same as that obtained from the average performed for the same molecule over time.

At very high fields, the first order approximation of the exponential in Eq. (1.29) may not be accurate. In these cases, the magnetic susceptibility is not field independent but slightly decreases with field. This results in a nonlinear field dependence of magnetization (saturation effect). The magnetic susceptibility per molecule is described by the Brillouin function:

$$\chi = \frac{\mu_0 \mu_B g_e}{2B_0}\left[(2S+1)\coth\left(\frac{(2S+1)\mu_B g_e B_0}{2kT}\right) - \coth\left(\frac{\mu_B g_e B_0}{2kT}\right)\right]. \tag{1.37}$$

The field dependence of χ for $S = 1/2$ systems is shown in Fig. 1.17. The decrease with respect to the Curie value becomes appreciable for fields larger than 20 T, values that are reached today by high field NMR spectrometers. The lower the temperature, the larger is the decrease of χ.

All the above treatment holds for a single S manifold. If there is some orbital contribution to the magnetic moment, it cannot be neglected. All the calculations should be repeated by using Hamiltonian Eq. (1.20) for a generic direction k and, by keeping in mind [Eq. (1.27)] that $\chi_M = \mu_0 N_A \langle \mu \rangle / B_0$

$$\chi_{M_{kk}} = -\frac{\mu_0 N_A \mu_B}{B_0} \frac{\sum_i \langle \phi_i | L_{kk} + g_e S_{kk} | \phi_i \rangle e^{-E_{ikk}/(kT)}}{\sum_i e^{-E_{ikk}/(kT)}} \tag{1.38}$$

where the sum is over all the levels of the S manifold now containing the orbital part. E_{ikk} is the Zeeman energy and may also contain ZFS effects. In the spin-Hamiltonian formalism, neglecting ZFS effects

$$E_{ikk} = g_{kk}\mu_B M_S B_0$$
$$\langle \phi | L_{kk} + g_e S_{kk} | \phi \rangle = g_{kk} M_S \tag{1.39}$$

and thus, if the exponential is approximated to the first order,

$$\chi_{M_{kk}} = \mu_0 N_A \mu_B^2 g_{kk}^2 \frac{S(S+1)}{3kT} \tag{1.40}$$

where g_{kk} is now different from g_e [Eq. (1.21)]. χ_M is now a tensorial quantity.

Eq. (1.38) is called the Van Vleck equation. In that form the Zeeman operator operates only to first order. Indeed, we should include the second order Zeeman term, which allows the interaction between the ground S multiplet (ϕ_i functions) with all the excited ones (ϕ_j functions):

$$\sum_{j \neq i} \frac{\mu_B^2 \left| \langle \phi_i | L_{kk} + g_e S_{kk} | \phi_j \rangle \right|^2}{E_i^0 - E_j^0} \tag{1.41}$$

where E_i^0 and E_j^0 are the energies at zero magnetic field ($E = E^0 + \mu_B(\boldsymbol{L} + g_e\boldsymbol{S}) \cdot \boldsymbol{B_0}$). The complete Van Vleck equation, including the population, is

$$\chi_{M_{kk}} = \mu_0 N_A \mu_B^2 \frac{\sum_i \left[\frac{\left| \langle \phi_i | L_{kk} + g_e S_{kk} | \phi_i \rangle \right|^2}{kT} - 2\sum_{j \neq i} \frac{\left| \langle \phi_i | L_{kk} + g_e S_{kk} | \phi_j \rangle \right|^2}{E_i^0 - E_j^0} \right] e^{-E_i^0/(kT)}}{\sum_i e^{-E_i^0/(kT)}} \tag{1.42}$$

A consequence of the anisotropy of the magnetic susceptibility is that the induced magnetic moment per particle $<\mu>$ is not isotropic but depends on the direction k with respect to the direction of $\boldsymbol{B}_0$, since, from Eqs. (1.27) and (1.36),

$$\langle \mu \rangle_{kk} = \frac{\chi_{kk} B_0}{\mu_0}. \tag{1.43}$$

Therefore, by representing the magnetic susceptibility as a tensor, the average induced magnetic moment depends on the orientation of the $\boldsymbol{\chi}$ tensor according to the following equation

$$\langle \mu \rangle = \frac{\boldsymbol{\chi} \cdot \boldsymbol{B_0}}{\mu_0} \tag{1.44}$$

Interestingly, the $\boldsymbol{<\mu>}$ vector is not parallel to $\boldsymbol{B}_0$ any more, unless $\boldsymbol{B}_0$ is oriented along one main direction of the $\boldsymbol{\chi}$ tensor, i.e., along one of the principal axes defining the reference frame in which $\boldsymbol{\chi}$ is diagonal.

Also diamagnetic systems, not containing unpaired electrons, can have a magnetic susceptibility, χ^{dia} [Eqs. (1.26) and (1.27)]. χ^{dia} is usually independent of H_0 (like the paramagnetic χ calculated above), negative (whereas χ is positive) and independent on temperature (at variance with χ). The diamagnetic susceptibility is due to the interaction of the magnetic field with the motions of the electrons in their orbitals. It is roughly proportional to the molecular weight, so that it is negligible for small complexes, but can be of the same order of the paramagnetic susceptibility in proteins. Also the diamagnetic susceptibility can be anisotropic due to the anisotropy of the electron currents, for instance in the presence of aromatic rings. Its presence must thus be taken into account when considering self-orientation effects (Chapter 3), especially in heme proteins or when multiple aromatic planes are staked together, as in the nucleic acids. In summary, the total molecular magnetic susceptibility is provided by the tensorial sum between the paramagnetic and the diamagnetic susceptibility:

$$\chi^{\text{mol}} = \chi + \chi^{\text{dia}}. \tag{1.45}$$

1.7 THE NUCLEAR MAGNETIC RESONANCE EXPERIMENT

The small difference in population among the M_I or M_S levels allows for the magnetic resonance experiments. From now on we will focus on the nuclear magnetic resonance experiment, but little would be changed if we dealt with EPR.

1.7.1 THE PULSE EXPERIMENT AND DEFINITION OF T_1 AND T_2

When a sample containing magnetically active nuclei is inserted into a static magnetic field, the spin magnetic moments start precessing at the Larmor frequency [Eq. (1.12)]. In the absence of the magnetic field, each spin is in a superposition state, i.e., a quantum mixture of "spin up" and "spin down" states with all possible combination coefficients, so that its magnetic moment can be randomly oriented in any direction and the macroscopic magnetization is zero. The static magnetic field imparts a bias towards orientations of the magnetic moments parallel to the magnetic field. As a result, a nonzero macroscopic magnetization is attained, as already described in Section 1.6 for electrons. Therefore, the nuclear magnetic moments of an ensemble of spins ½ will be a little more than 50% precessing about z with $M_I = ½$ and a little less than 50% precessing about $-z$ with $M_I = -½$. This difference provides a magnetization along z.

The achievement of the equilibrium magnetization at a given magnetic field from nonequilibrium conditions occurs through exchange of energy between the spin system and the surrounding, called lattice, which causes random fluctuations of spin orientations (as described in more detail in Chapter 4). The process is described by an exponential time constant known as longitudinal relaxation time, T_1, or, historically, as spin-lattice relaxation time (indicating that the energy of the spin bath is changed through interaction with the lattice). The transverse relaxation time, T_2, or spin-spin relaxation time, is the exponential time constant describing the achievement of the equilibrium of the magnetization component in the plane perpendicular to z. T_1^{-1} and T_2^{-1} are the corresponding *rate* constants, also indicated as R_1 and R_2.

In order to observe an NMR signal we need to perturb the equilibrium state. For practical reasons, the perturbation consists of driving the spins toward a plane perpendicular to the field. This is achieved by applying a magnetic field rotating with a frequency [Eqs. (1.12) and (1.16)] such that

$$h\nu = \hbar\omega = \hbar|\gamma_I| B_0 = \mu_N |g_I| B_0. \tag{1.46}$$

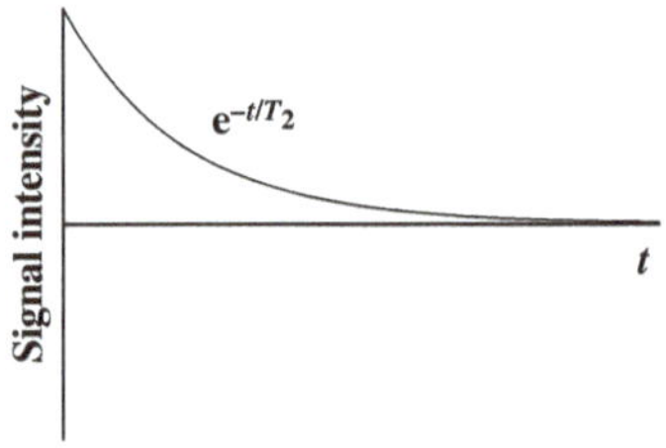

FIGURE 1.18 The FID detected on resonance with the precession frequency of the signal of interest.

The frequency ν is in the radiofrequency (r.f.) range; for the proton it is 1000 MHz at 23.5 T. If we send such a radiation for a time short with respect to T_1 or T_2, but with enough power to affect the spin system, we say that we send a pulse. The magnetization vector starts precessing about the r.f. field. The precession will continue with time, at least for times shorter than T_1 and T_2, with an angular velocity proportional to the r.f. field, i.e., to the power of the pulse. The precession angle or, better, rotation angle after a given time will thus depend on the product of power and time, i.e., on the energy of the pulse. After a given time, let us say when the rotation is 90 degrees, we stop the pulse and we let the system return to equilibrium. The coil, which transmitted the pulse, reveals now the disappearance of the magnetization in the *xy* plane. We say that the coil reveals the free induction decay (FID). The process is exponential with time constant T_2 (Fig. 1.18). The cosine Fourier transform of this exponential decay is

$$\int_0^\infty e^{-t/T_2}\cos(\omega t)\,\mathrm{d}t = \frac{T_2}{1+\omega^2 T_2^2}. \tag{1.47}$$

The frequency function is a Lorentzian with linewidth at half height $\Delta\nu_{1/2} = (\pi T_2)^{-1}(= R_2/\pi)$. R_2 is a measure of the uncertainty of the energy levels, which gives the linewidth in every spectroscopy. The uncertainty principle, according to which the uncertainty in energy of a level is inversely proportional to the lifetime, tells us that T_2 is a measure of the lifetime of the energy levels.

If the frequency of the pulse is different from the resonating frequency, nothing changes as long as the pulse contains the latter frequency as a component. It can be shown that a pulse effectively covers an interval of frequencies of the order of the reciprocal of the pulse length, centered at its own frequency (carrier frequency). The FID now has the shape shown in Fig. 1.19; besides intensity and linewidth, it contains the information about the nuclear resonance frequency.

1.7.2 THE CHEMICAL SHIFT

The effective magnetic field, which a nuclear spin senses when placed in an external magnetic field, is the sum of several contributions, whose nature is not discussed here except for that due to the interaction with unpaired electron(s). For example, the paired electrons will decrease the external magnetic field, since they experience an induced magnetic field which opposes to the external magnetic field.

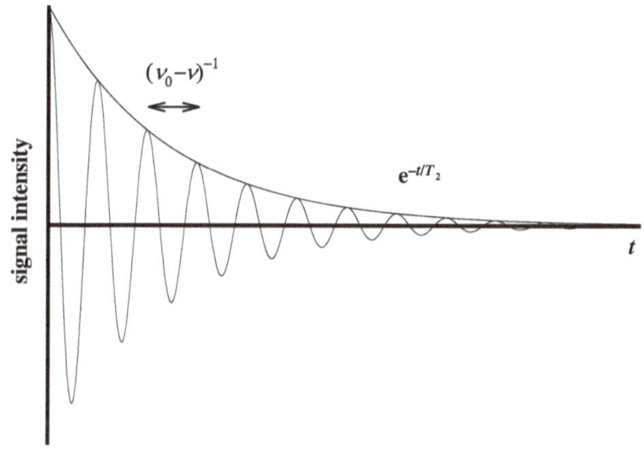

FIGURE 1.19 The FID detected off resonance with respect to the precession frequency of the signal of interest. The FID has the shape of a damped oscillation, where the time separation between adjacent maxima equals the reciprocal of the difference $\nu_0 - \nu$.

This is the same phenomenon which gives rise to bulk diamagnetism. Therefore, all nuclei in a diamagnetic molecule will experience a magnetic field smaller than free nuclei. Since free nuclei are not a practical standard, nuclei in convenient chemical compounds are used as standards. Typically, tetramethylsilane (TMS) is used in proton NMR because its protons are surrounded by a relatively large amount of paired electrons and essentially are at an extreme of the range. The effective magnetic field sensed by such protons is

$$B = B_0(1-\sigma_{\mathrm{TMS}}) \tag{1.48}$$

where σ is the so-called shielding constant and $B_0\sigma$ is the induced magnetic field which opposes to B_0 [equal to $-\mu_0 M$ of Eq. (1.2)].

Actually, the nucleus senses a different field in different directions around it. Therefore, the shift is a tensorial quantity. Upon free rotation in solution, the tensorial quantity is averaged to its trace, σ. The proton of $CHCl_3$ will experience a smaller shielding constant because chlorine, being quite electronegative, will attract the electrons and remove them from the proton. So, the proton of $CHCl_3$ will sense a larger effective magnetic field that the protons of TMS when placed in the same magnetic field (Fig. 1.20).

For fixed B_0, the frequency needed for the transition will be larger for $CHCl_3$ than for TMS. The difference in frequency is called chemical shift with respect to TMS. In this case the shift is positive:

$$\Delta\nu = \nu_{\mathrm{CHCl_3}} - \nu_{\mathrm{TMS}}. \tag{1.49}$$

We sometimes say that the $CHCl_3$ proton resonates downfield (at a smaller external magnetic field) with respect to TMS, which in turn resonates upfield (at a larger external magnetic field) if we imagine to keep the frequency fixed and to vary the magnetic field.

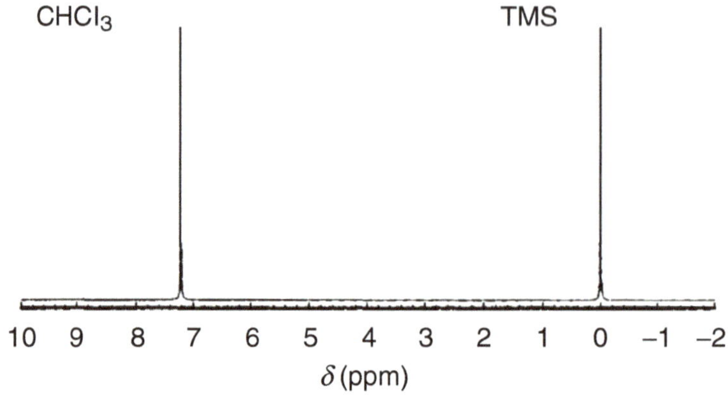

FIGURE 1.20 NMR spectrum of a mixture of $CHCl_3$ and TMS taken at 600 MHz proton Larmor frequency. By taking the chemical shift of TMS as zero, the chemical shift of $CHCl_3$ is 7.28 ppm. Shifts to the left of TMS are taken as positive and called down field. At the chosen magnetic field, the chemical shift of $CHCl_3$ is 4326 Hz.

Under the above definition, the chemical shift is expressed in frequency units (Hz) and depends on the external magnetic field, which then needs to be specified. It may be convenient to express the chemical shift as a pure number, i.e., divided by the frequency of the standard:

$$\frac{\Delta \nu}{\nu_0} = \delta = \frac{\nu_{\mathrm{CHCl_3}} - \nu_{\mathrm{TMS}}}{\nu_{\mathrm{TMS}}} = 7.28 \times 10^{-6} = 7.28\,\mathrm{ppm} \tag{1.50}$$

where ν_0 is the frequency of the spectrometer magnetic field referred to the standard and ppm means parts per million. The symbol δ will be used for chemical shift throughout the book, and all chemical shift equations will be expressed in δ.

The magnetic moments of unpaired electrons will also have a preference for being aligned along the external magnetic field, in the absence of other restrictions. Therefore, their magnetic moment will sum up and increase the effective magnetic field sensed by a nucleus. The interaction between a magnetic nucleus and unpaired electrons is called hyperfine coupling, after Fermi's account of the hyperfine splitting of the lines in the atomic spectra due to the coupling between the proton and the unpaired electron [2]. The shift in this case is positive (i.e., downfield: since there is the contribution of the electron dipole, a smaller external field is needed to bring the nucleus to resonance). The effect of the presence of unpaired electrons on the nuclear properties will be the subject of this book and of course will be discussed in detail, hopefully in a clear, concise but exhaustive way.

In terms of the Hamiltonian, such hyperfine coupling is represented as

$$H = \boldsymbol{I} \cdot \boldsymbol{A} \cdot \boldsymbol{S} \tag{1.51}$$

where $\boldsymbol{A}$ is the hyperfine coupling tensor. It is a tensor because it has different values depending on the orientation of the molecular frame within an external magnetic field, which defines the laboratory z-axis (Chapter 2).

1.7.3 SOMETHING MORE ABOUT RELAXATION RATES

T_1 and T_2 have already been defined in Section 1.7.1. Since their understanding is fundamental to any approach to NMR experiments and theory, we repeat here the definitions by using a different wording. When an ensemble of independent and equivalent spins at equilibrium in a magnetic field is perturbed, for example, by irradiation at the right frequency or by suddenly changing the magnetic field, the system is not at equilibrium any longer. For simplicity we like to consider the spins independent, i.e., not interacting one another. This assumption is convenient to define T_1 but unrealistic, as we will see in the following chapters. The spins are also assumed to experience the same chemical shift. If we refer to the Boltzmann law, which accounts for the levels' population, we can say that, after a perturbation, the spin temperature is different from the lattice temperature. For lattice we mean the environment of the nuclei, which is assumed to have an infinite heat capacity.

After the perturbation, the system tends to reach equilibrium again. We assume that the return to equilibrium is a stochastic process, i.e., each step toward equilibrium is random and uncorrelated to other steps. This implies that the spins are not related to one another. Under these conditions, the return to equilibrium is a first order kinetic process. The decay is then given by

$$F(t) = e^{-Rt} \tag{1.52}$$

where t is the time and R is a rate constant, as anticipated in Section 1.7.1.

The magnetic field has cylindrical symmetry, i.e., the physical properties of a system do not change upon rotation about the field direction, and there is no difference between x and y directions. If the measurement of the return to equilibrium is performed along the B_0 direction, i.e., the return to equilibrium of M_z is followed, the rate constant is said to be longitudinal and is indicated as R_1. The magnetization reaches equilibrium through transitions between pairs of levels differing by ΔM_I or $\Delta M_S = \pm 1$. Such transitions are induced, like in the magnetic resonance experiments, by oscillating magnetic fields available within the lattice. Matter, even in the condensed phase, experiences continuous movements whose average kinetic energy is proportional to $kT/2$ for each degree of freedom. Matter contains magnetic dipoles, electric charges, etc., which are capable of originating magnetic fields upon sudden reorientations or sudden movements, which can be seen as short pulses. Longitudinal relaxation occurs through energy exchange with the lattice until the equilibrium population of the levels is obtained and the spin temperature equals the lattice temperature. We recall that the lattice is assumed to have an infinite heat capacity.

In a typical experiment, called inversion recovery, M_z is initially inverted by applying a 180 degrees pulse. Its recovery along the z-axis is given by

$$M_z(t) = M_z(\infty) - 2M_z(\infty)e^{-R_1 t} \tag{1.53}$$

where $M_z(\infty)$ is the equilibrium value of M_z. The value of $M_z(t)$ at various times t is sampled by applying a 90 degrees pulse and measuring the intensity of the detected signal. R_1 can then be extracted from a fitting of the data to an exponential recovery.

If the rate of reaching equilibrium is measured orthogonally to the magnetic field, i.e., by sampling the projection of the magnetization in the xy plane obtained, for instance, after a 90 degrees pulse, a different rate constant is obtained, which is called the transverse relaxation rate and indicated as R_2.

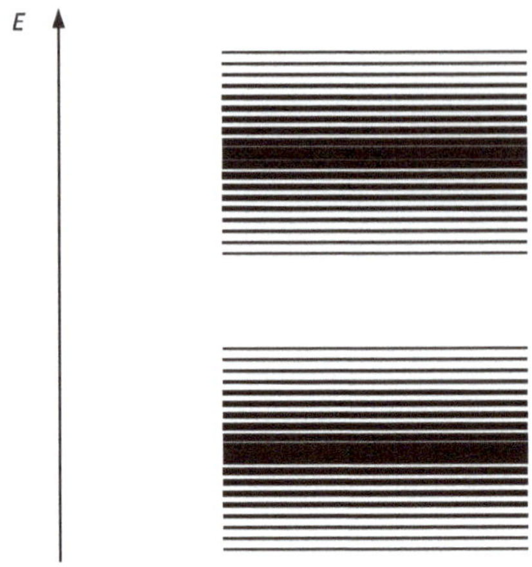

FIGURE 1.21 The spreading of energy levels as due to an effective fluctuating magnetic field along z.

In a pulsed experiment, the rate of reaching equilibrium is equal to the rate of disappearance of magnetization in the xy plane, and is proportional to the linewidth at half height:

$$\pi \Delta \nu_{1/2} = R_2. \tag{1.54}$$

All the mechanisms which contribute to R_1 contribute also to R_2, because the reestablishment of the equilibrium population leaves zero magnetization in the xy plane. Conversely, the dephasing, or fanning out, of the single spin components in the xy plane lowers the in plane magnetization without affecting the magnetization along z, and therefore it contributes only to R_2. Therefore

$$R_2 > R_1 \quad \text{or} \quad R_2 = R_1 + c \tag{1.55}$$

where c is the cumulative rate of all processes which lead to the disappearance of the magnetization in the xy plane besides those which lead to the reestablishment of Boltzmann equilibrium. Typically, local magnetic fields oscillating along B_0 at any frequency close to zero produce oscillations in the actual magnetic field and increase the linewidths; in other words, such fields $\boldsymbol{b}_z$ spread the spin moments (Fig. 1.21) through the vectorial interaction with the xy magnetization:

$$\boldsymbol{b}_z \times \boldsymbol{M}_{xy}.$$

The dephasing of the single spins leads to the equilibrium value of the magnetization in the xy plane, which is zero. We have already seen that the simultaneous change of orientation of two spins contributes only to R_2. Besides measuring the linewidth, typical experiments for measuring R_2 are the so-called spin–echo experiments. In their simplest form, a 90 degrees pulse is applied first, followed by a 180 degrees pulse after a variable time t. The intensity of the signal is measured after another time t following the 180 degrees pulse.

In another type of experiment, the magnetization is kept locked in the xy plane. If we use as reference frame a rotating coordinate system with z'coinciding with z and $x'$and y'rotating at the ω frequency, the magnetization vector, which precesses about the r.f. field, processes about a vector which

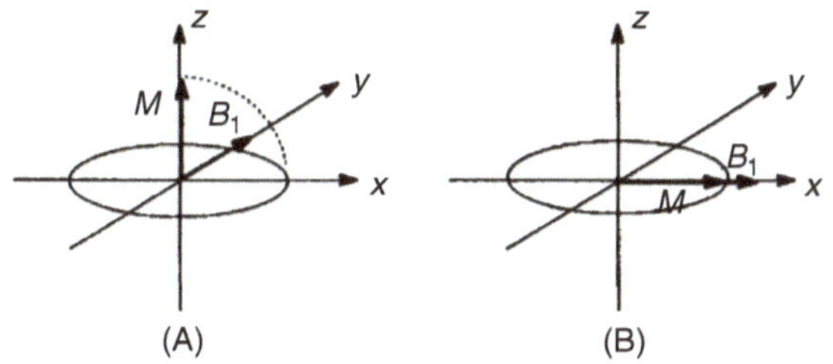

FIGURE 1.22 In a spin-locking experiment the magnetization is initially brought along the x-axis of the rotating frame (A), and then *locked* along that axis (B), by proper reorientation of the r.f. field B_1.

is constant in this frame, because the r.f. field also rotates at frequency ω. A 90 degrees pulse with B_1 along y' (in the rotating frame) thus sends the magnetization along x'. Then B_1 is aligned along x' (Fig. 1.22). The field B_1 originated by the transmitter is the only field sensed by the magnetization in the rotating frame. Then the magnetization precesses along B_1. At equilibrium the magnetization will return along z. The time constant for the kinetic process of returning to equilibrium, i.e., for the disappearance of the magnetization along y when a spin-locking field B_1 is operating, is called $T_{1\rho}$ and the corresponding rate constant is called $R_{1\rho}$.

The presence of unpaired electrons will activate new relaxation pathways, and we will have a nuclear R_i (i = 1 or 2) enhancement which is indicated as R_{iM}. Similarly, $R_{1\rho}$ also increases. This will be treated in some detail in Chapter 4.

We should end this section by recalling that, while the concept of relaxation is very general, the concept of relaxation rates is much less general, as it implies exponential return to equilibrium of magnetization. Exponential relaxation can only occur when the spin system is weakly coupled to the lattice, and the latter can be assumed to have infinite heat capacity. However, in the case of nuclei, very often the lattice is also constituted by other nuclear spin systems that have a limited heat capacity, of the same order as that of the spin system under investigation. *In the case of electron spins, sometimes the coupling with the lattice is too strong, and nonexponential electron relaxation may also occur.* These concepts will be further developed in Chapters 4 and 12. Despite these drawbacks, the concept of relaxation rates is still very useful and is thoroughly used, especially when dealing with nuclear relaxation rate enhancements due to unpaired electrons, which as we will see, are often truly exponential contributions.

GENERAL REFERENCES

For the physical concepts we have mainly referred to B.I. Bleaney, B. Bleaney, Electricity and Magnetism, third ed., Oxford University Press, London, 1976; M.H. Levitt, Spin Dynamics: Basics of Nuclear Magnetic Resonance, second ed., Wiley, Chichester, 2008.

There are numerous books on FT-NMR. The reference book is: R.R. Ernst, G. Bodenhausen, A. Wokaun, Principles of Nuclear Magnetic Resonance in One and Two Dimensions, Oxford University Press, Oxford, 1987. We may also indicate: J. Keeler, Understanding NMR Spectroscopy, second ed., Wiley, Chichester, 2010; P.J. Hore, J.A. Jones, S. Wimperis, NMR: The Toolkit, second ed., Oxford University Press, Oxford, 2015.

Previous books on NMR of paramagnetic molecules are: G.N. La Mar, W.DeW. Horrocks, Jr., R.H. Holm (Eds.), NMR of Paramagnetic Molecules, Academic Press, New York, 1973. I. Bertini, C. Luchinat, NMR of Paramagnetic Molecules in Biological Systems, Benjamin-Cummings, Menlo Park, 1986; J. Kowalewski, L. Maler, Nuclear Spin Relaxation in Liquids: Theory, Experiments, and Applications, Taylor & Francis, New York, 2006.

REFERENCES

[1] Koenig SH, Prodell AG, Kusch P. Phys. Rev. 1952;88:191.
[2] Fermi E. Z. Phys. 1930;60:320.

CHAPTER

THE HYPERFINE SHIFT

2

This chapter recalls the principles of the hyperfine coupling between electrons and nuclei in terms of energy and deals with its consequences on chemical shift. The equations for contact and pseudocontact shifts are derived and illustrated in a pictorial way. Their physical/chemical backgrounds are discussed as well as their limits of validity. The mechanisms of spin delocalization are illustrated.

2.1 NUCLEAR HYPERFINE SHIFT AND RELAXATION

How does the nucleus sense the electron? As mentioned in Section 1.7.3, nuclear and electron spins continuously change their levels in a magnetic field but the population distribution remains constant at a given temperature in the absence of perturbations. The change of level occurs through mechanisms which are different or have different degrees of efficiency for the two sets of spins, with the result that electrons relax orders of magnitude faster than nuclei. As a consequence, the nucleus in each of its M_I energy levels sees one electron rapidly changing its orientation among the M_S levels. The nucleus therefore senses an average magnetic moment of the electron, which would be zero if the populations of the M_S levels were equal. As this is not so in a magnetic field, the nucleus senses the average magnetic moment $\langle\mu\rangle$, which is just proportional to $\langle S_z\rangle$ (Section 1.6). It should be recalled that $\langle S_z\rangle$ is negative [Eq. (1.31)] due to the slight excess of population in the $M_S = -½$ state.

The average electron induced magnetic moment affects the energy of the nuclear $|I, M_I\rangle$ levels, i.e., affects the energy of the nucleus, because it represents a contribution to the overall magnetic field sensed by the nucleus. Such electron–nucleus interaction is represented by Eq. (1.51) and follows the treatment of the interaction between magnetic moments discussed in the previous chapter. In particular, there will be a dipolar interaction energy which averages zero in solution [Eq. (2.27)] or averages a value different from zero when the electron induced magnetic moment changes intensity with orientation as described in Fig. 1.14B and C [Eq. (2.46)]. A fraction of electron can sit on the resonating nucleus and so provide a direct shielding constant (Section 2.1.1). This contribution is called contact contribution.

If the interaction energy fluctuates, then nuclear relaxation enhancements occur (Chapter 4). Such relaxation enhancements are proportional to the average squared interaction energy

$$R_{1,2} \propto \langle E^2 \rangle.$$

Upon rotation the average electron induced magnetic moment causes nuclear relaxation. This is called Curie relaxation (Section 4.6). However, as it will be seen in Chapter 4, the full electron magnetic moment causes nuclear relaxation through dipolar and contact mechanisms by jumping over its various Zeeman levels. The electron jumps over the Zeeman levels with rates equal to the electronic R_1 value, and such jumps give also rise to nuclear relaxation.

NMR of Paramagnetic Molecules. http://dx.doi.org/10.1016/B978-0-444-63436-8.00002-8

The present chapter deals with the hyperfine shifts, which are only due to the average electron induced magnetic moment and therefore are related to $\langle S_z \rangle$. Chapter 3 will deal with self-orientation effects in high magnetic field and Chapter 4 will deal with nuclear hyperfine relaxation which, as anticipated earlier, depends on both average electron induced magnetic moment (Curie relaxation) and on the full electron magnetic moment (dipolar and contact relaxation).

2.1.1 THE SPIN DENSITY

The unpaired electron has the complication that it is not localized on a single point but, in general, is delocalized on the entire molecule. So, in every point of space where the molecular orbital (MO) containing the unpaired electron has a nonzero value, the average electron magnetic moment sensed by the nucleus is different from zero and is proportional to $\langle S_z \rangle$ times the fraction of unpaired electron present at that point (Section 2.2.2.2). Such a fraction is called spin density ρ, which for a single electron is given by the square of its wavefunction at that point.

The situation is further complicated by the fact that the presence of an unpaired electron in a MO polarizes the paired core electrons. This is called spin polarization. The MO containing the paired electrons is modified by the presence of the unpaired electron in another MO, in such a way that the electron with the spin aligned with the unpaired electron will have a slight preference to occupy the region of space of its MO which is closer to the unpaired electron itself (this is a manifestation of Hund's rule). Conversely, the other electron with spin antiparallel to the unpaired electron will have a slight preference to occupy regions of its MO far from the unpaired electron. This is illustrated in Fig. 2.1 for the case of two atomic orbitals.

At this point we recall that the unpaired electron may have some probability to sit just on the nucleus. Type s orbitals have maximal electron density on the nucleus, as their wavefunction is of the type $\exp(-r)$, where r is the electron–nucleus distance. Therefore, if the unpaired electron occupies an s orbital, or an

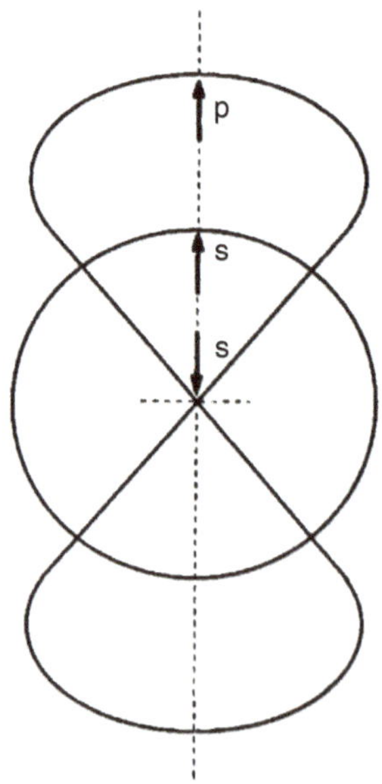

FIGURE 2.1 The unpaired electron in a 2p orbital affects the distribution of the two paired electrons in a 1s orbital.

MO containing an s orbital, there will be a finite probability that the electron resides on that nucleus. The amount of unpaired electrons residing on the nucleus is the spin density ρ at the nucleus. The spin density at the nucleus or in any other point is, in general, not just the positive quantity arising from the direct delocalization explained above, but also contains contributions due to spin polarization of paired electrons. The latter mechanism accounts for the presence of spin density on nuclei when the unpaired electron occupies a d orbital, which, like the p orbital, has a node at the nucleus. The unpaired electron on a p (or d) orbital can spin-polarize the electron pair occupying any s orbital. As shown in Fig. 2.1, the electron of a 1s pair with the same spin as the unpaired electron in a p or d orbital will preferentially be in the outer part of the 1s orbital, whereas the electron with reverse spin will preferentially be in the inner part of the 1s orbital. The two electrons of the pair will now have different probabilities to be at the nucleus. The difference provides the spin density at the nucleus. This contribution has opposite sign with respect to the contribution from the spin density in the orbital containing the unpaired electron. Since the unpaired electron has excess spin value $M_S = -½$ (*positive spin density*), the spin density at the nucleus arises from slight excess spin value $M_S = ½$, and is thus negative. Note, therefore, that although the unpaired electron always preferentially aligns with its spin along the external magnetic field, spin polarization may induce on the nucleus negative spin density as if the electron were polarized opposite to the magnetic field. Spin polarization is a local phenomenon and the g value of the spin density in an s orbital is rigorously 2.0023.

2.2 THE MAGNETIC NUCLEUS–UNPAIRED ELECTRON INTERACTION: THE HYPERFINE SHIFT

The average induced electron magnetic moment gives rise to a contribution to the chemical shift which is called hyperfine shift. Experimentally it is defined as the difference between the chemical shift in a paramagnetic system and that in an analogous diamagnetic system. It is customary to separately consider the effect of the spin density on the resonating nucleus from that outside it. The former part of the interaction is called contact or Fermi contact interaction. The latter is called dipolar interaction because it is a through space interaction which can be described by the dipolar interaction of two magnetic dipoles. The evaluation of the dipolar interaction requires, in principle, evaluating the integral all over the space of the interaction energy between the resonating nucleus and the spin density on each point of the space. In practice the problem can often be simplified.

2.2.1 THE FERMI CONTACT COUPLING

The contact shift is given by an additional magnetic field generated at the nucleus by the electron magnetic moment located at the nucleus itself. The latter magnetic moment originates from the spin density at the nucleus, weighted by $\langle S_z \rangle$. For an isolated electron spin, $\langle S_z \rangle$ is always negative (as long as the spin–orbit coupling is a small perturbation) (Chapter 1). In contrast, spin density can be either positive or negative because of spin polarization. For each s orbital the spin density at the nucleus, ρ, is given by

$$\rho = \psi_{-1/2}(0)^2 - \psi_{1/2}(0)^2 \tag{2.1}$$

where $\psi(0)$ is the value of the MO wavefunction at zero distance from the nucleus for –½ and ½ spins. Therefore, if the excess spin value is –½, the contribution to the magnetic field will be positive

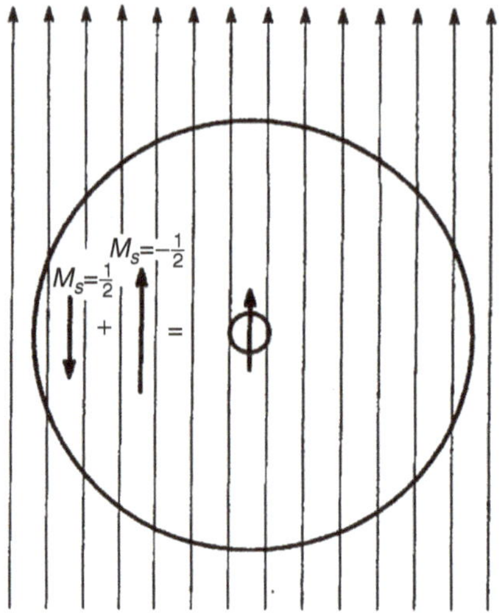

FIGURE 2.2 Excess –½ electron spin density contributes a magnetic field at the nucleus that adds to the external magnetic field.

(Fig. 2.2) as for a free electron and the chemical shift in frequency will be positive. The reverse holds for +½ spin. The spin density (when normalized to one electron) is proportional to what is called the contact coupling constant A, which expresses how much the nucleus and the electron sense each other:

$$A = \frac{2\mu_0}{3}\hbar\gamma_I g_e \mu_B \rho. \tag{2.2}$$

Since the s orbitals do not have orbital magnetic moment, $g = g_e = 2.0023$. A is an energy, which in SI units is expressed in joule. Sometimes it is convenient to have it expressed in frequency units, i.e., in hertz or in radians per second, and then we have to divide A by h or by $\hbar$ ($\hbar = h/2\pi$).[1] The general relationship between energy in joule and frequency—or angular frequency—is

$$E = h\nu = h\frac{\omega}{2\pi} = \hbar\omega. \tag{2.3}$$

In order to express the energy in wavenumbers, we should remember that $E = hc/\lambda$, and then E is proportional to the inverse of a length.

In quantum mechanical terms, the Hamiltonian for the contact interaction is

$$H = A\boldsymbol{I}\cdot\boldsymbol{S} \tag{2.4}$$

and the energy is obtained by evaluating the matrix elements $\langle \psi_{S,M_S,I,M_I} | H | \psi_{S,M_S,I,M_I} \rangle$. In the high field approximation ($g_e\mu_B B_0 \gg A$), which usually holds for high field NMR experiments, the quantization axis of both the nuclear and the electron spins is along the external magnetic field (for the latter, only if spin-orbit coupling is negligible, see later). Therefore, the energy of the contact interaction from Hamiltonian Eq. (2.4) can be written as

[1]In magnetic resonance there often is a factor 2π which complicates life. The frequency which we refer to is the angular frequency or the Larmor precession frequency. Such frequency ω is such that $\omega = 2\pi\nu$ [Eq. (1.46)], where ν is the frequency in hertz or s^{-1}, and ω is expressed in radians per second.

$$E^{\text{con}} = AM_I \langle S_z \rangle = -AM_I \frac{g_e \mu_B S(S+1)B_0}{3kT} \quad (2.5)$$

where we have used the definition of $\langle S_z \rangle$ given in Eq. (1.31). The contact shift is obtained by dividing the contact coupling energy by the nuclear Zeeman energy:

$$\delta^{\text{con}} = -\frac{E^{\text{con}}}{\hbar \gamma_I M_I B_0} = -\frac{A}{\hbar \gamma_I B_0} \langle S_z \rangle \quad (2.6)$$

and thus [1]

$$\delta^{\text{con}} = \frac{A}{\hbar} \frac{g_e \mu_B S(S+1)}{3\gamma_I kT}. \quad (2.7)$$

This equation holds under the same conditions of validity of Curie equation [Eq. (1.34)], i.e., for a single electron in an orbital well separated from the excited ones. It holds also in the case of multiple unpaired electrons in different MO, in the absence of sizable spin-orbit coupling effects (in these cases A is an average of the contact coupling constant of each electron, see later).

No nuclear parameter is contained in the expression of the contact shift. The γ_I appearing in both Eqs. (2.6) and (2.7) cancel the γ_I contained in A [Eq. (2.2)]. Eq. (2.6) tells us that the fractional variation in frequency is due to the presence of unpaired spin density, which determines A. The nuclear I value of the resonating nucleus does not appear in Eq. (2.7) because the frequency is referred to any $\Delta M_I = 1$ transition.

We should note that if $g = g_e$, the contact shift is isotropic (independent of orientation). If g is different from g_e and anisotropic (Section 1.4), then the contact shift is also anisotropic. The anisotropy of the shift is due to the fact that: (1) the energy spreading of the Zeeman levels is different for each orientation (Fig. 1.14), and therefore the value of $\langle S_z \rangle$ becomes orientation dependent; and (2) the values of $\langle S, M_S | S_z | S, M_S \rangle$ of Eq. (1.31) are orientation dependent as the result of efficient spin–orbit coupling. On the contrary, the contact coupling constant A is a constant whose value does not depend on the molecular orientation.

Let us suppose now that we have a solid with all molecules aligned with one another (Fig. 2.3) and that we perform the NMR experiment on a single crystal. If g of the S manifold equals g_e, the contact

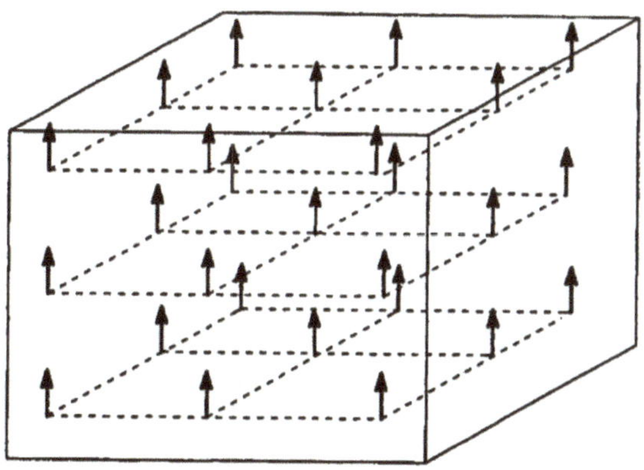

FIGURE 2.3 Ideal crystal with all molecules iso-oriented.

shift contribution will be independent of the crystal orientation in the magnetic field. If, however, g has a different value in any k direction, then spin–orbit coupling is not negligible and the contact shift will be orientation dependent. Specific calculations are needed. If, however, we arbitrarily neglect the anisotropy of $\langle \psi | S_z | \psi \rangle$, the following equation can be written

$$\delta_{kk}^{\text{con}} = \frac{A}{\hbar} \frac{g_{kk} \mu_B S(S+1)}{3\gamma_I kT}. \tag{2.8}$$

When the solid is dissolved in a liquid, the rotational rate of the molecules is fast with respect to the difference in hyperfine shift due to the electron Zeeman anisotropy, and the contact shift will be an average:

$$\delta_{\text{sol}}^{\text{con}} = \frac{A}{\hbar} \frac{\bar{g} \mu_B S(S+1)}{3\gamma_I kT}. \tag{2.9}$$

Typical values for hyperfine shift anisotropies are much smaller than the molecular reorientation rate τ_r^{-1} (Section 4.2) for all practical cases.

A correct equation for a singly populated S manifold is

$$\delta_{\text{sol}}^{\text{con}} = \frac{1}{\mu_0} \frac{A}{\hbar} \frac{1}{3\gamma_I \mu_B} \left(\frac{\chi_{xx}}{g_{xx}} + \frac{\chi_{yy}}{g_{yy}} + \frac{\chi_{zz}}{g_{zz}} \right) \tag{2.10}$$

where χ_{ii} and g_{ii} are the principal components of the paramagnetic susceptibility [$\boldsymbol{\chi} = \chi_M / N_A$, Eq. (1.36)] and $\mathbf{g}$ tensors. The reader can find a justification of it by comparing Eq. (2.10) with Eq. (2.9) through Eq. (1.40). Of course Eq. (2.10) is hardly used, as the principal χ values (and g values) are generally unknown.

We note here that Eq. (2.9) holds for a single electron in an orbital which is well separated by any other excited level. In the case of multiple unpaired electrons in different molecular orbitals, Eq. (2.9) still may hold in the absence of strong spin–orbit coupling effects, but the interpretation of the hyperfine constant becomes complicated: the overall hyperfine coupling is the sum of the hyperfine coupling for each molecular orbital. Indeed, each metal orbital which contains an unpaired electron is involved in a molecular orbital and provides a contribution to the total ρ for the various nuclei. The experimental data, however, provides through Eq. (2.9) the sum of the A values and therefore the sum of ρ. In order to make the spin density or contact constants comparable for different systems independent of the value of S, i.e., independent of the number of electrons, the values of ρ and A are normalized to one electron, i.e., they are divided by the number of electrons which is just $2S$. Eq. (2.2) becomes

$$A = \frac{\mu_0}{3S} \hbar \gamma_I g_e \mu_B \sum_i \rho_i. \tag{2.11}$$

A is now the average contact coupling per electron.

When there are multiple electrons, ZFS occurs, and the Boltzmann population of the levels may be taken into account. Then Eq. (2.10) should be used. As examples, we report the expressions for the Fermi contact shift for systems with axially symmetric distorsions and orbitally nondegenerate ground states characterized by $S = 1$, as for nickel(II) octahedral complexes, $S = 3/2$, as for chromium(III) octahedral complexes and cobalt(II) tetrahedral complexes, $S = 5/2$, as for weak field tetragonal complexes of manganese(II) and iron(III), all cases for which the orbital angular momentum is relatively small (Chapter 7). The zero field split energy levels, for $\mathbf{B}_0$ along z, can be represented as a function of the M_S values as represented in Fig. 1.15, where D is the ZFS parameter. The principal susceptibility values can then be calculated through the Van Vleck equation, [Eq. (1.38)], and, as long as $2D$ is larger than the perturbation due to an external magnetic field, they take the form:

$$S=1:\ \chi_{zz}=\mu_0\frac{2g_{\parallel}^2\mu_B^2}{kT}\left(\frac{e^{-X}}{1+2e^{-X}}\right);\quad X=\frac{D}{kT} \tag{2.12}$$

$$\chi_{xx}=\chi_{yy}=\mu_0\frac{2g_{\perp}^2\mu_B^2}{kT}\left(\frac{1-e^{-X}}{1+2e^{-X}}\right). \tag{2.13}$$

$$S=3/2:\ \chi_{zz}=\mu_0\frac{g_{\parallel}^2\mu_B^2}{4kT}\left(\frac{1+9e^{-2X}}{1+e^{-2X}}\right);\quad X=\frac{D}{kT} \tag{2.14}$$

$$\chi_{xx}=\chi_{yy}=\mu_0\frac{g_{\perp}^2\mu_B^2}{kT}\left(\frac{1+\frac{3}{4X}(1-e^{-2X})}{1+e^{-2X}}\right). \tag{2.15}$$

$$S=5/2:\ \chi_{zz}=\mu_0\frac{g_{\parallel}^2\mu_B^2}{4kT}\left(\frac{1+9e^{-2X}+25e^{-6X}}{1+e^{-2X}+e^{-6X}}\right);\quad X=\frac{D}{kT} \tag{2.16}$$

$$\chi_{xx}=\chi_{yy}=\mu_0\frac{g_{\perp}^2\mu_B^2}{4kT}\left(\frac{9+\frac{8}{X}-\frac{11}{2X}e^{-2X}-\frac{5}{2X}e^{-6X}}{1+e^{-2X}+e^{-6X}}\right). \tag{2.17}$$

By substituting into Eq. (2.10), and taking $e^{-X} = 1 - X$ ($D \ll kT$), the following expressions are obtained [2]:

$$S=1:\ \delta^{\text{con}}=\left(\frac{A}{\hbar}\right)\frac{2\bar{g}\mu_B}{3\gamma_I kT}\left[1-\frac{(g_{\parallel}-g_{\perp})D}{9\bar{g}kT}\right]. \tag{2.18}$$

$$S=3/2:\ \delta^{\text{con}}=\left(\frac{A}{\hbar}\right)\frac{5\bar{g}\mu_B}{4\gamma_I kT}\left[1-\frac{4(g_{\parallel}-g_{\perp})D}{15\bar{g}kT}\right]. \tag{2.19}$$

$$S=5/2:\ \delta^{\text{con}}=\left(\frac{A}{\hbar}\right)\frac{35\bar{g}\mu_B}{12\gamma_I kT}\left[1-\frac{32(g_{\parallel}-g_{\perp})D}{45\bar{g}kT}\right]. \tag{2.20}$$

These expressions contain one term that depends on T^{-1} and another that depends on T^{-2}, the latter term being a consequence of the Boltzmann population of the zero field split levels. It should be noted that the g values are not the experimental values obtained by EPR but the molecular g values, which can be obtained through a proper analysis of the experimental EPR spectra. For $S = {}^5/_2$ and excited states far elevated in energy, the molecular g values are close to the free electron, g_e, value and D is relatively small; therefore, the term in T^{-2} can be neglected. Concluding, the use of Eq. (2.9) instead of Eq. (2.10) may not introduce serious errors provided that empty orbitals are far in energy for the cases under considerations. An example of the temperature dependence of methyl proton NMR signals in a paramagnetic cobalt(II) cyclen-derivative [3] is reported in Fig. 2.4, The temperature dependence appears linear within the range of the measurements, but a linear fit would provide an intercept, i.e., a shift at infinite temperature, very far from zero ppm. This indicates that contributions with a T^{-2} dependence should also be present (Fig. 2.4).

Another case is that of one unpaired electron in a metal orbital in the presence of 1–2 empty metal orbitals close in energy. Spin–orbit coupling mixes the various orbitals so that the unpaired electron resides in an orbital which is a linear combination of all orbitals available. A typical case is that of low spin octahedral iron(III), where there is a single unpaired electron but d_{xy}, d_{xz}, and d_{yz} are close in energy. Here the g values are largely anisotropic (from $g_{max} \approx 3$ to $g_{min} \approx 1$) and the excited states may be partially populated, so that Eq. (2.9) may be an inadequate mean of describing the real case, especially as far as the temperature dependence is concerned. In fact, each excited level has a different hyperfine coupling constant and contributes according to its population, which can be estimated from the Boltzmann distribution [4].

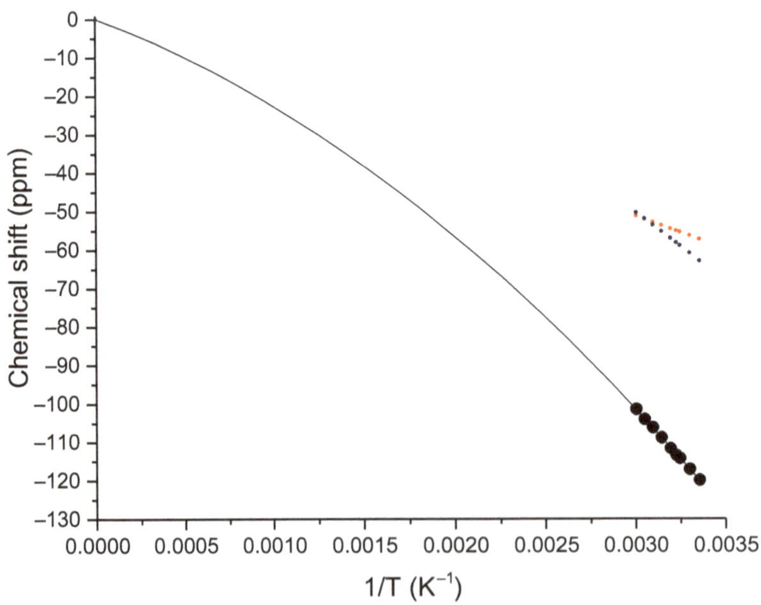

FIGURE 2.4 Temperature dependence of the methyl proton NMR signals in a paramagnetic cobalt(II) cyclen-derivative [3] at 500 MHz. Experimental data are shown as black symbols. The black line is the best fit curve against the function $\delta = BT^{-1} + CT^{-2}$. Red and blue symbols indicate the contributions from the T^{-1} and the T^{-2} terms, respectively.

It is now obvious that the case of $S > ½$ and low lying excited states can hardly be analyzed by means of Eq. (2.9) but requires a specific treatment which takes into full account the effects of spin–orbit and Zeeman interactions. It should be said that these treatments are rare in the literature, as they require a lot of skill and wisdom in choosing the right parameters.

The general treatment must indeed consider the presence of several electrons in the presence of orbitals close in energy. Spin-orbit coupling mixes the various orbitals so that electrons reside in orbitals which are linear combination of all orbitals available. All thermally populated electronic states must be considered. It can be shown that the contact shift in solution is obtained by averaging over the values along the three principal directions of the $\mathbf{g}$ ($\boldsymbol{\chi}$) tensor

$$\delta^{\text{con}} = \frac{\delta_{xx}^{\text{con}} + \delta_{yy}^{\text{con}} + \delta_{zz}^{\text{con}}}{3} \tag{2.21}$$

where

$$\begin{aligned}\delta_{kk}^{\text{con}} = \frac{-\mu_B}{\hbar\gamma_I kT\sum_i \exp(-E_i^0/kT)}\Bigg(&\sum_i \exp(-E_i^0/kT) < \phi_i|L_{kk} + g_e S_{kk}|\phi_i >< \phi_i|A_{ii}S_{kk}|\phi_i > + \\ &-kT\sum_{i\neq j}\frac{\exp(-E_i^0/kT) - \exp(-E_j^0/kT)}{E_i^0 - E_j^0} < \phi_i|L_{kk} + g_e S_{kk}|\phi_j >< \phi_j|A_{ij}S_{kk}|\phi_i >\Bigg)\end{aligned} \tag{2.22}$$

and $kk = xx$, yy, and zz [2].

The constants A_{ij} may or may not depend on the orbitals ϕ_i and ϕ_j [4]. They represent the proportionality constants which relate the orbital spin density contributions to the spin density at the nucleus for the ground and excited states.

Eqs. (2.21) and (2.22) indicate that the contact shift measured in solution results from the rotational average of contributions which may differ for different directions of the magnetic field in the molecular frame. Neglecting anisotropy of the contact shift can induce errors especially in partially oriented systems (Chapter 3), where anisotropy is not averaged out. For low-spin ruthenium(III) and iron(III) complexes, for instance, where the $\mathbf{g}$ tensor is strongly anisotropic and there are low-lying excited states, Eqs. (2.21) and (2.22) must be applied [5,6]. In the case of heme proteins, like cytochrome *c*, cytochrome b_5, and cyanometmyoglobin, Eqs. (2.21) and (2.22) can be used to calculate the contact shifts from the knowledge of the experimental g and $\Delta\chi$ values, which provided the energies of the three lowest orbital levels [4]. $[Fe(CN)_6]^{3-}$ and $[Mn(CN)_6]^{3-}$ can be studied in detail by using ^{13}C and ^{15}N solid state NMR spectroscopy (Chapter 5). The analysis of the NMR signal patterns revealed considerable dipolar shielding anisotropies, which can be ascribed to nonspherical distributions of the spin densities about the C and N atoms [7].

2.2.2 The dipolar coupling

This contribution to the shift is quite difficult to evaluate, because in general the spin density distribution all over the space is not known. The approach to this problem should be stepwise. We first consider that the unpaired electron is localized on the metal nucleus in a paramagnetic metal complex. We refer to this as to the metal-centered point–dipole approximation. Surely this contribution will always be present and often dominant. Then we will discuss the consequences of relaxing this condition. Even in the metal-centered approximation several cases should be discussed.

2.2.2.1 Metal-centered point–dipole approximation

The dipolar energy related to the dipole–dipole interaction [Eq. (1.1)] between a nucleus with magnetic moment $\boldsymbol{\mu}_1 = \hbar\gamma_I \boldsymbol{I}$ and the average induced electron magnetic moment $\boldsymbol{\mu}_2 = <\boldsymbol{\mu}>$ (Section 1.1) is

$$E = -\frac{\mu_0}{4\pi}\left[\frac{3(\hbar\gamma_I \boldsymbol{I}\cdot\boldsymbol{r})(<\mu>\cdot\boldsymbol{r})}{r^5} - \frac{\hbar\gamma_I \boldsymbol{I}\cdot<\mu>}{r^3}\right]. \tag{2.23}$$

Using Eq. (1.44), describing the relationship between $<\boldsymbol{\mu}>$ and the paramagnetic susceptibility tensor per molecule, $\boldsymbol{\chi} = \boldsymbol{\chi}_M/N_A$, and recalling that the nuclear spin is quantized along the direction of the external magnetic field, κ, Eq. (2.23) becomes:

$$E^{\text{dip}} = -\frac{\hbar\gamma_I B_0}{4\pi r^5} M_I \boldsymbol{\kappa}\cdot[3\boldsymbol{r}(\boldsymbol{r}\cdot\boldsymbol{\chi}) - r^2\boldsymbol{\chi}]\cdot\boldsymbol{\kappa} \tag{2.24}$$

If the external magnetic field is aligned with the z-axis and both $\boldsymbol{\mu}_1$ and $<\boldsymbol{\mu}>$ are oriented along this axis, the dipolar interaction energy between the nuclear magnetic moment and the average induced electron magnetic moment, according to Eq. (1.4), is

$$E^{\text{dip}} = -\frac{\mu_0}{4\pi}\hbar\gamma_I M_I \langle\mu\rangle \frac{1}{r^3}(3\cos^2\gamma - 1) \tag{2.25}$$

where γ is the angle between the metal–nucleus vector and the external magnetic field, which is coincident with the molecular χ_{zz} direction (Fig. 2.5). From Eq. (1.27), (1.33), and (1.34), we can write E^{dip} in terms of the z component of the paramagnetic susceptibility tensor $\boldsymbol{\chi}$

$$E^{\text{dip}} = -\frac{B_0}{4\pi r^3}\hbar\gamma_I M_I \chi_{zz}(3\cos^2\gamma - 1). \tag{2.26}$$

The dipolar interaction energy averages zero for isotropic $\boldsymbol{\chi}$. This can be easily verified by averaging Eq. (2.26) over all orientations, as

$$\int_{-1}^{1}(3\cos^2 y - 1)\,\mathrm{d}\cos y = 0. \tag{2.27}$$

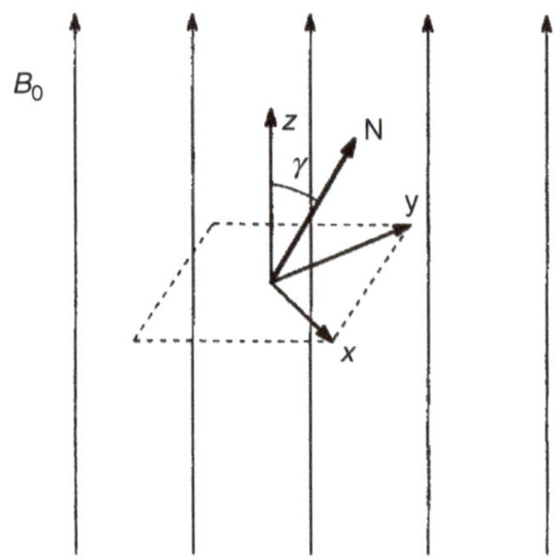

FIGURE 2.5 A molecule possessing magnetic anisotropy, with the z-axis oriented along the external magnetic field and a nucleus having a metal–nucleus vector at an angle γ with the external magnetic field.

From Eq. (2.24), a dipolar shielding tensor $\boldsymbol{\sigma}$ can be defined, so that the dipolar energy is

$$E^{\mathrm{dip}} = \hbar\gamma_I B_0 M_I \kappa \cdot \sigma \cdot \kappa = -\frac{\hbar\gamma_I B_0 M_I}{4\pi r^5}\begin{pmatrix}\kappa_x\\ \kappa_y\\ \kappa_z\end{pmatrix}^T \begin{pmatrix}(3x^2-r^2) & 3xy & 3xz\\ 3xy & (3y^2-r^2) & 3yz\\ 3xz & 3yz & (3z^2-r^2)\end{pmatrix}\begin{pmatrix}\chi_{xx} & \chi_{xy} & \chi_{xz}\\ \chi_{xy} & \chi_{yy} & \chi_{yz}\\ \chi_{xz} & \chi_{yz} & \chi_{zz}\end{pmatrix}\begin{pmatrix}\kappa_x\\ \kappa_y\\ \kappa_z\end{pmatrix} \tag{2.28}$$

From the dipolar interaction energy, the dipolar shift can be obtained by evaluating from Eq. (2.28) ΔE^{dip} between two states differing by $\Delta M_I = \pm 1$ and dividing it by the nuclear Zeeman energy $\hbar\gamma_I B_0$ (Appendix IV):

$$\delta^{\mathrm{dip}} = -\frac{\Delta E^{\mathrm{dip}}}{\hbar\gamma_I B_0} = \frac{1}{4\pi r^5}\begin{pmatrix}\kappa_x\\ \kappa_y\\ \kappa_z\end{pmatrix}^T \begin{pmatrix}(3x^2-r^2) & 3xy & 3xz\\ 3xy & (3y^2-r^2) & 3yz\\ 3xz & 3yz & (3z^2-r^2)\end{pmatrix}\begin{pmatrix}\chi_{xx} & \chi_{xy} & \chi_{xz}\\ \chi_{xy} & \chi_{yy} & \chi_{yz}\\ \chi_{xz} & \chi_{yz} & \chi_{zz}\end{pmatrix}\begin{pmatrix}\kappa_x\\ \kappa_y\\ \kappa_z\end{pmatrix} \tag{2.29}$$

With the magnetic field along a generic κ direction described by the spherical angles α and β in the frame where the χ tensor is diagonal (α being the angle between the external magnetic field and the z-axis of the tensor), (Fig. 2.6), for an axial χ tensor, the dipolar shift thus results

$$\begin{aligned}\delta^{\mathrm{dip}} = \frac{1}{4\pi r^3}\Big[&\chi_{\parallel}\cos^2\alpha(3\cos^2\theta-1)+\chi_{\perp}\sin^2\alpha\left(3\sin^2\theta\cos^2(\varphi-\beta)-1\right)\\ &+\frac{3}{4}(\chi_{\parallel}+\chi_{\perp})\sin 2\alpha\sin 2\theta\cos(\varphi-\beta)\Big]\end{aligned} \tag{2.30}$$

where the angles θ and φ are spherical angles describing the position of the nucleus in the χ frame, i.e., θ is the angle between the metal–nucleus vector $\boldsymbol{r}$ and the z-axis of the χ tensor, and φ is an angle related to the projection of $\boldsymbol{r}$ on the xy plane.

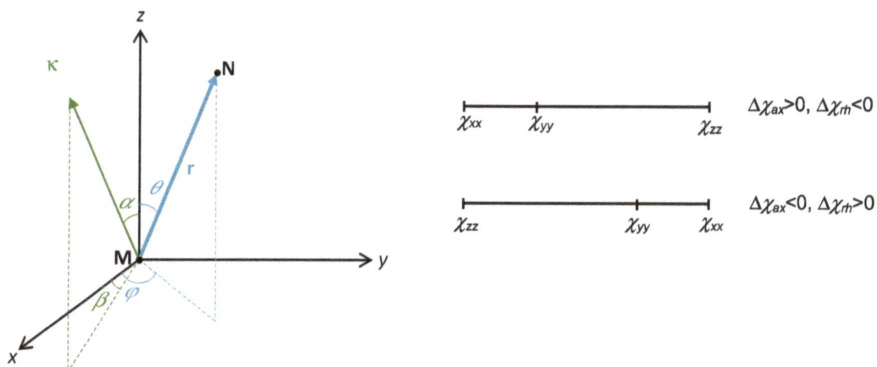

FIGURE 2.6 In a molecule possessing magnetic anisotropy a reference frame is defined by the main direction of the anisotropy tensor. α is the angle of the external field direction κ with the z-axis; β is the angle from the x-axis of the projection of the external field direction on the xy plane; θ is the angle between the metal–nucleus vector $\boldsymbol{r}$ and the z-axis; φ is the angle from the x-axis of the projection of $\mathbf{r}$ on the xy plane (M, metal; N, nucleus).

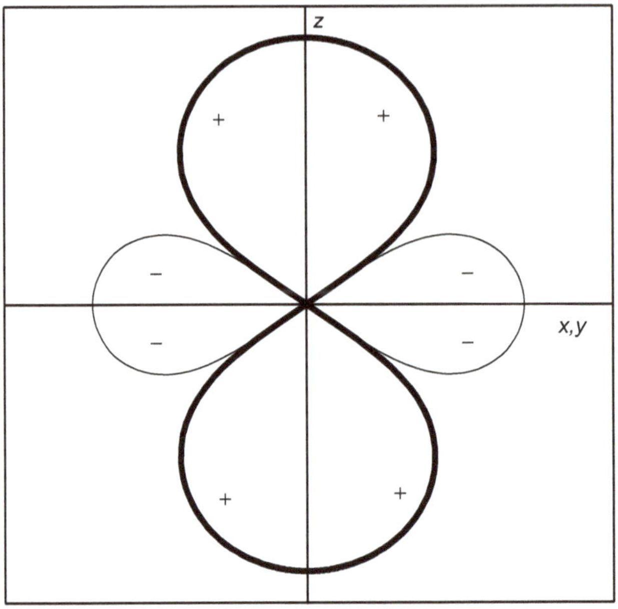

FIGURE 2.7 Angular dependence of the pseudocontact shift for an axial system, shown as a surface of constant absolute value of δ^{pcs}. In the example δ^{pcs} is positive along the *z*-axis and negative in the *xy* plane. The three-dimensional shape of the surface is similar to the representation of a d_{z^2} orbital.

This is indeed what is expected in ^1H ENDOR spectroscopy in single crystals of isooriented molecules and in single crystal static solid state NMR (Section 5.2.3). By integration of Eq. (2.30) over all orientations of the magnetic field, mimicking the effect of the molecular rotation in solution, the following equation is obtained (Appendix IV):

$$\delta^{pcs} = \frac{1}{12\pi r^3}(\chi_{\parallel} - \chi_{\perp})(3\cos^2\theta - 1). \tag{2.31}$$

This contribution to the shift is isotropic because it is already averaged out over all the orientations. Then it is similar to the contact shift[2], and is therefore called *pseudocontact shift,* δ^{pcs}. This shift is also referred to as "isotropic dipolar shift" or, in the older literature, also as "dipolar shift." Although of course it is legitimate to call it dipolar shift, the "isotropic" qualification should be used to distinguish it from the dipolar shift in the solid state, that is significantly larger and orientation dependent (Chapter 5). To avoid confusion, the term pseudocontact shift is preferable. Note that, when χ is isotropic, the pseudocontact shift is zero [Eq. (2.31)], but a dipolar shift is observed in the solid state [Eq. (2.30) and Eq. (IV.1)].

[2]Actually, to its, often dominant, isotropic component.

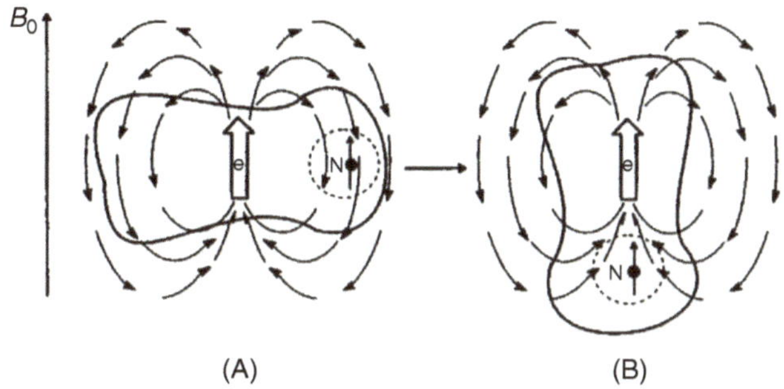

FIGURE 2.8 A nucleus N immersed in a dipolar magnetic field of an electron spin, the latter aligned along the external magnetic field, experiences an extra field that subtracts (A) or adds (B) to the external field.

The angular dependence of the pseudocontact shift is shown in Fig. 2.7. For $\theta = 54.74$ degrees, the pseudocontact shift is zero. If $\chi_{\parallel} > \chi_{\perp}$, then the pseudocontact shift is positive for θ values less than 54.74 degrees.

There is also an intuitive way of seeing the effect of an isotropic electron magnetic moment on a nucleus. Fig. 2.8 illustrates that, indeed, the dipolar magnetic field experienced by the nucleus for different orientations of the metal–nucleus vector with respect to the magnetic field changes sign, being positive when the vector is along the field and negative when the vector is perpendicular to the field. Therefore, one can accept that the average over all the orientations is zero. However, if the induced magnetic field changes intensity with the molecular orientation because χ is anisotropic, complete cancellation upon rotation does not occur any longer. It should be remembered that the anisotropy of χ (Sections 1.4 and 1.6) is due to orbital contributions to the magnetic moments.

In order to determine the pseudocontact shifts for the case of a rhombic χ tensor, an orientational average of Eq. (2.29) should be performed. This can be done by taking the average of the values calculated along three principal directions e_x, e_y, e_z, and therefore, the pseudocontact shift is [8,9]

$$\delta^{\text{pcs}} = -\frac{\Delta E^{\text{dip}}}{\hbar\gamma_I B_0} = -\frac{1}{3}(e_x \cdot \sigma \cdot e_x + e_y \cdot \sigma \cdot e_y + e_z \cdot \sigma \cdot e_z). \tag{2.32}$$

This expression, valid in any reference frame (with the metal ion in the origin), can be written as

$$\delta^{\text{pcs}} = -\frac{1}{3}Tr(\sigma) = \frac{1}{12\pi r^5} Tr\left[\begin{pmatrix} (3x^2 - r^2) & 3xy & 3xz \\ 3xy & (3y^2 - r^2) & 3yz \\ 3xz & 3yz & (3z^2 - r^2) \end{pmatrix}\begin{pmatrix} \chi_{xx} & \chi_{xy} & \chi_{xz} \\ \chi_{xy} & \chi_{yy} & \chi_{yz} \\ \chi_{xz} & \chi_{yz} & \chi_{zz} \end{pmatrix}\right]. \tag{2.33}$$

From Eq. (2.33), we obtain

$$\delta^{\text{pcs}} = \frac{1}{12\pi r^5}\left[(3x^2 - r^2)\chi_{xx} + 3xy\chi_{xy} + 3xz\chi_{xz} + 3xy\chi_{xy} + (3y^2 - r^2)\chi_{yy} + 3yz\chi_{yz} + 3xz\chi_{xz} + 3yz\chi_{yz} + (3z^2 - r^2)\chi_{zz}\right] \tag{2.34}$$

and thus

$$\delta^{pcs} = \frac{1}{4\pi r^5}\left[(3z^2 - r^2)\chi_{zz}/3 + (3x^2 - r^2)\chi_{xx}/3 + (3y^2 - r^2)\chi_{yy}/3 + 2xy\chi_{xy} + 2xz\chi_{xz} + 2yz\chi_{yz}\right] \quad (2.35)$$

and

$$\delta^{pcs} = \frac{1}{4\pi r^5}\left[z^2\chi_{zz} + x^2\chi_{xx} + y^2\chi_{yy} - r^2(\chi_{zz} + \chi_{xx} + \chi_{yy})/3 + 2xy\chi_{xy} + 2xz\chi_{xz} + 2yz\chi_{yz}\right] \quad (2.36)$$

which can also be easily rewritten using the direction cosines of the metal–nucleus vector $l = x/r$, $m = y/r$, $n = z/r$.

The equation for the pseudocontact shift can be written in a variety of ways, depending on the reference frame and on the coordinates used to describe the position of atoms (Cartesian, spherical or direction cosines). The most common forms, obtained from Eq. (2.35) (Appendix IV), valid for a generic reference frame with the metal ion in the origin, are reported in Eqs. (2.37)–(2.39). In Cartesian coordinates ($\bar{\chi} = (\chi_{xx} + \chi_{yy} + \chi_{zz})/3$):

$$\delta^{pcs} = \frac{1}{4\pi r^3}\left[(\chi_{zz} - \bar{\chi})\frac{2z^2 - x^2 - y^2}{2r^2} + (\chi_{xx} - \chi_{yy})\frac{x^2 - y^2}{2r^2} + \chi_{xy}\frac{2xy}{r^2} + \chi_{xz}\frac{2xz}{r^2} + \chi_{yz}\frac{2yz}{r^2}\right] \quad (2.37)$$

in spherical coordinates:

$$\delta^{pcs} = \frac{1}{4\pi r^3}\left[(\chi_{zz} - \bar{\chi})\frac{3\cos^2\theta - 1}{2} + (\chi_{xx} - \chi_{yy})\frac{\sin^2\theta\cos 2\varphi}{2} + \chi_{xy}\sin^2\theta\sin 2\varphi + \chi_{xz}\sin 2\theta\cos\varphi + \chi_{yz}\sin 2\theta\sin\varphi\right] \quad (2.38)$$

or using the director cosines of the metal–nucleus vector:

$$\delta^{pcs} = \frac{1}{4\pi r^3}\left[(\chi_{zz} - \bar{\chi})\frac{3n^2 - 1}{2} + (\chi_{xx} - \chi_{yy})\frac{l^2 - m^2}{2} + 2\chi_{xy}lm + 2\chi_{xz}ln + 2\chi_{yz}mn\right]. \quad (2.39)$$

2.2.2.1.1 Pseudocontact shift equation in the principal frame of χ

In case the reference frame is chosen to coincide with the main directions of the χ tensor, and $\tilde{\chi}_{xx}$, $\tilde{\chi}_{yy}$, and $\tilde{\chi}_{zz}$ are the principal components of the paramagnetic susceptibility tensor, the dipolar shielding tensor is [10,11]

$$\sigma = -\frac{1}{4\pi r^5}\begin{pmatrix} (3x^2 - r^2)\tilde{\chi}_{xx} & 3xy\tilde{\chi}_{yy} & 3xz\tilde{\chi}_{zz} \\ 3xy\tilde{\chi}_{xx} & (3y^2 - r^2)\tilde{\chi}_{yy} & 3yz\tilde{\chi}_{zz} \\ 3xz\tilde{\chi}_{xx} & 3yz\tilde{\chi}_{yy} & (3z^2 - r^2)\tilde{\chi}_{zz} \end{pmatrix} \quad (2.40)$$

where x, y, and z are the coordinates of the nucleus with respect to that principal frame, with the metal ion at the origin. As it is clear from Eq. (2.32), the rank zero component of the shielding tensor provides the orientationally-averaged pseudocontact shift, reported in Eq. (2.41):

$$\delta^{pcs} = \frac{1}{12\pi r^3}\frac{(3x^2 - r^2)\tilde{\chi}_{xx} + (3y^2 - r^2)\tilde{\chi}_{yy} + (3z^2 - r^2)\tilde{\chi}_{zz}}{r^2}. \quad (2.41)$$

Again, the isotropic part of $\boldsymbol{\chi}$ provides no contribution (as it is clear from the fact that $3x^2+3y^2+3z^2-3r^2=0$). Eq. (2.41) can be rewritten using the direction cosines of the metal–nucleus vector $l=x/r$, $m=y/r$, $n=z/r$,

$$\delta^{pcs}=\frac{1}{12\pi r^3}\left[\tilde{\chi}_{xx}\left(3l^2-1\right)+\tilde{\chi}_{yy}\left(3m^2-1\right)+\tilde{\chi}_{zz}\left(3n^2-1\right)\right] \tag{2.42}$$

or, since $\tilde{\chi}_{xx}+\tilde{\chi}_{yy}+\tilde{\chi}_{zz}=3\overline{\chi}=3\chi^{iso}$,

$$\delta^{pcs}=\frac{1}{4\pi}\frac{1}{r^3}\left[\tilde{\chi}_{xx}l^2+\tilde{\chi}_{yy}m^2+\tilde{\chi}_{zz}n^2-\overline{\chi}\right]. \tag{2.43}$$

These equations contain all three principal components of the $\boldsymbol{\chi}$ tensor. As $\boldsymbol{\chi}$ is generally unknown and only its anisotropy is relevant for pseudocontact shifts, functional forms that depend only on differences among the magnetic susceptibility components may be preferable. To this end, it may be convenient to define $\Delta\tilde{\chi}=\tilde{\chi}-\chi^{\text{iso}}$, so that Eq. (2.43) becomes

$$\delta^{pcs}=\frac{1}{4\pi}\frac{1}{r^3}\left[\Delta\tilde{\chi}_{xx}l^2+\Delta\tilde{\chi}_{yy}m^2+\Delta\tilde{\chi}_{zz}n^2\right]. \tag{2.44}$$

Eq. (2.41) can also be conveniently written in spherical coordinates, where the angles θ and φ define the position of the atom with respect to the principal axes of the paramagnetic susceptibility tensor centered on the metal ion (Fig. 2.6), θ being the angle between the metal–nucleus vector and the z-axis of the tensor, and φ being the angle between the projection of the metal–nucleus vector on the xy plane and the x-axis. The result is

$$\delta^{pcs}=\frac{1}{12\pi r^3}\left[(3\sin^2\theta\cos^2\varphi-1)\tilde{\chi}_{xx}+(3\sin^2\theta\sin^2\varphi-1)\tilde{\chi}_{yy}+(3\cos^2\theta-1)\tilde{\chi}_{zz}\right] \tag{2.45}$$

and thus, after some rearrangements (Appendix IV),

$$\delta^{pcs}=\frac{1}{12\pi}\frac{1}{r^3}\left[\Delta\chi_{ax}(3\cos^2\theta-1)+\tfrac{3}{2}\Delta\chi_{rh}\sin^2\theta\cos 2\varphi\right] \tag{2.46}$$

where $\Delta\chi_{ax}$ and $\Delta\chi_{rh}$ are defined as the axial and rhombic anisotropy parameters of the paramagnetic susceptibility tensor $\boldsymbol{\chi}$ of the metal,

$$\begin{aligned}\Delta\chi_{ax}&=\tilde{\chi}_{zz}-\frac{\tilde{\chi}_{xx}+\tilde{\chi}_{yy}}{2}=\frac{3}{2}\left(\tilde{\chi}_{zz}-\overline{\chi}\right)\\ \Delta\chi_{rh}&=\tilde{\chi}_{xx}-\tilde{\chi}_{yy}\end{aligned} \tag{2.47}$$

and $\overline{\chi}$ is the trace of $\boldsymbol{\chi}$ divided by three. Again, Eq. (2.46) shows that indeed the pseudocontact shifts do not depend on the total susceptibility value (i.e., on the trace of the tensor) but on its anisotropies only. The principal axes of the $\boldsymbol{\chi}$ tensor are conventionally defined as x-, y-, and z-axis in such a way that $|\Delta\chi_{rh}|\le\frac{2}{3}|\Delta\chi_{ax}|$ and the sign of $\Delta\chi_{rh}$ is opposite with respect to the sign of $\Delta\chi_{ax}$. This implies that the axes are defined so that the z-axis indicates the direction along which the magnetic susceptibility is most different with respect to those along the two perpendicular directions (it is the largest if $\Delta\chi_{ax}$ is positive, the smallest if $\Delta\chi_{ax}$ is negative), and the x axis indicates the direction with the smallest (if $\Delta\chi_{ax}$ is positive) or largest (if $\Delta\chi_{ax}$ is negative) magnetic susceptibility (Fig. 2.6).

Eq. (2.46) is the most common form used for the analysis of pseudocontact shifts. Other forms that can be useful [together with the already reported Eqs. (2.41)–(2.44)], according to the type of numerical calculations that are needed, are obtained for the Cartesian coordinates (obtained from Eq. (2.46) using the relationship $\cos 2\varphi = \cos^2\varphi - \sin^2\varphi$)

$$\delta^{\text{pcs}} = \frac{1}{12\pi}\frac{1}{r^5}\left(\Delta\chi_{ax}(3z^2 - r^2) + \frac{3}{2}\Delta\chi_{rh}(x^2 - y^2)\right) = \frac{1}{12\pi r^3}\left(\Delta\chi_{ax}\frac{2z^2 - x^2 - y^2}{r^2} + \frac{3}{2}\Delta\chi_{rh}\frac{x^2 - y^2}{r^2}\right). \quad (2.48)$$

Eq. (2.46) can also be expressed in terms of the direction cosines l, m, and n of the metal–nucleus vector with respect to the principal directions of the χ-tensor (Fig. 2.6):

$$\delta^{\text{pcs}} = \frac{1}{12\pi r^3}\left\{\Delta\chi_{ax}(3n^2 - 1) + \frac{3}{2}\Delta\chi_{rh}\left(l^2 - m^2\right)\right\}. \quad (2.49)$$

In summary, the equations for the pseudocontact shifts depend on (1) the atomic coordinates (in Cartesian coordinates, spherical coordinates or direction cosines form) and (2) the $\Delta\chi$ tensor ($\Delta\chi = \chi - \chi^{\text{iso}}$) parameters. The latter are 5 parameters: $\Delta\chi_{ax}$, $\Delta\chi_{rh}$, and the three Euler angles defining the orientation of the χ tensor in a different frame, if the equations valid in the principal frame of χ are used. If the equations valid for a generic frame are used, the 5 parameters are $\Delta\chi_{zz}$, $\Delta\chi_{xx} - \Delta\chi_{yy}$, $\Delta\chi_{xy}$, $\Delta\chi_{xz}$, and $\Delta\chi_{yz}$ (we recall that, by definition, the trace of $\Delta\chi$ is zero, and thus $\Delta\chi_{xx} + \Delta\chi_{yy} + \Delta\chi_{zz} = 0$).

Pseudocontact shifts can thus be always predicted from known coordinates and tensor components; however, the principal directions and components of the χ tensor are seldom available (Section 2.8). On the other hand, the traceless part of the χ tensor can be calculated from known coordinates and pseudocontact shifts, provided these are known for at least 5 nuclei (or at least 8 if the metal coordinates are also unknown). This can be done using Eq. (2.38), through a linear minimization in the 5 independent components of the χ tensor, or using Eq. (2.46), through a best fit of $\Delta\chi_{ax}$, $\Delta\chi_{rh}$, and the three Euler angles describing the orientation of the χ tensor. In a generic frame, the anisotropic part of the χ tensor can be obtained from the transpose of the rotation matrix described by the Euler angles (α,β,γ) times the anisotropic part of the diagonal $\tilde{\chi}$ tensor (i.e., the paramagnetic susceptibility anisotropy tensor in its main frame) times the rotation matrix:

$$\chi - \chi^{\text{iso}} = \mathbf{R}(\alpha,\beta,\gamma)^T \begin{pmatrix} -\Delta\chi_{ax}/3 + \Delta\chi_{rh}/2 & 0 & 0 \\ 0 & -\Delta\chi_{ax}/3 - \Delta\chi_{rh}/2 & 0 \\ 0 & 0 & 2\Delta\chi_{ax}/3 \end{pmatrix} \mathbf{R}(\alpha,\beta,\gamma). \quad (2.50)$$

On the contrary, atomic coordinates cannot be calculated from known pseudocontact shifts (measured in the presence of a paramagnetic metal ion) and tensor components. This can be easily deduced from the fact that the functional form of the pseudocontact shift equation can provide multiple solutions for the nuclear coordinates, with the consequence that obtaining structural information from pseudocontact shifts is not a straightforward task (Chapter 9).

Pseudocontact shifts can be still evaluated by expressing the principal paramagnetic susceptibility values as a function of the principal g values, in analogy with Eq. (1.40):

$$\chi_{kk} = \mu_0\mu_B^2 g_{kk}^2 \frac{S(S+1)}{3kT}. \quad (2.51)$$

As seen in Section 1.6, this approximation holds when the spin multiplet ground state is well isolated from excited electronic states, and ZFS is negligible. If $S = ½$, the g values can easily be measured through EPR spectroscopy and the g directions can be determined by single-crystal EPR measurements. When the latter measurements are not available, sometimes the principal g directions can be guessed from the symmetry of the molecule (if any), and an independent estimate of the pseudocontact shift can still be attempted.

By substituting the molecular χ values with the expression in Eq. (2.51), Eq. (2.31) and (2.46) become, respectively

$$\delta^{pcs} = \frac{\mu_0}{4\pi}\frac{\mu_B^2 S(S+1)}{9kT}(g_\parallel^2 - g_\perp^2)\frac{1}{r^3}(3\cos^2\theta - 1) \tag{2.52}$$

$$\delta^{pcs} = \frac{\mu_0}{4\pi}\frac{\mu_B^2 S(S+1)}{18kT}\frac{1}{r^3}\left\{\left[2g_{zz}^2 - \left(g_{xx}^2 + g_{yy}^2\right)\right](3\cos^2\theta - 1) + 3\left(g_{xx}^2 - g_{yy}^2\right)\sin^2\theta\cos 2\varphi\right\} \tag{2.53}$$

where g_{xx}, g_{yy} ($g_\perp$ in axial symmetry), and g_{zz} ($g_\parallel$ in axial symmetry) are the molecular g values. As in the case of contact shift, the pseudocontact shift does not contain parameters of the resonating nucleus. At variance with contact shifts, the pseudocontact shifts can be predicted exactly even if more than one S multiplet is populated, provided that the five χ tensor parameters are known.

Note that Eqs. (2.52) and (2.53) [and, likewise, Eqs. (2.31) and (2.46)] can be written in a way analogous to Eq. (2.9):

$$\delta^{pcs} = \frac{A^{pc}}{\hbar}\frac{\bar{g}\mu_B S(S+1)}{3\gamma_I kT} \tag{2.54}$$

where A^{pc} from, for example, Eq. (2.52) is given by

$$A^{pc} = \frac{\mu_0}{12\pi}\hbar\gamma_I\bar{g}\mu_B\left(\frac{g_\parallel^2}{\bar{g}^2} - \frac{g_\perp^2}{\bar{g}^2}\right)\frac{1}{r^3}(3\cos^2\theta - 1). \tag{2.55}$$

A of Eqs. (2.2) or (2.4), together with A^{pc} of Eq. (2.55), provide orientational average values of the hyperfine coupling tensor of Eq. (1.51).

When there are multiple electrons and ZFS occurs in orbitally nondegenerate states, the principal paramagnetic susceptibility values can be calculated through the Van Vleck equation [Eq. (1.42)], as long as $2D$ is larger than the Zeeman energy, and are given by Eqs. (2.12)–(2.17). Note that most of the magnetic anisotropy is determined by D, as $g_\parallel$ and $g_\perp$ are similar. By substituting the paramagnetic susceptibility values in Eq. (2.31) with Eqs. (2.12)–(2.17), and taking $e^X = 1 + X$ ($D \ll kT$), the following equations are obtained [2]:

$$S = 1: \delta^{pcs} = \frac{\mu_0}{4\pi}\frac{2\mu_B^2(g_\parallel^2 - g_\perp^2)}{9kT}\frac{3\cos^2\theta - 1}{r^3}\left[1 - \frac{(g_\parallel^2 + \frac{1}{2}g_\perp^2)}{3(g_\parallel^2 - g_\perp^2)}\frac{D}{kT}\right] \tag{2.56}$$

$$S = 3/2: \delta^{pcs} = \frac{\mu_0}{4\pi}\frac{5\mu_B^2(g_\parallel^2 - g_\perp^2)}{12kT}\frac{3\cos^2\theta - 1}{r^3}\left[1 - \frac{4(g_\parallel^2 + \frac{1}{2}g_\perp^2)}{5(g_\parallel^2 - g_\perp^2)}\frac{D}{kT}\right] \tag{2.57}$$

$$S = 5/2: \delta^{pcs} = \frac{\mu_0}{4\pi}\frac{35\mu_B^2(g_\parallel^2 - g_\perp^2)}{36kT}\frac{3\cos^2\theta - 1}{r^3}\left[1 - \frac{32(g_\parallel^2 + \frac{1}{2}g_\perp^2)}{15(g_\parallel^2 - g_\perp^2)}\frac{D}{kT}\right]. \tag{2.58}$$

These equations give the sum of the paramagnetic susceptibility contributions of each state i in the Van Vleck equation, weighted by the population according to the Boltzmann distribution. In contrast to the case of the Fermi contact shift [Eqs. (2.18)–(2.20)], the term depending on T^{-2} is already sizable for relatively small values of D/kT, since $g_{\parallel}^2 - g_{\perp}^2$ is small [2]. The term in T^{-1}, which depends on g anisotropy, is always small.

Scheme 2.1 summarizes the basic concepts related to the hyperfine coupling: it is composed of two terms, called contact and dipolar, depending on whether the electron spin is delocalized onto the resonating nucleus through the electronic orbitals or it is interacting through space with the nuclear spin, respectively. Both contact and dipolar terms can be described by tensors (the contact term depends on the g/χ tensor, the dipolar term is described by the dipolar shielding tensor σ). The isotropic averages of these tensors result in the contact shift and pseudocontact shifts, respectively, which can be measured in solution or in the solid state under magic angle spinning (MAS), and depend only on the traces of the tensors. Spinning side bands arising under MAS in solid state NMR (Section 5.1.2) contain the information on the anisotropy of the tensors.

2.2.2.2 Ligand-centered contributions. The first improvement of the dipole–dipole approximation is introduced by the consideration that the unpaired electron(s) is at least spread within the cage formed by the donor atoms. In the case of 4f electrons this does not produce a significant effect, whereas a 3d electron appears as a magnetic point–dipole only to nuclei that are relatively far away from the metal. An estimate of the lower limit of this distance is 700 pm [8]. In order to properly evaluate the pseudocontact term at shorter distances it is necessary to consider the appropriate MOs and to apply the dipolar interaction Hamiltonian to these wavefunctions. Attempts at this evaluation assuming a series

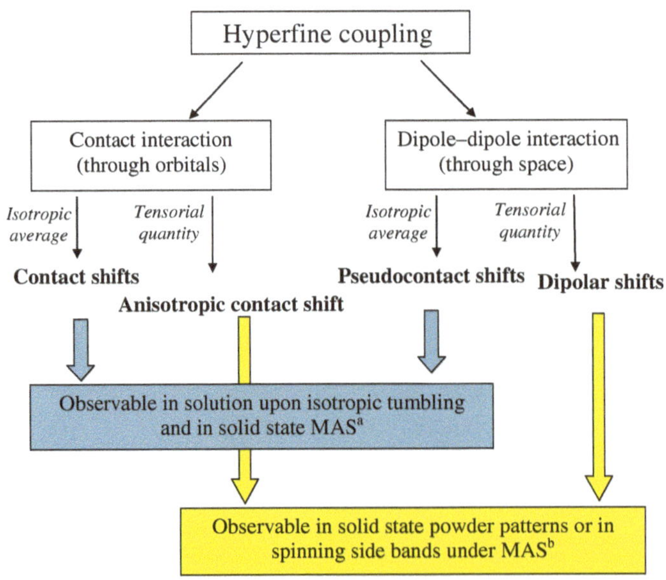

SCHEME 2.1 [a]Anisotropic effects can be partially reintroduced in solution by anisotropic tumbling (Chapter 3), and in the solid state by interference between radiofrequency pulses and MAS (Chapter 5).[b]The anisotropic terms also affect nuclear relaxation in solution (Chapter 4).

of point multipoles instead of a simple point–dipole, with the aim of mimicking the electron distribution within the metal atomic orbitals, have been partially successful [9,10,12–15]. These considerations provide considerable corrections on nuclei close to the paramagnetic center. Nuclei far from the paramagnetic center still see the electron as localized at the metal.

Because of the spatial delocalization of the electron spin density, the dipolar interaction can thus be expressed semiclassically as the coupling of the nuclear magnetic dipole moment to a continuous microscopic magnetization $M(r)$ due to the unpaired electron spin density; in the point dipole approximation, $M(r)$ is taken as a discrete sum over pure magnetic dipoles located at several specific distances r_i with high electron density [15]. Excellent agreement with experimental data is obtained for instance for the ^2H paramagnetic shift of $CuCl_2 \cdot 2D_2O$ in the solid state by positioning fractional point dipoles at the highest electron densities of each lobe of the singly occupied molecular orbital (SOMO), which interacts with the ligand p orbitals in a sigma-like fashion [15].

Finally, if the electron is considered to be delocalized on a large part of the molecule as the result of π delocalization, the problem is further complicated. Pseudocontact shift is also provided by spin density in a p orbital which has an orbital contribution to the magnetic moment different from zero. This is very relevant for heteronuclei like ^{15}N or ^{13}C which may bear unpaired spin density on p_z orbitals. The picture is of the type of Fig. 2.1 where, besides spin polarization, dipolar interactions occur. For example, fractions of unpaired electrons delocalized onto nitrogen p orbitals in $Fe(CN)_6^{3-}$ are predicted to affect seriously the overall ^{14}N hyperfine shifts [16]. Therefore, the analysis of hyperfine shifts for heteronuclei requires ligand-centered pseudocontact shifts to be taken into account. Hydrogen nuclei have no accessible p orbitals and hence any ligand-centered effect can only arise from unpaired spin density on p_π orbitals of neighboring atoms. Thus, the effects are expected to be smaller, although not necessarily negligible.

To treat the case of a nonpoint electron spin density distribution $\rho(\boldsymbol{r})$, it has been recently shown [17] that we need to solve the equation

$$\nabla^2 \delta^{\text{pcs}} = -\frac{1}{3} Tr\left(\frac{\partial^2 \rho(\boldsymbol{r})}{\partial \boldsymbol{r}\, \partial \boldsymbol{r}^T} \cdot \chi \right)$$

where the second-order partial derivatives of $\rho(\boldsymbol{r})$ with respect to position vector $\boldsymbol{r}$ and its transpose $\boldsymbol{r}^T$ provide the Hessian matrix of the electron spin density, Tr indicates the trace of the vector and ∇^2 indicates the Laplace operator, corresponding to the sum of the second partial derivatives with respect to the each spatial coordinate. The solution of this equation can be expressed in terms of spherical harmonics, analogously to Poisson's equation. This expression can be used to calculate the pseudocontact shifts at the nuclear positions from the electron probability density distribution $\rho(\boldsymbol{r})$, the susceptibility anisotropy tensor $\boldsymbol{\chi}$ and the nuclear coordinates. It has been proposed that this expression can also be used to recover the electron spin density distribution and the susceptibility anisotropy tensor from nuclear coordinates and measured pseudocontact shifts through a minimization procedure [17].

2.3 SHIFT AND SPIN PATTERNS FOR PROTONS AND DEUTERONS IN SOLUTION

From the equations given so far, it is difficult to predict any behavior of chemical compounds from the hyperfine shift point of view. Still, any effort is welcome in providing rules and expectations under certain conditions. We will see here that this is somewhat possible.

Chemical bonds involving protons and deuterons are very similar to one another. Therefore, since the contact and pseudocontact shifts are independent of the nuclear parameters, the shifts are essentially the same. Both nuclei have the characteristic that only 1s orbitals are involved. Therefore, protons and deuterons are discussed together. Sometimes, the use of ^{2}H NMR can be appropriate because it has a smaller linewidth than ^{1}H (Section 4.10).

Let us first try to discuss the contact contribution to the shift for protons and deuterons. As we shall see, there are conditions under which the pseudocontact shifts are small and negligible, or can be determined separately and subtracted from the hyperfine shifts. Furthermore, in the case of a monodentate ligand which exchanges rapidly from bound to free, the average pseudocontact shift in the absence of geometric constraints is zero.

The contact coupling constant is proportional to the spin density normalized to one electron. The proportionality constant depends on the magnetic properties of the two interacting particles. The following equation relates the contact hyperfine coupling with spin density (Eq. 2.11)

$$\frac{A}{h} = \frac{K\rho}{2S} \tag{2.59}$$

where $\rho/(2S)$ is the unpaired electron spin density normalized to one electron. For the hydrogen atom such a constant K is experimentally found to be equal to 1.42×10^{9} Hz [11].[3] In hydrogen and hydrogen isotopes (deuterium, tritium) there is only one pathway to generate contact spin density at the nucleus, i.e., through the 1s orbital, and no other orbital contributes to the contact or pseudocontact shift. The relationship between the constants K for the hydrogen isotopes {by recalling that A contains γ_I [Eq. (2.2)] and that $\Sigma_k\, \rho_k$ is the same for all isotopes} is given by ($k = 2, 3$)

$$K(^{k}\mathrm{H}) = \frac{\gamma(^{k}\mathrm{H})}{\gamma(^{1}\mathrm{H})} K(^{1}\mathrm{H}). \tag{2.60}$$

Let us now consider an organic radical, for example, an alkane radical. The unpaired electron will be in a MO. If the alkane radical is negative, the electron will be in the *lowest unoccupied* MO (LUMO), if positive or neutral in the *highest occupied* MO (HOMO). The amount of unpaired electron directly delocalized at the hydrogen nucleus will be proportional to the coefficient representing its 1s orbital in the MO containing the unpaired electron. Furthermore, we may have spin polarization from the fraction of an unpaired electron in a p orbital of a carbon or nitrogen atom to the hydrogen 1s orbital. Typical examples of spin polarization will be discussed later.

When the unpaired electron in the molecule resides on a metal ion, spin delocalization on the ligand can occur through different mechanisms, and different signs of spin density can arise. Positive contributions to the spin density on the ligand arise from the MO(s) involving partially filled metal orbitals. The HOMO of the ligand which is involved in the MO together with the metal orbital bears a fraction of unpaired electron. Negative contributions occur through spin polarization from unpaired electrons in the atomic orbitals of the metal and doubly occupied MOs of the metal ligand complex.

The variety of spin density distributions on the same ligand by changing the metal ions depends on the different weights of the two above contributions. Spin density distributions can also change if σ and π mechanisms contribute to different extents.

---

[3]The constant K is also theoretically calculated to be 1.42×10^{9} Hz by using Eqs. (2.59) and (2.11): $K = \frac{2\mu_0}{6\pi}\gamma_I g_e \mu_B \psi(0)^2$, where $\psi(0)^2 = \frac{4}{4\pi}\left(\frac{1}{a_0}\right)^3$ and $a_0 = 53$ pm.

2.3.1 METAL ION-WATER INTERACTIONS

Let us now consider the mechanism of spin delocalization in the water molecule when bound to a paramagnetic metal ion. Table 2.1 reports the proton *A/h* values for some hexaaqua metal complexes and the vanadyl pentaaqua complex. They have been determined from hyperfine shifts, although sometimes they have been determined as absolute values from relaxation studies (Chapter 4). Spin density has been evaluated through Eq. (2.59). The proton hyperfine shifts, assumed to be mainly contact in origin as a result of the averaged high symmetry, are downfield and the hyperfine coupling constants are positive. With the possible exceptions of Ti^{3+}, Fe^{3+}, and Mn^{2+}, it has been suggested that the hydrogen spin density, which is also listed in Table 2.1, tends to correlate with the number of unpaired t_{2g} electrons. The t_{2g} orbitals have the correct symmetry for π bonding. Unpaired electrons in the e_g orbitals, which have the correct symmetry for σ bonding, do not affect the value of ρ, and indeed the experimental proton hyperfine coupling constants in Cu^{2+} (d^9) and Ni^{2+} (d^8) complexes, having no unpaired electrons in t_{2g} orbitals, are very small (Chapter 7 and Table 7.1). It appears, therefore, that unpaired π electrons of the metal ions delocalize onto the H_2O moiety. One fully occupied water MO of σ type is capable of directly overlapping with the t_{2g} 3d orbital set of the metal. Thus the unpaired t_{2g} electron is capable of directly producing spin density on the protons. A further π interaction with the metal t_{2g} set arises from a nonbonding p orbital of oxygen, which would place spin density on the protons only through spin polarization. The sign of the observed shifts shows that the former mechanism is dominant [18]. The

Table 2.1 Hyperfine Coupling Constants[a] of Water Nuclei in Aqua Complexes of Common Metal Ions

Metal Ion	Electronic Configuration	2*S*	n[b]	*A/h* (MHz)		References
				^1H[c]	^{17}O	
Ti^{3+}	d^1	1	1	4.5 (3.2×10^{-3})[d]	4.4[d]	[19,20]
VO^{2+}	d^1	1	1	2.1 (1.4×10^{-3})[d]	3.8[d]	[21,22]
V^{2+}	d^3	3	3	2.0 (4.2×10^{-3})		[23]
Cr^{3+}	d^3	3	3	2.0 (4.2×10^{-3})		[23–25]
Mn^{2+}	d^5	5	3	0.6–1.0 ((2.2–$3.5) \times 10^{-3}$)	5.3	[23,26,27]
Fe^{3+}	d^5	5	3	0.4–1.3 ((1.4–$4.6) \times 10^{-3}$)		[23,28]
Fe^{2+}	d^6	4	2		9.4	[26]
Co^{2+}	d^7	3	1	0.4 (1.1×10^{-3})	14.8–15.7	[18,26,29]
Ni^{2+}	d^8	2	0	0.2 (3×10^{-4})	20.4	[18,30]
Cu^{2+}	d^9	1	0	0.06–0.19 (4.2×10^{-5}–1.3×10^{-4})	50.0	[18,31]
Ln^{3+}	f^1–f^6, f^8–f^{13}				0.6–0.8	[32]

[a] *Contact for protons; contact + ligand centered pseudocontact for oxygen. Positive signs indicate downfield shifts for* ^1H *and upfield shifts for* ^{17}O *signals.*
[b] *Number of unpaired electrons in* t_{2g} *orbitals for high-spin complexes with* O_h *symmetry.*
[c] ρ_H *values in parentheses, estimated from Eq. (2.59).*
[d] *Absolute values (obtained from relaxation data).*

^{17}O data (Table 2.1) show large hyperfine coupling constants when unpaired electrons are located in the e_g orbitals, which have the correct symmetry for σ bonding. The data for $Cu(H_2O)_6^{2+}$ and $Ni(H_2O)_6^{2+}$ demonstrate this clearly. It is obvious that direct delocalization through metal–oxygen σ bonding involving oxygen s orbitals gives rise to quite sizable spin density on the oxygen nucleus.

If two unpaired electrons populate the e_g orbitals, with the t_{2g} orbitals fully occupied (Table 7.1), there is an excess of $M_S = -½$ occupancy (positive spin density). Such –½ spin will be transmitted on the ligand through a MO which is a linear combination of the d orbitals and the ligand HOMO. The contribution to the ligand HOMO of atomic orbitals of the other ligand atoms decreases dramatically with the number of bonds. It follows that the spin density transmitted through this mechanism will rapidly decrease to zero as the number of chemical bonds between the metal and the resonating proton increases [33,34]. The positive shift (in frequency) is due to –½ spin density [Eqs. (2.6) and (2.59)], which in turn is the consequence of direct transfer with the same sign from d orbitals to ligand orbitals.

A second mechanism involving spin polarization may also be operative, in which spin polarization yields alternate positive and negative unpaired spin density along the ligand backbone. It is possible that spin polarization and direct delocalization mechanisms are simultaneously operative. For instance, the contact shifts are upfield for protons of ammonia coordinated to Cu^{2+}, Ni^{2+}, and Co^{2+} [35] for the mechanism discussed above; the contact shifts are downfield for Mn^{2+} owing to a dominant mechanism of direct transfer of spin density from t_{2g} orbitals to 1s hydrogen orbitals [35].

If the ligand has a σ MO system which has the correct symmetry to overlap with the $d_{x^2-y^2}$ and d_{z^2} orbitals, the shift pattern is predominantly of σ-type. However, spin polarization can induce ½ spin density in the π system. Once some unpaired spin density is in a p_π orbital, it spin-polarizes the electrons of an adjacent σ bond, let's say between carbon and hydrogen, thus producing a further mechanism for transferring spin density on the hydrogen. The proton *A/h* value from this mechanism is

$$\frac{A}{h} = Q_{\mathrm{CH}}^{\mathrm{H}} \frac{\rho_{\mathrm{C}}^{\pi}}{2S} \tag{2.61}$$

analogously to Eq. (2.59) for the free hydrogen atom, but the unpaired spin density ρ_{C}^{π} is now that residing on the neighboring carbon p_π orbital and the proportionality constant $Q_{\mathrm{CH}}^{\mathrm{H}}$ is much smaller. This makes the shift pattern to be α-H ≫ γ-H > β-H in the case of the hexa(pyridine) nickel(II) complex [36–40]. The constant $Q_{\mathrm{CH}}^{\mathrm{H}}$ is negative and of the order of −70 MHz (Section 2.4 and Table 2.2). Therefore, we can say that the spin density on the proton has opposite sign of that on the attached sp^2 carbon (Fig. 2.9).

Table 2.2 *Q* Values (MHz) for Fragments Involved in π Electron Spin Delocalization Systems[a]

$Q_{\mathrm{CH}}^{\mathrm{H}}$	$Q_{\mathrm{CH}}^{\mathrm{H}}$	$Q_{\mathrm{CC'}}^{\mathrm{C}}$	$Q_{\mathrm{C'C}}^{\mathrm{C}}$	$\langle Q_{\mathrm{CCH_3}}^{\mathrm{H}} \rangle$	$Q_{\mathrm{CF}}^{\mathrm{F}}$	$Q_{\mathrm{FC}}^{\mathrm{F}}$	References
+54.6	–65.8	+40.3	–39.0				[41]
	–75.6	+53.2					[42]
				+ 75			[43]
					–410	+2370	[44]

[a]*The notation $Q_{\mathrm{XY}}^{\mathrm{Y}}$ means the McConnell constant for a nucleus Y attached to an atom X in sp^2 configuration and bearing unpaired electron spin density on its p_z orbital. The notation $Q_{\mathrm{XY}}^{\mathrm{X}}$ means the McConnell constant for the nucleus of atom X in the same moiety.*

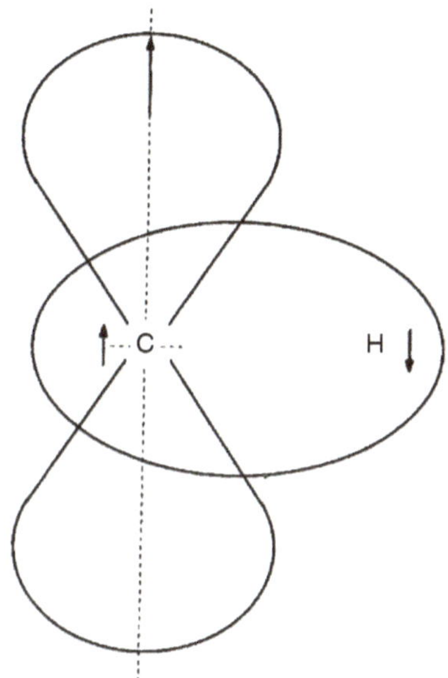

FIGURE 2.9 The spin density on protons has the opposite sign of that on the attached sp^2 carbon.

2.4 PROTON HYPERFINE COUPLING AND CONFORMATION

The contact hyperfine shift contains information on conformational arrangements because it is somehow related to chemical bonds. The information, however, is generally hidden, with the exception of few cases. For example, if we have a CH moiety of an sp^3 carbon bound to an sp^2 carbon which bears spin density in the p_z orbital, a spin density of the same sign is transferred to the methyl protons. A mechanism to originate spin density at the methyl proton is the direct overlap between the sp^2 carbon p_z and the 1s hydrogen orbital [45–48] (Fig. 2.10). Such overlap depends on the dihedral angle θ between the p_z(axis)–C–C plane and the C–C–H plane (Fig. 2.10). The relationship is described by the general Karplus equation [49]

$$\frac{A}{h} = (a\cos^2\theta + b\cos\theta + c)\frac{\rho_C^\pi}{2S} \tag{2.62}$$

where ρ_C^π is the spin density on the p_z orbital and *a*, *b*, and *c* are constants. For the present moiety, *b* equals zero. The constant *a* represents the maximal direct overlap of the 1s hydrogen orbital with the carbon p_z orbital, which occurs when $\theta = 0$ (Fig. 2.10). The constant *c* represents the θ-independent transfer of spin density through the C–C–H σ bonds, which is the only mechanism available when $\theta = 90$ degrees. The latter mechanism is less efficient than the former, and *c* is much smaller than *a* [45]. The constant *a* is about 140 MHz [45,49]. When the CH moiety belongs to a CH_3 group and the group is free to rotate, an average over all positions should be calculated. The average value of $\cos^2\theta$ is ½, so that $a\langle\cos^2\theta\rangle \cong 70$ MHz. Experimentally, a value of $a\langle\cos^2\theta\rangle + c$ of $\cong$75 MHz is found [45]. In other words, if there is a positive (–½) spin density on the p_z orbital of the sp^2 carbon atom, a positive spin density is present on average on a proton of a CH_3 group, whereas it would be negative if the proton were directly attached to the sp^2 carbon atom (Fig. 2.9). In general, we can say that *A*/*h* is always given

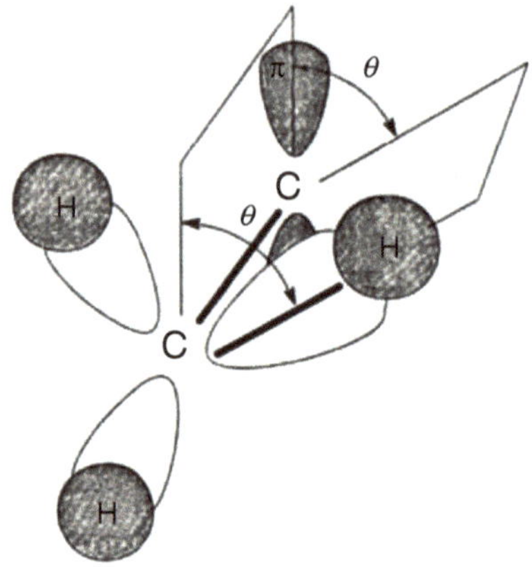

FIGURE 2.10 CH moiety of an sp^3 carbon bound to an sp^2 carbon, illustrating the definition of the dihedral angle θ.

by an equation such as Eq. (2.59), where the K value (which equals 1420 MHz for the isolated hydrogen atom) is scaled down to a constant Q, called McConnell constant, which depends on the geometry of the system. As we have seen, the constant Q is ≈ -70 MHz for a proton directly attached to an sp^2 carbon [Eq. (2.61)] or $\approx +75$ MHz for a freely rotating CH_3 group. Some other Q values are collected in Table 2.2 [41–44].

This reasoning induces one to think that, even in a moiety of the type M–D–C–H, where D is the donor atom, the CH proton experiences a contact coupling depending on the orientation of the D–C–H plane (Fig. 2.11). If the spin density is along the M–D bond (Fig. 2.11A), then the coupling depends on

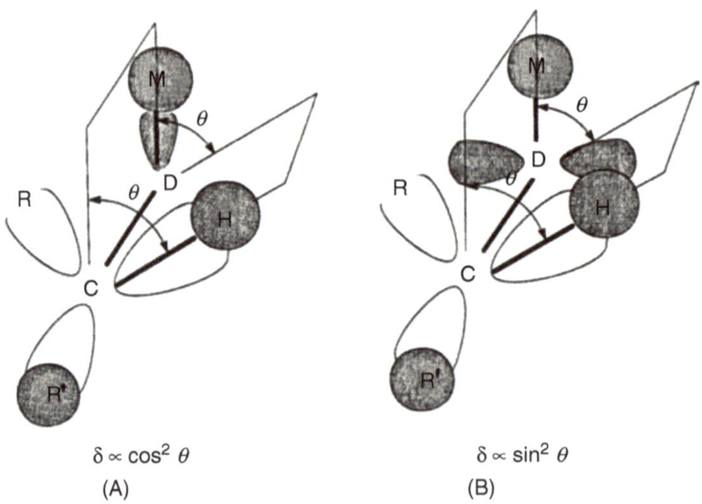

FIGURE 2.11 Dihedral angle θ between the M–D–C plane and the D–C–H plane of M–D–C–H moieties. The case of the spin density being along the M–D bond is shown in (A) and the case of the spin density being in a p orbital of the D atom orthogonal to the M–D bond is shown in (B).

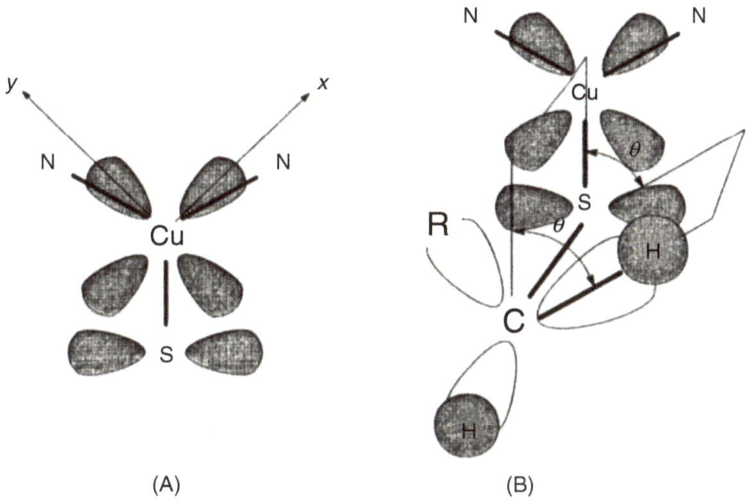

FIGURE 2.12 Arrangement of the donor atoms in the equatorial plane of copper and of the copper $d_{x^2-y^2}$ and sulfur p orbitals (A). The dihedral angle θ between the Cu–S–C plane and one of the S–C–H planes is depicted in (B).

Adapted from Ref. [52].

the dihedral angle between the M–D–C and D–C–H plane in a fashion similar to Eq. (2.62), where b may not be negligible [33], θ being now defined with respect to the M–D bond rather than with respect to a nonbonding p_z orbital as in Fig. 2.10. This is the case when D is an amino nitrogen [33–35]. If the spin density is in a p orbital of the D atom orthogonal to the M–D bond (Fig. 2.11B), a dependence on $\sin^2\theta$ is more appropriate. This seems to be the case for *S* as donor atom [50].

In the case of oxidized ferredoxins, a class of proteins containing $Fe_4S_4^{2+}$ clusters (Chapter 11), it has indeed been found that the contact shifts of the βCH_2 protons of the cysteines coordinated to the iron atoms (Fig. 11.19) depend on θ as

$$\delta = a\sin^2\theta + b\cos\theta + c \quad a \cong 10.3 \quad b \cong -2.2 \quad c \cong 3.9 \tag{2.63}$$

The relative values of the c and a constants give an idea of the importance of the two processes of electron delocalization, through M–D σ and D π bonds, the latter being dominant.

In the case of blue proteins, a class of proteins containing type 1 copper(II) (Section 7.3.1.2), there is a plane with copper(II), two histidine nitrogens and one cysteine sulfur, with the x and y directions defined as in Fig. 2.12A. The $d_{x^2-y^2}$ orbital contains the unpaired electron. The sulfur donor atom has a π orbital which overlaps the $d_{x^2-y^2}$ orbital (Fig. 2.12B). Large downfield shifts are observed (Chapter 7), and again the angular dependence of the shifts of the CH_2 protons is expected to be mainly of $\sin^2$ type [51].

2.5 THE ORIGIN OF THE SHIFTS IN HETERONUCLEI

The predictions of the proton shifts is difficult; the prediction is even more difficult in the case of heteroatoms. In the case of σ spin density like in a aliphatic amine (Table 2.3), the ^{13}C Fermi contact shift for the α-carbon of the aliphatic amine is upfield (A/h is negative) because of predominant spin polarization

Table 2.3 Estimated *A/h* Values (MHz) for ^{1}H, ^{13}C, and ^{14}N Nuclei of Nitrogen- and Sulfur-Containing Ligands of Nickel(II)

Ligand								References
Ammonia			N^1H_3 −1.5					[35]
Ethylamine		^{14}N	N^1H_2	$^{13}CH_2$	C^1H_2	$^{13}CH_3$	C^1H_3	[34]
		14.4	−2.64	−0.70	1.02	1.80	0.22	
$(bpy)Ni(xan\text{-}R)_2$	R	α-C^1H	β-C^1H	γ-C^1H	δ-C^1H	ε-C^1H	ζ-C^1H	[53]
	ethyl	−0.03	0.03					
	propyl	−0.03	0.04	0.02				
	butyl	−0.03	0.04	0.01	0.003			
	pentyl	−0.03	0.04	0.01	0	0.001		
	hexyl	−0.03	0.04	0.01	0	0	0	
	cyclo-	−0.23	0.09	0.005	0.001			
	hexyl		0.01	0.001				
	R	α-^{13}CH	β-^{13}CH	γ-^{13}CH	δ-^{13}CH	ε-^{13}CH	ζ-^{13}CH	
	ethyl	0.20	0.18					
	propyl	0.18	0.17	0.02				
	butyl	0.18	0.17	0.02	0.02			
	pentyl	0.18	0.17	0.02	0.02	0.004		
	hexyl	0.18	0.16	0.03	0.02	0.003	0.002	
	cyclohexyl	0.15	0.04	0.02	0.01			

bpy, bipyridyl; xan-R, 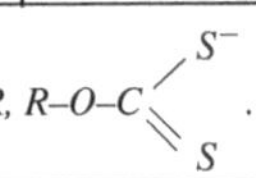

effects, whereas that for other carbon atoms is downfield, and rapidly attenuates with the number of bonds. A sizable downfield shift is experienced by the ^{14}N nucleus when nitrogen is a donor atom.

More hints are available for aromatic moieties. If we consider three sp^2 carbon atoms C′–C–C″ in a delocalized moiety (e.g., in an aromatic ring), and we assume that unpaired spin density resides in the p_z orbitals, we have the following contributions to the contact interaction at the central carbon atom C [54]:

- spin polarization of the 1s orbital from the p_z orbital of the same atom C;
- spin polarization of the 2s orbital from the p_z orbital of the same atom C, through the three σ bonds in which the 2s orbital is involved; and
- spin polarization of the 2s orbital from the p_z orbitals of the neighboring carbon atoms C′ and C″, through their σ bonds to atom C.

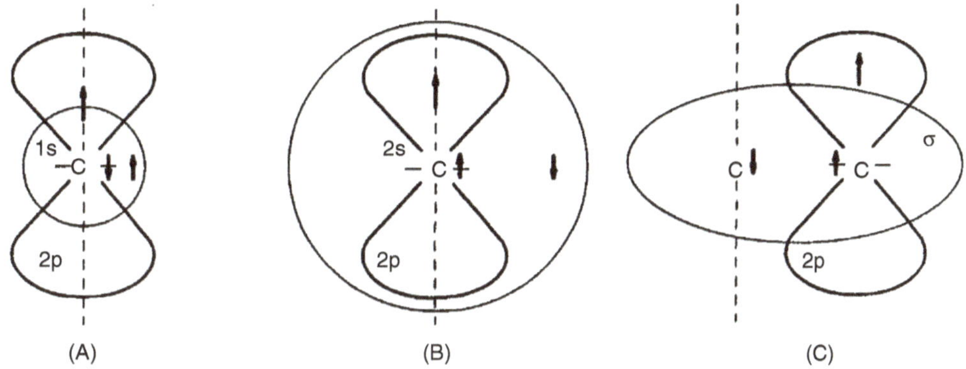

FIGURE 2.13 Spin polarization mechanisms arising from unpaired spin density on a $2p_z$ orbital, illustrated for a ^{13}C nucleus. (A) Polarization through a 1s orbital; (B) polarization through a 2s orbital; (C) polarization on a nearby carbon through a σ bond.

The first contribution is proportional to the spin density on the p_z orbital, ρ_C^π, through a negative proportionality constant, S^C, which has been estimated to be –35.5 MHz. Therefore, the resulting spin density at the C nucleus is opposite in sign to that of the p_z orbital (Fig. 2.13A).

The second contribution is again proportional to ρ_C^π and arises from the spin polarization of the σ bonds involving C′, C″, and the H atom bound to C. The proportionality constants, $Q_{CC'}^C$, $Q_{CC''}^C$, and Q_{CH}^C are of the same type of Q_{CH}^H in Eq. (2.61). They are all positive (Table 2.2), and therefore the resulting spin density at the C nucleus through the involvement of its 2s orbital in the three σ bonds is of the same sign as that of the p_z orbital (Fig. 2.13B).

At variance with the first two, the third contribution is proportional to the p_z spin density on the neighboring carbon atoms, $\rho_{C'}^\pi$ and $\rho_{C''}^\pi$. The proportionality constants, $Q_{CC'}^C$, and $Q_{CC''}^C$ are negative (Table 2.2) and the sign of the resulting spin density at the C nucleus is opposite to that on the p_z orbitals of the neighboring carbons (Fig. 2.13C).

The hyperfine coupling constant on the carbon atom C is therefore given by:

$$\frac{A}{h} = (S^C + Q_{CC'}^C + Q_{CC''}^C + Q_{CH}^C)\rho_C^\pi + Q_{C'C}^C \rho_{C'}^\pi + Q_{C''C}^C \rho_{C''}^\pi. \tag{2.64}$$

From the foregoing analysis, it appears that there are several terms making a significant contribution to the hyperfine coupling constant. As they are both positive and negative and of similar magnitude (Table 2.2), the final value of the hyperfine coupling constant is therefore the result of considerable cancellation among the various contributions. In Eq. (2.64) the three spin densities ρ_C^π, $\rho_{C'}^\pi$, and $\rho_{C''}^\pi$ are unknown. It appears that, even considering the shift of the proton attached to the carbon, which provides an independent estimate of ρ_C^π through Eq. (2.61), the number of parameters defining the shift of the heteronucleus is larger than the two experimental shifts, unless the shifts of the protons attached to the neighboring carbons are also considered. Similar considerations hold if the central atom is an sp^2 nitrogen. A deeper insight can be achieved by using experimental data on nuclear relaxation (Section 4.10).

Table 2.4 Estimated *A/h* Values (MHz) for ^{1}H, ^{13}C, and ^{14}N Nuclei of Pyridine and γ-Picoline Axially bis-Coordinated to Nickel(II) bis-Acetylacetonate

Ligand		*A/h* (MHz)	References
Pyridine	^{14}N	+18	[58]
Pyridine	α-^{13}C	−0.3	[55,59]
Pyridine	β-^{13}C	+0.7	[55,59]
Pyridine	γ-^{13}C	−0.1	[55,59]
γ-Picoline	γ-$^{13}CH_3$	+0.1	[59]
Pyridine	α-^{1}H	+0.9	
Pyridine	β-^{1}H	+0.3	
Pyridine	γ-^{1}H	+0.35	
γ-Picoline	γ-$C^{1}H_3$	−0.1	[59]

^{13}C data are available on several ligands. In Tables 2.4 and 2.5 the relative ^{13}C hyperfine shifts of pyridine and pyridine-*N*-oxide are reported as an example [55,56]. In both cases it appears that there is alternation in the shifts, which would indicate π spin delocalization. However, the ratio between the shifts of protons and carbons at each position are different from one another, and differ from what would be expected on the basis of a simple spin polarization mechanism. It is concluded that more than one mechanism is operative.

The analysis of the solid state NMR spectra of $[Fe(CN)_6]^{3-}$ and $[Mn(CN)_6]^{3-}$ indicates that the sign of the shift, as well as of its anisotropy, is opposite between ^{13}C e ^{15}N, and this suggests a different origin for the contact shift for the two nuclei: whereas ^{15}N sees a direct delocalization of the electron spin density in a π orbital orthogonal to the bond axis, the ^{13}C nucleus senses a spin polarization effect from a σ orbital [7].

Data are also available on ^{13}C and ^{15}N of cyanide bound to a heme moiety. Cyanide is the most common reactant of oxidized heme moieties [60]. All the monocyano and dicyano derivatives of isolated porphyrins and hemoproteins are low spin and contain either diamagnetic iron(II) or paramagnetic ($S = ½$) iron(III). ^{13}C and ^{15}N NMR spectra have been reported for the latter systems and, as shown in Tables 2.6 and 2.7, dramatic paramagnetic effects have been measured for both nuclei: the ^{13}C hyperfine shifts range from −2000 to −2400 ppm upfield [61] whereas the ^{15}N shifts range from 600 to

Table 2.5 Hyperfine Shifts (ppm) of Ring Nuclei in Octahedral Complexes of Ni^{2+} With Aniline and Pyridine-*N*-Oxide

Ligand		*Ortho*	*Meta*	*Para*	References
Aniline	^{1}H	−6.9	+3.8	−7.6	[57]
Py-NO	^{1}H	−13	+11	−17	
	^{13}C	+166	−65	+90	[56]

Table 2.6 ^{13}C Chemical Shifts (ppm from TMS) for Coordinated Cyanide in Low Spin Iron(III) Porphyrins and Hemoproteins [61]

Iron Species	Porphyrin 3,8-R Group	^{13}C Chemical Shift
Fe protoporphyrin IX	$-CH=CH_2$	−2393
Hemin c	$-CH(CH_3)SCH_2-CH(NH_2)COO^-$	−2300
Fe 3,8-disulfonate-deuteroporphyrin IX	$-SO_3^-$	−2167
Fe 5,10,15,20 tetra(4-carboxyphenyl)-porphyrin		−1968
(C_2H_5NC)-myoglobin (diamagnetic)		171.3

Table 2.7 ^{15}N Chemical Shifts (ppm from $^{15}NO_3^-$) for Coordinated Cyanide in Low Spin Iron(III) Porphyrins and Hemoproteins

Iron Species	Solvent	^{15}N Chemical Shift	References
Hemin	Me_2SO	732	
	Py	696	
	CH_3OD	506	[62]
	$CH_3OD + D_2O$ (1:1)	480	
	H_2O	448	
(4-AcPy)hemin	4-AcPy + D_2O	945	[62]
(Py)hemin	Py + D_2O	989	[62]
(3,5-Lu)hemin	3,5-Lu + D_2O	1070	[62]
$(1-CH_3Im)$hemin	$1-CH_3Im + D_2O$	926	[62]
Myoglobin	D_2O	931–948[a]	[62]
Cytochrome *c*	D_2O	841	[63]
Hemoglobin	D_2O	975 (α)	[64]
		1047 (β)	
Horseradish peroxidase	D_2O	576	[65]
Cytochrome *c* peroxidase	D_2O	587	[65]
Lignin peroxidase	D_2O	608	[65]
Manganese peroxidase	D_2O	639	[65]

Abbreviations: AcPy, acetylpyridine; Py, pyridine; Lu, lutidine; Im, imidazole.
[a]Depending on pH.

1200 ppm downfield [62–65]. The spin delocalization would be dominantly spin polarization through an Fe–C π bond, which may also account for the opposite sign of carbon and nitrogen shifts. Possible metal-centered pseudocontact contributions estimated from the known metal–nucleus distance and magnetic anisotropy values are of the order of 100–300 ppm downfield. Theoretical studies [16] have estimated sizable ligand-centered pseudocontact contributions to the shifts of nitrogen in $Fe(CN)_6^{3-}$.

Although the magnitude of the shifts is only slightly sensitive to the variations in the nature of the heme and to its contacts with the surrounding protein, it is strongly dependent on the identity of the trans ligand. For example, the ^{15}N signal in protohemin cyanide shifts from 945 to 989 to 1070 ppm downfield on passing from 4-acetylpyridine to pyridine to 3,5-lutidine as the *trans* ligand [63]. Both ^{13}C and ^{15}N hyperfine shifts are also very sensitive to changes in the solvent. Aprotic solvents usually give the largest absolute values, which decrease by as much as 200 ppm on passing from DMSO to methanol to water [62,63]. In general, the axial ligand field strength of cyanide, and hence the extent of unpaired spin delocalization, decreases with the increasing hydrogen bonding capabilities of the solvent.

^{15}N NMR data are available for cyanometmyoglobin, cyanocytochrome *c*, cyanomethemoglobin, and some peroxidases (Table 2.7) [62–65]. The large observable differences may be related to the different coordination strength of the *trans* histidine and/or to the different hydrophobicity of the CN binding site. The difference in the ^{15}N chemical shift (about 70 ppm) of the α and β subunits in hemoglobin (Table 2.7) [64] illustrates the sensitivity of the cyanide nuclei to the protein environment around the axial positions.

For ^{19}F Fermi contact shift of fluorine bound to sp^2 carbon atoms, spin density on the nucleus arises from spin polarization by ρ_C^π, as for analogous CC′ moieties, and from spin polarization by ρ_F^π, which occurs via direct delocalization through C–F π bonding (Fig. 2.14). The hyperfine coupling is therefore

$$\frac{A}{h} = Q_{CF}^{F}\rho_C^\pi + Q_{FC}^{F}\rho_F^\pi \tag{2.65}$$

where Q_{CF}^{F} is analogous to Q_{CH}^{H} or $Q_{C'C}^{C}$ and is negative, and Q_{FC}^{F}, which is the polarization effect of ρ_F^π on the C–F σ bond, is positive by analogy to $Q_{CC'}^{C}$ (Table 2.2).

^{31}P contact shifts in ligands coordinating through P–O moieties in oxovanadium(IV), cobalt(II), and nickel(II) complexes are downfield and of the order of 10^3 ppm [66–68]. Smaller downfield shifts are

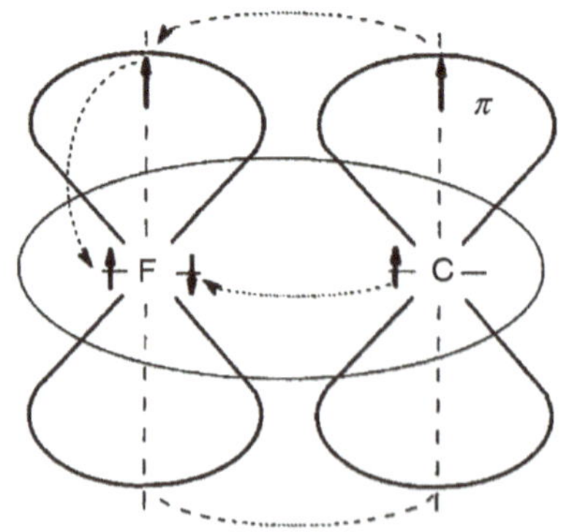

FIGURE 2.14 Spin polarization (through σ bonding) and direct delocalization (through π bonding) mechanisms for unpaired spin density transfer from a sp^2 carbon to a fluorine nucleus.

observed for nickel-coordinated P–S moieties [69]. Direct P–M coordination gives rise to even larger hyperfine coupling than in P–O moieties. In low spin cobalt(II) complexes, *A/h* values of the order of 10^2 MHz are found [70].

2.6 WHEN IS METAL-CENTERED PSEUDOCONTACT SHIFT EXPECTED?

Pseudocontact shift is expected every time there are energy levels close to the ground state. This causes orbital contributions to the ground state, and such contributions are orientation dependent. Therefore, the paramagnetic susceptibility tensor is anisotropic. Anisotropy of the paramagnetic susceptibility tensors arises also from sizable ZFS of the *S* manifold (Section 1.4).

Six-coordinated high spin cobalt(II) in pure octahedral symmetry has a $^4T_{2g}$ ground state which is triply degenerate (Section 7.3.7.2). Under spin–orbit coupling and low symmetry components up to six Kramers doublets can be obtained (Fig. 2.15) within 1000 cm^{-1}. The orbital contribution to the paramagnetic susceptibility is sizable, and the orbital contribution is intrinsically anisotropic. The pseudocontact shifts estimated [71] for a tetragonal complex are reported in Table 2.8. Octahedral nickel(II) complexes have a $^3A_{2g}$ ground state which is orbitally nondegenerate, with the first excited state at about 10,000 cm^{-1}. So, to a first approximation, its effects on the paramagnetic susceptibility tensor are modest.

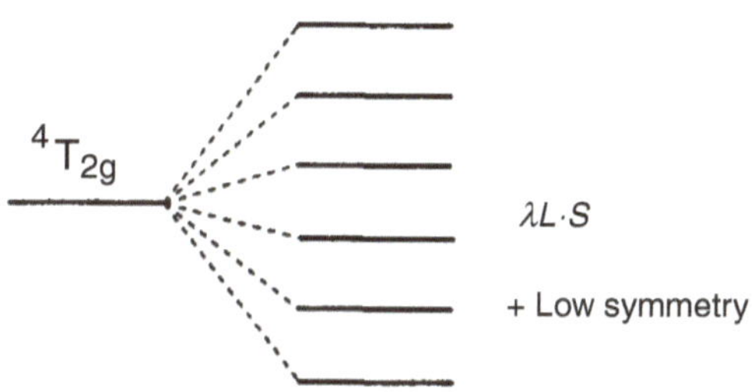

FIGURE 2.15 A ^{4}T state is split into six Kramers' doublets by spin orbit coupling and low symmetry components.

Table 2.8 Hyperfine (δ^{hyp}) and Pseudocontact (δ^{pcs}) Shifts[a] of the ^{1}H and ^{13}C Nuclei of Pyridine bis-Coordinated to Cobalt(II) bis-Acetylacetonate [71]

Atom	δ^{hyp} (ppm)	δ^{pcs} (ppm)
α-H	+32.9	–39.5
β-H	+5.0	–18.1
γ-H	–9.4	–15.6
α-C	–199	–92.5
β-C	+229	–35.7
γ-C	–73.8	–28.3

[a] *Pseudocontact shifts are estimated using single crystal magnetic susceptibility data.*

It is foreseeable that pseudocontact shifts occur in low spin iron(III) porphyrins which in octahedral symmetry would have an orbitally triply degenerate ground state. Therefore, g and χ values in iron porphyrins are highly anisotropic (Section 7.3.5.2) [72,73].

A more detailed description of the size of the pseudocontact shifts obtained for transition metal ions and lanthanoids is reported in Chapters 7 and 8, respectively.

2.7 ATTEMPTS TO SEPARATE CONTACT AND PSEUDOCONTACT SHIFTS

As a result of the nature of the pseudocontact shifts, a knowledge of the metal paramagnetic susceptibility tensor provides a method for their independent evaluation, if it is assumed that the shifts are metal-centered dipolar in origin. In principle, therefore, by performing paramagnetic susceptibility measurements on single crystals with several orientations of the magnetic field, it is possible to determine the magnetic anisotropy and the principal axes of the metal magnetic tensor.[4] By knowing the geometrical coordinates, the pseudocontact shifts can be easily predicted using Eq. (2.46). Such a procedure assumes that the magnetic anisotropy and the molecular axes are the same in the solid state and in solution. This is somewhat unjustified. An example of this procedure is shown in Table 2.8.

The procedure based on the direct use of the $\mathbf{g}$ tensor anisotropy and Eq. (2.53) is quite common for $S = ½$ systems, since g values from frozen solutions are easily obtainable. In this case, both the second order Zeeman contributions and possibly the effects of temperature on the g values are neglected. Furthermore, the directions of the molecular axes are arbitrarily assumed unless single-crystal data are available. Attempts are available in the literature regarding low spin cobalt(II) [74] and copper(II) [51].

Other semiquantitative methods for evaluating the metal-centered pseudocontact contributions to the shifts are based on the pattern predicted for a series of metal ions of the pseudocontact shifts assuming axial symmetry. Eq. (2.31), which is valid for the axial case, can be rewritten as

$$\delta^{pcs} = G_{r,\theta} \cdot D_{x} \qquad (2.66)$$

where $G_{r,\theta}$ is the geometric factor which can be calculated from the structure and D_{χ} is the term containing the paramagnetic susceptibility. Since the latter term is constant for a given molecule, the ratios between the pseudocontact shifts should be the same as the ratios between the $G_{r,\theta}$ values. When this equation is obeyed, it can be concluded that the observed hyperfine shifts are mainly pseudocontact in origin. In other cases this same property allows the pseudocontact contribution on all nuclei to be estimated by assuming that at a certain nucleus position, the contact contribution is negligible. Horrocks has developed a variant [75] which is based on a comparison between the proton hyperfine shift pattern for a molecule where the shifts are only contact in origin, and an analogous molecule whose proton shifts have to be separated into their contributions. The spin delocalization mechanisms have to be the same, as it is possibly the case for nickel(II) and cobalt(II) pairs.

Examples of contributions for some metal ions of contact and pseudocontact shifts are reported in Section 7.3. The case of lanthanoids will be discussed in Section 8.4. Protocols to separate and analyze hyperfine shifts in paramagnetic metalloproteins are presented in Section 9.3.

[4]The diamagnetic contribution, if not negligible, should be appropriately subtracted.

2.8 PREDICTION OF HYPERFINE SHIFTS

The ab initio calculation of hyperfine shifts from a molecular structure is a quite cumbersome problem which started to be addressed only recently. It requires the evaluation of the contributions to the shift from all thermally accessible energy levels of the $2S + 1$ electron spin states and from ZFS. This passes through the calculation of the anisotropic $\boldsymbol{g}$ and $\boldsymbol{A}$ (hyperfine coupling) tensors and of their product, which gives a variety of terms (nine terms for $S = 1/2$) identifying the various contributions to the hyperfine shifts [76]. Three terms identify the isotropic contact shift (with separated contributions from the deviation of the isotropic g-value from g_e, and from the pseudocontact term to the isotropic hyperfine coupling constant), one term the isotropic pseudocontact shift, and five terms the anisotropic contributions. The treatment can be extended to $S > ½$ systems with ZFS corrections applied a posteriori, as validated from the analysis of the ^1H and ^{13}C hyperfine shifts calculated for paramagnetic metallocenes ($(C_5H_5)_2M$, M = V, Cr, Mn, Co, Ni) [77].

In the case of $S > 1/2$, Pennanen and Vaara [78] proposed an NMR theory including the a priori treatment of ZFS effects for the ab initio calculation of hyperfine shifts for arbitrary electronic ground state degeneracy and systems in the weak spin-orbit coupling limit. Equations describing the hyperfine shifts were then proposed as a function of the $\boldsymbol{g}$, $\boldsymbol{A}$, and $\boldsymbol{D}$ (ZFS) tensors for systems with a ZFS much smaller than the energy separating the ground state from the excited states, and assuming that excited states are not thermally accessible [79,80]. Furthermore, Van den Heuvel and Soncini [81] proposed a more general theory which applies to any degree of electronic degeneracy, and arbitrarily strongly spin-orbit coupling limit, as that arising in strong spin-orbit coupled lanthanoids.

Density-functional theory (DFT) calculations are often performed to determine the EPR parameters ($\boldsymbol{g}$, $\boldsymbol{A}$, and $\boldsymbol{D}$), and they have been the subject of substantial development over the past decades. An accurate treatment of spin-orbit coupling is particularly critical for quantitative $\boldsymbol{g}$-tensor calculations [82,83]. DFT methods can actually be profitably used to calculate the hyperfine coupling $\boldsymbol{A}$ tensor, but have been shown to be inaccurate in determining the ZFS and, sometimes, the $\boldsymbol{g}$ tensor [84–86]. A higher accuracy is achieved using ab initio molecular orbital-based or wave function theories (WFT) at the complete active space self-consistent field (CASSCF) [87] and n-electron valence state perturbation theory 2 (NEVPT2) [88] levels. Using such approach, the ^1H hyperfine shifts for cobalt(II) pyrazolylborate complexes, for instance, are predicted in good agreement with the experimental data [80].

Neglecting ZFS, contact shifts can be calculated with a good accuracy after geometry optimization of a structural model with DFT at the B3LYP/cc-pVTZ level [89]. DFT calculations of hyperfine coupling constants, g values and orbital shieldings can in fact be used to calculate the contact shifts using Eq. (2.7) or Eq. (2.8). A good agreement with the experimental NMR data is obtained, for instance, for nitroxide radicals ($S = 1/2$, no ZFS), where the pseudocontact shifts can be neglected, and the measured ^1H and ^{13}C hyperfine shifts have values up to ±30 ppm and ±1500 ppm, respectively [76,89]. An overall agreement between DFT predictions and experimental data is also observed for some transition metal ions complexes, as in the case of chromium(III), manganese(III), iron(III), and copper(II) acetylacetonates, with the metal ions in the high spin states (Cr: $S = 3/2$, Mn: $S = 2$, Fe: $S = 5/2$, Cu: $S = 1/2$), as expected in the presence of a weak ligand in an octahedral field [89]. A correct trend of the hyperfine shifts is calculated for several paramagnetic ruthenium(III) complexes, with fast electronic relaxation ($10^{-12} - 10^{-11}$ s) in the $S = 1/2$ electronic state, although some deviances with respect to the experimental values are observed in this case [90]. An improved agreement with the experimental data in the hyperfine shifts of these complexes is achieved using a fully relativistic approach [91]. Notably, a good agreement between calculated and experimental shifts is obtained also for paramagnetic iron-sulfur proteins, as in the case of iron(III) rubredoxin [92].

DFT calculations of contact shift can be useful to recover the overall shape of the NMR spectra for certain assumed geometries of the metal ligands and spin states of the paramagnetic metals, and thus to validate the structural models as well as to determine the spin states, S. In fact, since the NMR spectral features are remarkably different according to the spin state, the comparison of the experimental NMR spectrum with the spectra calculated for all spin states can allow for a reliable identification of the correct spin state [93]. Furthermore, the comparison of the positions of calculated and experimental shifts may result useful for the assignment of most peaks. For instance, the major features of the NMR spectra calculated for iron(II), iron(III), and iron(IV) complexes with given molecular structures can allow for a structural characterization and the determination of the spin state of these systems [93,94].

Calculation of the molecular orbital electron density may also provide a way to predict the pseudocontact shift values by considering that electron spin density is spatially delocalized at positions associated with high electron density. The dipolar coupling with the nucleus is then expressed, in the point dipole approximation, as a discrete sum on the contributions from the dipole–dipole interactions between the electron magnetic dipoles located in these positions and the nuclear magnetic dipole (Section 2.2.2.2) [15].

REFERENCES

[1] McConnell HM, Chesnut DB. J. Chem. Phys. 1958;28:107.
[2] Kurland RJ, McGarvey BR. J. Magn. Reson. 1970;2:286.
[3] Tsitovich PB, Cox JM, Benedict JB, Morrow JR. Inorg. Chem. 2016;55:700.
[4] Bertini I, Luchinat C, Parigi G. Eur. J. Inorg. Chem. 2000;2473.
[5] McGarvey BR, Batista NC, Bezerra CWB, Schultz MS, Franco DW. Inorg. Chem. 1998;37:2865.
[6] McGarvey BR. Coord. Chem. Rev. 1998;170:75.
[7] Köhler FH, Lescouëzec R. Angew. Chem. Int. Ed. 2004;43:2571.
[8] Golding RM, Stubbs LC. J. Magn. Reson. 1979;33:627.
[9] Riley JP, Raynes WT. Mol. Phys. 1977;33:619.
[10] Stiles PJ. Mol. Phys. 1975;29:1271.
[11] Beringer R, Heald MA. Phys. Rev. 1954;95:1474.
[12] Buckingham AD, Stiles PJ. Mol. Phys. 1972;24:99.
[13] Stiles PJ. Mol. Phys. 1974;27:501.
[14] Golding RM, Pascual RO, Stubbs LC. Mol. Phys. 1976;31:1933.
[15] Walder BJ, Dey KK, Davis MC, Baltisberger JH, Grandinetti PJ. J. Chem. Phys. 2015;142:014201.
[16] Golding RM, Pascual RO, Vrbancich J. J. Mol. Phys. 1976;31:731.
[17] Charnock GTP, Kuprov I. Phys. Chem. Chem. Phys. 2014;16:20184.
[18] Luz Z, Shulman RG. J. Chem. Phys. 1965;43:3750.
[19] Bertini I, Luchinat C, Xia Z. Inorg. Chem. 1992;31:3152.
[20] Chmelnick AM, Fiat D. J. Chem. Phys. 1969;51:4238.
[21] Bertini I, Luchinat C, Xia Z. J. Magn. Reson. 1992;99:235.
[22] Reuben J, Fiat D. Inorg. Chem. 1967;6:579.
[23] Bloembergen N, Morgan LO. J. Chem. Phys. 1961;34:842.
[24] Luz Z, Shulman RG. J. Chem. Phys. 1965;43:3750.
[25] Bertini I, Fragai M, Luchinat C, Parigi G. Inorg. Chem. 2001;40:4030.

[26] Ducommun Y, Newman KE, Merbach A. Inorg. Chem. 1980;19:3696.
[27] Bertini I, Briganti F, Luchinat C, Xia Z. J. Magn. Reson. 1993;101:198.
[28] Bertini I, Capozzi F, Luchinat C, Xia Z. J. Phys. Chem. 1993;97:1134.
[29] Swift TJ, Connick RE. J. Chem. Phys. 1962;37:307.
[30] Ducommun Y, Earl WL, Merbach AE. Inorg. Chem. 1979;18:2754.
[31] Lewis WB, Alei M, Morgan LO. J. Chem. Phys. 1966;44:2409.
[32] Reuben J, Fiat D. J. Chem. Phys. 1969;51:4909.
[33] Fitzgerald RJ, Drago RS. J. Am. Chem. Soc. 1967;89:2879.
[34] Quaegebeur JP, Chachaty C, Yasukawa T. Mol. Phys. 1979;37:409.
[35] Wayland BB, Rice WL. Inorg. Chem. 1967;6:2270.
[36] Happe JA, Ward RL. J. Chem. Phys. 1963;39:1211.
[37] Kluiber RW, Horrocks WD Jr. Inorg. Chem. 1967;6:166.
[38] Holm RH, Everett GW, Horrocks WD Jr. J. Am. Chem. Soc. 1966;88:1071.
[39] Zhidomirov GM, Schastnev PV, Chuvylkin ND. J. Struct. Chem. 1970;11:458.
[40] Cramer RE, Drago RS. J. Am. Chem. Soc. 1970;92:66.
[41] Karplus M, Fraenkel GK. J. Chem. Phys. 1961;35:1312.
[42] Strom ET, Underwood GR, Jurkowitz D. Mol. Phys. 1972;24:901.
[43] Derbishire W. Mol. Phys. 1962;5:225.
[44] Eaton DR, Josey AD, Phillips WD, Benson RE. Mol. Phys. 1962;5:407.
[45] Stone EW, Maki AH. J. Chem. Phys. 1962;37:1326.
[46] McLachlan AD. Mol. Phys. 1958;1:233.
[47] Heller C, McConnell HM. J. Chem. Phys. 1960;32:1535.
[48] Lykos PG. J. Chem. Phys. 1960;32:625.
[49] Karplus M. J. Am. Chem. Soc. 1963;85:2870.
[50] Bertini I, Capozzi F, Luchinat C, Piccioli M, Vila AJ. J. Am. Chem. Soc. 1994;116:651.
[51] Bertini I, Ciurli S, Dikiy A, Gasanov R, Luchinat C, Martini G, Safarov N. J. Am. Chem. Soc. 1999;121:2037.
[52] Holm RH, Kennepohl P, Solomon EI. Chem. Rev. 1996;96:2239.
[53] Sheppard SK, Eaton GR, Eaton SS. Inorg. Chem. 1989;28:4496.
[54] McConnell HM. Proc. Natl. Acad. Sci. USA 1972;69:335.
[55] Hayamizu K, Murata M, Yamamoto O. Bull. Chem. Soc. Jpn. 1975;48:1842.
[56] Bertini I, Gatteschi D, Scozzafava A. Inorg. Chim. Acta 1972;6:185.
[57] Bertini I, Luchinat C, Scozzafava A. Inorg. Chim. Acta 1976;19:201.
[58] Ito S, Yano Y. Bull. Chem. Soc. Jpn. 1984;57:2824.
[59] Doddrell DM, Roberts JD. J. Am. Chem. Soc. 1970;92:6839.
[60] Goff HM. In: Lever ABP, Gray HB, editors. Iron Porphyrins. Reading, MA: Addison-Wesley; 1983. p. 237.
[61] Goff HM. J. Am. Chem. Soc. 1977;99:7723.
[62] Morishima I, Inubushi T, Neya S, Ogawa S, Yonezawa T. Biochem. Biophys. Res. Commun. 1977;78:739.
[63] Morishima I, Inubushi T. J. Am. Chem. Soc. 1978;100:3568.
[64] Behere DV, Gonzales-Vergara E, Goff HM. Biochem. Biophys. Res. Commun. 1985;131:607.
[65] Banci L, Bertini I, Kuan I-C, Tien M, Turano P, Vila AJ. Biochemistry 1993;32:13483.
[66] Pan H, Gerstein BC. J. Magn. Reson. 1991;92:618.
[67] McLaughlin AC, Grathwohl C, Richards RE. J. Magn. Reson. 1978;31:283.
[68] Boubel J-C, Delpuech J-J, Peguy A. J. Chem. Soc. Dalton Trans. 1978;1506.
[69] Bhasin CP, Srivastava G, Mehrotra RC. Inorg. Chim. Acta 1987;128:69.
[70] Labauze G, Raynor JB. J. Chem. Soc. Dalton Trans. 1981;590.
[71] Horrocks WD Jr, Hall DD. Inorg. Chem. 1971;10:2368.
[72] Horrocks WD Jr, Greenberg ES. Biochim. Biophys. Acta 1973;322:38.
[73] Trehwella J, Wright PE. Biochim. Biophys. Acta 1980;625:202.

[74] Shirazi A, Goff HM. Inorg. Chem. 1982;21:3420.
[75] Horrocks WD Jr. Inorg. Chem. 1970;9:670.
[76] Kaupp M, Köhler FH. Coord. Chem. Rev. 2009;253:2376.
[77] Hrobárik P, Revialkine R, Arbuznikov AV, Malkina OL, Malkin VG, Köhler FH, Kaupp M. J. Chem. Phys. 2007;126:024107.
[78] Pennanen TO, Vaara J. Phys. Rev. Lett. 2008;100:133002.
[79] Soncini A, Van den Heuvel W. J. Chem. Phys. 2013;138:021103.
[80] Rouf SA, Mareš J, Vaara J. J. Chem. Theory Comput. 2015;11:1683.
[81] Van den Heuvel W, Soncini A. Phys. Rev. Lett. 2012;109:073001.
[82] Neese F. J. Chem. Phys. 2007;127:164112.
[83] Malkina OL, Vaara J, Schimmelpfennig B, Munzarová M, Malkin VG, Kaupp M. J. Am. Chem. Soc. 2000;122:9206.
[84] Dlugopolska K, Ruman T, Danilczuk M, Pogocki D. Appl. Magn. Reson. 2008;35:271.
[85] Schmitt S, Jost P, van Wüllen C. J. Chem. Phys. 2011;134:194113.
[86] Kubica A, Kowalewski J, Kruk D, Odelius M. J. Chem. Phys. 2013;138:064304.
[87] Roos BO, Taylor PR, Siegbahn PEM. Chem. Phys. 1980;48:157.
[88] Angeli C, Borini S, Cimiraglia R. Theor. Chem. Acc. 2004;111:352.
[89] Rastrelli F, Bagno A. Chem. Eur. J. 2009;15:7990.
[90] Rastrelli F, Bagno A. Magn. Reson. Chem. 2010;48:S132–41.
[91] Komorovsky S, Repisky M, Ruud K, Malkina OL, Malkin VG. J. Phys. Chem. A 2013;117:14209.
[92] Wilkens SJ, Xia B, Weinhold F, Markley JL, Westler WM. J. Am. Chem. Soc. 1998;120:4806.
[93] Borgogno A, Rastrelli F, Bagno A. Dalton Trans. 2014;43:9486.
[94] Borgogno A, Rastrelli F, Bagno A. Chem. Eur. J. 2015;21:12960.

CHAPTER

THE EFFECT OF PARTIAL ORIENTATION: RESIDUAL DIPOLAR COUPLINGS

3

Sharp proton NMR lines are associated to short electron relaxation times, which in turn depend on the availability of low lying excited states (Chapter 4). Such systems also provide magnetic susceptibility anisotropy and pseudocontact shifts (Chapter 2). Magnetic anisotropy induces partial orientation (self-orientation) in high magnetic fields. Under these circumstances, the dipolar coupling energies average to values different from zero and residual dipolar couplings arise. Hyperfine shifts are also expected to change under partial orientation. The correction of the pseudocontact shifts in system partially oriented due to magnetic susceptibility anisotropy is negligible with respect to the isotropic values, but may become detectable at the highest magnetic fields available today.

3.1 PARTIAL ORIENTATION EFFECTS ON NUCLEUS-NUCLEUS DIPOLAR COUPLING

The network of the interactions among nuclear magnetic moments, between magnetic moments of nuclei and unpaired electrons, and between nuclear magnetic moments and the field generated by electron currents in orbitals, is responsible for the prohibitively broad NMR spectra of static samples in the solid state (Chapter 5). In solution, the fast isotropic reorientations of molecules average out the contributions corresponding to different nuclear positions with respect to the external magnetic field. This allows for a substantial simplification of the NMR spectrum, leaving only sharp lines. In diamagnetic systems the latter are centered at an average value of the chemical shielding interactions; in paramagnetic systems they are shifted with respect to this average value by the pseudocontact shift described in Chapter 2 (Table 3.1). In both cases they are split by the scalar coupling with covalently bound spins (^1J splitting).

This simplification of the spectra implies a substantial loss of relevant structural and dynamical information. This information can be partially recovered by making anisotropic the distribution of the molecular orientations. Anisotropic reorientations can be achieved either when molecules interact with strongly aligned objects or when the molecules themselves have some preferred orientations in a high magnetic field because their magnetic susceptibility is anisotropic. In the latter case, partial self-alignment of the molecules occurs. Many structured nucleic acid fragments have this feature, as well as several molecules containing paramagnetic metal ions.

Partial alignment, either due to external media or to self-alignment, reintroduces some effects related to the anisotropy of the nuclear interactions (Table 3.1). The most widely known and exploited effect is the so-called residual dipolar coupling, measured as a perturbation to the J-splitting (Section 12.4) observed for alignments as small as 1/1000, with respect to the value measured for unaligned sample (Fig. 3.1). Another effect related to paramagnetism is the residual dipolar shift, discussed in Section 3.5.

NMR of Paramagnetic Molecules. http://dx.doi.org/10.1016/B978-0-444-63436-8.00003-X

Table 3.1 Summary of Phenomena Involving Nuclear Spins and Their Manifestations in Isotropic and Anisotropic Solutions

Effect	Due to Interaction of the Nuclear Spin with	Isotropic Average	Result in the Presence of Partial Alignment
Chemical shielding	Electron currents in orbitals	Isotropic chemical shift	Residual chemical shift anisotropy [1]
Dipolar shielding	Average electron magnetic moment	Pseudocontact shift	Residual dipolar shift [2]
Quadrupolar coupling	Electric field gradients	(average to zero)	Residual quadrupolar coupling [3]
Dipole–dipole coupling	Magnetic moments of the neighboring nuclei	(average to zero)	Residual dipolar coupling [4]

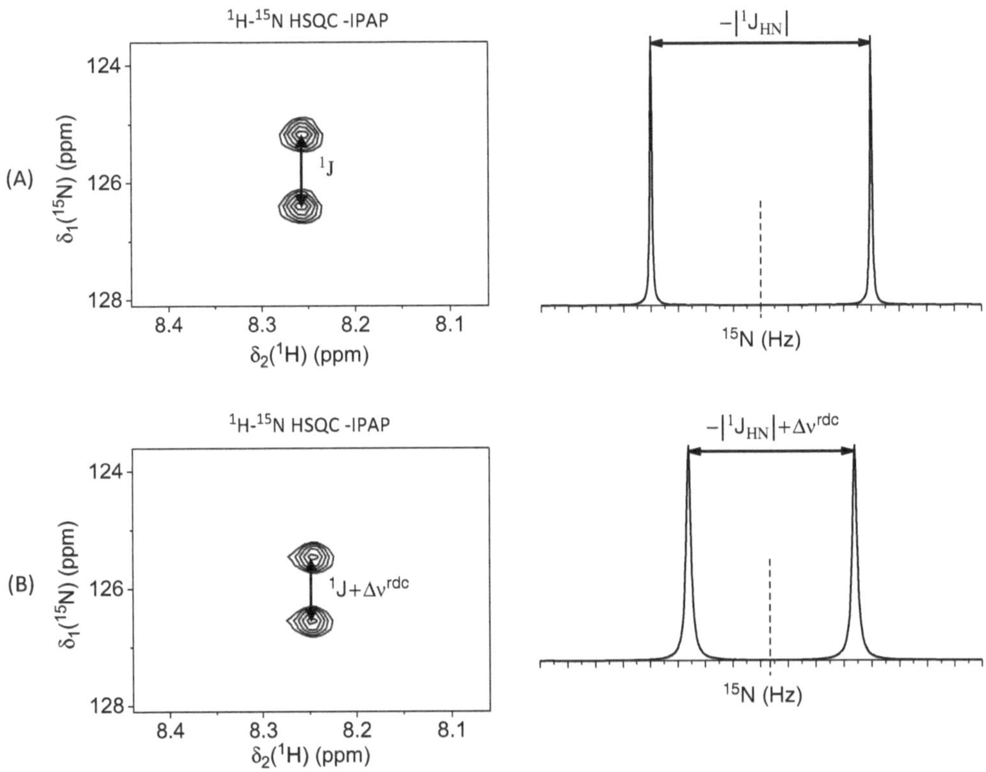

FIGURE 3.1 Coupled peaks in ^{1}H-^{15}N IPAP HSQC experiments for (A) a diamagnetic sample and (B) a paramagnetic sample with sizable magnetic susceptibility anisotropy. The difference in the frequency distance between the two peaks is due to the (positive) residual dipolar coupling; the mean of the peak frequencies is shifted due to the pseudocontact shift.

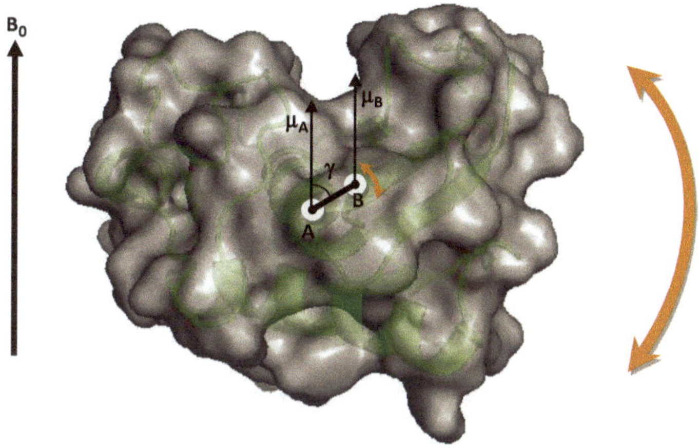

FIGURE 3.2 The angle γ describes the orientation of the vector connecting the two coupled nuclei A and B with respect to the external magnetic field, along which the magnetic moments of the nuclei are oriented. The angle γ changes due to molecular reorientation in solution and, if present, due to internal molecular dynamics.

Partial alignment of a molecule causes the occurrence of residual dipolar couplings because when not all orientations have the same probability, the dipole-dipole interaction between nuclear magnetic moments does not average to zero. From Eqs. (1.1), (1.4), and $\boldsymbol{\mu} = \hbar\gamma\boldsymbol{I}$, the residual dipolar coupling for the coupled nuclei A and B, at distance r_{AB}, is (Fig. 3.2)

$$\Delta\nu^{\mathrm{rdc}} = \frac{< E^{\mathrm{dip}} >}{h} = -\frac{\mu_0\hbar\gamma_A\gamma_B}{8\pi^2 r_{AB}^3}\langle 3\cos^2\gamma - 1\rangle \neq 0 \tag{3.1}$$

where γ is the angle between the direction of $\boldsymbol{B}_0$ and that of the AB vector (i.e., of the vector connecting the coupled nuclei A and B). The equation describing the residual dipolar couplings can be written as a function of the Saupe tensor (or order tensor) $\boldsymbol{S}$ [5], which describes the degree of alignment of a molecule [2],

$$\Delta\nu^{\mathrm{rdc}} = -\frac{\mu_0\hbar\gamma_A\gamma_B}{8\pi^2 r_{AB}^3}\left[S_{zz}\left(3\cos^2\alpha - 1\right) + \left(S_{xx} - S_{yy}\right)\sin^2\alpha\cos 2\beta\right] \tag{3.2}$$

where α is the angle between the AB vector and the z-axis of the $\boldsymbol{S}$ tensor, β is the angle which describes the position of the projection of the AB vector on the xy plane of the $\boldsymbol{S}$ tensor, relative to the x-axis, when x, y, and z are the principal directions of the traceless order tensor $\boldsymbol{S}$, and

$$S_{ii} = \frac{1}{2}\int_0^{\pi}\left(3\cos^2\theta_{i\zeta} - 1\right)P(\theta_{i\zeta})\sin\theta_{i\zeta}\,d\theta_{i\zeta} \tag{3.3}$$

where $\theta_{i\zeta}$ is the angle between the i direction and the magnetic field direction, and $P(\theta_{i\zeta})$ is the probability for that orientation. According to the typical convention, the axes of the reference frame are defined so that the magnitudes of the principal components are $|S_{zz}| > |S_{yy}| > |S_{xx}|$. As shown in Appendix II [Eq. (II.3)], when $P(\theta_{i\zeta})$ is constant, $S_{ii} = 0$ and thus $\Delta\nu^{\mathrm{rdc}} = 0$.

Eq. (3.2) is often written [6] as

$$\Delta\nu^{\text{rdc}} = -\frac{\mu_0 \hbar \gamma_A \gamma_B S_{\text{LS}}}{8\pi^2 r_{AB}^3} A_{ax} \left[(3\cos^2\alpha - 1) + \frac{3}{2} R \sin^2\alpha \cos 2\beta \right] \tag{3.4}$$

where the alignment (traceless) tensor $\boldsymbol{A}$ is related to the Saupe matrix through the relationship $\boldsymbol{A} = \frac{2}{3}\boldsymbol{S}$, so that, since $S_{xx} + S_{yy} + S_{zz} = 0$,

$$A_{ax} = \frac{2}{3}\left(S_{zz} - \frac{S_{xx} + S_{yy}}{2} \right) = S_{zz}$$
$$A_{rh} = \frac{2}{3}\left(S_{xx} - S_{yy} \right) \tag{3.5}$$

and $R = \frac{A_{rh}}{A_{ax}}$. In Eq. (3.4) S_{LS} is the generalized Lipari-Szabo order parameter [7], not to be confused with the Saupe matrix. S_{LS} ($0 \le S_{\text{LS}} \le 1$) accounts for the effect of internal motions with correlation times shorter than the molecular reorientation time. It corresponds to the spin relaxation order parameter, but it is sensitive to motions extending up to the millisecond time scale [8]. In fact, the averaging in Eq. (3.1) extends to times as long as 10^{-2} s because *J*-splitting values are of the order of 1–100 Hz; therefore, the orientation of the AB vector with respect to the magnetic field depends on the superposition of the internal molecular motions up to the milliseconds time scale and of the orientational distribution of the molecule as a rigid object (described by the $\boldsymbol{A}$, or $\boldsymbol{S}$, tensors). Equations are also available for the case of internal motions not axially symmetric about the mean orientation, so that residual dipolar couplings depend on the generalized order parameters $S_{zz,int}$ and $S_{xx,int} - S_{yy,int}$ instead of a single S_{LS} Lipari-Szabo parameter [9]. The case of extensive internal molecular mobility will be discussed in Section 9.6.

3.2 THE MAGNETIC FIELD INDUCED ORDER

As seen in Section 1.6, the magnetic susceptibility describes the proportionality between the induced magnetic moment in a molecule and the applied magnetic field [Eq. (1.27)], i.e., the extent of the magnetization of a substance in a magnetic field. It is represented by a tensor, indicated as $\boldsymbol{\chi}^{\text{mol}}$ [Eq. (1.45)], because $<\boldsymbol{\mu}>$ can be different depending on the orientation of the molecule with respect to the magnetic field. Anisotropy in the magnetic susceptibility tensor of a molecule causes partial orientation of the molecule itself.

In diamagnetic molecules the anisotropy of $<\boldsymbol{\mu}>$, and thus of $\boldsymbol{\chi}^{\text{dia}}$, is due to the interaction of the magnetic field with the motion of the electrons in their orbitals (contributions from paramagnetic excited states should also be considered). It is mainly related to the anisotropy of the electron currents, for instance obtained in the presence of aromatic rings. It is usually rather small in biomolecules, except for instance in the case of heme proteins or when multiple aromatic planes are stacked together, as in the case of double stranded nucleic acids.

Even if the local contributions to the overall susceptibility are small, they can become very relevant when they are summed over several repetitions in one or two dimensions. This is the case of bicelles and filamentous phages: such macroscopic objects can align completely in a magnetic field. Thus, they provide an anisotropic environment, which induces partial alignment also on the biomolecules in solution.

In the presence of a paramagnetic metal, the unpaired electron spin(s) provide a paramagnetic contribution [$\boldsymbol{\chi}$, see Eq. (1.36)] to the $\boldsymbol{\chi}^{\mathrm{mol}}$ tensor ($\boldsymbol{\chi}^{\mathrm{mol}} = \boldsymbol{\chi}^{\mathrm{dia}} + \boldsymbol{\chi}$). As described in Section 1.6, in electron configurations with a nonzero orbital magnetic moment, the average induced electron magnetic moment $<\boldsymbol{\mu}>$ is anisotropic. In brief, in paramagnetic systems, an anisotropic $\boldsymbol{\chi}$ tensor arises due to the orbital contribution to the electron magnetic moment, and this induces partial self-orientation of the molecule.

In all cases where the molecular magnetic susceptibility tensor, $\boldsymbol{\chi}^{\mathrm{mol}}$, is anisotropic, since the average magnetic moment per particle $<\boldsymbol{\mu}>$ induced by a magnetic field $\boldsymbol{B}_0$ is defined as

$$<\mu> = \frac{\boldsymbol{\chi}^{\mathrm{mol}} \cdot \boldsymbol{B}_0}{\mu_0}, \tag{3.6}$$

the change in the energy of the system resulting from the placement of the molecule in the magnetic field

$$E = -\frac{\boldsymbol{B}_0 \cdot <\mu>}{2}, \tag{3.7}$$

is anisotropic, depending on the orientation and magnitude of $<\boldsymbol{\mu}>$ with respect to $\boldsymbol{B}_0$. The different energy for the different molecular orientations relative to $\boldsymbol{B}_0$ causes a partial alignment favoring the lowest energy states (Fig. 3.3). This alignment can be expressed by the components of the order tensor [see Eq. (3.3)]

$$S_{ii} = \frac{\int \frac{3\cos^2\alpha - 1}{2} \exp[-E(\alpha,\beta)/(kT)] \mathrm{d}\cos\alpha\, \mathrm{d}\beta}{\int \exp[-E(\alpha,\beta)/(kT)] \mathrm{d}\cos\alpha\, \mathrm{d}\beta} \tag{3.8}$$

where α and β are the spherical angles describing the direction of $\boldsymbol{B}_0$ in the frame where the $\boldsymbol{\chi}$ tensor is diagonal. When the anisotropic energy $E(\alpha, \beta)$, described by Eq. (3.7), is much smaller for a subset

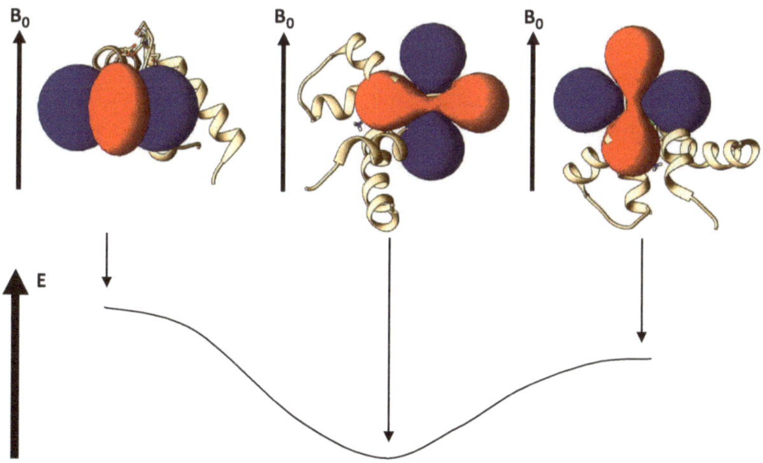

FIGURE 3.3 The energy of the system differs depending on the molecular orientations with respect to the external magnetic field due to anisotropy of the magnetic susceptibility tensor. This causes partial self-orientation of the molecule, and thus occurrence of self-orientation residual dipolar couplings.

of orientations than for all others, so much so that the latter are not thermally accessible, the molecular alignment is complete: this is the case of bicelles and other orienting devices that are totally aligned by the magnetic field. On the other hand, if the differences in energy among the different orientations are small, the alignment is only partial. Therefore,

$$S_{ii} = \frac{\int \frac{3\cos^2\alpha - 1}{2} \exp\left[\frac{B_0^2}{2\mu_0 kT}\left(\chi_{ii}^{\text{mol}}\cos^2\alpha + \chi_{jj}^{\text{mol}}\sin^2\alpha\cos^2\beta + \chi_{kk}^{\text{mol}}\sin^2\alpha\sin^2\beta\right)\right] \mathrm{d}\cos\alpha\,\mathrm{d}\beta}{\int \exp\left[\frac{B_0^2}{2\mu_0 kT}\left(\chi_{ii}^{\text{mol}}\cos^2\alpha + \chi_{jj}^{\text{mol}}\sin^2\alpha\cos^2\beta + \chi_{kk}^{\text{mol}}\sin^2\alpha\sin^2\beta\right)\right] \mathrm{d}\cos\alpha\,\mathrm{d}\beta} \tag{3.9}$$

as obtained from Eqs. (3.6)–(3.8), with $\boldsymbol{\chi}^{\text{mol}}$ representing the whole molecular magnetic susceptibility tensor, defined in Eq. (1.45), and thus containing both the diamagnetic and the paramagnetic contributions. The exponential can be approximated to first order, so that

$$S_{ii} = \frac{B_0^2}{15\mu_0 kT}\left(\chi_{ii} - \frac{\chi_{jj} + \chi_{kk}}{2}\right) = \frac{3}{2}\frac{B_0^2}{15\mu_0 kT}\left(\chi_{ii}^{\text{mol}} - \bar{\chi}^{\text{mol}}\right) \tag{3.10}$$

where

$$\bar{\chi}^{\text{mol}} = \frac{\chi_{xx}^{\text{mol}} + \chi_{yy}^{\text{mol}} + \chi_{zz}^{\text{mol}}}{3}. \tag{3.11}$$

3.3 PARAMAGNETIC METAL IONS AND RESIDUAL DIPOLAR COUPLINGS

Substituting the S_{xx}, S_{yy}, and S_{zz} components provided by Eq. (3.10) into Eq. (3.2), the equation for self-orientation residual dipolar couplings (in Hz) is obtained:

$$\Delta\nu^{\text{rdc}} = -\frac{1}{4\pi}\frac{B_0^2}{15kT}\frac{\gamma_A\gamma_B\hbar}{2\pi r_{AB}^3}\left[\Delta\chi_{ax}^{\text{mol}}(3\cos^2\alpha - 1) + \frac{3}{2}\Delta\chi_{rh}^{\text{mol}}\sin^2\alpha\cos 2\beta\right] \tag{3.12}$$

where α is the angle between the AB vector and the z-axis of the $\boldsymbol{\chi}^{\text{mol}}$ tensor, β is the angle which describes the position of the projection of the AB vector on the xy plane of the $\boldsymbol{\chi}^{\text{mol}}$ tensor, relative to the x-axis (Fig. 3.4), and

$$\begin{aligned} \Delta\chi_{ax}^{\text{mol}} &= \chi_{zz}^{\text{mol}} - \frac{\chi_{xx}^{\text{mol}} + \chi_{yy}^{\text{mol}}}{2} = \frac{3}{2}\left(\chi_{zz}^{\text{mol}} - \bar{\chi}^{\text{mol}}\right) \\ \Delta\chi_{rh}^{\text{mol}} &= \chi_{xx}^{\text{mol}} - \chi_{yy}^{\text{mol}} \end{aligned} \tag{3.13}$$

analogously to the definition of axial and rhombic anisotropies provided in Eq. (2.47). As in that case, the conventional definition of the x-, y-, and z-axis of the $\boldsymbol{\chi}^{\text{mol}}$ tensor is such that $|\Delta\chi_{rh}^{\text{mol}}| \le \frac{2}{3}|\Delta\chi_{ax}^{\text{mol}}|$. To be noted that whereas pseudocontact shifts depend on the anisotropy of the magnetic susceptibility tensor of the paramagnetic metal, $\boldsymbol{\chi}$, only, residual dipolar couplings depend on the anisotropy of the whole molecular magnetic susceptibility tensor, including the diamagnetic contribution. Of course, when the magnetic susceptibility tensor is isotropic, ($\chi_{xx}^{\text{mol}} = \chi_{yy}^{\text{mol}} = \chi_{zz}^{\text{mol}} = \bar{\chi}^{\text{mol}}$) the residual dipolar couplings are zero. Therefore, it is common to express the residual dipolar couplings (as also the pseudocontact

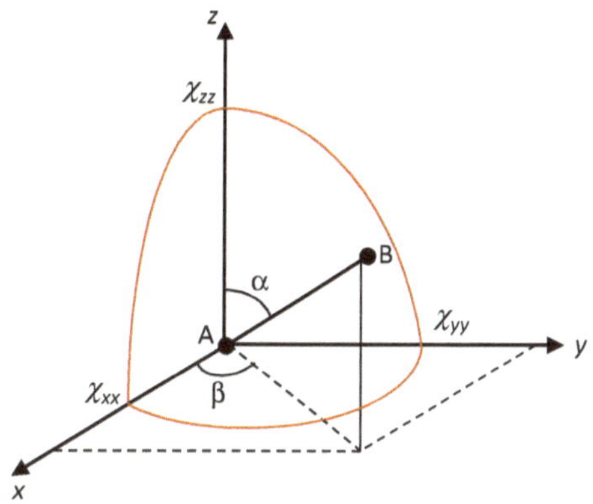

FIGURE 3.4 Angles α and β defining the orientation of the coupled nuclei vector in the principal frame of the magnetic susceptibility anisotropy tensor. The x-, y-, and z-axes are defined so that $|\chi_{xx}| < |\chi_{yy}| < |\chi_{zz}|$.

shifts) as a function of the magnetic susceptibility anisotropy tensor ($\Delta\boldsymbol{\chi}^{\text{mol}}$), defined as the difference between $\boldsymbol{\chi}^{\text{mol}}$ and $\bar{\chi}^{\text{mol}}$.

When the difference is calculated between the splitting of coupled nuclei observed for the paramagnetic molecule and a diamagnetic analog (Section 12.4), the contribution from the paramagnetic metal can be isolated and the equation to be used to analyze the residual dipolar coupling data becomes

$$\Delta\nu^{\text{rdc}} = -\frac{1}{4\pi}\frac{B_0^2}{15kT}\frac{\gamma_A\gamma_B\hbar S_{\text{LS}}}{2\pi r_{AB}^3}\left[\Delta\chi_{ax}(3\cos^2\alpha-1)+\frac{3}{2}\Delta\chi_{rh}\sin^2\alpha\cos 2\beta\right] \tag{3.14}$$

where $\Delta\chi_{ax}$ and $\Delta\chi_{rh}$ coincide with those defined as in Eq. (2.47) and the generalized Lipari-Szabo order parameter S_{LS} is introduced [analogously to Eq. (3.4)].

The equation describing self-orientation residual dipolar couplings is thus just a specialized case of the Eq. (3.4) used, when orientation is induced by external orienting media. For both external and internal orienting media, the residual dipolar couplings globally depend on five parameters, the two anisotropy values (A_{ax} and R for residual dipolar couplings induced by external orienting media, and $\Delta\chi_{ax}$ and $\Delta\chi_{rh}$ for self-orientation residual dipolar couplings) and the three angles defining the orientation of the tensor principal components. Once these five parameters have been determined, the residual dipolar coupling of any specific nuclear pair depends on the orientation of the nuclear pair vector in the common reference frame where the tensor is diagonal.

Eq. (3.14) describes the residual dipolar couplings in the frame where the $\boldsymbol{\chi}$ tensor is diagonal, in the sense that the angles α and β are spherical coordinates providing the orientation of the AB vector in such frame. Residual dipolar couplings can be written in the same frame in Cartesian coordinates (Section 2.2.2.1)

$$\Delta\nu^{\text{rdc}} = -\frac{1}{4\pi}\frac{B_0^2}{15kT}\frac{\gamma_A\gamma_B\hbar S_{\text{LS}}}{2\pi r_{AB}^3}\left[\Delta\chi_{ax}\frac{2z_{AB}^2-x_{AB}^2-y_{AB}^2}{r_{AB}^2}+\frac{3}{2}\Delta\chi_{rh}\frac{x_{AB}^2-y_{AB}^2}{r_{AB}^2}\right] \tag{3.15}$$

where $x_{AB}^2=(x_A-x_B)^2$, $y_{AB}^2=(y_A-y_B)^2$, and $z_{AB}^2=(z_A-z_B)^2$, or using the direction cosines (cos φ_i are the direction cosines relating a vector to the $i = x, y, z$ axes of the reference frame)

$$\Delta\nu^{\mathrm{rdc}} = -\frac{3}{4\pi}\frac{B_0^2}{15kT}\frac{\gamma_A\gamma_B\hbar S_{\mathrm{LS}}}{2\pi r_{AB}^3}\sum_i \chi_{ii}\cos^2\varphi_i \tag{3.16}$$

where the index i runs on the three principal directions of the χ tensor and φ_i are the angles between the AB vector and each of these principal directions.

Analogously to pseudocontact shifts, also for the residual dipolar couplings the equations can be written in any reference frame, different from that defined by the principal directions of the magnetic susceptibility anisotropy tensor, in spherical coordinates

$$\begin{aligned}\Delta\nu^{\mathrm{rdc}} = -\frac{3}{4\pi}\frac{B_0^2}{15kT}\frac{\gamma_A\gamma_B\hbar S_{\mathrm{LS}}}{2\pi r_{AB}^3}\Bigg[&\chi_{zz}\frac{(3\cos^2\alpha-1)}{2}+(\chi_{xx}-\chi_{yy})\frac{\sin^2\alpha\cos 2\beta}{2}\\ &+\chi_{xy}\sin^2\alpha\sin 2\beta+\chi_{xz}\sin 2\alpha\cos\beta+\chi_{yz}\sin 2\alpha\sin\beta\Bigg]\end{aligned} \tag{3.17}$$

where α and β are the angles between the AB vector and the z-axis of the chosen reference frame and between the projection of the AB vector on the xy plane and the x-axis of the chosen frame, respectively; in Cartesian coordinates

$$\begin{aligned}\Delta\nu^{\mathrm{rdc}} = -\frac{3}{4\pi}\frac{B_0^2}{15kT}\frac{\gamma_A\gamma_B\hbar S_{\mathrm{LS}}}{2\pi r_{AB}^3}\Bigg[&\chi_{zz}\frac{2z_{AB}^2-x_{AB}^2-y_{AB}^2}{2r_{AB}^2}+(\chi_{xx}-\chi_{yy})\frac{x_{AB}^2-y_{AB}^2}{2r_{AB}^2}\\ &+\chi_{xy}\frac{2x_{AB}y_{AB}}{r_{AB}^2}+\chi_{xz}\frac{2x_{AB}z_{AB}}{r_{AB}^2}+\chi_{yz}\frac{2y_{AB}z_{AB}}{r_{AB}^2}\Bigg]\end{aligned} \tag{3.18}$$

or using the direction cosines

$$\Delta\nu^{\mathrm{rdc}} = -\frac{3}{4\pi}\frac{B_0^2}{15kT}\frac{\gamma_A\gamma_B\hbar S_{\mathrm{LS}}}{2\pi r_{AB}^3}\sum_{i,j}\chi_{ij}\cos\varphi_i\cos\varphi_j \tag{3.19}$$

where the indices i and j run over the three axes x, y, and z, and φ_i are the angles between the AB vector and each of the three axes. In these cases, the χ tensor is not diagonal, and the five unknowns are the 5 components defining such symmetric tensor. The sixth component (one of the three diagonal elements, for instance χ_{xx}) can be set to the value making the tensor traceless (i.e., $\chi_{xx} = -\chi_{yy} - \chi_{zz}$) because, as it should be clear from Eq. (3.14) and the definition of $\Delta\chi_{ax}$ and $\Delta\chi_{rh}$, the residual dipolar couplings do not depend on the trace of the χ tensor.

From Eqs. (3.16) and (3.19), the residual dipolar couplings can also be written as

$$\Delta\nu^{\mathrm{rdc}} = -\frac{3}{4\pi}\frac{B_0^2}{15kT}\frac{\gamma_A\gamma_B\hbar S_{\mathrm{LS}}}{2\pi r_{AB}^5}\boldsymbol{P}^T\chi\boldsymbol{P} \tag{3.20}$$

where $\boldsymbol{P}$ represents the coupled nuclei vector with components x_{AB}, y_{AB}, and z_{AB}, written in the same frame of the χ tensor (and $\boldsymbol{P}^T$ is the transposed vector).

Despite the same functional form of the equations describing pseudocontact shifts [Eq. (2.46)] and residual dipolar couplings [Eqs. (3.12) and (3.14)], the spherical coordinates are differently defined: the angles in the equation for the pseudocontact shifts describe the position of the nuclei in the frame defined by the χ tensor, with origin on the paramagnetic metal ion position (Fig. 2.6), while those in the equations for the residual dipolar couplings describe the orientation of the vector connecting the coupled nuclei (Fig. 3.4) in the frame defined by the magnetic susceptibility anisotropy tensor (either $\boldsymbol{\chi}^{\mathrm{mol}}$ or $\boldsymbol{\chi}$, depending on the measurement procedure, see Section 3.4); the distance corresponds to the metal-nuclear distance (r) for pseudocontact shifts and to the fixed distance between coupled nuclei (r_{AB}) for residual dipolar couplings. Therefore, the residual dipolar couplings do not depend at all on the distance of the observed nuclei from the paramagnetic center, as well as on the position of each nucleus with respect to the $\boldsymbol{\chi}$ tensor. Furthermore, the residual dipolar couplings increase quadratically with the applied magnetic field and depend on the γ of the coupled nuclei.

As for pseudocontact shifts, residual dipolar couplings can be predicted by knowing nuclear coordinates and tensor components. On the other hand, the tensor components can be calculated if at least 5 residual dipolar couplings are known (in rigid systems), because the anisotropy (traceless part) of the $\boldsymbol{\chi}$ tensor is determined by 5 terms ($\Delta\chi_{ax}$, $\Delta\chi_{rh}$ and three angles defining the reference frame where the tensor is diagonal, or, alternatively, χ_{yy}, χ_{zz}, χ_{xy}, χ_{xz}, and χ_{yz}). On the contrary, there is a large degeneracy in the nuclear coordinates that can be determined from known residual dipolar couplings and tensor components (Figs. 3.5 and 3.6 and Section 9.4).

From Eq. (3.19), it is possible to write a residual dipolar coupling, after normalization to a prefactor, as a product between one vector containing the direction cosines ($\boldsymbol{v}$) of the AB vector with respect to the coordinate axes and one vector containing the 5 independent components of the $\boldsymbol{\chi}$ tensor:

$$\Delta\nu^{\mathrm{rdc,norm}} = \frac{\Delta\nu^{rdc}}{-\dfrac{3}{4\pi}\dfrac{B_0^2}{15kT}\dfrac{\gamma_A\gamma_B\hbar S_{\mathrm{LS}}}{2\pi r_{AB}^3}} =$$

$$= \left(\cos^2\varphi_y - \cos^2\varphi_x \;\; 2\cos\varphi_x\cos\varphi_y \;\; 2\cos\varphi_x\cos\varphi_z \;\; 2\cos\varphi_y\cos\varphi_z \;\; \cos^2\varphi_z - \cos^2\varphi_x\right)\begin{pmatrix}\chi_{yy}\\ \chi_{xy}\\ \chi_{xz}\\ \chi_{yz}\\ \chi_{zz}\end{pmatrix} =$$

$$= \boldsymbol{v}\begin{pmatrix}\chi_{yy}\\ \chi_{xy}\\ \chi_{xz}\\ \chi_{yz}\\ \chi_{zz}\end{pmatrix}. \qquad (3.21)$$

The vector containing the components of the $\boldsymbol{\chi}$ tensor is the same for all nuclear pairs, whereas the vector $\boldsymbol{v}$ depends on the orientation of the pair in the reference frame. Therefore, if residual dipolar couplings are measured for several pairs ($n \geq 5$), it is possible to write the vector ($\boldsymbol{D}$) containing the normalized residual dipolar couplings as a product between an $n \times 5$ matrix, where the rows are the 5-component $\boldsymbol{v}_i$ vectors, and the vector containing the components of the $\boldsymbol{\chi}$ tensor:

$$\boldsymbol{D} = \begin{pmatrix}\Delta\nu_1^{\mathrm{rdc,norm}}\\ \Delta\nu_2^{\mathrm{rdc,norm}}\\ \vdots\\ \Delta\nu_n^{\mathrm{rdc,norm}}\end{pmatrix} = \begin{pmatrix}\boldsymbol{v}_1\\ \boldsymbol{v}_2\\ \vdots\\ \boldsymbol{v}_n\end{pmatrix}\begin{pmatrix}\chi_{yy}\\ \chi_{xy}\\ \chi_{xz}\\ \chi_{yz}\\ \chi_{zz}\end{pmatrix} = \boldsymbol{V}\boldsymbol{\chi}. \qquad (3.22)$$

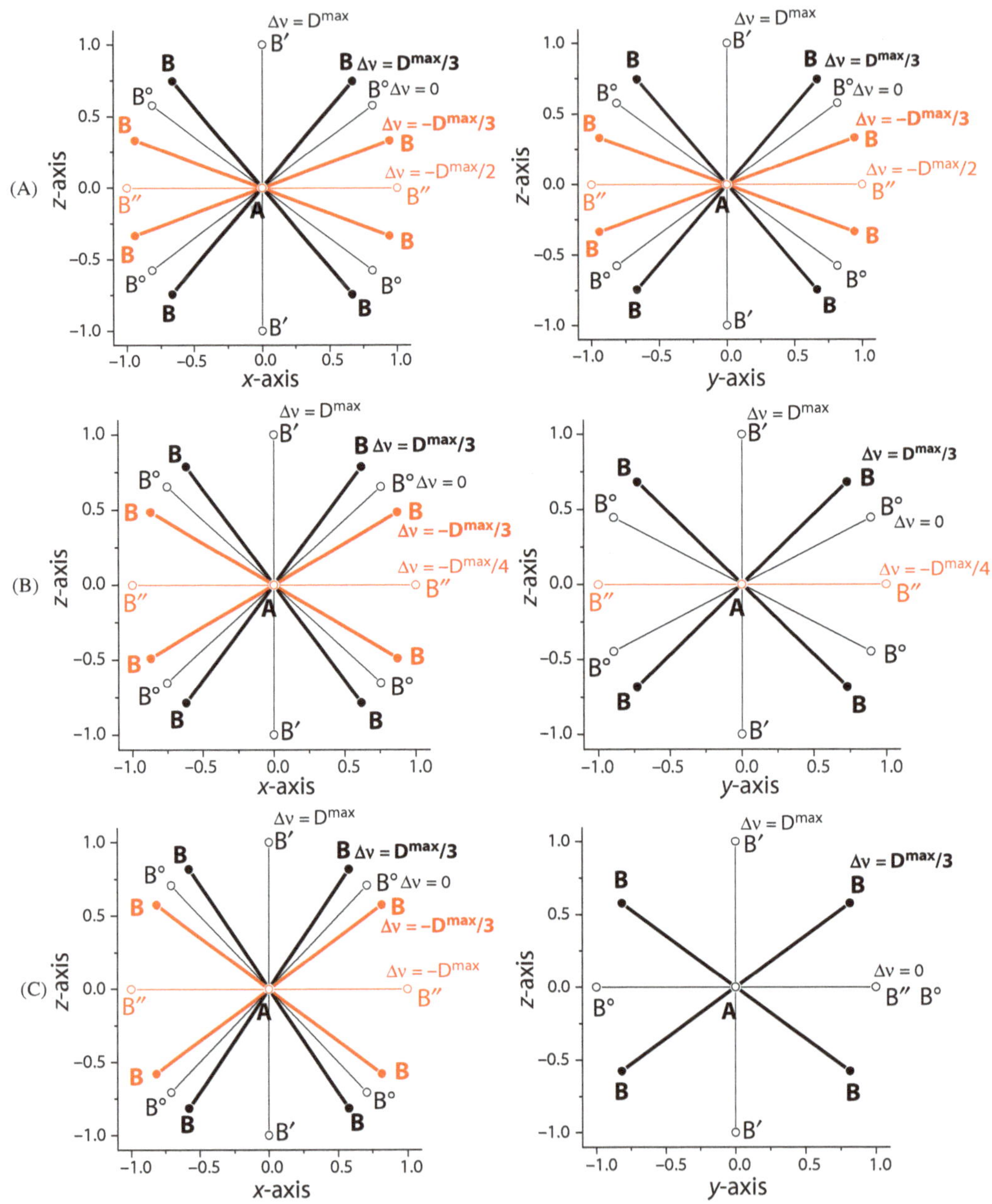

FIGURE 3.5 Possible angular positions within the *xz* and *yz* planes of a nucleus B coupled to the nucleus A positioned in the origin of the main frame of a magnetic susceptibility tensor with $\Delta\chi_{rh} = 0$ (A), $|\Delta\chi_{rh}| = |\Delta\chi_{ax}|/3$ (B), $|\Delta\chi_{rh}| = 2|\Delta\chi_{ax}|/3$ (C), and with residual dipolar coupling Δν equal to 1/3 of the maximum value, D^{max}. The value D^{max} is experienced when the nucleus is along the *z*-axis (B'). B° indicate the nuclear positions with residual dipolar coupling equal to zero; the value of the residual dipolar coupling in the plane perpendicular to the *z*-axis depends on the $\Delta\chi_{rh}$ value and on the angular position. The positions of the nucleus B with residual dipolar coupling equal to $-D^{max}/3$ are also shown, in red.

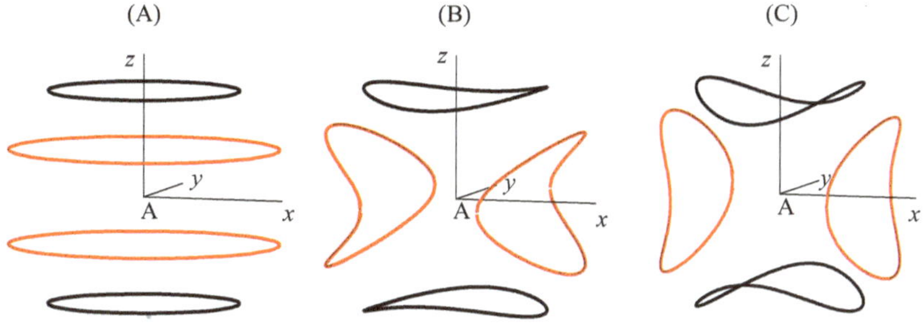

FIGURE 3.6 Possible positions of a nucleus B (lines in black) coupled to the nucleus A positioned in the origin of the main frame of a magnetic susceptibility tensor with $\Delta\chi_{rh} = 0$ (A), $|\Delta\chi_{rh}| = |\Delta\chi_{ax}|/3$ (B), $|\Delta\chi_{rh}| = 2|\Delta\chi_{ax}|/3$ (C) and with residual dipolar coupling $\Delta\nu$ equal to 1/3 of the maximum value, D^{max}, experienced when the nucleus is along the z-axis. The positions of the nucleus B with residual dipolar coupling equal to $-D^{max}/3$ are also shown, in red.

If the molecular structure is known, the $\boldsymbol{v}_i$ vectors can be calculated, so that it is possible to determine the $\boldsymbol{\chi}$ tensor by determining the Moore–Penrose inverse of the $\boldsymbol{V}$ matrix:

$$\boldsymbol{\chi} = \boldsymbol{V}^{-1}\boldsymbol{D} \tag{3.23}$$

using the singular value decomposition (SVD) method [10]. Alternatively, the 5 independent components of the $\boldsymbol{\chi}$ tensor can be determined through best fit calculations, performed in order to minimize the quadratic difference between experimental and calculated residual dipolar couplings, again with the elements of the $\boldsymbol{V}$ matrix [or with the nuclear coordinates if Eq. (3.18) is used] calculated from the known molecular structure. Due to the presence of experimental errors, the accuracy of the approach requires a number of experimental residual dipolar couplings substantially larger than five and well distributed within the range of values permitted by the tensor.

3.4 RESIDUAL DIPOLAR COUPLINGS IN THE PRESENCE OF DIAMAGNETIC AND PARAMAGNETIC ANISOTROPY

We have seen that in paramagnetic molecules, residual dipolar couplings must be calculated by taking into account the presence of both a paramagnetic and a diamagnetic contribution to the magnetic susceptibility anisotropy (when the diamagnetic susceptibility anisotropy is not negligible with respect to paramagnetic susceptibility). If a diamagnetic analog of the investigated molecule is available, the corresponding residual dipolar couplings depend only on the diamagnetic contribution to the magnetic susceptibility anisotropy.

In this case, residual dipolar couplings can be obtained by subtracting the experimental ^1J couplings measured for a diamagnetic analog from the corresponding values measured for the paramagnetic sample (assuming that cross-correlations effects are negligible, see Section 4.9). In this way the residual dipolar couplings depend on the paramagnetic $\boldsymbol{\chi}$ tensor only, i.e., on the same tensor responsible for the pseudocontact shifts. Alternatively, the differences between residual dipolar couplings at two different magnetic fields (B_{01} and B_{02}) can be obtained by subtracting the ^1J couplings measured at these fields, and in this case we obtain

$$\Delta(\Delta\nu)^{\text{rdc}} = -\frac{1}{4\pi}\frac{B_{01}^2 - B_{02}^2}{15kT}\frac{\gamma_A\gamma_B\hbar}{2\pi r_{AB}^3}\left[\Delta\chi_{ax}^{\text{mol}}(3\cos^2\alpha - 1) + \frac{3}{2}\Delta\chi_{rh}^{\text{mol}}\sin^2\alpha\cos 2\beta\right]. \tag{3.24}$$

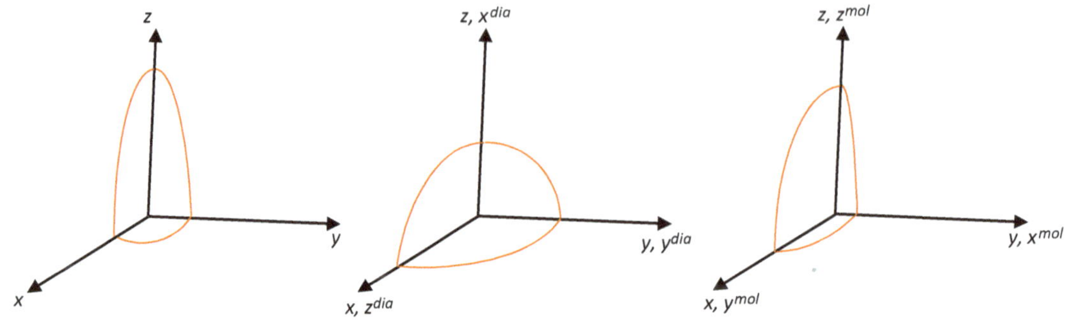

$$\Delta\chi + \Delta\chi^{dia} = \Delta\chi^{mol}$$

$$\begin{pmatrix} -2 & 0 & 0 \\ 0 & -2 & 0 \\ 0 & 0 & 4 \end{pmatrix} + \begin{pmatrix} 2 & 0 & 0 \\ 0 & -1 & 0 \\ 0 & 0 & -1 \end{pmatrix} = \begin{pmatrix} 0 & 0 & 0 \\ 0 & -3 & 0 \\ 0 & 0 & 3 \end{pmatrix}$$

$$\Delta\chi_{ax} = 6 \qquad \Delta\chi_{ax}^{dia} = 3 \qquad \Delta\chi_{ax}^{mol} = 4.5$$

$$\Delta\chi_{rh} = 0 \qquad \Delta\chi_{rh}^{dia} = 0 \qquad \Delta\chi_{rh}^{mol} = -3$$

FIGURE 3.7 Observed residual dipolar couplings depend on the tensorial sum of the diamagnetic and the paramagnetic susceptibility anisotropy tensors

These values thus depend on the overall $\boldsymbol{\chi}^{\mathbf{mol}}$ tensor. Since the diamagnetic and paramagnetic susceptibility tensors are diagonal, in general, in different frames, the residual dipolar couplings measured in paramagnetic samples depend on their tensorial sum (Fig. 3.7) [11].

As an example, in cytochrome b_5 [11], a heme protein containing an iron ion, ^{1}H-^{15}N ^1J couplings have been measured at two different fields (500 and 800 MHz) for the protein in both the oxidized (paramagnetic, low spin iron(III)) and reduced (diamagnetic, low spin iron(II)) states. From the analysis of the difference between the ^1J couplings at the two fields measured for the oxidized protein against the protein structure, the overall $\boldsymbol{\chi}^{\mathbf{mol}}$ tensor can be determined. The corresponding anisotropy values are $\Delta\chi_{ax} = 2.20 \pm 0.05 \times 10^{-32}$ m^3 and $\Delta\chi_{rh} = -1.34 \pm 0.04 \times 10^{-32}$ m^3. If the same analysis is performed for the reduced protein, the $\boldsymbol{\chi}^{\mathbf{dia}}$ tensor can be determined, in this case with anisotropy values $\Delta\chi_{ax} = -0.8 \pm 0.1 \times 10^{-32}$ m^3 and $\Delta\chi_{rh} = 0.1 \pm 0.3 \times 10^{-32}$ m^3. The paramagnetic $\boldsymbol{\chi}$ tensor can be evaluated from the pseudocontact shifts for the protein in the oxidized form, and its anisotropy values are $\Delta\chi_{ax} = 2.8 \pm 0.1 \times 10^{-32}$ m^3 and $\Delta\chi_{rh} = -1.1 \pm 0.2 \times 10^{-32}$ m^3, in good agreement with what expected from the tensorial sum $\boldsymbol{\chi}^{\mathbf{mol}} = \boldsymbol{\chi}^{\mathbf{dia}} + \boldsymbol{\chi}$. A similar analysis is available in the literature for oxidized and reduced cytochrome b_{562} from *E. coli* [12]. In this case, the diamagnetic and paramagnetic contributions to the molecular magnetic susceptibility anisotropy are of the same magnitude, but the *z*-axis of $\boldsymbol{\chi}^{\mathbf{dia}}$ tensor is almost perpendicular to the *z*-axis of the $\boldsymbol{\chi}$ tensor. The resulting diamagnetic, paramagnetic and total magnetic susceptibility anisotropy values are $\Delta\chi_{ax} = 1.6 \pm 0.1, -2.0 \pm 0.4, 1.8 \pm 0.2 \times 10^{-32}$ m^3, respectively, and $\Delta\chi_{rh} = -0.4 \pm 0.3, 0.5 \pm 0.3, -0.4 \pm 0.2 \times 10^{-32}$ m^3, respectively. Again, the tensors sum up as expected.

3.5 PARTIAL ORIENTATION EFFECTS ON THE DIPOLAR AND CONTACT SHIFTS

In Chapter 2 it is shown that the anisotropy of χ is responsible for pseudocontact shifts. We have seen in the previous sections that the anisotropy of χ can also induce partial orientation. The effect of partial orientation was however not taken into account when the equations for the pseudocontact shifts, as well as for the contact shifts, were derived. Therefore, Eqs. (2.34)–(2.39), (2.46), (2.48), and (2.49), provided to calculate the pseudocontact shifts, may need to be corrected for the fact that magnetic anisotropy induces partial orientation. Changes in contact shifts [Eq. (2.7)] may as well be expected in all cases that $<S_z>$ is not isotropic (Section 2.2.1) [13].

3.5.1 PARTIAL ORIENTATION EFFECTS ON THE PSEUDOCONTACT SHIFTS

Pseudocontact shifts occur when the metal magnetic susceptibility is anisotropic and they are derived (Chapter 2) under the assumption that every orientation of the molecule is equally probable. However, the magnetic susceptibility tensor axes are fixed within the molecule and the magnetic anisotropy introduces a dependence on orientation of the interaction energy between the paramagnetic ion and the magnetic field. If the energy of the interaction between the molecule and the magnetic field is orientation dependent, it follows that not all molecular orientations are equally probable (Fig. 3.3). As a consequence, changes in the observed pseudocontact shifts should be expected. Actually, it is not the metal magnetic susceptibility, the only cause of the orientation dependence of the molecular energy, but the overall molecular magnetic anisotropy, which includes the diamagnetic part [Eqs. (3.6) and (3.7)]. As we have seen, the latter is likely to increase with the size of the molecule, and it has been indeed found relevant in biological macromolecules.

We have seen in Section 2.2.2.1 that the contribution of the pseudocontact shift to the overall nuclear chemical shift can be calculated from the dipolar shielding tensor [Eqs. (2.28), (2.29), and (2.32)] as

$$\delta^{\mathrm{pcs}} = -\frac{1}{3}\left(e_x \cdot \sigma \cdot e_x + e_y \cdot \sigma \cdot e_y + e_z \cdot \sigma \cdot e_z\right) = -\frac{1}{3}Tr(\sigma) \tag{3.25}$$

where e_x, e_y, and e_z are the three principal directions of a reference frame. In the presence of partial orientation, the chemical shifts average to values, which depend on the components of the shielding tensor $\boldsymbol{\sigma}$ along the main directions of the order tensor $\boldsymbol{S}$ [14]:

$$\delta^{\mathrm{pcs}} = -\left[\frac{Tr(\sigma)}{3} + \frac{2}{3}\left(\sigma_{xx}S_{xx} + \sigma_{yy}S_{yy} + \sigma_{zz}S_{zz}\right)\right]. \tag{3.26}$$

Therefore, if $\chi^{\mathrm{mol}} \cong \chi$, by substituting the $\boldsymbol{S}$ components defined in Eq. (3.10) into Eq. (3.26) (the χ and $\boldsymbol{S}$ tensors are diagonal in the same frame) Eq. (3.27) is obtained

$$\delta^{\mathrm{pcs}} = -\frac{Tr(\sigma)}{3} - \frac{2}{3}\frac{B_0^2}{15\mu_0 kT}\left[\sigma_{xx}\left(\chi_{xx} - \frac{\chi_{yy}+\chi_{zz}}{2}\right) + \sigma_{yy}\left(\chi_{yy} - \frac{\chi_{xx}+\chi_{zz}}{2}\right) + \sigma_{zz}\left(\chi_{zz} - \frac{\chi_{xx}+\chi_{yy}}{2}\right)\right]. \tag{3.27}$$

Since $\sigma_{xx} = -(3x^2 - r^2)\chi_{xx}/(4\pi r^5)$, $\sigma_{yy} = -(3y^2 - r^2)\chi_{yy}/(4\pi r^5)$, and $\sigma_{zz} = -(3z^2 - r^2)\chi_{zz}/(4\pi r^5)$ [x, y, and z are the coordinates of the nucleus in the frame where $\boldsymbol{\chi}$ is diagonal, see Eq. (2.40)], using the relationships $\cos^2\varphi = (\cos 2\varphi + 1)/2$, $3\sin^2\theta\cos^2\varphi - 1 = (3\sin^2\theta\cos 2\varphi)/2 - (3\cos^2\theta - 1)/2$, $\sin^2\varphi = -(\cos 2\varphi - 1)/2$, $3\sin^2\theta\sin^2\varphi - 1 = -(3\sin^2\theta\cos 2\varphi)/2 - (3\cos^2\theta - 1)/2$, the equation for the pseudocontact shifts can be determined taking into account the effect of partial orientation due to the anisotropy of the $\boldsymbol{\chi}$ tensor (to be compared with Eq. (2.46) describing the pseudocontact shifts as determined for the case of isotropic reorientation),

$$\delta^{\mathrm{pcs}} = \frac{1}{12\pi r^3}\left\{\Delta\chi_{ax}\left(3\cos^2\theta - 1\right)\left[1 + \frac{B_0^2}{15\mu_0 kT}\left(\left(2\chi_{zz} + \frac{\chi_{xx} + \chi_{yy}}{2}\right) - \frac{3}{4}\frac{\Delta\chi_{rh}^2}{\Delta\chi_{ax}}\right)\right]\right.$$
$$\left. + \frac{3}{2}\Delta\chi_{rh}\sin^2\theta\cos 2\varphi\left(1 + \frac{B_0^2}{15\mu_0 kT}\left(2\chi_{xx} + 2\chi_{yy} - \chi_{zz}\right)\right)\right\} \tag{3.28}$$

where θ is the angle between the metal–nucleus vector $\boldsymbol{r}$ and the z-axis of the diagonal $\boldsymbol{\chi}$ tensor, and φ is an angle related to the projection of $\boldsymbol{r}$ on the xy plane of the tensor. For systems with axial symmetry ($\chi_{xx} = \chi_{yy} = \chi_{\perp}$) Eq. (3.28) becomes [15]

$$\delta^{\mathrm{pcs}} = \frac{(3\cos^2\theta - 1)(\chi_{\parallel} - \chi_{\perp})}{12\pi r^3}\left(1 + \frac{B_0^2}{15\mu_0 kT}(2\chi_{\parallel} + \chi_{\perp})\right). \tag{3.29}$$

This correction of the pseudocontact shifts due to partial orientation is however minor and can be neglected in most practical applications [15]. It is predicted to increase with the square of the magnetic field. Up to magnetic fields of the order of 20 T, this effect is hard to observe in practice because it is of the same order and of opposite sign with respect to the difference in pseudocontact shift calculated using the field independent first order approximation [Curie law, Eq. (1.34)], instead of the exact field dependent value [Brillouin function, Eq. (1.37)] of the magnetic susceptibility. In fact, if the magnetic susceptibility is field dependent, as described by the Brillouin equation [Eq. (1.37)], because at high magnetic fields the electronic Zeeman energy is not much smaller than kT (Section 1.6) [16], pseudocontact shifts [described by Eq. (2.46)] become field dependent also in the absence of partial orientation [17], as well as contact shifts [Eq. (2.10)] [15].

The difference in population among the electron Zeeman levels can differ from what predicted from the Curie law even more than predicted by the Brillouin equation in the presence of ZFS effects. Thus, the consequent alteration of the pseudocontact shifts can be even larger than that due to Eq. (3.28), but at the same time it is a small fraction of the pseudocontact shift. Quantitative estimates have not been attempted. On the other hand, when a paramagnetic molecule is oriented mechanically by using orienting devices such as liquid crystals, a large and measurable change in hyperfine shift can be observed [18].

Another effect to be considered in the analysis of the chemical shifts of self-oriented molecules is the residual anisotropic chemical shift (RACS), due to the anisotropy in the orientational sampling of the chemical shifts, which results from the partial alignment of the molecule. It thus depends on the chemical shift anisotropy (CSA) tensor [19], defined from the difference between the chemical shielding tensor (Chapter 5) and its isotropic value (equal to the trace of the shielding tensor divided by 3). The residual anisotropic chemical shifts are described by the equation

$$\delta^{\mathrm{racs}} = \frac{B_0^2}{15\mu_0 kT}\sum_{i,j}\delta_{ii}^{CSA}\cos^2\theta_{ij}\Delta\chi_{jj} \tag{3.30}$$

where the indices i and j run over the three axes x, y and z, and θ_{ij} are the angles of the principal axes of the magnetic susceptibility anisotropy tensor, $\boldsymbol{\Delta\chi}$, with respect to the principal axes of the CSA tensor, $\boldsymbol{\delta}^{\mathrm{CSA}}$. This effect can be evaluated from the knowledge of the CSA tensor, and thus from the molecular structure [20], and its evaluation can be important for a correct determination of the pseudocontact shifts. Residual anisotropic chemical shifts can be significant at high magnetic fields for nuclei with large CSA tensors, like amide ^{15}N and carbonyl/aromatic ^{13}C nuclei. They are usually negligible for protons.

3.5.2 PARTIAL ORIENTATION EFFECTS ON THE CONTACT SHIFTS

In the presence of partial orientation, either due to anisotropy of the susceptibility tensor or to external orienting agents, and of an orientation dependent $<S_z>$ [Eq. (2.6)], the contact shifts average to values that depend on the components of S_{ii} along the main directions of the order tensor $\boldsymbol{S}$ [analogously to Eq. (3.26) for pseudocontact shifts]:

$$\delta^{\mathrm{con}} = \left(\frac{1}{3}+\frac{2}{3}S_{xx}\right)\tilde{\delta}^{\mathrm{con}}_{xx} + \left(\frac{1}{3}+\frac{2}{3}S_{yy}\right)\tilde{\delta}^{\mathrm{con}}_{yy} + \left(\frac{1}{3}+\frac{2}{3}S_{zz}\right)\tilde{\delta}^{\mathrm{con}}_{zz}. \quad (3.31)$$

However, since the frame where the order tensor $\boldsymbol{S}$ is diagonal can be different from the principal directions of the $\boldsymbol{g}$ tensor, the axes of the $\boldsymbol{S}$ frame will not coincide with the directions along which the contact shift values are calculated, $\delta^{\mathrm{con}}_{xx}$, $\delta^{\mathrm{con}}_{yy}$, and $\delta^{\mathrm{con}}_{zz}$ [see Eq. (2.8)]. Therefore $\tilde{\delta}^{\mathrm{con}}_{\mathrm{xx}}$, $\tilde{\delta}^{\mathrm{con}}_{\mathrm{yy}}$, and $\tilde{\delta}^{\mathrm{con}}_{zz}$ in Eq. (3.31) must be calculated by projecting the values of $\delta^{\mathrm{con}}_{xx}$, $\delta^{\mathrm{con}}_{yy}$, and $\delta^{\mathrm{con}}_{zz}$ along the axes defining the frame where $\boldsymbol{S}$ is diagonal

$$\tilde{\delta}^{\mathrm{con}}_{ii} = \delta^{\mathrm{con}}_{xx}\sin^2\alpha\cos^2\beta + \delta^{\mathrm{con}}_{yy}\sin^2\alpha\sin^2\beta + \delta^{\mathrm{con}}_{zz}\cos^2\alpha \quad (3.32)$$

where α and β are the spherical angles defining the position of the i axis of the $\boldsymbol{S}$ tensor in the frame of the $\boldsymbol{g}$ tensor [21]. Partial orientation effects on contact shifts due to molecular self-orientation are usually not taken into account because considered a minor perturbation with respect to the precision of the experimental data.

GENERAL REFERENCES

Prestegard JH, Al-Hashimi HM, Tolman JR. NMR structures of biomolecules using field oriented media and residual dipolar couplings. Q. Rev. Biophys. 2000;33:371–424.

Bertini I, Luchinat C, Parigi G. Magnetic susceptibility in paramagnetic NMR. Prog. Nucl. Magn. Reson. Spectrosc. 2002;40:249–73.

REFERENCES

[1] Boyd J, Redfield C. J. Am. Chem. Soc. 1999;121:7441.

[2] Bertini I, Luchinat C, Parigi G. Prog. Nucl. Magn. Reson. Spectrosc. 2002;40:249.

[3] Lohman JAB, Maclean C. Chem. Phys. Lett. 1978;58:483.

[4] Lohman JAB, Maclean C. Chem. Phys. 1978;35:269.

[5] Saupe A. Angew Chem. Int. (Ed. Engl.) 1968;7:97.

[6] Kramer F, Deshmukh MV, Kessler H, Glaser SJ. Concepts Magn. Reson. 2004;21A:10.
[7] Lipari G, Szabo A. J. Am. Chem. Soc. 1982;104:4546.
[8] Tolman JR, Flanagan JM, Kennedy MA, Prestegard JH. Nat. Struct. Biol. 1997;4:292.
[9] Prestegard JH, Al-Hashimi HM, Tolman JR. Q. Rev. Biophys. 2000;33:371.
[10] Losonczi JA, Andrec M, Fischer MW, Prestegard JH. J. Magn. Reson. 1999;138:334.
[11] Banci L, Bertini I, Huber JG, Luchinat C, Rosato A. J. Am. Chem. Soc. 1998;120:12903.
[12] Arnesano F, Banci L, Bertini I, van der Wetering K, Czisch M, Kaptein R. J. Biomol. NMR 2000;17:295.
[13] Kurland RJ, McGarvey BR. J. Magn. Reson. 1970;2:286.
[14] Jokisaari J. Encyclopedia of NMR. Chichester: John Wiley & Sons; 1996. pp 839.
[15] Bertini I, Felli IC, Luchinat C. J. Magn. Reson. 1998;134:360.
[16] Bleaney BI, Bleaney B. Electricity and Magnetism. Oxford: Oxford University Press; 1976.
[17] John M, Otting G. Chem. Phys. Chem. 2007;8:2309.
[18] Bertini I, Castellani F, Luchinat C, Martini G, Parigi G, Ristori S. J. Phys. Chem. 2000;104:10653.
[19] John M, Park AY, Pintacuda G, Dixon NE, Otting G. J. Am. Chem. Soc. 2005;127:17190.
[20] Schmitz C, Stanton-Cook MJ, Su XC, Otting G, Huber T. J. Biomol. NMR 2008;41:179.
[21] Bertini I, Luchinat C, Parigi G. Eur. J. Inorg. Chem. 2000;2473.

CHAPTER

RELAXATION

4

The sole presence of an electron spin causes nuclear relaxation. The correlation time for the electron nucleus interaction is presented as well as equations valid for dipolar and contact interactions. To do so, electron relaxation mechanisms need to be quickly reviewed. All the subtleties of nuclear relaxation enhancements are presented pictorially and quantitatively. The Overhauser effect is introduced and discussed. The nuclear Overhauser effect allows us to know which magnetic nuclei are close to other magnetic nuclei, thus providing information on their distances; the electron-nuclear Overhauser effect provides a mechanism for dynamic nuclear polarization (DNP).

4.1 INTRODUCTION

In paramagnetic molecules, the magnetic nucleus does not see unpaired electrons as localized, but as spin density distributed throughout space, with an integrated intensity equal to that of the total unpaired electrons. As a result, in every unit volume, the spin density will spend some time in the low Zeeman energy level(s) (negative M_S, Chapter 1) and some time (slightly less) in the upper energy level(s) (positive M_S). Changes in M_S values involve changes in the orientation of the electron magnetic moment. The time sharing of the levels occurs through electron relaxation. Electron relaxation thus provides fluctuating magnetic fields and causes nuclear relaxation. When reference is made to spin density at the resonating nucleus, nuclear relaxation is *contact* in origin. The remaining spin density on the molecule and the relative electron relaxation are sensed by the resonating nucleus through dipolar coupling. The consequent nuclear relaxation is *dipolar* in origin. Often, reference is made to metal-centered relaxation; in this case, the whole electron is considered localized on the metal. Dipolar relaxation due to spin density elsewhere is called ligand-centered relaxation. This will be discussed in Section 4.4.1.

Restricting ourselves to metal-centered nuclear relaxation, the fluctuation of the electron dipolar field at the nucleus due to electron relaxation can be easily visualized as depicted in Fig. 4.1A. We now want to mention other mechanisms, besides electron relaxation, which occur in solution and through which unpaired electrons cause nuclear relaxation. All of these mechanisms will be discussed in detail in this chapter.

1. Rotation of the molecular frame causes the nucleus to see the electron in different positions (Fig. 4.1B). If the rotation is faster than the electron relaxation time, on the rotational time scale the nucleus sees the electron with the same M_S value but on different positions in space. This random motion of the electron around the nucleus is again seen as a fluctuating magnetic field that causes nuclear relaxation through dipolar coupling.
2. In addition, the nucleus sees the induced electronic magnetic moment $\langle \mu \rangle$ [(Eq. (1.33)], which is aligned along the magnetic field as a result of the time average over the Zeeman levels. Upon rotation, the induced electronic magnetic field causes fluctuating magnetic fields sensed by the nucleus through space. This is another dipolar mechanism that can be also visualized by referring

NMR of Paramagnetic Molecules. http://dx.doi.org/10.1016/B978-0-444-63436-8.00004-1

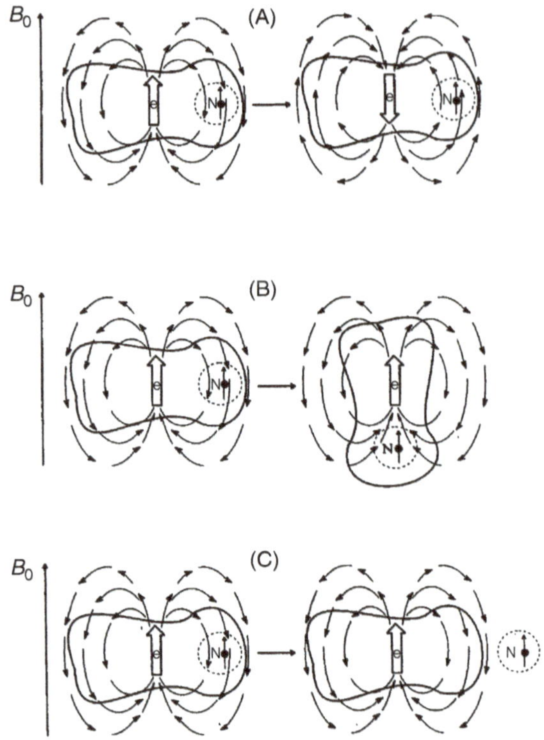

FIGURE 4.1 Pictorial representation of the motions causing nuclear relaxation: electron spin relaxation (A), molecular rotation (B) and chemical exchange (C). It can be seen that the electron dipolar field at the nucleus fluctuates with time in direction (A), intensity (C) or both (B).

to Fig. 4.1B, where the electronic vector is now the difference in electron population of the Zeeman levels and called Curie spin relaxation.

3. Finally, the approach and binding of a moiety containing the resonating nucleus to a group containing the unpaired electron, and the following detachment (chemical exchange), also causes fluctuating magnetic fields at the nucleus (Fig. 4.1C) through both contact and dipolar mechanisms. A limiting case of chemical exchange is the outer sphere interaction (Section 6.6).

Relaxation measurements provide a wealth of information both on the extent of the interaction between the resonating nuclei and the unpaired electrons, and on the time dependence of the parameters associated with the interaction. Whereas the dipolar coupling depends on the electron–nucleus distance, and therefore contains structural information, the contact contribution is related to the unpaired spin density on the various resonating nuclei and therefore to the topology (through chemical bonds) and the overall electronic structure of the molecule. The time-dependent phenomena associated with electron–nucleus interactions are related to the molecular system, and to the lifetimes of different chemical situations, for the resonating nucleus. Obtaining either structural or dynamic information, however, is only possible if an in-depth analysis of a series of experimental results provides sufficient data to characterize the system within the theoretical framework discussed in this chapter.

In real systems, each unpaired electron is delocalized through chemical bonds in the neighborhood of the metal, to an extent which depends on the chemical properties of the system. Therefore, the dipolar interaction with the resonating nucleus should be evaluated by integrating over all the points where there is a finite unpaired spin density. Approximate procedures will be given in Section 4.4.1. An account of the nuclear relaxation times based on this general approach requires details of the electronic structure of the system, including the unpaired spin densities on the various non-s orbitals of the molecule, which is difficult to treat in general terms.

4.2 THE CORRELATION TIME

Electrons switch between levels characterized by M_S values. Let us examine now an ensemble of n molecules, each with an unpaired electron, in a magnetic field at a given temperature. The bulk system is at constant energy but at the molecular level electrons move, molecules rotate, there are concerted atomic motions (vibrations) within the molecules and, in solution, molecular collisions. Is it possible to have information on these dynamics on a system which is at equilibrium? The answer is yes, through the *correlation function.* The correlation function is a product of the value of any time-dependent property at time zero with the value at time t, summed up to a large number n of particles. It is a function of time. In this case the property can be the M_S value of an unpaired electron and the particles are the molecules. The correlation function has its maximum value at $t = 0$; since each molecule has one unpaired electron, the product of the M_S value at time zero times the M_S value at time t ($t = 0$) is either ½ · ½ or – ½ · (–½), i.e., always equal to +¼. With time, some spins will change their orientation (Fig. 4.1A), so some particles will contribute by –¼. The value of the correlation function $C(t)$ decreases. It can be shown that the decay of the correlation function can be approximated by an exponential in the absence of constraints (Fig. 4.2).[1]

$$C(t) = \sum_i M_S(0)M_S(t) \approx \left(\sum_i \frac{1}{4}\right)\exp(-t/\tau_c). \tag{4.1}$$

From Eq. (4.1), the correlation time τ_c is defined as the time constant for which the correlation function exponentially decays to zero. At a time small compared with τ_c, $\exp(-t/\tau_c) \rightarrow 1$ and we expect that essentially all spins maintain their original value. Therefore, most of the products will be ¼. At times long with respect to τ_c, we expect that all spins have changed their orientations many times, so that on

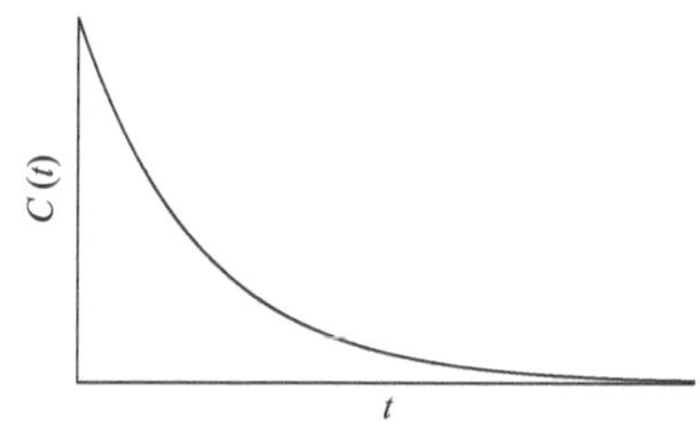

FIGURE 4.2 Exponential decay with time of the correlation function.

[1]Here, and in the following, only positive values of time t are considered. Otherwise, all equations of the type of Eq. (4.1) should contain the absolute value of time in the exponential decay term, $\exp(-|t|\tau_c)$.

the average half of the spins will result with the same M_S with respect to their original values, and the other half will have the other M_S value. Under these conditions, statistically half of the products will be ¼ and half –¼, and the summation over a large number n of spins will yield zero.

Strictly speaking, the time dependence of $C(t)$ is not exponential for $t \ll \tau_c$; by referring to the above example, after a time much shorter than τ_c but still larger than zero, it is likely that $C(t) = C(0)$. Indeed, an expansion of the initial part of $C(t)$ would show a flat region around zero. In other words, it must be that $\mathrm{d}C(0)/\mathrm{d}t = 0$, whereas for a true exponential $\mathrm{d}C(0)/\mathrm{d}t = -\left(\sum_i \frac{1}{4}\right)/\tau_c$. Only for longer times does $C(t)$ become effectively exponential. The small initial deviation is, however, irrelevant for our purposes. Real cases of strong deviations from an exponential behavior do exist, but we will often restrict our interest to exponential correlation functions. In any case, the correlation time of a certain time-dependent process can be defined as the integral of the correlation function of that process, independently of the actual law for the time dependence.

The Fourier transform of an exponential decay in the time domain, as seen in Section 1.7.1, is a Lorentzian in the frequency domain:

$$J(\omega) = \frac{\tau_c}{1+\omega^2\tau_c^2}. \tag{4.2}$$

This function (Fig. 4.3) has its inflection point at $|\omega|\tau_c = 1$. We say that the function undergoes a dispersion at $|\omega|\tau_c = 1$. This function provides the intensity of the various frequencies available in the lattice due to the fluctuations to which τ_c is related. It is also called noise, power, or spectral density function. The nucleus picks up the needed ω frequency for its relaxation. The probability of this to occur depends on the spectral density [i.e., on the value of function (4.2)] at that frequency.

Transverse nuclear relaxation can also occur when the local fields at the nucleus fluctuate slowly, i.e., with an ω frequency near zero (Section 4.4). Then the spectral density function takes the form

$$J(0) = \tau_c. \tag{4.3}$$

This is a so-called nondispersive term. This effect can be viewed as transient increases or decreases of the precession frequency due to local field fluctuations along z (the $\boldsymbol{B}_0$ direction). These changes in

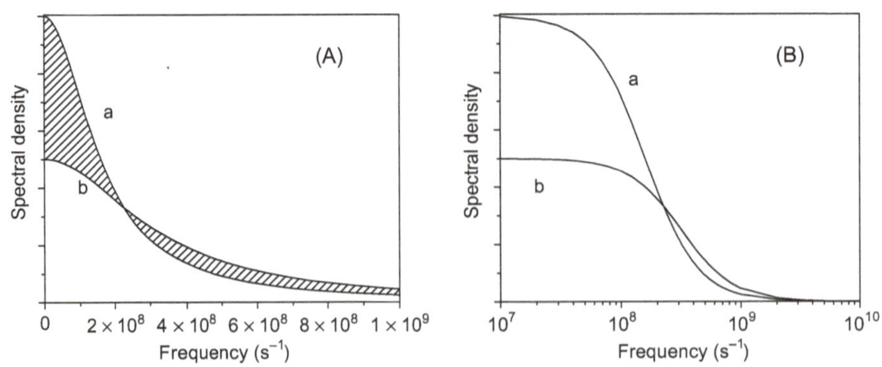

FIGURE 4.3 Spectral density function $J(\omega)$ as a function of frequency, linear scale (A) and log scale (B). The profiles are obtained for $\tau_c = 1 \times 10^{-9}$ s (a) and 5×10^{-10} s (b). The two dashed areas in (A) are equivalent, showing that the area under the a and b curves is the same. The inflection points occur at $\omega\tau_c = 1$.

the precession frequency do not appreciably affect the z-projection of the nuclear magnetic moment but "dephase" its projection in the xy plane. In cases where two nuclear spins relax each other, the sum and the difference of the two frequencies (zero and double quantum transitions) are needed besides ω, and then we will also have functions of the type

$$J(\omega_1 - \omega_2) = \frac{\tau_c}{1+(\omega_1 - \omega_2)^2 \tau_c^2}$$
$$J(\omega_1 + \omega_2) = \frac{\tau_c}{1+(\omega_1 + \omega_2)^2 \tau_c^2}. \quad (4.4)$$

Finally, local field fluctuations along z for the first spin can also be caused by fluctuations of the xy component of the second spin. This originates a term of the type of $J(0)$ but "offset" by the precessional frequency of the second spin, ω_2:

$$J(\omega_2) = \frac{\tau_c}{1+\omega_2^2 \tau_c^2}. \quad (4.5)$$

Analogous spectral densities will also be found in the treatment of electron relaxation [Eqs. (4.12) and (4.13)].

Once the correlation time is fixed, the value of the spectral density function at a given frequency can be determined. When τ_c is long, the spectral density is large in the small frequency range below τ_c^{-1}; when τ_c is short, we have a larger range of frequencies available but with a lower intensity (Fig. 4.3). Of course, τ_c refers to real physical mechanisms of fluctuations. We use the word fluctuation to indicate a movement which is random and unpredictable. We also use the words "uncorrelated" and "stochastic" for this behavior. The ω frequencies generally arise from sudden and short movements, which provide a range of frequencies just like in pulsed NMR (Section 1.7.1).

The movements capable of relaxing the nuclear spin that are of interest here are related to the presence of unpaired electrons, as has been discussed in Section 4.1. They are electron spin relaxation, molecular rotation, and chemical exchange. These correlation times are indicated as τ_e or T_{1e} (electronic relaxation correlation time), τ_r (rotational correlation time), and τ_M (exchange correlation time). All of them can modulate the dipolar coupling energy and therefore can cause nuclear relaxation. Each of them contributes to the decay of the correlation function. If these movements are independent of one another, then the correlation function decays according to the product

$$\exp(-t/\tau_e)\exp(-t/\tau_r)\exp(-t/\tau_M) = \exp\left[-\left(\tau_e^{-1} + \tau_r^{-1} + \tau_M^{-1}\right)t\right].$$

Therefore, the overall correlation time for dipolar coupling is such that

$$\left(\tau_c^{-1}\right)^{\text{dip}} = \tau_e^{-1} + \tau_r^{-1} + \tau_M^{-1} \quad (4.6)$$

i.e., its reciprocal is the sum of the single reciprocal correlation times. It often happens that only one dominates. In the case of relaxation by contact or isotropic coupling, only chemical exchange and electron relaxation can modulate the coupling. Thus

$$\left(\tau_c^{-1}\right)^{\text{con}} = \tau_e^{-1} + \tau_M^{-1}. \quad (4.7)$$

Sometimes distinction can be made between τ_{c1} and τ_{c2} if τ_e is specified as T_{1e} (longitudinal) or T_{2e} (transverse) electronic correlation time (Section 4.4).

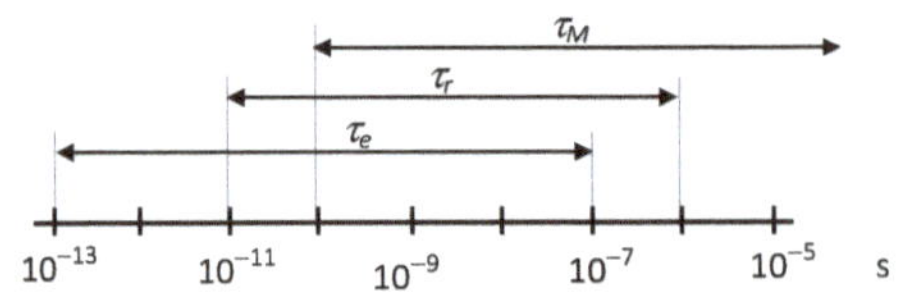

FIGURE 4.4 Ranges of typical values for τ_e, τ_r and τ_M as defined in Eq. (4.6).

Electronic relaxation times fall in the range 10^{-7}–10^{-13} s (Fig. 4.4), whereas the exchange time can be indefinitely long or as short as 10^{-10} s, depending on the chemical bond strength. As far as rotation is concerned, the rotational correlation time can be predicted for spherical rigid particles [1–3].

$$\tau_r = \frac{4\pi\eta a^3}{3kT} = \frac{\eta M}{dN_A kT} \tag{4.8}$$

where η is the viscosity (kg s^{-1} m^{-1}) of the solvent, a is the radius of the molecule, assumed spherical, M is the molar mass (kg mol^{-1}), d is the density (kg m^{-3}) of the molecule (usually taken equal to 10^3), and N_A is Avogadro's constant. Values for τ_r in water solutions at room temperature range from 3×10^{-11} s for hexaaqua metal complexes, to 10^{-8} s for small proteins ($M \approx 10$ kg mol^{-1}) to 10^{-6} s for larger macromolecules ($M \approx 10^3$ kg mol^{-1}) (Fig. 4.4). The value of τ_r may also be estimated from T_1 measured at several magnetic fields on a diamagnetic analog of the metal complexes of interest. In fact, in diamagnetic systems the nuclear relaxation times are usually determined by dipolar coupling with other nuclei in the same molecule, the correlation time for the interaction being the rotational correlation time of the molecule itself (in the absence of chemical exchange).

4.3 ELECTRON RELAXATION

From the foregoing discussion, it appears that electron relaxation may be important for NMR on paramagnetic substances because τ_e may be the correlation time. Indeed, it is always important. Let us say we have two paramagnetic ($S = ½$) compounds in water, one with τ_e of 10^{-8} s and the other with τ_e of 10^{-12} s. The size of each molecule is the same and such that the rotational correlation time is 10^{-10} s. For the former complex $\tau_c \approx \tau_r = 10^{-10}$ s and for the latter $\tau_c \approx \tau_e = 10^{-12}$ s. So, in one case nuclear relaxation depends on a correlation time which is due to rotation, whereas in the other it depends on the electron relaxation time. When the rotational mechanism determines the correlation time, and the latter happens to be long, nuclear relaxation rates will be invariably large and the NMR lines broad. As a consequence, we lose resolution and the capability of revealing connectivities among signals.

As will be seen soon, as a practical consequence, we can state that high resolution NMR can be fruitfully attempted for those systems for which τ_c is determined by τ_e: the shorter τ_e, the better. Detection of signals and exploitation of the NMR experiment is still convenient for diamagnetic moieties interacting with paramagnetic systems with long τ_e when in the presence of fast chemical exchange. Under these circumstances the linewidths of nuclei of the moiety which is in molar excess and exchanging with the moiety bound to the paramagnetic center are reduced by the molar fraction of the bound moiety, and still contain structural an dynamic information (Chapter 6).

In Table 4.1 estimates of the electronic relaxation times at high NMR fields are reported for some paramagnetic metal ions in solution and at room temperature. As τ_e values may be strongly field

Table 4.1 Electronic Relaxation Times of Some Common Paramagnetic Metal Ions and Nuclear Relaxation Rates for a Proton at 5 Å From the Metal, at 800 MHz ^{1}H Resonance Frequency, Due to Dipolar and Curie Relaxation, Estimated From Eqs. (4.17), (4.18), (4.30) and (4.31), With $\tau_c = \tau_e$

Metal Ion	Electronic Configuration	S	τ_e(s)	Small Molecules ($\tau_r = 10^{-10}$ s) $R^I_{1,2M}$	Large Molecules ($\tau_r = 10^{-8}$ s) R^I_{1M}	R^I_{2M}	References
Ti^{3+}	d^1	1/2	10^{-10}–10^{-11}	40–300	50–400	100–500	[4,5]
VO^{2+}	d^1	1/2	$\approx 10^{-8a}$	300–500	40	15,000	[6–8]
V^{3+}	d^2	1	$\approx 10^{-11}$	100–150	100	500	[9,10]
V^{2+}	d^3	3/2	$\approx 10^{-9}$	1,500–2,000	1000	15,000	[11–13]
Cr^{3+}	d^3	3/2	5×10^{-9}–5×10^{-10}	1,500–2,000	200–1500	10,000–60,000	[14–16]
Cr^{2+}	d^4	2	10^{-11}–10^{-12}	50–500	40–400	4,000	est. from [17]
Mn^{3+}	d^4	2	10^{-10}–10^{-11b}	300–2000	300–3000	4,000–8000	[10,16,18]
Mn^{2+}	d^5	5/2	$\approx 10^{-8c}$	4,000–6,000	400	200,000	[14,19]
Fe^{3+} (H.S.)	d^5	5/2	10^{-9}–10^{-10d}	500–6,000	500–3000	8,000–50,000	[16,20,21]
Fe^{3+} (L.S.)	d^5	1/2	10^{-11}–10^{-13}	2–50	1–60	50–150	[22,23]
Fe^{2+} (H.S.)	d^6, 5–6 coord.	2	10^{-12}–10^{-13}	50–150	10–50	4,000	[24], est. from [25]
	d^6, 4 coord.	2	$\approx 10^{-11}$	300–500	400	4,000	est. from [26]
Co^{2+} (H.S.)	d^7, 5–6 coord.	3/2	5×10^{-12}–10^{-13}	20–200	10–200	1,000–2,000	[27,28]
	d^7, 4 coord.	3/2	$\approx 10^{-11}$	200–300	200	2,000	[29]
Co^{2+} (L.S.)	d^7	1/2	10^{-9}–10^{-10}	200–400	200–400	500–3,000	[12]
Ni^{2+}	d^8, 5–6 coord.	1	$\approx 10^{-10e}$	600–700	1000	2,000	[30]
	d^8, 4 coord.	1	$\approx 10^{-12}$	20–30	20	400	[31,32]
Cu^{2+}	d^9	1/2	$(1–5) \times 10^{-9}$	300–500	40–200	3,000–20,000	[33]
Ru^{3+}	d^5	1/2	10^{-11}–10^{-12}	5–50	5–50	50–150	[10,34]
Re^{3+}	d^4	2	10^{-12}–10^{-13}	50–150	10–50	4,000	[35]

(Continued)

Table 4.1 Electronic Relaxation Times of Some Common Paramagnetic Metal Ions and Nuclear Relaxation Rates for a Proton at 5 Å From the Metal, at 800 MHz ^{1}H Resonance Frequency, Due to Dipolar and Curie Relaxation, Estimated From Eqs. (4.17), (4.18), (4.30) and (4.31), With $\tau_c = \tau_e$ *(cont.)*

Metal Ion	Electronic Configuration	S	τ_e(s)	Small Molecules ($\tau_r = 10^{-10}$ s)	Large Molecules ($\tau_r = 10^{-8}$ s)		References
				$R^I_{1,2M}$	R^I_{1M}	R^I_{2M}	
Gd^{3+}	f^7	7/2	10^{-8}–10^{-9}[f]	5,000–15,000	800–5000	100,000–400,000	[36–42]
<u>Ln^{3+}</u>		J					[43–46]
Ce^{3+}	f^1	5/2	4×10^{-13}–10^{-3}	6–12	3–7	200–300	
Pr^{3+}	f^2	4	3×10^{-13}–6×10^{-14}	16–30	4–13	1,000	
Nd^{3+}	f^3	9/2	8×10^{-13}–2×10^{-13}	20–40	10–20	1,100	
Sm^{3+}	f^5	5/2	2×10^{-13}–5×10^{-14}	0.2–0.6	0.2–0.6	3–4	
Eu^{2+}	f^7	7/2	5×10^{-14}–10^{-14}	300–400	20–30	30,000	
Tb^{3+}	f^8	6	8×10^{-13}–2×10^{-13}	800–1,000	100–200	60,000	
Dy^{3+}	f^9	15/2	10^{-12}–4×10^{-13}	1,100–1,500	170–300	80,000	
Ho^{3+}	f^{10}	8	8×10^{-13}–2×10^{-13}	1,000–1,500	150–250	80,000	
Er^{3+}	f^{11}	15/2	8×10^{-13}–3×10^{-13}	700–1,000	120–200	50,000	
Tm^{3+}	f^{12}	6	2×10^{-12}–5×10^{-13}	300–500	80–200	20,000	
Yb^{3+}	f^{13}	7/2	5×10^{-13}–2×10^{-13}	50–70	16–30	3,000	

Metal ions more suitable for high resolution NMR are underlined.
[a] *Field dependence of τ_e with a $T_{1e(0)}$ value of 5×10^{-10} s has also been reported [47].*
[b] *τ_e values in the 10^{-11}–10^{-12} s range have also been proposed [48].*
[c] *τ_e is strongly field dependent; $T_{1e(0)}$ values of 10^{-9}–10^{-10} s have been reported [49,50].*
[d] *τ_e is field dependent, $T_{1e(0)}$ values being highly variable and ranging from 10^{-9} [21] to less than 10^{-11} s [51].*
[e] *τ_e is strongly field dependent. $T_{1e(0)}$ values of 10^{-11}–10^{-12} s have been reported [52,53].*
[f] *τ_e is field dependent. $T_{1e(0)}$ values of 10^{-8}–4×10^{-10} s have been reported [38,39].*

dependent, the low field limiting values $T_{1e(0)}$, when available, are also reported. The table also reports the calculated linewidth at 800 MHz of a proton at 5 Å distance (in the presence of dipolar coupling, see later) for two values of τ_r. The underlined metal ions are those for which high resolution NMR is feasible and nD NMR (Chapter 12) can be attempted also for nuclei experiencing hyperfine shifts. In particular, lanthanoids(III) [except gadolinium(III)] and low spin iron(III) are particularly suitable for NMR. Then, tetrahedral nickel(II) and high spin six-coordinated cobalt(II) complexes have short electronic relaxation times. Manganese(III), low spin cobalt(II), and high spin heme-containing iron(III) compounds are at the borderline. Oxovanadium(IV), chromium(III), copper(II), manganese(II), and gadolinium(III) are not suitable for NMR experiments, except on nuclei in rapid exchange with diamagnetic species in excess. In Chapter 11, we will see that magnetic coupling in polymetallic systems may provide short effective electronic relaxation times independently of the electronic relaxation times of the monometallic system. We are now going to cursorily review the principles of electron relaxation in order to be able to account for the values of the electronic relaxation times or to predict them.

4.3.1 THE MAIN MECHANISMS FOR ELECTRON RELAXATION

Mechanisms analogous to those illustrated in Fig. 4.1 apply also to electron relaxation (see later). However, electrons have other more efficient relaxation mechanisms which overcome the former ones. They are based on the presence of spin-orbit coupling. Spin-orbit coupling arises from the coupling between the electron spin magnetic moment and the electronic orbital motion about the charged nucleus. Molecular motions modulate the orbital magnetic moment and then affect the electron spin. Several possible mechanisms for electron relaxation have been successfully identified, and equations have been derived for many of them [54,55].

Solid-state theories ascribe electron relaxation to the coupling of electronic spin transitions with transitions between lattice vibrational levels, or more generally with *phonons*. Disappearance (depopulation of a vibrational level) or creation (population of a vibrational level) of phonons modulate the orbital component of the electron magnetic moment.

The orbit–lattice interaction allows the electron to change energy level, which may imply changes in the electron spin state. It is customary to separate the coupling into direct processes [56], Orbach processes [57], and Raman processes [56,57], all of which are shown in Fig. 4.5. In the direct process, a phonon with the same energy as the electron spin transition between states with different M_S values is required. The direct process may be important only at about liquid helium temperature, where only low energy phonons are available.

At higher temperatures, two-phonon relaxation processes dominate because the energy of most phonons is much larger than the electron Zeeman energy. In the Orbach process, the interaction with a first phonon may determine a spin transition from an electron spin ground state with a given M_S to an electronic excited state, and a subsequent interaction with a second phonon may cause a second transition back to the electronic ground state but with different M_S. In summary, the electron–phonon coupling can provide the energy for jumps from the ground to the excited states and vice versa. Such jumps may involve spin changes. The Raman process is operative when there are no excited states available for electronic transitions; in the Raman process two phonons of high energy may simultaneously interact with the spin system so that their energy difference equals the electron Zeeman energy. The availability of the phonon spectrum for a given molecule may largely improve the temperature dependence of the electron relaxation rate due to the Raman process, as the number of phonons at high frequencies in the real spectrum may be much larger than predicted by the Debye spectrum used in the models [58].

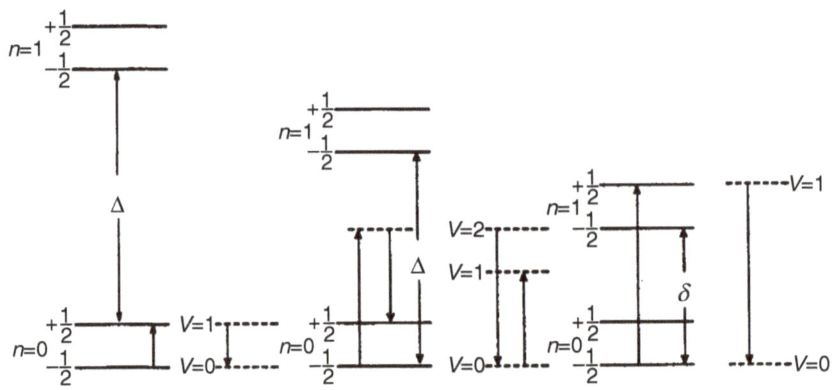

FIGURE 4.5 Lattice and spin transitions are coupled by (A) direct processes, (B) Raman processes, (C) Orbach processes. The proximity of the excited electronic state favors both Orbach and Raman processes. The electronic states are labeled with n, the lattice vibrational states are labeled with V. Δ and δ indicate energy separations with excited states coupled to the ground state by spin–orbit coupling.

Following the solid-state approach, equations have been derived [59,60] also for the electron spin relaxation of $S = ½$ ions in solution determined by the aforementioned processes. At 300 K, corresponding to a thermal energy of about 200 cm^{-1}, the most probable phonon energies range between 50 and 1000 cm^{-1}. Therefore, in the presence of low-lying electronic energy levels at about the energy of the phonons, the Orbach process can be very effective, and at room temperature it can be the fastest mechanism for electron relaxation. The electron relaxation time is described in the cases of Orbach-type mechanism by the Eq. (4.9) [6]:

$$T_{1e}^{-1} = \frac{C}{\delta^2 \tau}\left[\exp(\hbar\delta/kT) - 1\right]^{-1} \tag{4.9}$$

where $\hbar\delta$ is the excitation energy to the first excited states and τ is a mean correlation time for the intermolecular fluctuations, or for collisions with solvent molecules. No field dependence in the electron relaxation rate is accounted for by this relaxation mechanism.

In solution, besides phonons, collisions with solvent should be taken into consideration, whose correlation time is usually in the range 10^{-11} to 10^{-12} s. However, there is no satisfactory theory that unifies relaxation in the solid state and in solution. The reason for this is that the solid state theory was developed for low temperatures, while solution theories were developed for room temperature. The phonon description is a powerful one when phonons are few. By increasing temperature, the treatment becomes cumbersome, and it is more convenient to use stochastic theory (Section 4.2) instead of analyzing the vibrational transitions.

Two mechanisms proposed for $S = ½$ ions are typical of solutions and are determined by the rotation of the molecule. The first is the so-called spin-rotation mechanism [61]. This mechanism arises from the fact that, upon a sudden change in the rotational motion of a molecule, for example, after a collision, the molecular skeleton and the electron cloud may be slightly misplaced from one another, thus providing an instantaneous electric dipole moment. Therefore, there is a coupling between the spin angular momentum and the magnetic field generated by the instantaneous electric dipole moment, the latter related to the rotational angular momentum of the molecule; this coupling causes relaxation. The smaller the molecule, the larger the effect; for macromolecules, this mechanism is indeed negligible. The contribution to the relaxation rate of this mechanism is thus inversely proportional to that of the mechanisms modulated by τ_r, which increases with the size of the molecule. The effect is also proportional to the departure of $\overline{g}$ from g_e. The second kind of mechanism is due to fluctuations in electron Hamiltonian terms, due to Brownian motions. The electron Hamiltonian is given by

$$H = \mu_B \boldsymbol{L} \cdot \boldsymbol{B_0} + g_e \mu_B \boldsymbol{S} \cdot \boldsymbol{B_0} + \boldsymbol{S} \cdot \boldsymbol{D} \cdot \boldsymbol{S} + \boldsymbol{I} \cdot \boldsymbol{A} \cdot \boldsymbol{S} \tag{4.10}$$

where the first two terms represent the coupling between the orbital and the intrinsic magnetic moment with the external magnetic field, the third term the ZFS and the fourth, the hyperfine coupling between unpaired electron and metal nucleus. In the spin Hamiltonian formalism, the first two terms may be written as

$$\mu_B (\boldsymbol{L} + g_e \boldsymbol{S}) \cdot \boldsymbol{B_0} = \mu_B \boldsymbol{S} \cdot \boldsymbol{g} \cdot \boldsymbol{B_0} \tag{4.11}$$

where $\boldsymbol{g}$ is a tensor which contains the anisotropy introduced by the orbital angular operator (Section 1.4). Fluctuations of $\boldsymbol{g}, \boldsymbol{A}$, and $\boldsymbol{D}$ can cause electron relaxation. Stochastic fluctuations can arise from molecular reorientation (with correlation time τ_r) and/or from molecular distortions, for example, due to collisions (with correlation time τ_v) [62], the latter mechanism being usually dominant. If $\boldsymbol{g}$ or $\boldsymbol{A}$ are anisotropic, their values are orientation dependent, and are therefore modulated by rotation [62]. As already seen, electrons have an efficient relaxation mechanism with respect to nuclear relaxation also because of the presence of spin–orbit coupling, that arises from the coupling between electron spin magnetic moment and orbital momentum, due to its motion around the electrically charged nucleus. In $S > ½$ systems, the presence of spin–spin coupling must also be considered. In the spin Hamiltonian formalism all these terms are included in the ZFS Hamiltonian $\boldsymbol{S} \cdot \boldsymbol{D} \cdot \boldsymbol{S}$. Therefore, for ions with $S > ½$, relaxation mechanisms can arise from modulation of the quadratic ZFS in solution; these mechanisms have been discussed in detail by Bloembergen and Morgan [63] and Rubinstein et al. [14]. Deformations of the coordination polyhedron by collision with solvent molecules cause a transient ZFS which allows the coupling of rotation or distortional motion with spin transitions. Therefore, even in high symmetry systems without static ZFS, a transient ZFS can be originated by instantaneous deformations due to colliding solvent molecules. The following equations have been derived for R_{1e} $(= T_{1e}^{-1})$ and R_{2e} $(= T_{2e}^{-1})$ for $S = 1$

$$R_{1e} = \frac{2\Delta_t^2}{50}[4S(S+1)-3] \left(\frac{\tau_v}{1+\omega_S^2\tau_v^2} + \frac{4\tau_v}{1+4\omega_S^2\tau_v^2} \right) \tag{4.12}$$

$$R_{2e} = \frac{\Delta_t^2}{50}[4S(S+1)-3] \left(3\tau_v + \frac{5\tau_v}{1+\omega_S^2\tau_v^2} + \frac{2\tau_v}{1+4\omega_S^2\tau_v^2} \right) \tag{4.13}$$

where Δ_t^2 is the mean squared fluctuation of the ZFS and τ_v is the correlation time for the instantaneous distortions of the metal coordination polyhedron. Analogous equations also hold for the two electronic transitions ½ → –½ and ½ → ³⁄₂ of S = ³⁄₂ systems [14]. For S = ⁵⁄₂ three different relaxation times for each type should be considered [14]. Eqs. (4.12) and (4.13) are still good approximations for their effective averages. This is the so-called pseudorotational model. From Eqs. (4.12) and (4.13), the low-field limiting value of the electronic relaxation rate $R_{1e(0)}$ is given by: $R_{1e(0)} = (\Delta_t^2/5)[4S(S+1)-3]\tau_v$. Eqs. (4.12) and (4.13) have been derived in the absence of a static ZFS term in the electron Hamiltonian.

From the foregoing, the importance of the contribution of the various electronic relaxation mechanisms can be considered qualitatively for various cases. The least efficient mechanisms are the rotational mechanisms (**g**, **A** modulation) and spin rotation, which account for R_{1e} values up to 10^9 s^{-1} and are only detected in the absence of other mechanisms. Contributions from the modulation of the quadratic ZFS to the relaxation rates through solvent bombardment can range from 10^8 to 10^{11} s^{-1}, depending on the extent of the instantaneous geometrical distortions and spin–orbit coupling. It should be remembered that the extent of the splitting of an S manifold depends on the spin–orbit coupling constant, which increases with the atomic number and thus from left to right and from top to bottom in the periodic table. A new mechanism based on a generalization of the spin rotation mechanism, and which is valid when the spin–orbit coupling energy is larger than the energy separation from the first excited electronic state, has been proposed, which accounts for electron relaxation rates as large as 10^{11} s^{-1}, as for example, in low spin ruthenium(III) complexes [64,65]. The Orbach and Raman processes may be the only processes that can account for relaxation rates higher than 10^{11} s^{-1}. Although in solution there are no phonons, vibrational transitions within molecular clusters, and thus local, may allow for fast electron relaxation. The Orbach process at room temperature occurs when there are energy separations between ground and excited states in the range 100–1000 cm^{-1}.

Transition metal complexes are known to experience structural distortions whenever they can decrease the overall energy of the system (Jahn–Teller effect); this is the case of degenerate orbitals which are asymmetrically occupied: a distortion will occur to remove the degeneracy, and the lowest energy level(s) will be most populated. For symmetric complexes, transient geometry distortions occur because the asymmetric distribution of the electron density may actually induce transient distorsions in the system due to the greater electrostatic repulsion of the orbitals pointing to the ligands (dynamic Jahn-Teller). A strong dynamic Jahn-Teller effect happens in the case of octahedral high spin d^4, low spin d^7, and d^9 configurations, where e_g orbitals are unevenly occupied (a weaker effect occurs when t_{2g} orbitals are unevenly occupied, as in the case of octahedral d^1, d^2, low spin d^4, low spin d^5, high spin d^6, and high spin d^7 configurations, Chapter 7). A strong dynamic Jahn-Teller effect is for instance observed in the copper aqua ion (d^9). In the case of six-coordinated metal ion complexes elongation occurs along the three principal axes (Fig. 4.6). The occurrence of elongation is random with a correlation time of about 5×10^{-12} s. Elongation changes the hyperfine coupling constant between the copper nucleus and the unpaired electrons. This can be a further electron relaxation mechanism operative for the $Cu(OH_2)_6^{2+}$ complex [66–68].

The electron relaxation mechanisms operative for several different metal ions will be discussed in more detail in Chapters 7 and 8. For the particular case of the aqua ions, they are summarized in Table 7.3.

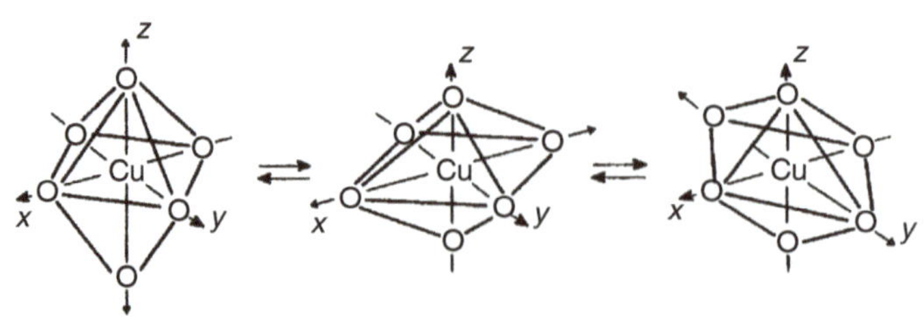

FIGURE 4.6 Elongations along the three principal axes of the Cu—O bonds in the hexaaqua copper(II) complex due to dynamic Jahn–Teller effects occur randomly with a correlation time of about 5×10^{-12} s.

4.4 NUCLEAR RELAXATION DUE TO DIPOLAR COUPLING WITH UNPAIRED ELECTRONS

From the classic description of the coupling between two point dipoles with magnetic moments μ_1 and μ_2, the energy of the interaction E^{dip} is proportional to their scalar product and to the inverse of the third power of their distance [Eq. (1.1)]. If the reciprocal orientation of the two vectors changes randomly with time, the fluctuating magnetic field produced by one at the location of the other may induce spin transitions on the other center, and thus provide a relaxation pathway. The interest here is limited to nuclear spin transitions induced by fluctuating magnetic fields originating from unpaired electron spins. It should be remembered that the electronic magnetic moment is orders of magnitude larger than that of nuclei; for example, the magnetic moment of the free electron is 658 times that of the proton (Section 1.2). The interaction between protons will anyway be present, even if usually much smaller, and it contributes to the diamagnetic relaxation.

While keeping in mind the general picture of nuclear relaxation in paramagnetic systems as described in Section 4.1, it is appropriate to consider first the simple case of dipolar coupling between two point-dipoles, as if the unpaired electrons were localized on the metal ion. The enhancement of the nuclear longitudinal relaxation rate R_{1M} due to dipolar coupling with unpaired electrons can be calculated starting from the Hamiltonian for the system:

$$H = g_e \mu_B \boldsymbol{S} \cdot \boldsymbol{B}_0 - \hbar \gamma_I \boldsymbol{I} \cdot \boldsymbol{B}_0 + \boldsymbol{I} \cdot \boldsymbol{A} \cdot \boldsymbol{S}. \tag{4.14}$$

This Hamiltonian is constituted by two time-independent terms, which refer to the Zeeman interactions for the electron (first term) and the nucleus ($\hbar\gamma_I = g_I \mu_N$, second term), as seen in Chapter 1, plus the interaction term between the electron and the nucleus spin [Eq. (1.51)]. The latter term fluctuates with time, and relaxation is due to its fluctuation.[2] As a result,[3] nuclear relaxation rates will

---

[2]Nuclear relaxation thus originates from the modulation of the hyperfine interaction

$$H_1(t) = \boldsymbol{I} \cdot \boldsymbol{A} \cdot \boldsymbol{S} = \frac{\mu_0}{4\pi} \hbar^2 \gamma_I \gamma_S \left(\frac{3(\boldsymbol{I} \cdot \boldsymbol{r})(\boldsymbol{S} \cdot \boldsymbol{r})}{r^5} - \frac{\boldsymbol{I} \cdot \boldsymbol{S}}{r^3} \right) + \frac{2}{3} \mu_0 g_e \mu_B g_I \mu_N \sum_i \delta(r)_i \boldsymbol{S} \cdot \boldsymbol{I}.$$

The first term provides the dipolar relaxation; the second, corresponding to the limit of the former for $r = 0$, provides the contact relaxation. δ indicates the Dirac δ operator and Σ_i represents the sum over all unpaired electrons.

[3]R_{1M} can be expressed in terms of the probabilities of all possible transitions involving different spin states. From the time dependent perturbation theory, if H_0 is the static Hamiltonian (first two terms in Eq. (4.14)) with eigenvalues $E_n = \hbar\omega_n$ and H_1 is the time dependent perturbing Hamiltonian (third term in Eq. (4.14)), which describes an interaction with energy $E_1(t)$ and whose explicit form will depend on the kind of interaction, the probability to have a transition from the m to the n state is proportional to

$$w_{mn} = \frac{1}{\hbar^2} \int_0^t \overline{E_{1(mn)}(t') E_{1(mn)}(t'')} e^{-i(\omega_m \ \omega_n)(t''-t')} \mathrm{d}(t''-t') + \mathrm{c.c.}$$

Since $H_1(t)$ is a 'stationary random perturbation', it is invariant under a change in the origin of the time, and the time (or ensemble) average will thus be

$$\overline{E_1(t')E_1(t'+t)} = \overline{E_1(0)^2} e^{-t/\tau_c}$$

where τ_c is the correlation time constant relative to the fluctuations. From the integration over the time, the spin density functions are thus obtained.

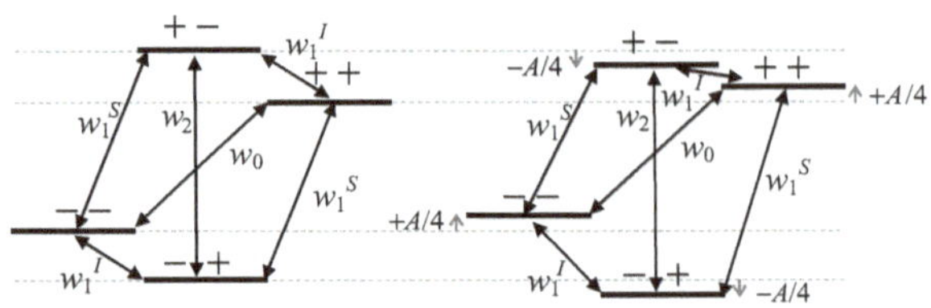

FIGURE 4.7 Energy levels, transition frequencies and transition probabilities per unit time (w_0, w_1 and w_2) in a magnetically coupled *S–I* system [(A) dipolar coupling; (B) contact coupling]. *A* is the contact hyperfine coupling constant (Appendix III). The order of the levels is for $g_I > 0$.

be proportional to the average of the square of this interaction energy and to the appropriate spectral density functions.[4]

$$R_{1M} \propto \langle (E_{\text{int}})^2 \rangle f(\omega, \tau_c). \tag{4.15}$$

Consider a single electron and a single nucleus separated by a distance r in a magnetic field B_0, and whose interaction is dipolar in origin. The $\mathbf{I} \cdot \mathbf{A} \cdot \mathbf{S}$ term is the dipolar coupling between the two (which average zero in solution in the absence of magnetic anisotropy). The energy levels and the transition frequencies ω are shown in Fig. 4.7. Each transition has associated a relative transition probability per unit time, w_0 for ω_0, w_1^S for ω_1^S, w_1^I for ω_1^I, w_2 for ω_2. The nuclear longitudinal relaxation rate, which describes the decay of the nuclear spin magnetization, will be expressed in terms of the probabilities of the transitions involving nuclear spin flipping (Appendix V):

$$R_{1M} = \frac{1}{10}\left(\frac{\mu_0}{4\pi}\right)^2 \frac{\hbar^2\gamma_I^2\gamma_S^2}{r^6}\left[\frac{\tau_c}{1+(\omega_I-\omega_S)^2\tau_c^2}+\frac{3\tau_c}{1+\omega_I^2\tau_c^2}+\frac{6\tau_c}{1+(\omega_I+\omega_S)^2\tau_c^2}\right] \tag{4.16}$$

which is the Solomon equation [69]. The three terms in parentheses are proportional to the transition probabilities and contain the spectral density functions $f(\omega, \tau_c)$ already discussed [Eqs. (4.2) and (4.4)]. The frequencies needed for the nuclear transitions are: ω_I (proton spin changes, electron spin remains the same), $\omega_I + \omega_S$ (proton and electron spins increase or decrease their energy), and $\omega_I - \omega_S$ (proton spin energy level increases and electron spin energy level decreases, or opposite). This equation can be generalized [70,71] to systems with $S > ½$

$$R_{1M} = \frac{2}{15}\left(\frac{\mu_0}{4\pi}\right)^2 \frac{\gamma_I^2 g_e^2 \mu_B^2 S(S+1)}{r^6}\left[\frac{\tau_c}{1+(\omega_I-\omega_S)^2\tau_c^2}+\frac{3\tau_c}{1+\omega_I^2\tau_c^2}+\frac{6\tau_c}{1+(\omega_I+\omega_S)^2\tau_c^2}\right]. \tag{4.17}$$

[4]We recall here that the time average of the interaction energy E_{dip} between two point-dipoles is zero (Chapter 2 and Appendix II), while the time average of its square is different from zero. The average dipolar shift in solution (pseudocontact shift) depends on the presence of magnetic anisotropy, which is necessary to give a nonzero average of E_{dip}; dipolar relaxation, on the contrary, will occur independently of the presence of magnetic anisotropy.

Note that Eq. (4.17) contains S but not I; R_{1M} is independent of I because it refers to $\Delta M_I = 1$ transitions [69,71]. J is needed instead of S for lanthanoids. Analogously [69–71], an equation can be derived for R_{2M}:

$$R_{2M} = \frac{1}{15}\left(\frac{\mu_0}{4\pi}\right)^2 \frac{\gamma_I^2 g_e^2 \mu_B^2 S(S+1)}{r^6}\left[4\tau_c + \frac{\tau_c}{1+(\omega_I-\omega_S)^2\tau_c^2} + \frac{3\tau_c}{1+\omega_I^2\tau_c^2} + \frac{6\tau_c}{1+(\omega_I+\omega_S)^2\tau_c^2} + \frac{6\tau_c}{1+\omega_S^2\tau_c^2}\right]. \quad (4.18)$$

R_{2M} differs from R_{1M} principally for the first and last terms. It is well known in NMR that frequencies near zero contribute to R_2 but not to R_1 (first term). This term originates from a $J(0)$ term [Eq. (4.3)], as described in Section 4.2. The last term is present because R_{2M} must also contain the probabilities of transitions connecting states with the same M_I but different M_S. When ω_S and ω_I are much smaller than τ_c^{-1} (fast motion limit) the denominators of the fractions between brackets become unity and the Solomon Eqs. (4.17) and (4.18) reduce to

$$R_{1M} = R_{2M} = \frac{4}{3}\left(\frac{\mu_0}{4\pi}\right)^2 \frac{\gamma_I^2 g_e^2 \mu_B^2 S(S+1)}{r^6}\tau_c. \quad (4.19)$$

At a magnetic field of 2.35 T, the proton transition frequency (Larmor frequency) is: $\nu_I = 100$ MHz; $|\omega_I| = 6.28 \times 10^8$ rad s^{-1}; $|\omega_S| = 4.13 \times 10^{11}$ rad s^{-1}. The ratio $|\omega_S|/|\omega_I|$ is 658 if the nucleus is a proton, and higher for other nuclei (except tritium, Appendix I). Therefore, the values $|\omega_I \pm \omega_S|$ in Eqs. (4.17) and (4.18) can be approximated by $|\omega_S|$.

It should be recalled that the longitudinal relaxation process requires energy exchange with the lattice. In this case, the lattice is represented by the electron spin system, which exchanges energy with the nuclear spin system. Every process that causes R_1 relaxation also contributes to R_2 relaxation. Spin–spin flip–flop transitions not involving energy exchange do contribute to R_2 only. In the fast motion limit, the coupling with the lattice has the same effect on the magnetization along the z axis and on that in the xy plane, and $R_1 = R_2$. However, outside this limit, the two functions $f_1(\omega, \tau_c)$ for R_{1M} and $f_2(\omega, \tau_c)/2$ for R_{2M}, corresponding to the expressions in parentheses in Eqs. (4.17) and (4.18), diverge, leading to $R_{1M} < R_{2M}$. The function $f_1(\omega, \tau_c)$ has its maximum value when the rate constant τ_c^{-1} equals the nuclear Larmor precession frequency $|\omega_I|$ (Fig. 4.8), which makes the coupling of the electron and

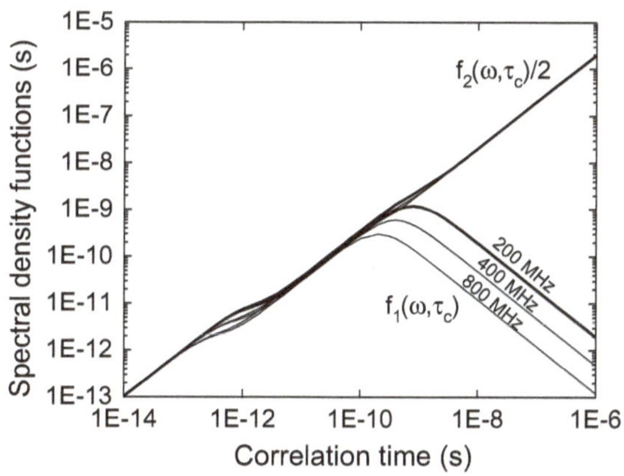

FIGURE 4.8 Plot of $f_1(\omega, \tau_c)$ and $\frac{1}{2}f_2(\omega, \tau_c)$ of Eqs. (4.17) and (4.18) against τ_c for various proton Larmor frequencies (MHz).

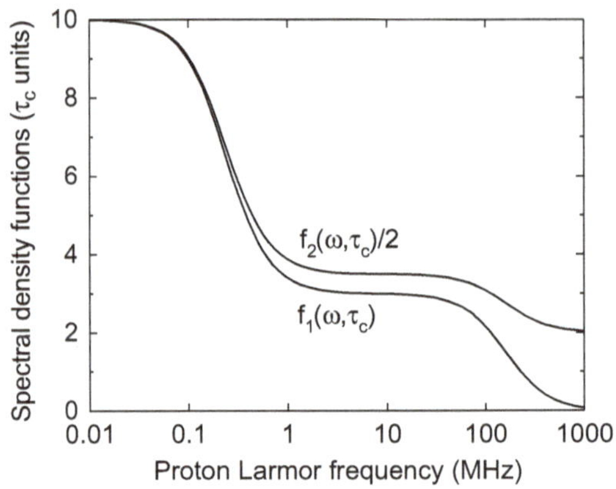

FIGURE 4.9 Plot of $f_1(\omega, \tau_c)$ and $\frac{1}{2}f_2(\omega, \tau_c)$ of Eqs. (4.17) and (4.18) as a function of magnetic field (expressed as proton Larmor frequency; log scale). $\tau_c = 1 \times 10^{-9}$ s

nuclear spins most efficient. The function $f_2(\omega, \tau_c)/2$ increases monotonically with τ_c and differs from $f_1(\omega, \tau_c)$ for $\tau_c > |\omega_I|^{-1}$ (Fig. 4.8) because of the frequency-independent term $4\tau_c$.

R_{1M} and R_{2M} values also depend on the applied magnetic field, as ω_I and ω_S are determined by it ($|\omega_I| = \gamma_I B_0$; $|\omega_S| = \gamma_S B_0 = 658|\omega_I|$ for protons). The magnetic fields currently used range from that of the Earth (25–65 μT) to 23.5 T ($|\omega_I|/2\pi = 1000$ MHz for protons). The field dependence of $f_1(\omega, \tau_c)$ (Fig. 4.9) shows two plateaus whose heights are in a 10/3 ratio and depend on the value of τ_c. The τ_c parameter is considered constant within the magnetic field range used, but this may not always be so [Eqs. (4.12) and (4.13), Section 4.3]. The inflection point of the first dispersion is halfway between the two plateaus and corresponds to a magnetic field value at which $|\omega_S| = \tau_c^{-1}$; the inflection point of the second dispersion is halfway between the second plateau and zero, and corresponds to $|\omega_I| = \tau_c^{-1}$. Profiles of the type of that shown in Fig. 4.9 can be obtained experimentally as described in Section 12.5, and are called nuclear magnetic relaxation dispersions (NMRD). Examples of experimental NMRD profiles will be found in Chapters 7, 8 and 10.

The field dependence of $f_2(\omega, \tau_c)/2$ is similar to that of $f_1(\omega, \tau_c)$ (Fig. 4.9), except that the relative height of the two plateaus is 20/7 and at high field the function levels off at a finite value that is one-fifth of the low field limit, because of the field independent term $4\tau_c$.

In the foregoing discussion, the electron–nucleus spin system was assumed to be rigidly held within a molecule isotropically rotating in solution. If the molecule cannot be treated as a sphere, its motion is, in general, anisotropic and three different correlation times should be considered. Furthermore, if the molecule is not rigid, the correlation times for rotations about single bonds should also be taken into account. For the case of a metal ion rigidly held to a macromolecule, but with a coordinated water molecule freely rotating about the M—O bond, the effect of neglecting the latter rotation on the calculation of R_{1M} for the protons is estimated to be about 20% [72]. The exact magnitude of the effect is a function of the angle that the metal–proton vector makes with the M—O bond.

When τ_c is significantly affected by τ_e, then it can be magnetic field dependent, since electron relaxation can often be described by equations similar to those for nuclear relaxation [see, e.g., Eqs. (4.12) and (4.13)]. Owing to this dependence, longitudinal (T_{1e}) and transverse (T_{2e}) electronic relaxation times may not be equal and, strictly speaking, when electron relaxation is the dominant correlation time, Eqs. (4.17) and (4.18) should be written as

$$R_{1M} = \frac{2}{15}\left(\frac{\mu_0}{4\pi}\right)^2 \frac{\gamma_I^2 g_e^2 \mu_B^2 S(S+1)}{r^6}\left(\frac{7T_{2e}}{1+\omega_S^2 T_{2e}^2} + \frac{3T_{1e}}{1+\omega_I^2 T_{1e}^2}\right) \quad (4.20)$$

$$R_{2M} = \frac{1}{15}\left(\frac{\mu_0}{4\pi}\right)^2 \frac{\gamma_I^2 g_e^2 \mu_B^2 S(S+1)}{r^6}\left(4T_{1e} + \frac{13T_{2e}}{1+\omega_S^2 T_{2e}^2} + \frac{3T_{1e}}{1+\omega_I^2 T_{1e}^2}\right) \quad (4.21)$$

where the terms containing ω_S, $\omega_I + \omega_S$ and $\omega_I - \omega_S$ have been collected for simplicity, since $|\omega_I| \ll |\omega_S|$. The longitudinal relaxation time $T_{1\rho}$ in the rotating frame is also shortened through coupling with unpaired electrons, according to the following equation [71]:

$$R_{1\rho M} = \frac{1}{15}\left(\frac{\mu_0}{4\pi}\right)^2 \frac{\gamma_I^2 g_e^2 \mu_B^2 S(S+1)}{r^6}\left(\frac{4T_{1e}}{1+\omega_1^2 T_{1e}^2} + \frac{13T_{2e}}{1+\omega_S^2 T_{2e}^2} + \frac{3T_{1e}}{1+\omega_I^2 T_{1e}^2}\right). \quad (4.22)$$

It differs from the equation for R_{2M} for the substitution of the nondispersive term with a $\omega_1 \tau_e$ dependent term, where ω_1 is the nuclear Larmor frequency in the B_1 field. Since the latter frequency is always such that $|\omega_1|\tau_c \ll 1$, $R_{1\rho M}$ can be safely taken equal to R_{2M}. This also holds for the other nuclear relaxation mechanisms discussed in the following sections.

In the low field (or fast motion) limit, in the absence of chemical exchange, and for $T_{2e} = T_{1e}$, Eqs. (4.20)–(4.22) reduce to

$$R_{1M} = R_{2M} = R_{1\rho M} = \frac{4}{3}\left(\frac{\mu_0}{4\pi}\right)^2 \frac{\gamma_I^2 g_e^2 \mu_B^2 S(S+1)}{r^6} T_{1e} \quad (4.23)$$

justifying the qualitative statement that the shorter the electronic relaxation times the smaller the paramagnetic effects on nuclear relaxation.

4.4.1 GENERALIZED DIPOLAR COUPLING

As anticipated in Sections 2.2.2 and 4.1, the unpaired electrons should not be considered as point-dipoles centered on the metal ion. They are at the least delocalized over the atomic orbitals of the metal ion itself. The effect of the deviation from the point–dipole approximation under these conditions is estimated to be negligible for nuclei already 3–4 Å away [73]. Electron delocalization onto the ligands, however, may heavily affect the overall relaxation phenomena. In this case the experimental R_{iM} may be higher than expected, and the ratios between the R_{iM} values of different nuclei do not follow the sixth power of the ratios between metal to nucleus distances. In the case of hexaaqua metal complexes, the point–dipole approximation provides shorter distances than observed in the solid state (Table 4.2) for both ^1H and ^{17}O. This implies spin density delocalization on the oxygen atom. Ab initio calculations of R_{1M} have been performed for both ^1H and ^{17}O nuclei in a series of hexaaqua complexes (Table 4.2). The calculated metal nucleus distances in the assumption of a purely metal-centered dipolar relaxation mechanism are sizably smaller than the crystallographic values for ^{17}O, and the difference dramatically increases from $3d^5$ to $3d^9$ metal ions [74]. The differences for protons are quite smaller [74].

R_{1M} calculations in the presence of ligand-centered contributions are possible for metal complexes with ligands having dominant π spin density delocalization mechanisms. With certain approximations,

Table 4.2 Crystallographic Values of Metal–Hydrogen and Metal–Oxygen Distances in Hexaaqua Complexes of Divalent 3d Metal Ions[a] Compared with Calculated Effective Distances

	Metal Ions				
Measurements	**Mn**	**Fe**	**Co**	**Ni**	**Cu**
^1H data					
r_{cryst} (pm)	290.6	290.6	278.1	275.5	270.4
r_{eff} (pm)	290.0	289.5	276.3	273.1	268.4
$r^6_{cryst} \cdot \langle 1/r^3 \rangle^2_{eff}$ [b]	1.01	1.02	1.04	1.05	1.05
^{17}O data					
r_{cryst} (pm)	222.0	222.0	209.0	206.1	201.0
r_{eff} (pm)	219.8	211.7	190.5	174.9	172.3
$r^6_{cryst} \cdot \langle 1/r^3 \rangle^2_{eff}$ [b]	1.06	1.33	1.74	2.67	2.52

[a]*From [74]. M–H distances were calculated using $r_{O\text{–}H}$ = 95.7 pm and HÔH=104.52°.*
[b]*The effective distances account for the experimental T_{1M}^{-1} by considering the electronic distribution predicted from ab initio MO calculations. $r^6_{cryst} \cdot \langle 1/r^3 \rangle^2_{eff}$ is the predicted enhancement coefficient of T_{1M}^{-1} with respect to a pure metal-centered dipolar relaxation mechanism.*

the relaxation rates of protons and carbon atoms in sp^2 CH moieties can be expressed [75–77] as the sum of a metal-centered term (given by the Solomon equation), a ligand-centered term and a cross term. The ligand-centered contributions will be proportional to the spin density ρ_{2p_z} in the carbon 2p$_z$ orbital, and to the square of the reciprocal third power of the average nucleus-2p$_z$ electron distance. Equations of the type of Eq. (4.24) have been proposed [77]:

$$R_{1M} = \frac{2}{15}\left(\frac{\mu_0}{4\pi}\right)^2 \gamma_I^2 g_e^2 \mu_B^2 S(S+1)\left[\rho_M^2 r_M^{-6} + \frac{4}{25}\rho_{2p_z}^2 \left\langle r_{2p_z}^{-3}\right\rangle^2 - \frac{2}{5}\rho_{2p_z}\left\langle r_{2p_z}^{-3}\right\rangle \rho_M r_M^{-3}\right] f_1(\omega,\tau_c) \tag{4.24}$$

where $\langle r_{2p_z}^{-3} \rangle$ is calculated to be 5.3×10^{-7} pm^{-3} for protons and 3.2×10^{-6} pm^{-3} for carbon atoms [75,76]. In general, ρ_{2p_z} can be evaluated from the contact part of the proton hyperfine shift. This can be an easy-to-follow procedure for estimating ligand-centered contributions to the relaxation rates for π delocalized spin density [77].

4.5 NUCLEAR RELAXATION DUE TO CONTACT COUPLING WITH UNPAIRED ELECTRONS

In the case of relaxation due to the contact term, the transition probabilities as defined in Fig. 4.7, where the splitting of the levels is now due to the contact contribution, are found to be

$$w_1^I = w_1^S = w_2 = 0. \tag{4.25}$$

This is because any scalar relaxation occurs through a flip-flop mechanism (Fig. 4.7)[5];

$$w_0 = \frac{1}{2}\left(\frac{A}{\hbar}\right)^2 \frac{\tau_c}{1+(\omega_S - \omega_I)^2\tau_c^2}. \tag{4.26}$$

Here the correlation time is τ_c^{con} of Eq. (4.7) and A is defined in Eq. (2.2). This differs from τ_c^{dip} of Eq. (4.6) because of the absence of the contribution from the rotational correlation time since the contact interaction, by its nature, is not altered by reorientation of the molecules.[6] In the absence of chemical exchange phenomena ($\tau_M^{-1} \ll \tau_e^{-1}$), the correlation time equals the electronic relaxation time.

The enhancement of the longitudinal contact relaxation rate, after extension to the general case $S > ½$, is given by the Bloembergen equation [71,78]

$$R_{1M} = \frac{2}{3}S(S+1)\left(\frac{A}{\hbar}\right)^2 \frac{T_{2e}}{1+\omega_S^2 T_{2e}^2} \tag{4.27}$$

where T_{2e} is the actual parameter determining τ_c. The equation for the transverse contact relaxation rate is [71,78]

$$R_{2M} = \frac{1}{3}S(S+1)\left(\frac{A}{\hbar}\right)^2 \left(T_{1e} + \frac{T_{2e}}{1+\omega_S^2 T_{2e}^2}\right). \tag{4.28}$$

The term $S(S + 1)(A/\hbar)^2$ is proportional to the square of the coupling energy between an electron spin vector S and the nuclear spin vector I.

In the fast motion limit, and for $T_{1e} = T_{2e}$, Eqs. (4.27) and (4.28) reduce to

$$R_{1M} = R_{2M} = \frac{2}{3}S(S+1)\left(\frac{A}{\hbar}\right)^2 T_{1e}. \tag{4.29}$$

As in the case of dipolar relaxation, the presence of a field-independent term in the equation for R_{2M} causes the two nuclear relaxation values to diverge for $\omega_S^2 T_{2e}^2 > 1$ (Figs. 4.10 and 4.11).

4.6 CURIE NUCLEAR SPIN RELAXATION

In deriving the dipolar and contact relaxation contributions due to the presence of unpaired electrons, as shown in Eqs. (4.20), (4.21), (4.27), and (4.28), the small difference in the population of the electron spin levels according to the Boltzmann distribution has been neglected. Such a difference accounts for

[5]In fact,

$$H_1^{cont} = A\boldsymbol{I}\cdot\boldsymbol{S} = A(I_x S_x + I_y S_y + I_z S_z) = \frac{A}{2}(S_+ I_- + S_- I_+ + 2S_z I_z),$$

(Appendix III) where $S_\pm$ and $I_\pm$ are the usual flip up and down spin operators. Therefore, contact relaxation contains only the $|\omega_I - \omega_S|$ term, since the transitions $|\omega_I + \omega_S|$ (corresponding to S_+I_+, S_-I_-), ω_S (S_+I_z, S_-I_z) and ω_I (S_zI_+, S_zI_-) do not contribute, at variance with dipolar relaxation.

[6]$\langle S_z\rangle$ can be orientation dependent in the case of anisotropic $\boldsymbol{g}$ and/or $\boldsymbol{\chi}$ tensors, but the contact coupling constant A does not depend on the molecular orientation.

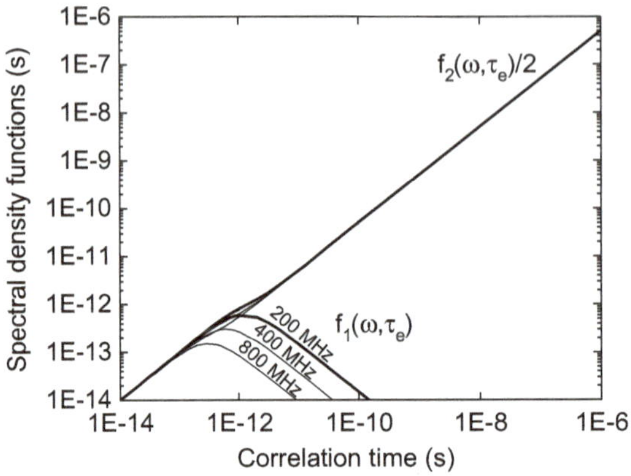

FIGURE 4.10 Plot of $f_1(\omega, \tau_e)$ and $\frac{1}{2}f_2(\omega, \tau_e)$ of Eqs. (4.27) and (4.28) against τ_e for various proton Larmor frequencies (MHz).

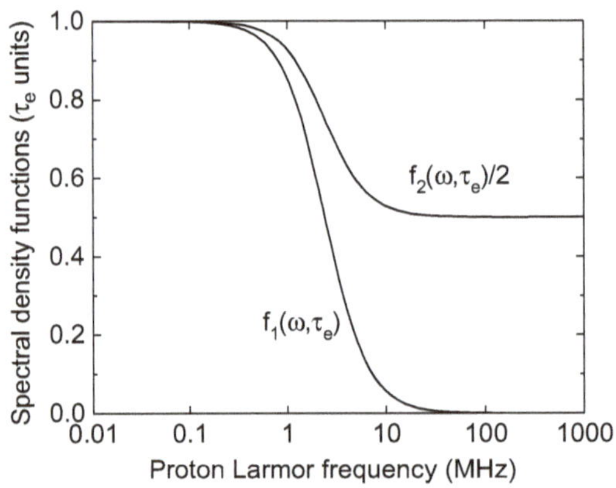

FIGURE 4.11 Plot of $f_1(\omega, \tau_e)$ and $\frac{1}{2}f_2(\omega, \tau_e)$ of Eqs. (4.27) and (4.28) as a function of magnetic field (expressed as proton Larmor frequency; log scale). $\tau_e = 1 \times 10^{-10}$ s.

the time-averaged magnetic moment of the molecule $\langle\mu\rangle$ [Eq. (1.33)], which is related to the $\langle S_z\rangle$ value defined in Eq. (1.31) for a simplified case. The interaction of the nuclear spins with this static magnetic moment related to $\langle S_z\rangle$ provides a further relaxation contribution. Such interaction, of course, cannot be modulated by electron relaxation since $\langle S_z\rangle$ is already an average over the electron spin states. The correlation time for the coupling is only determined by τ_r (or possibly by τ_M). This relaxation mechanism is usually called magnetic susceptibility relaxation or Curie spin relaxation, to reflect its relationship with the magnetic susceptibility of a sample (Curie law).

The dipolar contributions to R_{1M} and R_{2M} provided by this mechanism are [71,79,80]

$$\begin{aligned} R_{1M} &= \frac{2}{5}\left(\frac{\mu_0}{4\pi}\right)^2 \frac{\gamma_I^2 g_e^2 \mu_B^2}{r^6} \langle S_z \rangle^2 \frac{3\tau_r}{1+\omega_I^2\tau_r^2} \\ &= \frac{2}{5}\left(\frac{\mu_0}{4\pi}\right)^2 \frac{\omega_I^2 g_e^4 \mu_B^4 S^2(S+1)^2}{(3kT)^2 r^6} \frac{3\tau_r}{1+\omega_I^2\tau_r^2} \end{aligned} \tag{4.30}$$

$$\begin{aligned} R_{2M} &= \frac{1}{5}\left(\frac{\mu_0}{4\pi}\right)^2 \frac{\gamma_I^2 g_e^2 \mu_B^2}{r^6} \langle S_z \rangle^2 \left(4\tau_r + \frac{3\tau_r}{1+\omega_I^2\tau_r^2}\right) \\ &= \frac{1}{5}\left(\frac{\mu_0}{4\pi}\right)^2 \frac{\omega_I^2 g_e^4 \mu_B^4 S^2(S+1)^2}{(3kT)^2 r^6} \left(4\tau_r + \frac{3\tau_r}{1+\omega_I^2\tau_r^2}\right). \end{aligned} \tag{4.31}$$

In the case of lanthanoids $\langle J_z \rangle^2$ should replace $\langle S_z \rangle^2$. Although the value of $\langle S_z \rangle^2$ is much smaller than the value of $S(S+1)/3$, which appears in the corresponding Solomon equations [Eqs. (4.20) and (4.21)], Curie spin relaxation may be significant when the dipolar coupling described by the Solomon equations is governed by the electronic relaxation times; that is, when the latter are much smaller than the rotational correlation times. Furthermore, since $\langle S_z \rangle^2$ depends on the square of the external magnetic field, Curie spin relaxation increases with the square of the nuclear Larmor frequency ω_I—see Eqs. (4.30) and (4.31) and Fig. 4.12. For example, when $\tau_r = 5 \times 10^{-11}$ s, $\tau_e = 5 \times 10^{-13}$ s and $B_0 = 14$ T, Curie spin relaxation can be sizably larger than the Solomon contribution to both R_1 and R_2. This is the case of aqualanthanoid(III) ions except gadolinium(III) [43]. It is also quite common for Curie relaxation to dominate R_2 in macromolecules. For example, when $\tau_r = 10^{-7}$ s, $\tau_c = T_{1e} = 10^{-12}$ s, and $B_0 = 6$ T, Curie spin relaxation for R_{2M} is one order of magnitude larger than the Solomon contribution. Under these conditions, Curie spin relaxation does not significantly contribute to R_{1M}; at high magnetic field ($\omega_I^2\tau_r^2 \gg 1$) the dependence on ω_I^2 in the numerator is canceled out. R_{1M} levels off at a value that is often small compared with the dipolar contribution.

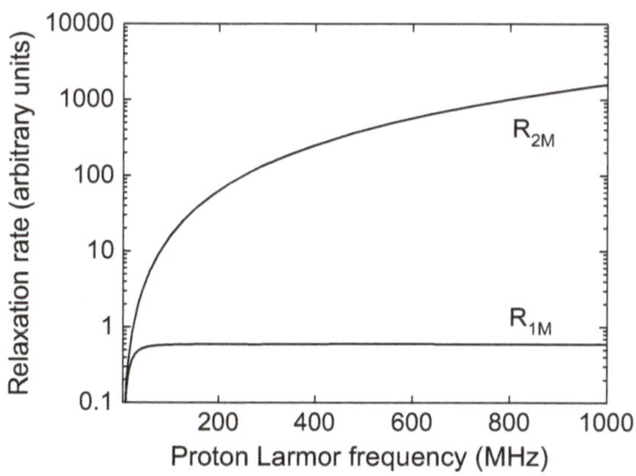

FIGURE 4.12 Field dependence of the Curie spin relaxation contributions to R_{1M} and R_{2M} (arbitrary scale); $\tau_r = 1 \times 10^{-8}$ s.

The occurrence of this particular relaxation mechanism can be recognized through the field and temperature dependence of the linewidths. In the absence of other effects, the line broadening is proportional to the square of the magnetic field (Fig. 4.12) and to η/T^3, the latter dependence arising from the $1/T^2$ dependence of $\langle S_z \rangle^2$ [Eqs. (4.30) and (4.31)] and the η/T dependence of τ_r [Eq. (4.8)].

In principle, there may also be a Curie relaxation contribution of the contact-type whenever there is chemical exchange or intramolecular rotation to modulate the coupling. The contribution to R_{2M} would then be

$$R_{2M} = \frac{4}{3}\left(\frac{A}{\hbar}\right)^2 \langle S_z \rangle^2 \tau_M = \frac{4}{3}(\Delta\omega_I)^2 \tau_M \tag{4.32}$$

where $\Delta\omega_I$ is the observed contact shift ($\Delta\omega_I = 2\pi\Delta\nu_I$).

The relaxation equations discussed here and in Section 4.4 and 4.5 take a different form in the case of lanthanoids and actinoids (Section 8.5). For these systems, in fact, the J quantum number substitutes the S quantum number and g_J substitutes g_e.

4.7 FURTHER ELECTRONIC EFFECTS ON NUCLEAR RELAXATION

4.7.1 THE EFFECT OF *g* ANISOTROPY AND OF THE SPLITTING OF THE *S* MANIFOLD AT ZERO MAGNETIC FIELD

The electronic term which is the first term in the Hamiltonian written in Eq. (4.14) and used to derive the Solomon and Bloembergen equations [Eqs. (4.17), (4.18), (4.20), (4.21), (4.27), and (4.28)] may be inappropriate in many cases, since the electron energy levels may be strongly affected by the presence of ZFS or hyperfine coupling with the metal nucleus. Therefore, the electron static Hamiltonian to be solved to find the ω_S values, i.e., all electron energy transitions, and their probabilities, will be, in general,

$$H_0 = \mu_B \boldsymbol{S}\cdot\boldsymbol{g}\cdot\boldsymbol{B}_0 + \boldsymbol{S}\cdot\boldsymbol{D}\cdot\boldsymbol{S} + \boldsymbol{I}\cdot\boldsymbol{A}\cdot\boldsymbol{S}. \tag{4.33}$$

Let us discuss first the case in which only the first term is present. In the Solomon and Bloembergen equations for R_i ($i = 1, 2$) there is the ω_S parameter at the denominator of a Lorentzian function. Up to now ω_S has been taken equal to that of the free electron. However, in the presence of orbital contributions, the Zeeman splitting of the M_S levels changes its value and ω_S equals $|\gamma_S|B_0$ or $(g/\hbar)\mu_B B_0$. When g is anisotropic (Fig. 1.14), the value of ω_S is different from that of the free electron and is orientation dependent. The principal consequence is that another parameter (at least) is needed, i.e., the θ angle between the metal–nucleus vector and the z direction of the $\boldsymbol{g}$ tensor (Section 1.4). A second consequence is that the ω_S fluctuations in solution must be taken into account when integrating over all the orientations. Appropriate equations for nuclear relaxation have been derived for both the cases in which rotation is faster [81,82] or slower [83,84] than the electronic relaxation time. In practical cases, the deviations from the Solomon profile are within 10–20% (e.g., Fig. 4.13).

When there is splitting of the S manifold at zero magnetic field [second term in Eq. (4.33)], a major effect is a reduction of the amplitude of the $\omega_S\tau_c$ dispersion, because the energy of most electronic spin transitions increases (Fig. 1.15) and can be much larger than the Zeeman energy and larger than $\hbar\tau_c^{-1}$, so that for these transitions $\omega_S\tau_c \gg 1$. We should distinguish between systems with half integer and integer S

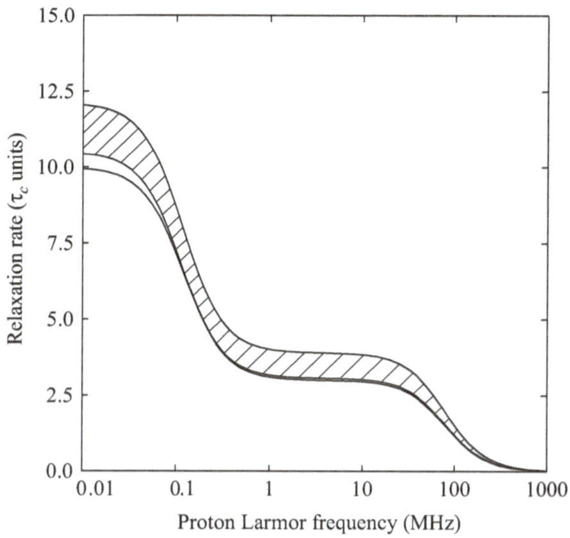

FIGURE 4.13 Plot of the spectral density functions for dipolar relaxation in the presence of an axially symmetric g tensor. Conditions: $g_{\parallel} = 2.3$, $g_{\perp} = 2.0$, $\tau_c = 2 \times 10^{-9}$ s, $\theta = 0$ degrees (upper curve) and $\theta = 90$ degrees (middle curve) compared with the Solomon behavior (lower curve).

values. In the former case we always have an $M_S = -\frac{1}{2} \rightarrow \frac{1}{2}$ transition with energy of the order of $g\mu_B B_0$. The $M_S = \frac{1}{2} \rightarrow \frac{3}{2}$ transition, for example, may contain the term D (e.g., Fig. 1.15). If D is much greater than the Zeeman energy, as is often the case, $|\omega_S|$ is much larger than in the case of the $-\frac{1}{2} \rightarrow \frac{1}{2}$ transition. Then the condition $|\omega_S|\tau_c \gg 1$ holds and the Lorentzian function has a negligible value. Such electronic transitions do not contribute to nuclear relaxation. It should also be mentioned that this holds when $D \gg \hbar\tau_c^{-1}$. Neglecting the ZFS may introduce an error of about a factor two in R_{1M} (Fig. 4.14C). In the case of integer S, all transitions contain D, so that the $\omega_S\tau_c$ dispersion may be totally quenched (Fig. 4.14A,B). Further complications arise due to the effect of rhombicity of ZFS, described by the parameter E (Section 1.4). E can vary in the range $-1/3 \leq E/D \leq 1/3$. We only note here that rhombicity causes a drop to zero of the relaxation rate for integer S systems, whereas its effect is minor for half-integer S systems.

In any case, the presence of ZFS may cause the occurrence of a further dispersion in the plot of relaxation rate as a function of proton Larmor frequency, corresponding to the transition from the dominant ZFS limit to the dominant Zeeman limit[7] (Fig. 4.14).

---

[7]The electron Hamiltonian is in fact given by

$$H = g_e \mu_B \boldsymbol{S} \cdot \boldsymbol{B}_0 + \boldsymbol{S} \cdot \boldsymbol{D} \cdot \boldsymbol{S}$$

where $\boldsymbol{D}$ is the traceless ZFS tensor. The axial and rhombic components of ZFS, D and E (Section 1.4), are defined as

$$D = D_z - \frac{D_x + D_y}{2} \quad \text{and} \quad E = \frac{D_x - D_y}{2}.$$

In the low field limit the ZFS term (the second in the above equation) is dominant, while in the high field limit the Zeeman term (the first in the above equation) is dominant.

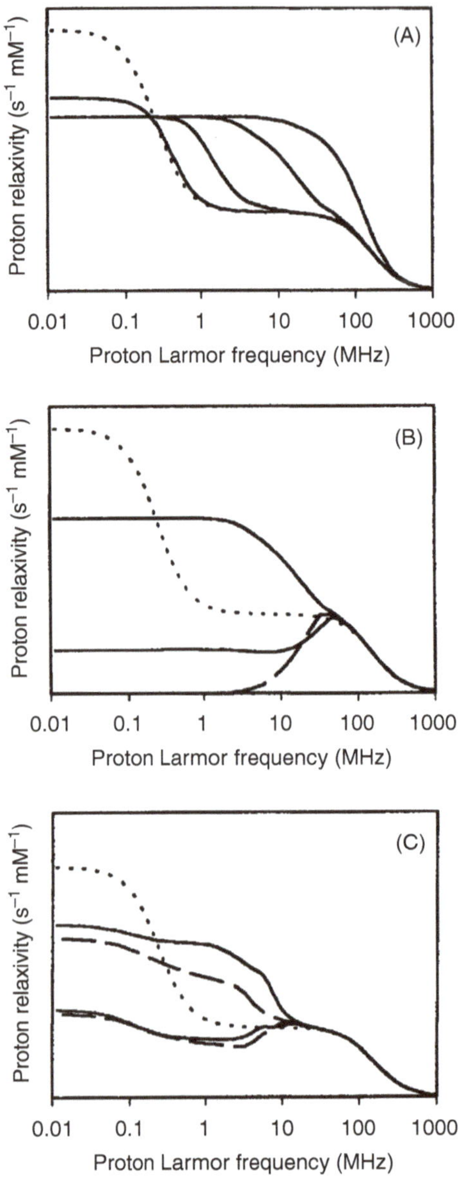

FIGURE 4.14 Plot of the nuclear longitudinal dipolar relaxation rate in the presence of ZFS, for (A) $S = 1$, $\theta = 0$, $D = 0.01, 0.1, 1, 10\ \text{cm}^{-1}$, $E = 0$ (axial case, solid lines); (B) $S = 1$, $D + E = 1\ \text{cm}^{-1}$, $E = 0$, $\theta = 0$ (upper solid line), $\theta = 90$ degrees (lower solid line) and $E/D = 1/3$, $\theta = 0$ and 90 degrees (dashed line); (C) $S = 5/2$, $D + E = 0.1\ \text{cm}^{-1}$, $E = 0$, $\theta = 0$ (upper solid line), $\theta = 90$ degrees (lower solid line) and $E/D = 1/3$, $\theta = 0$ and 90 degrees (dashed lines); (D) $I = 3/2$, $\theta = 0$, $A_x = A_y = A_z = 200, 100, 50, 25 \times 10^{-4}\ \text{cm}^{-1}$ (solid lines in order of increasing low field relaxivity); (E) $I = 3/2$, $\theta = 0$, $A_z = 200, 100, 50, 25 \times 10^{-4}\ \text{cm}^{-1}$, $A_x = A_y = 0$; (F) $I = 3/2$, $\theta = 90$ degrees, $A_z = 200, 100, 50, 25 \times 10^{-4}\ \text{cm}^{-1}$, $A_x = A_y = 0$. The dotted line shows the Solomon profile. τ_c is 10^{-9} s in all cases.

A further effect of static ZFS when it influences nuclear relaxation is that again the latter depends on an angle (with axial ZFS) or two angles (with rhombic ZFS), which take into account the position of the resonating nucleus within the $\boldsymbol{D}$ tensor frame.

Another mechanism to provide splitting of the S manifold is the hyperfine coupling between the unpaired electron and the metal nucleus. For example, at zero magnetic field an $S = ½$ $I = ½$ system gives two sets of levels of degeneracy 3 and 1, separated by A, where A is the metal–nucleus-unpaired-electron hyperfine coupling. Again, the change in the energy levels due to this mechanism can affect the transition probabilities. The effect is sizable, however, when $A > \hbar\tau_c^{-1}$, and at low magnetic fields where $A > g\mu_B B_0$. In high resolution NMR, the latter condition is practically never met.

As seen in the presence of static ZFS, the effect on the field dependence of nuclear relaxation can be very different depending on the values of $A_{\parallel}$, $A_{\perp}$ and on the position of the nucleus with respect to the molecular frame defined by the $\boldsymbol{A}$ tensor, which is determined by an angle (if axial) or two angles (if rhombic). Again, the general effect is that of reducing the relaxation rate at all fields with a Zeeman energy smaller than the hyperfine coupling energy. We note that, at variance with ZFS, the hyperfine coupling with the metal nuclear spin affects the relaxivity profiles also in the case of I and S equal to 1/2 and for isotropic systems ($A_{\parallel} = A_{\perp}$) [84].

The theoretical approach to the description of R_{1M} in the presence of static ZFS or hyperfine coupling with the metal nucleus gives analytical solutions that are exact within certain limiting experimental conditions [29]. Numerical solutions [85,86], and computer programs [87] for the more general cases are also available.

4.7.2 FIELD DEPENDENCE OF ELECTRON RELAXATION

In case the electron relaxation is influencing the correlation time for nuclear relaxation, according to Eqs. (4.6) and (4.7), the field dependence of the proton relaxation rate also reflects the field dependence of τ_e, if present. In general, the latter is described by Eqs. (4.12) and (4.13), in the absence of splitting of the S manifold at zero magnetic field. It can be seen that τ_e is constant until the $\omega_S\tau_v$ dispersion occurs. Since τ_e must be larger than τ_v to stay within the range of validity of all the mentioned equations (Redfield limit, Section 4.13), as τ_v is the correlation time for electron relaxation, it results that the electron relaxation time starts to increase after the occurrence of the $\omega_S\tau_e$ dispersion. As far as τ_e increases, R_{1M} increases until the $\omega_I\tau_e$ dispersion occurs [Eq. (4.20)]. Therefore, the overall shape of the profile obtained by plotting the relaxation rate as a function of proton Larmor frequency results in a peak in the high field region, between the $\omega_S\tau_e$ dispersion and the $\omega_I\tau_e$ dispersion. Fig. 4.15A and B shows that with increasing the value of τ_v, both the value of the maximum of the peak increases and the peak moves toward lower frequencies. The increase in relaxation rate at high fields may be even larger for R_{2M} in case of the presence of contact relaxation (Fig. 4.15C). As shown by Eq. (4.28), R_{2M} in fact increases with the electron relaxation time, that, in presence of a field dependence as described by Eq. (4.12) goes to infinity. The transverse relaxation rate at high fields would thus be limited by exchange processes only, if present, that contribute to τ_c through the exchange time τ_M (Eq. (4.7)). Exchange phenomena are described in more detail in Chapter 6. Some examples where these behaviors have been observed will be reported in Chapters 7 and 10.

The presence of static ZFS can affect the description of the field dependence of the electron relaxation time, because it changes the electron energy levels. This implies a change in the nuclear spin transition probabilities as well as in the electron spin relaxation processes. Several approaches have

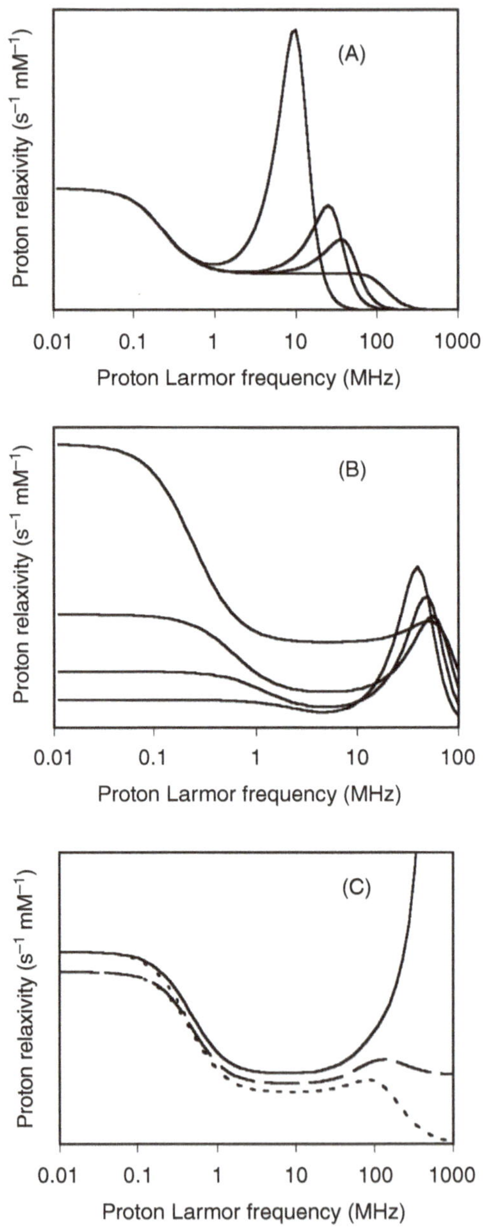

FIGURE 4.15 (A) Plot of the nuclear longitudinal dipolar relaxation rate in the presence of a field dependent electron relaxation time ($S = \frac{5}{2}$) for $T_{1e(0)} = 10^{-9}$ s and $\tau_v = 1, 5, 10, 50 \times 10^{-12}$ s. The amplitude of the peak that appears in the high field part of the profile increases with increasing τ_v. (B) Plot of the longitudinal dipolar relaxation rate in the presence of a field dependent electron relaxation time ($S = \frac{5}{2}$) for $\Delta_t = 0.047$ cm^{-1} and $\tau_v = 2, 5, 10, 20 \times 10^{-12}$ s. (C) Plot of the longitudinal relaxation rate in the presence of a field dependent electron relaxation time with $A/h = 0$ and 1 MHz (dotted lines) and of the transverse relaxation rate with $A/h = 0$ (dashed line) and 1 MHz (solid line). Conditions: $S = \frac{5}{2}$, $\Delta_t = 0.047$ cm^{-1}, $\tau_v = 2 \times 10^{-12}$ s ($T_{1e(0)} = 10^{-9}$ s) and $\tau_r = 10^{-9}$ s.

been proposed to take into account the effect of static ZFS on the electron relaxation time and its field dependence, when the latter is ascribed to a modulation of transient ZFS [87–94].

4.8 A COMPARISON OF DIPOLAR, CONTACT, AND CURIE NUCLEAR SPIN RELAXATION

Once R_{1M} and R_{2M} have been measured, it is useful to try to understand the relative weight of dipolar, contact, and Curie spin contributions to the overall relaxation effect. Indeed, each of the three contributions is independently capable of providing valuable information whereas the whole value may not. The dipolar and Curie relaxation mechanisms—Eqs. (4.17), (4.18), (4.30), and (4.31)—provide experimental values that are related to the distance of the nucleus from the paramagnetic metal and τ_c or τ_r, assuming that ligand-centered effects can be neglected. Measurements at various magnetic fields lead, in principle, to an estimation of these parameters. The field dependence of the dipolar contribution can be observed in any region of the accessible magnetic field range that is from 0.001 to 1000 MHz, whereas Curie relaxation can usually be observed only at relatively high fields. Even at a single magnetic field value, measurements of dipolar R_{1M} contributions on various nuclei of the same moiety give information on their relative distance from the metal, as long as τ_c is the same. This is also true for R_{2M} values when they are determined by dipolar and/or Curie contributions.

The presence of contact relaxation indicates that a given moiety is covalently bound to a paramagnetic metal ion and provides an estimate of the absolute value of A [Eqs. (4.27) and (4.28)]. Sometimes the contact coupling constant can be evaluated by chemical shift measurements, and it is therefore possible to predict whether the contact relaxation contributions to R_{1M}, R_{2M}, or both, are negligible or sizable.

Unlike hyperfine isotropic shifts, which often contain pseudocontact and contact contributions of the same order of magnitude, relaxation rates can often be recognized to be dominated by only one of the possible contributions. In addition, whereas contact and pseudocontact shifts may happen to have different signs, thereby making their separation more uncertain, relaxation contributions are obviously always positive and additive.

Some qualitative guidelines can be given to make an a priori estimate of the relative weight of dipolar, contact, and Curie relaxation contributions. Consider first the fast motion limit where $R_{1M} = R_{2M}$ and none of the frequency-dependent terms is dispersed. The equations take the simple form already noted:

$$R_{1M} = R_{2M} = \frac{4}{3}\left(\frac{\mu_0}{4\pi}\right)^2 \frac{\gamma_I^2 g_e^2 \mu_B^2 S(S+1)}{r^6}\tau_c^{\mathrm{dip}} + \frac{2}{3}S(S+1)\left(\frac{A}{\hbar}\right)^2 \tau_c^{\mathrm{con}}. \tag{4.34}$$

Under these conditions the Curie spin contribution is always negligible (Section 4.6). If $\tau_c^{\mathrm{dip}} = \tau_c^{\mathrm{con}}$ it is only necessary to compare the following two expressions:

$$\frac{4}{3}\frac{\gamma_I^2}{r^6}8.6\times10^{-61} \quad \text{and} \quad \frac{2}{3}\left(\frac{A}{h}\right)^2$$

where 8.6×10^{-61} is $(\mu_0/4\pi)^2 g_e^2 \mu_B^2$ (T^2 m^6). For hydrogen nuclei, γ_I^2 is 7.16×10^{16} rad^2 s^{-2} T^{-2}, whereas $(A/\hbar)^2$ does not usually exceed 10^{12} rad^2 s^{-2}, and is often found to be much lower. Thus, for metal–nucleus distances of 500 pm or smaller, the dipolar term is largely dominant. The value of the dipolar term drops dramatically with increasing r owing to its r^{-6} dependence, and for larger distances it could become smaller than the contact term. However, the $(A/\hbar)^2$ value is also qualitatively expected to decrease with increasing distance from the paramagnetic center. Therefore, unless some efficient unpaired spin delocalization pathway is operative, the dipolar term in proton relaxation is usually dominant for most systems of chemical interest.

The situation is different when the rotational correlation time is shorter than τ_e and therefore dominates the overall correlation time τ_c in the dipolar term. In this case, the relative importance of the dipolar term is decreased by a factor τ_e/τ_r, which can be as large as 10^2–10^3 for small complexes with rotational correlation times of 10^{-10} to 10^{-11} s and electronic relaxation times of 10^{-8} to 10^{-9} s, for example, Cu^{2+}, Mn^{2+}, and VO^{2+}. In macromolecular complexes, rotational correlation times are much larger, and situations of this type do not occur.

Outside the fast motion limit the relative weight of contact and dipolar interactions on R_{1M} and R_{2M} may also be different. The following considerations are particularly relevant to proton relaxation. Curie contributions to R_{2M} can be sizable. By comparing Eqs. (4.21) and (4.28) on the one hand and Eq. (4.31) on the other, it can be noted that they contain terms of the type

$$S(S+1)\tau_e \quad \text{and} \quad \omega_I^2 S^2 (S+1)^2 \tau_r$$

respectively. Therefore, it is expected that Curie contributions will be comparatively higher, the higher the field, the higher S, and the higher the τ_r/τ_e ratio.

Once the Curie contribution to R_{2M} is estimated and subtracted, the contribution of contact and dipolar interactions can be estimated by examining the correlation time dependence of the paramagnetic relaxation depicted in Figs. 4.8 and 4.10. It appears that the maximum for R_{1M} occurs at $|\omega_I|\tau_c^{\mathrm{dip}} \approx 1$ in the dipolar term and at $|\omega_S|\tau_c^{\mathrm{con}} \approx 1$ in the contact term. Taking for simplicity $\tau_c^{\mathrm{dip}} = \tau_c^{\mathrm{con}}$, this means that in the intermediate situation where $|\omega_S|\tau_c^{\mathrm{dip}} > 1 > |\omega_I|\tau_c^{\mathrm{con}}$ the relative importance of the contact term is even smaller than that estimated in the fast motion limit. The equation for R_{2M} has nondispersive terms in both the dipolar and contact contributions (accounting for one-fifth and one-half of the total effect measured in the fast motion limit respectively), and therefore the conclusions drawn in the fast motion limit are still qualitatively correct.

A comparison of R_{1M} and R_{2M} values may thus be useful to evaluate the occurrence of relaxation by contact interactions. Taking again $\tau_c^{\mathrm{dip}} = \tau_c^{\mathrm{con}} = \tau_c$, and keeping in mind the τ_c (or field) dependence of R_{1M} and R_{2M} as given by Eqs. (4.27) and (4.28), the R_{1M}/R_{2M} ratios are expected to be as reported in Table 4.3. When the estimate of τ_c is such that the intermediate situation ($|\omega_S|\tau_c > 1 > |\omega_I|\tau_c$) occurs, dominant dipolar relaxation will still give R_{1M}/R_{2M} ratios close to unity, whereas dominant contact

Table 4.3 Ratios Between R_{1M} and R_{2M} for Dipolar and Contact Contributions in the Various Motional Regimes

	$\omega_S\tau_c < 1$	$\omega_S\tau_c > 1 > \omega_I\tau_c$	$\omega_I\tau_c > 1$
R_{1M}/R_{2M} (dipolar)	1	6/7	$\ll 1$
R_{1M}/R_{2M} (contact)	1	$\ll 1$	$\ll 1$

relaxation will give $R_{2M} \gg R_{1M}$. In the latter case, no information is obtained concerning the mechanism controlling R_{1M}; however, an idea of what happens to R_{1M} can be perceived by the following procedure:

- use R_{2M} and a reasonable estimate for τ_c^{con} to calculate $(A/\hbar)^2$, as if R_{2M} were completely determined by contact relaxation;
- use the values of τ_c^{con} and $(A/\hbar)^2$ in the contact equation for R_{1M} to estimate the upper limit of the contact contribution of the longitudinal relaxation;
- compare this value with the experimental value of R_{1M} to check whether the contact contribution is negligible.

When electron relaxation has a field dependence as described by Eq. (4.12), contact relaxation can determine very large values of R_{2M}, as shown by Eq. (4.28), while the contribution from dipolar relaxation can be negligible if $\tau_c^{dip} = \tau_r$.

4.9 CROSS CORRELATION EFFECTS

When a nucleus can relax through two different types of interactions modulated by the same motion, a cross-correlated relaxation effect can arise resulting from the interference between the two interactions. In fact, since relaxation depends on the square of the energy related to all interactions involving the considered nuclear spin, in the presence of two types of interactions, say A and B, relaxation can be described as the sum of three contributions, two of them depending on the squared energies of each interaction, $A^2 + B^2$, and a third one corresponding to the cross-term related to the product between the energy terms of the two interactions, 2AB.

One of these cross correlation effects, occurring in diamagnetic systems, is that between the dipole–dipole interaction and the chemical shielding anisotropy, CSA [95], which can lead to differential line-broadening of spin-spin coupled doublet components [96,97] and to relaxation-allowed coherence transfer [98] (Section 12.3.2).

In paramagnetic systems, when the electron spin relaxation is fast with respect to the rotational time, the electron spin is decoupled from the nuclear spins and there is no cross correlation with the dipole–dipole interaction between nuclei; however, a dipole–dipole interaction remains between the nuclear spin and the static averaged electron magnetic moment, which is modulated by the molecular reorientation time to cause Curie spin relaxation (Section 4.6). The interference between this interaction and the dipole–dipole interaction between nuclei can actually provide cross correlated relaxation. ^{1}H COSY cross peaks in paramagnetic proteins often arise from cross correlation between interproton dipolar coupling and Curie interaction rather than from true scalar couplings [99]. This paramagnetic cross-correlation is conceptually identical to the dipole dipole-CSA cross-correlation [95].

In the presence of dipole-dipole and Curie interactions, the energy of the system is provided by the sum of the two terms. As discussed before, since relaxation is proportional to the square of the total energy, in addition to the Curie and dipolar relaxation terms, a cross term appears as the double product of the two energy terms. The cross term either decreases or increases the rates calculated from the Curie and dipolar relaxation terms, depending on S being in the 1/2 or –1/2 spin state. This causes a difference in nuclear linewidths [100]. Following Goldman [95], the difference in the linewidth (in Hz) of, for example, the two proton dimension components of the HN doublet in HSQC spectra of ^{15}N enriched paramagnetic proteins, calculated in the assumption of isotropic χ tensor, is given by:

$$\Delta\nu_{1/2} = -\frac{2}{15\pi}\frac{\mu_0}{4\pi}\frac{B_0\gamma_{\mathrm{H}}^2\gamma_{\mathrm{N}}\hbar}{r_{\mathrm{HN}}^3}(\sigma_{\parallel}-\sigma_{\perp})\frac{3\cos^2\theta_{\mathrm{MHN}}-1}{2}\left(4\tau_c+\frac{3\tau_c}{1+\omega_I^2\tau_c^2}\right) \tag{4.35}$$

where

$$\sigma_{\parallel}-\sigma_{\perp} = -\frac{\mu_0}{4\pi}\mu_B^2 g_e^2\frac{S(S+1)}{r^3kT} = -\frac{3\chi_{\mathrm{iso}}}{4\pi r^3} \tag{4.36}$$

as obtained by diagonalization of the traceless dipolar shielding tensor [Eq. (2.40)].

Therefore [99–101],

$$\begin{aligned}\Delta\nu_{1/2} &= \frac{2}{15\pi}\left(\frac{\mu_0}{4\pi}\right)^2\frac{B_0\gamma_{\mathrm{H}}^2\gamma_{\mathrm{N}}\mu_B^2 g_e^2\hbar S(S+1)}{r^3 r_{\mathrm{HN}}^3 kT}\frac{3\cos^2\theta_{\mathrm{MHN}}-1}{2}\left(4\tau_c+\frac{3\tau_c}{1+\omega_I^2\tau_c^2}\right)\\ &= \frac{\mu_0}{4\pi}\frac{B_0\gamma_{\mathrm{H}}^2\gamma_{\mathrm{N}}\hbar\chi}{10\pi^2 r^3 r_{\mathrm{HN}}^3}\frac{3\cos^2\theta_{\mathrm{MHN}}-1}{2}\left(4\tau_c+\frac{3\tau_c}{1+\omega_I^2\tau_c^2}\right).\end{aligned} \tag{4.37}$$

The angle θ_{MHN} is that between the HN axis and the H-metal axis (Fig. 4.16), r is the H-metal distance and τ_c is the correlation time modulating both relaxation processes. Eq. (4.37) can be written in Cartesian coordinates in the form:

$$\Delta\nu_{1/2} = \frac{\mu_0}{4\pi}\frac{B_0\gamma_{\mathrm{H}}^2\gamma_{\mathrm{N}}\hbar\chi}{10\pi^2}\frac{3(\boldsymbol{r}\cdot\boldsymbol{r}_{\mathrm{HN}})^2-r^2r_{\mathrm{HN}}^2}{2r^5r_{\mathrm{HN}}^5}\left(4\tau_c+\frac{3\tau_c}{1+\omega_I^2\tau_c^2}\right). \tag{4.38}$$

The calculation of cross-correlated relaxation caused by the interference of nuclear dipole-dipole and the Curie spin interaction can also be generalized to treat the case of anisotropic magnetic susceptibility [102].

Cross-correlated relaxation between dipole–dipole and Curie interactions can be experimentally determined only if the cross-correlated relaxation between dipole–dipole and CSA interactions is known. The latter can be measured in analogous diamagnetic compounds, so that dipolar-Curie cross correlation can be obtained by subtraction (Section 12.3.2). This implies the assumption that the former is not affected by the presence of the paramagnetic metal ion. This assumption may not be necessarily valid.

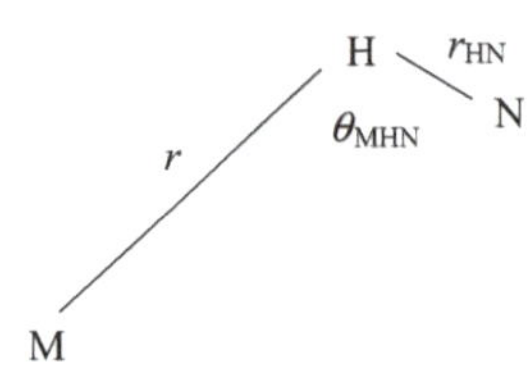

FIGURE 4.16 Cross correlation effects between dipole–dipole and Curie interactions, for the nucleus H of the HN doublet, depend on the angle θ_{MHN} between the H-N direction and the H-metal direction.

Dipole–dipole-Curie cross-correlated relaxation rates were measured for the amide protons of the iron protein cytochrome *c*' and of the cerium-substituted calcium binding protein calbindin D_{9k}, and resulted in good agreement with the values predicted from Eq. (4.37) [102–105].

The imaginary part of the same spectral density function as in Eq. (4.37) causes a dynamic frequency shift [70], which manifests itself as a change in the position of the individual lines of a multiplet, each line being shifted by an equal amount in opposite direction. The magnitude of this additional shift is given by [101]

$$\Delta\nu_{\text{DFS}} = \frac{2}{5\pi}\left(\frac{\mu_0}{4\pi}\right)^2 \frac{B_0\gamma_{\text{H}}^2\gamma_{\text{N}}g_e^2\mu_B^2\hbar S(S+1)}{r_{\text{HN}}^3 r^3 kT}\frac{3\cos^2\theta_{\text{MHN}}-1}{2}\frac{\omega_I\tau_c^2}{1+\omega_I^2\tau_c^2} \tag{4.39}$$

For large correlation times and/or at high magnetic fields, the dynamic frequency shift is independent on the value of the correlation time, on the nucleus observed, and on the magnetic field, as $\frac{B_0\gamma_{\text{H}}^2\gamma_{\text{N}}\omega_I\tau_c^2}{1+\omega_I^2\tau_c^2} \approx \gamma_{\text{H}}\gamma_{\text{N}}$. The absence of a field dependence can be used to separate the dynamic frequency shift from the paramagnetism-induced residual dipolar coupling, as the latter depends on the second power of the magnetic field (Section 3.3). Dynamic frequency shifts are expected to be too small to be used as structural restraints [104,106,107] (Chapter 9). In fact, they are smaller than 2 Hz for an amide proton at 3 Å from a metal ion with $S = 1/2$, as well as for an amide proton at 10 Å from the terbium(III) ion ($J = 6$, $g_J = 3/2$, Chapter 8) which, even at this distance, is experiencing a paramagnetic line broadening of several hundreds of Hz.

Another cross correlation effect is due to the interference between Curie and CSA relaxation [108]. The effect of this cross correlation is that of increasing or decreasing the nuclear relaxation rates depending on the relative orientation of the principal axes of the CSA and the dipolar shielding [Eq. (2.40)] tensors. Its magnitude also depends on the product of the anisotropies of the CSA and of the dipolar shielding tensors, and is usually very small, unless the Curie spin is so large that paramagnetic relaxation is predominantly due to the Curie rather than the dipolar contribution.

4.10 NUCLEAR PARAMETERS AND RELAXATION

For nuclei other than protons, the magnetic moment is smaller (with the exception of ^{3}H). The nuclei with $I > ½$ have a nuclear electric quadrupole moment which is an efficient source of nuclear relaxation by itself. Neglecting this contribution to relaxation, we may say that any dipolar coupling with the unpaired electron is smaller than in the case of the proton because the magnetic moment is smaller. Since the latter is proportional to γ_I and relaxation depends on the square of the coupling energy, R_i (i = 1, 2) depends on γ_I^2. As a consequence, the linewidths decrease, for example, on passing from ^{1}H to ^{15}N nuclei because the dipolar interaction with unpaired electrons is 1/100 that of ^{1}H when the two nuclei are at the same distance from the paramagnetic center. On the one hand, it is a pity that we soon lose such a wealth of information on the nucleus–electron coupling! On the other hand, this property can be exploited in the case of metal ions with slow electron relaxation that cause broad ^{1}H lines. However, ^{2}H lines are much sharper. Of course, when the compounds are large in molecular size, quadrupolar relaxation becomes dominant and ^{2}H spectroscopy may not be convenient.

As far as contact contributions are concerned, the nuclear γ_I parameter is contained in A [Eq. (2.2)] and therefore the same γ_I^2 dependence as in dipolar relaxation is introduced in contact relaxation.

Again, heteronuclei are expected to be less relaxed owing to their smaller γ_I. However, heteronuclei can be directly coordinated to the paramagnetic metal ion. In this case the spin density on the nucleus can be very large and thus A can be very large compared with the proton case. Values of $(A/\hbar)^2$ as large as 10^{17} rad^2 s^{-2} can be obtained for directly coordinated nuclei like ^{17}O, ^{19}F, ^{14}N, and ^{15}N. In such cases, contact interaction may easily be the dominant mechanism for nuclear relaxation, especially for R_{2M}. In the case of imidazole complexes, even the noncoordinated nitrogen is quite broad. In this case, contact and/or ligand-centered effects provide efficient nuclear relaxation even for a nucleus at 4–4.5 Å from the metal [109].

4.11 THE EFFECT OF TEMPERATURE ON THE ELECTRON-NUCLEUS SPIN INTERACTION

No explicit temperature dependence is included in the equations for R_{1M} and R_{2M}, except for cases where Curie spin relaxation is the dominant term (Section 4.6). In the latter case, Curie paramagnetism has a T^{-1} dependence and therefore relaxation depends on T^{-2}. The effect of temperature on linewidths determined by Curie relaxation is dramatic also because of the τ_r dependence on temperature, as shown in Eq. (4.8). All the correlation times modulating the electron–nucleus coupling, either contact or dipolar, are generally temperature dependent, although in different ways, and their variation will therefore be reflected in the values of R_{1M} and R_{2M}.

In the limits of validity for the Solomon and Bloembergen equations, the correlation times for contact and dipolar relaxation are given by Eqs. (4.6) and (4.7) respectively. The exchange time τ_M is rarely short enough to dominate in Eqs. (4.6) and (4.7); however, when this is the case, a strong temperature dependence of the nuclear relaxation times is expected, because the variation of the exchange rate with temperature is generally exponential [110]. When the rotational correlation time is the dominant term, the effect of temperature would be anticipated to be less, since, as already seen, τ_r may be approximated by the Stokes-Einstein Eq. (4.8). The effect is enhanced, however, by the change in viscosity η of the solvent; in the case of water, η decreases as much as 2.7 times from 0°C to 40°C.[8] Qualitatively, both τ_r and τ_M decrease with increasing temperature; this causes a decrease in the nuclear relaxation rates in the nondispersive regions and an increase in the second half of the ω_I dispersive region (Fig. 4.17).

The temperature dependence of τ_e, including its sign, may be a matter of extensive speculation, mainly because the origin of the electron–spin relaxation mechanism in each particular case may be different, and in general is not known in detail. When equations of the type such as Eqs. (4.12) and (4.13) hold, and by assuming that the correlation time for electron relaxation τ_v has an exponential temperature dependence, τ_e would be expected to increase with increasing temperature in the nondispersive region ($|\omega_S|\tau_v \ll 1$) and to decrease in the dispersive region (high magnetic field). In fact, from Eqs. (4.12) and (4.13), τ_e^{-1} is proportional to $\tau_v/(1+\omega_S^2\tau_v^2)$, i.e., proportional to τ_v as long as $|\omega_S|\tau_v \ll 1$, and to τ_v^{-1} when $|\omega_S|\tau_v \gg 1$. However, this picture may be further complicated by a possible temperature dependence of the Δ_t^2 term in Eqs. (4.12) and (4.13), making it difficult to predict the overall behavior of τ_e. In addition, different electronic relaxation mechanisms may be operative at different

[8]The transport properties are also well modeled by exponential laws, but the energy is generally smaller than for chemical exchange.

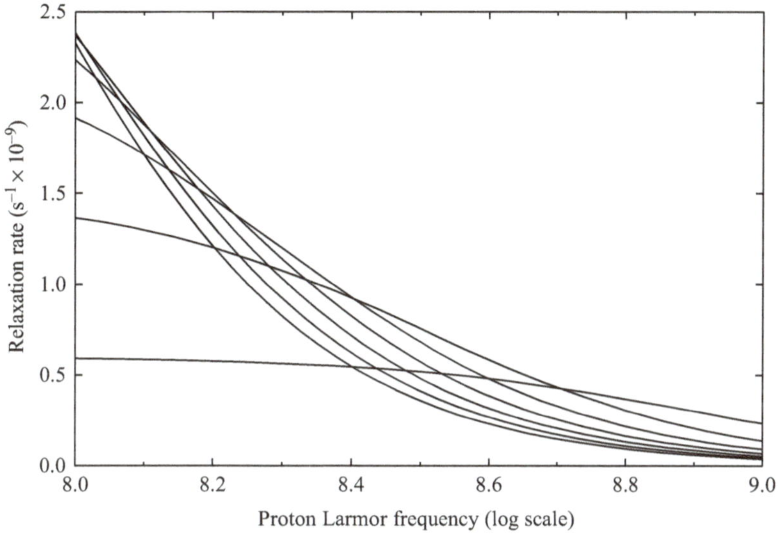

FIGURE 4.17 Plot of $f_1(\omega, \tau_c)$ of Eq. (4.17) as a function of the proton Larmor frequency in the high frequency region. The curves are calculated for τ_c values decreasing from 2×10^{-9} s (upper curve on the left) in steps of 3×10^{-10} s.

temperatures, although the possibility that the switching from one mechanism to another occurs in a few tens of degrees around room temperature is rather unlikely.

In the coupled metal systems that will be discussed in Chapter 11, the effect of temperature is generally more complicated. Besides the effects described above, the overall temperature dependence will obviously also depend on the sign and magnitude of the exchange coupling constant *J*. The reader is referred to Chapter 11 for more details.

The effects of chemical exchange on the NMR parameters, and hence the influence of temperature through modulation of the exchange process, will be considered in Section 6.3.

4.12 STABLE FREE RADICALS

Free radicals have electronic relaxation times of about 10^{-7}–10^{-6} s at room temperature, which are the longest among paramagnetic compounds [54,55]. Electron relaxation is governed by several mechanisms (spin-rotation, nucleus-electron dipole-dipole interaction, spin diffusion, etc.), the contribution of which in different conditions can be studied by analyzing the increase in the electron relaxation times observed with increasing the reorientation time of the radical [111,112]. Electron relaxation times are field dependent, as they increase with increasing field, although an opposite trend may be observed above 1 T [113]. Whereas at low radical concentrations ($<10^{-3}$ M), the electronic relaxation time is concentration independent, at high concentrations T_{1e} decreases for the increasing effect of intermolecular dipole-dipole relaxation process [114]. The correlation time for the NMR experiment when the radical is free in solution is the rotational correlation time.

Small molecules are already at the limit of detectability as far as proton NMR is concerned, whereas they are easily detected through ^{2}H NMR. If other factors concur favorably (interactions between radicals at high concentration, tendency to aggregate, availability of low-lying excited states), the proton NMR signals can also be detected for radicals. Nevertheless, it is, in general, a hard task to perform high resolution studies on radicals.

When the radical interacts rigidly with a macromolecule its rotational correlation time increases. For very large macromolecules, the correlation time for the NMR experiment eventually becomes the electronic relaxation time. If the bound radical still has motional freedom, this motion may be faster than the rotational time of the whole molecule and thus determine the correlation time. Of course, if the system is in chemical exchange, the exchange time may also be shorter than the electronic relaxation time and so represents the limiting NMR correlation time. An immobilized radical produces a very large broadening effect on the NMR line of nearby nuclei, which is probably the largest effect that a paramagnetic species can cause.

For organic radicals like trityl and TEMPONE, the temperature dependence of electron relaxation indicates that the latter is dominated by the Raman process from about 20 to 100 K, where it is several orders of magnitude longer than at room temperature [115]. The Raman process increases with increasing g anisotropy and decreasing rigidity of the molecule.

Radicals are increasingly used as relaxation agents (Chapter 9) and polarizing agents in dynamic nuclear polarization (DNP) experiments (Sections 4.15 and 5.3) both in solution and in the solid state.

4.13 REDFIELD LIMIT AND BEYOND

The relaxation equation derived so far for electrons and nuclei share a common assumption usually called the perturbation regime or Redfield limit [116]. The assumption allows the use of the linear response theory; it states that the energy of the coupling between the spin and the lattice, $E_{\text{int}}/\hbar$ (in frequency units), whose modulation is responsible for the spin relaxation, must be smaller than the inverse of the correlation time, τ_c, for the modulation of the coupling itself, $E_{\text{int}}/\hbar \ll \tau_c^{-1}$. The relaxation equations that hold in the Redfield limit have the general form (in the low field or extreme narrowing limit)

$$T_1^{-1} = \left(\frac{E_{\text{int}}}{\hbar}\right)^2 \tau_c \tag{4.40}$$

[Eq. (4.15)] in addition to a proportionality coefficient which is of the order of a few units. As $E_{\text{int}}/\hbar \ll \tau_c^{-1}$ must hold, we can substitute τ_c^{-1} for $E_{\text{int}}/\hbar$ in Eq. (4.40) and obtain

$$T_1^{-1} << \tau_c^{-1} \tag{4.41}$$

i.e., the T_1 arising from the interaction can never be as short as the correlation time with which the interaction is modulated. This is another way of expressing the Redfield limit. Furthermore, from Eq. (4.40) we obtain

$$T_1^{-1} << \frac{E_{\text{int}}}{\hbar} \tag{4.42}$$

which is another way of defining the Redfield limit, i.e., the energy of the spin–lattice coupling (in frequency units) is always larger than the relaxation rate enhancement induced on the spin by the coupling.

The fact that equations are not available when this limit does not hold (strong coupling regime) should not automatically mean that a real case cannot exist in which the interaction energy is larger than τ_c^{-1}. It only means that the consequences of this fact according to the standard relaxation equations (i.e., the correlation time is longer than T_1) do not make physical sense, because it is not conceivable that a modulation can induce a change in a physical property (the orientation of a spin) at a rate higher than the modulation rate itself. Therefore, what happens must be that, if we imagine that in a system initially within the Redfield limit we can increase the spin–lattice interaction energy continuously, we should have an increase in T_1^{-1} that initially is proportional to the square of $E_{int}/\hbar$, and then becomes slower and slower as T_1^{-1} approaches $E_{int}/\hbar$ (and simultaneously τ_c^{-1}). The limiting value should probably be of the same order as $E_{int}/\hbar$ and τ_c^{-1}.

In practice nuclear spin–lattice relaxation is always within the Redfield limit, i.e., the interaction energy with the lattice is always much smaller than τ_c^{-1}. This is true even with paramagnetic systems, where the nuclear spin–lattice interaction energy is often much larger than usual. On the other hand, it is not obvious that electrons are always in the Redfield limit [91]. When electrons are outside the Redfield limit, although nuclear relaxation is in the Redfield limit, it is not easy to define a correlation time because the electron relaxation time may not be known and might even be undefined. Outside the Redfield limit, in fact, electron relaxation may not be described by an exponential and therefore a relaxation time may not be defined. This problem was theoretically solved by treating the electron spin and the lattice to which it is so strongly coupled as a whole, using the Liouville superoperator formalism [88,89,117,118]. The nuclear spin is coupled to this *generalized lattice* containing both the electron and its lattice. No electron relaxation time is defined, or any nucleus–lattice correlation time. The only motion of the generalized lattice considered is rotation and it is described by a rotational diffusion equation. This approach has been applied to nickel(II) systems, where the electron relaxation, outside the Redfield limit, is mainly caused by transient ZFS of the $S = 1$ manifold (Section 4.3.1).

4.14 THE NUCLEAR OVERHAUSER EFFECT

The comprehension of the nuclear Overhauser effect [119,120] (NOE) is of fundamental importance in dealing with nuclear relaxation [121,122]. Especially in paramagnetic compounds, it still represents a most powerful technique to detect dipolar connectivities; finally it allows the comprehension of two-dimensional spectroscopies based on dipolar coupling. The NOE is the fractional variation of the intensity of a signal when another signal is selectively saturated.

The first application to paramagnetic compounds appeared in 1983 on signals with T_1 of the order of 50 ms: they were aliphatic residues in the active cavity of met-myoglobin cyanide which contains low spin iron(III) [123]. Soon it became apparent that larger the molecular size, the larger the reorientational correlation time and the larger the NOEs. It was possible then to measure NOEs between signals with T_1 shorter than 10 ms with linewidths of several hundred Hertz (at half-height) as long as the distances were small, as for geminal protons. This increased the number of successful applications of the NOE technique in the field of metalloproteins. In the meantime, improvements were made in the difference spectra techniques, which permit the recording of the FID when a given signal is saturated and the subtraction of the FID when it is not; this is then repeated a number of times and the differences summed up. The baseline remains flat and small NOEs can be detected even under a large envelope of signals. When no NOE is observed, from the signal to noise ratio of the difference spectrum we can

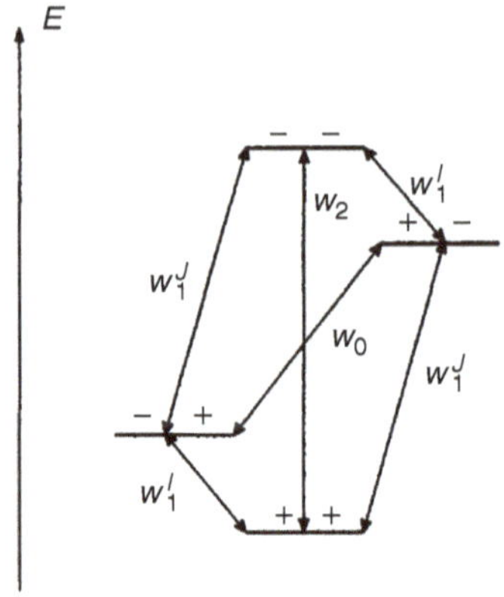

FIGURE 4.18 Spin transitions and transition probabilities in a two unlike spin system experiencing dipolar coupling. The + and – refer to the signs of M_I and M_J for the two nuclear spins. The two nuclei are assumed to have positive magnetogyric ratios.

estimate an upper limit for NOE intensities, and therefore a lower limit for the efficiency of the dipolar coupling between the proton of the saturated signal and any other proton nearby.

The NOE is the result of the transfer of spin population (called polarization transfer) between two dipole coupled nuclei I and J, i.e., a change of the $\langle I_z \rangle$ and/or $\langle J_z \rangle$ values due to the coupling. The NOE depends on the square of the magnitude of each nuclear magnetic moment (see later); therefore its application is essentially limited to proton NMR. In protein NMR the ^{15}N–^{1}H NOE in peptide NH groups is measured to learn about the N–H reorientational correlation time.

In order to understand NOEs, it is convenient to refer to an energy diagram of the type shown in Fig. 4.18, which shows the energies of two dipole coupled nuclei with $I = ½$, $J = ½$. In case the magnetogyric ratios γ_I and γ_J are positive, as in the example, the ground state is that with $M_I = M_J = +½$, and the most excited level has $M_I = M_J = -½$. At intermediate energies there are the two levels, $M_I = -½$, $M_J = +½$, and $M_I = +½$, $M_J = -½$. The dipolar coupling energy averages zero by rotation [Eq. (2.27)]. Therefore, the energy levels of Fig. 4.18 do not depend on the extent of the spin–spin dipolar coupling. This figure is similar to Fig. 4.7A, where the dipolar coupling between an electron spin and a nuclear spin was depicted. The NMR spectrum consists of two lines at energies $\hbar\gamma_I B_0$ and $\hbar\gamma_J B_0$. When the equilibrium population of two levels is altered by providing the proper frequency to the system, for example, the frequency corresponding to the w_1^I transition, the population of all levels will be perturbed. Since the intensity of a signal in NMR depends on the difference in population between the involved levels, selective irradiation of one signal causes a change in the intensity in the other signal. As already defined, the fractional variation of intensity is just the NOE.

4.14.1 STEADY STATE NOE

The steady state NOE is the fractional variation η_I in the integrated NMR signal intensity of a nuclear spin I when another spin J is saturated for enough time to allow the system to reach a new steady state equilibrium. In order to fully understand the phenomenon it is convenient to refer first to a saturation time long in comparison with the T_1 of the signal on which NOE is going to be measured; for short irradiation times we will then deal with NOE as a function of time $\eta_I(t)$. We now express the intensity of a signal from spin I as the expectation value of I_z, $\langle I_z \rangle$, as already defined (Section 1.6). The larger the difference in population between the M_I levels, the larger $\langle I_z \rangle$. After saturating signal J with a proper soft pulse, we have made the population of the + + level equal to that of the + – level, as well as that of the – + level equal to that of the – – level (Fig. 4.18). If the saturation lasts for a long enough time, the difference in population between + + and – + levels, and between the + – and – – levels, reaches a steady state condition and $\langle I_z(t) \rangle$ reaches a new value that is different from the original equilibrium value $\langle I_z(\infty) \rangle$. The NOE $\eta_{I(J)}$ (which indicates that NOE is observed on signal I when J is saturated), will be given by

$$\eta_{I(J)} = \frac{\langle I_z(t) \rangle - \langle I_z(\infty) \rangle}{\langle I_z(\infty) \rangle}. \qquad (4.43)$$

It can be shown (Appendix VII) that the following relationship holds:

$$\langle I_z(t) \rangle = \langle I_z(\infty) \rangle + [\sigma_{I(J)}/\rho_{I(J)}]\langle J_z(\infty) \rangle. \qquad (4.44)$$

In order to define $\sigma_{I(J)}$ and $\rho_{I(J)}$, it is convenient to refer to Fig. 4.18 and to define w_1^I and w_1^J as the transition probabilities between two states involving a single quantum transition either of spin I or J; w_0 is the zero quantum transition and corresponds to the $- + \rightarrow + -$ transition and vice versa; w_2 corresponds to the $- - \rightarrow + +$ transition and vice versa, which is a double quantum transition. $\rho_{I(J)}$ is defined as

$$\rho_{I(J)} = w_0 + 2w_1^I + w_2. \qquad (4.45)$$

It represents the total probability for nucleus I to change its spin component along z in a coupled two spin system.

$\sigma_{I(J)}$ ($= \sigma_{J(I)}$ when I and J have the same spin multiplicity) is called cross-relaxation rate, and is given by

$$\sigma_{I(J)} = \sigma_{J(I)} = w_2 - w_0. \qquad (4.46)$$

The transitions corresponding to w_2 and w_0 involve simultaneous changes in spin state of both nuclei. The difference between w_2 and w_0 tells us how the variation in population of J affects the equilibrium of I. In other words, one can think in terms of transfer of spin population from *J to I* when J is saturated. The larger the cross relaxation, the larger the dipolar coupling. The NOE on I is proportional to the cross relaxation from J and inversely proportional to the capability of I to return to equilibrium once its equilibrium is perturbed through cross relaxation.

The transition probabilities depend on the mean squared interaction energy relative to the mechanism which causes the transition, times the value of the spectral density at the required frequencies [Eq. (4.15)]. The square of the dipolar interaction energy is, as usual [Eq. (1.4) and Appendix V], proportional to $\langle \mu_1 \cdot \mu_2/r^3 \rangle^2$, where μ_1 and μ_2 are the magnetic moments of the two spins. The actual equations are

$$\rho_{I(J)} = \left(\frac{\mu_0}{4\mu}\right)^2 \frac{2\hbar^2\gamma_I^2\gamma_J^2 J(J+1)}{15r_{IJ}^6}\left[\frac{\tau_c}{1+(\omega_I-\omega_J)^2\tau_c^2} + \frac{3\tau_c}{1+\omega_I^2\tau_c^2} + \frac{6\tau_c}{(1+(\omega_I+\omega_J)^2\tau_c^2}\right] \qquad (4.47)$$

and

$$\sigma_{I(J)} = \left(\frac{\mu_0}{4\mu}\right)^2 \frac{2\hbar^2\gamma_I^2\gamma_J^2 J(J+1)}{15r_{IJ}^6}\left[\frac{6\tau_c}{1+(\omega_I+\omega_J)^2\tau_c^2} - \frac{\tau_c}{(1+(\omega_I-\omega_J)^2\tau_c^2}\right]. \tag{4.48}$$

The correlation time is the reorientational correlation time. This is generally the rotational correlation time unless internal motions of a group in a molecule are faster. In the latter case, pairs of nuclei in the same molecule can have different reorientational correlation times.

Note the analogy between Eq. (4.47) and Eq. (4.17): indeed, $\rho_{I(J)}$ is a longitudinal relaxation time of spin I due to the interaction with spin J upon reciprocal reorientation. The point is that it cannot be measured, except approximately, through a T_1 measurement because T_1 is defined in the absence of links with other nuclear spins, i.e., in the absence of cross relaxation.

By combining Eqs. (4.43) and (4.44), we obtain

$$\eta_{I(J)} = [\sigma_{I(J)}/\rho_{I(J)}]\langle J_z(\infty)\rangle / \langle I_z(\infty)\rangle. \tag{4.49}$$

Since $\langle I_z(\infty)\rangle$ is proportional to $I(I+1)\gamma_I$, and $\langle J_z(\infty)\rangle$ is proportional to $J(J+1)\gamma_J$ (Eqs. (1.31)), in the general case of heteronuclear spins we have

$$\eta_{I(J)} = \frac{\sigma_{I(J)}}{\rho_{I(J)}}\frac{J(J+1)\gamma_J}{I(I+1)\gamma_I}. \tag{4.50}$$

The relative values of I, J, γ_I, and γ_J are important for determining the size and the sign of heteronuclear NOE; obviously, in homonuclear NOE they cancel each other and Eq. (4.49) takes the simple form

$$\eta_{I(J)} = \frac{\sigma_{I(J)}}{\rho_{I(J)}}. \tag{4.51}$$

Cross relaxation can occur in principle also between nuclei coupled by a time-dependent scalar interaction. In this case only w_0 can contribute to it (Section 4.5). The correlation time for the reorientation is either chemical exchange or the relaxation rate of nucleus J if nucleus I is observed. The correlation time is generally too long to make the experiment successful.

4.14.2 STEADY STATE NOE IN REAL LIFE

The first consideration by looking at Eqs. (4.47), (4.48), and (4.49) is that NOE does not contain any structural information within the present scheme. Indeed, both $\rho_{I(J)}$ and $\sigma_{I(J)}$ contain the same r_{IJ} parameter which cancels out in $\eta_{I(J)}$. In practice, the two-spins scheme is never a good description of any real system. The nucleus I, on which NOE is going to be observed, is always coupled to other nuclei. When dealing with hyperfine coupled nuclei, often the major source of relaxation is not the coupling with J but the coupling with the unpaired electrons. The expression for $\eta_{I(J)}$ then is (Appendix VII)

$$\eta_{I(J)} = \frac{\sigma_{I(J)}}{\rho_{I(J)} + \rho_{I(\text{other})} + R_{1M}^I} = \frac{\sigma_{I(J)}}{\rho_I} \tag{4.52}$$

where $\rho_{I(J)}$, as defined before, is the contribution to longitudinal relaxation of nucleus I due to the coupling with nucleus J, $\rho_{I(\text{other})}$ includes all the couplings with other nuclei responsible for relaxation of nucleus I with the exception of the coupling with nucleus J, and R^I_{1M} is the paramagnetic contribution to the longitudinal relaxation rate of nucleus I arising from the various mechanisms described in Sections 4.4–4.6. The sum of all these relaxation rates is the total rate ρ_I.

When the denominator in Eq. (4.52) is dominated by the paramagnetic contribution to nuclear relaxation ($\rho_I \cong R^I_{1M}$), the following simplified equation holds

$$\eta_{I(J)} = \sigma_{I(J)}/R^I_{1M} = \sigma_{I(J)}T^I_{1M}. \tag{4.53}$$

In Eq. (4.53) the r^{-6} parameter is not canceled out, because $\sigma_{I(J)}$ depends on r_{I1}^{-6} but R^I_{1M} does not. When the nucleus–electron coupling is dipolar in origin, R^I_{1M} depends on r^{-6} where r is the distance between the resonating nucleus and the unpaired electron [Eq. (4.20)]. This partially holds when R_{1M} is not dominating.

Even if not important in principle, it is important in practice to underline that ρ_I is not a time constant for an exponential process; the process would be exponential only in the absence of cross relaxation. In paramagnetic compounds the total relaxation may be dominated by R^I_{1M} and the total time dependence approaches an exponential behavior [124]. Therefore, ρ_I is the measured T_1^{-1} of the nucleus (either from a selective or a nonselective experiment). When R^I_{1M} is not dominating, after a selective excitation of I, the initial return to equilibrium can be assumed to be exponential and its rate constant, $T^{-1}_{1\text{sel}}$ can be taken as ρ_I.

Of course, the larger the denominator in Eq. (4.52), the smaller the NOE. That is why NOE in paramagnetic complexes has developed so late: the coupling with unpaired electrons makes the relaxation very effective and the NOE very small. In geminal protons belonging to macromolecules ($\tau_r \approx 10^{-8}$ s) with ρ_I of, say, 100 s^{-1} due to paramagnetic contributions, the NOE is about 20% at 500 MHz. If the distance between protons increases to 3 Å the NOE drops to 0.8%. In small complexes τ_r is typically 10–100 times shorter than in the above example and the NOEs have intensities smaller by a similar amount. The problem would be hopeless if we could not gain one to two orders of magnitude in sample concentration. It is therefore important to have a strict control of the temperature (remember that the hyperfine shift is temperature dependent) during the hours of summation of the difference FIDs (Section 12.3), and to increase the signal to noise ratio until the percent differences can be measured with a precision of two figures by selecting a high enough solute concentration and number of scans.

When measuring steady state NOE, the effect measured on saturating I or on saturating J may be different because ρ_I and ρ_J are in general different. We say that steady state NOE is not symmetric. It follows that larger NOEs, and then more favorable cases, occur when the signal with larger ρ is saturated.

4.14.3 SELECTIVE AND NONSELECTIVE T_1

It is intuitive that the NOE will be larger the larger the cross relaxation, and that the NOE will be smaller the larger ρ_I. It may be appropriate here to discuss in a more quantitative way the response of a dipole-coupled two spin system to selective and nonselective experiments, and the relationship between the results of these experiments and the quantity of interest, ρ_I (Section 12.1.5). When measuring T_1 of a signal one can excite (for instance, invert) all the signals simultaneously (nonselective experiment). If the nuclei are dipole coupled, then they will return to equilibrium also by exchanging polarization through cross relaxation. Alternatively, only one signal is inverted (selective experiment) and its return to equilibrium measured. Cross relaxation will be different from the previous case. In both cases, the return to equilibrium is not exponential in the presence of cross relaxation. If this is the case, the measurement of T_1 is approximate, and even the definition of T_1 may be invalid. This is because the other nuclei are part of the lattice but this lattice does not possess the infinite heat capacity required to have an exponential behavior of the magnetization. In fact, cross relaxation is capable of affecting the spin state populations of the other nuclei.

We start from the expression for $\mathrm{d}\langle I_z\rangle/\mathrm{d}t$ derived in Appendix VII [Eq. (VII.13)] and write the corresponding expression for $\mathrm{d}\langle J_z\rangle/\mathrm{d}t$:

$$\begin{aligned}\frac{\mathrm{d}\langle I_z(t)\rangle}{\mathrm{d}t} &= -\left[\langle I_z(t)\rangle - \langle I_z(\infty)\rangle\right]\rho_I - \left[\langle J_z(t)\rangle - \langle J_z(\infty)\rangle\right]\sigma_{I(J)} \\ \frac{\mathrm{d}\langle J_z(t)\rangle}{\mathrm{d}t} &= -\left[\langle J_z(t)\rangle - \langle J_z(\infty)\rangle\right]\rho_J - \left[\langle I_z(t)\rangle - \langle I_z(\infty)\rangle\right]\sigma_{J(I)}\end{aligned} \tag{4.54}$$

where we have substituted $\rho_{I(J)}$ and $\rho_{J(J)}$ with the corresponding total rates ρ_I and ρ_J [Eq. (4.52)]. The two equations (analogous to those for the case of chemical exchange[9], Section 6.3.4) constitute a system of coupled differential equations, whose general solutions are

$$\begin{aligned}\langle I_z(t)\rangle &= \langle I_z(\infty)\rangle + C_1\exp(-\lambda_1 t) + C_2\exp(-\lambda_2 t) \\ \langle J_z(t)\rangle &= \langle J_z(\infty)\rangle + \frac{\lambda_1 - \rho_I}{\sigma_{J(I)}}C_1\exp(-\lambda_1 t) + \frac{\lambda_2 - \rho_I}{\sigma_{J(I)}}C_2\exp(-\lambda_2 t)\end{aligned} \tag{4.55}$$

where $\lambda_{1,2} = \rho \pm D$, and

$$\rho = \frac{1}{2}(\rho_I + \rho_J) \quad D = \frac{1}{2}\left[(\rho_I - \rho_J)^2 + 4\sigma_{I(J)}\sigma_{J(I)}\right]^{1/2}. \tag{4.56}$$

The explicit expressions for C_1 and C_2 can be derived for many particular cases [121], including the selective and nonselective inversion recovery experiments of interest here. For the selective case

$$C_1 = \frac{2(\lambda_2 - \rho_I)\langle I_z(\infty)\rangle}{\lambda_1 - \lambda_2} \quad C_2 = -\frac{2(\lambda_1 - \rho_I)\langle I_z(\infty)\rangle}{\lambda_1 - \lambda_2} \tag{4.57}$$

and for the nonselective experiment

$$\begin{aligned}C_1 &= 2\langle I_z(\infty)\rangle\frac{\lambda_2 - \rho_I}{\lambda_1 - \lambda_2} - 2\langle J_z(\infty)\rangle\frac{\sigma_{J(I)}}{\lambda_1 - \lambda_2} \\ C_2 &= -2\langle I_z(\infty)\rangle\frac{\lambda_1 - \rho_I}{\lambda_1 - \lambda_2} + 2\langle J_z(\infty)\rangle\frac{\sigma_{J(I)}}{\lambda_1 - \lambda_2}.\end{aligned} \tag{4.58}$$

Calculated magnetization recovery profiles are reported in Fig. 12.10. It is clear that in no case is the recovery exponential.

[9]The magnetization, M_z^I, along z is proportional to $\langle I_z\rangle$, as reported in Section 1.6. Therefore Eqs. (4.50), (4.55), (4.60), and (4.61) can be rewritten by substituting $\langle I_z\rangle$ with M_z^I and $\langle J_z\rangle$ with M_z^J in case I and J are nuclei of the same nuclear species.

4.14.4 STEADY STATE NOE IN PARAMAGNETIC COMPOUNDS

By combining Eqs. (4.47) and (4.48) it appears that there is an NOE dependence on the magnetic field and on the reorientational correlation time. In Fig. 4.19A, the NOE dependence on the external magnetic field is shown for τ_r of 10 ns and for two protons at 1.8 Å, by taking $\rho_{I(\text{other})} = 0$ and $R^I_{1M} = 0$, 50 and 200 s^{-1}. The dependence on τ_r under the same conditions is shown at 800 MHz (Fig. 4.19B). From Fig. 4.19A, it appears that at low $\omega\tau$ values (fast motion limit) and $R^I_{1M} = 0$ the maximal homonuclear NOE is 0.5, whereas at large values of $\omega\tau$ (slow motion limit) the maximal NOE is −1. Of course, these are limiting values in the absence of R^I_{1M} (and $\rho_{I(\text{other})}$): the hyperfine coupled relaxation decreases the above absolute values, but leaves the dependence unaltered. The NOE is larger at larger magnetic fields and/or when rotation is slowed down (Fig. 4.19B). It is important to note that, for very long τ_r values, NOE values close to −1 could in principle be achieved even in the presence of paramagnetic relaxation;

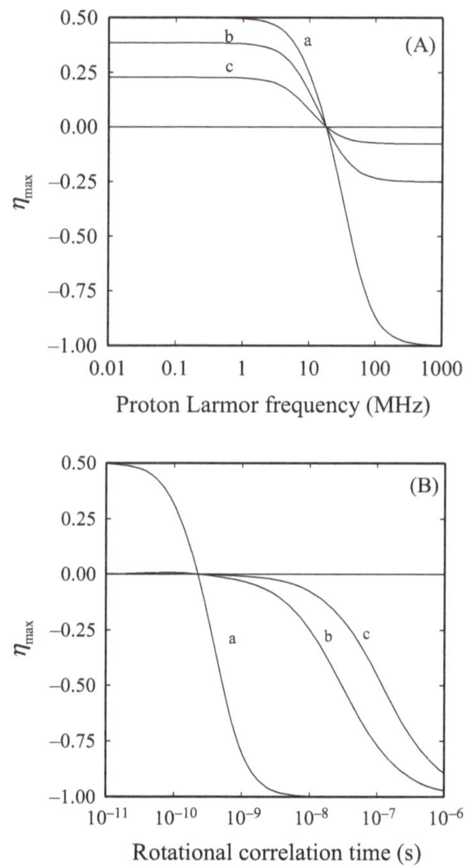

FIGURE 4.19 NOE values as a function of (A) magnetic field for $\tau_r = 10$ ns and (B) τ_r at a field of 800 MHz, for two hydrogen nuclei at 1.8 Å distance ($\rho_{I(J)}$) = 23.6 s^{-1} in the fast motion and 2.36 s^{-1} in the slow motion regime). Curves (a), (b) and (c) refer to a field-independent $R^I_{1M} = 0$, 50 and 200 s^{-1}, respectively.

this is because $\sigma_{I(J)}$ increases with τ_r whereas the denominator of Eq. (4.52) increases less because R_{1M}^I does not depend on τ_r but rather on τ_e. Sometimes, with small proteins, it is even convenient to add ethylene glycol to their water solutions in order to increase viscosity and then τ_r [125,126].

4.15 OVERHAUSER DNP

Overhauser DNP is the DNP mechanism active in solution to increase the difference in population between the nuclear spin states, and thus to increase the NMR signal intensity (Section 5.3 for the other DNP mechanisms operating in the solid state). To be operative, the sample must contain paramagnetic molecules (called polarizing agents), usually radicals, which have a larger difference in population (polarization) between the spin states with respect to nuclei, due to their larger γ (Chapter 1). After microwave irradiation at the electron resonance transition frequency, the electron spin polarization is transferred to adjacent nuclei by the hyperfine interaction coupling electron and nuclear spins. In the presence of exchange conditions (Chapter 6), the polarization is then transferred to all solvent nuclei.

In the presence of hyperfine coupling between nucleus I and unpaired electron(s) S, the rate of variation with time of the nuclear magnetization along the magnetic field direction, M_z^I, after a perturbation from its Boltzmann equilibrium value, $M_z^I(\infty)$, can be written analogously to the case of nuclear dipole-nuclear dipole interaction, discussed in Section 4.14 (Appendix VII) [69]. It depends on the transition probabilities among the different proton-electron spin levels (Fig. 4.7) and on the relaxation processes not involving the electron spin, $R_{1\text{dia}}^I$:

$$\frac{\mathrm{d}M_z^I(t)}{\mathrm{d}t} = -\left(w_0 + 2w_1^I + w_2 + R_{1\text{dia}}^I\right)\left(M_z^I(t) - M_z^I(\infty)\right) - (w_2 - w_0)\left(M_z^S(t) - M_z^S(\infty)\right) \tag{4.59}$$

where M_z^S is the electron magnetization along the z direction. At steady state conditions $\left(\frac{\mathrm{d}M_z^I(t)}{\mathrm{d}t} = 0\right)$ under continuous microwave irradiation of the allowed electron spin transition, the magnetization $M_{z(\text{ss})}^I$ becomes

$$M_{z(\text{ss})}^I = M_z^I(\infty) - \frac{w_2 - w_0}{w_0 + 2w_1^I + w_2 + R_{1\text{dia}}^I}\left(M_{z(\text{ss})}^S - M_z^S(\infty)\right) \tag{4.60}$$

which can be rewritten as

$$\varepsilon = \frac{M_{z(\text{ss})}^I - M_z^I(\infty)}{M_z^I(\infty)} = \frac{w_2 - w_0}{w_0 + 2w_1^I + w_2}\,\frac{w_0 + 2w_1^I + w_2}{w_0 + 2w_1^I + w_2 + R_{1\text{dia}}^I}\,\frac{M_z^S(\infty) - M_{z(\text{ss})}^S}{M_z^S(\infty)}\,\frac{M_z^S(\infty)}{M_z^I(\infty)} = -\xi fs\left|\frac{\gamma_S}{\gamma_I}\right| \tag{4.61}$$

that corresponds to the Overhauser enhancement ε [127]. In this expression, the three fractions represent the coupling factor ξ, the leakage factor f, and the saturation factor s, respectively.

The enhancement ε of the nuclear spin magnetization of solvent nuclei is thus given by the ratio of the magnetogyric ratios of electron and nuclear spin multiplied by the coupling factor ξ, the leakage factor f and the saturation factor s. The nuclear spin magnetization of solvent nuclei is in fact proportional to that of the nuclei bound to the paramagnetic molecule through the proportionality constant f_M (fast exchange conditions, Chapter 6).

The coupling factor ξ describes the efficiency of the cross-relaxation processes and is given by:

$$\xi = \frac{\sigma_{I(S)}}{\rho_{I(S)}} = \frac{w_2 - w_0}{w_0 + 2w_1^I + w_2} = \frac{w_2 - w_0}{R_{1M}^I} \tag{4.62}$$

where R_{1M}^I is the paramagnetic contribution to the longitudinal relaxation rate of nucleus I, arising from the hyperfine interaction with unpaired electron(s).

The leakage factor f accounts for the paramagnetic contribution to the spin relaxation of solvent nuclei, $R_{1\mathrm{p}}^I$ (Section 6.3.2), with respect to the total relaxation rate, $R_{1\mathrm{p}}^I + R_{1\mathrm{dia}}^I$, comprising the diamagnetic contributions due to the couplings to all other nuclei:

$$f = \frac{w_0 + 2w_1^I + w_2}{w_0 + 2w_1^I + w_2 + R_{1\mathrm{dia}}^I} = \frac{R_{1\mathrm{p}}^I}{R_{1\mathrm{p}}^I + R_{1\mathrm{dia}}^I} = 1 - \frac{T_{1I}}{T_{1I}^0} \tag{4.63}$$

where T_{1I} and T_{1I}^0 are the longitudinal nuclear relaxation times in the presence and absence of paramagnetic species, respectively. The leakage factor approaches 1 when the diamagnetic contribution to the relaxation rate is negligible with respect to the paramagnetic contribution.[10]

The saturation factor s describes the efficiency of the microwave pumping. For a single homogeneous EPR line [70]:

$$s = \frac{M_z^S(\infty) - M_{z(\mathrm{ss})}^S}{M_z^S(\infty)} = \frac{\gamma_e^2 B_1^2 T_{1e} T_{2e}}{1 + \gamma_e^2 B_1^2 T_{1e} T_{2e}}, \tag{4.64}$$

where B_1 is the magnetic field strength of the microwave in the rotating frame. If the electron transitions are completely saturated, $M_{z(\mathrm{ss})}^S = 0$ and the saturation factor is 1.

From the transition probabilities calculated for the electron-nucleus dipole–dipole interaction (Appendix VII) [69,128]

$$w_0 = kJ(\omega_I - \omega_S, \tau_c) \tag{4.65}$$

$$w_1^I = \frac{3}{2} kJ(\omega_I, \tau_c) \tag{4.66}$$

$$w_2 = 6kJ(\omega_I + \omega_S, \tau_c) \tag{4.67}$$

where $J(\omega, \tau)$ is the Lorentzian spectral density function, τ_c is the correlation time for the dipolar interaction [Eq. (4.6)], and k is proportional to the squared dipole–dipole energy,

$$k = \frac{2}{15}\left(\frac{\mu_0}{4\pi}\right)^2 \frac{\gamma_I^2 g_e^2 \mu_B^2 S(S+1)}{r^6} \tag{4.68}$$

the coupling factor results

$$\xi = \frac{5kJ(\omega_S, \tau_c)}{R_{1M}^I} \tag{4.69}$$

[10]Note that with this definition of f, a large leakage of magnetization results in a low f value, whereas a very small leakage provides f values close to 1.

which, using the Eq. (4.17), here rewritten as Eq. (4.70)

$$R_{1M}^{I} = k\left(7J(\omega_S, \tau_c) + 3J(\omega_I, \tau_c)\right), \tag{4.70}$$

can be rearranged in the following form

$$\xi = \frac{5}{7}\left[1 - \frac{3kJ(\omega_I, \tau_c)}{R_{1M}^{I}}\right]. \tag{4.71}$$

As shown in Section 4.4, the low field value of R_{1M}^{I} is $10k\tau_c$; after the ω_S dispersion, it approaches the value of $3kJ(\omega_I, \tau_c)$. Eq. (4.71) shows that for a purely dipolar interaction, the coupling factor is a positive quantity which can take the maximum value of 0.5, and decreases with increasing applied magnetic field. The largest value of 0.5 is for $R_{1M}^{I} = 10k\tau_c$; it approaches zero for $R_{1M}^{I} \approx 3kJ(\omega_I, \tau_c)$. As shown in Fig. 4.20A, at a given field, the larger the correlation time, the smaller is the coupling factor. This is related to its direct dependence on the $J(\omega_S, \tau_c)$ spectral density function [Eq. (4.69)], and thus to the occurrence of its dispersion.

In the presence of a contact coupling constant, A, a further term is present in w_0 (Section 4.5) equal to

$$w_0^{\text{con}} = R_{1M}^{\text{con}} = \frac{2}{3}S(S+1)\left(\frac{A}{\hbar}\right)^2 J(\omega_I - \omega_S, \tau_{\text{con}}) \tag{4.72}$$

where τ_{con} is the correlation time for the contact interaction [Eq. (4.7)].
In this case, $R_{1M}^{I} = R_{1M}^{\text{dip}} + R_{1M}^{\text{con}}$ and the coupling factor thus becomes

$$\xi = \frac{5kJ(\omega_S, \tau_c) - R_{1M}^{\text{con}}}{R_{1M}^{I}} \tag{4.73}$$

which can be rearranged in the following form [129]

$$\xi = \frac{5}{7}\left[1 - \frac{3kJ(\omega_I, \tau_c)}{R_{1M}^{I}}\right] - \frac{12}{7}\frac{R_{1M}^{\text{con}}}{R_{1M}^{I}} \tag{4.74}$$

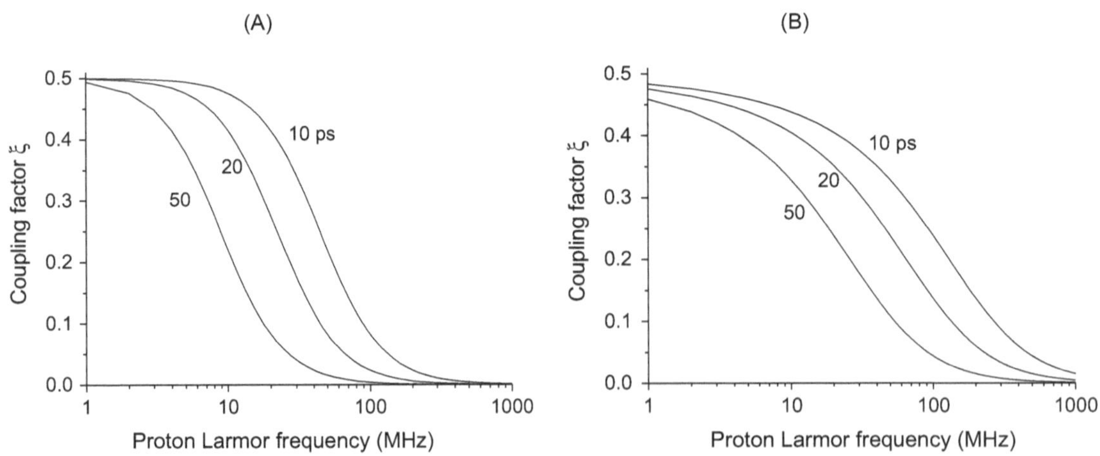

FIGURE 4.20 Field dependence of the coupling factor for correlation times of 10, 20 or 50 ps, in the case of (A) solvent molecules bound to the paramagnetic molecules or (B) diffusing solvent molecules.

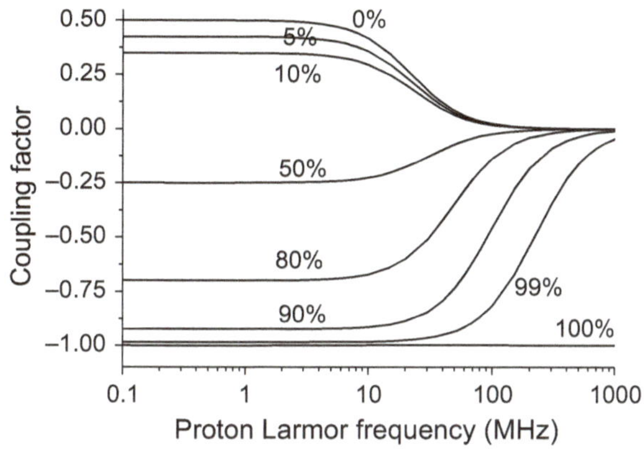

FIGURE 4.21 Field dependence of the coupling factor for different percent of the low field contact relaxation rate with respect to the total (dipolar plus contact) relaxation rate. The calculations are performed for a correlation time of 20 ps.

Eq. (4.74) shows that for a purely contact interaction ξ is always -1. If both contact and dipolar couplings are present, ξ can then range from 0.5 to -1. Fig. 4.21 shows the values of ξ as a function of the magnetic field for different ratios of the low field contact and dipolar contributions. When contact relaxation is dominant, very large DNP enhancements may result even at high magnetic fields [130]. On the contrary, small contact contributions reduce the absolute value of the coupling factor.

Fig. 4.22 shows the coupling factors calculated from relaxation measurements performed for manganese(II) aqua ion (Section 8.3.5.4) [131]. The coupling factor at low field is large and negative (around -0.5 at 25°C), as expected from the dominant contribution to relaxation of the contact interaction. It becomes close to 0.5 with increasing the magnetic field up to the point that contact relaxation is basically dispersed and dipolar relaxation is not. Further increase of the magnetic field causes a fast decrease of the coupling factor as a consequence of the occurrence of the dipolar dispersion.

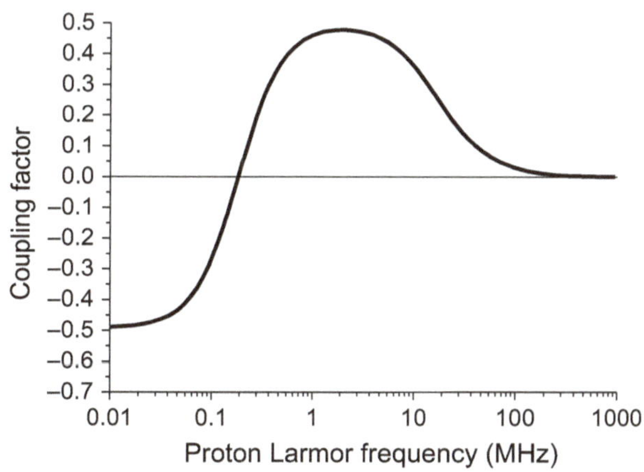

FIGURE 4.22 Coupling factor at 25°C calculated from the relaxation profile of a water solution with manganese(II) aqua ions (Fig. 8.45, Section 8.3.5.4).

Adapted from Ref. [131].

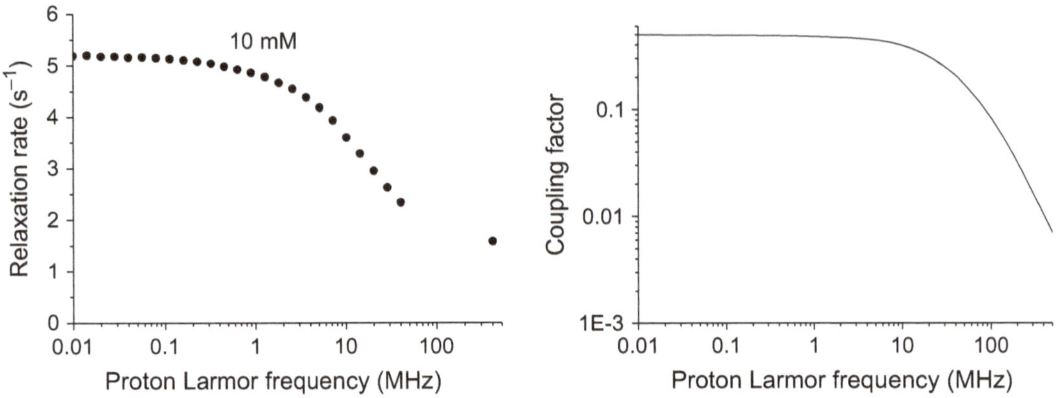

FIGURE 4.23 Solvent water relaxation rates and coupling factors for solutions of TEMPONE 10 mM at 298 K.

Adapted from Ref. [129].

In the presence of paramagnetic nuclear relaxation rates determined from solvent molecules diffusing around the paramagnetic molecules, instead of bound to them (Section 6.6), the constant k and the form of the spectral density functions in Eqs. (4.65)–(4.67), (4.69)–(4.71), and (4.73)–(4.74) are somewhat different [Eqs. (6.48)–(6.49)], providing the R^{I}_{1p} as reported in Eq. (6.46). Fig. 4.20B shows the values of ξ as a function of the magnetic field, as calculated from the Freed diffusion model for some values of the diffusional correlation time (Section 6.6) [132]. In this case the decrease of the coupling factor for large fields occurs more slowly than for bound solvent molecules. The dependence of ξ on the correlation time, at a given field, can be used to estimate the correlation time for the coupling factor determined from DNP measurements. The correlation time can in turn provide useful information on the mobility of the molecule at the site where the paramagnetic center is placed [133]. Finally, if diffusion occurs on a time scale of the order of, or longer than, the molecular reorientation time, contributions from both diffusional solvent molecules and bound solvent molecules should be considered.

DNP is usually performed using as polarizing agents nitroxide radicals, which are characterized by sharp EPR lines, fast rotation, fast diffusion, and favourable distribution of the unpaired electron. These features make the nitroxide radicals ideally suited for solution DNP.

The above equations can be used to estimate the coupling factor from the field dependence of the relaxation rate. In the absence of contact contributions, the $3kJ(\omega_I, \tau_c)$ term in Eq. (4.71) (in the absence of g-anisotropy, ZFS and hyperfine coupling to the metal nucleus) is in fact 3/10 of the low field relaxation and R^{I}_{1M} is the paramagnetic relaxation at the field of interest. A good agreement between the relaxation-derived ξ and the experimental DNP enhancements is obtained for solutions of the radicals TEMPOL and TEMPONE (Fig. 4.23). In these cases, the most significant contributions to the relaxation rate and to the $3kJ(\omega_I, \tau)$ term are provided by translational diffusion, but, although no water molecule is expected to bind the radical, the effect of molecular reorientation during solvent diffusion should also be taken into account (Section 6.6). In water solution at 298 K, at 15 MHz the coupling factor ξ is 0.35 [129,134]; at 400 MHz it decreases to 0.008 [135]. When the temperature increases, also the coupling factor increases as a consequence of the shorter correlation times.

REFERENCES

[1] Stokes G. Trans. Cambridge Philos. Soc. 1956;9:5.
[2] Einstein A. Investigations on the Theory of the Brownian MovementNew York: Dover; 1956.
[3] Debye P. Polar MoleculesNew York: Dover; 1929.
[4] Wilson RC, Myers RJ. J. Chem. Phys. 1976;64:2208.
[5] Bertini I, Luchinat C, Xia Z. Inorg. Chem. 1992;31:3152.
[6] Kivelson D. J. Chem. Phys. 1966;45:1324.
[7] McCain DC, Myers RJ. J. Phys. Chem. 1967;71:192.
[8] Rogers RN, Pake GE. J. Phys. Chem. 1960;33:1107.
[9] Chmelnick AM, Fiat D. J. Magn. Reson. 1972;8:325.
[10] Eaton DR. J. Am. Chem. Soc. 1965;87:3097.
[11] McPherson GL, Koch RC, Stucky GD. J. Chem. Phys. 1974;60:1424.
[12] Bencini A, Gatteschi D. In: Melson GA, Figgis BN, editors. Transition Metal Chemistry. New York and Basel: Marcel Dekker, INC; 1982.
[13] Goodman BA, Raynor JB. In: Emeleus HJ, Sharpe AG, editors. Advances in Inorganic Chemistry and Radiochemistry. New York: Academic Press; 1970.
[14] Rubinstein M, Baram A, Luz Z. Mol. Phys. 1971;20:67.
[15] McGarvey BR. J. Phys. Chem. 1957;61:1232.
[16] La Mar GN, Walker FA. J. Am. Chem. Soc. 1973;95:1782.
[17] La Mar GN, Van Hecke GR. J. Am. Chem. Soc. 1969;91:3442.
[18] Koenig SH, Brown RD III, Spiller M. Magn. Reson. Med. 1987;4:252.
[19] Grant M, Dodgen HW, Hunt JP. Inorg. Chem. 1971;10:71.
[20] Bertini I, Capozzi F, Luchinat C, Xia Z. J. Phys. Chem. 1993;97:1134.
[21] Bertini I, Galas O, Luchinat C, Messori L, Parigi G. J. Phys. Chem. 1995;99:14217.
[22] Yu LP, La Mar GN, Rajarathnam K. J. Am. Chem. Soc. 1990;112:9527.
[23] Banci L, Bertini I, Marconi S, Pierattelli R, Sligar SG. J. Am. Chem. Soc. 1994;116:4866.
[24] Ducommun Y, Newman KE, Merbach A. Inorg. Chem. 1980;19:3696.
[25] Banci L, Bertini I, Marconi S, Pierattelli R. Eur. J. Biochem. 1993;215:431.
[26] Werth MT, Kurtz DM Jr, Moura I, LeGall J. J. Am. Chem. Soc. 1987;109:273.
[27] Chmelnick AM, Fiat D. J. Chem. Phys. 1967;47:3986.
[28] Banci L, Bertini I, Luchinat C. Inorg. Chim. Acta 1985;100:173.
[29] Bertini I, Luchinat C, Mancini M, Spina G. J. Magn. Reson. 1984;59:213.
[30] Chachaty C, Forchioni A, Verlet J, Ronfard-Haret JC. Chem. Phys. Lett. 1974;29:436.
[31] Ming L-J, Banci L, Luchinat C, Bertini I, Valentine JS. Inorg. Chem. 1988;27:4458.
[32] La Mar GN. J. Am. Chem. Soc. 1965;87:3567.
[33] Banci L, Bertini I, Luchinat C. Magn. Reson. Rev. 1986;11:1.
[34] Steiner UE, Bürβner D. Z. Phys. Chem. N. F. 1990;169:159.
[35] Paulo A, Domingos A, Pires di Matos A, Santos I, Carvalho MFNN, Pombiero AJL. Inorg. Chem. 1994;33:4729.
[36] Vishnevskaya GP, Kozyrev BM. J. Struct. Chem. 1966;7:20.
[37] Hudson A, Lewis JWE. Trans. Faraday Soc. 1970;66:1297.
[38] Koenig SH. Magn. Reson. Med. 1991;22:183.
[39] Hernandez G, Tweedle M, Bryant RG. Inorg. Chem. 1990;29:5110.
[40] Aime S, Barbero L, Botta M. Magn. Res. Imaging 1991;9:843.
[41] Powell DH, Dhubhghaill OMN, Pubanz D, Helm L, Lebedev YS, Schlaepfer W, Merbach AE. J. Am. Chem. Soc. 1996;118:9333.

[42] Mastarone DJ, Harrison VSR, Eckermann AL, Parigi G, Luchinat C, Meade TJ. J. Am. Chem. Soc. 2011;133:5329.
[43] Bertini I, Capozzi F, Luchinat C, Nicastro G, Xia Z. J. Phys. Chem. 1993;101:198.
[44] Misra SK, Orhun U. Solid State Commun. 1987;63:867.
[45] Malhotra VM, Buckmaster HA, Dixon JM. J. Phys. C: Solid State Phys. 1980;13:3921.
[46] Aime S, Botta M, Ermondi G. Inorg. Chem. 1992;31:4291.
[47] Bertini I, Luchinat C, Xia Z. J. Magn. Reson. 1992;99:235.
[48] Luz Z, Fiat D. J. Chem. Phys. 1967;46:469.
[49] Koenig SH, Brown RD III. J. Magn. Reson. 1985;61:426.
[50] Bertini I, Briganti F, Luchinat C, Xia Z. J. Magn. Reson. 1993;101:198.
[51] Bloch JB, Navon G. J. Inorg. Nucl. Chem. 1980;42:693.
[52] Friedman HL, Holz M, Hertz HG. J. Chem. Phys. 1979;70:3369.
[53] Kowalewski J, Larsson T, Westlund PO. J. Magn. Reson. 1987;74:56.
[54] Bertini I, Luchinat C, Martini G. In: Poole CP, editor. Handbook of Electron Spin Resonance. New York: American Institute of Physics; 1994. p. 51.
[55] Bertini I, Luchinat C, Martini G. In: Poole CP, editor. Handbook of Electron Spin Resonance. New York: American Institute of Physics; 1994. p. 79.
[56] Van Vleck JH. Phys. Rev. 1940;57:426.
[57] Orbach R. Proc. R. Soc. London, Ser. A 1961;264:458.
[58] Hoffmann SK, Lijewski S. J. Magn. Reson. 2015;252:49.
[59] Kivelson D. J. Chem. Phys. 1960;33:1094.
[60] Al'tshuler SA, Valiev KA. Sov. Phys. 1959;35:661.
[61] Atkins PW, Kivelson D. J. Chem. Phys. 1966;44:169.
[62] Bruno GV, Harrington JK, Eastman MP. J. Phys. Chem. 1977;81:1111.
[63] Bloembergen N, Morgan LO. J. Chem. Phys. 1961;34:842.
[64] Serebrennikov YA, Steiner UE. J. Chem. Phys. 1994;100:7508.
[65] Steiner UE, Serebrennikov YA. J. Chem. Phys. 1994;100:7503.
[66] Powell DH, Helm L, Merbach AE. J. Chem. Phys. 1991;95:9258.
[67] Freed JH, Kooser RG. J. Chem. Phys. 1968;49:4715.
[68] Kooser RG, Volland WV, Freed JH. J. Chem. Phys. 1969;50:5243.
[69] Solomon I. Phys. Rev. 1955;99:559.
[70] Abragam A. The Principles of Nuclear MagnetismOxford: Oxford University Press; 1961.
[71] Koenig SH. J. Magn. Reson. 1982;47:441.
[72] Woessner DE. J. Chem. Phys. 1962;37:647.
[73] Golding RM, Pascual RO, McGarvey BR. J. Magn. Reson. 1982;46:30.
[74] Nordenskiöld L, Laaksonen A, Kowalewski J. J. Am. Chem. Soc. 1982;104:379.
[75] Gottlieb HPW, Barfield M, Doddrell DM. J. Chem. Phys. 1977;67:3785.
[76] Doddrell DM, Healy PC, Bendall MR. J. Magn. Reson. 1978;29:163.
[77] Mispelter J, Momenteau M, Lhoste J-M. J. Chem. Soc. Dalton Trans. 1981;1729.
[78] Bloembergen N. J. Chem. Phys. 1957;27:575.
[79] Guéron M. J. Magn. Reson. 1975;19:58.
[80] Vega AJ, Fiat D. Mol. Phys. 1976;31:347.
[81] Sternlicht H. J. Chem. Phys. 1965;42:2250.
[82] Vasavada KV, Nageswara Rao BD. J. Magn. Reson. 1989;81:275.
[83] Bertini I, Luchinat C, Vasavada KV. J. Magn. Reson. 1990;89:243.
[84] Bertini I, Briganti F, Luchinat C, Mancini M, Spina G. J. Magn. Reson. 1985;63:41.
[85] Banci L, Bertini I, Briganti F, Luchinat C. J. Magn. Reson. 1986;66:58.
[86] Sharp RR. J. Chem. Phys. 1993;98:2507.

[87] Bertini I, Galas O, Luchinat C, Parigi G. J. Magn. Reson. Ser. A 1995;113:151.
[88] Bertini I, Kowalewski J, Luchinat C, Nilsson T, Parigi G. J. Chem. Phys. 1999;111:5795.
[89] Kruk D, Nilsson T, Kowalewski J. Phys. Chem. Chem. Phys. 2001;3:4907.
[90] Rast S, Fries PH, Belorizky E, Borel A, Helm L, Merbach AE. J. Chem. Phys. 2001;115:7554.
[91] Fries PH, Belorizky E. J. Chem. Phys. 2007;126:204503.
[92] Larsson T, Westlund PO, Kowalewski J, Koenig SH. J. Chem. Phys. 1994;101:1116.
[93] Nilsson T, Kowalewski J. J. Magn. Reson. 2000;146:345.
[94] Kowalewski J, Kruk D, Parigi G. Adv. Inorg. Chem. 2005;57:41.
[95] Goldman M. J. Magn. Reson. 1984;60:437.
[96] Farrar TC, Quintero-Arcaya RA. Chem. Phys. Lett. 1985;122:41.
[97] Farrar TC, Quintero-Arcaya RA. J. Phys. Chem. 1987;91:3224.
[98] Wimperis S, Bodenhausen G. Mol. Phys. 1989;66:897.
[99] Bertini I, Luchinat C, Tarchi D. Chem. Phys. Lett. 1993;203:445.
[100] Bertini I, Luchinat C, Piccioli M, Tarchi D. Concepts Magn. Reson. 1994;6:307.
[101] Ghose R, Prestegard JH. J. Magn. Reson. 1997;128:138.
[102] Bertini I, Kowalewski J, Luchinat C, Parigi G. J. Magn. Reson. 2001;152:103.
[103] Boisbouvier J, Gans P, Blackledge M, Brutscher B, Marion D. J. Am. Chem. Soc. 1999;121:7700.
[104] Hus JC, Marion D, Blackledge M. J. Mol. Biol. 2000;298:927.
[105] Bertini I, Cavallaro G, Cosenza M, Kümmerle R, Luchinat C, Piccioli M, Poggi L. J. Biomol. NMR 2002;23:115.
[106] Desvaux H, Gochin M. Mol. Phys. 1999;96:1317.
[107] Volkman BF, Wilkens SJ, Lee AL, Xia B, Westler WM, Berger R, Markley JL. J. Am. Chem. Soc. 1999;121:4677.
[108] Pintacuda G, Kaikkonen A, Otting G. J. Magn. Reson. 2004;171:233.
[109] Yamamoto Y, Nanai N, Inoue Y, Chujo R. J. Chem. Soc. Chem. Commun. 1989;1419.
[110] Basolo F., Pearson R.G. Mechanisms of inorganic reactions. A study of metal complexes in solutions, Wiley, New York, 1963.
[111] Robinson BH, Haas DA, Mailer C. Science 1994;263:490.
[112] Biller JR, Elajaili H, Meyer V, Rosen GM, Eaton SS, Eaton GR. J. Magn. Reson. 2013;236:47.
[113] Froncisz W, Camenisch TG, Ratke JJ, Anderson JR, Subczynski WK, Strangeway RA, Sidabras JW, Hyde JS. J. Magn. Reson. 2008;193:297.
[114] Koptyug IV, Bossmann SH, Turro NJ. J. Am. Chem. Soc. 1996;118:1435.
[115] Zhou Y, Bowler BE, Eaton GR, Eaton SS. J. Magn. Reson. 1999;139:165.
[116] Redfield AG. Res. Dev. 1957;1:19.
[117] Benetis N, Kowalewski J, Nordenskiöld L, Wennerström H, Westlund PO. Mol. Phys. 1983;48:329.
[118] Benetis N, Kowalewski J, Nordenskiöld L, Wennerström H, Westlund PO. Mol. Phys. 1983;50:515.
[119] Overhauser AW. Phys. Rev. 1953;89:689.
[120] Overhauser AW. Phys. Rev. 1953;92:411.
[121] Noggle JH, Schirmer RE. The Nuclear Overhauser EffectNew York: Academic Press; 1971.
[122] Neuhaus D, Williamson M. The Nuclear Overhauser Effect in Structural and Conformational AnalysisNew York: VCH; 1989.
[123] Johnson RD, Ramaprasad S, La Mar GN. J. Am. Chem. Soc. 1983;105:7205.
[124] Granot J. J. Magn. Reson. 1982;49:257.
[125] Dugad LB, La Mar GN, Unger SW. J. Am. Chem. Soc. 1990;112:1386.
[126] Bertini I, Luchinat C, Messori L, Vasak M. Eur. J. Biochem. 1993;211:235.
[127] Hausser KH, Stehlik D. Adv. Magn. Reson. 1968;3:79.
[128] Bloembergen N. J. Chem. Phys. 1957;27:572.
[129] Bennati M, Luchinat C, Parigi G, Türke M-T. Phys. Chem. Chem. Phys. 2010;12:5902.

[130] Loening NM, Rosay M, Weis V, Griffin RG. J. Am. Chem. Soc. 2002;124:8808.
[131] Luchinat C, Parigi G, Ravera E. J. Biomol. NMR 2014;58:239.
[132] Hwang LP, Freed JH. J. Chem. Phys. 1975;63:4017.
[133] Franck JM, Pavlova A, Scott JA, Han S. Prog. Nucl. Magn. Reson. Spectrosc. 2013;74:33.
[134] Höfer P, Parigi G, Luchinat C, Carl P, Guthausen G, Reese M, Carlomagno T, Griesinger C, Bennati M. J. Am. Chem. Soc. 2008;130:3254.
[135] Neugebauer P, Krummenacker JG, Denysenkov VP, Parigi G, Luchinat C, Prisner TF. Phys. Chem. Chem. Phys. 2013;15:6049.

CHAPTER

HIGH RESOLUTION SOLID-STATE NMR IN PARAMAGNETIC MOLECULES

5

Whether a molecule is in solution or in a solid, its nuclear interactions are the same. But free reorientation in solution induces an averaging which sizably simplify the spectra. This chapter is thus intended to show how the anisotropy of the magnetic interactions affects the spectra, how it can be averaged also in the solid state, and to experimental approach NMR of samples in the solid state.

5.1 NUCLEAR INTERACTIONS IN SOLID-STATE NMR: STATIC AND MAS

5.1.1 THE SPATIAL DEPENDENCE OF NUCLEAR INTERACTIONS

As already extensively discussed in Chapter 1, in the presence of a magnetic field the nuclear spin magnetic energy levels are mainly determined by the interaction between the nuclear magnetic moments and the external magnetic field. This picture is complicated by the interactions of the nuclear magnetic moments with (1) the additional field originating from the electron currents in orbitals (chemical shielding), (2) the electric field gradients (quadrupolar coupling), (3) the magnetic moments of neighboring nuclei (dipole–dipole and scalar couplings), and finally, (4) the magnetic moment of unpaired electrons (dipolar shielding and contact shielding), which is the subject of this book. All these interactions have a characteristic spatial dependency, as they depend on spatially-dependent molecular properties [(1), (2), and (4)] and/or on the orientation of the vector connecting the two nuclei (or the nucleus and the electron) with respect to the external magnetic field [(3) and (4)], as summarized in Table 5.1.

It follows that for each set of equivalent nuclei in a molecule, the position of the signal and its multiplicity depend on the orientation of the molecule in the field. In the more realistic scenario of the molecule being embedded in a single crystal, with all molecules equally oriented in the crystal lattice, the positions and multiplicities of all the signals for each set of nonequivalent nuclei have a different dependency on the orientation of the molecule with respect to the crystal and on the orientation of the crystal with respect to the field.

The situation is exemplified in Fig. 5.1 (chemical shielding) and Fig. 5.2 (dipole–dipole coupling), where the position of the crystallite is changed with respect to the magnetic field. For simplicity, both the chemical shielding and the dipole–dipole coupling tensors are assumed to be aligned with the principal axis of the crystal.

The situation of a crystalline powder follows intuitively, since all the orientations of the crystallites are equally possible: as a result, all the frequencies and multiplicities appear at the same time, as exemplified in Fig. 5.1D, Fig. 5.2D, and Fig. 5.3.

NMR of Paramagnetic Molecules. http://dx.doi.org/10.1016/B978-0-444-63436-8.00005-3

Table 5.1 Spatial Dependence of Nuclear Interactions and Their Rotational Average Under MAS

Interaction	Orientation Dependence	Averages to	Notes
Chemical shielding	$3\cos^2\theta - 1$	Chemical shift	
Dipolar shielding	$3\cos^2\theta - 1$	Pseudocontact shift	see Eq. (5.6)
Heteronuclear dipole–dipole	$3\cos^2\theta - 1$	0	
Nuclear quadrupole (1st order)	$3\cos^2\theta - 1$	0	Analogous to zero field splitting for electrons (Section 1.4 and Eq. (1.24))
Nuclear quadrupole (2nd order)	$35\cos^4\theta - 30\cos^2\theta + 3$	Not averaged by MAS, but scaled to −7/18 of the static value, isotropic value is called quadrupolar shift	See above
Contact shielding	$3\cos^2\theta - 1$—anisotropic only when caused by spin polarization	Contact shift	

5.1.2 MAGIC ANGLE SPINNING (MAS)

Fast isotropic reorientation of molecules in solution average out the anisotropic part of these interactions, yielding a substantially simplified NMR spectrum, consisting of sharp lines centered at the average value of the chemical shielding interactions and split by the scalar coupling with covalently bound spins. Also in the solid state sharp spectra can be obtained. This is, for instance, the case of single crystals: as already mentioned above, all sets of equivalent nuclei for a given orientation of the molecule in the lattice and of the lattice in the field will experience a single shift, a single value of dipole–dipole coupling and quadrupole, if present; thus the spectrum intuitively consists of several sharp lines. However, it is impractical to interpret single crystal spectra when several resonances are present, because of several nonequivalent sets of nuclei and/or because more than one orientation of the molecule in the unit cell. For this reason a more general approach is sought for: a common workaround for obtaining a "solution-like" spectrum of a powdered solid is resorting to the so-called magic angle spinning (MAS) [1–3]. In this technique, the sample is spun at a given frequency about an axis that forms an angle of 54.7356° with the applied magnetic field (Fig. 5.4).

For any anisotropic interaction having a $3\cos^2\theta - 1$ spatial dependency, the average value can be calculated to be [3]:

$$\langle 3\cos^2\theta - 1\rangle = \frac{1}{2}(3\cos^2\theta_R - 1)(3\cos^2\beta - 1) \tag{5.1}$$

where θ is the orientation of the z-axis of the principal axis frame with respect to the external magnetic field, θ_R is the angle at which the rotation is performed and β is the orientation of the z-axis of the principal axis frame with respect to the tilt of the rotor axis, i.e., the difference between θ and θ_R (Fig. 5.4). This average equals zero for $\theta_R = 54.7356°$, and as long as the spinning frequency is at least 3–4 times larger than the frequency span of the anisotropic interaction that needs to be cancelled. As will be discussed after, for

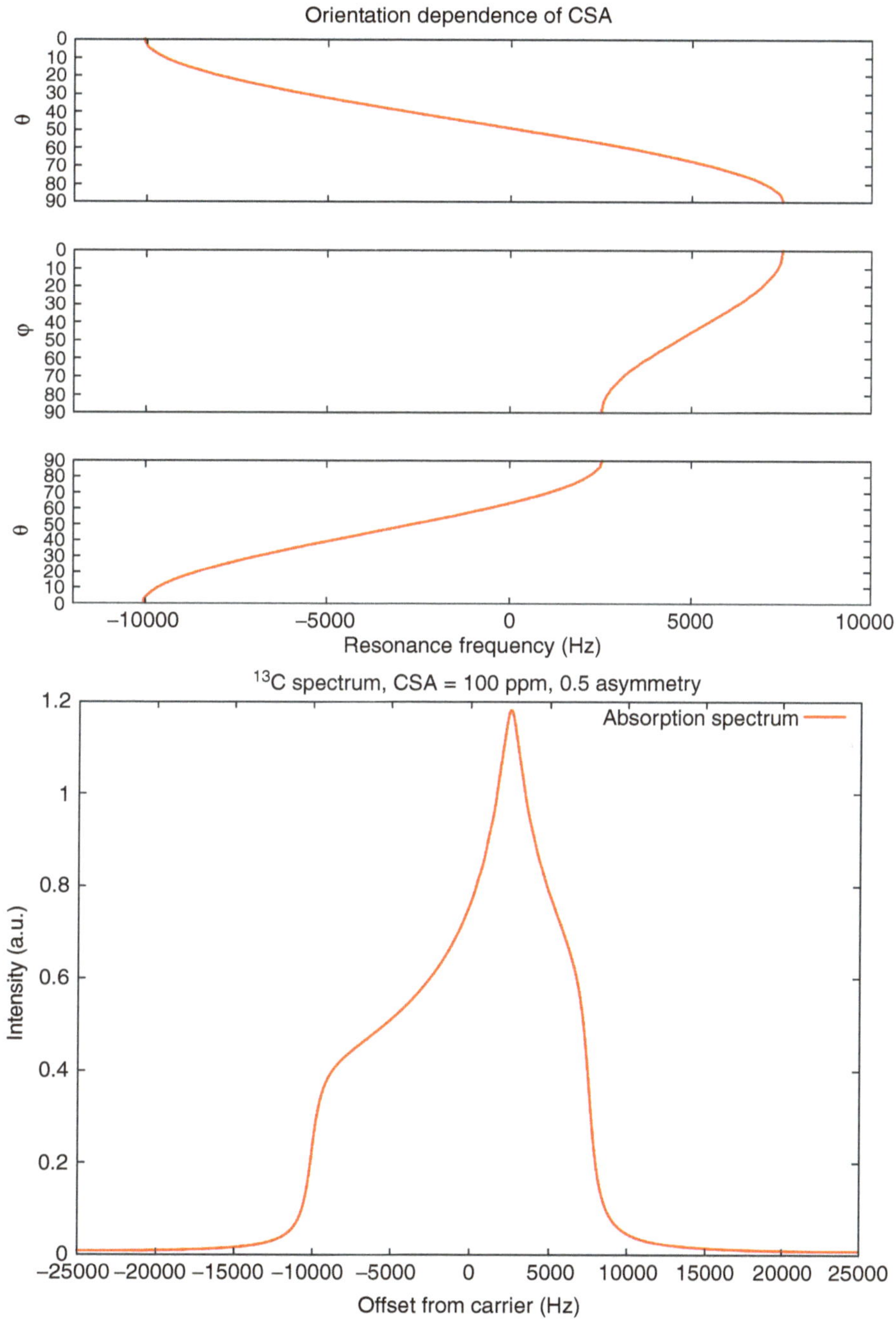

FIGURE 5.1 Orientation dependence of the chemical shielding anisotropy in a single crystal. In panel (A) the crystallite is rotated from the position where the principal axis frame of the CSA tensor is aligned with the field to the position where it is perpendicular to the field; in panel (B) the rotation is accomplished around the magnetic field, spanning 90° in total; in panel (C) the crystallite is rotated around an axis perpendicular to the field back to the starting position. Panel (D) represents the full powder pattern.

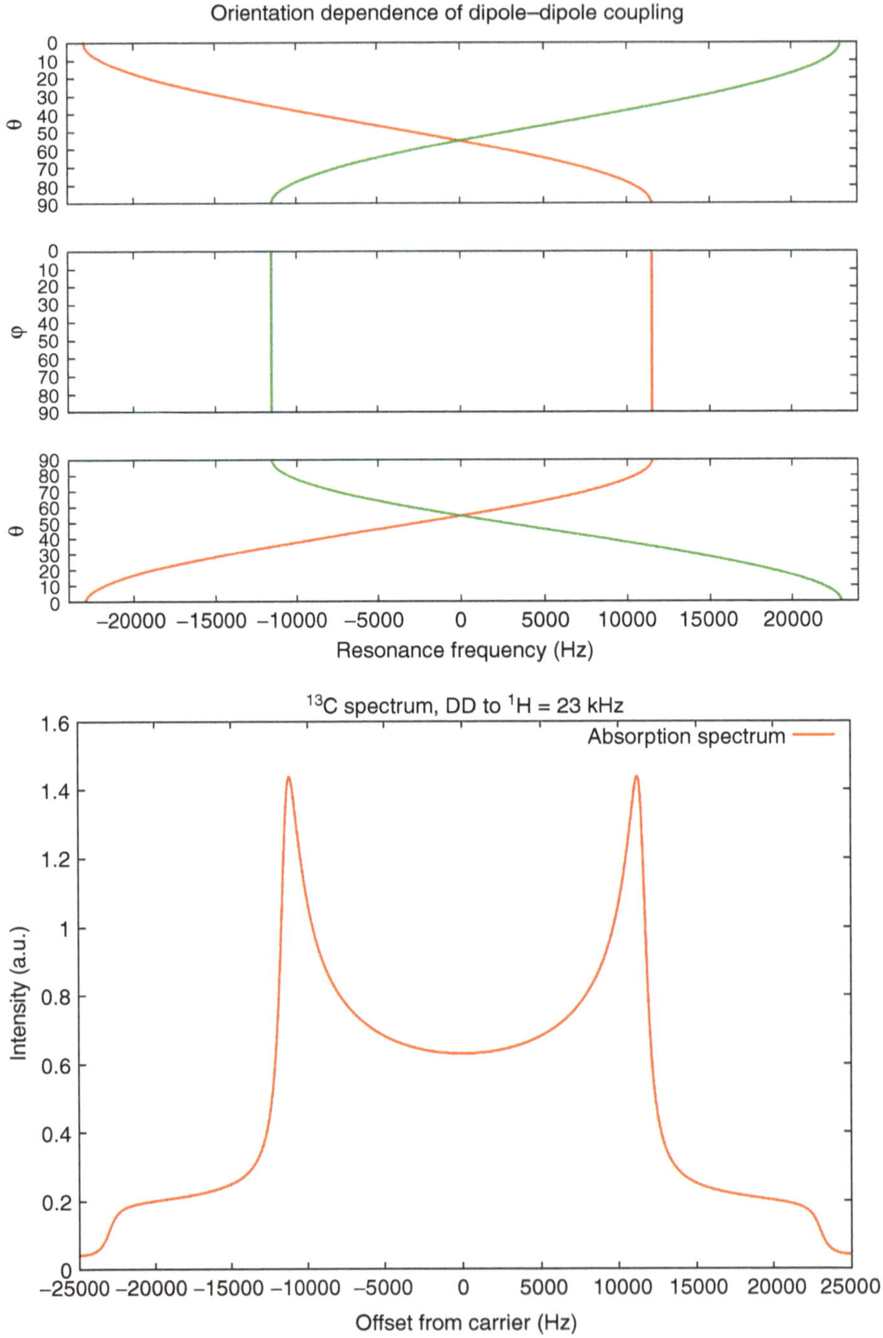

FIGURE 5.2 Orientation dependence of the dipole–dipole coupling in a single crystal. The movements of the crystallite are the same as in Fig. 5.1, in the starting position the vector between the two nuclei is aligned with the field. Panel (D) represents the full powder pattern.

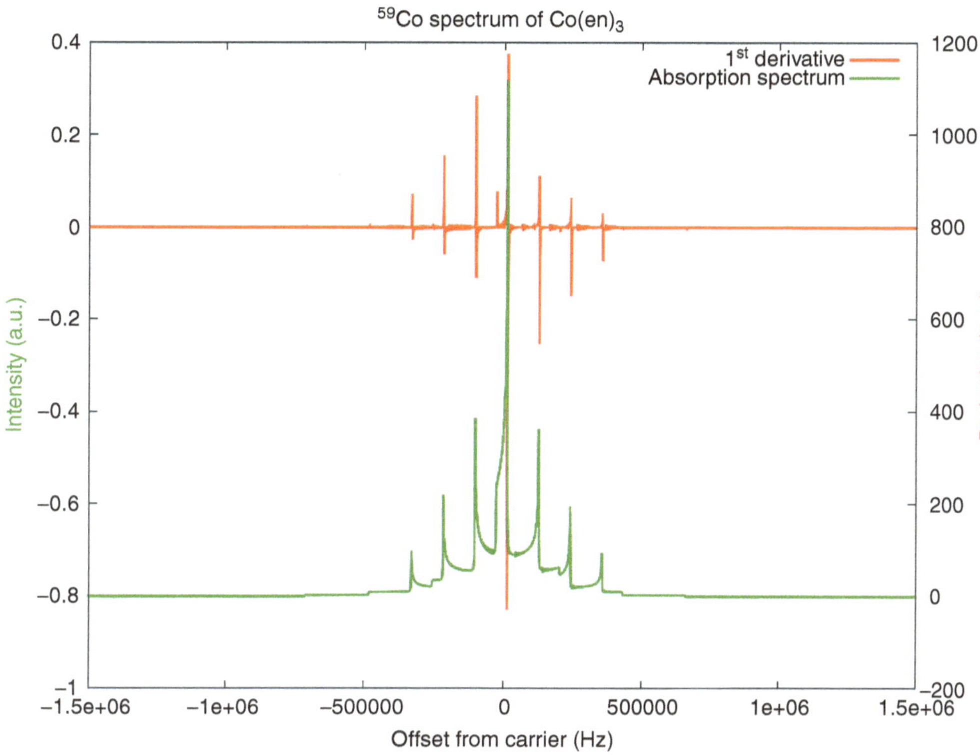

FIGURE 5.3 ^{59}Co powder pattern for tris-ethylenediamine-cobalt(III) trichloride, simulated from experimental quadrupolar coupling parameters [49]. Top trace in red indicates represents the derivative spectrum.

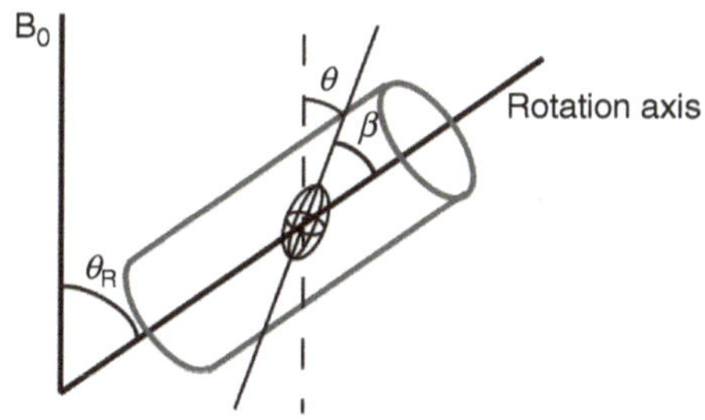

FIGURE 5.4 Magic angle spinning setup. Angle θ represent the inclination of the principal axes frame of a nuclear interaction tensor with respect to the applied magnetic field; θ_R is the tilt angle of the rotor (gray cylinder) and β is the angle that brings the principal axis of the nuclear interaction tensor along the rotation axis.

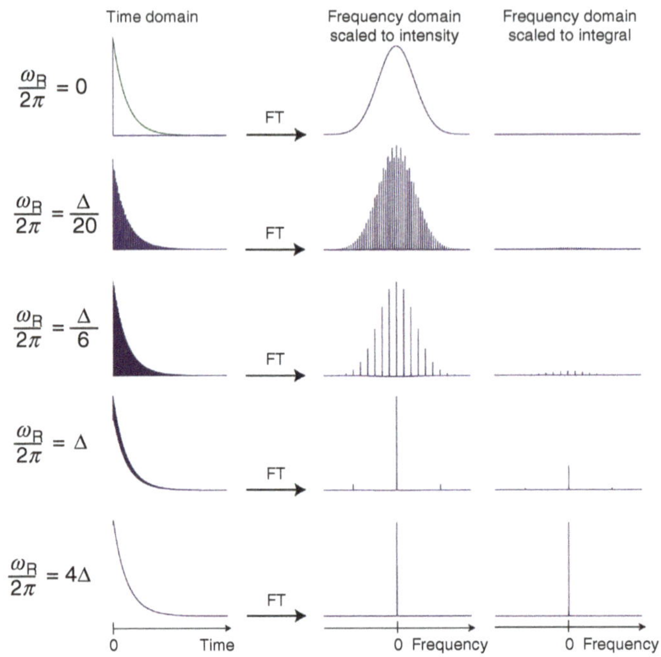

FIGURE 5.5 Formation of rotational echoes and spinning sidebands.

lower spinning frequencies, the anisotropic part of the interaction acquires a time-dependence that causes the signal to split over a manifold of spinning sidebands, each separated by the rotation frequency. A hand waving argument for this is the following: when a powdered sample is excited, spins belonging to different crystallites evolve differently, depending on their orientation with respect to the field. If the sample is rotating, the evolution frequency of each spin will vary, causing a dephasing of the signal. However, when the rotor comes back to the starting position, all the spins will have again all the evolution frequencies that they had in the beginning; in this way the signals are rephased, reaching again the starting intensity (neglecting transverse relaxation). This effect is known as rotational echo (Fig. 5.5).

A quantitative description of MAS requires a more complicated mathematical treatment. In general, the aforementioned NMR interactions can be described as the coupling of two vectors via a second order Cartesian tensor. The description of the evolution of the system under space and spin operators can be conveniently treated separating the spin (T) and space (A) components of the Hamiltonian, by the use of the so-called "spherical tensors" [4]. Each component can be split into a linear combination of spherical basis tensors:

$$H = \Sigma_{k=0}^{2} \Sigma_{q=-k}^{k} (-1)^q A_{kq} T_{k-q} \tag{5.2}$$

Spherical tensors transform under rotation according to simple rules, which are useful to understand some of the following steps. A more exhaustive discussion can be found in Duer [5]. After such decomposition, only the terms that commute with the Zeeman Hamiltonian are preserved (secular approximation), so that:

$$H = A_{00}T_{00} - A_{10}T_{10} + A_{20}T_{20} \tag{5.3}$$

Table 5.2 Wigner Rotation Matrix Elements

$d^{(2)}_{00}(\theta_R)$	$\frac{1}{2}(3\cos^2\theta_R - 1)$
$d^{(2)}_{01}(\theta_R) = d^{(2)}_{0-1}(\theta_R)$	$-\sqrt{\frac{3}{2}}\sin\theta_R\cos\theta_R$
$d^{(2)}_{02}(\theta_R) = d^{(2)}_{0-2}(\theta_R)$	$-\sqrt{\frac{3}{8}}\sin^2\theta_R$

The anisotropies of NMR-relevant interactions are described by traceless symmetric second rank tensors, thus the term A_{10} in Eq. (5.3) is dropped. The term A_{00} represents the isotropic part, and the term A_{20} represents the anisotropic component.

Moving from laboratory frame to the rotor frame (Fig. 5.4) [6], the Hamiltonian is transformed into:

$$H = A_{00}T_{00} + T_{20}\sum_{p=-2}^{2} A_{2p}e^{-ip\omega_R t}d^{(2)}_{0p}(\theta_R) \tag{5.4}$$

where the $d^{(2)}_{0p}(\theta_R)$ are the Wigner rotation matrix elements reported in Table 5.2.

The only time dependence of the Hamiltonian Eq. (5.4) is retained in the terms A_{2p}, oscillating at ω_R and $2\omega_R$. As already mentioned, the term A_{20} shown in Eq. (5.1) vanishes for $\theta_R = 54.7356°$.

This means that it is possible to decouple some interactions by mechanical rotation at the magic angle and also accounts for the formation of rotational echoes as described earlier.

5.1.3 HOMONUCLEAR COUPLING AND ITS IMPLICATIONS FOR HETERONUCLEAR DECOUPLING

The effect of MAS is that of showing the isotropic frequency of all signals in the spectrum of a powder sample. However, when dealing with proton-dense solids, MAS alone does not suffice in efficiently removing the heteronuclear coupling, the reason for this behavior being summarized in the following paragraphs.

Let us consider the case of an isolated homonuclear spin couple: the effect of dipole–dipole coupling acting between them is analogous to what has been described for the contact coupling, with the diagonal terms of the Hamiltonian that shifts the energy of all the states by a constant factor. The off-diagonal terms, containing the transverse spin operator components, cause a mixing of the $|\alpha\beta\rangle$ and $|\beta\alpha\rangle$ states, which is stronger when the two states are closer in energy. Analogously to the two spin case, when a third spin is introduced, further states are created, and only $|\alpha\alpha\alpha\rangle$ and $|\beta\beta\beta\rangle$ are eigenstates of the dipole–dipole coupling to the third spin; this means that the remaining possible states of the spin system evolve under the influence of the dipolar coupling. The only, unlikely, exception is when the three $|\alpha_2\beta\rangle$ states (irrespectively of which of the spins is in the β state) (as well as the three $|\beta_2\alpha\rangle$ states, irrespectively of which of the spins is in the α state) are exactly degenerate (identical spins), then the states that are formed are still eigenstates of the dipole–dipole coupling. The situation for N spins follows from the above: N manifolds will be created with increasing quasi-degeneracy from 1 for all $|\alpha_N\rangle$ or $|\beta_N\rangle$ (which are eigenstates of the dipolar coupling and thus will not be broadened), N for the states where a single spin is reversed, N(N–1) for states where two spins are reversed and so on. Broadening by the time-dependent fluctuation of the remaining states increases with the total number of states in the manifold. The dipole–dipole coupled multispin system behaves as a time-dependent mixture of nearly degenerate Zeeman functions and there is a continuous exchange of longitudinal magnetization between the spins of the dipole–dipole coupled system, which will be referred to as "dipolar network" from now on. Only the total z component of the N spin system is conserved. The effect of MAS is that of modulating the dipole–dipole coupling (Table 5.1). This induces a periodicity in the interaction, and thus breaks up each manifold into spinning sidebands [7,8]. In general, not all spins are magnetically equivalent and, as a result, the dipole–dipole interactions differ and the dipole–dipole interactions of all pairs do not commute. This results in further broadening of the sideband pattern [5]. When now considering the coupling of an X nucleus to a proton in a proton-dense sample, one needs to take into account the coupling of the X spin to the proton dipolar network, which yields an incoherent contribution, thus a homogeneous broadening of the line. To avoid this, either large spinning frequencies are applied (> 40 kHz for protein samples), or strong radiofrequency irradiation is required to maintain the saturation of the whole proton bath [9]. At higher spinning rates, lower power decoupling can be applied, because the effective homonuclear dipole–dipole coupling is reduced, leading to reduced fluctuations of the time-dependent states of the spin system and lower power can effectively saturate the relevant protons, in turn yielding improved H-X decoupling. This goes up to the point that decoupling can be applied only to remove the effect of J-couplings as done in solution NMR. This is actually more relevant for paramagnetic systems: the mixing of the states with intermediate total angular momentum is stronger the closer the original states are in energy: intuitively, a large hyperfine interaction will reduce the energy overlap and will yield more efficient decoupling for the paramagnetic system as compared to its diamagnetic analog [10].

Of note, when the decoupling field corresponds to a multiple of the rotation frequency, the effect of decoupling vanishes and the dipole–dipole interaction is reestablished (this is called recoupling, Section 5.1.4) [11].

5.1.4 RECOUPLING

There is a large amount of information contained in the anisotropic interactions, and the spectral simplification allowed for by MAS comes at the price of losing such information (Section 5.2.3). However, there are strategies to reintroduce such information: MAS modulates in a fixed manner the spatial part of the Hamiltonian in Eq. (5.4), but all Hamiltonians also have a spin part, which can be manipulated by radiofrequency pulses. Thus one can imagine for instance that the effect of MAS can be reverted by applying rotor-synchronized inversion pulses (Sections 12.3.6 and 12.3.8). This approach can be schematized by using as an example the RFDR experiment (Section 12.3.6), in which the recoupling is used to obtain cross-peaks between neighboring spins in a two dimensional homonuclear spectrum. The RFDR can be depicted as follows: MAS modulates the homonuclear dipole–dipole interaction with a $e^{-i\omega_r t}$ dependence. The interaction thus cycles between positive and negative values, and the overall effect is that the dipole–dipole interaction is averaged to zero. Now, let's imagine that an (infinitely short) inversion pulse is given at half of a cycle when the interaction (assumed starting from a positive value) reaches the zero point. Then, another positive-value half-cycle will follow. The average value will now be smaller than the full interaction, but still different from zero, allowing for recoupling of the interaction [12]. This same strategy can be applied to reintroduce CSA and heteronuclear dipole–dipole coupling as well (Section 12.3.8), and tailored pulse sequences are available to reintroduce selectively some terms. An exhaustive description of recoupling methods is given by Levitt in [13].

5.2 NMR SPECTRA OF PARAMAGNETIC COMPLEXES IN THE SOLID STATE

5.2.1 THE EFFECT OF PARAMAGNETIC CENTERS

5.2.1.1 The Curie-spin interaction

As in the case of solution NMR, the solid-state NMR spectra of paramagnetic substances shows both the diamagnetic and the paramagnetic contributions and, in general, the considerations made for the spatially-dependent diamagnetic interactions will hold true also for the paramagnetic ones (Table 5.1).

It is instructive to analyze the case of the dipole–dipole coupling between the nucleus and the average induced magnetic moment, named the "Curie spin," which has been discussed in detail in Chapter 2, to show how the interaction is sensed by nuclei framed in a solid matrix. The Hamiltonian [Eq. (2.24)] for the dipolar interaction between the nucleus I with magnetic moment $\mu_1 = \hbar\gamma_I \boldsymbol{I}$ and the average induced electron magnetic moment $\mu_2 = \langle\mu\rangle$, since $\boldsymbol{I}$ is quantized along the direction $\boldsymbol{k}$ of $\boldsymbol{B}_0$ ($\boldsymbol{I} = I_k\boldsymbol{k}$), is

$$H = -\frac{\mu_0}{4\pi}\left[\frac{3(\hbar\gamma_I \boldsymbol{I}\cdot\boldsymbol{r})(\langle\mu\rangle\cdot\boldsymbol{r})}{r^5} - \frac{\hbar\gamma_I \boldsymbol{I}\cdot\langle\mu\rangle}{r^3}\right] = -\frac{\hbar\gamma_I B_0}{4\pi r^5} I_k\boldsymbol{k}\ \left[3\boldsymbol{r}(\boldsymbol{r}\cdot\chi) - r^2\chi\right]\boldsymbol{k} \tag{2.24}$$

where $\boldsymbol{\chi}$ is the magnetic susceptibility tensor per molecule. We can then define a dipolar shielding (second rank) tensor [Eq. (2.28)]

$$\sigma = 3\boldsymbol{r}(\boldsymbol{r}\cdot\chi) - r^2\chi, \tag{5.5}$$

as a physical effect that changes both the direction and the intensity of the magnetic field as sensed by the nuclei. Hamiltonian Eq. (2.24) can thus be rewritten as:

$$H = -\frac{\hbar\gamma_I}{4\pi r^5}\boldsymbol{I}\cdot\sigma\cdot\boldsymbol{B}, \tag{5.6}$$

which can be compared with the chemical shielding Hamiltonian:

$$H = \hbar\gamma_I \boldsymbol{I}\cdot\boldsymbol{\delta}\cdot\boldsymbol{B}, \tag{5.7}$$

where $\boldsymbol{\delta}$ denotes the chemical shielding tensor. From this comparison it is clear that the dipolar shielding Hamiltonian has the same symmetry with respect to spin and space operations as the chemical shielding Hamiltonian [4,14]. The trace of the dipolar interaction is the pseudocontact shift, which is zero in the case of isotropic magnetic susceptibility (Fig. 5.6).

In solution, molecular tumbling causes random sampling of all the possible shielding values, and this provides a relaxation mechanism [15–17]. On the contrary, since the random molecular tumbling is abolished in a solid, neither interaction contributes sizably to relaxation (we assume the absence of local fluctuations for the time being, see later) [18,19]. This means that the quality of solid-state NMR spectra does not depend on the size of the system under investigation. This feature makes solid-state NMR particularly attractive for large paramagnetic biomolecules.

However, it is important to recall that the magnetic susceptibility of a paramagnetic center, in particular that of extremely anisotropic lanthanoids, can be modulated by small fluctuations of the coordination environment [20,21]. On the basis of this, it can be postulated that such fluctuations may reintroduce Curie-spin relaxation in solids, although in this case, it is not dependent on the molecular tumbling time but on the characteristic timescale of the distortion of the coordination environment.

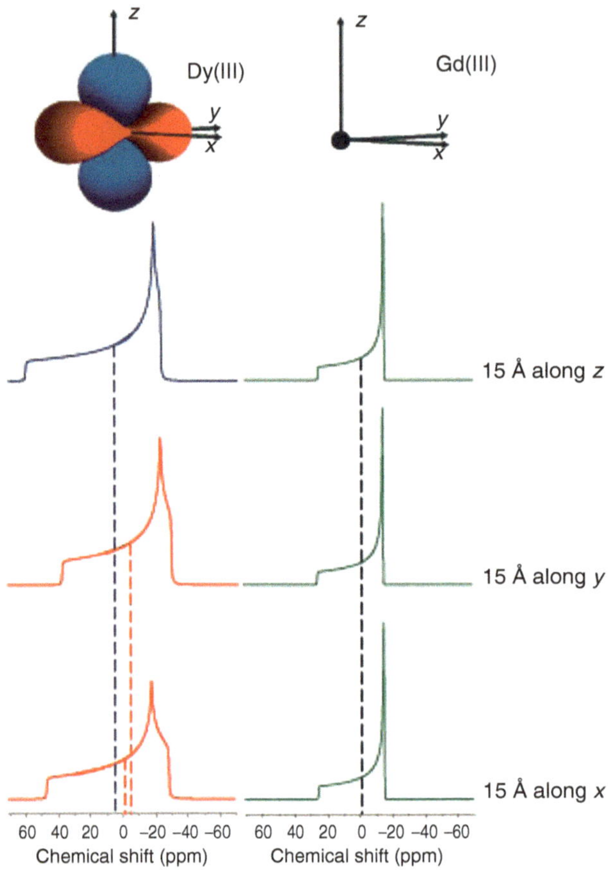

FIGURE 5.6 Powder patterns of dipolar shielding anisotropy on nuclei located at 15 Å distance from the metal center, in the case of Dy(III) and Gd(III). The magnitude of the magnetic susceptibility is calculated with Eq. (2.29), the anisotropy of Dy is simulated on the basis of the experimentally determined values in Ref. [72] and the powder patterns are calculated with Simpson [73,74] using the zcw28656 crystal file, Available from: http://nmr.au.dk/downloads/.

Reproduced from Ref. [75].

As in solution NMR, contact shift is present in solids. The contact shift is proportional to the spin density at the nucleus (in an s-orbital) and, as such, is mainly isotropic. In the presence of anisotropy of $\boldsymbol{g}$ or of the magnetic susceptibility, also the contact shift can be anisotropic. In addition to this, spin density in p, d, or f orbitals can produce a coupling to the nucleus. Such interaction can be described via a second-rank tensor [22,23]:

$$A_{2,0} = \alpha \frac{\mu_0}{4\pi} \hbar g_e \mu_B \gamma_I \left\langle \frac{1}{r_{orb}^3} \right\rangle \rho_{orb} \tag{5.8}$$

that has the same transformation properties with respect to space operations as the Curie-spin interaction, and thus cannot be separated. In this equation, α and r_{orb} are specific for each molecule and reflect the averaging of the electron position over its wavefunction. ρ_{orb} is the fraction of spin density in a given orbital/orbital set.

The two contributions can be separated only in case the parameters relevant to the Curie-spin interaction can be obtained from data related to nuclei at large distance from the metal (i.e., from pseudocontact shift on further nuclei, as also described in Section 9.3) or if the crystal field parameters that determine the $\Delta\chi$ values are known.

5.2.1.2 Paramagnetic relaxation in solids

Paramagnetic centers cause relaxation in solids as well as in liquids. Modulation of the interactions is only possible through electron relaxation, as the chemical exchange and reorientational motions are largely abolished and exert negligible modulation of the local field generated by the electrons. In this picture, the main contributions to relaxation are dipolar (Section 4.4) and contact (Section 4.5) mechanisms.

The dipolar (or Solomon) mechanism is represented by the dipolar interaction between electronic and nuclear spins, which randomly fluctuates about a null average, modulated mainly by electron relaxation. The contact (or Bloembergen) mechanism occurs as the contact interaction between the electronic and nuclear spins is also randomly fluctuating. As already mentioned the other mechanism available in solution (i.e., the Curie-spin interaction, Section 4.6) does not give rise to relaxation in solids, as it is only modulated by exchange and reorientations that are abolished in solids.

As already observed for the solution-state case, the paramagnetic centers can be divided into two classes: on the one side there are NO radicals, and metal ions such as VO^{2+}, Cr^{3+}, Mn^{2+}, Cu^{2+}, and Gd^{3+} possessing isotropic (or nearly isotropic) magnetic susceptibility and sizable electron spin relaxation times that mainly affect the surrounding nuclei by dipolar relaxation; on the other side there are the metal ions with fast relaxing electron spins, resulting in smaller perturbation of the nuclear spin relaxation. In particular, since Curie-spin relaxation is absent, the lanthanoid ions possessing large magnetic susceptibilities are less experimentally prohibitive.

5.2.1.3 Paramagnetic-assisted condensed data collection

It is worthy to recall some key features of the NMR spectra that are directly related to the presence of paramagnetic centers both in the solid state and in solution and consider their importance in the case of solid-state NMR: (1) relaxation is in general faster and (2) the paramagnetic interactions dramatically change the resonance frequencies of the nuclei. The effect of (2) is double:

1. the peaks may be so far shifted that the broadening due to inefficient decoupling does not affect the resolution (for ^{13}C nuclei);
2. shift of the signals reduces the homonuclear coupling drastically, leading to a breakup of the ^{1}H dipolar network. This, for instance, allows for direct ^{1}H detection and also mitigates the need for high-power decoupling for nuclei with low magnetogyric ratio γ, such as ^{13}C and ^{15}N.

Low-power decoupling (or no decoupling at all) and/or ^{1}H detection, combined with the fast recycling allowed for by the enhanced relaxation, gives a natural increase of the signal-to-noise per unit time in the study of (almost) any paramagnetic system.

5.2.2 PROTON-ENHANCED SPECTROSCOPY

As already discussed, while in liquids direct magnetic dipole–dipole interactions between nuclear spins are averaged to zero due to rapid rotational diffusion, no such averaging motions exist in solids with the natural consequence that nuclear dipole–dipole coupling remains to broaden the spectra by thousands of kHz. Thus, ever since the pioneering studies on solids, methods have been sought to mitigate or erase such broadening.

Heteronuclear decoupling can be accomplished by MAS and by tailored radiofrequency decoupling. In the case of low γ nuclei, homonuclear decoupling naturally occurs either because of dilution of the spins, or it results from the increased separation of the energy levels for increased static field whereas the dipolar coupling remains the same, or because of both things at the same time. MAS contributes to remove the remaining couplings. As suggested before, in the case of paramagnetic systems also hyperfine shifts concur in reducing the overlap between the energy levels. For these reasons low γ nuclei are still preferentially detected in the solid state.

The removal of homonuclear dipole–dipole broadening comes at the price of reduced sensitivity, and often in prohibitively long longitudinal relaxation times. For these reasons, that are in common with solution NMR too, it is common to use high γ nuclei as a source of polarization.

By far, the most used approach to low γ nuclear polarization is proton-enhanced nuclear induction spectroscopy, now indicated as cross-polarization, introduced by Waugh and coworkers [24]. The process that leads to cross-polarization can be depicted, following the thermodynamic treatment described in reference [25], as follows: the aim is to transfer energy from one oscillator that has a lower spin temperature (the high γ nucleus) to another that has a higher spin temperature (the low γ nucleus). It is clear that each oscillator has its own characteristic frequency, i.e., the Larmor frequency at the selected field. Transfer of energy between the two oscillators would cause a change in the total energy of the system and is thus not allowed. As we have already seen in Section 1.7.3, a spin-locking field causes the spins to precess at some frequency around the resonant field. Now, intuitively, the power of the two spin-locking fields on I and S spins could be chosen so as to have the same precession frequency for both kind of spins, i.e., the two oscillators have now the same characteristic frequency. This condition is known as Hartmann-Hahn [26] matching:

$$\omega_I = \omega_S. \tag{5.9}$$

If there is a coupling active between the two spins, for instance the dipolar coupling, any flip-flop transition is energy conserving, and a spin-thermal contact will be established: the energy will be transferred from one oscillator to the other [27].

Under MAS, the coupling becomes time-dependent, thus the effectiveness of magnetization transfer between the two spin baths also becomes time-dependent. Vega and coworkers have given a quantitative explanation of the spin dynamics [28]. The complete description is based on the available spin states in a system with a single low γ spin coupled to N abundant, high γ spins all coupled between them, and is beyond the scope of this book. A simple picture can be derived anyway: the time-dependent homonuclear HH dipole–dipole coupling breaks up each manifold with total spin M for the abundant species in a set of n levels separated by ω_R. Each manifold can be associated to the low-abundance spin in its α or β state, yielding two manifolds separated by ω_{1S}. Of course the manifold at $M+1$ and $M-1$ are separated from M by ω_{1I}. Now, Eq. (5.4) suggests that the heteronuclear interaction between I and S will couple the nth rotational level with total I-spin equal to M and the S-spin in α state to the $n+k$th state with total I spin equal to $M-1$ and S-spin in the β state, with $k = -2, -1, 1, 2$. From this it follows that the original Hartmann-Hahn matching condition can be relaxed so as to fulfill the following relationship:

$$\omega_I = \omega_S \pm k\omega_R \tag{5.10}$$

The effect of the dependency on the frequency of MAS is more evident at higher spinning rates, where all the matching conditions can be more effectively separated (Fig. 5.7) [29]. The result of

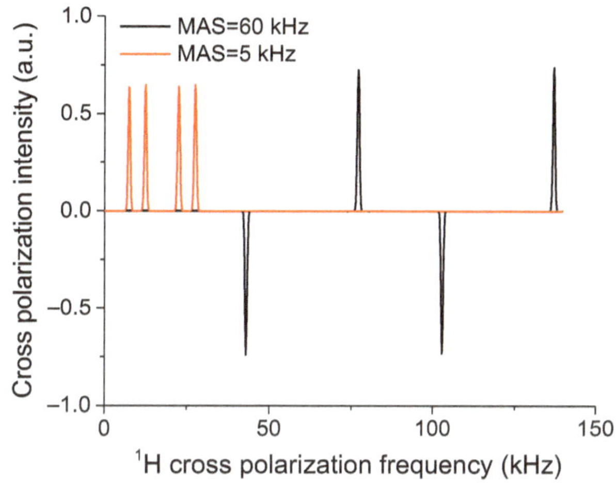

FIGURE 5.7 Separation of matching conditions on passing from 5 kHz to 60 kHz magic angle spinning, simulated using Simpson.

Adapted from Refs. [73,74].

Eq. (5.10) is derived in the frame of perturbation theory of the homonuclear coupling between the abundant spin I, i.e., for very high spinning frequencies. When the interaction between the I spins is strong, the requirement for k to be -2, -1, 1, or 2 is lifted.

It is to be noted that cross-polarization can be effective between any pair of dipole–dipole coupled nuclei, i.e., it does not require the ^{13}C nucleus to be bound to a ^{1}H. Furthermore during the time course of the cross-polarization experiment, the magnetization is usually spread to nearby nuclei by spin diffusion. The outcome of this is that magnetization is spread through all the X-nucleus spectrum, regardless of the degree of substitution of the specific nucleus. It is still important to recall that the rates and efficiencies of cross-polarization still depend on the degree of substitution, especially at high field and high spinning rates, and selective transfers can still be achieved under peculiar matching conditions [29,30]. In the earlier discussion of cross-polarization in both spinning and static solids, we have assumed that cross-polarization proceeds to full equilibrium between the I and S spin reservoirs, omitting any consideration on the time dependence of the process. Intuitively, the time-limiting process will be relaxation of the magnetization to equilibrium, occurring during the spin-locking. We have seen (Section 1.7.3) that this corresponds to $T_{1\rho}$ relaxation. The increase of the cross-polarization efficiency is, instead, related to a time constant sometimes referred to as T_{IS}, which is specific for specific functional groups. For a rigid lattice, T_{IS} is calculated to be inversely proportional to the extent of the HX interaction and directly proportional to the square root of the interactions in the HH dipolar network: cross-polarization builds up faster, the stronger the HX interactions and the weaker the HH interactions [25,31,32]. As a side remark, the strength of the dipolar network is smaller when large hyperfine shifts are present, as it is the case of paramagnetic compounds, thus the matching conditions become narrower, but the CP efficiency might even be increased.

Still, one final consideration is required. As mentioned above, the overall efficiency of cross polarization relies on broadband spin-locking and critically depends on the lifetime of the spin-locked magnetization [33]. Thus, when paramagnetic relaxation is extremely fast and the signal is spread over hundreds of kHz, CP might be substituted by approaches based on transferred-echo-double-resonance

(TEDOR) or dipolar-INEPT or variations thereof (Chapter 12) [33–35], although CP is the method of choice for starting any session of solid-state NMR data acquisition.

5.2.3 MEASUREMENT OF PARAMAGNETIC INTERACTIONS AND THEIR ANISOTROPY

The experimental problem can be separated into two "components," the first being the large spectral width to be covered, the second being the expected relaxation properties of the system (Chapters 7 and 12). It is in principle rather easy to obtain a meaningful set of experiments on a paramagnetic system, either via excitation and detection of ^{1}H nuclei, proton-enhanced heteronuclear experiments on ligand nuclei or on metal countercations (^{23}Na), and even detection of the NMR-active metal center (^{51}V, ^{59}Co). Whereas the hyperfine interactions tend to spread the isotropic signal over several hundreds of part per million and the dipolar part tends to broaden the signals again over several hundreds of part per million, even in the absence of a net isotropic shift, still paramagnetic relaxation in solids may be more forgiving, and large shifts help in reducing nuclear–nuclear interactions. In the case of many heavy nuclei, large paramagnetic shifts and anisotropies are also compounded with large chemical shielding anisotropies and/or with large quadrupolar patterns. In addition, there might be several nuclei sensing different environments, raising the need of distinguishing the patterns of the different sites. Such paramagnetic systems with large anisotropies at several sites can be studied in different ways, which will be again discussed and exemplified in Chapters 7 and 12. Among those, it is instructive to consider at this point the following approaches: (1) single crystal spectroscopy, (2) isotropic/anisotropic separation, and (3) single-crystallite selective excitation. These experimental approaches are especially well-suited for all those systems which possess a very large dispersion of the hyperfine shifts but mild relaxation effects, i.e., the spectra show very well dispersed and sharp resonances. These properties tend to be related to one another (Chapter 7).

Single crystals—A single crystal spectrum shows a sharp line (or multiplet) for all the sets of equivalent nuclei in a molecule, for each molecular orientation in the unit cell of the crystal. All the signals of course appear together in the spectrum, and thus need to be well resolved from one another (Section 5.1). Upon rotation of the crystal (usually about each of the three crystallographic axes), the spatial dependence of the signals is revealed, as long as the signal can be followed through the angular range (Figs. 5.1 and 5.2). The angular dependence of the position and multiplicity of each resonance allows for the description of all the spatial-dependent components of the Hamiltonian [36]. The main drawbacks of the method are the need of goniometric probeheads plus, since the lines need to be well resolved, there is a need for large, high-purity crystals of the compounds of interest.

Isotropic/anisotropic separation—As opposed to the single-crystal spectroscopy, this method and the following apply to powdered samples, where all the frequencies appear simultaneously. The basic idea of isotropic/anisotropic separation is to phase-modulate the spinning sidebands in the indirect dimension so that they cancel out in the 2D processing [37,38] (Section 12.1.4). For this reason, each component of the spectrum can be separated into its own spinning-sideband pattern, allowing for recovery, by a relatively simple 2D experiment, of the isotropic information and the spin interactions that are encoded in the powder pattern. The main drawback of this method is that the full powder pattern must be excited and refocused over the timespan of one or more rotor periods, requiring high-power irradiation, low fields, or both. Pintacuda and coworkers have developed a version of this experiment that relies on short, high-power adiabatic pulses that have been shown to refocus efficiently very wide spectral widths [39]. Also transverse relaxation may strongly affect the outcome of the experiment, as a significant fraction of the magnetization can be lost during the indirect detection period.

Single-crystallite excitation—At variance with the previous methodology, no high-power pulses are required in this case, and there is a significant advantage in using high fields [40,41]. This method relies on the use of trains of phase-modulated selective pulses centered at the spinning sidebands. These pulses have the overall effect of causing efficient excitation of different subsets of crystallites over the whole powder pattern, as different crystallites contribute differently to the intensity of each sideband. In this case the drawback is that each selective pulse is rather long (>1 ms) and, when relaxation starts occurring already at the time of the pulse, the efficiency of the experiment is reduced: this imposes a lower limit on the $T_{1\rho}$ of the nuclei to be detected (Section 1.7.3).

5.2.4 DESCRIPTION OF CRYSTAL LATTICES (COMPLEXES AND PROTEINS)

The packing of molecules in crystals makes it possible for each nucleus to interact with several paramagnetic centers, also from neighboring molecules, experiencing a significant intermolecular contribution (Fig. 5.8). To a good approximation, each of the surrounding paramagnetic centers contributes independently to the local field on a given nucleus, and the total electron–nucleus dipole–dipole interactions are given by the sum over all the crystal neighbors (Section 9.8).

Protons are an attractive target for paramagnetic crystals, because they are 100% abundant, have (almost) the highest possible magnetogyric ratio γ, and their CSA is usually negligible, making it easier to detect the dipolar shielding anisotropy with minor distortion. In the case of small paramagnetic complexes, Kervern and coworkers [42] developed a protocol for crystal structure determination via the combined use of pseudocontact shifts and of the determination of full dipolar shielding anisotropy tensor, (if the space group and crystal lattice parameters are known):

1. create maps at a given confidence level, which sample the metal position in the cell, the orientation and anisotropy of the χ tensor;
2. generate a molecular model inside each of the maps, and position the ligands;
3. score the structures against the experimental data.

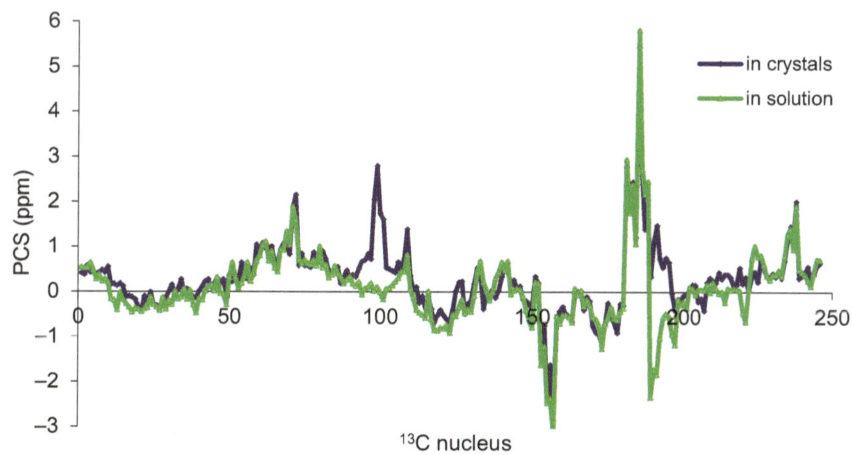

FIGURE 5.8 Comparison of the ^{13}C pseudocontact shifts observed for solid and liquid Co(II)-bound catalytic domain of MMP12 (Chapter 12).

In the case of biomolecules, in particular of proteins, pseudocontact shifts have a further added value: to measure a pseudocontact shift only requires to know the assignment of a single nucleus; for this reason the pseudocontact shifts are immediately available for all assigned NMR signals, while many distance restraints are initially ambiguous, and can be introduced only after a first emerging structural family is calculated and used to resolve ambiguities (Chapter 9) [43]. Since pseudocontact shifts have both an intramolecular and an intermolecular contribution (Fig. 5.8), a structural refinement that takes onto account both contributions will yield at the same time: (1) the structure of the protein and (2) the crystal packing, once the cell parameters are available from powder X-ray diffraction spectra.

5.3 DYNAMIC NUCLEAR POLARIZATION IN THE SOLID STATE

NMR is a fairly insensitive technique. Low γ nuclei can receive magnetization from higher γ nuclei, as brilliantly demonstrated since the 1960s by the Waugh's and Freeman's groups in solids and liquids respectively [24,44]. Thanks to the brilliant intuition of Overhauser and the experimental application by Carver and Slichter in the 1950s [45,46], we know that magnetization can be transferred to nuclei also from the spin with the highest magnetogyric ratio: the electron. This is known as dynamic nuclear polarization (DNP, Chapter 4). In order to work, DNP requires the following: (1) one unpaired electron (polarizing agent), (2) one nucleus coupled to the unpaired electron and (3) microwave irradiation of the EPR spectrum to drive transitions (see later). The electron spin needs to be slowly relaxing, so that the EPR transitions can be efficiently saturated, and the EPR line must be either much smaller than the nuclear Larmor frequency (Section 5.3.1) or slightly smaller than twice the nuclear Larmor frequency (Section 5.3.1) [47]. Polarizing agents are thus usually stable organic free radicals, such as amine-oxyls or tri-arylmethyl radicals, as these molecules tend to have long relaxation times and sharp EPR resonances (Section 4.12). The use of some gadolinium(III), manganese(II) [48], and chromium(III) [49] with very limited g and D anisotropies has been reported and it is foreseen that their use might become more relevant at higher field given the fact that their linewidth will not increase (sizeably) with field, and relaxation will be reduced.

For the theoretical description of DNP in the solid state, two mechanisms are generally invoked: a two spin process known as solid effect (SE, Section 5.3.1) [50] and a three spin process known as cross effect (CE, Section 5.3.2) [51]. Recently, a third mechanism, the Overhauser effect, strictly analogous to what observed in solution and in conducting solids, was shown to be present also in the solid state [52]. This latter mechanism is described in Section 4.15.

The different mechanisms have different frequency dependencies (sometimes referred to as DNP spectrum). As such, the analysis of the frequency-dependent DNP spectrum provides information on which mechanism prevails under different conditions. The full theoretical treatment of the frequency dependence of the DNP spectrum requires a very profound quantomechanical treatment and, at the time of writing, is at the frontier of the DNP research. In this context, the elegant treatment developed by Vega and coworkers provides an almost pictorial, yet physically sound, access to the underlying phenomena and will be used to guide the reader through this section [53,54].

The main idea behind the simplest formulation of the model is that each mechanism provides a specific lineshape of the DNP spectrum, and each spin packet in the EPR line simply dictates the intensity of the shape around a given frequency, i.e., the DNP spectrum is the convolution of the mechanism-specific shape with the EPR lineshape.

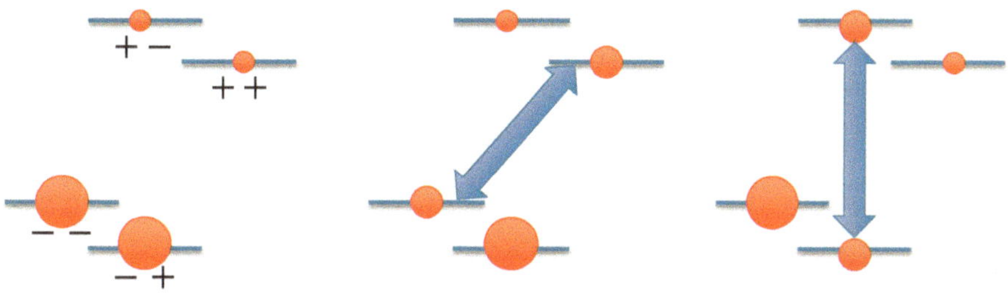

FIGURE 5.9 Energy levels of an electron-nucleus spin system involved in solid-effect DNP. The different size of the red circles represents the different population of the levels.

5.3.1 SOLID EFFECT

Solid effect is the occurrence of a forbidden transition (either zero- or double-quantum transition) in an electron-nucleus system, according to the scheme reported in Fig. 5.9.

This means that the DNP efficiency is highest at two frequencies:

$$\omega_{MW} = \omega_e \pm \omega_n \tag{5.11}$$

When the double quantum transition is irradiated, the enhancement is positive (Fig. 5.9), and when the zero-quantum transition is irradiated, it is negative.

A simplified view suggests that the typical lineshape of SE comprises of two peaks, with opposite sign and separated by $2\omega_n$ (Fig. 5.10A), to be convoluted with the EPR lineshape of the radical (Fig. 5.10B) [53].

This also means that there can be DNP intensity outside the EPR spectrum. Since SE is the irradiation of a forbidden transition, it requires high power microwave irradiation to occur efficiently.

The forbidden transition can be driven because the electron and nuclear states are not pure: the states are mixed by the pseudosecular part of the hyperfine interaction. This same physical picture is also at the basis of the ESEEM experiment as described by Mims [55].

Let us assume that the system can be described by the following Hamiltonian [Eq. (4.14)]:

$$H = \hbar(\omega_S S_z - \omega_I I_z) + \boldsymbol{I} \cdot \boldsymbol{A} \cdot \boldsymbol{S} \tag{5.12}$$

where $\boldsymbol{A}$ is the hyperfine coupling tensor expressed in the laboratory frame. For the sake of simplicity let us also assume that g_e is isotropic, that the nucleus does not experience any isotropic contact interaction, and that the dipole–dipole interaction can be approximated by a point–dipole interaction. The dipole–dipole interaction can be written in the laboratory frame as

$$H_{\text{dip}} = \boldsymbol{I} \cdot \boldsymbol{A} \cdot \boldsymbol{S} \approx S_z(A_{zx} I_x + A_{zy} I_y + A_{zz} I_z) \tag{5.13}$$

where the terms in S_x and S_y have been dropped as they precess at too high frequency to be sensed by the nuclei. Eq. (5.13) is then transformed by rotation of $\varphi = \arctan\left(\frac{A_{zy}}{A_{zx}}\right)$ of the laboratory frame around the z-axis to yield

$$H_{\text{dip}} = S_z(BI_x + AI_z), \tag{5.14}$$

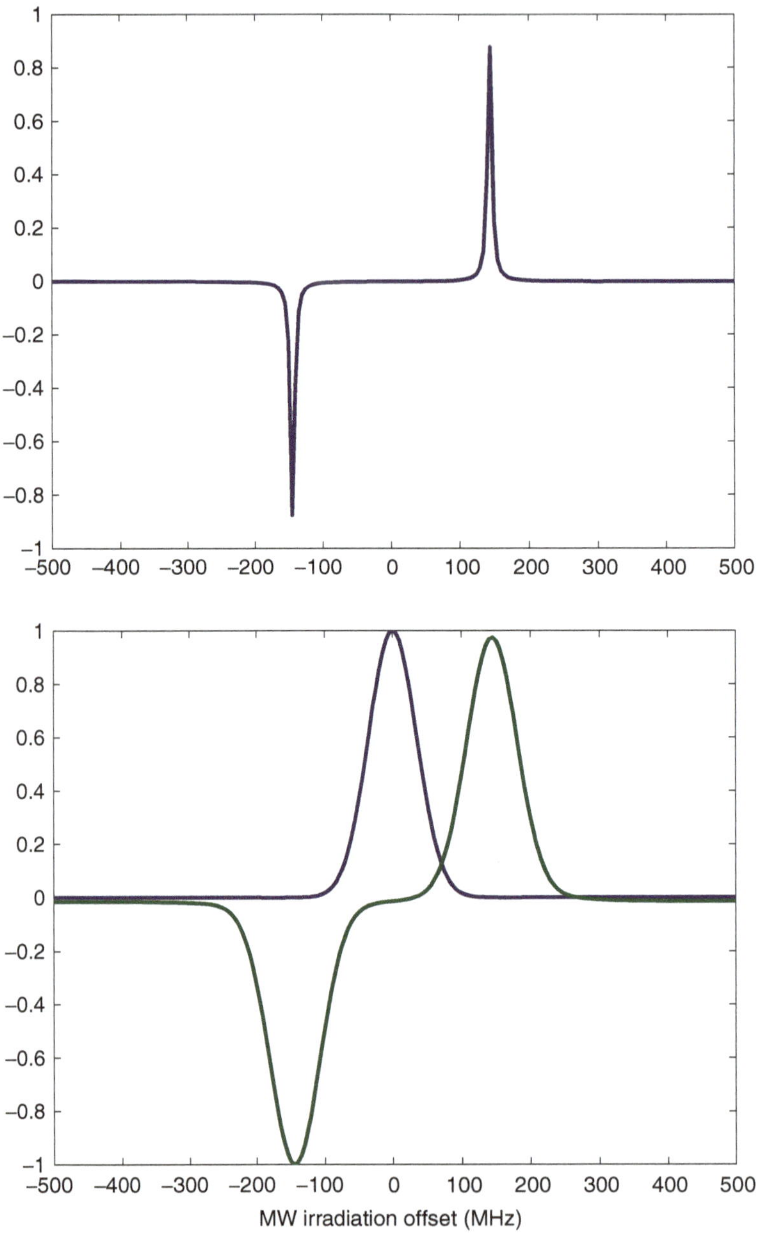

FIGURE 5.10 Characteristic lineshape (DNP spectrum) of the SE and result of its convolution with the EPR lineshape, calculated according to Ref. [53]. The zero point corresponds to matching of the center of the EPR line. The line is calculated with T_{1n} = 1000 s, T_{1e} = 500 ms, T_{2e}= 600 µs, ω_n = 144 MHz, B = 3 MHz and the microwave field is assumed to have 500 kHz nutation frequency.

where $A = A_{zz}$ and $B = \sqrt{A_{zx}^2 + A_{zy}^2}$. The presence of the anisotropic term in I_x causes the nuclear spin not to be quantized along the magnetic field, thus the nuclear spin Hamiltonian in the laboratory frame has nonnegligible off-diagonal components:

$$H_{\text{dip}} = \begin{pmatrix} \frac{\omega_S}{2} - \frac{\omega_I}{2} + \frac{A}{4} & \frac{B}{4} & 0 & 0 \\ \frac{B}{4} & \frac{\omega_S}{2} + \frac{\omega_I}{2} - \frac{A}{4} & 0 & 0 \\ 0 & 0 & -\frac{\omega_S}{2} - \frac{\omega_I}{2} - \frac{A}{4} & -\frac{B}{4} \\ 0 & 0 & -\frac{B}{4} & -\frac{\omega_S}{2} + \frac{\omega_I}{2} + \frac{A}{4} \end{pmatrix} \tag{5.15}$$

Diagonalization of the matrix yields the following eigenvectors:

$$|1\rangle = \cos\frac{\eta_\alpha}{2}|+,-\rangle + \sin\frac{\eta_\alpha}{2}|+,+\rangle$$

$$|2\rangle = \cos\frac{\eta_\alpha}{2}|+,+\rangle + \sin\frac{\eta_\alpha}{2}|+,-\rangle$$

$$|3\rangle = \cos\frac{\eta_\beta}{2}|-,-\rangle + \sin\frac{\eta_\beta}{2}|-,+\rangle$$

$$|4\rangle = \cos\frac{\eta_\beta}{2}|-,+\rangle + \sin\frac{\eta_\beta}{2}|-,-\rangle$$

Where $\eta_\alpha = \arctan\frac{-B}{A + 2\omega_n}$ and $\eta_\beta = \arctan\frac{-B}{A - 2\omega_n}$ are called "branching angles" and indicate the tilt of the quantization axis for I with respect to the z-axis, whether the electron is in α or β. It has been derived that "forbidden" two spin transitions are indeed allowed with $\sin^2\frac{\eta_\alpha - \eta_b}{2}$ probability, i.e., with a probability that decreases for increased magnetic field. As far as the transition frequencies are concerned, neglecting the effect of B on the transition frequencies, one gets $\omega_S \pm \frac{A}{2}$ as frequency for the allowed transitions (1–3 and 2–4) and $\omega_S \pm \frac{A}{2} \mp \omega_I$ for the "forbidden" transitions.

5.3.2 CROSS EFFECT IN STATIC AND ROTATING SOLIDS, COMPARISON OF DNP EFFICIENCIES AND RADICAL DESIGN

Cross effect is the transfer of polarization from a pair of coupled electrons to their surrounding nuclei, by irradiation of an allowed transition of the electron-electron-nucleus system. It occurs when one spin packet of one electron spectrum is irradiated and has another spin packet at ω_I distance in frequency, and the two electrons are dipole–dipole coupled, as per the scheme described in Fig. 5.11: the polarization transfer is maximized when the energy of the level with different M_S for the two electrons (for instance, +1/2, − 1/2) and with M_I = +1/2 is equal to the energy of the level

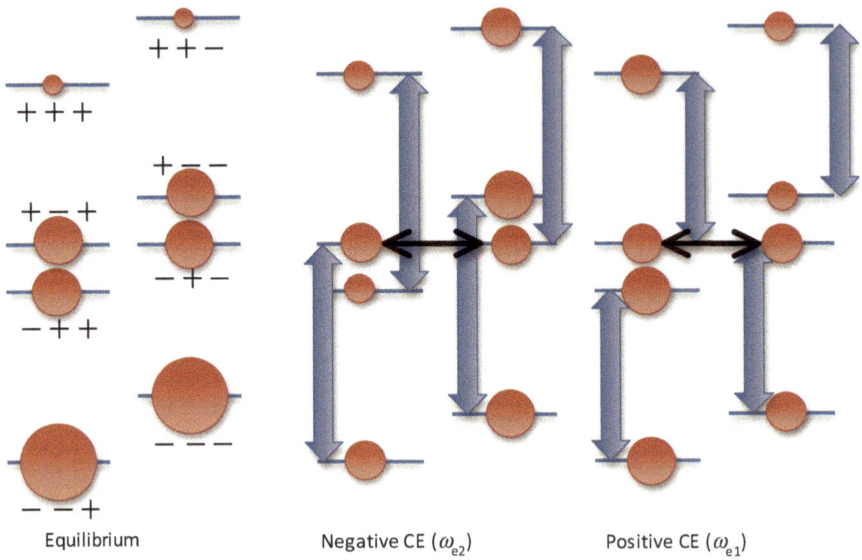

FIGURE 5.11 Energy levels of an electron-electron-nucleus spin system involved in cross-effect DNP. The different size of the red circles represents the different population of the levels.

with exchanged M_S for the two electrons (for instance, −1/2, + 1/2) and with $M_I = -1/2$. Therefore, the matching condition is that

$$|\omega_{S1} - \omega_{S2}| = \omega_I, \tag{5.16}$$

where ω_{S1} and ω_{S2} are the Larmor frequencies of the dipolar coupled electrons S_1 and S_2.

The corresponding lineshape is described in Fig. 5.12A. Finally, convolution of such lineshape with the EPR spectrum of the radical again yields the characteristic DNP spectrum (Fig. 5.12B) [54]. A complete treatment includes the contribution of off-resonance saturation effects of the EPR spectrum [56–58], MW irradiation does not need to directly saturate one of the transitions of the CE electron pairs, but it is enough that it saturates the transitions of other electrons which then indirectly depolarize the CE electron pairs via electron spectral diffusion (eSD) [57]. The qualitative picture of the process does not change.

There is another view of DNP which is based on thermodynamic arguments, somewhat similarly to what is described for the H-X contact during cross-polarization: for DNP it is assumed that the microwave field, together with a coupling within the electron and nuclear thermal baths allows for establishment of a common spin temperature. This mechanism is referred to as thermal mixing (TM) [59–61]. Of note, the quantum modeling of CE, including the application of a large microwave power allows for a substantial cooling of the nuclear and dipolar energy bath, with some heating of the electron Zeeman bath, along the lines of the TM mixing description. The possibility of reproducing such distribution of the energy levels somewhat lifts the need for the introduction of the TM model [62]. In practice, any experimental DNP spectrum can be deconvoluted in the SE and the CE contributions, and the interplay

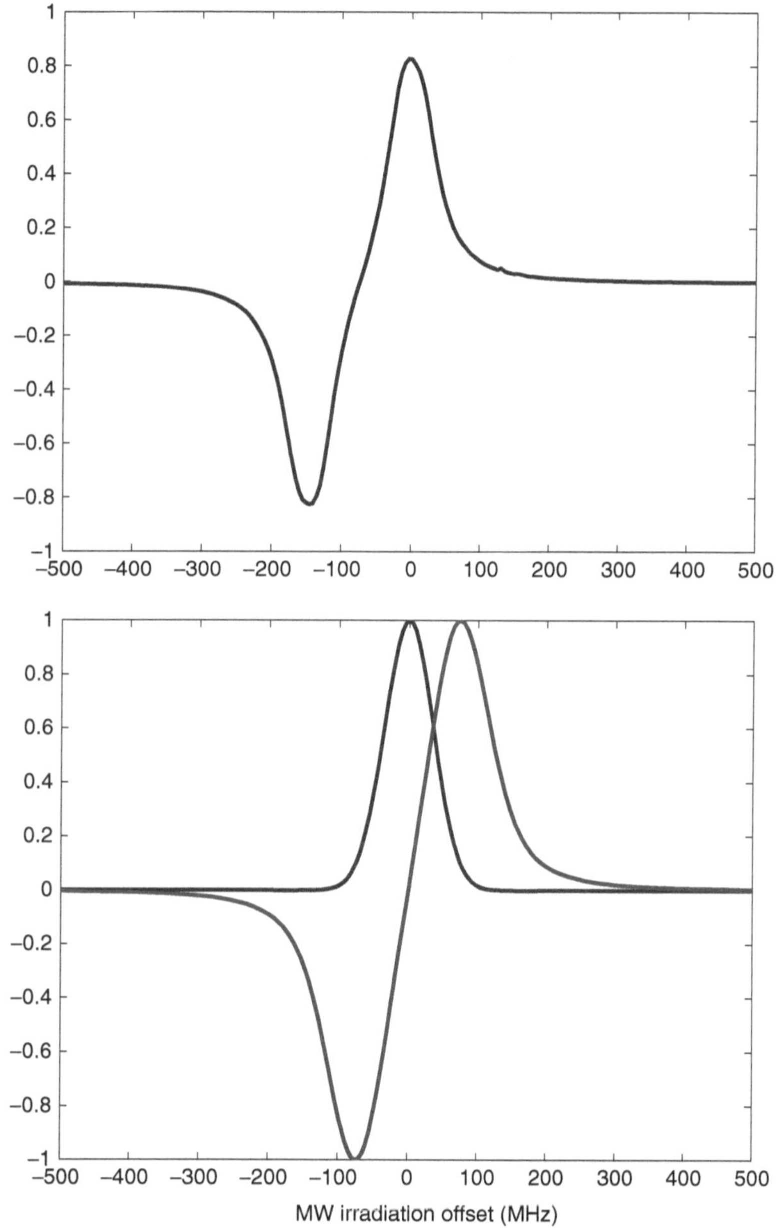

FIGURE 5.12 Characteristic lineshape (DNP spectrum) of the CE and result of its convolution with the EPR lineshape, calculated according to Ref. [54]. The zero point corresponds to matching of the center of the EPR line. The line is calculated with T_{1n} = 1000 s, T_{2e} = 500 ms, T_{2e} = 600 μs, ω_n = 144 MHz, B = 3 MHz, D (electron-electron dipolar coupling) = 12 MHz and the microwave field is assumed to have 500 kHz nutation frequency.

of SE and CE is an interesting tool for the understanding of the ongoing phenomena: at low radical concentration, CE is impaired by the low number of couples at the right distance and right orientation; on the other hand, SE relies on the irradiation of a forbidden transition and, as such, requires higher power than CE. This is also apparent in the intrinsic linewidths of the two mechanisms, which are sizably different: the intrinsic linewidth represents the transition probability as a function of the MW irradiation offset, and for SE it is scaled down by a factor $\sin^2\frac{\eta_\alpha-\eta_b}{2}\approx\sin^2\frac{B}{2\omega_n}$. When B is 3 MHz and ω_n = 144 MHz the scaling factor is about 10^{-4}.

This said, we can summarize which process is dominating at under different experimental conditions:

1. Dielectric solid, liquid nitrogen temperature, high concentration of radical, the radical EPR spectrum is wider than the ω_n frequency—CE dominates, SE is present.
2. Dielectric solid, liquid temperature, low concentration of radical, the radical EPR spectrum is wider than the ω_n frequency—SE dominates, CE is present.
3. Dielectric solid, liquid temperature, low concentration of radical, the radical EPR spectrum is narrower than the ω_n frequency—SE dominates, CE is absent.

In rotating solids, the matching of the CE occurs several times over a rotor period. When two energy levels are coming closer. However, levels do not cross (i.e., there is an adiabatic flip-flop or flip-flop-flip transition), because of perturbation from the electron-electron coupling, the hyperfine interaction and the microwave irradiation [63,64]. We should consider the following situations: (1) avoided crossing of one electron with microwave irradiation, (2) three spin avoided crossing, (3) electron-electron avoided crossing. The effect of (1) is that when the electron's resonant frequency crosses the microwave irradiation, a change in electron population may occur, provided the microwave irradiation is fast enough to cause a transition when the electron is crossing the exact frequency. The effect of (2) is the cross-effect itself. The effect of (3) is that of equalizing the electron population and, since cross-effect relies on the difference in population between electron spin packets separated by ω_n, this is detrimental for DNP.

The discussion earlier on can assist in deriving some rules to be followed for the design of polarizing agents. The first important factor to be considered is the field at which the measurement will be carried out, which determines the linewidth of the radical that should be used. The second key determinant is the temperature, which determines the electron relaxation times: lower electron relaxation rates are requested to obtain saturation and to achieve a substantial population difference between different spin packets of the EPR spectrum. It has been proposed that a delocalization of the low-frequency vibrational motions over a large number of atoms rather than on the atoms where the electron mostly resides is beneficial for increasing relaxation times [65], this has been translated into the "rule of thumb" that the electron relaxation times increase with increased radical molecular weight [66]. Of course, fast local fluctuations of the magnetic field as provided, for instance, by methyl group rotation concur to decrease the electron relaxation times [67], so either methyl groups are substituted by bulkier moieties or deuteration is applied [68]. The last key determinant is the radical distribution through the sample [69]. For CE DNP, the radicals need to come close to one another, but with the appropriate orientation, otherwise they only do affect each other relaxation properties. This is particularly true for CE under MAS, which is decreased when nonproductive electron-electron avoided crossings occur. So, in order to obtain a larger set of active electron-electron couples, one better rely on chemical grafting of two or more radical moieties on the same molecule rather than trusting on the stochastic distribution of the radical in solution. This has led to the development of the so-called biradicals, which are the most commonly used chemicals for high field DNP [68,70,71].

REFERENCES

[1] Andrew ER, Bradbury A, Eades RG. Nature 1958;182:1659.
[2] Andrew ER, Bradbury A, Eades RG. Nature 1959;183:1802.
[3] Lowe IJ. Phys. Rev. Lett. 1959;2:285.
[4] Matti Mariq M, Waugh JS. J. Chem. Phys. 1979;70:3300.
[5] Solid-State NMR Spectroscopy Principles and ApplicationsOxford: Blackwell Science; 2002.
[6] Sun BQ, Pines A. Appl. Magn. Reson. 1993;5:43.
[7] Leskes M, Thakur RS, Madhu PK, Kurur ND, Vega S. J. Chem. Phys. 2007;127:024501.
[8] Leskes M, Madhu PK, Vega S. Prog. Nucl. Magn. Reson. Spectrosc. 2010;57:345.
[9] Kotecha M, Wickramasinghe NP, Ishii Y. Magn. Reson. Chem. 2007;45(Suppl. 1):S221–30.
[10] Wickramasinghe NP, Parthasarathy S, Jones CR, Bhardwaj C, Long F, Kotecha M, Mehboob S, Fung LWM, Past J, Samoson A, Ishii Y. Nat. Methods 2009;6:215.
[11] Takegoshi K, Nakamura S, Terao T. Chem. Phys. Lett. 2001;344:631.
[12] Sodickson DK, Levitt MH, Vega S, Griffin RG. J. Chem. Phys. 1993;98:6742.
[13] Levitt MH. In: Encyclopedia of Magnetic Resonance, Grant DM, Harris RK, editors. Chichester 165, 2002.
[14] Grandinetti PJ, Ash JT, Trease NM. Prog. Nucl. Magn. Reson. Spectrosc. 2011;59:121.
[15] Abragam A. The Principles of Nuclear MagnetismOxford: Oxford University Press; 1961.
[16] Gueron M. J. Magn. Reson. 1975;19:58.
[17] Vega AJ, Fiat D. Mol. Phys. 1976;31:347.
[18] Haeberlen U, Waugh JS. Phys. Rev. 1969;185:420.
[19] Kervern G, Steuernagel S, Engelke F, Pintacuda G, Emsley L. J. Am. Chem. Soc. 2007;129:14118.
[20] Cucinotta G, Perfetti M, Luzon J, Etienne M, Car PE, Caneschi A, Calvez G, Bernot K, Sessoli R. Angew. Chem. Int. Edit. 2012;51:1606.
[21] Boulon ME, Cucinotta G, Luzon J, Degl'Innocenti C, Perfetti M, Bernot K, Calvez G, Caneschi A, Sessoli R. Angew. Chem. 2013;125:368.
[22] Morton JR, Preston KF. J. Magn. Reson. 1978;30:577.
[23] Heise H, Kohler FH, Xie X. J. Magn. Reson. 2001;150:198.
[24] Pines A, Gibby MG, Waugh JS. J. Chem. Phys. 1972;56:1776.
[25] Pines A, Gibby MG, Waugh JS. J. Chem. Phys. 1973;59:569.
[26] Hartmann SR, Hahn EL. Phys. Rev. 1962;128:2042.
[27] Marx R, Glaser SJ. J. Magn. Reson. 2003;164:338.
[28] Marks D, Vega S. J. Magn. Reson. A 1996;118:157.
[29] Laage S, Marchetti A, Sein J, Pierattelli R, Sass HJ, Grzesiek S, Lesage A, Pintacuda G, Emsley L. J. Am. Chem. Soc. 2008;130:17216.
[30] Demers JP, Vijayan V, Becker S, Lange A. J. Magn. Reson. 2010;205:216.
[31] Kolodziejski W, Klinowski J. Chem. Rev. 2002;102:613.
[32] Van Vleck JH. Phys. Rev. 1948;74:1168.
[33] Kervern G, Pintacuda G, Emsley L. Chem. Phys. Lett. 2007;435:157.
[34] Wickramasinghe NP, Ishii Y. J. Magn. Reson. 2006;181:233.
[35] Swamy SKK, Karczmarska A, Makowska-Janusik M, Kassiba A, Dittmer J. Chem. Phys. Chem. 2013;14:1864.

[36] Vosegaard T, Hald E, Langer V, Skov HJ, Daugaard P, Bildsoe H, Jakobsen HJ. J. Magn. Reson. 1998;135:126.
[37] Hung I, Gan Z. J. Magn. Reson. 2010;204:150.
[38] Walder BJ, Dey KK, Kaseman DC, Baltisberger JH, Grandinetti PJ. J. Chem. Phys. 2013;138:174203.
[39] Pell AJ, Clement RJ, Grey CP, Emsley L, Pintacuda G. J. Chem. Phys. 2013;138:114201.
[40] Pell AJ, Kervern G, Emsley L, Deschamps M, Massiot D, Grandinetti PJ, Pintacuda G. J. Chem. Phys. 2011;134:024117.
[41] Pell AJ, Pintacuda G, Emsley L. J. Chem. Phys. 2011;135:144201.
[42] Kervern G, D'Aléo A, Toupet L, Maury O, Emsley L, Pintacuda G. Angew. Chem. Int. Ed. Engl. 2009;48:3082.
[43] Luchinat C, Parigi G, Ravera E, Rinaldelli M. J. Am. Chem. Soc. 2012;134:5006.
[44] Morris GA, Freeman R. J. Am. Chem. Soc. 1979;101:760.
[45] Overhauser AW. Phys. Rev. 1953;89:689.
[46] Carver TR, Slichter CP. Phys. Rev. 1953;92:212.
[47] Ni QZ, Daviso E, Can TV, Markhasin E, Jawla SK, Swager TM, Temkin RJ, Herzfeld J, Griffin RG. Acc. Chem. Res. 2013;46:1933.
[48] Corzilius B, Smith AA, Barnes AB, Luchinat C, Bertini I, Griffin RG. J. Am. Chem. Soc. 2011;133:5648.
[49] Corzilius B, Michaelis VK, Penzel S, Ravera E, Smith A, Luchinat C, Griffin RG. J. Am. Chem. Soc. 2014;136:11716.
[50] Jeffries JD. Phys. Rev. 1957;106:164.
[51] Hwang CF, Hill DA. Phys. Rev. Lett. 1967;18:110.
[52] Can TV, Caporini MA, Mentink-Vigier F, Corzilius B, Walish JJ, Rosay M, Maas WE, Baldus M, Vega S, Swager TM, Griffin RG. J. Chem. Phys. 2014;141:064202.
[53] Hovav Y, Feintuch A, Vega S. J. Magn. Reson. 2010;207:176.
[54] Hovav Y, Feintuch A, Vega S. J. Magn. Reson. 2012;214:29.
[55] Mims WB. Phys. Rev. B 1972;5:2409.
[56] Hovav Y, Shimon D, Kaminker I, Feintuch A, Goldfarb D, Vega S. Phys. Chem. Chem. Phys. 2015;17:6053.
[57] Hovav Y, Kaminker I, Shimon D, Feintuch A, Goldfarb D, Vega S. Phys. Chem. Chem. Phys. 2015;17:226.
[58] Ravera E, Shimon D, Feintuch A, Goldfarb D, Vega S, Flori A, Menichetti L, Luchinat C, Parigi G. Phys. Chem. Chem. Phys. 2015;17:26969.
[59] Provotorov BN. Sov. Phys. JETP 1962;14:1126.
[60] Abragam A, Borghini MIn: Gorter CJ, editor. Progress in Low Temperature Physics, vol. 4. Amsterdam: North-Holland Publishing Company; 1964. p. 384.
[61] Borghini M. Phys. Rev. Lett. 1968;20:419.
[62] Hovav Y, Feintuch A, Vega S. Phys. Chem. Chem. Phys. 2013;15:188.
[63] Mentink-Vigier F, Akbey U, Hovav Y, Vega S, Oschkinat H, Feintuch A. J. Magn. Reson. 2012;224:13.
[64] Thurber KR, Tycko R. J. Chem. Phys. 2012;137:084508.
[65] Sato H, Kathirvelu V, Fielding A, Blinco JP, Micallef AS, Bottle SE, Eaton SS, Eaton GR. Mol. Phys. 2007;105:2137.
[66] Sauvee C, Rosay M, Casano G, Aussenac F, Weber RT, Ouari O, Tordo P. Angew. Chem. Int. Ed. 2013;52:10858.
[67] Zecevic A, Eaton GR, Eaton SS, Lindgren M. Mol. Phys. 1998;95:1255.
[68] Kubicki DJ, Casano G, Schwarzwälder M, Abel S, Sauvee C, Ganesan K, Yulikov M, Rossini AJ, Jeschke G, Coperet C, Lesage A, Tordo P, Ouari O, Emsley L. Chem. Sci. 2016;7:550.
[69] Banerjee D, Shimon D, Feintuch A, Vega S, Goldfarb D. J. Magn. Reson. 2013;230:212.
[70] Mathies G, Caporini MA, Michaelis VK, Liu Y, Hu KN, Mance D, Zweier JL, Rosay M, Baldus M, Griffin RG. Angew. Chem. Int. Ed. Engl. 2015;54:11770.
[71] Song C, Hu KN, Joo CG, Swager TM, Griffin RG. J. Am. Chem. Soc. 2006;128:11385.
[72] Bertini I, Janik MBL, Lee Y-M, Luchinat C, Rosato A. J. Am. Chem. Soc. 2001;123:4181.
[73] Bak M, Rasmussen JT, Nielsen NC. J. Magn. Reson. 2000;147:296.
[74] Tosner Z, Andersen R, Stevensson B, Eden M, Nielsen NC, Vosegaard T. J. Magn. Reson. 2014;246:79.
[75] Bhaumik A, Luchinat C, Parigi G, Ravera E, Rinaldelli M. Cryst. Eng. Comm. 2013;15:8639.

CHAPTER

CHEMICAL EXCHANGE, CHEMICAL EQUILIBRIA, AND DYNAMICS

6

Chemical exchange is capable of affecting the NMR parameters both in terms of shifts and relaxation. The matter is complex and the various detailed treatments may be laborious. Here the exchange of a nucleus between two sites is presented both qualitatively and quantitatively. The exchange time can be measured through saturation transfer experiments. Some thermodynamic parameters are obtained from the NMR parameters by varying the concentration of the species at equilibrium. In this chapter, the relaxation due to outer-sphere interactions is introduced as well as the chemical shifts due to bulk paramagnetic effects, which are used to measure the paramagnetism of molecules in solution.

6.1 INTRODUCTION

If a chemical species containing the investigated nucleus undergoes chemical exchange among different chemical environments, the nuclear NMR parameters are affected in several ways. A full treatment of chemical exchange is rather cumbersome, both from the points of view of shift and of relaxation properties, and is beyond the scope of this book. It is worth noting that even in the simple case of exchange between two sites, the general solution implies nonexponentiality of relaxation. Fortunately, there are some limiting cases which are much easier to treat and for which simple equations can be given. These limiting cases also often hold when one of the chemical environments is paramagnetic.

In general, in the presence of chemical exchange, the nuclei of interest will spend a fraction of time residing in each of n different chemical environments. The probability P_i of finding a chosen nucleus in a particular site i, will be equal to the molar fraction f_i of nuclei in that environment:

$$P_i = f_i = \frac{N_i}{\sum_{j=1}^{n} N_j} \tag{6.1}$$

where N_i is the number of nuclei in the ith site, $\Sigma_j N_j$ is the total number of nuclei in the sample, and $\Sigma_i P_i = 1$ by definition.

Here, the only approximation is to neglect the time spent by a nucleus *in moving* from one environment to another. For example, consider a 0.1 mM water solution of a $Ni(CH_3NH_2)_6^{2+}$ complex, in the presence of 0.1 M excess CH_3NH_2, assuming that the complex is fully formed under these conditions. If there is chemical exchange between bound and free methylamine, the methyl protons will spend

$$P_{bound} = \frac{6\times10^{-4}}{10^{-1}+6\times10^{-4}} \approx 6\times10^{-3}$$

NMR of Paramagnetic Molecules. http://dx.doi.org/10.1016/B978-0-444-63436-8.00006-5

of the time in the nickel-bound state and

$$P_{\text{free}} = \frac{10^{-1}}{10^{-1} + 6\times10^{-4}} \approx 1$$

of the time in the bulk solution.

The NH_2 protons can also exchange with water molecules. The probability of finding such a proton in each of the three environments will be

$$P_{\text{bound}} = \frac{6\times10^{-4}}{55.5 + 10^{-1} + 6\times10^{-4}} \approx 1.1\times10^{-5}$$

$$P_{\text{free}} = \frac{10^{-1}}{55.5 + 10^{-1} + 6\times10^{-4}} \approx 1.8\times10^{-3}$$

$$P_{\text{H}_2\text{O}} = \frac{55.5}{55.5 + 10^{-1} + 6\times10^{-4}} \approx 1$$

where 55.5 is the approximate molarity of water. Therefore, the exchangeable protons will only spend about 1/100,000 (i.e., P_{bound}) of their time in a bound position.

The probabilities, or molar fractions, are equal to the fraction of time a nucleus will spend in a particular environment when observed for a suitably long period of time. They do not reveal *how long* the nucleus will stay, on average, in each particular position; in other words, the average lifetime of the nucleus in each environment cannot be calculated from the molar fractions. As will be seen, the chemical lifetime or its reciprocal, the chemical dissociation rate, are crucial parameters governing the influence of chemical exchange on NMR parameters.

6.2 A PICTORIAL VIEW OF CHEMICAL EXCHANGE

Consider, for simplicity, the exchange of a nucleus with spin I between two sites, A and B:

$$I^{\text{A}} \underset{\tau_{\text{B}}^{-1}}{\overset{\tau_{\text{A}}^{-1}}{\rightleftharpoons}} I^{\text{B}}$$

Although the dissociation rate constants τ_{A}^{-1} and τ_{B}^{-1} are not known, it is known that, at equilibrium, $\tau_{\text{A}}/\tau_{\text{B}} = P_{\text{A}}/P_{\text{B}}$. We will give a full description of the effect of variation of τ_{A} (and τ_{B}) on the nuclear NMR parameters and describe the relevant equations involved [1,2].

The nuclei in the A site will resonate at a Larmor precession frequency ω_{A}, and those in the B site will resonate at a frequency ω_{B}. Under very slow chemical exchange conditions, two signals will be observed, centered at frequency ω_{A} and ω_{B}, with intensities proportional to P_{A} and P_{B} respectively. When $P_{\text{A}} = P_{\text{B}}$, the two signals will be of equal intensity and $\tau_{\text{A}}^{-1} = \tau_{\text{B}}^{-1} = \tau_M^{-1}$ [Fig. 6.1A(1)]. If $R_{2\text{A}} = R_{2\text{B}}$ the two signals have the same linewidth, given by $R_{2\text{A}}/\pi = R_{2\text{B}}/\pi$. We will also assume that $R_{1\text{A}} = R_{1\text{B}}$. When the exchange rate increases, i.e., the lifetime in each chemical environment decreases, the uncertainty principle states that the uncertainty in the energy of the nucleus in each of the two environments must increase. The energy uncertainty is reflected in an increase of signal linewidth. The effect becomes appreciable when τ_M becomes comparable with, or shorter than $T_{2\text{A}}$, $T_{2\text{B}}$. Then the effective T_2 for the signals becomes close to τ_M [Fig. 6.1A(2–4)]. In contrast, T_1 is only affected by the efficiency of coupling of the spin system with the lattice in order to undergo spin transitions. Therefore, if these mechanisms are equal in each environment ($R_{1A} = R_{1B}$), longitudinal relaxation will not be affected by exchange between the two sites.

When the exchange rate is further increased to the point where it is about equal to the difference $\Delta\omega$ in resonance frequency of the two environments, the linewidth is also of the same magnitude. Therefore, the two signals are as broad as their separation and a single broad resonance is observed, extending from ω_A to ω_B [Fig. 6.1A(5)].

When the exchange rate becomes faster than $\Delta\omega$, then the single resonance, centered at $(\omega_A + \omega_B)/2$, becomes sharper and sharper. From the NMR point of view, the nuclei are now experiencing an average environment and as the exchange time decreases further, the linewidth eventually returns to the value of the separate signals in the absence of exchange [Fig. 6.1A(6–9)].

It has been shown that by increasing τ_M^{-1} the resonance frequency of each signal passes from ω_A or ω_B to the average value $(\omega_A + \omega_B)/2$. This transition is not smooth, and the two signals begin to collapse only for τ_M^{-1} values very close to $\Delta\omega$.

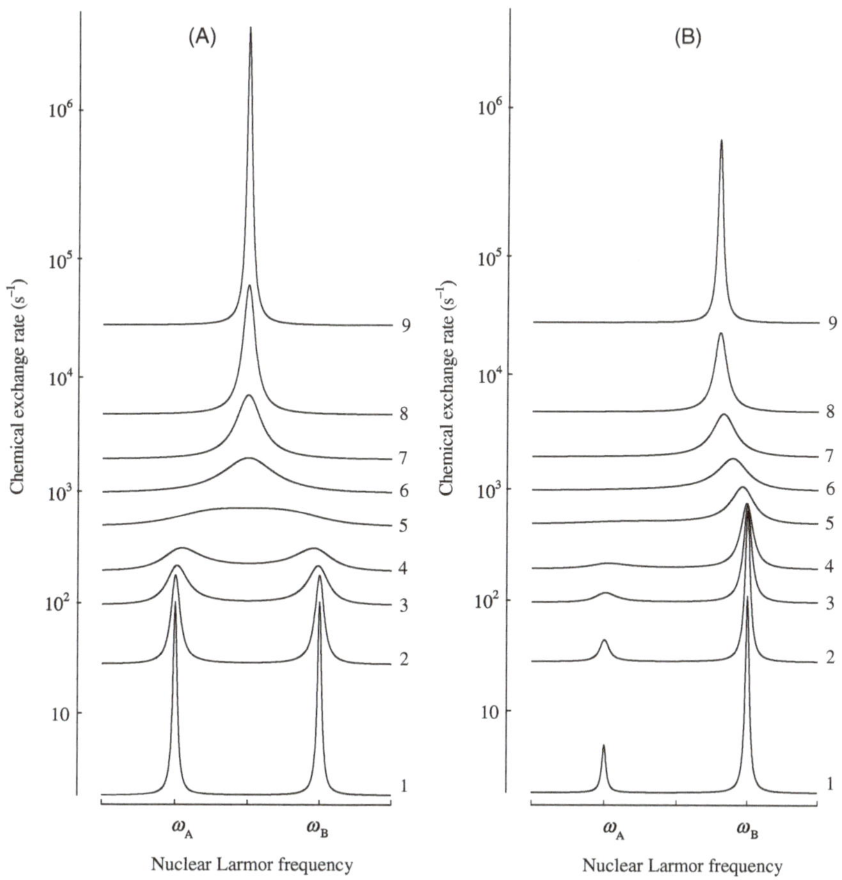

FIGURE 6.1 Calculated NMR spectra of a system constituted by a nuclear species in chemical exchange between two sites, A and B, as a function of the chemical exchange rate, τ_M^{-1}. The chemical shift separation between A and B signals is $\Delta\omega = 1000$ rad s^{-1}. The various situations are: (A) $R_{2A} = R_{2B}$, $P_A = P_B$; (B) $R_{2A} = R_{2B}$, $P_A < P_B$.

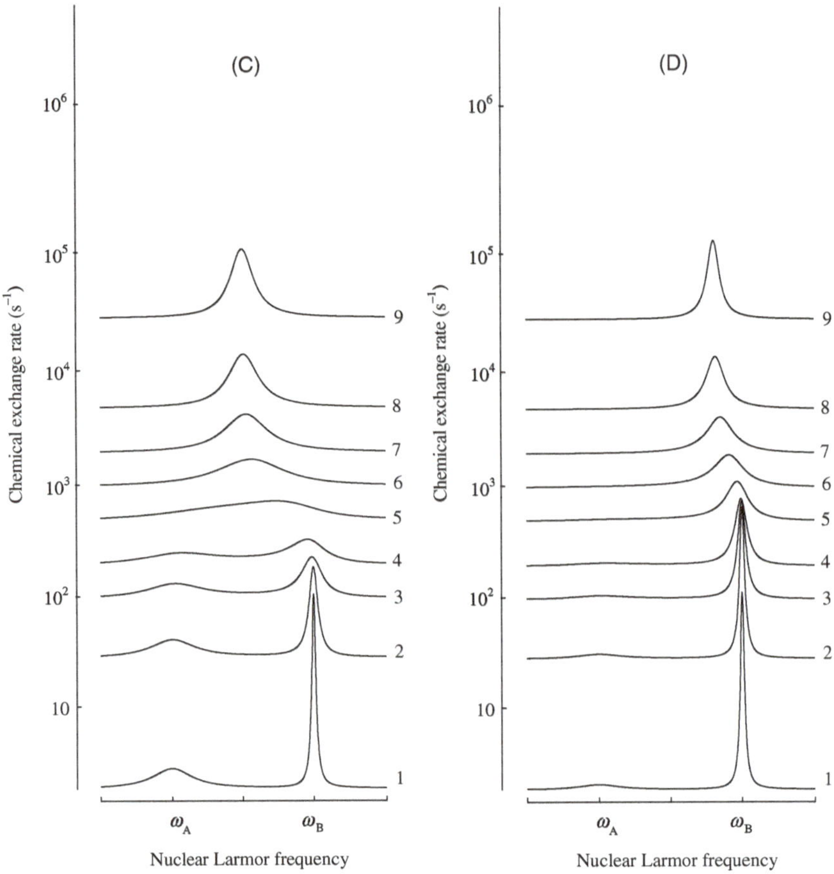

FIGURE 6.1 *(cont.)* (C) $R_{2A} > R_{2B}$, $P_A = P_B$; (D) $R_{2A} > R_{2B}$, $P_A < P_B$; and no exchange (1); slow exchange (2–4); coalescence (5); fast exchange (6–9).

The situation $\tau_M^{-1} = \Delta\omega/\sqrt{2}$ [Fig. 6.1A(5)] represents signal coalescence, i.e., the borderline between the so-called slow- and fast-exchange regions. Thus, slow- or fast-exchange rates can only be defined with respect to the difference in Larmor frequency $\Delta\omega$, and therefore are dependent on the strength of the external magnetic field and on the nuclear species involved. For instance, exchange between two sites for a given nuclear species can be fast at 200 MHz and slow at 800 MHz; likewise, for a given field, exchange may be fast for deuterons and slow for protons in the same chemical environment.

6.3 NMR PARAMETERS IN THE PRESENCE OF EXCHANGE

6.3.1 EXACT SOLUTIONS FOR TWO-SITE EXCHANGE

Having introduced the concept of exchange in a pictorial way, we treat now the more general case of exchange between two sites of different population and different intrinsic relaxation times. Under such conditions relaxation is not a single exponential but rather a double exponential process, where the individual rate constants are given by Eqs. (6.2) and (6.3). The chemical shifts are given by Eq. (6.4) [3,4]:

$$R_{1\text{exch}} = A_1 \pm \left[A_1^2 - \left(\tau_{1\text{A}}^{-1}\tau_{1\text{B}}^{-1} - \tau_\text{A}^{-1}\tau_\text{B}^{-1}\right)\right]^{1/2} \tag{6.2}$$

$$R_{2\text{exch}} = A_2 \pm \left[\frac{G + (G^2 + H^2)^{1/2}}{2}\right]^{1/2} \tag{6.3}$$

$$\omega_\text{exch} = \frac{\omega_\text{A} + \omega_\text{B}}{2} \pm \left[\frac{-G + (G^2 + H^2)^{1/2}}{2}\right]^{1/2} \tag{6.4}$$

where

$$\begin{aligned}
A_1 &= \frac{\tau_{1\text{A}}^{-1} + \tau_{1\text{B}}^{-1}}{2} \\
A_2 &= \frac{\tau_{2\text{A}}^{-1} + \tau_{2\text{B}}^{-1}}{2} \\
G &= \frac{\left(\tau_{2\text{A}}^{-1} - \tau_{2\text{B}}^{-1}\right)^2}{4} + \tau_\text{A}^{-1}\tau_\text{B}^{-1} - \frac{\Delta\omega^2}{4} \\
H &= \frac{\tau_{2\text{A}}^{-1} - \tau_{2\text{B}}^{-1}}{4}\Delta\omega \\
\tau_{1\text{A}}^{-1} &= R_{1\text{A}} + \tau_\text{A}^{-1} \\
\tau_{1\text{B}}^{-1} &= R_{1\text{B}} + \tau_\text{B}^{-1} \\
\tau_{2\text{A}}^{-1} &= R_{2\text{A}} + \tau_\text{A}^{-1} \\
\tau_{2\text{B}}^{-1} &= R_{2\text{B}} + \tau_\text{B}^{-1} \\
\tau_\text{AB}^{-1} &= \tau_\text{A}^{-1} + \tau_\text{B}^{-1}.
\end{aligned}$$

Fig. 6.1B–D illustrates the situation where the populations are unequal (B and D) and the relaxation times are unequal (C and D). It can be seen that when the populations are different, the average signal sits at the weighted average position and has a linewidth which is the weighted average of the two linewidths. The linewidths of the various intermediate exchange situations can be appreciated in the plots of Fig. 6.1B–D.

In the special case of $\tau_\text{A} = \tau_\text{B} = \tau$, $R_{1\text{A}} = R_{1\text{B}} = R_1$, and $R_{2\text{A}} = R_{2\text{B}} = R_2$, the above quantities reduce to:

$$\begin{aligned}
A_1 &= R_1 + \tau^{-1} \\
A_2 &= R_2 + \tau^{-1} \\
G &= \tau^{-2} - \Delta\omega^2/4 \\
H &= 0
\end{aligned}$$

and Eqs. (6.2)–(6.4) become

$$R_{1\text{exch}} = R_1 = \text{constant} \tag{6.5}$$

$$R_{2\text{exch}} = R_2 + \tau^{-1} \pm \left(\frac{G + |G|}{2}\right)^{1/2} \tag{6.6}$$

$$\omega_\text{exch} = \frac{\omega_\text{A} + \omega_\text{B}}{2} \pm \left(\frac{-G + |G|}{2}\right)^{1/2}. \tag{6.7}$$

Eq. (6.6) accounts for the linewidths of the signals reported in Fig. 6.1A.

6.3.2 EXCHANGE OF EXCESS METAL LIGANDS

When the paramagnetic site is the least populated, and the difference in population is very large, Eqs. (6.2)–(6.4) can be simplified [3–6]. For example, the chemical shift for the signal in the diamagnetic site evolves as follows as a function of τ_M

$$\Delta\omega_{\rm p} = \omega_{\rm p} - \omega_{\rm d} = \frac{f_M}{\tau_M^2}\frac{\Delta\omega_M}{\left(R_{2M}+\tau_M^{-1}\right)^2+\left(\Delta\omega_M\right)^2} \tag{6.8}$$

where $\omega_{\rm p}$ is the experimental shift, $\omega_{\rm d}$ is the chemical shift of the diamagnetic site, and $\Delta\omega_M$ is the difference in chemical shift between the metal site and the diamagnetic site. This paramagnetic contribution to the shift in the metal site is given by the equations described in Section 2.2.[1] R_{2M} is the transverse nuclear relaxation rate enhancement in the paramagnetic site (Sections 4.4–4.6). The behavior is illustrated in Fig. 6.2. For $\tau_M^{-1} \gg R_{2M}, \Delta\omega_M$, Eq. (6.8) becomes

$$\Delta\omega_{\rm p} = f_M\Delta\omega_M \tag{6.9}$$

i.e., the signal sits at the weighted average position between $\omega_{\rm d}$ and ω_M.

In analogy with the chemical shift, the relaxation rates of the bulk nuclei will be termed $R_{1\rm d}$ and $R_{2\rm d}$. In the presence of chemical exchange with nuclei in the paramagnetic site, such rates will be enhanced by an amount $R_{1\rm p}$ or $R_{2\rm p}$, and the measured values will be

$$R_{1\,\rm meas} = R_{1\rm d} + R_{1\rm p} \quad R_{2\,\rm meas} = R_{2\rm d} + R_{2\rm p}$$

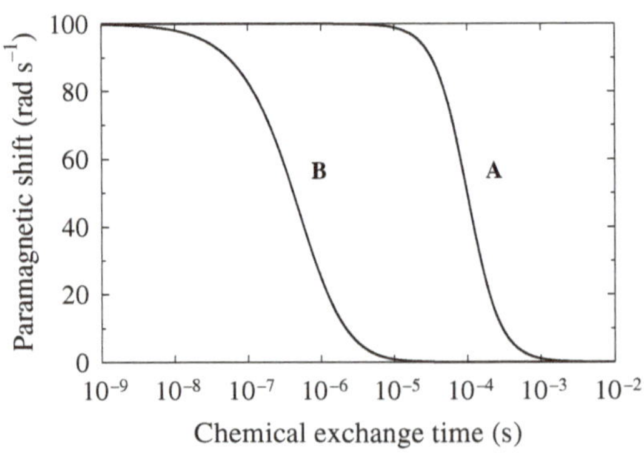

FIGURE 6.2 Paramagnetic contribution to the chemical shift $\Delta\omega_{\rm p}$ as a function of the exchange rate of the nucleus from the paramagnetic site τ_M. Conditions: $\Delta\omega_M = 10^4$ rad s^{-1}, $f_M = 10^{-2}$, $R_{2M} = 10^2$ (A) or 10^6 (B) s^{-1}.

[1]Note that the equations in Section 2.2 report the paramagnetic shifts ($\delta^{\rm con}$, $\delta^{\rm pcs}$, etc.) in ppm. To obtain $\Delta\omega_M$ (in units of rad s^{-1}) one should multiply the shift in ppm by the nuclear Larmor frequency (in MHz), and further multiply by 2π. For example, a proton experiencing $\delta^{\rm con} = 3.75$ ppm yields $\Delta\omega_M = 3.75 \times 500 \times 2\pi = 11{,}775$ rad s^{-1} when measured in a 500 MHz spectrometer, whereas a deuteron in the same chemical environment yields $\Delta\omega_M = 3.75 \times 76.75 \times 2\pi = 1807.5$ rad s^{-1}, where 76.75 MHz is the deuterium Larmor frequency in a magnetic field corresponding to a 500 MHz proton Larmor frequency.

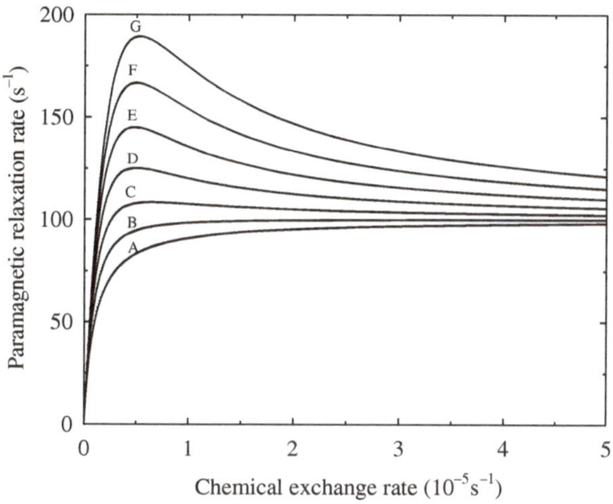

FIGURE 6.3 Paramagnetic contributions to the nuclear relaxation rates R_{1p} (A) and R_{2p} (A–G) as a function of the exchange rate of the nucleus τ_M^{-1} from the paramagnetic site. Conditions: $R_{1M} = R_{2M} = 10^4$ s^{-1}, $f_M = 10^{-2}$, $\Delta\omega_M = 0$ ((A) coincident with R_{1p}), 1×10^4 (B), 1.5×10^4 (C), 2×10^4 (D), 2.5×10^4 (E), 3×10^4 (F), 3.5×10^4 (G) rad s^{-1}.

The longitudinal relaxation rate enhancement R_{1p} goes asymptotically from zero to $f_M\, R_{1M}$ with increasing τ_M^{-1} (Fig. 6.3A), according to the following equation:

$$R_{1p} = f_M R_{1M} \frac{\tau_M^{-1}}{\left(R_{1M} + \tau_M^{-1}\right)^2} = f_M (T_{1M} + \tau_M)^{-1} \tag{6.10}$$

where R_{1M} is the nuclear longitudinal relaxation enhancement in the paramagnetic site (Sections 4.4–4.6). τ_M thus acts as a limiting factor for the propagation of the effect from the paramagnetic site to the bulk nuclei. In other words, in the absence of exchange there is no paramagnetic contribution to the relaxation rate of the bulk nuclei, whereas in the fast-exchange limit ($\tau_M^{-1} \ll R_{1M}$) the measured relaxation rate is the weighted average between the diamagnetic R_{1d} and the paramagnetic R_{1M} rates:

$$R_{1\ \mathrm{meas}} = R_{1d} + f_M R_{1M}$$

where the molar fraction of bound nuclei f_M is the weighting factor. It should be remembered that, as in the case of the paramagnetic shift, R_{1M} can be very large compared to R_{1d}, so that even for small f_M values a noticeable relaxation rate enhancement may be measured.

The transverse relaxation rate enhancement R_{2p} could be treated in the same way if it were not for the difference in chemical shift $\Delta\omega_M$ between the paramagnetic and the diamagnetic species. As has already been shown, the difference in chemical shift causes a line broadening—an increasing in R_{2M}—even when the exchange is between two diamagnetic sites. Furthermore, the effect is expected to be much larger in this case since much larger chemical shift differences are involved. The general formula relating R_{2p} to $\Delta\omega_M$ and τ_M^{-1} of the bulk nuclei is [5,6]

$$R_{2p} = \frac{f_M}{\tau_M} \frac{R_{2M}^2 + R_{2M}\tau_M^{-1} + (\Delta\omega_M)^2}{\left(R_{2M} + \tau_M^{-1}\right)^2 + (\Delta\omega_M)^2}. \tag{6.11}$$

Fig. 6.3 shows the variation of R_{2p} as a function of τ_M^{-1} for different values of $\Delta\omega_M$.

Note that in the absence of exchange ($\tau_M^{-1} = 0$), $R_{2p} = 0$, and for very fast exchange ($\tau_M^{-1} \gg R_{2M}$, $(\Delta\omega_M)^2/R_{2M}$), $R_{2p} = f_M R_{2M}$, as found for the longitudinal relaxation rate. If $\tau_M^{-1} \gg R_{2M}$, R_{1M} and $\Delta\omega_M$, $R_{2p} = f_M(R_{2M} + (\Delta\omega_M)^2\tau_M)$ and $R_{1p} = f_M R_{1M}$; thus, if $R_{2M} - R_{1M} \ll (\Delta\omega_M)^2\tau_M$, $R_{2p} - R_{1p} = f_M(\Delta\omega_M)^2\tau_M$. The latter expression can be used to determine τ_M from the values of the experimentally available R_{1p}, R_{2p}, and $\Delta\omega_M$ [7].

In the intermediate region, where τ_M^{-1} is comparable with $\Delta\omega_M$, R_{2p} may be larger than R_{1p} even if $R_{2M} = R_{1M}$ because of exchange broadening effects.

If little or no isotropic shift is present ($\Delta\omega_M \ll R_{2M}$), Eq. (6.11) reduces to

$$R_{2p} = f_M R_{2M} \frac{\tau_M^{-1}}{R_{2M} + \tau_M^{-1}} = f_M (T_{2M} + \tau_M)^{-1} \tag{6.12}$$

which has the same form as Eq. (6.10) for R_{1p}.

In the very fast exchange regions, the observed paramagnetic effects on the chemical shift $\Delta\omega_p$ and the relaxation rates R_{1p}, R_{2p}, are simply proportional to the full paramagnetic effects in the metal binding site, as summarized:

$$\Delta\omega_p = f_M \Delta\omega_M \quad R_{1p} = f_M R_{1M} \quad R_{2p} = f_M R_{2M}. \tag{6.13}$$

Therefore, if the molar fraction of bound ligand f_M is known, the full paramagnetic effects can be calculated. Of course, if no exchange takes place, the ligand in the bulk solution is completely insensitive to the presence of the paramagnetic metal ion.

In the intermediate situation the measured parameters are complicated functions of the full effects experienced in the paramagnetic site ($\Delta\omega_M$, R_{1M}, R_{2M}) and the exchange rate τ_M^{-1}. The τ_M^{-1} dependence of such functions has been summarized in Figs. 6.2–6.4. There are, however, particular regions in which the measured parameters are simpler functions of τ_M^{-1}. These regions correspond to cases in which Eqs. (6.8), (6.10), and (6.11) can be approximated by

$$\textit{Case I}: \Delta\omega_p = \frac{f_M}{\tau_M^2 \Delta\omega_M} \quad \left(\Delta\omega_M \gg R_{2M}, \tau_M^{-1}\right). \tag{6.14}$$

Case I occurs when the difference in shift between the paramagnetic and the diamagnetic site, in angular frequency units, is large compared with both transverse relaxation and the exchange rate. In this case the diamagnetic line is only slightly shifted, and the shift is inversely proportional to the paramagnetic shift.

$$\textit{Case II}: \Delta\omega_p = \frac{f_M \Delta\omega_M}{\tau_M^2 R_{2M}} \quad \left(R_{2M} \gg \tau_M^{-1}, \Delta\omega_M\right). \tag{6.15}$$

Case II occurs when the transverse relaxation rate is faster than both exchange rate and paramagnetic shift. Here the observed shift is directly proportional to the paramagnetic shift.

$$\textit{Case III}: R_{1p} = f_M \tau_M^{-1} \quad \left(R_{1M} \gg \tau_M^{-1}\right) \tag{6.16}$$

$$\textit{Case IV}: R_{2p} = f_M \tau_M^{-1} \quad \left(R_{2M} \gg \tau_M^{-1}\right). \tag{6.17}$$

Cases III and IV represent the slow exchange limits for longitudinal (III) and transverse (IV) relaxation enhancements. The observed enhancements are proportional to the exchange rate, independently of the values of R_{1M} and R_{2M}.

$$Case\,V: R_{2\mathrm{p}} = f_M\,(\Delta\omega_M)^2\,\tau_M \quad (\tau_M^{-1} \gg (\Delta\omega_M)^2\,\tau_M \gg R_{2M}). \tag{6.18}$$

Case V is an intermediate exchange case that can be encountered for transverse relaxation. Here, the dependence on τ_M is opposite to that in Case IV. Case V occurs in the presence of relatively large paramagnetic shift, as long as $(\Delta\omega_M)^2$ is small compared to τ_M^{-2} but large compared to $R_{2M}\tau_M^{-1}$.

In Cases III and IV, $R_{1\mathrm{p}}$ and $R_{2\mathrm{p}}$ are a direct measure of the exchange rate τ_M^{-1}.

Eqs. (6.11)–(6.18) also hold for $R_{1\rho\mathrm{p}}$ by simply substituting $R_{1\rho M}$ for R_{2M}.

6.3.3 TEMPERATURE AND EXCHANGE

The measurement of the exchange time τ_M may provide useful kinetic information on the system. Kinetic parameters for the dissociation process may be obtained by performing relaxation measurements as a function of temperature. If it is assumed that the dissociation of the ligand from the paramagnetic site is a first order kinetic process, the dissociation rate constant τ_M^{-1} is given by the Eyring relationship

$$\tau_M^{-1} = \frac{kT}{h}\exp\left(\frac{-\Delta G^{\neq}}{RT}\right) \tag{6.19}$$

where $\Delta G^{\neq}$ is the free energy of activation for the dissociation process.

In the normal range of temperatures used in NMR experiments, the major source of variation of τ_M^{-1} with temperature is contained in the exponential part. In other words, a plot of log τ_M against $1/T$ (Arrhenius plot) will give a fairly straight line. Given the linear relationship between τ_M^{-1} and the relaxation rates, straight lines will also be obtained by plotting log $R_{1\mathrm{p}}$ or log $R_{2\mathrm{p}}$ (Cases III and IV) against $1/T$ ($\propto$ log τ_M) (Fig. 6.4). Note that a straight line of opposite slope is obtained in Case V.

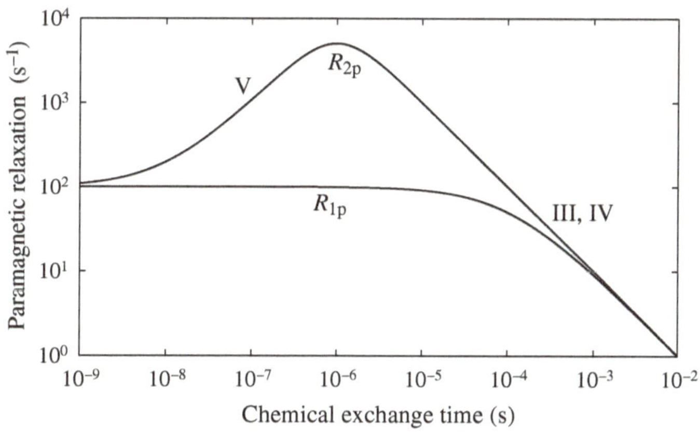

FIGURE 6.4 Linear dependence of $R_{1\mathrm{p}}$ and $R_{2\mathrm{p}}$ (log scale) on τ_M (log scale) in the exchange regions corresponding to Cases III–V. Conditions: $R_{1,2M} = 10^4$ s^{-1}, $f_M = 10^{-2}$, and $\Delta\omega_M = 10^6$ rad s^{-1}.

The appropriate equations can then be found by simply taking the log of expressions III, IV, and V, with τ_M given by Eq. (6.19) (and remembering that $\Delta G = \Delta H - T\Delta S$). From Eqs. (6.16) and (6.17), we thus obtain

$$\log R_{1p} = \log R_{2p} = \log\left(f_M \frac{kT}{h}\right) - \frac{\Delta H^{\neq}}{RT} + \frac{\Delta S^{\neq}}{R} \cong \text{const} - \frac{\Delta H^{\neq}}{RT} \tag{6.20}$$

while from Eq. (6.18) we obtain:

$$\log R_{2p} = \log\left(f_M \Delta\omega_M^2 \frac{h}{kT}\right) + \frac{\Delta H^{\neq}}{RT} - \frac{\Delta S^{\neq}}{R} \cong \text{const}' + \frac{\Delta H^{\neq}}{RT} \tag{6.21}$$

where $\Delta H^{\neq}$ and $\Delta S^{\neq}$ are the enthalpy and entropy of activation for the dissociation process.

Temperature dependent experiments are often performed just to check whether R_{1p} in Eq. (6.10) is in the fast exchange or in the exchange limited region. A strong increase of R_{1p} with temperature is a clear indication of τ_M being the limiting factor.

6.3.4 SATURATION TRANSFER

As already mentioned (Section 1.7.3), when a nuclear spin system experiences fluctuating magnetic fields, relaxation occurs. If these fields are produced by other nuclear spins, there is a reciprocal influence that leads to a complicated picture (Section 12.1.5). However, if these fluctuating magnetic fields are produced by particles which belong to the lattice, i.e., their energy is negligibly affected by nuclear relaxation (e.g., unpaired electrons), the description of the effects is much simpler (Chapter 4) and the return to equilibrium after a perturbation is exponential. Under these circumstances we are going to analyze the effect of chemical exchange.

We have already defined the equilibrium magnetization of a spin I in a given magnetic field B_0 as $M_z(\infty)$, where the (∞) refers to the fact that the sample must have been exposed to the field for time sufficiently long for equilibrium magnetization to be virtually achieved (Section 1.7.3). After any perturbation from equilibrium of the nuclear spin system such that, at time zero after the perturbation, $M_z(0) \neq M_z(\infty)$, the system will tend to return to equilibrium with a simple rate law of the type

$$\frac{\mathrm{d}M_z(t)}{\mathrm{d}t} = -R_1\left[M_z(t) - M_z(\infty)\right] \tag{6.22}$$

which integrates to an exponential magnetization recovery

$$M_z(t) - M_z(\infty) = \left[M_z(0) - M_z(\infty)\right]\exp(-R_1 t) \tag{6.23}$$

where R_1 being the rate constant for the longitudinal relaxation process. In the case of an inversion recovery experiment, $M_z(0) = -M_z(\infty)$, and Eq. (6.23) reduces to

$$M_z(t) = M_z(\infty) - 2M_z(\infty)\exp(-R_1 t) \tag{6.24}$$

which is Eq. (1.53).

The rate equations for a spin system in chemical exchange between two sites

$$I^{A} \underset{k_{-1}}{\overset{k_1}{\rightleftarrows}} I^{B}$$

(where $k_1 = \tau_A^{-1}, k_{-1} = \tau_B^{-1}$), are given by[2]

$$\frac{dM_z^A(t)}{dt} = -R_1^A\left[M_z^A(t) - M_z^A(\infty)\right] - k_1\left[M_z^A(t) - M_z^A(\infty)\right] + k_{-1}\left[M_z^B(t) - M_z^B(\infty)\right] \tag{6.25}$$

$$\frac{dM_z^B(t)}{dt} = -R_1^B\left[M_z^B(t) - M_z^B(\infty)\right] - k_{-1}\left[M_z^B(t) - M_z^B(\infty)\right] + k_1\left[M_z^A(t) - M_z^A(\infty)\right]. \tag{6.26}$$

As anticipated in Section 6.1, solution of these two coupled differential equations gives biexponential behavior, therefore preventing the definition of R_1 for each site (although the two rate constants for the two components are known [Eq. (6.2)]. However, Eqs. (6.25) and (6.26) explicitly contain the rate constants for the exchange process, k_1 and k_{-1}, suggesting that they can be used to obtain information on the exchange dynamics through some kind of perturbation on the equilibrium populations of the spins [8]. Suppose that the signal I^A is saturated by applying a weak r.f. on the A resonance for a reasonably long period of time. Then, the steady state intensity of signal B can be obtained from Eq. (6.26) by setting $dM_z^B(t)/dt$ and $M_z^A(t)$ to zero:

$$0 = -\left(R_1^B + k_{-1}\right)\left[M_z^B(t) - M_z^B(\infty)\right] - k_1 M_z^A(\infty) \tag{6.27}$$

from which we obtain an expression for the fractional change in intensity of signal B upon saturation of A:

$$\frac{M_z^B(t) - M_z^B(\infty)}{M_z^B(\infty)} = \frac{M_z^A(\infty)}{M_z^B(\infty)} \frac{-k_1}{R_1^B + k_{-1}}. \tag{6.28}$$

Therefore, by knowing the longitudinal relaxation rate constant of the nucleus in the B site R_1^B in the absence of exchange and by measuring the fractional change in intensity of signal B (called saturation transfer), the rate constants can be obtained. We recall that the two rate constants are related by the fractional populations of the two sites (Section 6.2), in turn proportional to the equilibrium intensities of the two signals

$$k_1\, M_z^A(\infty) = k_{-1}\, M_z^B(\infty) \tag{6.29}$$

thus, Eq. (6.28) can be further simplified as

$$\frac{M_z^B(t) - M_z^B(\infty)}{M_z^B(\infty)} = \frac{-k_{-1}}{R_1^B + k_{-1}}. \tag{6.30}$$

These equations hold for any population distribution of the two sites. Eq. (6.30) is the equation to be used when a saturation transfer experiment is going to be planned.

[2]The transfer of spin population between sites A and B can be written as magnetization transfer through Eqs. (1.26), (1.27), and (1.33).

From the functional form of Eq. (6.30) it is easy to predict the behavior of saturation transfer as a function of the exchange rate. When the rate constant for the B → A transformation (k_{-1}) is much smaller than the longitudinal relaxation rate of the nucleus in the B site (R_1^{B}), the saturation transfer tends to zero. When the rate constant is much higher, the saturation transfer tends to −1, i.e., there is a total transfer of magnetization to the B site when the A site is kept saturated. Note that these conditions are referred to as fast exchange, even if the exchange is still slow with respect to the chemical shift separation. Fast exchange conditions on the relaxation time scale are often reached before the signals are actually coalesced, so that there is still a reasonable range of exchange rates for which saturation transfer can be observed.

Throughout this section it has been assumed that relaxation in the A and B sites is intrinsically exponential. Warnings about this assumption have been made elsewhere (Section 12.1.5).

The methods to detect and analyze saturation transfer in 2D experiments are described in Section 12.3.1.

6.4 EQUILIBRIUM CONSTANTS

6.4.1 NMR OF METAL LIGANDS

Consider a metal complex *CM*, where *C* is a multidentate ligand that leaves an empty coordination position on the metal, in the presence of a monodentate ligand *L*. *CM* could also be a metalloenzyme interacting with a substrate or an inhibitor *L*. The paramagnetic effects observed on a nucleus of *L* can then be used to obtain information on its dissociation constant:

$$K = \frac{[CM][L]}{[CML]} = \frac{(C_{CM} - [CML])(C_L - [CML])}{[CML]} \tag{6.31}$$

where C_{CM} and C_L are the total concentrations of all the metal-containing and ligand-containing species respectively, and $[CML]$ is the equilibrium concentration of the adduct.

Under fast exchange conditions, the molar fraction of bound ligand f_M can be expressed in terms of Eqs. (6.13) and (6.31) as

$$\begin{aligned} f_M &= \frac{[CML]}{C_L} = \frac{\Delta\omega_{\mathrm{p}}}{\Delta\omega_M} = \frac{R_{1\mathrm{p}}}{R_{1M}} = \frac{R_{2\mathrm{p}}}{R_{2M}} \\ &= \frac{K + C_{CM} + C_L - \left[(K + C_{CM} + C_L)^2 - 4C_{CM}C_L\right]^{1/2}}{2C_L}. \end{aligned} \tag{6.32}$$

Therefore, measurements of either $\Delta\omega_{\mathrm{p}}$, $R_{1\mathrm{p}}$, or $R_{2\mathrm{p}}$ at various concentrations of C_L and/or C_{CM} allow a two-parameter fitting of the data through Eq. (6.32) in terms of K and $\Delta\omega_M$, R_{1M}, or R_{2M}.

If the experimental conditions are such that $[CML]$ is always much smaller than C_L (i.e., large excess of *L* with respect to *CM*), then Eq. (6.31) becomes

$$K = \frac{(C_{CM} - [CML])C_L}{[CML]} \tag{6.33}$$

and Eq. (6.32) then becomes

$$f_M = \frac{[CML]}{C_L} = \frac{C_{CM}}{K + C_L}. \tag{6.34}$$

Expressing f_M in the form of Eq. (6.13) gives

$$C_L = C_{CM}\frac{\Delta\omega_M}{\Delta\omega_p} - K = C_{CM}\frac{R_{1M}}{R_{1p}} - K = C_{CM}\frac{R_{2M}}{R_{2p}} - K. \tag{6.35}$$

For constant C_{CM}, a plot of $1/\Delta\omega_p$, $1/R_{1p}$ (= T_{1p}), or $1/R_{2p}$ (= T_{2p}) against C_L, gives a straight line; $\Delta\omega_M$, R_{1M}, or R_{2M} can then be obtained from the slope, and K can be found from the intercept on the y-axis.

Eq. (6.35), like Eq. (6.32) is valid when the ligand is in fast exchange; it is also valid under the exchange-limited conditions, which are described in Eqs. (6.8), (6.10), and (6.11), as long as τ_M is not dependent on C_L. In this case, K can still be obtained, in addition to the limit values of R_{1p}, R_{2p}, or $\Delta\omega_p$ at $f_M = 1$.

6.4.2 NMR OF WATER PROTONS (THE ENHANCEMENT FACTOR)

If one or more ligands L in large excess interact with a metal ion in a metal complex CM in the presence of free metal ions M in solution, then the exchange of ligand L among three sites should be considered. A typical case is when L is a coordinating solvent molecule, for example, water. The molar fraction of water nuclei is given by

$$f_M = \frac{p[M]}{C_{H_2O}} \quad f_{CM} = \frac{q[CM]}{C_{H_2O}} \quad f_{\text{bulk}} = \frac{[H_2O]}{C_{H_2O}} \approx 1 \tag{6.36}$$

where p and q are the numbers of water molecules interacting with free and bound metal ions respectively. The water proton R_{1p} is given by the contribution of the metal, $(R_{1p})_M$, and of the complex $(R_{1p})_{CM}$, both of which can be expressed through Eq. (6.10):

$$\begin{aligned} R_{1p} &= f_M R_{1M}\frac{\tau_M^{-1}}{R_{1M} + \tau_M^{-1}} + f_{CM} R_{1CM}\frac{\tau_{CM}^{-1}}{R_{1CM} + \tau_{CM}^{-1}} \\ &= \frac{p[M]}{C_{H_2O}} R_{1M}\frac{\tau_M^{-1}}{R_{1M} + \tau_M^{-1}} + \frac{q[CM]}{C_{H_2O}} R_{1CM}\frac{\tau_{CM}^{-1}}{R_{1CM} + \tau_{CM}^{-1}}. \end{aligned} \tag{6.37}$$

The two limiting cases of the metal completely free ($K \to \infty$) or completely bound ($K \to 0$), where K is the dissociation constant of the CM complex, would give

$$\begin{aligned} R_{1p}(\infty) &= \frac{pC_M}{C_{H_2O}} R_{1M}\frac{\tau_M^{-1}}{R_{1M} + \tau_M^{-1}} \\ R_{1p}(0) &= \frac{qC_M}{C_{H_2O}} R_{1CM}\frac{\tau_{CM}^{-1}}{R_{1CM} + \tau_{CM}^{-1}} \end{aligned} \tag{6.38}$$

where C_M is the total metal concentration. The first case can be simulated by simply not adding any ligand C to the solution. It is then customary to define an enhancement factor as [9]

$$\varepsilon = \frac{R_{1\mathrm{p}}}{R_{1\mathrm{p}}(\infty)} \tag{6.39}$$

where $R_{1\mathrm{p}}$ is measured in a solution containing both the metal ion and the ligand C, and $R_{1\mathrm{p}}(\infty)$ is obtained from a solution containing the same concentration of metal ion but in the absence of the ligand C. From Eqs. (6.37) and (6.38)

$$\begin{aligned}\varepsilon &= \frac{[M]}{C_M} + \frac{q[CM]}{pC_M}\frac{R_{1CM}\tau_{CM}^{-1}}{R_{1M}\tau_M^{-1}}\frac{R_{1M}+\tau_M^{-1}}{R_{1CM}+\tau_{CM}^{-1}} \\ &= \frac{[M]}{C_M} + \frac{q[CM]}{pC_M}\frac{T_{1M}+\tau_M}{T_{1CM}+\tau_{CM}} = \frac{[M]}{C_M} + \frac{[CM]}{C_M}\varepsilon_0 \\ &= f_\mathrm{f} + f_\mathrm{b}\varepsilon_0 = 1 + f_\mathrm{b}(\varepsilon_0 - 1)\end{aligned} \tag{6.40}$$

where f_f and f_b are the molar fractions of free and bound metal ion and

$$\varepsilon_0 = \frac{q}{p}\frac{T_{1M}+\tau_M}{T_{1CM}+\tau_{CM}} = \frac{R_{1\mathrm{p}}(0)}{R_{1\mathrm{p}}(\infty)}.$$

When the CM complex is fully formed, $[M] = 0$ and $[CM] = C_M$; therefore, from Eq. (6.40), $\varepsilon = \varepsilon_0$. ε_0 *is thus defined as the enhancement factor measured in a solution where all the metal is complexed.* Since usually $q < p$, ε_0 should be smaller than unity if the intrinsic nuclear relaxation times are the same in the metal complex and in the aquaion. However, as often $T_{1CM} < T_{1M}$ owing to a longer correlation time τ_c in the complex (Chapter 4), ε_0 can be larger than unity. This is particularly true when C is a macromolecule (e.g., a protein) and M is a metal ion with long electronic relaxation times.

As it appears in Eq. (6.40), ε_0 is defined only in terms of the molar fraction of bound metal ion; that is, independently of the actual concentrations. By using an equation for f_b analogous to Eq. (6.32), the enhancement factor can be expressed as:

$$\varepsilon = 1 + \frac{K + C_M + C_C - \left[\left(K + C_M + C_C\right)^2 - 4C_M C_C\right]^{1/2}}{2C_M}(\varepsilon_0 - 1). \tag{6.41}$$

Both K and ε_0 can be obtained through a two-parameter fitting of the ε data obtained at various C_C and/or C_M concentrations.

Eq. (6.41) is valid irrespective of the rate at which water exchanges from the two paramagnetic sites. In fact, the enhancement factor is defined in terms of the quantities $T_{1M} + \tau_M$ and $T_{1CM} + \tau_{CM}$, which are likely to be constant and, in particular, independent of the concentration of the various species in solution. Furthermore, when the exchanging ligand is the solvent, as in the above examples, its concentration is virtually constant under any circumstances.

6.5 τ_M AS CORRELATION TIME

Let us consider a ligand nucleus (e.g., a proton in a water molecule) and its interaction with a paramagnetic center (e.g., a metal ion) to which the ligand is bound. In the absence of exchange, the hyperfine shifts and relaxation rates are given by the equations developed in Chapters 2 and 4. In the case of fast exchange of ligands (e.g., bulk water), we have seen in Section 6.3.2, that the shift and relaxation rates are the weighted averages between the free and bound ligands. This means that *exchange does not alter the shift and relaxing capabilities of the metal site.* Whereas this is always true as far as the shift is concerned, it may not be true for relaxation rates. We have seen that nuclear relaxation in the bound species is proportional to the average squared interaction energy (either dipolar or contact) and to a function of the correlation time that modulates the interaction. We have also seen that the correlation time for the dipolar interaction is given by the reciprocal of the sum of the electronic relaxation rate, the reorientation correlation rate, and the chemical exchange rate, whereas in the case of contact relaxation it is given by the reciprocal of the sum of the electronic relaxation rate and the chemical exchange rate [Eqs. (4.6) and (4.7)]. We have learned that relaxation arises from the modulation of the interaction energy by whatever random process. If chemical exchange is present, clearly the interaction energy can be modulated by exchange, because the interaction (both dipolar and contact) is lost upon ligand detachment. In the limit situation where τ_M becomes shorter than τ_e (for contact) or shorter than both τ_e and τ_r (for dipolar relaxation) then it becomes the correlation time [Eqs. (4.6) and (4.7)]. Of course, the R_{1M} and R_{2M} profiles still maintain the same shape as shown in Figs. 4.9 and 4.11.

6.6 BEYOND THE CONCEPT OF BINDING SITE: OUTER SPHERE RELAXATION

So far we have assumed that the molecule bearing the nucleus under investigation spends a finite (although small) time in a well-defined binding site, and the rest of the time in the bulk solution at a distance from the metal which may be considered infinite. In this process, we have considered negligible the time spent in approaching and leaving the binding site (i.e., spent at a finite metal–nucleus distance). As τ_M becomes shorter and shorter, it will eventually reach the point where it is comparable with, or shorter than, the diffusional correlation time of the molecule. Under these conditions, the metal–nucleus interaction during the approach and departure of the ligand becomes a substantial fraction of the total interaction. A limit situation can be reached where a binding site no longer exists and the metal–nucleus interaction is only exerted by random encounters between the molecules, regulated only by diffusion processes. The correlation time for the latter, termed τ_D, then becomes an important parameter. In the only presence of diffusion-controlled interactions, contact shifts no longer exist, and dipolar shifts average to zero if the approach to the ion can occur in every direction. Therefore, hyperfine shifts are zero (although an effect on the shifts can still be detected, Section 6.8), whereas relaxation enhancements do not drop to zero. This situation goes under the name of "outer sphere" relaxation. At variance with the correlation functions encountered so far, the correlation function for molecular diffusion is not exponential [9,10]. Furthermore, its form depends on the assumed diffusional model.

The diffusional correlation time τ_D depends on the size of both the metal and the ligand-containing moieties, according to their diffusion coefficients, D_M and D_L, and on the minimal distance that can be achieved between the ligand and the metal ion, called *distance of closest approach*, d [9,10].

$$\tau_D = \frac{d^2}{D_M + D_L}. \tag{6.42}$$

In turn, the diffusion coefficients are defined by assuming that the molecules behave as rigid spheres in a medium of viscosity η:

$$D_M = \frac{kT}{6\pi a_M \eta} \quad D_L = \frac{kT}{6\pi a_L \eta} \tag{6.43}$$

where a_M and a_L are the radii of the metal-containing and ligand-containing molecules. Note that, when one of the two molecules has a much larger size than the other, its diffusion coefficient is much smaller than the other, and does not contribute appreciably to the denominator of Eq. (6.42). In other words, the diffusional correlation time is only dependent on the size of the small molecule.

In outer sphere relaxation two limiting situations may thus occur, depending on whether the electronic relaxation time τ_e is shorter or longer than the diffusional correlation time τ_D. Since the metal ion and the interacting nucleus are not held together in the same molecular framework, the rotational correlation time τ_r is ineffective in modulating the electron–nucleus interaction and needs not be considered further, τ_D is typically in the range 10^{-9}–10^{-11} s and seldom reaches 10^{-12} s (for instance, water molecules in water have $\tau_D \approx 3 \times 10^{-12}$ s). Thus, τ_e can still be the dominant correlation time when the metal ion undergoes fast electron relaxation. In such a case, one can figure out that, for each metal–ligand distance between d and infinity, the metal–ligand system can be considered as frozen on the time interval over which the electron undergoes many transitions between the spin levels. Therefore, R_{1p} and R_{2p} can be evaluated by simply integrating the Solomon equations [Eqs. (4.20) and (4.21)] over the distance range between d and infinity [9–11]:

$$R_{1p} = \frac{2}{15}\left(\frac{\mu_0}{4\pi}\right)^2 1000 N_A [M] \frac{4}{3}\pi \frac{\gamma_I^2 g_e^2 \mu_B^2 S(S+1)}{d^3}\left(\frac{7T_{2e}}{1+\omega_S^2 T_{2e}^2} + \frac{3T_{1e}}{1+\omega_I^2 T_{1e}^2}\right) \tag{6.44}$$

$$R_{2p} = \frac{1}{15}\left(\frac{\mu_0}{4\pi}\right)^2 1000 N_A [M] \frac{4}{3}\pi \frac{\gamma_I^2 g_e^2 \mu_B^2 S(S+1)}{d^3}\left(4T_{1e} + \frac{13T_{2e}}{1+\omega_S^2 T_{2e}^2} + \frac{3T_{1e}}{1+\omega_I^2 T_{1e}^2}\right) \tag{6.45}$$

where N_A is the Avogadro constant and $[M]$ is the concentration of the molecule bearing the paramagnetic center expressed in mol dm^{-3}. Note that only the paramagnetic enhancements R_{1p} and R_{2p} can be evaluated, since R_{1M} and R_{2M} cannot be defined. Also, at variance with normal chemical exchange situations, R_{1p} and R_{2p} do not depend on ligand concentration but only on the concentration of the metal-containing species. Indeed, when no binding site exists, each nucleus may interact with more than one metal at a time, independently of the concentration of nuclei; R_{1p} and R_{2p} are therefore only proportional to the number of metal ions per unit volume, which is expressed by $N_A[M] \times 1000$ in SI units.

In the diffusion-controlled regime different equations should be derived, taking into account that the interaction energy is now modulated by fluctuations in r between d and infinity. In this case, the kind of integration to be performed depends on the model assumed for the diffusional behavior of the system.

According to one of the most commonly used models for diffusion [12,13], the following equations have been derived when τ_D is the dominant correlation time:

$$R_{1p} = \frac{32}{405}\pi\left(\frac{\mu_0}{4\pi}\right)^2 \frac{1000N_A[M]\gamma_I^2 g_e^2 \mu_B^2 S(S+1)}{d(D_M + D_L)}\{7J^{tr}(\omega_S) + 3J^{tr}(\omega_I)\} \quad (6.46)$$

$$R_{2p} = \frac{16}{405}\pi\left(\frac{\mu_0}{4\pi}\right)^2 \frac{1000N_A[M]\gamma_I^2 g_e^2 \mu_B^2 S(S+1)}{d(D_M + D_L)}\{4J^{tr}(0) + 13J^{tr}(\omega_S) + 3J^{tr}(\omega_I)\} \quad (6.47)$$

where the spectral density functions are given by

$$J^{tr}(\omega) = \frac{1 + 5z/8 + z^2/8}{1 + z + z^2/2 + z^3/6 + 4z^4/81 + z^5/81 + z^6/648} \quad (6.48)$$

with

$$z = (2\omega\tau_D)^{1/2}. \quad (6.49)$$

It should be noted that, because the interaction energy is averaged in a different way with respect to the case of shortest τ_e discussed earlier, the equations look very different. In particular, the $J^{tr}(\omega)$ do not have the usual Lorentzian form (Fig. 6.5). In the case of τ_D and τ_e having comparable values [14]:

$$z = \sqrt{2\left(\omega\tau_D + \frac{\tau_D}{\tau_e}\right)}. \quad (6.50)$$

Calculations were also performed for electron spin Hamiltonian dominated by zero field splitting [15–17].

In the earlier model it is assumed that both the unpaired electron(s) and the dipole–dipole coupled nucleus are in the center of spherical molecules. If this is not the case, and the rotational correlation

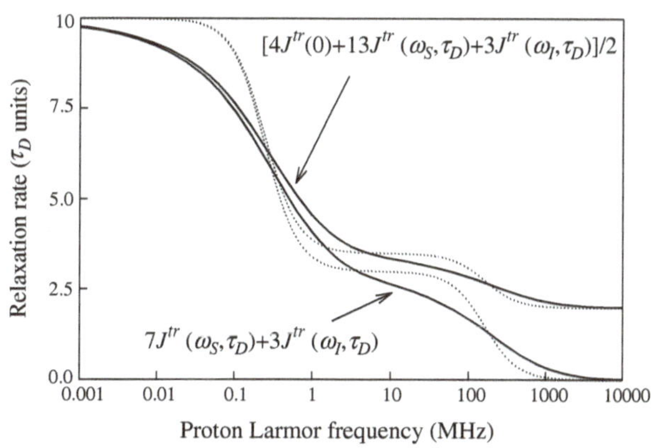

FIGURE 6.5 Plot of the spectral density functions of Eqs. (6.46) and (6.47) as a function of magnetic field (expressed as proton Larmor frequency; log scale). $\tau_D = 2 \times 10^{-9}$ s. The Solomon profiles obtained for $\tau_c = 2 \times 10^{-9}$ s are also reported (*dotted lines*) for comparison purposes.

time τ_r is not much longer than τ_D, relaxation is also affected by molecular rotation of both interacting spins, so that the spectral density function [described by Eq. (6.48)] must contain both translational and rotational contributions. This is referred to as the eccentricity effect. A rigorous mathematical description of the translational-rotational modulations of the intermolecular coupling is rather cumbersome [18,19]. A decomposition approximation is proposed, based on the treatment of the translational and rotational effects as independent of each other. A Lorentzian spectral density function, or a Cole-Davidson function[3] in the case of high viscosity liquids [20–23], is added to the translational spectral density, with a phenomenological parameter e which reflects the rotational influence on the modulation of the dipole–dipole interaction (for spins placed in the molecular centers $e = 0$):

$$R_{1p} = \frac{32}{405}\pi\left(\frac{\mu_0}{4\pi}\right)^2 \frac{1000N_A[M]\gamma_I^2 g_e^2 \mu_B^2 S(S+1)}{d^3} \times\left\{7\left[\tau_D J^{tr}(\omega_S)+e\frac{\tau_r}{1+\omega_S^2\tau_r^2}\right]+3\left[\tau_D J^{tr}(\omega_I)+e\frac{\tau_r}{1+\omega_I^2\tau_r^2}\right]\right\}. \tag{6.51}$$

When the interacting spins are not placed in the center of spherical molecules, the distance of closest approach d is different depending on the direction along which the solvent molecules approach the paramagnetic complex. When used to analyze experimental data, the above equations thus provide d values representing weighted averages resulting from the contributions coming from the different collision distances along the different directions (also Section 6.7).

Finally, for very short electron relaxation times ($\tau_e \ll \tau_D$), as in lanthanoids except gadolinium(III) (Chapter 8), an additional contribution to the outer-sphere relaxation can be provided by the Curie spin interaction modulated by τ_D [24]:

$$R_{1p} = \frac{32}{405}\pi\left(\frac{\mu_0}{4\pi}\right)^2 \frac{1000N_A[M]\omega_I^2 g_e^4 \mu_B^4 S^2(S+1)^2}{3k^2T^2 d\left(D_M+D_L\right)}\left\{7J^{tr}(\omega_S)+3J^{tr}(\omega_I)\right\} \tag{6.52}$$

where for lanthanoids (except gadolinium(III)) S and g_e should be replaced by J and g_J, respectively (Chapter 8).

6.7 SOLVENT RELAXIVITY AND ITS INFORMATION CONTENT

The presence of a paramagnetic complex in solution changes the relaxation rate of solvent nuclei as a result of (1) chemical exchange of the solvent molecules coordinated at a fixed distance to the paramagnetic metal or closeby (inner-sphere relaxation), and (2) solvent free diffusion around the paramagnetic

[3]The Cole-Davidson spectral density, $J^{CD}(\omega)$, is used to describe the rotational dynamics of high viscosity liquids:

$$J^{CD}(\omega) = \frac{\sin\left[\beta \arctan(\omega\tau_{CD})\right]}{\omega(1+\omega^2\tau_{CD}^2)^{\beta/2}}$$

The correlation time τ_{CD} is related to τ_r according to the relationship: $\tau_r = \beta\tau_{CD}$. The parameter β reflects the deviation from the Lorentzian function; for $\beta = 1$, $J^{CD}(\omega)$ simplifies to the Lorentzian function.

center (outer-sphere relaxation). From Eqs. (6.10) and (6.46), the paramagnetic enhancement to the longitudinal relaxation rate R_{1p} of a solvent nuclei is provided by:

$$R_{1p} = f_M (R_{1M}^{-1} + \tau_M)^{-1} + \frac{32}{405}\pi \left(\frac{\mu_0}{4\pi}\right)^2 \frac{1000 N_A [M] \gamma_I^2 g_e^2 \mu_B^2 S(S+1)}{d(D_M + D_L)} \{7J^{tr}(\omega_S) + 3J^{tr}(\omega_I)\} \tag{6.53}$$

where the first and the second terms report the inner-sphere and outer-sphere contributions, respectively. R_{1M}, the relaxation rate of the coordinated nuclei, is provided by the sum of dipolar relaxation, described by Eqs. (4.17) (in the absence of static ZFS and hyperfine coupling between unpaired electron(s) and metal nucleus, Section 4.7.1), contact relaxation, described by Eq. (4.27), and Curie spin relaxation, described by Eq. (4.30). In the absence of exchange (τ_M approaching infinite), there is no inner-sphere contribution to R_{1p}. In the presence of fast exchange ($\tau_M \ll T_{1M}$), on the contrary, the inner-sphere contribution is proportional to the relaxation rates of the coordinated nuclei. When $\tau_M \approx T_{1M}$ (intermediate exchange), the solvent exchange acts as a limiting factor for the propagation of the paramagnetic effect from the nuclei in bound position to the solvent nuclei. This makes the relaxation rates always smaller than for an analogous system in fast exchange conditions. Furthermore, all dispersions in the relaxivity profile are less evident, the ratios between the relaxation rates before and after the dispersions being smaller than those calculated for T_{1M}.

The concentration of the paramagnetic species $[M]$ is contained in the inner-sphere term within the molar fraction of bound nuclei f_M. As a result, R_{1p} has a linear dependence on the concentration of the paramagnetic molecule. It is thus convenient to define the relaxivity of a paramagnetic molecule as the paramagnetic contribution to the relaxation rates in the presence of 1 mM concentration of the paramagnetic species.

In summary, ^{1}H relaxivity depends on (1) structural parameters, i.e., on the number and the distance of the protons close, at a fixed distance, to the paramagnetic ion, and on the distance of closest approach between the paramagnetic molecules and the other proton-containing molecules freely diffusing in solution, (2) dynamic parameters, i.e., the lifetime of the protons around the paramagnetic ion (τ_M), the reorientation time of the molecule (τ_r), the extent and the correlation time of local motions of the moieties to which the protons belong (motions which can also modulate the dipole–dipole interaction analogously to τ_r), the diffusional correlation time (τ_D), and (3) electronic parameters, i.e., the electron spin quantum number, the electron relaxation times and their field dependence, the contact coupling constant A, and the ZFS parameters (Sections 1.4 and 4.7). In the presence of a nuclear spin quantum number larger than zero, also the hyperfine coupling between electron(s) and nuclear spins can be important (Section 4.7.1).

The nuclear magnetic relaxation dispersion (NMRD) is a technique based on the measurement of the solvent molecule relaxation rates in magnetic fields of intensity varying of several orders of magnitude (Section 12.5.2), from frequencies as low as 0.01 MHz of proton Larmor frequency up to the highest frequencies of the high resolution NMR spectrometers. From the measurement of the NMRD profiles of paramagnetic complexes, usually performed in water solutions, we can thus determine the solvent water proton relaxivity. Since the latter depends on all the parameters listed above, it represents a precious source of information for the characterization of structure, dynamics, and electronic properties of the paramagnetic complex.

The analysis of the relaxivity profiles is commonly performed using Eq. (6.53) and the equations reported in Sections 4.3–4.6 (called, all together, Solomon-Bloembergen-Morgan model) for the calculation of R_{1M}. As described in Section 4.7.1, this approach can fail in the presence of static ZFS and/or hyperfine coupling between electrons and metal nuclear spins. In these cases, more complicated approaches should be applied [16,25,26]. In the so called slow rotation limit ($\tau_r \gg \tau_e$), the (modified) Florence NMRD program [27–30] can be used. The program provides the ^{1}H relaxivity profiles due to contact and dipolar contributions including the effects, for any metal nucleus spin quantum number and any electron spin quantum number, of both static and transient ZFS, of the hyperfine coupling of any rhombicity between electron spin and metal nuclear spin, and of any ***g*** tensor anisotropy.

Due to the large number of parameters on which the relaxivity depends, it is often convenient to acquire relaxation profiles at different temperatures. This permits to include some constraints on the temperature dependence of the different correlation times, as described in Section 10.4.2, and to exclude any temperature dependence for other parameters like the distance between bound nuclei and paramagnetic metal, the distance of closest approach, the ZFS parameters, the hyperfine coupling with the metal nucleus, the constant of contact interaction, and, possibly, Δ_t [Eq. (4.12)]. As a result, T_{1M} may either decrease or increase with increasing temperature, depending on the lifetime τ_M, on the magnetic field and on the correlation time which modulates the relaxation process. If the correlation time decreases with increasing temperature, the low field R_{1p} decreases when $\tau_M \ll T_{1M}$ (fast exchange limit). On the other hand, since τ_M always decreases with increasing temperature, R_{1p} increases if $\tau_M \gg T_{1M}$ [slow exchange limit, Eq. (6.53)] and if the outer-sphere contribution is much smaller than the inner-sphere contribution. The outer-sphere contribution always decreases with increasing temperature.

Regarding the outer-sphere term, when the paramagnetic center is not located at the center of a spherical molecule and Eq. (6.53) is used to analyze the experimental relaxation profile, a distance of closest approach *d* shorter than the value expected for the distance between intermolecular centers is often obtained, with a diffusion coefficient representing an upper bound limit of the correct diffusion coefficient value [31]. If the directions along which the solvent molecules can approach the paramagnetic center are restricted within a solid angle, much smaller than spherical, as for instance in the case of a paramagnetic metal ion located on the periphery of a macromolecule, the outer-sphere contribution should be appropriately scaled down. For instance, if the water molecules can diffuse until a short distance of closest approach only in a semisphere, the outer-sphere term should be divided by two.

6.8 BULK SUSCEPTIBILITY SHIFT

In the diffusion limit, the hyperfine shift, defined as the shift difference between a paramagnetic and a diamagnetic environment both measured with respect to the same internal standard, is zero. However, the absolute shifts of all nuclei in the paramagnetic solution are all offset by the same extent with respect to an analogous diamagnetic solution because of the bulk paramagnetism of the sample. As the effect is the same for all nuclei in the sample, when chemical shifts are reported relative to an internal reference signal the effect is therefore canceled. However, it is possible to measure this effect, which is proportional to the magnetic susceptibility of the paramagnetic solute and to its concentration, by referring the shifts to an *external* reference signal. The experimental procedure is referred to as the Evans method [32], and constitutes one of the best ways of measuring solute magnetic susceptibilities in solution around room temperature.

The contribution to the absolute chemical shift from the magnetic susceptibility of a sample is given by the following equation [33]:

$$\delta_\chi = m\left(\tfrac{1}{3}-\alpha\right)\left[\chi^{\text{para}} + \chi^{\text{dia}} - \chi^{\text{solv}}\left(1+\frac{\rho_{\text{solv}}-\rho_{\text{soltn}}}{m}\right)\right] \tag{6.54}$$

where m is the mass of solute per unit volume, α is a demagnetization factor, χ^{para} and χ^{dia} are the paramagnetic and diamagnetic contributions to the mass susceptibility of the solute, χ^{solv} is the mass susceptibility of the solvent, and ρ_{solv} and ρ_{soltn} are the densities of the solvent and the solution respectively. χ^{dia} is negative and usually smaller than χ^{para} in absolute value, unless when dealing with large macromolecules. The demagnetization factor α depends on the geometry of the sample and on its orientation relative to the external magnetic field. For a spherical sample $\alpha = ⅓$ and the susceptibility effect on the chemical shift vanishes. For cylindrical samples, $\alpha = ½$ if the magnetic field is perpendicular to the cylinder's axis (as in most electromagnets) and $\alpha = 0$ if the magnetic field is parallel to the cylinder axis (as is usual in superconducting magnets). The quantity ⅓ – α is therefore equal to –⅙ in the former case and ⅓ in the latter; that is, opposite in sign and double in magnitude. Therefore, a high field cryomagnet would be preferable both because of α and higher resolution.

The experimental setup consists of two coaxial tubes (Fig. 6.6) [34], one of which, for example, the inner one, containing a solution of an inert probe substance and the paramagnetic solute, and the other containing a solution of only the inert probe substance in the same solvent. The shifts of the probe

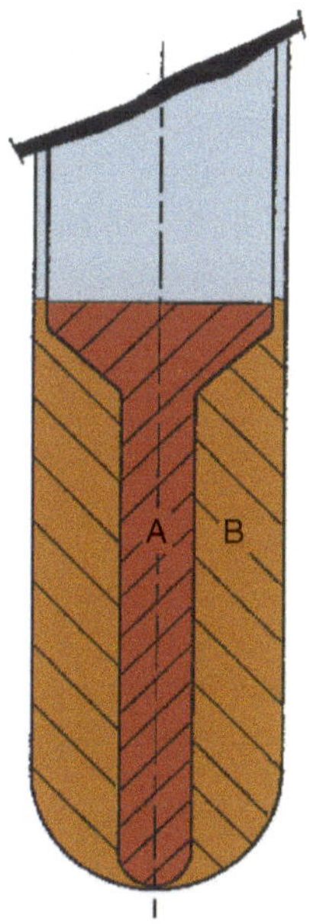

FIGURE 6.6 Coaxial NMR tubes for the measurement of magnetic susceptibility of a paramagnetic solute. Solution A contains the paramagnetic solute and an inert probe substance. Solution B contains the probe substance in the same solvent. The measurement is then repeated substituting the paramagnetic solute in A with a diamagnetic analog of the same concentration.

substance differ in the two solutions, and two different signals are observed. Their chemical shift separation is measured, and the experiment repeated with the same solution in the outer tube and a diamagnetic analog in the inner tube. The chemical shift separation of the two signals is again measured. The difference between the two values, if the concentrations of the paramagnetic and diamagnetic solutes are the same, is directly related to χ^{para}

$$\Delta\delta = m\left(\tfrac{1}{3}-\alpha\right)\chi^{para}. \tag{6.55}$$

Note that the terms containing the solvent susceptibility in Eq. (6.54) cancel because the two solutions have the same density. Eq. (6.55) can also be rewritten as

$$\Delta\delta = 1000M\left(\tfrac{1}{3}-\alpha\right)\chi_M^{para} \tag{6.56}$$

where $1000M$ is the concentration (mol m^{-3}) of the paramagnetic solute and χ_M^{para} is the paramagnetic contribution to the molar susceptibility, which is directly related to the effective magnetic moment of the paramagnetic center (Chapter 1):

$$\mu_{eff}^2 = \frac{3kT}{N_A\mu_0}\chi_M^{para}. \tag{6.57}$$

Typical values of χ_M^{para} range from 2×10^{-8} to 3×10^{-7} m^3 mol^{-1}, so that for a millimolar solution in a superconducting magnet ($1000M = 1$ mol m^{-3}, $\alpha = 0$), $\Delta\delta$ goes from 0.007 to 0.1 ppm, that is from 6 to 90 Hz on a 900 MHz instrument.

Possible sources of error may arise if the probe is not completely inert. In fact, even a very weak interaction of the probe with the paramagnetic center may cause specific effects on the chemical shift, which may substantially alter the measurement, given the relatively small effects being measured. The best approach is to repeat the measurements using different probes. Another source of error, for very large macromolecules, lies in the sizable χ^{dia} term, whose absolute value may be of the same order of magnitude or even larger than χ^{para}, making the subtraction more critical.

REFERENCES

[1] Gutowsky HS, McCall DM, Slichter CP. J. Chem. Phys. 1953;21:279.
[2] Gutowsky HS, Holm CH. J. Chem. Phys. 1956;25:1228.
[3] McLaughlin AC, Leigh JS Jr. J. Magn. Reson. 1973;9:296.
[4] Leigh JS Jr. J. Magn. Reson. 1971;4:308.
[5] Swift TJ, Connick RE. J. Chem. Phys. 1962;37:307.
[6] Swift TJ. In: La Mar GN, Horrocks WD Jr, Holm RH, editors. NMR of Paramagnetic Molecules. New York: Academic Press; 1973. p. 53.
[7] Cossy C, Helm L, Merbach AE. Inorg. Chem. 1988;27:1973.
[8] Johnston ER. Concepts Magn. Reson. 1995;7:219.
[9] Dwek RA. Nuclear Magnetic Resonance in Biochemistry: Applications to Enzyme SystemsLondon: Oxford University Press; 1973.
[10] Abragam A. The Principles of Nuclear MagnetismOxford: Oxford University Press; 1961.
[11] Luz Z, Meiboom S. J. Chem. Phys. 1964;40:2686.
[12] Hwang LP, Freed JH. J. Chem. Phys. 1975;63:4017.

[13] Polnaszek CF, Bryant RG. J. Chem. Phys. 1978;68:4034.
[14] Freed JH. J. Chem. Phys. 1978;68:4034.
[15] Bayburt T, Sharp RR. J. Chem. Phys. 1990;92:5892.
[16] Kowalewski J, Kruk D, Parigi G. Adv. Inorg. Chem. 2005;57:41.
[17] Kruk D, Nilsson T, Kowalewski J. Mol. Phys. 2001;99:1435.
[18] Ayant Y, Belorizky E, Fries PH, Rosset J. J. Phys. 1977;38:325.
[19] Albrand JP, Taieb MC, Fries PH, Belorizky E. J. Chem. Phys. 1983;78:5809.
[20] Neugebauer P, Krummenacker JG, Denysenkov VP, Parigi G, Luchinat C, Prisner TF. Phys. Chem. Chem. Phys. 2013;15:6049.
[21] Meier R, Kruk D, Gmeiner J, Rössler EA. J. Chem. Phys. 2012;136:034508.
[22] Kruk D, Korpaa A, Kowalewski J, Rössler EA, Mosciski J. J. Chem. Phys. 2012;137:044512.
[23] Kruk D, Hoffmann SK, Goslar J, Lijewski S, Kubica-Misztal A, Korpaa A, Oglodek I, Kowalewski J, Rössler EA, Mosciski J. J. Chem. Phys. 2013;139:244502.
[24] Vigouroux C, Belorizky E, Fries PH. Eur. Phys. J. D 1999;5:243.
[25] Larsson T, Westlund PO, Kowalewski J, Koenig SH. J. Chem. Phys. 1994;101:1116.
[26] Nilsson T, Kowalewski J. J. Magn. Reson. 2000;146:345.
[27] Kruk D, Nilsson T, Kowalewski J. Phys. Chem. Chem. Phys. 2001;3:4907.
[28] Nilsson T, Kowalewski J. Mol. Phys. 2000;98:1617.
[29] Bertini I, Kowalewski J, Luchinat C, Nilsson T, Parigi G. J. Chem. Phys. 1999;111:5795.
[30] Bertini I, Galas O, Luchinat C, Parigi G. J. Magn. Reson. Ser. A 1995;113:151.
[31] Polnaszek CF, Bryant RG. J. Chem. Phys. 1984;81:4038.
[32] Evans DF. J. Chem. Soc. 1959;2003.
[33] Dickinson WC. Phys. Rev. 1951;81:717.
[34] Phillips WD, Poe M. Methods Enzymol. 1972;24:304.

CHAPTER

TRANSITION METAL IONS: SHIFT AND RELAXATION

7

This chapter aims at providing the reader with guidelines for the interpretation of the spectra of compounds with some common metal ions. Here, only mononuclear complexes are considered. When dealing with the NMR spectra of a paramagnetic compound in solution, one has first of all to figure out the electron relaxation times (and electron relaxation mechanisms) and the electron–nucleus correlation time; then has to guess nuclear relaxation, in order to set the experiments. Only then structural and dynamic information from the nuclear relaxation properties and from the hyperfine shifts can be safely gained. The experience of the authors—and of many others—will be shared with the reader in the following sections, without any ambition of being comprehensive or reviewing the field.

7.1 AN OVERVIEW OF THE ELECTRONIC PROPERTIES OF TRANSITION METAL IONS

7.1.1 INTRODUCTORY REMARKS ON THE ELECTRONIC STRUCTURE OF METAL IONS

The introductory comments in this section are intended to: (1) help the reader who is not familiar with the chemistry of transition metal ions to understand the relationship between the electronic ground state of the metal and the NMR properties of the interacting nuclei; and (2) to provide a uniform vocabulary for the phenomena associated with the electronic properties of metal ions.

The present chapter discusses 3d metal ions, and extension to 4d and 5d elements does not require, in principle, any new concept. Discussion of f elements will be given in Chapter 8.

Electronic $3d^n$ configurations give rise to several free ion terms depending on the occupancy of the five d orbitals. Such terms arise because of all the possible combinations, C, of the M_L orbital quantum number and the M_S spin quantum number for each electron. C can be found as a binomial coefficient expressing the possibilities for a single d electron to occupy 10 spin-orbitals:

$$C = \frac{10!}{n!(10-n)!}$$

The C microstates are grouped into subsets of degenerate microstates, which define the free ion terms. The sets are indicated as ^{2S+1}L, where S is the total multielectron spin quantum number ($S = \sum M_S$) and L is the total orbital quantum number ($L = \sum M_L$); it is to be noted that L is given as a letter, according to the following convention:

L	0	1	2	3	4	5	6
Symbol	S	P	D	F	G	H	I

NMR of Paramagnetic Molecules. http://dx.doi.org/10.1016/B978-0-444-63436-8.00008-9

Table 7.1 Free Ion Configuration and Population of Electronic d Orbitals in Octahedral Symmetry

Electronic Configuration	M_L of Occupied Orbitals	L	S	Free Ion Configuration	Ligand Field O_h (Octahedral)	
d^1	2	2	1/2	^2D	e_g – – t_{2g} ↑ – –	^2T
d^2	2,1	3	1	^3F	– – ↑ ↑ –	^3T
d^3	2,1,0	3	3/2	^4F	– – ↑ ↑ ↑	^4A
d^4	2,1,0,−1	2	2	^5D	↑ – ↑ ↑ ↑	^5E
d^5	2,1,0,−1,−2	0	5/2	^6S	↑ ↑ – – ↑ ↑ ↑ ⇃↾ ⇃↾ ↑ HS LS (S = 1/2)	^6A ^2T
d^6	2,2,1,0,−1,−2	2	2	^5D	↑ ↑ ⇃↾ ↑ ↑	^5T
d^7	2,2,1,1,0,−1,−2	3	3/2	^4F	↑ ↑ ↑ – ⇃↾ ⇃↾ ↑ ⇃↾ ⇃↾ ⇃↾ HS LS (S = 1/2)	^4T ^2E
d^8	2,2,1,1,0,0,−1,−2	3	1	^3F	↑ ↑ ⇃↾ ⇃↾ ⇃↾	^3A
d^9	2,2,1,1,0,0,−1,−1,−2	2	1/2	^2D	⇃↾ ↑ ⇃↾ ⇃↾ ⇃↾	^2E

The total number of degenerate states of the free ion term is given by the spin multiplicity multiplied by the orbital multiplicity $(2S + 1)(2L + 1)$. The ground term is of most interest and corresponds to the combination of largest S and L values. Examples are reported in Table 7.1 ("Free ion configuration" column) for the different electronic configurations.

In metal complexes, the degeneracy of the five d orbitals is reduced due to the surrounding electric charge distribution. The electrons in the ligands are in fact closer to some of the d-orbitals and farther from others. This causes a loss of degeneracy because the electrons in d orbitals closer to the ligands have a higher energy than those farther away. In octahedral symmetry (O_h symmetry) three orbitals have energy lower that the other two. The degeneracy of the states is then determined taking into account the number of electrons in the d orbitals. The symbols that are used to identify the degeneracy are A and B for nondegenerate, E for doubly degenerate, and T for triply degenerate [1,2].

1. If the three orbitals lower in energy in O_h symmetry are populated by one electron, three degenerate states are possible, according to the three possible positions of the electron in the three levels (T symmetry).
2. Analogously, if there are two electrons in the three low energy orbitals, still there are three possible states (T symmetry).

3. On the contrary, if there are three electrons, there is only a nondegenerate ground state (A symmetry).
4. If there are four electrons, and the separation between the lower and the higher sets of orbitals is smaller than the spin pairing energy (high spin configuration), one of the four electrons occupies the higher set, and the ground state is doubly degenerate (E symmetry); on the contrary, if the separation between the lower and the higher sets of orbitals is larger than the spin pairing energy (low spin configuration), the ground state is triply degenerate (T symmetry).
5. If there are five electrons, there is only a nondegenerate ground state (A symmetry) if the system is high spin, and three degenerate ground states (T symmetry) if the system is low spin.

The cases from six to nine electrons are reported in Table 7.1.

Similar considerations can be made for other idealized symmetries, like tetrahedral (T_d), square planar (D_{4h}), square pyramidal (C_{4v}), trigonal bipyramidal (D_{3h}), etc. The energy levels of the d orbitals in these idealized symmetries are shown in Table 7.2. In octahedral symmetry the five d orbitals are split in three degenerate lower energy orbitals, symmetric with respect to inversion and asymmetric with respect to the C_2 axis that is perpendicular to the principal C_4 axis (T_{2g} orbitals, Table 7.1), and two degenerate higher energy orbitals, symmetric with respect to inversion (E_g orbitals).

The degeneracy of the states in the ligand-field is further removed by real symmetries, which are distorted and not idealized. For instance, in low spin d^5 electronic configuration (5 electrons in the d orbitals, with total $S = 1/2$), in distorted octahedral coordination (Table 7.1), the unpaired electron is in the lowest orbital after filling the two other T_{2g} orbitals with lowest energy with two electrons each. The first excited state is obtained by promoting one electron so that the unpaired electron resides in the second lowest orbital. Furthermore, the coupling between the magnetic moment operator associated to the electron spin and orbital angular momenta, that is, the spin–orbit coupling operator—accounts for the splitting of the S manifold. Theory also expects that the geometry of the chromophore is such that every spin-degeneracy of the ground state is removed, except the double degeneracy associated with half-integer S (Jahn-Teller theorem, Section 4.3.1). The double degeneracy is a consequence of the symmetry properties of angular momenta (Kramers doublets). The combined effects of small low-symmetry components and spin–orbit coupling on the ground level originating in the chosen idealized symmetry give rise to a set of levels, relatively close in energy, whose energy separation depends on the zero field splitting (ZFS, Section 1.4). The splitting of the degenerate states in low-symmetry may be of the order of kT and higher, as it can range up to several thousands of wave numbers (cm^{-1}).

Sometimes the departure from an idealized geometry is so large that the orbitally nondegenerate ground level is separated by thousands of wave numbers from the first excited state. For example, elongated tetragonal copper(II) complexes have a $^2B_{1g}$ ground level separated by several thousands of wave numbers from the $^2A_{1g}$ level (both arising from the splitting of a degenerate E_g level in octahedral field). The energy separation in this case is not referred to as ZFS. When the system is orbitally nondegenerate even in the idealized symmetry (A and B symmetries), the zero-field splitting of the S manifold for $S \geq 1$ ranges from zero to few tens of wave numbers (cm^{-1}).

7.1.2 ELECTRON RELAXATION FOR THE DIFFERENT METAL IONS

The electron relaxation time of a metal ion largely depends on the availability of low-lying excited states, which make spin–orbit coupling efficient. Typically, T_{1e} values are longer than 10^{-11} s when the energy of the excited states is much larger than that of the ground state, whereas they are shorter than 10^{-11} s

Table 7.2 Common d-Orbital Splitting Patterns in Octahedral (O_h), Tetrahedral (T_d), and Tetragonal (D_{4h} or C_{4v}) Symmetries

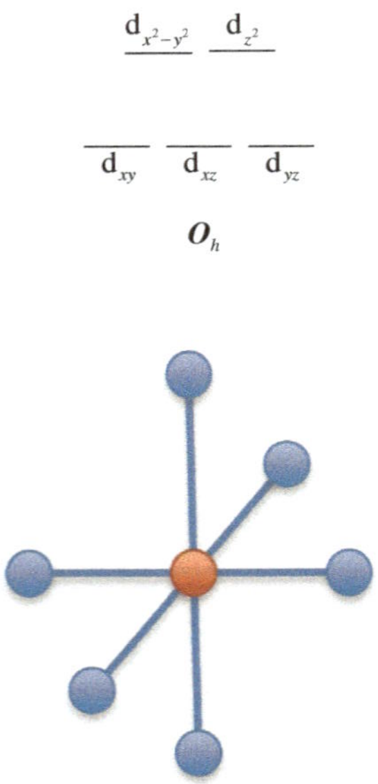

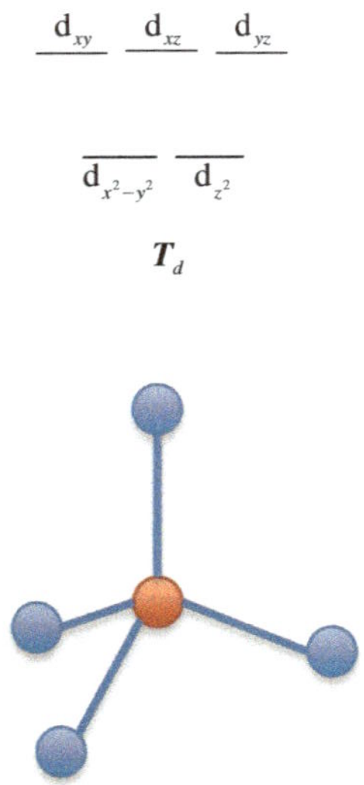

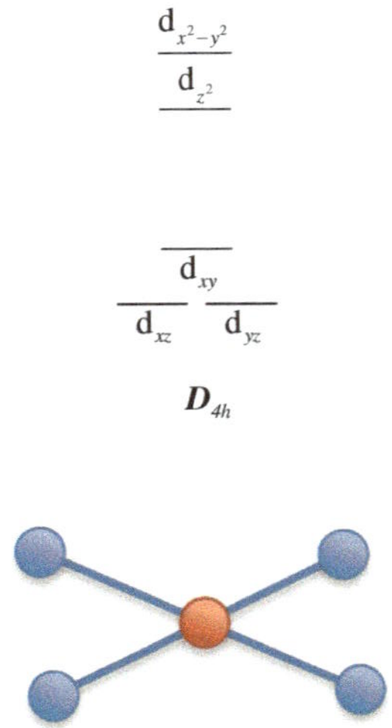

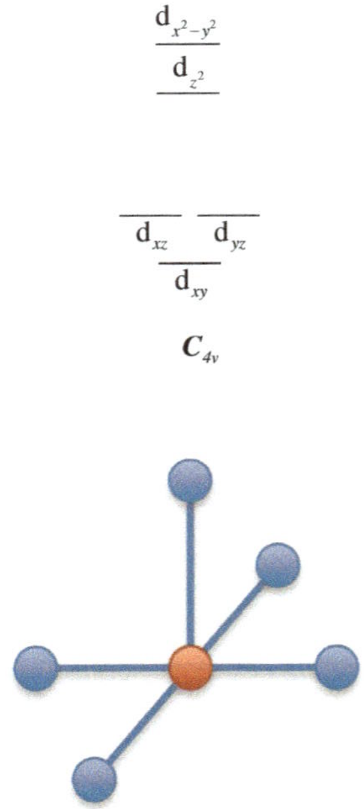

when the differences in energy of the excited states with respect to the ground state are comparable to kT. Therefore, T_{1e} can also change depending on the molecular geometry and on the chelating groups.

Metal ions with $S = ½$ have no ZFS. They can be divided into three classes, according to the spin–orbit interaction of the paramagnetic center:

1. Complexes with the ground state well separated from the excited states. They usually have very long electron relaxation time ($\approx 10^{-7}$ s), like in radicals.
2. Complexes with excited states far above the ground state in energy and where the spin-orbit coupling causes anisotropy in the hyperfine coupling to the metal nucleus (Section 4.7.1) and in the $\boldsymbol{g}$ tensors. In these cases, electron relaxation occurs through modulation of the $\boldsymbol{A}$ and $\boldsymbol{g}$ anisotropy. The electron relaxation times are typically relatively long (10^{-8}–10^{-10} s at room temperature) and field independent. Examples are Cu^{2+} and VO^{2+} aqua ions.
3. Complexes with low-lying excited states, with ground levels deriving from an orbitally degenerate ground level in idealized symmetry, allowing the Orbach mechanism to operate efficiently: typical examples are Ti^{3+}, with $T_{1e} \approx 10^{-11}$ s, and low spin Fe^{3+}, with $T_{1e} \approx 10^{-12}$ s.

In metal ions with $S > 1/2$, the electronic configuration of the paramagnetic center can dramatically alter the relaxation properties. Complexes with a nondegenerate ground state (A/B symmetry), where the excited electronic levels are high in energy, typically have: (1) nearly isotropic $\boldsymbol{g}$ tensor, with value similar to that of free electrons, (2) isotropic $\boldsymbol{A}$ tensor, (3) small transient ZFS. In these complexes the most efficient electron relaxation mechanism is often the modulation of transient ZFS with a correlation time independent of τ_r and ascribed to the collisions with the solvent molecules (responsible of the deformation of the coordination polyhedron causing transient ZFS). The electron relaxation is usually field dependent with $T_{1e} \approx 10^{-9}$–10^{-11} s at low field. Typical examples are Mn^{2+}, high spin Fe^{3+}, Cr^{3+}, Ni^{2+} (in O_h symmetry), Gd^{3+} aqua ions. In complexes where a static ZFS is present (as for Mn^{2+} or Fe^{3+} complexes), also modulation of the latter with a correlation time related to τ_r is a further possible electron relaxation mechanism, which may coexist with the collisional mechanism. In such low symmetry systems, it is reasonable to expect the magnitude of transient ZFS is related to that of the static ZFS, as the former can be seen as a perturbation of the latter. As a consequence, systems with increasing static ZFS experience faster electron relaxation rates.

In metal ions with $S > 1/2$ and with nearly doubly degenerate or triply degenerate ground states (i.e., with idealized E or T symmetry) the Orbach-type mechanism can efficiently operate due to the low-lying energy levels. The electronic relaxation times are usually short, of the order of 10^{-12} s. Typical examples are high spin pseudooctahedral Co^{2+}, pseudotetrahedral nickel(II), high spin Fe^{2+}, lanthanoids(III) except Gd^{3+}. High spin pseudotetrahedral cobalt(II) and pseudooctahedral nickel(II) have excited states higher in energy than those in the pseudotetrahedral and pseudooctahedral analogs, so the Orbach and Raman relaxation mechanisms are expected to be relatively less efficient and probably comparable with the modulation of the quadratic ZFS.

In order to investigate electron and nuclear properties, it is convenient to study the interactions of a paramagnetic metal ion with water protons, by measuring longitudinal water proton relaxation rates as a function of magnetic field. Details on the technique used to perform relaxation measurements as a function of magnetic field (Field-cycling relaxometry) are reported in Section 12.5.2. This type of measurements, called nuclear magnetic relaxation dispersion (NMRD), permit the determination of the field dependence of R_{1M} (as depicted in Sections 4.4–4.6) by exploiting the chemical exchange of the bound water with the bulk water molecules according to Eq. (6.10) (Section 6.7).

Table 7.3 Summary of the ^{1}H NMRD Parameters for Some Aqua Ion Complexes at 298 K

Metal Ion	I	S	A/h (MHz)	$T_{1e(0)}$ (ps)[a]	τ_v (ps)	τ_c[b]	Main Electron Relaxation Mechanism
Cu(II)	3/2	½	0–0.2	3000 (type 1: 500)	–	τ_r	Raman, g, A-anisotropy, dynamic Jahn-Teller
VO(IV)	7/2	1/2	2.1	500	–	τ_r	A-anisotropy, dynamic Jahn-Teller
Ti(III)		1/2	4.5	40	–	τ_r, T_{1e}	Orbach
Ni(II)		1	0.2	3–10	2	T_{1e}	ZFS modulation
Cr(III)		3/2	2.0	400	2	τ_r	ZFS modulation
Mn(II)	5/2	5/2	0.7	3000	5	τ_r	ZFS modulation
Fe(III) HS		5/2	0.4	90	5	τ_r, T_{1e}	ZFS modulation
Fe(II) HS		2		≈1	–	T_{1e}	Orbach
Co(II)	7/2	3/2	0.4	3–6	–	T_{1e}	Orbach, ZFS modulation

[a] *See also Section 4.3.*
[b] *correlation time for dipolar relaxation*

A summary of the electron relaxatation parameters derived through ^{1}H NMRD for some aqua ion complexes at 298 K is reported in Table 7.3. These parameters will be discussed in detail in the next sections.

7.1.3 MAGNETIC SUSCEPTIBILITY ANISOTROPY FOR THE DIFFERENT METAL IONS

Pseudocontact shifts and paramagnetic residual dipolar couplings depend on the magnetic susceptibility anisotropy, which, on the other hand, can be estimated from the **g** anisotropy using Eq. (1.38), valid when the ground state is well isolated from excited electronic states and ZFS is negligible, or Eq. (1.40), which is generally valid. The **g** anisotropy can be interpreted as a deviation of **g** from the free electron g_e value along the different directions: this deviation increases with increasing the spin-orbit coupling and decreases with increasing the energy separation between the d orbitals [3]. In the presence of well separated excited states, the **g** anisotropy, and thus also the magnetic susceptibility anisotropy, is modest, whereas in the presence of low-lying excited states it can be quite large, and also the magnetic susceptibility anisotropy can increase up to 3×10^{-32} m^3 (for Ni^{2+}, Fe^{3+}) or even 7×10^{-32} m^3 (for high spin Co^{2+}). For the same reason, very large magnetic susceptibility anisotropies can arise in lanthanoids (depending on the g_J and J values, Chapter 8).

7.2 A QUICK SUMMARY OF FIRST ROW TRANSITION METAL IONS

We here briefly summarize the most common oxidation states of first row transition metal ions, along with their preferred geometry and subsequent magnetic properties. This summary is by no means intended to be exhaustive, and the reader is referred to manuals of inorganic chemistry for a more

Table 7.4 Electronic Configuration of the Main Paramagnetic Oxidation States of Transition Metal Ions

Ti(III)	V(IV)	Cr(III)	Mn(II)	Mn(III)	Fe(III)	Fe(II)	Co(II)	Ni(II)	Cu(II)
$3d^1$	$3d^1$	$3d^3$	$3d^5$	$3d^4$	$3d^5$	$3d^6$	$3d^7$	$3d^8$	$3d^9$

thorough description and further understanding. The electronic configuration of the main oxidation states of the paramagnetic transition metal ions are summarized in Table 7.4 for readers' convenience.

7.2.1 TITANIUM

Titanium compounds are known with oxidation numbers −1 to 4, the more stable being by far the +4 state. States −1, +2, and +3 are paramagnetic.

7.2.2 VANADIUM

Compounds of vanadium are known with oxidation states −3 to 5, the most common being +4 (d^1). Vanadium has 100% naturally abundant ^{51}V nuclear isotope, that has $I = 7/2$ and a substantial gyromagnetic ratio.

7.2.3 CHROMIUM

Chromium has a vast and varied coordination chemistry, and compounds are known with oxidation numbers ranging from −4 to +6, the most stable being the +3 and +6 states. Oxidation numbers +2 (d^4, not in trigonal bipyramidal geometry), +3 (d^3), +4 (d^2), and +5 (d^1) are paramagnetic. Less stable paramagnetic oxidation states are accessible to chromium: (1) chromium(II) has the same electronic structure as manganese(III), i.e. $3d^4$, $S = 2$, although the stability is much lower. As in the case of manganese(III), ZFS is of the order of 2–3 cm^{-1}; (2) chromium(V) is highly unstable, however complexes can be obtained through reduction of chromium(VI) in the presence of diols—these compounds have the same electronic configuration ($3d^1$) as oxovanadium(IV) and were used for dynamic nuclear polarization [4].

7.2.4 MANGANESE

Also manganese has a substantially vast coordination chemistry, with oxidation states ranging from −3 to 7. Common states are Mn^{2+} (d^5) and Mn^{3+} (d^4).

7.2.5 IRON

Iron compounds are known with positive oxidation numbers ranging from +2 to +6, the most stable being by far the +2 (d^6) and +3 (d^5) states. In complexes, iron(III) commonly gives rise to high ($S = 5/2$) or low ($S = 1/2$) spin states, depending on geometry and ligand field strength. Iron(II) tetrahedral or pseudotetrahedral complexes are always high spin ($S = 2$), octahedral or pseudooctahedral complexes may have either high spin ($S = 2$) or low spin ($S = 0$) ground states. Iron(II) complexes with $S = 1$ can be observed when the deviation from octahedral symmetry is large. Iron(IV) (d^4) is also not unusual, although much less common, than iron(II) and iron(III), when stabilized by a tight coordination environment.

7.2.6 COBALT

The most common cobalt oxidation numbers are +2 and +3. The first (d^7) is favored by low ligand field strength. As a consequence, cobalt(III) compounds, that are obtained with ligands with high ligand field strength, tend to be low-spin (diamagnetic), at variance with iron(II).

7.2.7 NICKEL

The most common oxidation number of nickel is +2 (d^8). As such, it strongly favors square planar complexes (diamagnetic). In case of steric constriction, tetrahedral (paramagnetic) or pentacoordinated (diamagnetic or paramagnetic) complexes are possible; with weak ligands, pseudooctahedral (paramagnetic) complexes are also possible. Nickel can also be found with oxidation number +3 (d^7), as for instance in metallocenium ions.

7.2.8 COPPER

Copper is usually found in +1, +2, and +3 oxidation states. Copper(II) (d^9) is by far the most common and stable. The +1 oxidation state being possible only in anaerobic condition with coordinating solvent and/or stabilized by crystal packing forces (e.g., copper(I) iodide) and has a strong tendency to disproportionate in copper(II) and copper(0). Copper(III) oxidation state is quite unusual, although it can be achieved by forcing the coordination environment to be square planar. As such, although copper(III) is a d^8 ion, it is intrinsically diamagnetic, as square planar nickel(II).

7.3 PROPERTIES OF PARAMAGNETIC IONS SORTED BY ELECTRON CONFIGURATION

7.3.1 d^1/d^9 (Cu^{2+}, Ti^{3+}, VO^{2+})

7.3.1.1 Copper(II), type 2

Copper(II) has a $3d^9$ electronic configuration, which gives rise to a ^2D free ion ground state (Fig. 7.1). In ideal octahedral symmetry the ground state is ^2E. However, pure octahedral and tetrahedral symmetries can never be observed because Jahn–Teller distortions (Section 4.3.1) remove the orbital degeneracy of the ground state. The separation of the electronic energy levels depends on the coordination number and stereochemistry, as well as on the nature of the ligands. In nonidealized geometry (often of D_{4h} or C_{4v} type) the orbitally nondegenerate ground state is separated by several thousands of wave numbers (cm^{-1}) from the first excited state. Therefore, copper(II) behaves as an isolated ground state and exhibits long electron relaxation time (of the order of 10^{-9} s), because electron relaxation mechanisms are relatively inefficient. Copper(II) complexes have thus relatively sharp EPR signals, and it is often possible to record these spectra at room temperature. The main electron relaxation mechanism is probably a Raman-type process and/or modulation of $\boldsymbol{g}$ and $\boldsymbol{A}$ anisotropy by molecular tumbling for small complexes with $\tau_r \approx 10^{-11}$ s [5], or other rotation-independent motions for macromolecules.

In six-coordinated complexes, due to dynamic Jahn–Teller effects, instantaneous elongations occur along the three coordination axes, yielding a nondegenerate ground state. Elongation also changes the hyperfine coupling constant between the copper nucleus and the unpaired electrons. It has been proposed that this random process, with a correlation time of about 5×10^{-12} s (mean lifetime of the

Conf.	L	S	Free ion conf.	Ligand field									
				Octahedral (O_h)		Tetrahedral (T_d)		Square planar (D_{4h})		Square pyramid. (C_{4v})		Trigonal bipyram. (D_{3h})	
d^9	2	1/2	^{2}D	↓↑ ↑ ↓↑ ↓↑ ↓↑	^{2}E	↓↑ ↓↑ ↑ ↓↑ ↓↑	^{2}T	↑ ↓↑ ↓↑ ↓↑ ↓↑	^{2}A	↑ ↓↑ ↓↑ ↓↑ ↓↑	^{2}A	↑ ↓↑ ↓↑ ↓↑ ↓↑	^{2}A

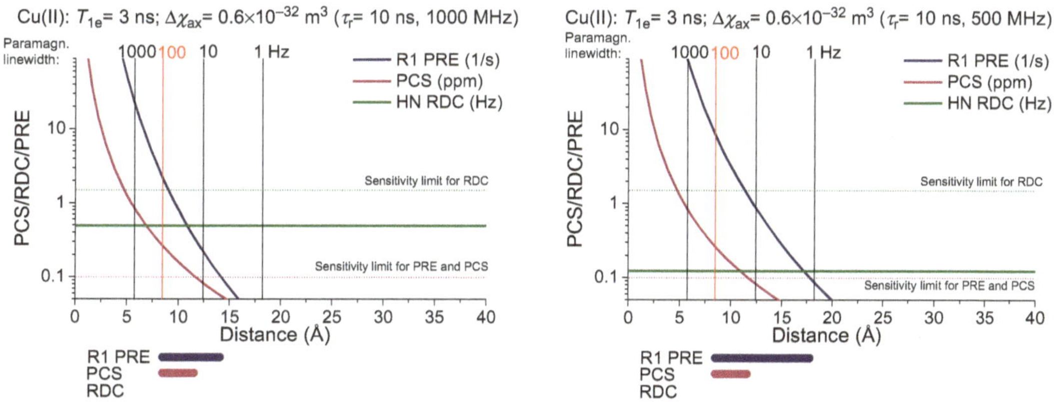

FIGURE 7.1 Population of the electronic d orbitals in different idealized symmetries in copper(II) complexes and ^{1}H R_1 paramagnetic relaxation enhancement (in s^{-1}, *blue line*), maximum pseudocontact shift (in ppm, *pink line*) and maximum HN residual dipolar coupling (in Hz, *green line*) as a function of the distance from the metal ion, for typical values of electron relaxation time and axial magnetic susceptibility anisotropy, for a molecule with τ_r of 10 ns, at 1000 MHz (left panel) or 500 MHz (right panel). The plots also shows the distances with ^{1}H paramagnetic linewidth of 1, 10, 100, and 1000 Hz. The bars under the plots indicate the range of distances with ^{1}H R_1 paramagnetic relaxation enhancement, pseudocontact shift and HN residual dipolar coupling values above the experimental sensitivity limits. The maximum pseudocontact shift is obtained for a nucleus on the z-axis of the magnetic susceptibility tensor; the maximum residual dipolar coupling is obtained for an amide vector parallel to the z-axis of the magnetic susceptibility tensor.

complex between hops), can actually be a further electron relaxation mechanism operative for symmetric complexes, as, for example, the copper aqua ion [6–8].

Due to the well separated excited states, the magnetic susceptibility anisotropy is modest. For copper proteins, estimates for $\Delta\chi_{ax}$ of about 6×10^{-33} m^3 were obtained [9,10]. Fig. 7.1 shows the typical maximum values of pseudocontact shifts and paramagnetic amide proton residual dipolar couplings expected as a function of the distance from the metal ion, at 500 or 1000 MHz, for a protein with reorientation time of 10 ns. The figure also shows the ^{1}H paramagnetic relaxation enhancement and linewidths calculated with an electron relaxation time of 3 ns. In the assumption that the detection limit of the nuclear peaks in standard NMR spectra corresponds to a paramagnetic linewidth of 100 Hz, relaxation enhancements

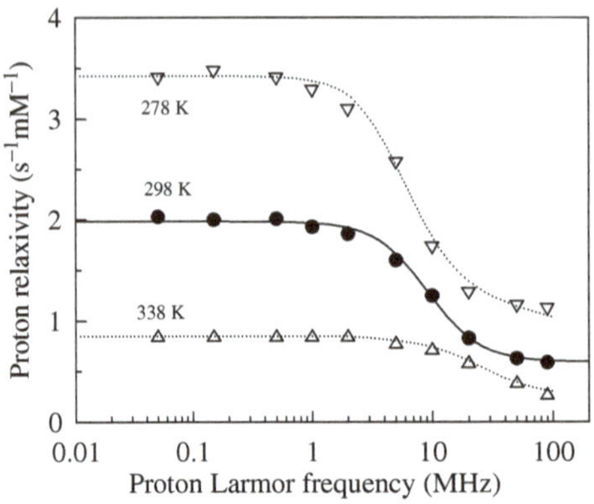

FIGURE 7.2 Water ^{1}H NMRD profiles for an aqueous solution of $Cu(H_2O)_6^{2+}$ at 298 K (•), 278 K (∇), and 338 K (Δ). The lines represent the best-fit curves obtained using the Solomon equation and $\tau_c = 2.6 \times 10^{-11}$ s (298 K), 4.0×10^{-11} s (278 K) and 0.95×10^{-11} s (338 K).

and pseudocontact shifts are calculated to be measurable for distances larger than about 8 Å. Since they decrease as r^{-6} and r^{-3} (r is the nuclear-metal distance), respectively, their range of observability is thus from about 8 Å to the distance corresponding to the sensitivity limit for these observables, which can be estimated around 0.1 s^{-1} and 0.1 ppm, respectively. Paramagnetic residual dipolar couplings are not observed because from the typical $\Delta\chi_{ax}$ values they cannot be larger than 1 Hz.

The water proton NMRD profile of copper(II) aqua ion at 298 K [11] (Fig. 7.2) is in good accordance with what expected from the point dipole–point dipole relaxation theory, as described by the Solomon equation [Eq. (4.17)]: only one dispersion appears, and the relaxivity values before and after the dispersion are in the 10:3 ratio. The best fitting procedure applied to a configuration of 12 water protons bound to the metal ion provides a distance between water protons and the paramagnetic center equal to 2.7 Å, and a correlation time of 3×10^{-11} s, which defines the position of the ω_S dispersion. The correlation time is determined by rotation as expected from the Stokes–Einstein equation [Eq. (4.8)]. The electron relaxation time is in fact expected to be one order of magnitude longer (Table 7.3). This also ensures the absence of any contact relaxation contribution as otherwise the corresponding dispersion would have been present in the NMRD profile. Actually, the unpaired electron sits in an e_g orbital, and no delocalization is thus expected on the water molecule. In fact, one fully occupied water molecular orbital of σ type directly overlaps through π bonding with the t_{2g} metal orbital set [12]. Profiles acquired at different temperature warrant that the system is in a fast exchange regime, i.e. $\tau_M << T_{1M}$, as proved by the decrease in relaxivity for increasing temperatures. Furthermore, at higher temperatures, the dispersion is centered at a lower frequency, in agreement with a faster reorientation correlation time. An alternative model for copper(II) aqua ion has been proposed [13], in which the ion is believed to be mostly five coordinated, and rapidly cycling between square pyramidal and trigonal bipyramidal geometries.

When the viscosity of the solution is increased by using ethyleneglycol or glycerol-water mixtures as solvent, the rotational correlation time increases. This determines: (1) higher relaxivity values at low frequencies; (2) a shift toward lower frequencies of the ω_S dispersion; (3) the appearance of a second dispersion (ascribed to the ω_I dispersion) at high fields. Fig. 7.3 shows the ^{1}H NMRD profiles of copper(II) in ethylene glycol at different temperatures [14]. The profile maintains the shape

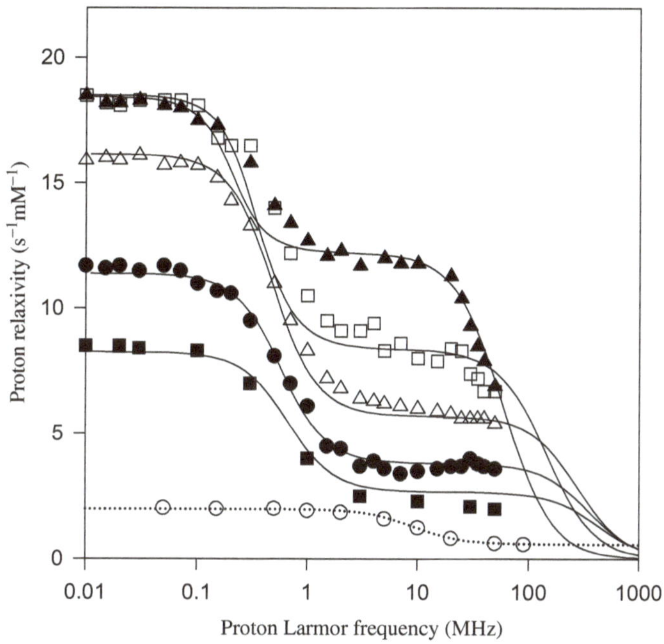

FIGURE 7.3 Solvent ^1H NMRD profiles for ethyleneglycol solutions of $Cu(ClO_4)_2 \cdot 6H_2O$ at 264 (▲), 278 (□), 288 (Δ), 298 (●), and 312 (■) K as compared to those of water solution at 298 K (○) [14]. The solid lines are best fit curves obtained using an isotropic *A* value.

predicted by the Solomon equation [Eq. (3.16)] until τ_r is increased so much that the correlation time is determined by the electron relaxation. In such condition, in fact, electron energy levels and their transition probability must be calculated by considering, besides the Zeeman interaction energy, also the hyperfine coupling between metal ion (copper(II) has $I = 3/2$) and the unpaired electron ($S = 1/2$). As shown in Section 4.7.1, the effect on the shape of the relaxivity profile can be quite dramatic, in the range of frequencies where the hyperfine coupling energy is larger than the Zeeman energy, i.e. $Ach \geq g_e\mu_B B_0$, where A is the electron-copper nucleus hyperfine coupling constant in cm^{-1}, and thus for proton Larmor frequency smaller than $Ac\gamma_I/\gamma_S$. The profiles were fit with an isotropic hyperfine constant of 0.0026 cm^{-1}. The presence of hyperfine coupling between metal ion and unpaired electron can be easily recognized by observing that the ratio in relaxation rate before and after the ω_S dispersion is different from the expected 10/3 value.

^1H NMRD profiles of copper(II) proteins should be always fit by considering the presence of hyperfine coupling between the unpaired electron and the metal nucleus. The shape of the ^1H NMRD profiles can be very different depending on the values of $A_{\parallel}$, $A_{\perp}$, and on the position of the protons with respect to the molecular frame defined by the hyperfine A tensor. The values of $A_{\parallel}$ and $A_{\perp}$ can be obtained by EPR measurements or directly from the relaxation profile. Typical $A_{\parallel}$ values for type 2 copper containing proteins at room temperature are between 120 and 190 $\times 10^{-4}$ cm^{-1}, as measured in copper bovine carbonic anhydrase II and its anion adducts, Cu_2Zn_2 superoxide dismutase and adducts, Cu_2 transferrin, Cu_2 alkaline phosphatase, Cu phthalate dioxygenase [15], whereas $A_{\perp}$ is usually very small, up to 40 $\times 10^{-4}$ cm^{-1}. The electron relaxation time has values in the 10^{-9}–10^{-8} s range, i.e. one

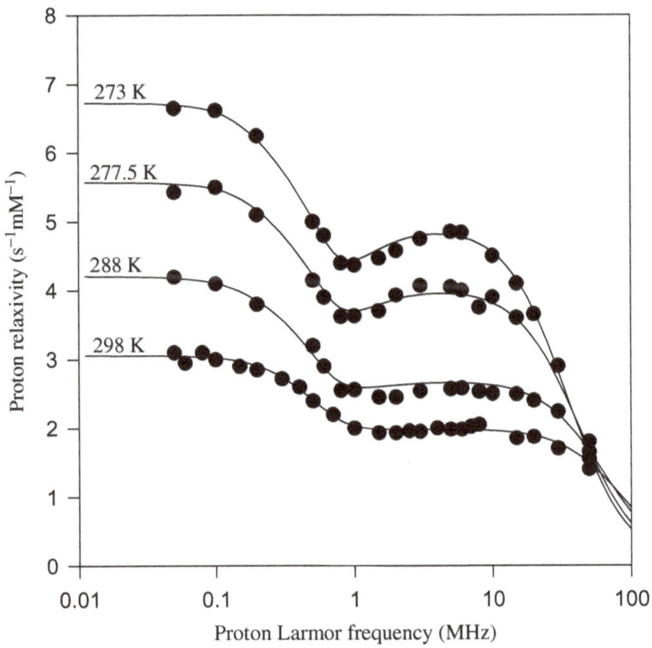

FIGURE 7.4 Water ^{1}H NMRD profiles for superoxide dismutase solutions at various temperatures [18]. The *solid lines* are best-fit curves obtained with the inclusion of the effect of hyperfine coupling with the metal nucleus.

Adapted from Ref. [19].

order of magnitude longer than the aqua ion, and is essentially independent of the reorientation time of the macromolecule and the viscosity of the solution. Therefore, rotation independent mechanisms have to be operative. We also find that T_{1e} decreases with increasing temperature. No field dependence in the electron relaxation time was ever found in the investigated region between 0.01 and 100 MHz of proton Larmor frequency, nor at 800 MHz when high resolution is achieved [10].

As an example, the ^{1}H NMRD profiles of the copper(II) protein superoxide dismutase are shown in Fig. 7.4 for different temperatures [16]. Copper(II) in this protein sits in a distorted tetragonal coordination environment. The profiles show the ω_S and ω_I dispersions, the rate between the relaxivity values before and after the ω_S dispersion being much different from 10:3. Low temperature profiles also show a relaxivity peak at about 5 MHz, which can be accounted for by considering the hyperfine coupling to the metal nuclear spin (Section 4.7.1). The fits indicate the presence of an axial water molecule, coordinated to the copper ion with a Cu–H distance of 3.2 Å, T_{1e} values ranging from 4.6×10^{-9} s at 273 K to 1.8×10^{-9} s at 298 K, and $A_{\parallel} = 0.0137$ cm^{-1} and $A_{\perp}$ negligible, as obtained from EPR measurements [17]. Best fit values for several copper proteins are given in Table 7.5. The distance r is reported for two water protons coordinated to the metal ion. If no exchangeable proton is present in the first coordination sphere, any paramagnetic effect would be due to second sphere and/or outer-sphere water molecules.

As a second example, the ^{1}H NMRD profiles of the protein copper(II)-CopC at 278, 288, and 303 K are shown in Fig. 7.5 [10]. The fit provided the correlation time τ_c equal to $0.109 \times 10^{-12}\, e^{3012/T}$ s (where T is the temperature), corresponding to 5×10^{-9} s at 278 K and 2×10^{-9} s at 303 K, $A_{\parallel}$ equal to 177×10^{-4} cm^{-1}, $A_{\perp} \approx 0$, $\theta = 42$ degrees and two water protons at 3.5 Å. Both electron relaxation time and reorientation time are expected to contribute to the value of τ_c, as τ_r is estimated around 5×10^{-9} s

Table 7.5 Best Fit Values of NMRD Parameters for Several Copper Proteins

Complex	T (K)	$r_{M\text{-}H}$[a] (Å)	$\tau_c \approx T_{1e}$ (ns)	$A_{\parallel}$ (cm^{-1})	$A_{\perp}$ (cm^{-1})	References
CuBCA II	298	2.8	1.9	0.0131	0.0020	[20,21]
CuBCA II + HCO_3^-	298	3.4	2.1	0.0131	0.0020	[20,21]
CuBCA II + N_3^-	298	2.7	2.6	0.0124	0.0040	[20,21]
CuBCA II + $C_2O_4^{2-}$	298	3.6	3.1	0.0150		[22]
Cu_2Zn_2SOD	298	3.2	1.8	0.0137	0.0040	[19]
	288	3.2	2.5	0.0137	0.0040	[19]
	278	3.2	3.8	0.0137	0.0040	[19]
	273	3.2	4.6	0.0137	0.0040	[19]
Cu_2Zn_2SOD + NCO^-	298	3.9	3.2	0.0158		[23]
Cu_2Zn_2SOD + NCS^-	298	3.6	2.4	0.0148		[23]
Cu_2Zn_2SOD + N_3^-	298	4.9	3.9	0.0157		[23]
Cu_2Zn_2SOD + CN^-	298	5.1	7.7	0.0188		[23]
Cu_2Cu_2 SOD	298	3.4	4.2	0.0143	0.0020	[24]
Cu_2E_2 SOD	298	3.2	3.6	0.0148	0.0020	[24]
Cu_2TRN	281	3.9	7.6	0.0167	0.0020	[25]
	298	3.7	5.7	0.0167	0.0020	[25]
	311	3.7	5.4	0.0167	0.0020	[25]
Cu_2AP	298	3.0	3.0	0.0164		[26]
Cu_2Mg_4AP + 2P_i	298	3.5	3.0	0.0129		[26]
Benzylamide oxidase	298	2.7	7.0	0.0165		[27]
CuPDO	298	3.4	5.4	0.0153	0.0027	[28]
CuPDO + phthalate	298	2.5	13	0.0168		[28]
Azurin (type 1)	293	5	0.8	0.0062		[29]
Azurin His46Gly (type 2)	293	3.2	15	0.0160		[29]

Abbreviations: BCA II, bovine carbonic anhydrase II; SOD, superoxide dismutase; TRN, transferrin; AP, alkaline phosphatase; P_i, inorganic phosphate; PDO, phthalate dioxygenase; E, empty.
[a]*Assuming two equivalent protons.*

at 303 K. It is worth noting that it was necessary to acquire the profiles at different temperature and to fit them simultaneously in order to obtain a unique fit. In fact, each profile can be fit independently with more than one set of parameters.

In copper(II) systems, a small (10–15%) g-anisotropy is always present. Calculations show that such modest g-anisotropy has negligible effect on the dispersion curves, thus affecting the resulting best fit parameters only slightly [19].

A copper(II) complex with square-pyramidal geometry and its ^{1}H NMR spectrum are shown in Fig. 7.6. The relatively broad signals, in the range from 20.4 to −13 ppm, as typical for mononuclear

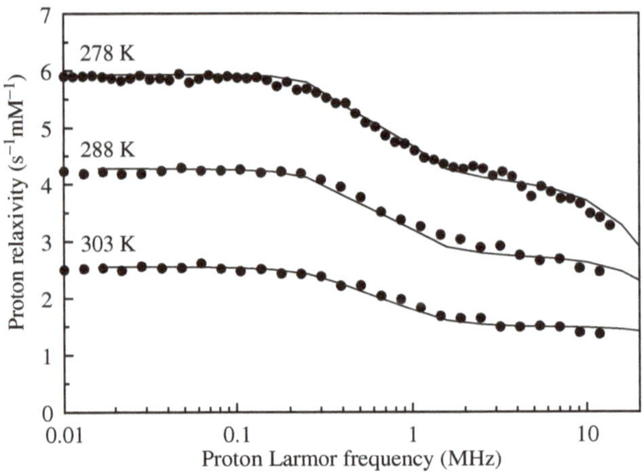

FIGURE 7.5 Water ^{1}H NMRD profiles for solutions of copper(II)-CopC obtained at three different temperatures.

Reproduced with permission from Ref. [10].

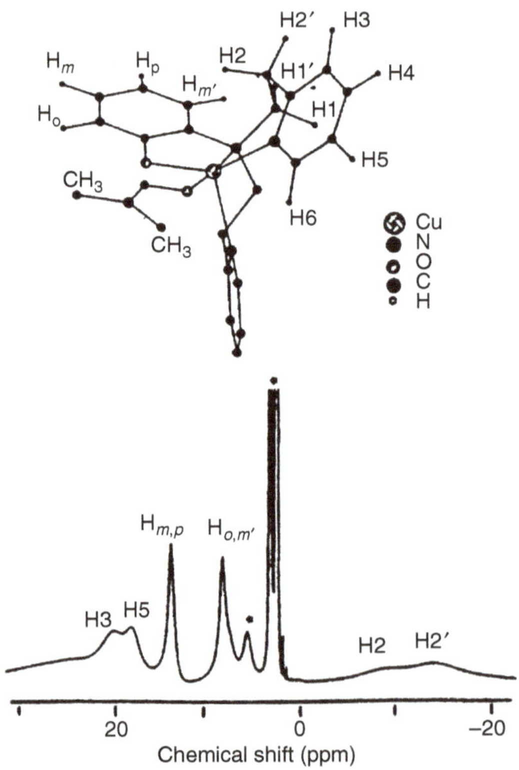

FIGURE 7.6 ^{1}H NMR spectrum at 400 MHz of a copper(II) complex (inset) illustrating the modest spectral resolution. The asterisk represents peaks due to DMF.

Adapted from Ref. [30].

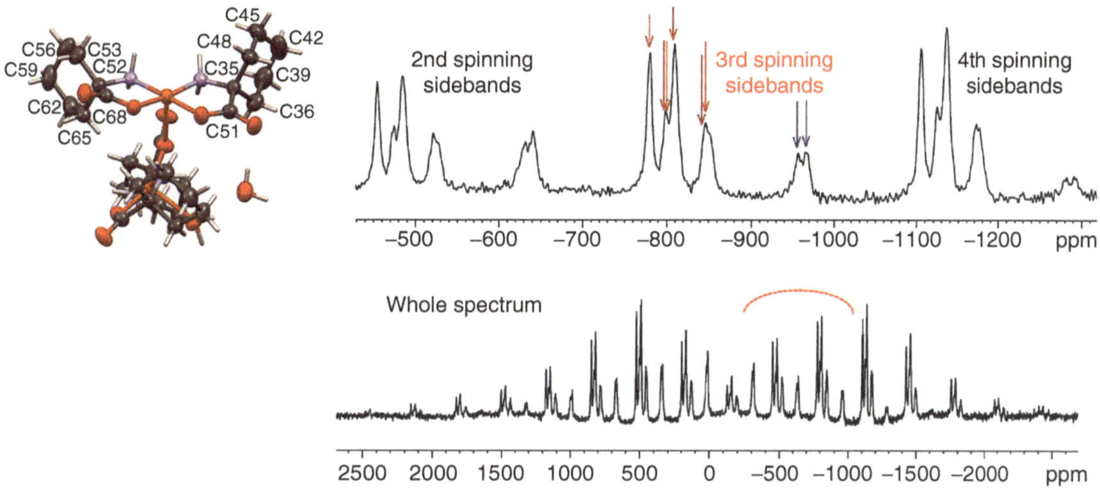

FIGURE 7.7 ^{2}H fast-MAS spectrum of the deuterated form of the copper complex shown in the inset (bis(1-amino(cyclo)hexane-1-carboxylato-κ^2N,O)copper(II); 9.38 T, MAS 20 kHz, temperature 300 K). Top: extension, 2nd to 4th low-frequency spinning sidebands. The *red and blue arrows* indicate the 3rd spinning sidebands of the ND and OD deuterons, respectively.

Reproduced with permission from Ref. [32].

copper(II) systems, are related to a correlation time of 4×10^{-11} s, which is the rotational time, as the electron relaxation time is about two orders of magnitude longer [30]. The reader is referred to Section 11.3.2 for a comparison with a dimeric copper species with similar ligands.

As mentioned in Section 12.3.8, a number of ^{2}H NMR investigations have been carried out historically on metal complexes because (1) the quadrupolar interaction does not provide relaxation in the solid state (Chapter 5), thus allowing for the detection of spectra with a good resolution, (2) the ^{2}H quadrupole interaction, which is about 200 kHz, is efficiently modulated by molecular motions occurring on a 10^{-3}–10^{-7} s time scale, thus providing an handle on such motions [31], and (3) the rather low gyromagnetic ratio of the deuteron makes it less susceptible to paramagnetic relaxation than ^{1}H NMR spectroscopy. As an example, the ^{2}H NMR spectrum of a deuterated copper(II) complex is shown in Fig. 7.7, with hyperfine shifted peaks mainly due to the contact interaction. In this case, a quadrupolar coupling constant close to 200 kHz indicates a rather rigid environment for deuterons [32]. Another brilliant example exploits two-dimensional shift-quadrupole correlation spectra (Fig. 7.8) to detect the reciprocal orientation of the quadrupolar tensor and the dipolar shielding tensor [Eq. (2.28)] [33].

The Cu_A site, common in biology (inset in Fig. 7.9), is dinuclear with two copper atoms bridged by the thiolate sulfurs of two cysteine ligands. One unpaired electron is delocalized over two metals, which are thus $Cu^{1.5+}$. The NMR spectra show narrow lines from the copper ligands (Fig. 7.9) [34,35], corresponding to an electron relaxation time of 10^{-11} s, as in Cu^{2+}–Cu^{2+} dimers (Section 11.3.2). However, in Cu_A there is no magnetic coupling between the two centers, as they contain only one unpaired electron just as an isolated Cu^{2+} ion. Electron relaxation of Cu_A may be fast because the orbital overlap between the two copper centers provides new relaxation mechanisms not available to a monomer (as Orbach or Raman relaxation).

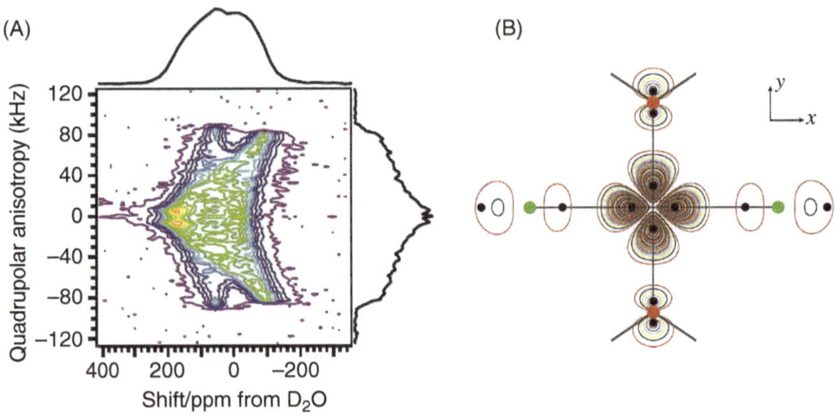

FIGURE 7.8 Shift/quadrupole correlation spectrum of $CuCl_2 \cdot 2D_2O$ (panel A). The dipolar shielding anisotropy could be fit breaking up the dipole of copper over the points with higher SOMO density represented in panel B—*green dots* indicate chloride ions, *red dots* represent the oxygen atoms of water molecules and the *black dots* indicate the placement of the partial dipoles.

Reproduced with permission from Ref. [33].

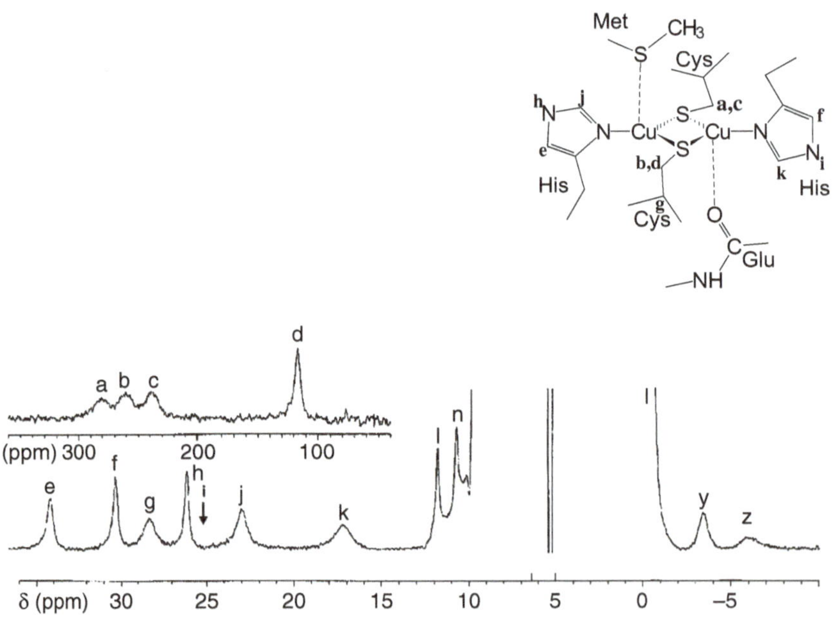

FIGURE 7.9 600 MHz ^1H NMR spectrum of water solutions of the *Thermus* Cu_A domain at pH 8 and 278 K. Signal i is observable at lower pH.

Adapted from Ref. [34].

7.3.1.2 Copper(II), type 1

When copper is bound to one sulfur atom of a cysteine and two nitrogens of two histidines, with one or two weakly bound axial ligands, in an essentially tetrahedrally distorted—trigonal ligand environment (type 1 copper proteins, Fig. 7.10), the excited levels are low in energy [36], and the T_{1e} values are reduced of about one order of magnitude with respect to type 2 copper, to about 5×10^{-10} s [37] (see azurin in Table 7.5). Examples are blue copper proteins, like ceruloplasmin and azurin, and copper(II) substituted liver alcohol dehydrogenase [29,38,39]. The estimated maximum pseudocontact shifts, amide proton paramagnetic residual dipolar couplings, ^{1}H paramagnetic relaxation enhancements and linewidths are shown in Fig. 7.11.

Proton relaxivity and electron relaxation are strictly related to the coordination environment, as shown in the study of the NMRD profiles of azurin (M_r 14000) and some of its mutants. The relaxivity of wild type azurin is very low (Fig. 7.12) due to a solvent protected copper site, with the closest water at a distance of more than 5 Å from the copper ion. The fit, performed taking into account the presence of hyperfine coupling with the metal nucleus ($A_{\parallel} = 62 \times 10^{-4}$ cm^{-1}), provided T_{1e} values of 8×10^{-10} s. Although the metal site in azurin is relatively inaccessible, several mutations of the copper ligands were performed to open it up to the solvent. The ^{1}H NMRD profiles indicate the presence of water coordination for the His117Gly (two water molecules) and His46Gly (one water molecule) mutants [29] (Fig. 7.12) and larger T_{1e} values, of 5×10^{-9} and 15×10^{-9} s, respectively. The increase in T_{1e} is made evident by the shift of the $\omega_I T_{1e}$ dispersion (between 10 and 100 MHz) to a lower Larmor frequency than that of the wild type azurin. The differences in water coordination thus change the coordination geometry of the metal ion, which seems to be the main factor contributing to the variations of the electron relaxation of copper. The different coordination environments range from the trigonal (type 1) copper site of the wild type azurin to the tetragonal (type 2) sites of the mutants: as the tetragonal character decreases, electron relaxation becomes significantly faster.

As far as NMR is concerned, the hyperfine coupled proton lines of copper(II) macromolecules are broad beyond detection, due to the long electronic relaxation times. Only partial signal assignments of oxidized amicyanin [41] and azurin [42], two blue copper proteins, were possible with standard techniques. The complete spectra of oxidized blue copper proteins plastocyanin, azurin and stellacyanin

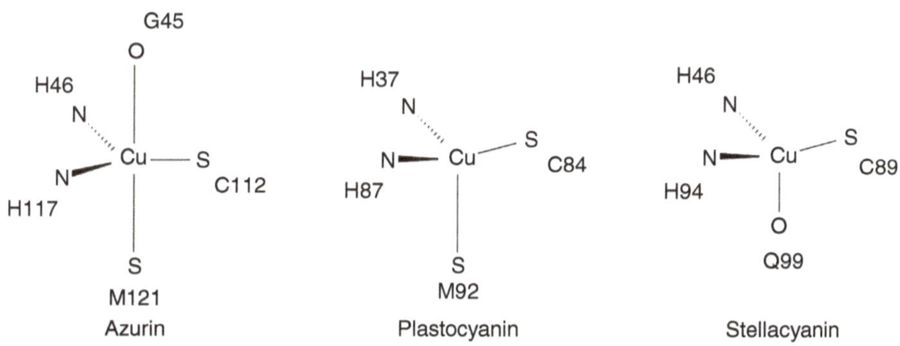

FIGURE 7.10 Schematic drawing of the active site of *P. aeruginosa* azurin, spinach plastocyanin and cucumber stellacyanin. *H*, histidine; *C*, cysteine; *M*, methionine; *G*, glycine; *Q*, glutamine.

Adapted from Ref. [40].

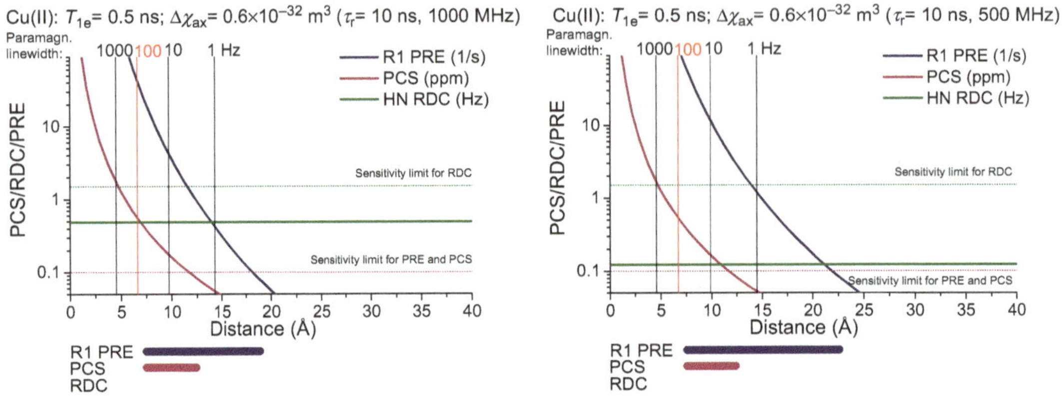

FIGURE 7.11 ^{1}H R_1 paramagnetic relaxation enhancement (in s^{-1}, *blue line*), maximum pseudocontact shift (in ppm, *pink line*), and maximum HN residual dipolar coupling (in Hz, *green line*) in type 1 copper(II) proteins as a function of the distance from the metal ion, for typical values of electron relaxation time and axial magnetic susceptibility anisotropy, for a molecule with τ_r of 10 ns, at 1000 MHz (left panel) or 500 MHz (right panel). The plots also shows the distances with ^{1}H paramagnetic linewidth of 1, 10, 100, and 1000 Hz. The bars under the plots indicate the range of distances with ^{1}H R_1 paramagnetic relaxation enhancement, pseudocontact shift and HN residual dipolar coupling values above the experimental sensitivity limits. The maximum pseudocontact shift is obtained for a nucleus on the *z*-axis of the magnetic susceptibility tensor; the maximum residual dipolar coupling is obtained for an amide vector parallel to the *z*-axis of the magnetic susceptibility tensor.

were assigned through saturation transfer with the reduced diamagnetic species (Table 7.6). To detect the β-CH_2 signals of the cysteine strongly bound to copper(II) in the trigonal plane, an audacious technique has been applied which is based on irradiation of regions where such signals are expected but not detected, since too broad, and the corresponding saturation transfer on the reduced species is observed. In this way, the protons of the copper(II)-coordinated cysteine have been located. The Cys84 β-CH_2 protons of plastocyanin were located at 650 and 489 ppm, with linewidths of 519 and

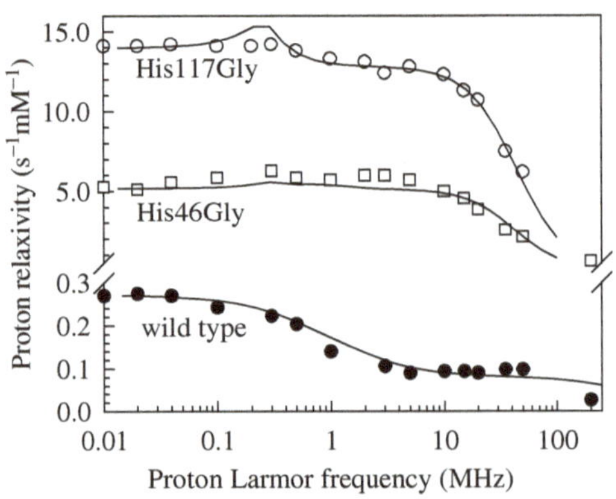

FIGURE 7.12 Paramagnetic enhancements to water ^{1}H NMRD profiles for solutions of wild type azurin (●), and its His117Gly (○) and His46Gly (□) derivatives at 278 K and pH 7.5.

Reproduced with permission from Ref. [29].

Table 7.6 Assignments of the Signals Corresponding to Copper-Ligands in Copper(II) and Copper(I) Azurin and Stellacyanin Recorded at 800 MHz

Assignment	δ_{OX}[a] (ppm)	δ_{RED} (ppm)	$\Delta\nu_{OX}$ (Hz)	A/h (MHz)
Azurin				
Hβ1/2 Cys-112	850	3.48	1.2×10^6	28/27
Hβ2/1 Cys-112	800	2.91	1.2×10^6	27/28
Hδ2 His-117	54.0	6.91	6000	1.61
Hδ2 His-46	49.1	5.92	5500	1.49
Hε1 His-117/46	46.7	6.78		1.45/1.02
Hε1 His-46/117	34.1	6.87		1.06/1.48
Hε2 His-117	27	11.69		
Hε2 His-46	26.9	11.46	1400	0.56
Hα Asn-47	19.9	4.69	1200	0.52
Hα Cys-112	–7.0	5.79		–0.38
NH Asn-47	–30	10.68		–1.3
Stellacyanin				
Hβ1/2 Cys-89	450	2.61	2.8×10^5	16/13
Hβ2/1 Cys-89	375	2.43	2.1×10^5	13/16
Hδ2 His-94/46	55.0	7.01	3800	1.77/1.51
Hδ2 His-46/94	48.0	7.10	3000	1.53/1.78
Hε1 His-94/46	41.2	7.60	3750	1.40/1.01
Hε1 His-46/94	29.8	7.42	2700	0.70/1.09
Hε2 His-46	26	10.10		0.62
Hα Asn-47	16.9	4.46	420	0.47
Hα Cys-89	–7.5	5.10		–0.41
NH Asn-47	–15	10.50		–0.8

[a]*The estimated errors are ±0.2 ppm for signals detected directly and ±10% for signals detected indirectly through saturation transfer.*

329 kHz [43]. Analogously, the hyperfine shifts and linewidths of the latter signals for oxidized azurin and stellacyanin were obtained, and it was observed that they differ dramatically from one protein to another: average hyperfine shifts of about 850, 600, and 400 ppm, and average linewidths of 1.2, 0.45, and 0.25 MHz are observed for azurin, plastocyanin, and stellacyanin, in that order (Fig. 7.13B). The contact hyperfine coupling constants for protons belonging to copper(II)-bound protein residues were calculated from the contact shifts, after correcting for the small pseudocontact contributions. The dependence of the contact shift on the Cu–S–C–H$^{\beta}$ dihedral angle is of the $\sin^2 \theta$ type, as expected for a spin density on β protons depending from an overlap between the sulfur p orbital and the 1s orbital of hydrogen [44]. The variations among the different proteins examined are interpreted as a

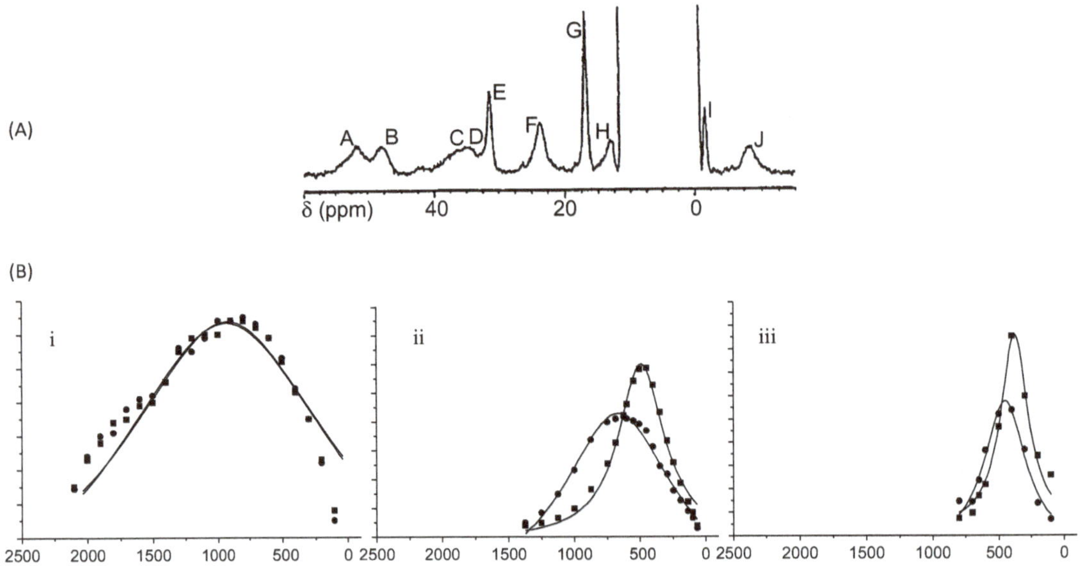

FIGURE 7.13 (A) ^{1}H NMR spectra at 298K of oxidized spinach plastocyanin at 800 MHz (adapted from [43]). (B) Far downfield region of the ^{1}H NMR spectra of oxidized (i) *P. aeruginosa* azurin, (ii) spinach plastocyanin and (iii) cucumber stellacyanin containing signals not observable in direct detection. The positions and line widths of the signals were obtained using saturation transfer experiments by plotting the intensity of the respective exchange connectivities with the reduced species as a function of the decoupler irradiation frequency.

Part (B): Adapted from [40].

measure of the out-of-plane displacement of the copper ion, related to the strength of the axial ligand(s), which increases on passing from azurin to plastocyanin to stellacyanin (Fig. 7.10). Passing from 600 to 800 MHz unambiguously indicates that the Curie contribution is negligible, as expected for small proteins containing $S = ½$ metal ions. The 800 MHz ^{1}H NMR spectrum of plastocyanin clearly shows eight downfield and two upfield hyperfine shifted signals, each of them accounting for one proton (Fig. 7.13A), with very short nuclear relaxation times. They were assigned either through saturation transfer or 1D NOE (Section 4.14).

The linewidths of the hyperfine shifted signals are determined by dipolar and contact contributions. The former depends on the reciprocal sixth power of the metal–nucleus distance, while the latter depends on the square of the hyperfine coupling constant, which in turn is proportional to the contact shift. Indeed, it was observed [43] that the linewidths of some signals (e.g., the β-CH_2 protons of the bound cysteine) are dominated by contact relaxation while some others (e.g., the bound His Hε1 signals) are dominated by dipolar relaxation. In any case, both dipolar and contact contributions are proportional to the electron longitudinal relaxation time T_{1e}. For signals with similar hyperfine shifts arising from protons at similar distance from the metal ion in the three proteins, the linewidths should hold the same ratio as the T_{1e} values. According to this reasoning, and referring to the histidine $H^{\delta 2}$ protons, it was qualitatively stated that T_{1e} is the shortest for plastocyanin. It then should be slightly longer in stellacyanin, and about two times longer in azurin [for which a value of 0.8 ns was estimated [29] (Table 7.5)].

7.3.1.3 Vanadium(IV)

Vanadium(IV) is a d^1 ion and thus, like for copper(II) that is a d^9 ion, ZFS modulation cannot be present. The electron relaxation times are expected to be long unless in the presence of low-lying excited states, and high resolution NMR is hardly performed. The oxovanadium(IV) aquaion is forced to adopt a C_{4v} rather than O_h symmetry, and thus the ground state is orbitally nondegenerate (^2B; antisymmetric with respect to the C_4 axis). Under these conditions, the first excited state is expected to be far in energy. Electron relaxation mechanisms have been proposed to be dynamic Jahn-Teller effect, modulation of hyperfine coupling with the metal nucleus, and spin-rotation. In the same coordination environment, electron relaxation in VO-proteins is about one order of magnitude slower than in type 2 copper(II) proteins, due to the stronger spin-orbit interaction of the latter ion (Section 1.4). The magnetic susceptibility anisotropy is expected to be even smaller than for copper(II). The maximum pseudocontact shifts and amide proton paramagnetic residual dipolar couplings and the ^1H paramagnetic relaxation enhancements and linewidths, shown in Fig. 7.14, are calculated for an electron relaxation time of 0.5 ns (for the aqua ion) and 20 ns (more common in low symmetry complexes and proteins), and $\Delta\chi_{ax} = 10^{-33}$ m^3.

The NMRD profiles of $VO(H_2O)_5^{2+}$ at different temperatures are shown in Fig. 7.15 [45]. The first dispersion in the profiles is ascribed to the contact relaxation and corresponds to an electron relaxation time of about 5×10^{-10} s (Table 7.3), the second to the dipolar relaxation and corresponds to the rotational correlation time of about 5×10^{-11} s. The value of the correlation time connected to the first dispersion cannot be ascribed to the presence of exchange, because the hydrogen exchange rate of the equatorial water molecules is much longer than 10^{-10} s. Therefore, it was concluded that this value corresponds to T_{1e} even if it is shorter than expected on the basis of the EPR spectra. With increasing temperature, the measurements indicate that the electron relaxation time increases, whereas the reorientation time decreases.

The protons of the four water molecules in the equatorial plane are found at a distance of 2.6 Å from the paramagnetic center, those of the fifth axial water at 2.9 Å and exchange much faster than the former. The constant of contact interaction for the equatorial water molecules is found, by fitting the NMRD data, equal to 2.1 MHz. A significant contribution for contact relaxation is actually expected because the unpaired electron sits in a T_{2g} orbital, which has the correct symmetry for directly overlapping the fully occupied water molecular orbitals of σ type [12].

By increasing viscosity, the longer value of τ_r moves the second dispersion toward lower frequency, at the same time increasing the relative contribution of the dipolar relaxation with respect to the contact relaxation [46]. The nonlinear increase of relaxation rate with τ_r at low fields provides evidence that τ_M is affecting the correlation time. When τ_r becomes longer than the electron relaxation time, the latter becomes the correlation time for nuclear relaxation.

Values of T_{1e} consistent with expectations are found for VO-protein compounds. In bis-oxovanadium(IV) transferrin, water protons are sensitive to the paramagnetic metal through second sphere water, the oxogroup occupying the only solvent-accessible coordination position. Furthermore, its EPR spectra are very sharp and characterized by a well-resolved hyperfine structure. The NMRD profile displays two dispersions (Fig. 7.16). The one at high frequency (ca. 10 MHz) is attributed to the ω_I dispersion, providing a value for τ_c of 2×10^{-8} s. Since the rotational time of the protein is of the order of $2–3 \times 10^{-7}$ s, τ_c is mainly determined by T_{1e}, with possible contributions by τ_M. The low field region is sizably affected by the presence of the hyperfine coupling between unpaired electrons and the $I = \frac{7}{2}$ metal nucleus. The constants of this interaction are $A_{\parallel} = 170 \times 10^{-4}$ and $A_{\perp} = 60 \times 10^{-4}$ cm^{-1} [25].

Conf.	L	S	Free ion conf.	Ligand field									
				Octahedral (O_h)		Tetrahedral (T_d)		Square planar (D_{4h})		Square pyramid. (C_{4v})		Trigonal bipyram. (D_{3h})	
d^1	2	1/2	^2D	– – ↑ – –	^2T	– – – ↑ –	^2E	– – – ↑ –	^2E	– – – – ↑	^2B	– – – ↑ –	^2E

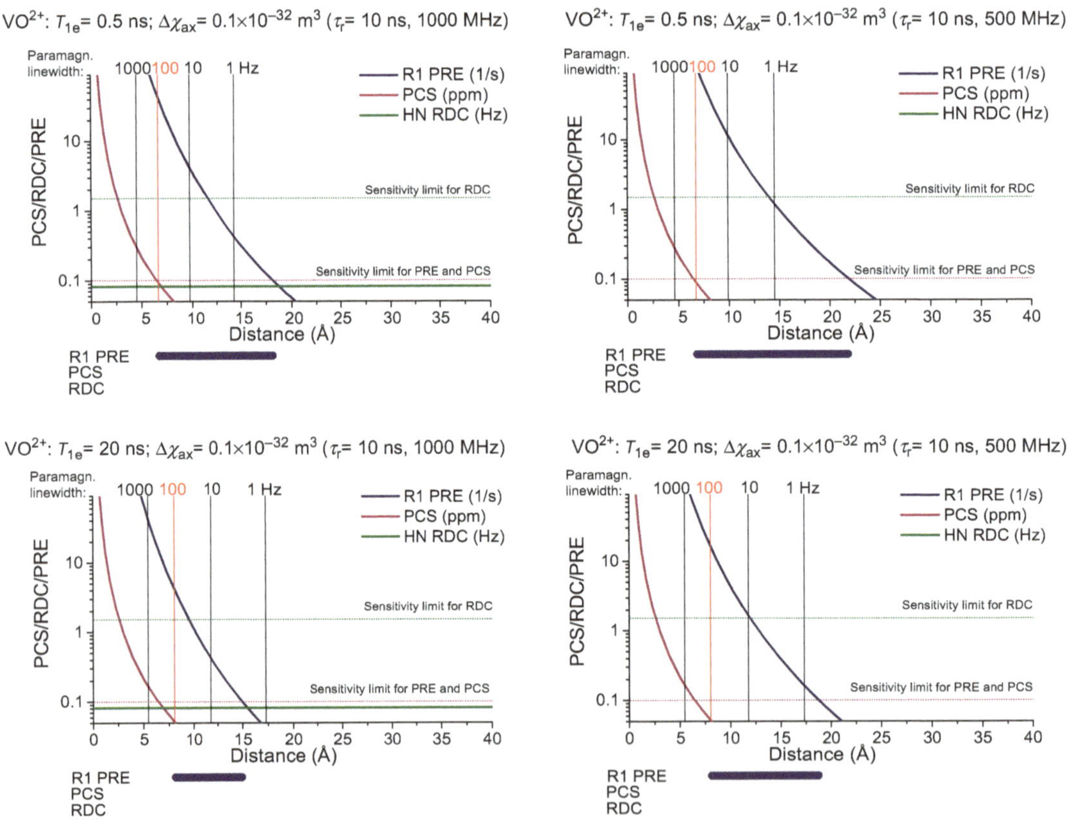

FIGURE 7.14 Population of the electronic d orbitals in different idealized symmetries in VO^{2+} complexes and ^1H R_1 paramagnetic relaxation enhancement (in s^{-1}, *blue line*), maximum pseudocontact shift (in ppm, *pink line*) and maximum HN residual dipolar coupling (in Hz, *green line*) as a function of the distance from the metal ion, for typical values of electron relaxation time and axial magnetic susceptibility anisotropy, for a molecule with τ_r of 10 ns, at 1000 MHz (left panel) or 500 MHz (right panel). The plots also shows the distances with ^1H paramagnetic linewidth of 1, 10, 100, and 1000 Hz. The bars under the plots indicate the range of distances with ^1H R_1 paramagnetic relaxation enhancement, pseudocontact shift and HN residual dipolar coupling values above the experimental sensitivity limits. The maximum pseudocontact shift is obtained for a nucleus on the z axis of the magnetic susceptibility tensor; the maximum residual dipolar coupling is obtained for an amide vector parallel to the z axis of the magnetic susceptibility tensor.

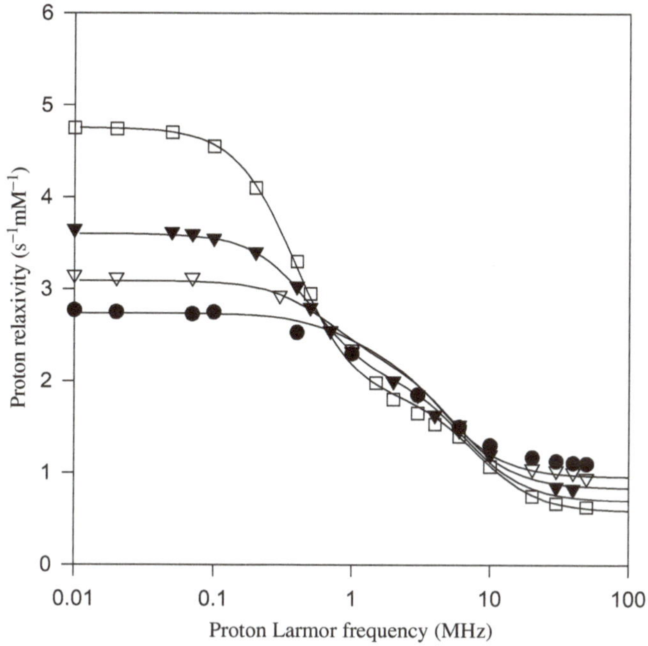

FIGURE 7.15 Water ^1H NMRD profiles for solutions of $VO(H_2O)_5^{2+}$ at (●) 278 K, (∇) 288 K, (▼) 298 K, and (□) 308 K. The *solid lines* represent the best-fit curves using the Solomon–Bloembergen–Morgan equations [Eqs. (4.12), (4.13), (4.17), and (4.27)].

Adapted from Ref. [45].

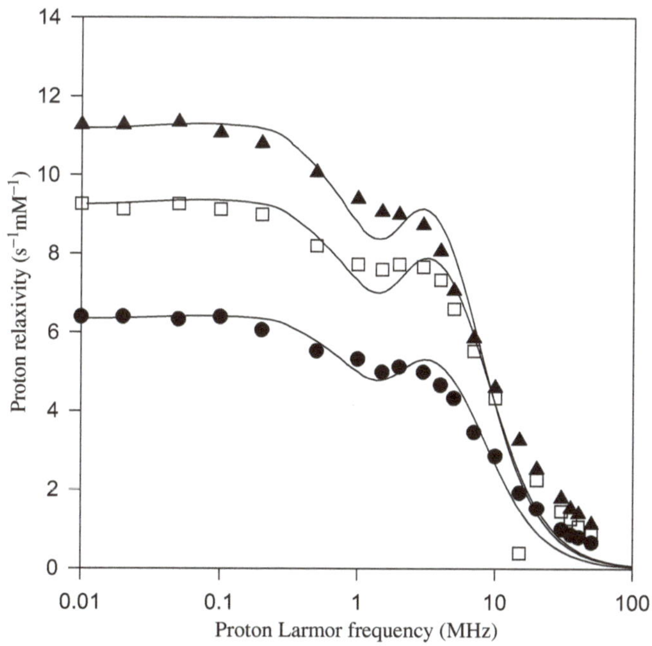

FIGURE 7.16 Water ^1H NMRD profiles for VO^{2+}-transferrin solutions at pH 8, for three temperatures: (▼) 281 K, (□) 298 K, and (●) 311 K.

Adapted from Ref. [25].

7.3.1.4 Titanium(III)

Titanium(III) is also a d^1 metal ion, and therefore is expected to have relatively long electron relaxation times. However, electron relaxation is somewhat faster than for copper(II) and oxovanadium(IV). Actually, the first excited state for titanium hexaaqua ion has been estimated to be around 2000 cm^{-1} higher than the ground state [47], while in the case of copper(II) and oxovanadium(IV), which have longer T_{1e} values, the first excited states are several thousands of cm^{-1}. Therefore, values for the electron relaxation time of 30 ps and $\Delta\chi_{ax} = 4 \times 10^{-33}$ m^3 are obtained; the corresponding maximum pseudocontact shifts and amide proton paramagnetic residual dipolar couplings and the ^1H paramagnetic relaxation enhancements and linewidths are shown in Fig. 7.17. The electron relaxation time is found to be field

Conf.	*L*	*S*	Free ion conf.	Ligand field: Octahedral (O_h)		Tetrahedral (T_d)		Square planar (D_{4h})		Square pyramid. (C_{4v})		Trigonal bipyram. (D_{3h})	
d^1	2	1/2	^2D	– – ↑ – –	^2T	– – – ↑ –	^2E	– – – ↑ –	^2E	– – – – ↑	^2B	– – – ↑ –	^2E

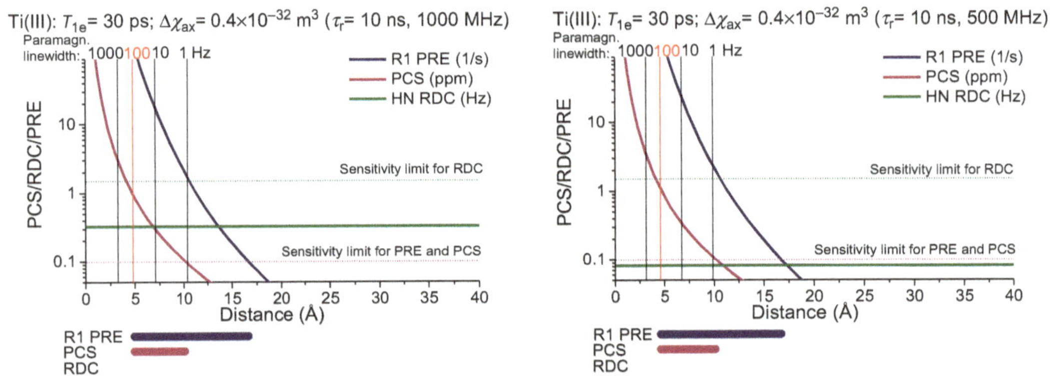

FIGURE 7.17 Population of the electronic d orbitals in different idealized symmetries in titanium(III) complexes and ^1H R_1 paramagnetic relaxation enhancement (in s^{-1}, *blue line*), maximum pseudocontact shift (in ppm, *pink line*) and maximum HN residual dipolar coupling (in Hz, *green line*) as a function of the distance from the metal ion, for typical values of electron relaxation time and axial magnetic susceptibility anisotropy, for a molecule with τ_r of 10 ns, at 1000 MHz (left panel) or 500 MHz (right panel). The plots also shows the distances with ^1H paramagnetic linewidth of 1, 10, 100, and 1000 Hz. The bars under the plots indicate the range of distances with ^1H R_1 paramagnetic relaxation enhancement, pseudocontact shift and HN residual dipolar coupling values above the experimental sensitivity limits. The maximum pseudocontact shift is obtained for a nucleus on the *z*-axis of the magnetic susceptibility tensor; the maximum residual dipolar coupling is obtained for an amide vector parallel to the *z*-axis of the magnetic susceptibility tensor.

independent up to 600 MHz and to be independent of τ_r, suggesting an Orbach relaxation mechanism (Table 7.3). As it happens for copper(II), the effective electron relaxation time decreases with increasing temperature. Such behavior is consistent with an Orbach-type mechanism [Eq. (4.9)], so that the electron relaxation time is linearly proportional to the mean correlation time for intermolecular fluctuations [48].

The ^{1}H NMRD profiles of water solution of $Ti(H_2O)_6^{3+}$ at 293 K is shown in Fig. 7.18. Only one dispersion is present, and the ratio between the relaxivity values before and after the dispersion is much larger than 10/3: this can be accounted for by considering the presence of contact and dipolar contributions to relaxivity with the same correlation time. The analysis of the profiles actually provide similar values of T_{1e} and τ_r, around 3×10^{-11} s, and a constant of contact interaction of 4.5 MHz [48], so that the contact and dipolar dispersions are overlapped. Twelve water protons are found to be at 2.62 Å from the metal ion. If a 10% outer-sphere contribution is subtracted from the data, the distance increases to 2.67 Å, which is a reasonably good value. The increase at high fields in the R_2 values cannot in this case be ascribed to the nondispersive term present in the contact relaxation equation, as in other cases (manganese(II), high spin iron(III)), because longitudinal relaxation measurements do not indicate field dependence in the electron relaxation time. Therefore, such increase can only be due to the chemical exchange contribution and indicates values for τ_M equal to 4.2×10^{-7} s and 1.2×10^{-7} s at 293 and 308 K, respectively [Eq. (6.11)]. Measurements on a solution containing 65% glycerol-d_8

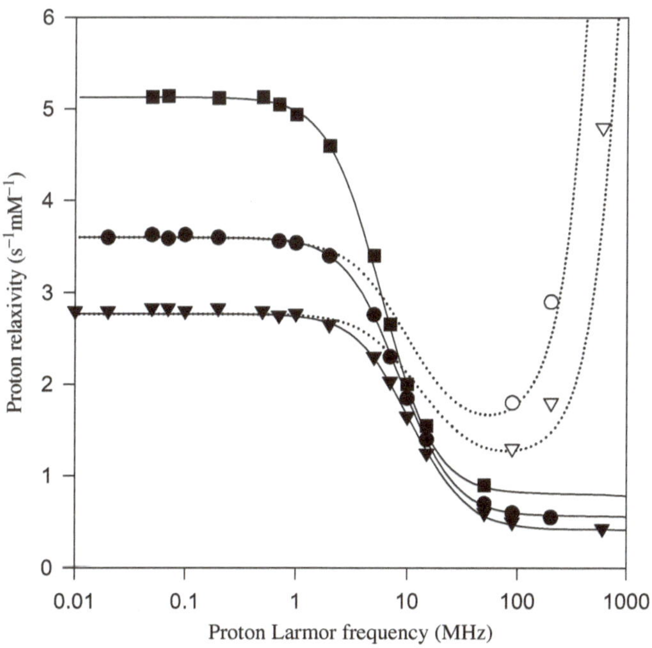

FIGURE 7.18 Water ^{1}H NMRD longitudinal profiles for $Ti(H_2O)_6^{3+}$ solutions at 278 K (■), 293 K (●), and 308 K (▼) and transverse profiles at 293 K (○) and 308 K (∇).

Adapted from Ref. [48].

indicate an electron relaxation time, six times longer than in pure water. Such an increase has been considered consistent with the increase in viscosity, being capable of affecting the efficiency of the Orbach mechanism [47,48].

In general, NMR spectra of titanium(III) complexes are expected to provide quite broad lines for hyperfine coupled protons.

7.3.2 d^2/d^8 (Ni^{2+},V^{3+})

7.3.2.1 Nickel(II)

Nickel(II) is a $3d^8$ ion and has two unpaired electrons (S = 1) when it is six coordinated or pseudo-tetrahedrally four coordinated. Tetrahedral nickel(II) has an orbitally triply degenerate ground state (^{3}T), as shown in Fig. 7.19. Due to the low-lying excited states, the Orbach-type processes are likely very efficient and short relaxation times, of the order of 10^{-12} s, are expected. Sharp proton NMR signals are thus obtained. Octahedral nickel(II) has an orbitally nondegenerate ground state (^{3}A). Pseudo-octahedral nickel(II) has excited states high in energy (about 10,000 cm^{-1} for the first excited state), so that Orbach-type relaxation mechanisms are expected to be inefficient, and the modulation of the ZFS by reorientation or collisions may become the dominant relaxation mechanism. Actually, pseudo-octahedral nickel(II) relaxes significantly slower than pseudo-tetrahedral nickel(II), and its typical electron relaxation time is of the order of 10^{-10} s. Typical values of the magnetic susceptibility anisotropy are in both cases around 2×10^{-32} m^3. The calculated maximum pseudocontact shifts and amide proton paramagnetic residual dipolar couplings and the ^{1}H paramagnetic relaxation enhancements and linewidths in the octahedral and tetrahedral symmetries are shown in Fig. 7.19. When in a planar four coordination, nickel(II) is always low spin, thus diamagnetic (S = 0). Five-coordinated nickel(II) complexes can either be high (S = 1) or low spin (S = 0) depending on the nature of the donor atoms.

Nickel(II) has integer spin quantum number: this accounts for the fact that very low relaxivity values, down to zero, at low field, can be due to the large rhombicity of the static ZFS. No dispersion is observed at low and medium fields, due to the fast electron relaxation time, and to the absence of the ω_S dispersion expected for all integer spin systems with static ZFS [49–52] (Section 4.7.1). The profiles are thus characterized by being flat and with low values of relaxivity up to about 50 MHz, when a sharp increase in relaxivity is detected, due to a field dependence of the electron relaxation time caused by fluctuations of the quadratic transient ZFS.

The proton relaxivity for the hexaaqua nickel(II) complex is shown in Fig. 7.20. From the increase noted at high fields (Table 7.3), a value of T_{1e} at low fields ($T_{1e(0)}$) around 3×10^{-12} s is calculated in the Solomon limit [Eqs. (4.12), (4.13), and (4.20)], or around 10^{-11} s if a ZFS of about 3 cm^{-1} [53] is taken into account (Section 4.7.1). The ω_I dispersion occurs outside the accessible frequency range. No contribution from contact relaxation is expected. Outer-sphere relaxation has been estimated to contribute about 10% of the total [53]. Nickel(II) systems are often outside the Redfield limit for electron relaxation (Section 4.13) and thus appropriate slow-motion theory should be applied. Relaxation data analyzed using the slow motion theory (Section 10.4.3) indicate $\tau_v \approx 2 \times 10^{-12}$ s, $\tau_r \approx 7 \times 10^{-12}$ s and Δ_t = 4–5 cm^{-1} at 324 K [54].

In water-glycerol solutions the relaxivity is sizably smaller (Fig. 7.20), indicating that the glycerol not only changes the solution viscosity, but may also enter the first coordination sphere of the metal ion, resulting in lower symmetry complexes, characterized by nonvanishing static ZFS of 2–6 cm^{-1} [54].

Conf.	L	S	Free ion conf.	Ligand field									
				Octahedral (O_h)		Tetrahedral (T_d)		Square planar (D_{4h})		Square pyramid. (C_{4v})		Trigonal bipyram. (D_{3h})	
d^8	3	1	^3F	↑ ↑ ↓↑ ↓↑ ↓↑	^3A	↓↑ ↑ ↑ ↓↑ ↓↑	^3T	– ↓↑ ↓↑ ↓↑↓↑	Diamagn.	↑ ↑ ↓↑ ↓↑ ↓↑	^3A	↑ ↓↑ ↑ ↓↑ ↓↑	^3E

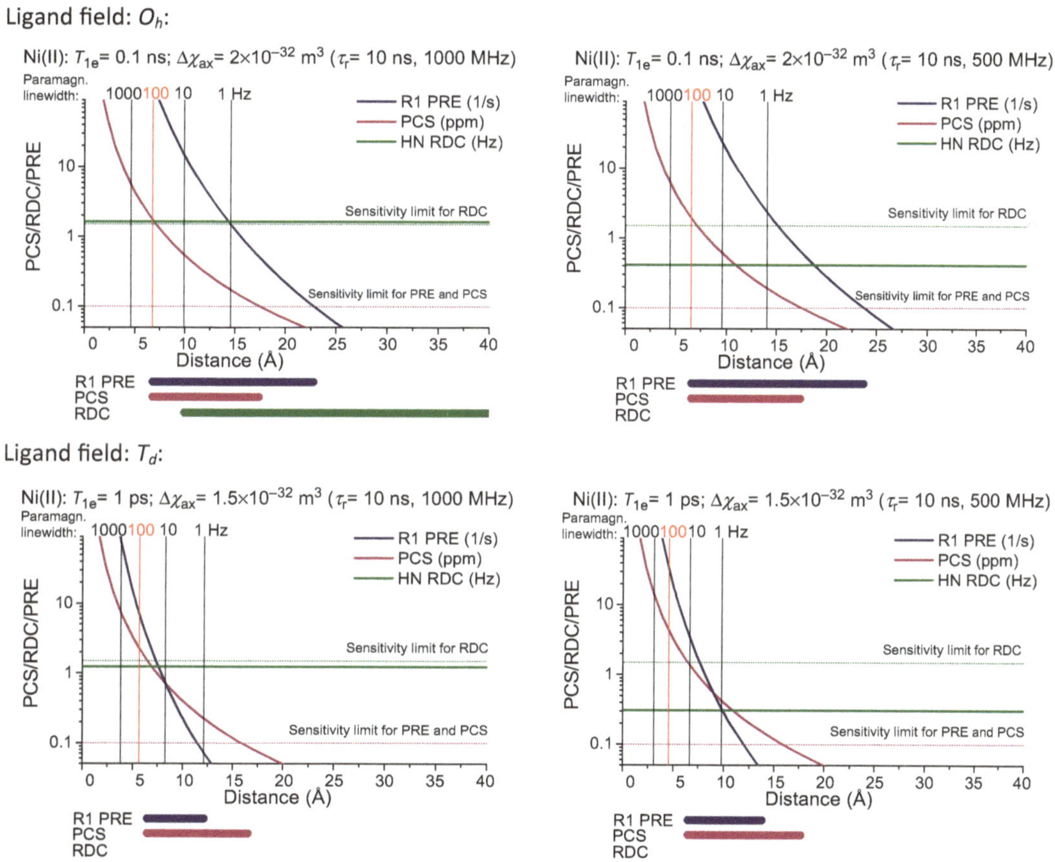

FIGURE 7.19 Population of the electronic d orbitals in different idealized symmetries in nickel(II) complexes and ^1H R_1 paramagnetic relaxation enhancement (in s^{-1}, *blue line*), maximum pseudocontact shift (in ppm, *pink line*) and maximum HN residual dipolar coupling (in Hz, *green line*) as a function of the distance from the metal ion, for typical values of electron relaxation time and axial magnetic susceptibility anisotropy, for a molecule with τ_r of 10 ns, at 1000 MHz (left panel) or 500 MHz (right panel). The plots also shows the distances with ^1H paramagnetic linewidth of 1, 10, 100, and 1000 Hz. The bars under the plots indicate the range of distances with ^1H R_1 paramagnetic relaxation enhancement, pseudocontact shift and HN residual dipolar coupling values above the experimental sensitivity limits. The maximum pseudocontact shift is obtained for a nucleus on the z-axis of the magnetic susceptibility tensor; the maximum residual dipolar coupling is obtained for an amide vector parallel to the z-axis of the magnetic susceptibility tensor.

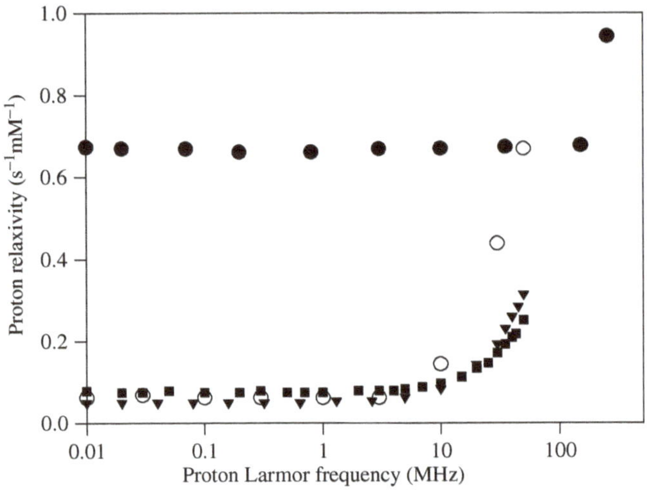

FIGURE 7.20 Solvent ^{1}H NMRD profiles for $Ni(H_2O)_6^{2+}$ solutions in water at 298 K (●), and in ethyleneglycol at 264 (▼) and 298 (■) K [14], and for nickel(II) bovine carbonic anhydrase II at pH 6.0 and 298 K (○).

Adapted from Ref. [50].

The ^{1}H NMRD profile of other nickel(II) complexes were fit according to the pseudorotational model, and D values of 5–6 cm^{-1} were obtained [55].

Water ^{1}H R_1 values have been measured for the hexa-coordinate nickel(II)-substituted bovine carbonic anhydrase II (M_r 30,000) [50,56] (Fig. 7.20): as in the aqua complex, the low field profile is flat, and a rise occurs at high field. Again, the absence of the ω_S dispersion is accounted for by the presence of a static ZFS, large with respect to the Zeeman splitting, and the increase in relaxivity is due to the field dependence of the electron relaxation. As for other metal ions, the correlation time, τ_v, for the modulation of the ZFS in proteins is longer than that in the aquaion, as indicated by the fact that the increase in proton relaxivity begins at lower fields.

As already discussed, a change in coordination number from six to four perturbs the electronic relaxation times of nickel(II) complexes, which becomes short due to Orbach relaxation. An excellent example of the sharp NMR signals observed for pseudotetrahedral nickel(II) complexes is the ^{1}H NMR spectrum of nickel(II)-substituted azurin [57,58]. Azurin is a small protein containing a single type 1 copper (Section 7.3.1.2) forced into distorted trigonal environment with weak oxygen and sulfur axial ligands, as shown in Fig. 7.10. The spectrum of the nickel(II) derivative shows several well resolved signals [57]. The two most downfield signals (Fig. 7.21) arise from the β-CH_2 protons of the equatorial cysteine, signal c from one of the γ-CH_2 protons of the axial methionine, signal e from one of the α-CH protons of the axial glycine, while the signals from the ring protons of the two equatorial histidines are all in the 30–70 ppm region.

7.3.2.2 Vanadium(III)

As vanadium(III) has a d^2 electron configuration, it has a triply degenerate ground state in octahedral symmetry. Therefore, pseudooctahedral vanadium(III), in analogy to pseudotetrahedral nickel(II), is expected to display short electronic relaxation times and large magnetic susceptibility anisotropy. Large pseudocontact shifts were consistently observed [59].

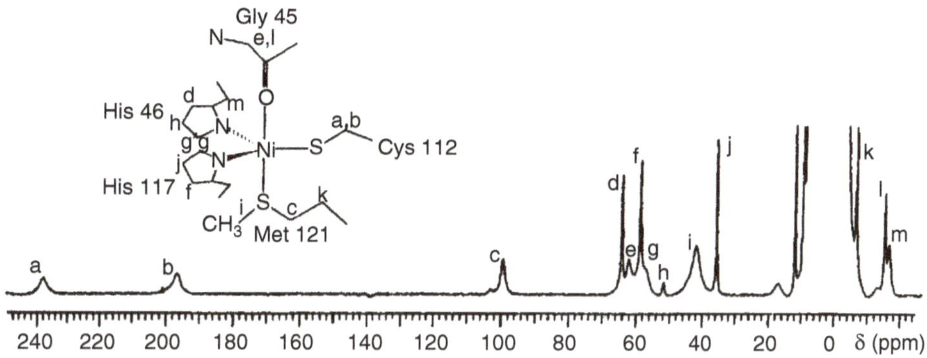

FIGURE 7.21 300 MHz ^{1}H NMR spectrum of nickel(II)-substituted azurin at pH 7.0 and 303 K, and a schematic drawing of the metal coordination polyhedron.

Adapted from Ref. [57].

7.3.3 d^3 (Cr^{3+})

Chromium(III) has a $3d^3$ electronic configuration, and is generally six coordinated (and remarkably co-ordinatively inert). In octahedral symmetry the free-ion ground state ^{4}F yields the ground orbital singlet $^{4}A_{2g}$ state (Fig. 7.22). The orbitally nondegenerate ground state of the chromium(III) ion makes Orbach relaxation inefficient and excludes fast electron relaxation, which is in fact of the order of 10^{-9}–10^{-10} s. Spin-orbit coupling in low symmetry splits the $^{4}A_{2g}$ ground state in two Kramers' doublets. The resulting ZFS is usually small, rarely exceeding 1 cm^{-1} [60]. The main electron relaxation mechanism is ascribed to modulation of transient ZFS (Table 7.3). The narrow electron line makes chromium(III) complexes suitable as polarizing agents in dynamic nuclear polarization [61] (Chapter 5). Basically almost no magnetic susceptibility anisotropy is expected. The calculated maximum pseudocontact shifts and amide proton paramagnetic residual dipolar couplings and the ^{1}H paramagnetic relaxation enhancements and linewidths in the octahedral symmetry are shown in Fig. 7.22 as a function of the nuclear distance from the ion.

The NMRD profiles for hexaaqua chromium(III) at pH 0 (to avoid formation of hydroxo-containing species) are shown in Fig. 7.23. The position of the first dispersion, in the 333 K profile, indicates a correlation time of 5×10^{-10} s. The high field dispersion corresponds to a correlation time around 3×10^{-11} s, and thus it is related to dipolar relaxation, modulated by the rotational correlation time τ_r. According to Stokes–Einstein law [Eq. (4.8)], this time increases with decreasing temperature, and correspondingly the position of the dispersion moves toward lower fields. The proton residence time τ_M is about 5×10^{-6} s, causing the occurrence of a slow exchanging regime, as clearly shown by the increase of solvent proton relaxivity at low fields with increasing the temperature. The occurrence of slow exchange hinders any increase in relaxivity below 300 K, thus explaining the disappearing of the low field dispersion in the low temperature profiles.

Since the correlation time related to the low field dispersion (5×10^{-10} s) is too long to be the reorientation correlation time and too fast to be the water proton lifetime, it has to be ascribed to the electron relaxation time (Table 7.3). Therefore, such dispersion must be due to contact relaxation. The constant of contact interaction, *A/h*, was found of about 2 MHz. The measurements also show that (1)

Conf.	L	S	Free ion conf.	Ligand field: Octahedral (O_h)		Tetrahedral (T_d)		Square planar (D_{4h})		Square pyramid. (C_{4v})		Trigonal bipyram. (D_{3h})	
d^3	3	3/2	^4F	– – ↑ ↑ ↑	^4A	↑ – – ↑ ↑	^4T	– – ↑ ↑ ↑	^4A	– – ↑ ↑ ↑	^4A	– ↑ – ↑ ↑	^4E

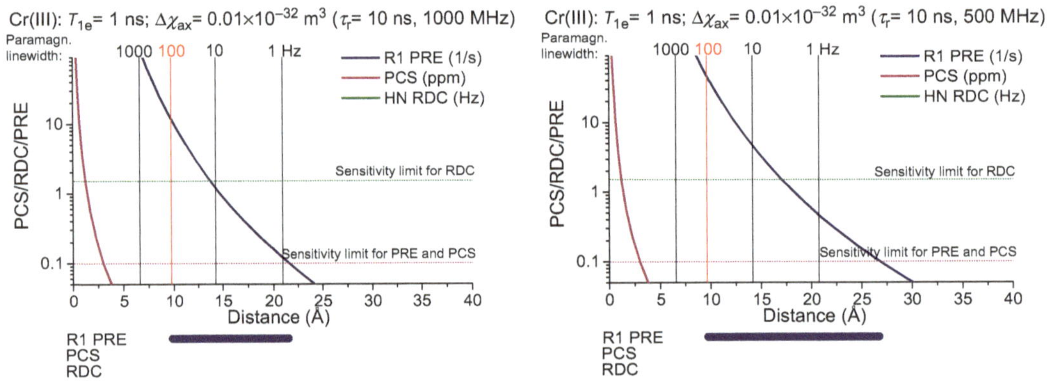

FIGURE 7.22 Population of the electronic d orbitals in different idealized symmetries in chromium(III) complexes and ^{1}H R_1 paramagnetic relaxation enhancement (in s^{-1}, *blue line*) and maximum pseudocontact shift (in ppm, *pink line*) as a function of the distance from the metal ion, for typical values of electron relaxation time and axial magnetic susceptibility anisotropy, for a molecule with τ_r of 10 ns, at 1000 MHz (left panel) or 500 MHz (right panel). The plots also shows the distances with ^{1}H paramagnetic linewidth of 1, 10, 100, and 1000 Hz. The bar under the plots indicates the range of distances with ^{1}H R_1 paramagnetic relaxation enhancement values above the experimental sensitivity limits. No paramagnetic residual dipolar couplings can be observed, due to the very small magnetic susceptibility anisotropy.

the effective electron relaxation time at low fields decreases with decreasing temperature and (2) the transverse relaxation rates increase sizably at high fields, thus indicating that the effective electron relaxation time is field dependent (Section 4.7.2). The occurrence of a field dependent electron relaxation time is confirmed by the longitudinal relaxation measurements in glycerol solution [62], as the typical high field relaxivity peak appears, with Δ_t of 0.11 cm^{-1} and τ_v of 7×10^{-12} s at 298 K.

Since the contribution to solvent proton relaxation of first sphere water molecules is reduced by the occurrence of a slow exchanging regime, second and outer sphere contributions can be detected [62]. In fact outer-sphere and/or second sphere protons contribute less than 5% of proton relaxivity for the highest temperature profile, and to about 30% for the lowest temperature profile. A simultaneous fit of longitudinal and transverse relaxation rates provides a second coordination sphere with

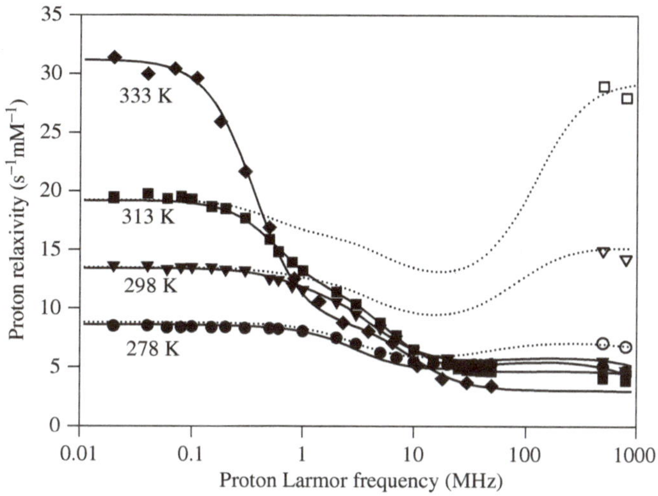

FIGURE 7.23 Water ^{1}H NMRD profiles for $Cr(H_2O)_6^{3+}$ solutions at pH 0 and 278 (●), 298 (▼), 313 (■), and 333 (♦) K. Solid symbols indicate R_1 measurements and open symbols R_2 measurements. The solid lines represent the best fit profiles of R_1; dashed lines indicate the best fit profiles of R_2.

Adapted from Ref. [62].

26 fast exchanging water protons at 4.5 Å from the metal ion [63], and a distance of closest approach for diffusing water molecules of about 6 Å. The minimization provided a metal-proton distance for the 6 coordinated water molecules of 2.71 Å, a transient ZFS of 0.11 cm^{-1}, a correlation time for electron relaxation of about 2×10^{-12} s, a reorientation time of about 70×10^{-12} s, and a lifetime of about 7×10^{-6} s, at room temperature. The reorientation time is about 2–3 times larger than expected for a hexaaqua ion. This is probably because the second sphere water molecules increase the apparent molar mass of the hydrated ion by about three times the molar mass of the hexaaqua chromium(III) ion if considered without second sphere water molecules. This would imply that the lifetime of second sphere water molecules is longer than the reorientation time. This suggests that second-sphere water molecules are less labile than in other metal(II) aqua ions, where the best-fit value of τ_r is much smaller than for the chromium(III) aqua ion, likely as a consequence of the different charge.

In chromium(III) complexes, proton hyperfine shifted signals are expected to be broad, possibly beyond detection. An example is provided by the case of the highly symmetric decamethylchromocenium ion (Fig. 7.24), which shows severely broadened resonances as compared to the corresponding diamagnetic cobalt(III) derivative [64]. Because of the negligible contribution of the cyclopentadienyl (Cp) orbitals to the molecular orbitals where the unpaired electrons are located, that are mainly 3d metal orbitals [65], there is a spin polarization effect on the Cp rings, with a resulting negative spin density for decamethylchromocenium.

Line broadening effects can also be exploited to monitor the binding of chromium(III), which is followed in proteins by the disappearance of signals [66–69]. ^{2}H NMR spectroscopy can be used in the study of small chromium(III) complexes to reduce the line broadening [70].

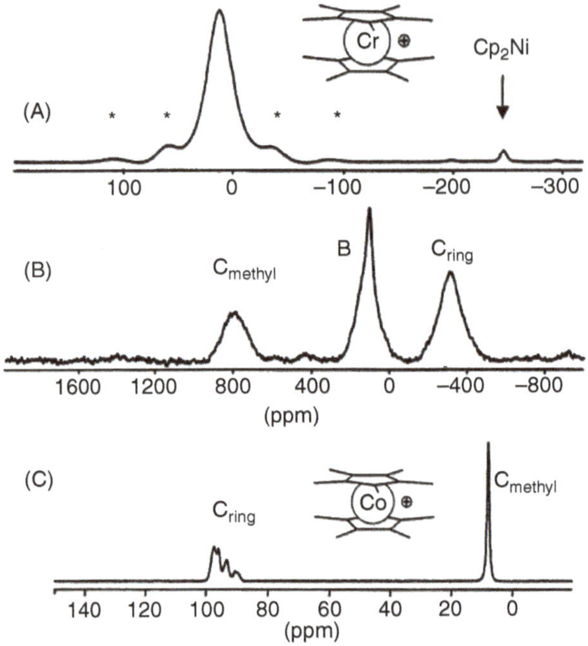

FIGURE 7.24 ^{1}H (A) and ^{13}C (B) MAS NMR spectra of $[(C_5Me_5)_2Cr]^+PF_6^-$ (Me, methyl, cation shown in the inset) at 300 MHz proton Larmor frequency. The position of the peaks with respect to the corresponding diamagnetic cobalt(III) compound (C) shows negative contact shifts. Spinning sidebands are marked by asterisks, B is the background signal of the probehead and Cp_2Ni is used as internal temperature standard.

Reproduced with permission from Ref. [64].

7.3.4 d^4 (Mn^{3+}, IRON(IV))

7.3.4.1 Manganese(III)

Manganese(III) is a d^4 ion and generally gives rise to high spin $S = 2$ compounds. Low symmetry and Jahn-Teller effects provide an orbitally nondegenerate ground state. ZFS is in general $\leq 1\ cm^{-1}$, except for porphyrin derivatives, where it is larger [71]. The porphyrin systems are tetradentate dianionic ligands and essentially planar, as reported in Fig. 7.25, and allows easy access of monodentate ligands to both the axial coordination positions [72]. The electron relaxation time in hexaaqua complexes is $\leq 10^{-11}$ s [73], and about 10^{-10} s in manganese(III)-porphyrins [46]. In the latter complexes the electron relaxation time is relatively longer than in the aqua ions because the C_{4v} symmetry causes the two more excited orbitals (which contain one electron only) to be not very close in energy (Fig. 7.26). The maximum pseudocontact shifts and amide proton paramagnetic residual dipolar couplings, and the ^{1}H paramagnetic relaxation enhancements and linewidths, calculated for an electron relaxation time of 50 ps and $\Delta\chi_{ax} = 5 \times 10^{-33}\ m^3$ are shown in Fig. 7.26.

The NMRD profiles of manganese(III) porphyrins (Fig. 7.27) are often characterized by a peak in relaxivity above about 2 MHz. This can be caused by a magnetic field dependent electron relaxation time. It was also suggested [74,75] that such peak can be ascribed to the effect of rhombic ZFS for

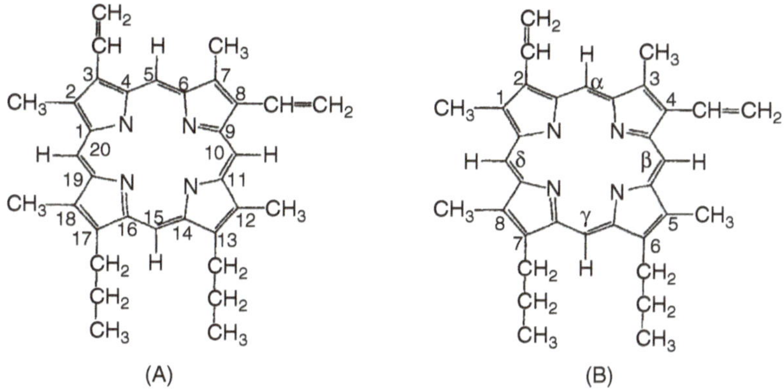

FIGURE 7.25 Labeling of the positions in the porphyrin ring. The porphyrin shown is protoporphyrin IX. (A) is the IUPAC recommendation, while (B) is still commonly used. The two numbering systems will be used in this book interchangeably.

Conf.	L	S	Free ion conf.	Ligand field: Octahedral (O_h)		Tetrahedral (T_d)		Square planar (D_{4h})		Square pyramid. (C_{4v})		Trigonal bipyram. (D_{3h})	
d^4	2	2	^5D	↑ – ↑ ↑ ↑	^5E	↑ ↑ – ↑ ↑	^5T	– ↑ ↑ ↑ ↑	^5A	– ↑ ↑ ↑ ↑	^5A	– ↑ ↑ ↑ ↑	^5A

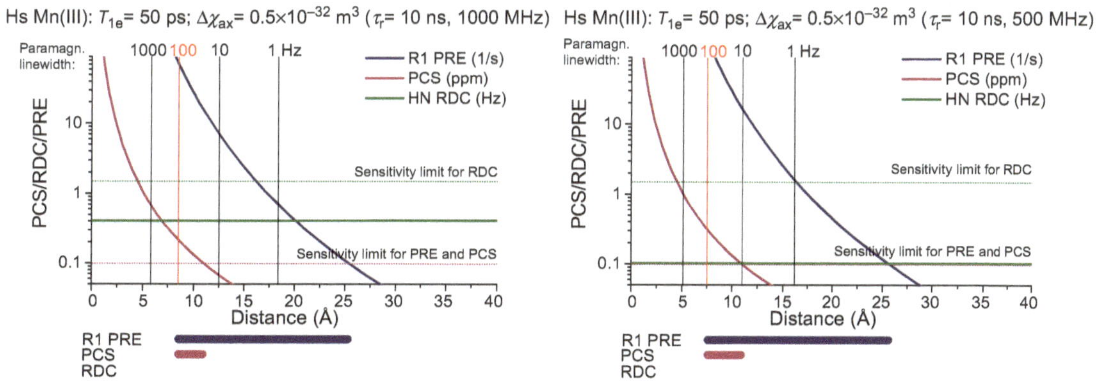

FIGURE 7.26 Population of the electronic d orbitals in different idealized symmetries in high spin manganese(III) complexes and ^{1}H R_1 paramagnetic relaxation enhancement (in s^{-1}, *blue line*), maximum pseudocontact shift (in ppm, *pink line*) and maximum HN residual dipolar coupling (in Hz, *green line*) as a function of the distance from the metal ion, for typical values of electron relaxation time and axial magnetic susceptibility anisotropy, for a molecule with τ_r of 10 ns, at 1000 MHz (left panel) or 500 MHz (right panel). The plots also shows the distances with ^{1}H paramagnetic linewidth of 1, 10, 100, and 1000 Hz. The bars under the plots indicate the range of distances with ^{1}H R_1 paramagnetic relaxation enhancement, pseudocontact shift and HN residual dipolar coupling values above the experimental sensitivity limits. The maximum pseudocontact shift is obtained for a nucleus on the z-axis of the magnetic susceptibility tensor; the maximum residual dipolar coupling is obtained for an amide vector parallel to the z-axis of the magnetic susceptibility tensor.

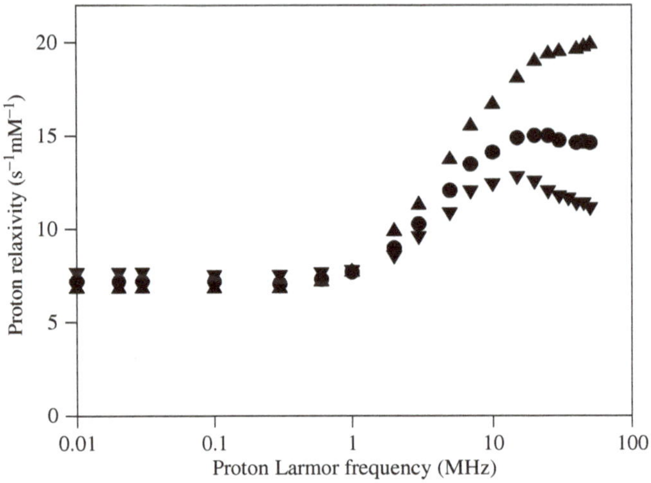

FIGURE 7.27 ^{1}H NMRD profiles of a water solution of the tetraphenylsulfanyl porphyrin ($TPPS_4$) manganese(III) complex at 278 (▲), 293 (●) and 308 (▼) K.

Adapted from Ref. [76].

integer spins. In this case, the increase in relaxivity would correspond to Zeeman energy of the order of the ZFS (Section 4.7.1), and no field dependence for electron relaxation would be needed.

^{1}H NMR spectra of manganese(III) porphyrin derivatives are available [46,77,78] in five- and six-coordinated adducts. The ^{1}H NMR spectrum in $CDCl_3$ of the five-coordinated tetra-*p*-tolyl porphynato manganese(III) (Table 7.7) chloride shows alternation of sign of the isotropic shifts of the *meta* and *para* protons and methyl substituents of the aryl moiety. These shifts, although very small, rule out the possibility of sizeable pseudocontact contributions. The large upfield shifts for the pyrrole protons, which compare with downfield shifts for α-CH_2 substituents, indicate π spin density on pyrrole rings. Predominance of σ spin density is proposed for the *meso* protons [77]. The π spin density distribution is consistent with porphyrin-to-metal π charge transfer as the delocalization mechanism.

7.3.4.2 Iron(IV)

Iron(IV) is much less common than iron(II) and iron(III). However, it is not unusual in biochemical oxidations catalyzed by metalloenzymes, many of which feature an iron(IV)–O catalytic intermediate, where the high oxidation state is stabilized by a tight coordination environment [80]. Iron(IV) nonheme complexes have been investigated as catalysts and as models of biological oxygenases. Depending on the ligands, such complexes may have ground state multiplicities of $S = 1$ or $S = 2$ [81]. A relatively stable complex is $[Fe^{IV}(O)(TMC)(CH_3CN)]^{2+}$ (TMC = 1,4,8,11-tetramethyl-1,4,8,11-tetraazacyclo-tetradecane) [82] (Fig. 7.28). The ^{1}H NMR spectrum of this complex can be interpreted in terms of $S = 1$ being the ground state, by comparison with the calculated spectra with $S = 1$ or $S = 2$ [83].

7.3.5 d^5 (Fe^{3+}, Mn^{2+})

7.3.5.1 Iron(III), high spin

High spin iron(III) complexes occur with some or all weak donor atoms. High spin iron(III) has one unpaired electron in each of the five d orbitals and every orbital can contribute to the overall spin density.

Table 7.7 Isotropic Shift Data (ppm) for Manganese(III) Porphyrins in $CDCl_3$ Solutions [77,78,79]

Complex	Temperature (K)	Pyrrole-H	Phenyl Protons				Propyl Protons			Ethyl Protons		Methyl Protons	*meso*-H
			o-H[a]	*m*-H	*p*-H	CH_3	α-CH_2	β-CH_2	γ-CH_3	α-CH_2	β-CH_3		
TPPMnCl	308	−30.3		+0.4	−0.4								
o-CH_3 TPPMnCl	308	−29.0		+0.5	−0.7	−0.4							
m-CH_3 TPPMnCl	308	−30.0		+0.5	−0.6	−0.22							
p-CH_3 TPPMnCl	308	−30.2		+0.6		+0.29							
T-*n*-PrPmnCl	308	−29.3					+4.7	−0.5	≈0				
OEPMnCl	308									+18.2	+0.7		+41.4[b]
EPMnCl	294									+18.5	+0.7	+31.7	−20.6
MPDMEMnCl	294									+17.0	+0.9	+31.9, +36.0	−20.5, −23.5

TPP, tetraphenylporphyrin; T-n-PrP, tetra-n-propylporphyrin; OEP, octaethylporphyrin; EP, etioporphyrin; MPDME, mesoporphyrin IX dimethylester.

[a] *Unresolved signal.*

[b] *Assignment controversial [77].*

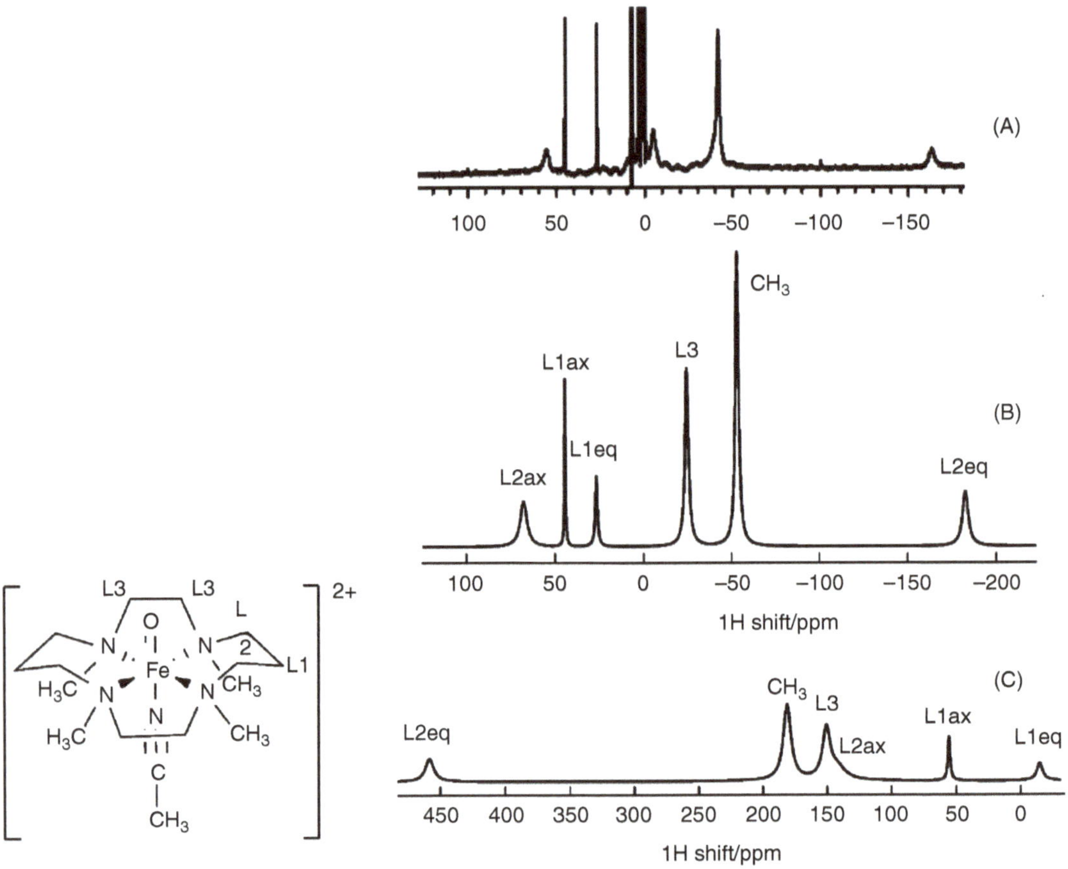

FIGURE 7.28 Molecular structure of $[Fe^{IV}(O)(TMC)(CH_3CH)]^{2+}$, experimental ^1H NMR spectrum (A), and calculated spectrum with $S = 1$ (B) or with $S = 2$ (C).

Reproduced with permission from Refs. [83,84].

The ground state is a sextuplet with an orbitally nondegenerate ground state (^6A). The orbitals and their occupancy in various symmetries are reported in Fig. 7.29. There are no excited levels with the same spin multiplicity, since moving one electron in an excited d orbital requires spin pairing, and thus electron relaxation is not efficient. Electron relaxation times have been reported in the range 10^{-11} s to 10^{-9} s.

As there are no excited states with the same multiplicity, spin orbit coupling can only occur to second order and is relatively weak, introducing a relatively small zero-field splitting [85]. Modulation of ZFS through solvent bombardment is expected to be the main electron relaxation mechanism. Therefore, the larger the ZFS, the faster the electron relaxation rate and the sharper the proton NMR lines [86]. When the zero field splitting, D, is small, as it may occur in quasi-symmetrical complexes, the effective electron relaxation time T_{1e} can be as large as 10^{-9} s [87] and $\Delta\chi_{ax}$ is as small as 6×10^{-33} m^3 [Eq. (2.58)]. D may be of the order of 10 cm^{-1} in porphyrin complexes with one axial ligand or with two weak ligands.

Conf.	L	S	Free ion conf.	Ligand field									
				Octahedral (O_h)		Tetrahedral (T_d)		Square planar (D_{4h})		Square pyramid. (C_{4v})		Trigonal bipyram. (D_{3h})	
d^5	0	5/2	^6S	↑ ↑ ↑ ↑ ↑	^6A	↑ ↑ ↑ ↑ ↑	^6A	↑ ↑ ↑ ↑ ↑	^6A	↑ ↑ ↑ ↑ ↑	^6A	↑ ↑ ↑ ↑ ↑	^6A

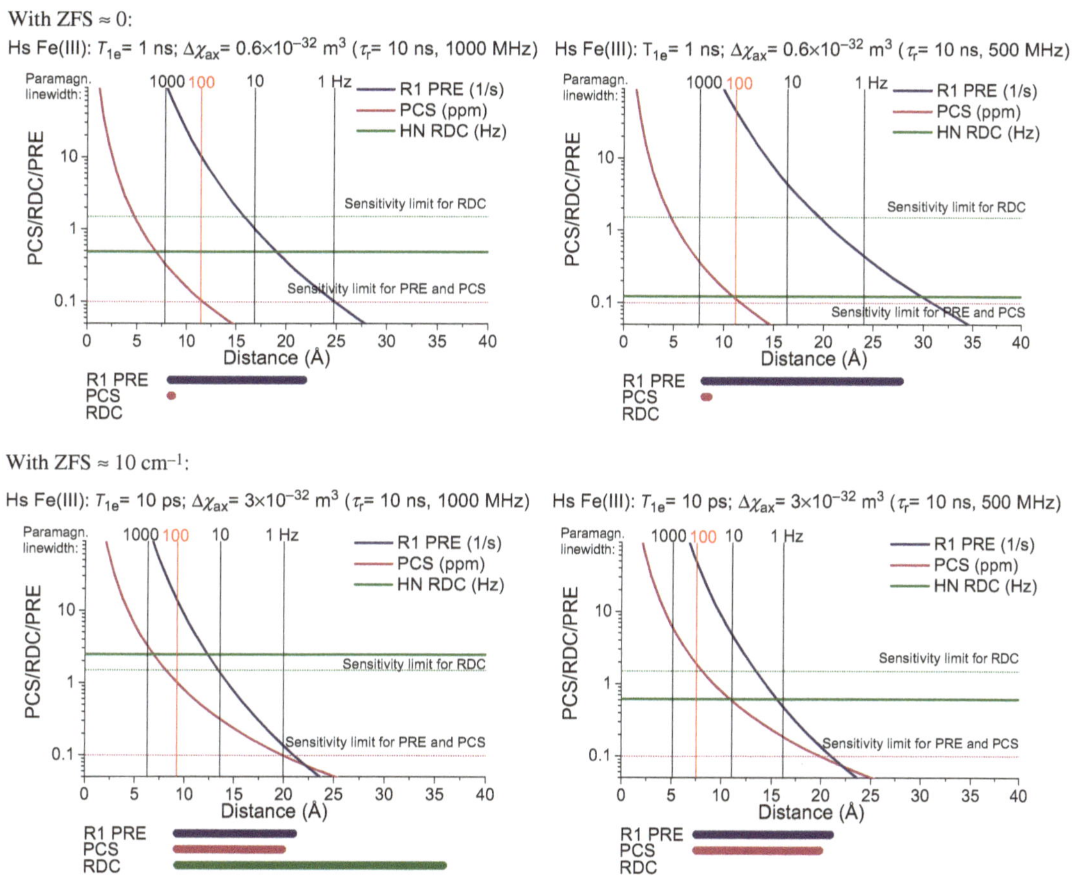

FIGURE 7.29 Population of the electronic d orbitals in different idealized symmetries in high spin iron(III) complexes and ^{1}H R_1 paramagnetic relaxation enhancement (in s^{-1}, *blue line*), maximum pseudocontact shift (in ppm, *pink line*) and maximum HN residual dipolar coupling (in Hz, *green line*) as a function of the distance from the metal ion, for typical values of electron relaxation time and axial magnetic susceptibility anisotropy, for a molecule with τ_r of 10 ns, at 1000 MHz (left panel) or 500 MHz (right panel). The plots also shows the distances with ^{1}H paramagnetic linewidth of 1, 10, 100, and 1000 Hz. The bars under the plots indicate the range of distances with ^{1}H R_1 paramagnetic relaxation enhancement, pseudocontact shift and HN residual dipolar coupling values above the experimental sensitivity limits. The maximum pseudocontact shift is obtained for a nucleus on the z-axis of the magnetic susceptibility tensor; the maximum residual dipolar coupling is obtained for an amide vector parallel to the z-axis of the magnetic susceptibility tensor.

In these cases, T_{1e} decreases to about 10^{-11} s, $\Delta\chi_{ax}$ increases to 3×10^{-32} m^3, and hyperfine coupled proton NMR signals can be detected. The maximum pseudocontact shifts and amide proton paramagnetic residual dipolar couplings and the ^{1}H paramagnetic relaxation enhancements and linewidths for these cases of very small and large ZFS are provided in Fig. 7.29.

7.3.5.1.1 Water proton relaxation

The ^{1}H NMRD profiles of water solutions of $Fe(H_2O)_6^{3+}$ in 1 M perchloric acid (to avoid formation of hydroxo-containing species) at 278, 288, 298, 308 K are shown in Fig. 7.30 [88]. Only one dispersion is displayed at about 7 MHz. It corresponds to a correlation time $\tau_c \approx 3 \times 10^{-11}$ s at 298 K. The small increase of R_{1p} above 20 MHz at the lowest temperatures makes evident that: (1) T_{1e} is field dependent; and (2) T_{1e} is influent in the determination of τ_c. Furthermore, since R_{1p} increases at low field with increasing the temperature, τ_M may contribute to the nuclear relaxation, according to Eq. (6.10). This agrees with a ratio between the values before and after the dispersion smaller than 10:3. From independent measurements it was possible to set $\tau_M = 3.8 \times 10^{-7}$ s at 298 K [12]. Both dipolar and contact relaxation are present, although only one dispersion is observed presumably because at low magnetic fields T_{1e} is similar to or smaller than τ_r, i.e. around 5×10^{-11}–10^{-10} s ($\Delta_t = 0.095$ cm^{-1}). The electron correlation time, τ_v, is estimated around 5×10^{-12} s at room temperature, as commonly found for small complexes in water solution (Table 7.3) and of the order expected for the mean lifetime between collisions with solvent molecules [89,90]. It is responsible for the high field increase of R_{1p} and R_{2p} (Section 4.7.2). The presence of contact relaxation is confirmed by the large increase of the R_{2p} values at high magnetic field, due to the field dependence of T_{1e} (Fig. 7.30). The constant of contact interaction, A/h, is found to be equal to 0.43 MHz, somewhat smaller than the value obtained from chemical shift measurements (1.2–1.3 MHz [12,88]). The best fit value of $r = 2.62$ Å is slightly shorter than

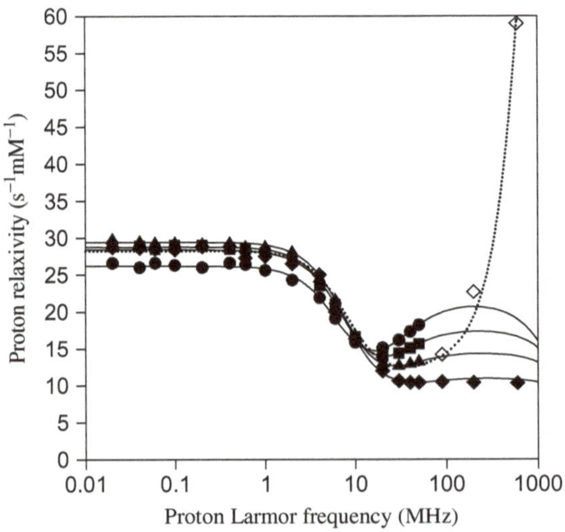

FIGURE 7.30 Water proton longitudinal relaxivity as a function of proton Larmor frequency (^{1}H NMRD profiles) for solutions of $Fe(OH_2)_6^{3+}$ at (●) 278 K, (■) 288 K, (▲) 298 K, (♦) 308 K. High field transverse relaxivity data at 308 K (◊) are also shown. The lines represent the best fit curves using the Solomon–Bloembergen–Morgan equations [Eqs. (4.12), (4.13), (4.17), (4.18), (4.27), and (4.28)].

Adapted from Ref. [88].

the distance evaluated with X-ray structure analysis. This could be due to the fact that outer-sphere relaxation has been neglected (it should contribute for about 1 s^{-1} mM^{-1} in the proton relaxivity [91]) as well as any contribution of the second coordination sphere. It may also be noted that τ_r for the water solution is somewhat longer than the value of about 3×10^{-11} s calculated by using the Stokes-Einstein hydrodynamic model [Eq. (4.8)]. This may suggest the presence of some second sphere waters. Water molecules in the second coordination sphere are in fact expected to be strongly hydrogen bonded to the inner shell of the coordinated water molecules. The ^{1}H NMRD profiles of iron(III) aqua ions decrease markedly in overall amplitude above pH 3 as a result of the formation and precipitation of a variety of hydroxides.

By increasing the viscosity through glycerol–water mixtures (Fig. 7.31), it is shown that the relative influence of τ_r on τ_c with respect to T_{1e} becomes lower and lower, and the hump in the high field region is thus more and more evident. Because the frequency at which R_{1p} begins to increase moves gradually to lower field with increasing the viscosity, also τ_v must increase with viscosity. By assuming that r, A, and Δ_t [Eqs. (4.12), (4.17), and (4.27)] are not affected by the presence of glycerol, the fit of the profile in Fig. 7.31, corresponding to a concentration of 60% of glycerol in solution, provides the values for $\tau_v = 1.4 \times 10^{-11}$ s (instead of 5.3×10^{-12} s, obtained for water solution) and $\tau_r = 2 \times 10^{-10}$ s (instead of 5.3×10^{-11} s), and a number of exchanging protons of 8 (instead of 12), which indicates that glycerol molecules replace water molecules in the coordination sphere. The increase in τ_v is a common feature observed in many other systems, like nickel(II) aqua ion, gadolinium(III) aqua ion, manganese(II) aqua ion, and in manganese(II)-proteins: the efficiency of the solvent bombardment in such systems is in fact reduced in the presence of glycerol. Some iron containing proteins seem to represent an exception to this behavior, despite their long τ_r. In fact, profiles have been acquired, for instance of methemoglobin [92] or catalase [93], where no field dependence of electron relaxation is evident up to 50 MHz.

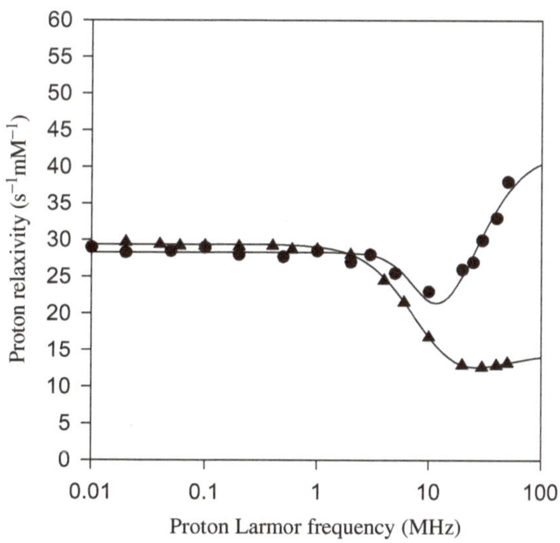

FIGURE 7.31 Water ^{1}H NMRD profiles of $Fe(OH_2)_6^{3+}$ at 298 K with (▲) pure water and (●) 60% glycerol. The lines represent the best fit curves using the Solomon–Bloembergen–Morgan equations [Eqs. (4.12), (4.13), (4.17), (4.18), (4.27), and (4.28)].

Adapted from Ref. [88].

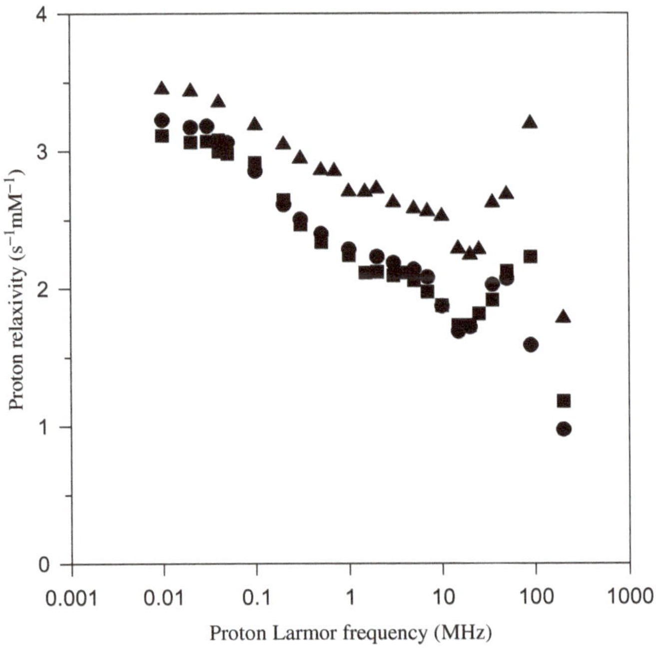

FIGURE 7.32 Paramagnetic contribution to the water ^{1}H NMRD profiles of diferric transferrin solutions at (●) 278 K, (■) 293 K, (▲) 308 K.

Adapted from Ref. [87].

A possible explanation could be a switch in the dominating electron relaxation mechanism between iron(III) aqua ion and iron(III) proteins. The large ZFS present in proteins may in fact favor other field independent electron relaxation mechanisms.

The ^{1}H NMRD profile of the diferric transferrin solution (Fig. 7.32) [87] is also instructive for the case of a macromolecule containing an iron(III) ion. The profile shows four inflections: the first is ascribed to the ω_S dispersion, the second one to the transition from the dominant ZFS limit to the dominant Zeeman limit (Section 4.7.1), the following increase is due to the field dependent electron relaxation time (Section 4.7.2) and finally the ω_I dispersion appears. The best fit analysis provides the presence of a rhombic ZFS with $D = 0.2$, $E/D = 1/3$, in accordance with EPR spectra [94]. The analysis suggests that two sets of electron relaxation times must be considered, of the order of 10^{-10} s at low fields, and of about 10^{-9} s at intermediate fields, before the $\omega_S\tau_v$ dispersion occurs. Eqs. (4.12) and (4.13) are thus inadequate to describe the field dependence of the electron relaxation over the whole range of frequencies due to the presence of static ZFS [95].

In oxidized rubredoxin, there is a pseudotetrahedral iron(III) coordinated to four cysteine sulfurs (Fig. 7.33). The water proton relaxivity profiles of oxidized rubredoxin at 283 and 298 K are shown in Fig. 7.34. They indicate that exchangeable water protons must be at distances larger than 4 Å from the iron atom to account for the measured low rates [96]. From the position of the dispersion, the electron relaxation time is estimated around 7×10^{-11} s at room temperature. As needed for slowly rotating systems with $S > 1/2$, the presence of a static ZFS (> 1.5 cm^{-1}) should be taken into account. In the C6S rubredoxin variant, where a hydroxide ion is coordinated to the metal ion, much higher relaxation

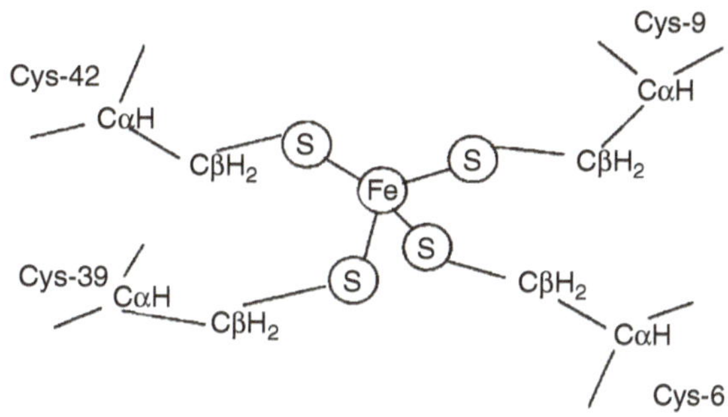

FIGURE 7.33 The iron core of rubredoxin from *Clostridium pasteurianum*.

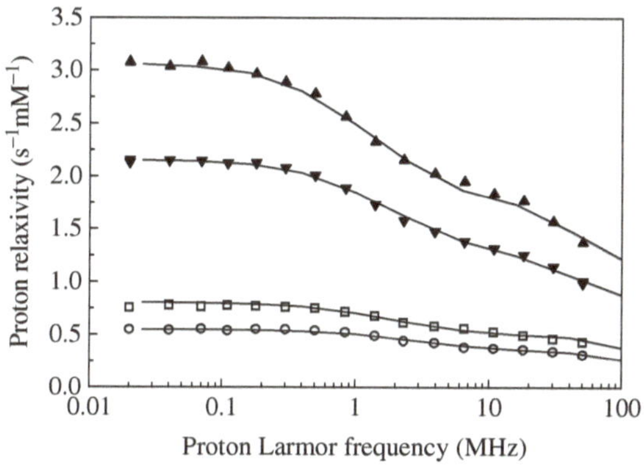

FIGURE 7.34 Paramagnetic enhancements to water ^{1}H NMRD profiles for solutions of wild type (open symbols) and C6S (filled symbols) rubredoxin at 298 K (○ and ▼, respectively) and 283 K (□ and ▲, respectively).

Reproduced with permission from Ref. [96].

rates were measured. This was ascribed to an increased hydration and to electron relaxation time of the order of 1.5×10^{-10} s at room temperature. The analysis excludes, in any case, the presence of first-sphere exchangeable protons, thus ruling out fast exchange of the coordinated hydroxide proton [96].

When water is directly coordinated to iron(III) in a macromolecule, slow exchange effects often quench the relaxivity [92,97]. The fluoride derivative of methemoglobin displays a clear example of a fast exchanging water molecule interacting with the fluoride, i.e. by H-bond (Fig. 7.35) [92,97]. In fact, while for methemoglobin the value of the relaxation rate is related to one water molecule coordinated to the paramagnetic center in slow exchange regime, besides the outer-sphere contribution, in fluoromethemoglobin the water molecule is replaced by fluoride, which is H-bonded to a water molecule. This second-sphere water molecule in fast exchange is responsible for the overall shape of the NMRD profile, according to the Solomon equation, providing a higher relaxivity than that of the directly coordinated but slowly exchanging water in methemoglobin.

7.3.5.1.2 High resolution NMR

From the above information it appears that high spin iron(III) is not an ideal metal ion to be studied by high resolution NMR because of the dramatic effects on linewidths due to effective electron relaxation

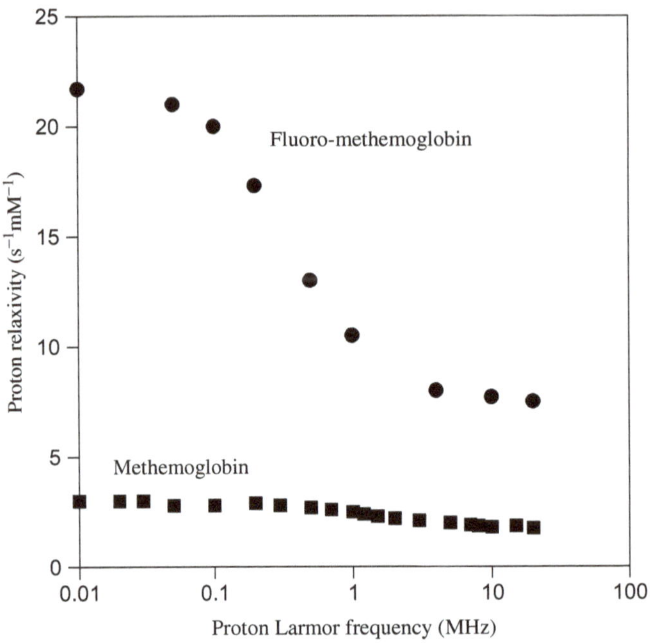

FIGURE 7.35 Water ^{1}H NMRD profiles for a solution of methemoglobin (■) and fluoro-methemoglobin (●) at 279 K. In the latter case, fast exchange is responsible for water proton relaxation enhancements which are quenched by slow exchange in the former case.

Adapted from Ref. [92].

times [as defined in Eqs. (4.12) and (4.13)] of the order of 10^{-10} s at low fields, and higher at high fields. On the other hand, in small complexes, τ_r may be the correlation time for dipole-dipole relaxation [98].

NMR studies of porphyrin-containing iron(III) complexes are very many owing to their importance in biological systems. In porphyrin systems (Fig. 7.25), the large tetragonal component due to strong equatorial ligands provides relatively large ZFS and, as mentioned, relatively short electron relaxation times. This fast electron correlation time contributes to the overall correlation time for dipole–dipole relaxation. In five-coordinated complexes with a halide in the apical position, the order of the D value is $I^- \geq Br^- > Cl^- > NCS^- > F^-$ [99]. Typical values range from a few wavenumbers to ca. 15 cm^{-1}. Spin–orbit coupling may also mix the $S = 5/2$ ground state with the excited $S = 3/2$ state [100]. This is known in the literature as quantum mechanical spin admixing, which is reported to be relevant either in five-coordinated complexes or six-coordinated with very weak apical ligands [101] (Section 7.3.5.3).

The occupancy of the d orbitals in high spin porphyrin compounds is of the type D_{4h}. Such occupancy accounts for both σ and π spin density distribution on the porphyrin ring. In fact, whereas the $d_{x^2-y^2}$ orbital has the correct symmetry to form σ bonds, the d_{xz} and d_{yz} have the correct symmetry to give rise to π bonds. Therefore, two mechanisms contribute to transfer unpaired electron spin on the resonating nucleus H: delocalization through σ bonds and through the π heme orbitals. The $d_{x^2-y^2}$ orbital has the correct symmetry to overlap with the σ bonds of the porphyrin and directly transmits spin density to the resonating protons. The unpaired electrons in the d_{xz} and d_{yz} orbitals delocalize along the pyrrole rings of the heme, through π bonds, and reach the peripheral heme carbon atoms. Thus, the 1s orbital of protons

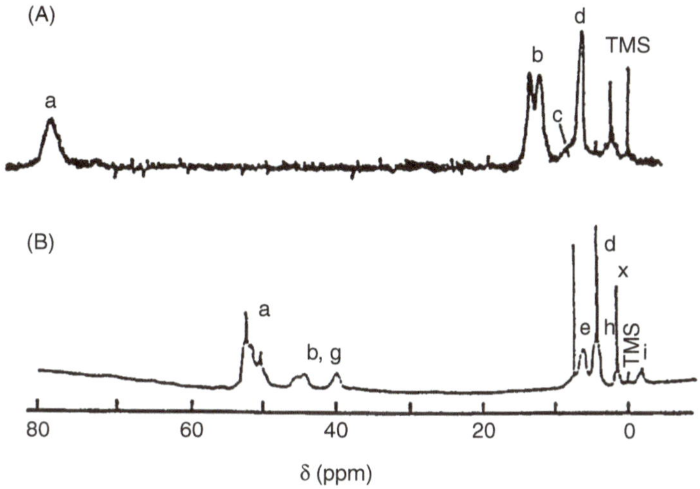

FIGURE 7.36 ^1H NMR spectra of high spin iron(III) porphyrins. (A) Fe TPP-Cl (no substituent at the pyrrole positions). Signal a refers to pyrrole protons; signals b, c, and d refer to the *meta, ortho,* and *para* phenyl protons respectively [102]. (B) Fe (protoporphyrin IX)-Cl (see Fig. 7.25A for the ligand). Signals a belong to the methyl groups, signals b and g to the 13,17 α-CH_2 and the 3,8 α-CH; signals d to the COOH; signals e to the 13,17 β-CH_2; signals h and i to the 3,8 β-CH *cis* and β-CH *trans* respectively.

Adapted from Ref. [103].

of CH_3 groups slightly overlaps with the p_π of the pyrrole carbons. This overlap depends on the dihedral angle between the pyrrole plane and the plane formed by the pyrrole carbon-side group carbon-attached proton (Section 2.4), but in the case of methyl protons it is averaged by their rotation around the C–C axis. Finally, through the d_{z^2} orbital, the unpaired electron delocalizes on the coordinated axial ligand.

The σ spin density transfer mechanism accounts for the large downfield shifts (Fig. 7.36) of signals 2, 3, 7, 8, 12, 13, 17, 18 (numbering as in Fig. 7.25A), or *pyrrole* signals. Consistently, CH_3 and CH_2 substituents in these positions are also downfield shifted. This is nicely shown in the proton NMR spectrum of FeTPPCl [102] (where TPP = 5,10,15,20-tetraphenylporphyrin) and of Fe(protoporphyrin IX dimethylester)Cl [103], as shown in Fig. 7.36. The protons of the phenyl groups at the 5, 10, 15, 20 (or *meso*) positions of FeTPPCl experience some alternating hyperfine shifts, which indicate that some π spin density is present on the phenyl rings as a result of a π spin density at the *meso* position. ^{13}C NMR data confirm the presence of π spin density of positive sign on the *meso* carbons [104]. *Meso* protons are generally downfield in six-coordinated complexes (about 50 ppm) [103] and upfield in five-coordinated complexes (Fig. 7.39) [105]. Such difference is ascribed to the coplanarity of the metal with the ligand in the former case which is lost in the latter [103]. A downfield pyrrole proton and methyl shifts, with the methyl shifts smaller, as found for the five-coordinate compounds, is indicative of a dominant σ delocalization. Contact shift arising from delocalized π spin density, on the other hand, yield upfield pyrrole proton and downfield methyl shifts. If both σ and π spin transfers take place, the added effect of the π mechanism on a dominant σ mechanism would be to decrease the downfield pyrrole-H and increase the downfield pyrrole-CH_3 shifts. Experimental data support the conclusion that π spin transfer is more important in six- than in five-coordinate complexes [103,106].

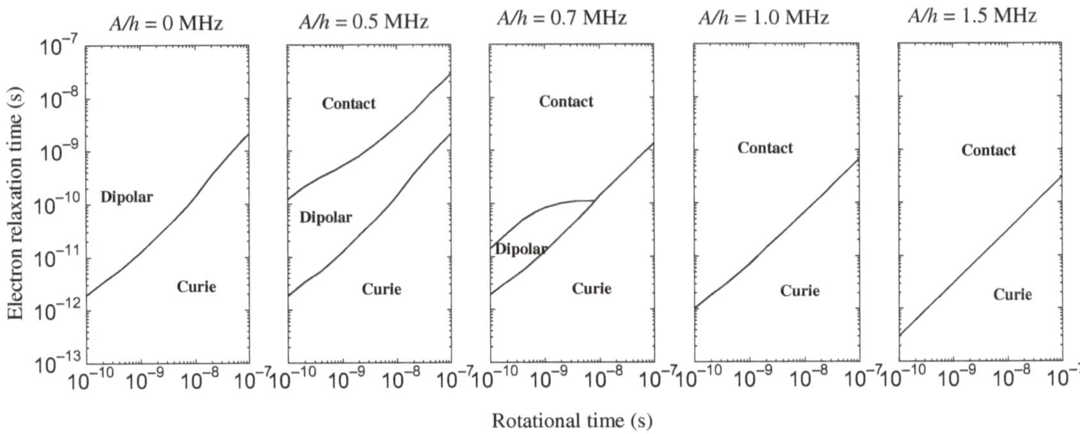

FIGURE 7.37 Predominance of dipolar, contact or Curie relaxation in signal linewidths at 800 MHz for different rotational and electron relaxation times, and for different constants for the contact interaction. Calculations have been performed for protons at 5 Å from a $S = 5/2$ ion.

In macromolecules, the correlation time for dipole-dipole relaxation is dominated by T_{1e}, and the linewidths depend dramatically on this value [107,108]. In turn, T_{1e} decreases with increasing the magnitude of ZFS [46]. When ZFS is large, as for instance in heme proteins, reasonably sharp lines can be obtained. However, the resolution is limited by the fact that, even when T_{1e} is shorter than about 10^{-10} s, Curie relaxation, which is always present, remains as the main source of linebroadening. Eqs. (4.31) and (4.28) show that Curie relaxation increases with increasing the rotational correlation time, and thus with the size of the protein, whereas the contact contribution to nuclear relaxation, present for iron coordinated amino acids, does not depend on it. On the other side, contact relaxation increases with increasing electron relaxation time, whereas Curie relaxation is independent on it. These features, sketched in Fig. 7.37, indicate that for relatively small values of τ_r and large values of T_{1e} the contact term is responsible for signal linewidth, and for relatively large values of τ_r and small values of T_{1e} Curie relaxation is responsible for signal linewidth. Dipolar relaxation may be important only in a small range between the earlier described regions. Examples, described in subsequent paragraph, are shown in Figs. 7.39 and 7.40.

The ^{1}H NMR spectra of ferricytochromes c' are typical of high spin iron(III). The iron atom is pentacoordinated, with four ligands provided by the nitrogen atoms of a porphyrin (Fig. 7.25) and the fifth ligand being a histidine residue exposed to the solvent. Being a cytochrome of c type, the heme moiety is covalently bound to two cysteinyl residues by means of thioether links (Fig. 7.38). The spectrum at pH 5.0 consists of four strongly downfield shifted signals each of intensity three (which are due to the four heme methyl groups) and eight downfield signals each of one proton intensity. These signals are due to the two α-CH protons establishing the heme thioether bridge, to the four α-CH_2 protons of the propionate heme side chains, and to the two β-CH_2 of the proximal histidine bound to the iron atom (Fig. 7.39) [109]. A sizable line broadening is observed, induced by Curie relaxation, since the protein is a dimer with M_r 28000. Broad upfield signals observed in the low pH species are attributable to *meso* protons, as expected for five-coordinated high spin iron(III). Cytochromes c' display pH-modulated

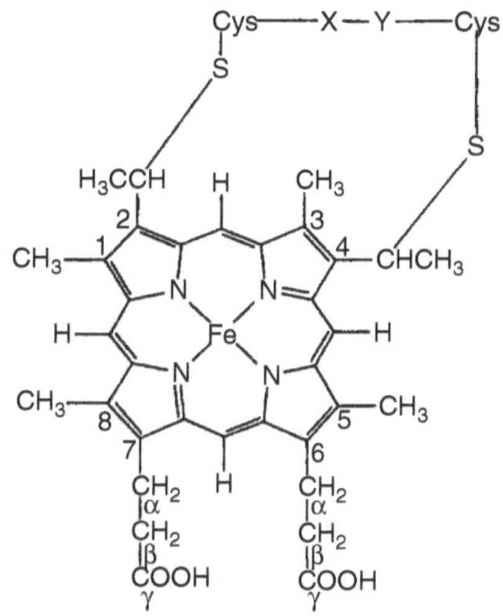

FIGURE 7.38 Schematic drawing of the heme moiety in cytochromes *c′* (labeling as in Fig. 7.25B).

transitions that can be monitored by means of NMR [109,110]. The spreading of well resolved peaks in the NMR spectrum in fact allows us to follow the changes in the hyperfine shifts by changing pH. The pair corresponding to the α-CH_2 propionate of carbon 7 experiences a dramatic pH dependence at low pH, in accordance with the fact that it belongs to a propionate and should be influenced by the presence of Glu 10, which is another ionizable group in the vicinity of the heme [109]. At high pH the histidine becomes a histidinato ion, thus providing a different shift pattern.

The ^{1}H NMR spectrum of met-myoglobin [111] is another example of high spin iron(III) (Fig. 7.40). The assignment of the signals experiencing large isotropic shifts is obtained from a combination of isotope labeling and NOEs (Sections 4.14 and 12.2) [105]: A to D are assigned to the ring methyl groups in the order 8, 5, 3, and 1; W, Y, and Z to three of the four *meso* protons, and all other signals to the other protons of the porphyrin ring. The shifts of the *meso* protons are indicative of six coordination. The shifts of the nuclei of the heme axial ligand, which often is an imidazole ring, are far downfield, indicating the predominance of σ spin delocalization. This is consistent with the presence of an unpaired electron in the d_{z^2} orbital.

In oxidized rubredoxin, the proton signals of the four iron(III) ligand cysteines (Fig. 7.33) are broad beyond detection. The $^{\beta}C^2H_2$ signals of ^{2}H labeled cysteines are located far downfield (300–900 ppm) (Fig. 7.41) [112], which correspond to hyperfine coupling constants of about 1–3 MHz for protons. Two of the four $^2H_\alpha$ signals appear downfield (180 and 150 ppm), while the other two appear upfield (−10 ppm, overlapped). Either dipolar or contact contributions of opposite sign as those operative on the $^{\beta}C^2H_2$ nuclei may be responsible for the upfield shifts [112].

7.3.5.2 Iron(III), low spin

Low spin iron(III) occurs with strong ligands and often with hexacoordination. Low spin iron(III) complexes, which have a ^{2}T configuration in octahedral symmetry, are characterized by low-lying excited states in pseudooctahedral symmetry. The electronic configuration of a generic low symmetry (C_2) low

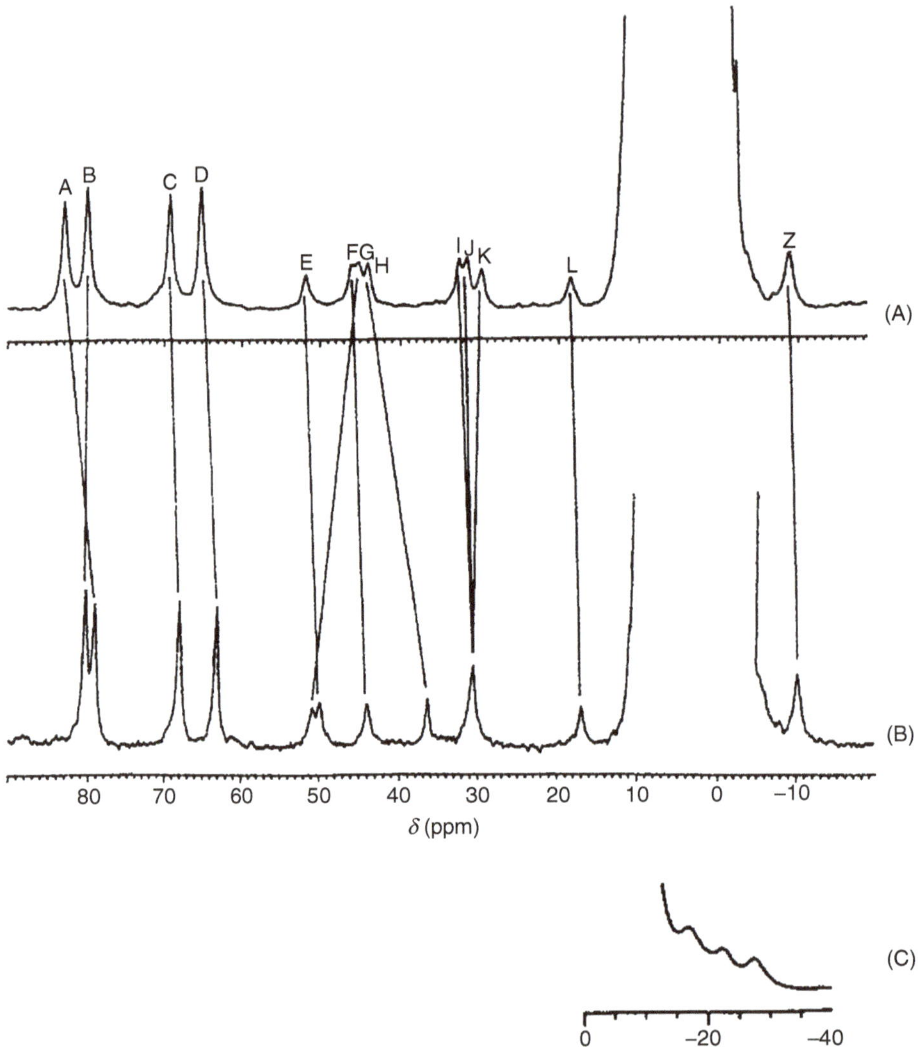

FIGURE 7.39 600 MHz ^{1}H NMR spectra of a five-coordinated iron(III) porphyrin, ferricytochrome *c′* from *R. gelatinosus* at 300K in H_2O at (A) pH 7.4 and (B) pH 4.0 . Signals have been assigned: A, B, C, D, 1-, 8-, 5-3-CH_3; E, 6-CHα′; F, 4-CHα; G, 7-CHα′; H, 7-CHα; I, 6-CHα; J, His β; K, His β′; L, 2-CHα. The upfield signals belonging to *meso* protons of cytochrome *c′* from *R. palustris*, at pH 5, are shown in (C) [106] (labeling as in Fig. 7.25B).

Part A and B: Adapted from Ref. [109].

spin iron(III) complex is shown in Fig. 7.42. The excited states are obtained by promoting one or more electrons of the lowest T_{2g} orbitals to the highest T_{2g} orbital, so that the unpaired electron resides in the second lowest or lowest orbitals. Because of the presence of low-lying excited states, Orbach processes are likely to be very efficient and short relaxation times are expected. Consistent with it, the *g* values of the ground state significantly deviate from 2 (typically, from 0.9 to 3.5) [85,113] and $\Delta\chi_{ax}$ of about

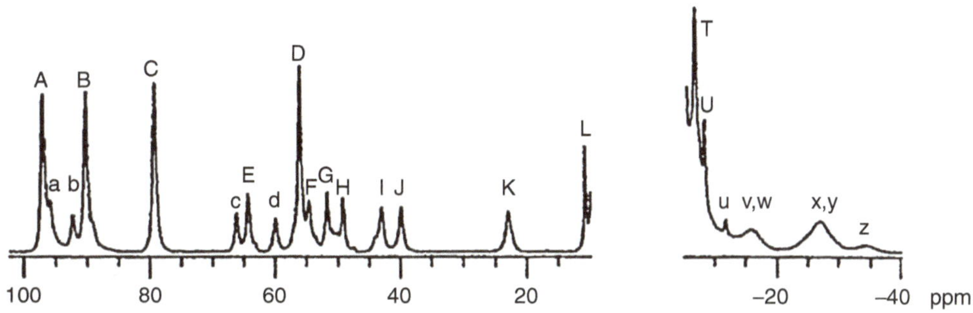

FIGURE 7.40 360 MHz ^{1}H NMR spectrum of *Aplysia* met-myoglobin in ^2H_2O at 298 K and pH 5.9. Chemical shifts are in ppm from DSS. Signal assignments: A, B, C, D, 8-, 5-, 3-, 1-CH_3; E, 7-CHα; F, 6-CHα; G, 4-CHα; H, 6-CHα′; I, 7-CHα′; J, 2-CHα; U, 4-CHβ; W, Y, Z, *meso* H; X, Val-E11 γ-CH_3. a–d methyl peaks correspond to the reversed heme orientation (labeling as in Fig. 7.25B).

Adapted from Ref. [105].

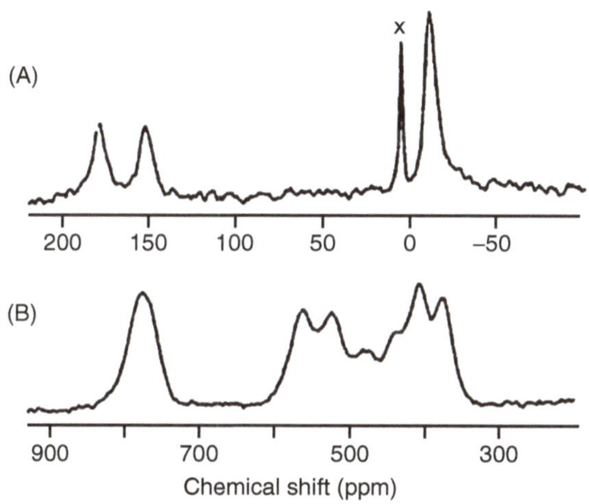

FIGURE 7.41 ^{2}H NMR spectra of oxidized [$^2H^{\alpha}$]Cys-labeled rubredoxin (A) and oxidized [$^2H^{\beta2,\beta3}$]Cys-labeled rubredoxin (B) at 308 K, pH 6.0. The peak labeled x arises from residual $^1H^2HO$.

Adapted from Ref. [112].

2.5×10^{-32} m^3 are observed (Fig. 7.42). The relatively large magnetic susceptibility anisotropy permits the measurements of paramagnetic residual dipolar couplings and of pseudocontact shifts in a wide range of distances.

A typical low spin iron(III) hexacoordinated complex is $Fe(CN)_6^{3-}$. Its NMRD in water indicates a $T_{1e} < 10^{-12}$ s. In heme complexes the effective T_{1e} is about 10^{-12} s when the apical ligands are His/His, His/Cys or His/CN. In the cases of His/OH$^-$ and Cys/H_2O T_{1e} is somewhat longer and the g anisotropy decreases.

In D_{4h} (tetragonal) symmetry there is only one unpaired electron in two degenerate orbitals of correct symmetry to give rise to π bonds with a π orbital of the porphyrin moiety [114]. The ^{1}H NMR

Conf.	L	S	Free ion conf.	Ligand field: Octahedral (O_h) (strong ligands)		Tetrahedral (T_d)		Square planar (D_{4h})		Square pyramid. (C_{4v})		Trigonal bipyram. (D_{3h})		C_2
d^5	0	1/2	^6S	– – ↓↑ ↓↑ ↑	^2T	↑ – – ↓↑ ↓↑	^2T	– – – ↑ ↓↑ ↓↑	^2A	– – – ↓↑ ↑ ↓↑	^2E	– ↑ _ ↓↑ ↓↑	^2E	– – ↑ ↓↑ ↓↑

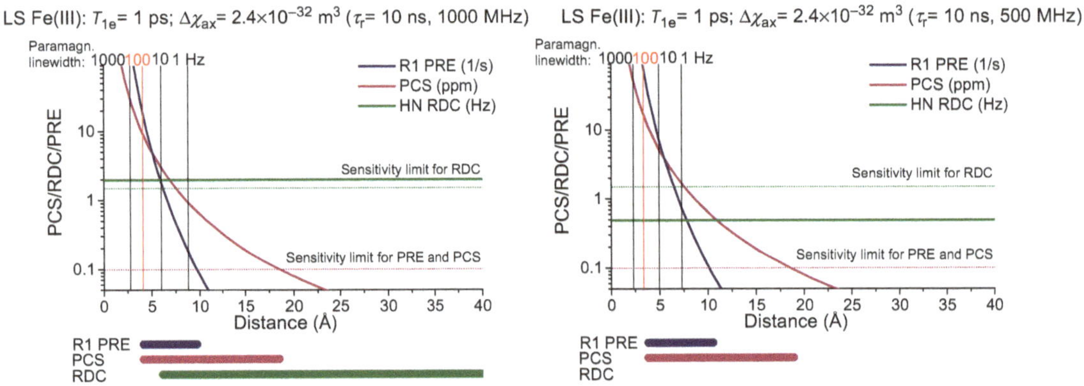

FIGURE 7.42 Population of the electronic d orbitals in different idealized symmetries in low spin iron(III) complexes and ^{1}H R_1 paramagnetic relaxation enhancement (in s^{-1}, *blue line*), maximum pseudocontact shift (in ppm, *pink line*) and maximum HN residual dipolar coupling (in Hz, *green line*) as a function of the distance from the metal ion, for typical values of electron relaxation time and axial magnetic susceptibility anisotropy, for a molecule with τ_r of 10 ns, at 1000 MHz (left panel) or 500 MHz (right panel). The plots also shows the distances with ^{1}H paramagnetic linewidth of 1, 10, 100, and 1000 Hz. The bars under the plots indicate the range of distances with ^{1}H R_1 paramagnetic relaxation enhancement, pseudocontact shift and HN residual dipolar coupling values above the experimental sensitivity limits. The maximum pseudocontact shift is obtained for a nucleus on the z-axis of the magnetic susceptibility tensor; the maximum residual dipolar coupling is obtained for an amide vector parallel to the z-axis of the magnetic susceptibility tensor.

spectrum of Fe(protoporphyrin IX)–imidazole–cyanide is reported in Fig. 7.43 [115]. The free rotation of the imidazole ring about the metal–nitrogen bond, which is fast on the NMR timescale, simulates a tetragonal symmetry as far as the chemical shifts are concerned [77]. The four methyls are all downfield, though to a quite smaller value than in the case of high spin iron(III) complexes. The pseudocontact shifts are larger than in the case of high spin iron(III). An estimate for bis-imidazole systems is reported in Table 7.8 [116,117].

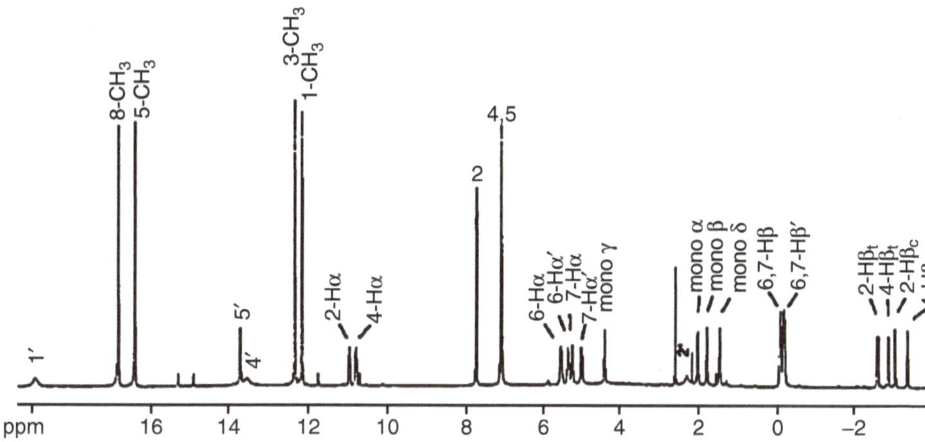

FIGURE 7.43 ^{1}H NMR spectrum of Fe(protoporphyrin IX)–imidazole–cyanide (labeling as in Fig. 7.25B).

Adapted from Ref. [115,118].

In proteins, the apical ligand is fixed and, whenever it is capable of π bonding, it does so with one metal orbital, thus removing the degeneracy of the d_{xz} and d_{yz} orbitals. This introduces large anisotropy in the peripheral substituents of the heme ring. As a consequence, in heme proteins, different pyrrole substituents may experience different unpaired electron delocalization depending on the orientation of the axial ligands. The splitting of the d_{xz} and d_{yz} orbitals can be as large as several times kT. When it is of the order of kT, the temperature dependence of the shifts for the heme protons is complicated because of the Boltzmann population of the excited level [119,120].

The NMR spectrum of the heme group of oxidized horse heart cytochrome *c* is shown in Fig. 7.44 [121,122]. Cytochrome *c* is a heme protein where the iron atom is hexacoordinated, with a histidine and a methionine as axial ligands. As common for low spin iron(III) systems, it presents relatively sharp signals and a narrow range of isotropic shifts. On the basis of their intensity, the four signals at about 34, 31, 10, and 7 ppm downfield are assigned to the ring methyl groups. A noticeable feature of this spectrum is the larger spreading of the ring methyl resonances with respect to isolated low spin iron(III) porphyrins. The pseudocontact shifts are relatively small for all but the *meso* positions, and therefore the spreading of the ring methyl shifts is indeed due to differences in the contact contributions, expected from the asymmetry of the unpaired electron spin density delocalization in the heme group induced by the axial ligand.

Pseudocontact contributions are absolutely dominant for protons many bonds away from the metal ion, and permit the determination of magnetic susceptibility anisotropies together with the principal direction of the χ tensor and a refinement of the solution structure. Once the correct magnetic susceptibility tensor is available, the contact and pseudocontact contributions to the hyperfine shifts of the heme moiety and of the iron ligands can be separated (Table 7.9). From the knowledge of the g and $\Delta\chi$ values, the contact shift of the heme methyl protons can also be estimated through a ligand field analysis together with the Kurland and McGarvey approach [123] and an angular dependence of the contact coupling constant [as provided by the following Eq. (7.2)], by taking into account the contributions of the various electronic levels [124].

Table 7.8 Separation of Contact and Pseudocontact Shifts in Low-Spin bis-imidazole Iron(III) Porphyrins [116,117] (labeling as in Fig. 7.25B)

Position	Hyperfine Shift (ppm)	Pseudocontact Shift (ppm)	Contact Shift (ppm)
Heme			
meso o-H[a]	−3.09	−3.09	0
meso m-H[a]	−1.49	−1.44	~0
meso p-H[a]	−1.37	−1.27	~0
meso p-CH_3[b]	−0.94	−0.94	0
pyrrole H[a,b]	−25.4	−5.8	−19.5
meso α-CH_2[c]	0.6	−4.5	5.1
meso H[d]	−7.0	−9.3	2.3
pyrrole α-CH_2[d]	2.0	−3.2	5.2
Axial imidazole			
1-H[a]	~2	−9.6	11.6
1-CH_3[e]	17.2	10.3	6.9
2-H[a]	−9.5	−28.0	18.5
2-CH_3[e]	~12	~12	~0
4-H[a]	9.7	−8.2	17.9
5-H[a]	4.0	−7.6	11.6
5-CH_3[e]	15.5	9.0	6.5

[a]*For Fe α,β,γ,δ-tetraphenylporphyrin*-$(Im)_2^+$.
[b]*For Fe α,β,γ,δ-tetratoluylporphyrin*-$(Im)_2^+$.
[c]*For Fe α,β,γ,δ-tetrapropylporphyrin*-$(Im)_2^+$.
[d]*For Fe 1,2,3,4,5,6,7,8-octaethylporphyrin*-$(Im)_2^+$.
[e]*For Fe α,β,γ,δ-tetraphenylporphyrin*-$(CH_3Im)_2^+$.

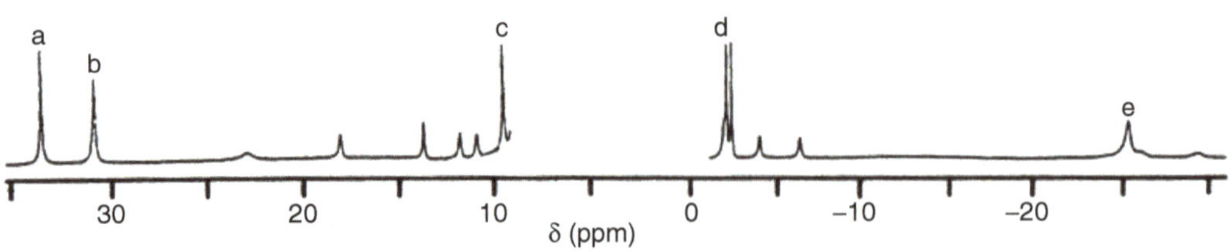

FIGURE 7.44 360 MHz ^{1}H NMR spectra of oxidized horse heart cytochrome *c*. The labeled signals are assigned to: *a*, 8-CH_3; *b*, 3-CH_3; *c*, 5-CH_3; *d*, thioether bridge 2-CH_3; *e*, axial methionine S-CH_3; the resonances at 7.4 ppm (1-CH_3) and 3.1 ppm (thioether bridge 4-CH_3) are not shown. Chemical shifts are in ppm from DSS (labeling as in Fig. 7.25B).

Adapted from Ref. [123].

Table 7.9 Separation of Contact and Pseudocontact Contributions to the Hyperfine Shift for Horse Heart Cytochrome *c* at 293 K [125] (labeling as in Fig. 7.25B)

		Chemical Shift				
Atom Name	**Residue Name**	**Oxidized Species (ppm)**	**Reduced Species (ppm)**	**Hyperfine Shift (ppm)**	**Calculated Pseudocontact Shift (ppm)**	**Contact Shift (ppm)**
Hα	Met	2.77	3.09	−0.32	−0.03 ± 0.30	−0.29
ε-CH_3	Met	−24.7	−3.30	−21.4	4.42 ± 1.4	−25.8
Hα	His	9.10	3.50	5.30	4.28 ± 0.04	1.04
Hδ2	His	24.6	0.14	24.5	25.7 ± 0.45	−1.20
Hε1	His	−25.7	0.51	−26.2	12.9 ± 0.64	−39.0
8-CH_3	heme	35.7	2.16	33.5	−1.86 ± 0.04	35.4
meso-δH	heme	2.11	9.06	−6.95	−9.06 ± 0.15	2.11
1-CH_3	heme	6.81	3.46	3.35	−4.38 ± 0.07	7.73
2-Hα	heme	−1.33	5.20	−6.53	−4.73 ± 0.05	−1.8
2-CH_3	heme	−2.63	1.46	−4.09	−3.09 ± 0.02	−1.0
meso-βH	heme	1.40	9.30	−7.90	−6.31 ±0.09	−1.6
3-CH_3	heme	32.8	3.84	29.0	−1.15 ±0.01	30.1
4-Hα	heme	2.09	6.30	−4.21	−2.70 ± 0.09	−1.51
4-CH_3	heme	3.05	2.57	0.48	−1.37 ± 0.02	1.85
meso-βH	heme	−0.92	9.61	−10.5	−10.63 ± 0.13	0.13
5-CH_3	heme	9.72	3.58	6.14	−4.86 ± 0.03	11.0
meso-γH	heme	7.50	9.64	−2.14	−6.25 ± 0.09	4.11

Homonuclear correlation spectroscopy (COSY) experiments (Section 12.3.2) substantiate the theoretical predictions, based on molecular orbital calculation, of the pattern of spin delocalization in the $3e_\pi$ orbitals of low-spin iron(III) complexes of unsymmetrically substituted tetraphenylporphyrins [126]. Furthermore, the correlations observed show that this π electron spin density distribution is differently modified by the electronic properties of a mono-*ortho*-substituted derivative, depending on the distribution of the electronic effect over both sets of pyrrole rings or only over the immediately adjacent pyrrole rings [126]. No NOESY cross peaks are detectable, consistently with expectations of small NOEs for relatively small molecules and effective paramagnetic relaxation [127].

In the case of His/CN systems, structurally pertinent to myoglobin cyanide, peroxidase cyanide, cytochrome *c* cyanide and the CN^- derivative of a cytochrome *c* mutant, where the axial ligand methionine is substituted with alanine (Ala80–cyt *c*–CN), the simple assignment of the four methyl protons, which is an almost trivial task, provides direct structural information on the axial ligands, which can be used for structural analysis in solution. The chemical shifts for each methyl proton at 298 K appear to

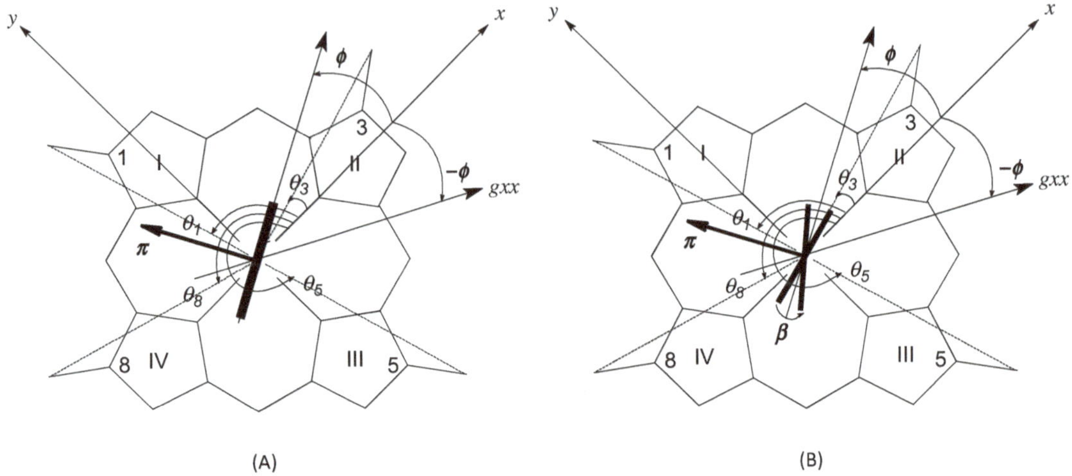

FIGURE 7.45 Schematic representation of the heme moiety. The x-axis is taken along the metal–pyrrole II direction. The θ_i angles for the four methyl groups are defined as the angles between the metal–methyl ith vector and the x-axis. The ϕ angle defines (A) the direction of the histidine ring plane, in histidine–cyanide systems; and (B) the direction of the bisector of the dihedral angle β formed by the two axial histidines, in bis-histidine systems.

be dependent on the angle ϕ between the metal–pyrrole II direction and the direction of the histidine π interaction according to the heuristic equation [128]:

$$\delta_i = a\sin^2(\theta_i - \phi) + b\cos^2(\theta_i + \phi) + c + k_i \qquad a \cong 18.4 \quad b \cong -0.8 \quad c \cong 6.1 \tag{7.1}$$

where θ_i is the angle between the metal–ith-methyl direction and the metal–pyrrole II axis (Fig. 7.45A) and k_i is a correction term to account for an average higher shift value of the protons of methyls 5 and 8 and an average lower shift value of the protons of methyls 1 and 3 ($k_i = \pm\, 0.7$ and ± 2.7 ppm for c-and b-type heme, respectively). Such equation finds its ground on the dependence on ϕ of contact and pseudocontact shifts for the methyl protons. The spin density on the pyrrole carbons determines the value of the contact shift of the methyl protons, which is maximal when the histidine plane forms an angle of 90 degrees with the Fe-carbon methyl line and zero when it points at the methyl group (i.e., the methyl is in the histidine plane). This may be expressed as:

$$\delta_i^{\mathrm{con}} \propto \sin^2(\theta_i - \phi) \tag{7.2}$$

From Eq. (2.46), the pseudocontact shift is proportional to a function of the type

$$\delta^{\mathrm{pcs}} \propto k\cos 2\Omega - 1$$

where k is the ratio between the rhombic and the axial susceptibility anisotropy, k being negative if $g_x < g_y$ as is always the case in the present systems. In these low spin heme systems, the g_x axis is rotated clockwise from a metal–pyrrole direction by an angle which is the same in magnitude but opposite in direction to the angle ϕ [129]. Therefore, $\Omega = \theta_i + \phi$, and the sum of contact and pseudocontact shift is consistent with the heuristic Eq. (7.1).

As an example, from the chemical shifts of methyl protons of the cyanide derivative of Met80Ala cytochrome *c*, an angle ϕ of 57 ± 9 degrees is found by using Eq. (7.1) for the orientation of the histidine plane with respect to the pyrrole II direction, to be compared to the value obtained from the NMR solution structure, equal to 49 degrees (Fig. 7.46A).

Analogously, in bis-histidine ferriheme proteins, cytochromes b_5 (Fig. 7.46B) [130–132], cytochromes c_3 [133] and cytochromes c_7 [134], the chemical shifts for each methyl proton at 298 K appear to be dependent on the ϕ and β structural angles according to the heuristic equation [128]:

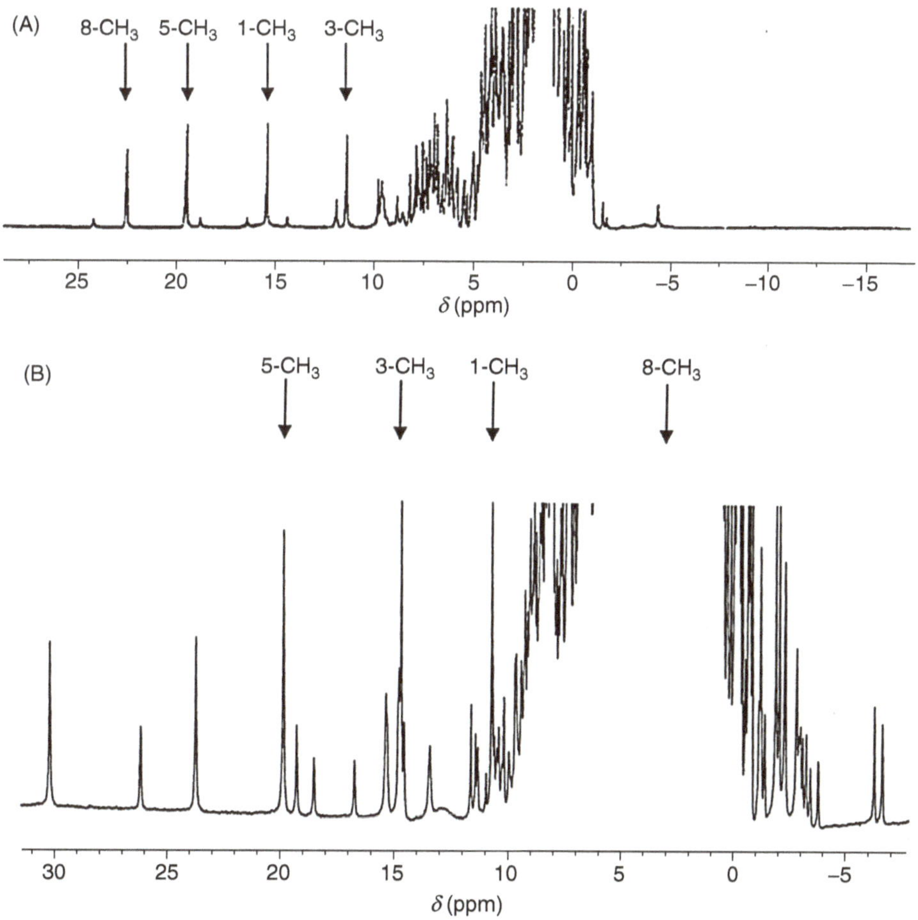

FIGURE 7.46 Hyperfine shifts of methyl protons in (A) Met80Ala cytochrome *c*–CN$^-$, and (B) cytochrome b_5. The former is a histidine–cyanide ferriheme protein, since the axial ligand methionine is substituted with alanine, the latter is a bis-histidine ferriheme protein (labeling as in Fig. 7.25B).

$$\begin{aligned}&\delta_i = \cos\beta[a\sin^2(\theta_i-\phi)+b\cos^2(\theta_i-\phi)+c]+d\sin\beta+k_i\\&a\cong 38.8\quad b\cong -10.5\quad c\cong -1.1\quad d\cong 9.4\end{aligned} \tag{7.3}$$

where β is the acute angle between the two histidine planes and ϕ is the angle between the bisector of the angle β and the metal–pyrrole II axis (Fig. 7.45B).

The $(d_{xz}d_{yz})^4d_{xy}^1$ state in porphyrins is often the excited state (Fig. 7.42) at about 1500 cm^{-1}. Weak axial ligands decrease its energy relative to the d_{xz}, d_{yz} orbitals. When there is distortion of the porphyrin ring due to bulky axial ligands (with weak σ donating and strong π accepting character), the $(d_{xz}d_{yz})^4d_{xy}^1$ state may become the ground state. In this case, the anisotropy of g is $g_\perp > 2 > g_\parallel$ and the spin delocalization which occurs through π orbitals involving d_{xz} and d_{yz} is small [135–138]. As a consequence, little hyperfine shifts are observed on the β-pyrrole positions. Conversely, large π spin delocalization is observed on the *meso* positions, probably due to spin delocalization from d_{xy} to the porphyrin ring by means of in plane π-bonding to the four porphyrin nitrogen atoms [135].

Exploitation of the paramagnetic properties of iron proteins permits not only to elucidate the structure of the system but also to get information on metal trafficking in cells [139] and on protein assembly [140]. Relevant examples in these respects are provided by iron-sulfur proteins and heme proteins.

7.3.5.3 Spin-admixed Fe^{3+}-P and high spin–low spin equilibria

High spin Fe^{3+}–PX compounds, where X is a very weak anion, have spin admixed ground states of $S = 5/2$ and $S = 3/2$ species. The pure $S = 5/2$ and $S = 3/2$ states are closely spaced [141], so spin–orbit coupling provides a mechanism for mixing them. Rather than creating the commonly observed thermal equilibrium between the two spin states (so called spin crossover), the selection rules of quantum mechanics and spin–orbit coupling allow the two states to mix and to create a new, discrete, admixed ground state. Such admixed $S = 3/2, 5/2$ states give rise to magnetic properties that lie along a continuum between the extremes of the pure $S = 5/2$ and $S = 3/2$ states [142]. The difference between the two spin states lies in the fact that the $d_{x^2-y^2}$ orbital is empty for $S = 3/2$ species. Since unpaired spin in the $d_{x^2-y^2}$ orbital is associated with predominant σ spin delocalization and downfield pyrrole proton isotropic shifts, whereas unpaired spin in d_{xz} and d_{yz} orbitals results in upfield pyrrole proton isotropic shifts through π spin delocalization, proton NMR resonances are quite sensitive to the $S = 5/2$ and $S = 3/2$ contributions in a spin-admixed complex. Therefore, in tetraphenylporphyrinate (TPP) complexes, ^{1}H NMR δ_{pyrrole} values shift dramatically upfield with increasing $S = 3/2$ character. High spin species such as FeCl(TPP) have large downfield shift (+80 ppm) of the eight pyrrole protons on the periphery of the porphyrin macrocycle, whereas species approaching pure intermediate spin have upfield shifts that can be as large as –62 ppm [142]. In addition, temperature changes affect the mixing of the two states with dramatic effects on the pyrrole proton shifts [101].

The spectrum of horseradish peroxidase (HRP) is reported in Fig. 7.47, together with the assignment, mainly obtained through the detection of proton–proton dipolar connectivities [143]. Methyl groups are shifted in the range 70–80 ppm downfield. The NH resonance of the histidine ring of the axial ligand occurs at around 100 ppm downfield, while the Hε1 and Hε2 resonances of the same histidine are too broad to be detected. The general pattern of the isotropic shifts is thus indicative of high spin iron(III); however, when the protein is substituted with deuterohemin, in which the two vinyl groups are substituted by pyrrole protons, the signals of the latter are not observed in the usual region around 60–80 ppm downfield, but most probably lie within the diamagnetic region. This has been taken to be consistent with the proposed admixture of $S = 5/2$ and $S = 3/2$ ground states [144]. This also

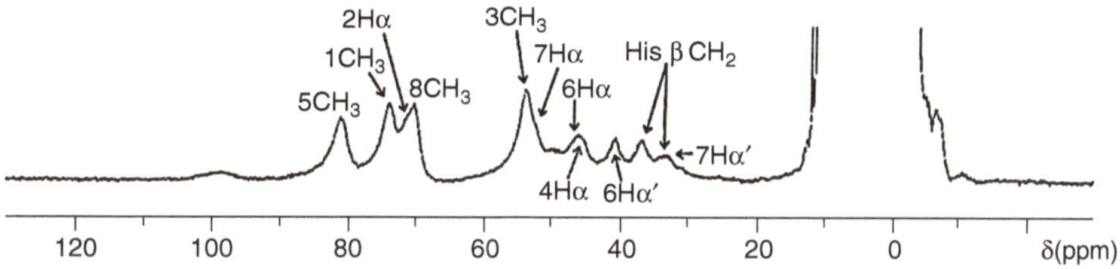

FIGURE 7.47 800 MHz ^{1}H NMR spectrum of horseradish peroxidase in 50 mM phosphate buffer. Shifts are in ppm from DSS (labeling as in Fig. 7.25B).

accounts for the fact that the magnetic moment (5.2 μ_B) is less than the spin-only value (5.92 μ_B) for $S = 5/2$, indicating a contribution from a $S < 5/2$ state [143].

Chemical equilibria between species with different spin states are common. When there is chemical equilibrium between $S = 5/2$ and $S = 1/2$, if the equilibrium is fast on the NMR time scale, i.e. the exchange rate between the two states is faster than the difference in resonance frequency of the two states, a weighted average shift is observed (Section 6.2). When the equilibrium is slow on the NMR time scale, saturation transfer can be observed if the exchange rate between the two states is slower than the difference in resonance frequency of the two states, but faster than the longitudinal relaxation rates of the nucleus in the two environments (Sections 6.3.4 and 12.3.1). As an example, saturation transfer experiments have been performed to assign heme proton signals in ferric low spin cyanide HRP [145,146]. In ferric high spin HRP large scalar interaction leads to the resolution of most heme signals. The interconversion between HRP and cyanide HRP in a sample containing both the forms can lead to magnetization transfer between resonances in the two states, due to CN^- exchange. The exchange rate is 2.5 s^{-1} at 328 K, which allows saturation transfer experiments to be performed, and allowed for the assignment of the heme signals of HRP through the assignment of the signals of cyanide HRP, upon saturation of the former. Cyanide HRP, on the other hand, possesses considerable magnetic anisotropy, and hence exhibits more favorable relaxation properties and it is easily assigned.

Also the ^{1}H NMR spectra of HasA, a heme carrier protein, show that iron(III) is in a nonpure spin-state [147]: the average chemical shift for heme methyl group (ca. 40 ppm) is, indeed, intermediate between typical values of a purely high-spin ($S = 5/2$) and those of a purely low-spin ($S = 1/2$) heme iron(III). This is confirmed by the temperature dependence of the chemical shifts of these signals which deviates from the linear dependence predicted by Eq. (2.7). Monodimensional ^{13}C spectra of holo ^{13}C-enriched HasA with unlabeled heme have permitted in this case to selectively observe the signals of the axial ligands, Y75 and H32, which are generally nondetectable in ^{1}H spectra of iron(III) heme proteins with $S > 1/2$ (Fig. 7.48). The temperature dependence of their chemical shifts indicates that when the temperature increases the shifts of the axial Y75 residue tend toward the diamagnetic values, and those of the H32 residue tend toward larger shifts (in absolute value). This indicates that the axial Y75 ligand is released upon increasing temperature, which leaves a pentacoordinated species or a six-coordinated species with a water molecule as the sixth ligand. In fact, upon breaking (or weakening) of the coordination bond between the iron and the Y residue, the nuclei of the latter experience no (or very little) contact shift and, therefore, their shifts will tend toward the diamagnetic values, as observed, and the signals of the H32 nuclei tend toward larger shifts, as expected for axial ligands when passing from a lower to a higher spin state. In monodimensional ^{13}C spectra of mutants

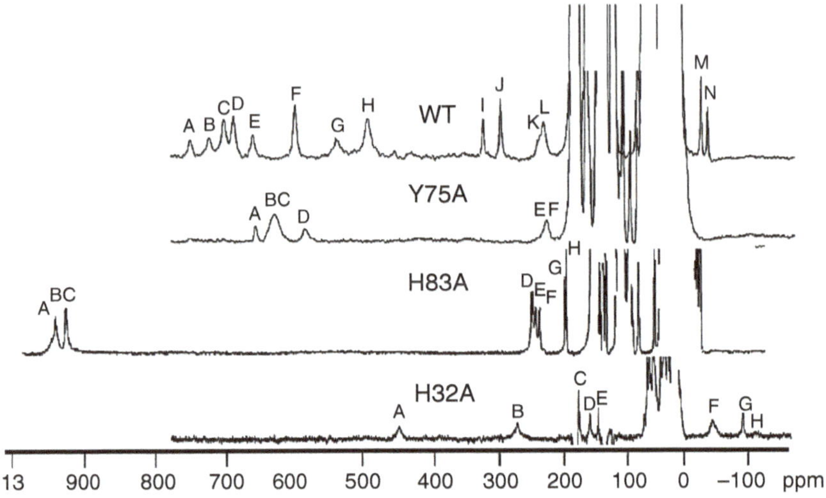

FIGURE 7.48 ^{13}C NMR spectra of wild type HasA and its mutants reflect the difference in the number of protein axial ligands (number of detectable signals) and spin state (chemical shift range).

Reproduced with permission from Ref. [139].

(Fig. 7.48), the pattern of hyperfine shifted resonances shows that H32 is the only axial ligand in Y75A and H83A (S = 5/2), whereas Y75 is the only axial ligand in H32A (S = 3/2) [139,148,149]. Of course, the presence of the paramagnetic heme iron(III) center in holoHasA causes the broadening beyond detection of the NMR signals of nuclei close to the metal, and hyperfine shifts for detectable nuclei. The appearance or disappearance of these paramagnetic effects can be exploited to monitor the transfer of the heme between proteins involved in heme transport networks [150,151]. Analogously, paramagnetic broadening due to the presence of iron(III) species permitted to identify the binding site of ferric species in ferritin [152].

7.3.5.4 Manganese(II)

Manganese(II) has a $3d^5$ electronic configuration, giving rise to an orbitally nondegenerate ^6S ground term and thus to high spin $S = 5/2$ compounds (Fig. 7.49), just like in the high spin iron(III) case already discussed. Because of the singly degenerate ground state (^6A), electron relaxation is not efficient. Degeneracy is only removed by ZFS due to second order spin–orbit coupling with excited states that arise from other free ion terms of lower multiplicity. This perturbation splits the spin degeneracy into three levels in a symmetry lower than cubic (Fig. 1.15C). Typical ZFS values for manganese(II) are in the range 0–1 cm^{-1} [60], i.e. they are generally smaller than that in high spin iron(III). This is because, compared to iron(III), the spin orbit coupling constant is smaller due to the smaller charge of the manganese(II) ion, and because excited states are closer in manganese(II) than in iron(III).

The most efficient electron relaxation mechanisms are bound to the zero field splitting modulation (Table 7.3), which may arise from rotation of the complex or, more probably, from distortions of the coordination sphere as a result of collisions with solvent molecules [Eqs. (4.12) and (4.13)]. Typical electron relaxation times at low fields are around 10^{-9}–10^{-10} s. The basically null g anisotropy as well as the small ZFS causes negligibly small $\Delta\chi_{ax}$ values. Fig. 7.49 shows the ^1H paramagnetic relaxation

Conf.	L	S	Free ion conf.	Ligand field: Octahedral (O_h)		Tetrahedral (T_d)		Square planar (D_{4h})		Square pyramid. (C_{4v})		Trigonal bipyram. (D_{3h})	
d^5	0	5/2	^6S	↑↑ ↑↑↑	^6A	↑↑↑ ↑↑	^6A	↑ ↑ ↑ ↑↑	^6A	↑ ↑ ↑↑ ↑	^6A	↑ ↑↑ ↑↑	^6A

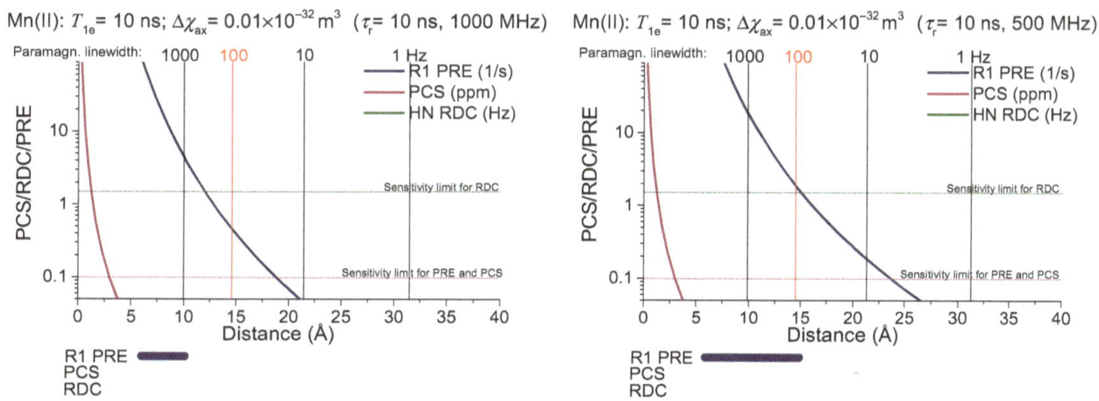

FIGURE 7.49 Population of the electronic d orbitals in different idealized symmetries in manganese(II) complexes and ^{1}H R_1 paramagnetic relaxation enhancement (in s^{-1}, *blue line*) and maximum pseudocontact shift (in ppm, *pink line*) as a function of the distance from the metal ion, for typical values of electron relaxation time and axial magnetic susceptibility anisotropy, for a molecule with τ_r of 10 ns, at 1000 MHz (left panel) or 500 MHz (right panel). The plots also shows the distances with ^{1}H paramagnetic linewidth of 1, 10, 100, and 1000 Hz. The bar under the plots indicate the range of distances with ^{1}H R_1 paramagnetic relaxation enhancement values above the experimental sensitivity limits. No paramagnetic residual dipolar couplings can be observed, due to the very small magnetic susceptibility anisotropy.

enhancement and linewidth at different distances from the metal ion, calculated for an electron relaxation time of 10 ns (at 500 and 1000 MHz).

The ^{1}H NMRD profiles of $Mn(H_2O)_6^{2+}$ in water solution show two dispersions in the 0.01–100 MHz range of proton Larmor frequency (Fig. 7.50). The first (at ca. 0.05 MHz, at 298 K) is attributed to the contact relaxation and the second (at ca. 7 MHz, at 298 K) to the dipolar relaxation. From a best fit analysis of the profiles, the electron relaxation time (around 3.5×10^{-9} s at low fields and 298 K), described by the parameters Δ_t (ranging between 0.015 and 0.03 cm^{-1}) and τ_v (ranging between 2 and 7×10^{-12} s), can be determined from the position of the first dispersion [153]; the reorientation correlation time $\tau_r = 3 \times 10^{-11}$ s is consistent with the position of the second dispersion and is in accordance with the value expected for hexaaquametal(II) complexes. The constant of contact interaction is about

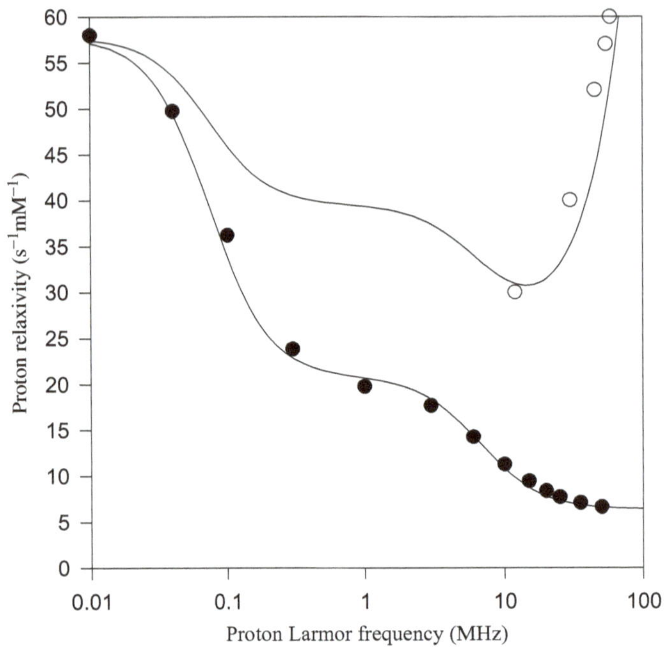

FIGURE 7.50 Water ^{1}H NMRD longitudinal (●) and transverse (○) profiles of $Mn(H_2O)_6^{2+}$ solutions at 298 K.

0.7 MHz (Table 7.3). The impressive increase of R_2 at high fields is due to the field dependence of the electron relaxation time and to the presence of a nondispersive T_{1e}–containing term in the equation for contact relaxation (Section 4.7.2). If it were not for the finite residence time, τ_M, of the water molecules in the coordination sphere, the increase in R_2 could continue as long as the electron relaxation time increases.

Fig. 7.51 reports the NMRD profiles of manganese(II) at different concentrations of d^8 glycerol and the best fit profiles (assuming $\Delta_t = 0.03$ cm^{-1}, $A/h = 0.64$ MHz, and $r = 2.78$ Å) [154]. The main features that can be observed by increasing viscosity are: (1) the relaxivity increases, (2) the contact dispersion progressively disappears, (3) the dipolar dispersion moves toward lower fields, (4) a peak appears at high fields. These features can be explained by an increase in the reorientation time, with τ_r increasing from 3×10^{-11} s for the water solution to 3×10^{-10} s for a solution containing 65% w/w of glycerol, at 288 K. At the same time the correlation time for electron relaxation, τ_v, increases (from 6×10^{-12} for the water solution to 5×10^{-11} s for a solution containing 65% w/w of glycerol, at 288 K), thus resulting in a decrease of the effective electron relaxation time at low fields, and thus of the contact contribution. At some point the correlation time for dipolar relaxation becomes the effective electron relaxation time, so that its field dependence produces the peak arising at high fields. Notably, both τ_r and τ_v increase linearly with viscosity, pointing out that modulation of the transient ZFS by collision with solvent is the dominant source of relaxation up to very high viscosity [154]. The fitting was obtained by keeping the number of bound waters free to change with the concentration of d^8 glycerol, because of the possible substitution of water by glycerol, and by taking into account the outer-sphere contribution, theoretically estimated around 5–10% of the total relaxivity [154].

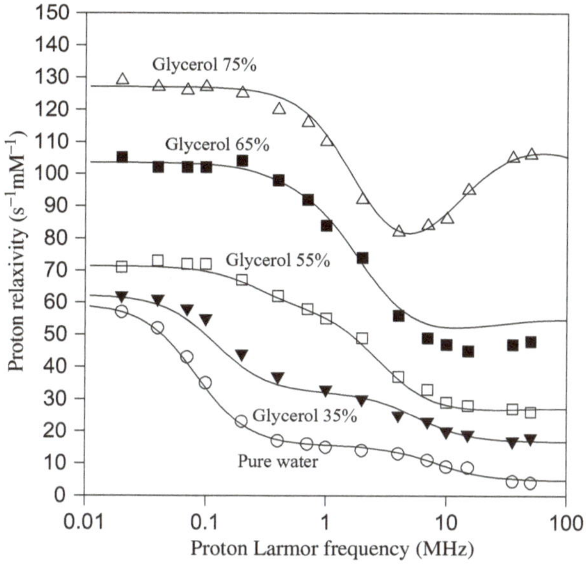

FIGURE 7.51 Water ^{1}H NMRD profiles for $Mn(H_2O)_6^{2+}$ solutions at 308 K in pure water (○) and with increasing amounts of d^8-glycerol: 35% (▼), 55% (□), 65% (■), 75% (Δ).

Adapted from Ref. [154].

In all manganese(II) proteins and in most complexes the contact interaction is found negligible. The ^{1}H NMRD profile of MnEDTA (Fig. 7.52), for instance, indicates the presence of the dipolar contribution only, and one water bound to the complex [155]. Actually the profile is very similar to that of GdDTPA (Chapter 10), and is provided by the sum of inner-sphere and outer-sphere contributions of the same order. The relaxation rate of manganese(II) complexes with DTPA is instead provided by outer-sphere relaxation only, since no water molecules are bound to the complex (Sections 6.5 and 6.6) [156].

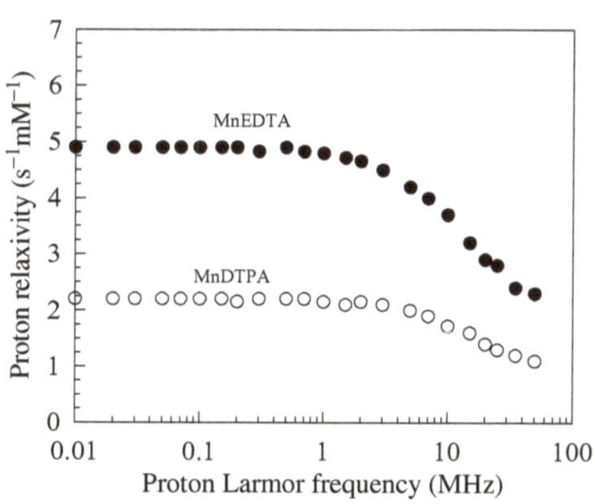

FIGURE 7.52 Water ^{1}H NMRD profiles for MnEDTA (●) and MnDTPA (○) solutions at 308 K.

Reproduced with permission from [157].

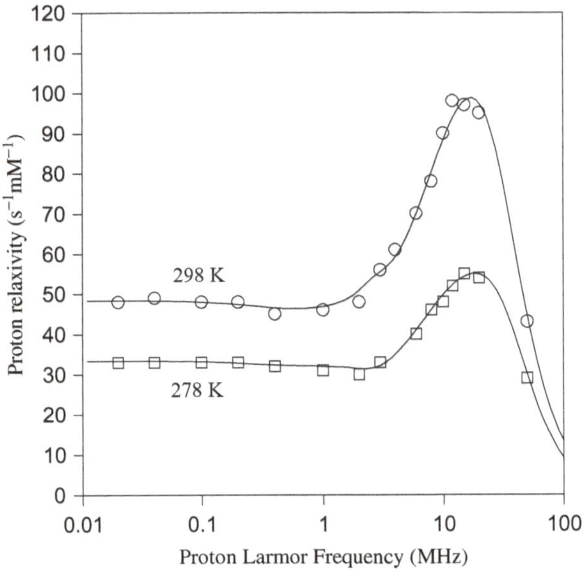

FIGURE 7.53 Water ^{1}H NMRD profiles for solutions of Mn^{2+} concanavalin A at 298 (○) and 278 (□) K [159]. The solid curves are calculated with $D = 0.04\ cm^{-1}$.

Fig. 7.53 shows the NMRD profiles obtained with manganese(II) bound to the protein concanavalin A. The profiles are basically flat up to 1 MHz, when a peak appears. The correlation time for dipolar relaxation is the electron relaxation time, being shorter, at low fields, than the reorientation correlation time. After the beginning of the ω_S dispersion, a peak arises due to the increase of the electron relaxation time and the final ω_I dispersion. Inclusion of ZFS is also required, as found from EPR measurements [158]. It is actually possible to fit the profile without considering the presence of static ZFS, but it results in wrong values (of the order of a factor 2–3) of τ_M, τ_r, τ_v, Δ_t, and, above all, of the hydration number. The main effect of static ZFS is in fact that of decreasing the proton relaxivity at low fields without altering the relaxivity at high fields (Section 4.7.1), thus affecting the best-fit values of all parameters. The fit performed with including ZFS with $D = 0.04\ cm^{-1}$ provides indication of 6–7 protons at 2.8 Å from the manganese(II) ion, with τ_v about 5×10^{-11} s and $T_{1e(0)}$ about 3×10^{-10} s at room temperature [158].

In Fig. 7.54 we report the ^{1}H NMRD profile of $[Mn(EDTA)(BOM)_2]^{2-}$ bound to human serum albumin (HSA) [160]. Again, the profiles are characterized by the following features: (1) the presence of a dispersion at about 0.2–1 MHz, due to the dipolar $\omega_S T_{2e}$ dispersion in the presence of static ZFS, (2) an increase in relaxivity with increasing the field from 1 to 10 MHz, due to the field dependence of T_{1e}, (3) a subsequent decrease in relaxivity due to the $\omega_I T_{1e}$ dispersion. The best-fit analysis performed taking into account the presence of a static ZFS with $D = 0.05\ cm^{-1}$ indicates that one fast exchanging water molecule is coordinated to the metal ion and that the electron relaxation is determined by the parameters $\Delta_t = 0.012\ cm^{-1}$ and $\tau_v = 4 \times 10^{-11}$ s [160,161]. This set of parameters can also account for the NMRD profile of $Mn(EDTA)(BOM)_2$ in water solution, if data are properly analyzed by taking into account fast rotation conditions [161]. To be noted that the fit of $Mn(EDTA)(BOM)_2$-HSA data

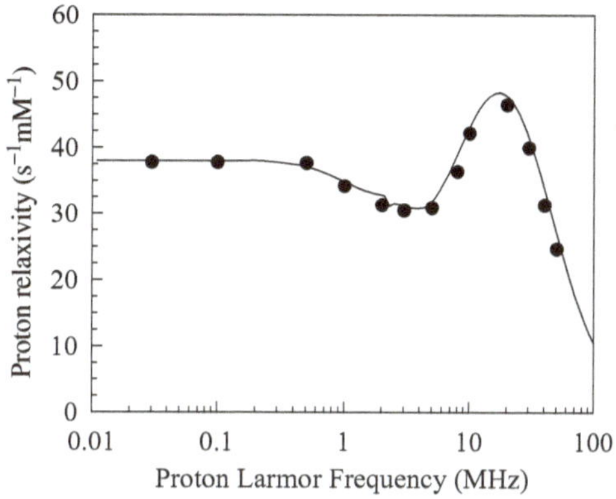

FIGURE 7.54 Water ^1H NMRD profiles for $Mn(EDTA)(BOM)_2$-HSA solutions at 298 K.

Adapted from Refs. [160,161].

performed with the Solomon-Bloembergen-Morgan (SBM) theory (Sections 6.7 and 10.2) would have provided somewhat wrong values [162].

Proton NMR spectra for manganese(II) systems are expected to have very broad signals, broader than for high spin iron(III) systems, since, the electronic configuration being the same, ZFS is smaller and thus the electron relaxation time is longer.

7.3.6 d^6 (Fe^{2+})

Iron(II) is a d^6 ion which can be high spin or low spin; high spin iron(II) is paramagnetic ($S = 2$), low spin iron(II) is diamagnetic in octahedral symmetry and paramagnetic in tetrahedral symmetry. High spin iron(II) complexes are obtained with weak or medium strength ligands. The electronic configuration are shown in Fig. 7.55. The electron relaxation times are rather short ($T_{1e} \approx 10^{-12}$ s) because the excited levels are close in energy for the same reasons as in the case of low spin iron(III), thus allowing the Orbach mechanism to be efficient. Consistently, the NMRD profile of $Fe(H_2O)_6^{2+}$, obtained from Mohr salt ($(NH_4)_2Fe(SO_4)_2 \cdot 6H_2O$), reported in Fig. 7.56, does not exhibit any dispersion below 50 MHz of proton Larmor frequency. The observed $\Delta\chi_{ax}$ values, of about 2×10^{-32} m^3, can provide relatively large pseudocontact shifts and paramagnetic residual dipolar couplings (Fig. 7.55).

Being $S = 2$ systems, a predominance of σ delocalization is expected as well as moderate π delocalization at the *meso* positions [163]. A spectrum is reported in Fig. 7.57 [163], where the pyrrole protons are downfield.

In proteins, the position of the axial histidine ring determines the inequivalence of protons in the porphyrin plane. Probably as a result of this, a variety of patterns is observed with either one or two of the four methyl substituents of protoporphyrin IX being downfield [86,109,165–167]. The orientation of the χ_{xx} and χ_{yy} axes in the porphyrin plane depends on the orientation of the axial π interaction, with a behavior analogous to the low spin iron(III) case, but with the *x*- and *y*-axes interchanged. In other words, when the π interaction is along a metal–pyrrole nitrogen direction, it defines the χ_{yy} and not the χ_{xx} direction. When the π interaction rotates clockwise away from the metal pyrrole direction by an angle α, it is χ_{yy} that rotates counterclockwise by α. Therefore, in analogy with the low spin iron(III) case, the pseudocontact shift can be calculated as a function of the orientation of the axial histidine for

Conf.	L	S	Free ion conf.	Ligand field: Octahedral (O_h)		Tetrahedral (T_d)		Square planar (D_{4h})		Square pyramid. (C_{4v})		Trigonal bipyram. (D_{3h})	
d^6	2	2	^5D	↑ ↑ ↓↑ ↑ ↑	^5T	↑ ↑ ↑ ↓↑ ↑	^5E	↑ ↑ ↑ ↓↑ ↑	^5E	↑ ↑ ↑ ↑ ↓↑	^5A	↑ ↑ ↑ ↓↑ ↑	^5E

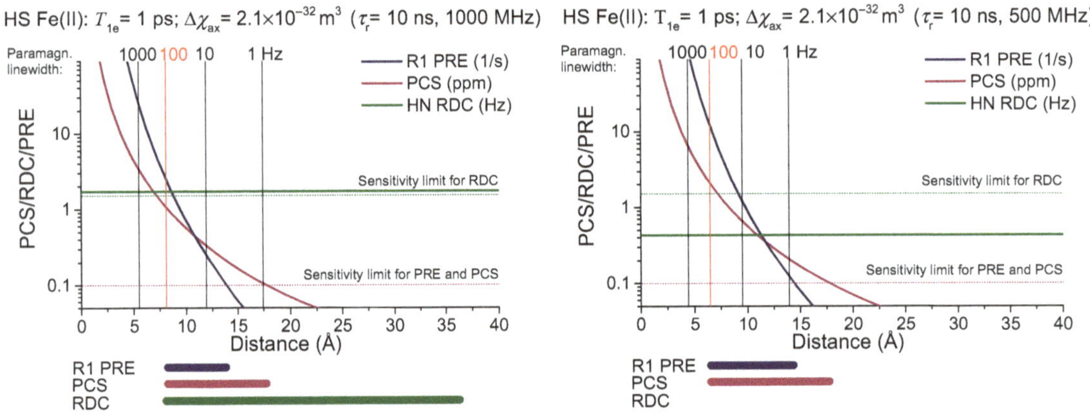

FIGURE 7.55 Population of the electronic d orbitals in different idealized symmetries in high spin iron(II) complexes and ^{1}H R_1 paramagnetic relaxation enhancement (in s^{-1}, *blue line*), maximum pseudocontact shift (in ppm, *pink line*) and maximum HN residual dipolar coupling (in Hz, *green line*) as a function of the distance from the metal ion, for typical values of electron relaxation time and axial magnetic susceptibility anisotropy, for a molecule with τ_r of 10 ns, at 1000 MHz (left panel) or 500 MHz (right panel). The plots also shows the distances with ^{1}H paramagnetic linewidth of 1, 10, 100, and 1000 Hz. The bars under the plots indicate the range of distances with ^{1}H R_1 paramagnetic relaxation enhancement, pseudocontact shift and HN residual dipolar coupling values above the experimental sensitivity limits. The maximum pseudocontact shift is obtained for a nucleus on the z-axis of the magnetic susceptibility tensor; the maximum residual dipolar coupling is obtained for an amide vector parallel to the z-axis of the magnetic susceptibility tensor.

all methyl protons, by using Eq. (2.46). The anisotropy of the contact shifts has been predicted to display the same dependence on the orientation of the axial π interaction, being maximal along the direction of the π interaction (i.e., perpendicular to the histidine plane) and zero at 90 degrees from the direction of the π interaction. The angle dependence is thus the same with respect to the low spin iron(III) case. Therefore, hyperfine shifts result to have a different angle dependence of that observed in low spin iron(III) systems, as the dependence is the same for the contact contribution and shifted by 90 degrees for the pseudocontact contribution. The overall effect is shown in Fig. 7.58. As an example, in reduced cytochrome c' the shift pattern of the methyls is three signals upfield and one downfield. This is due to

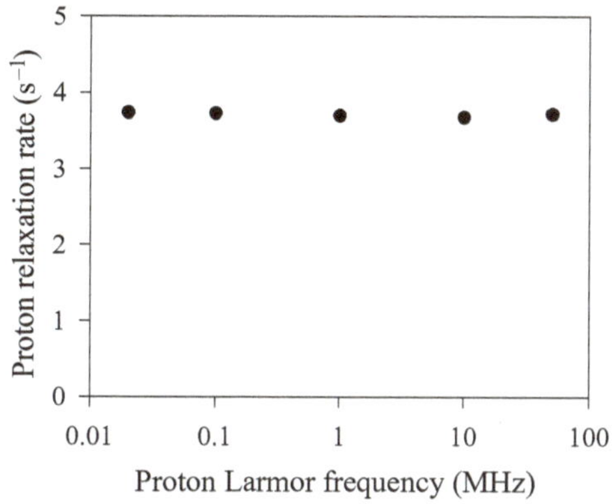

FIGURE 7.56 Water ^{1}H NRMD profiles for 10 mM solution of $(NH_4)_2Fe(SO_4)_2 \cdot 6H_2O$.

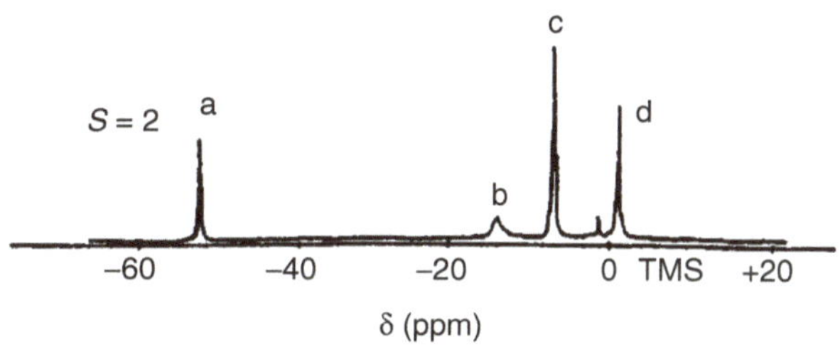

FIGURE 7.57 ^{1}H NMR spectrum of $S = 2$ Fe 5,10,15,20-tetraphenylporphyrin (2-CH_3Im). *a*, 1,2,3,4,5,6,7,8-H; *b*, 2-CH_3Im 4,5-H; *c*, phenyl *ortho-*, *meta-* and *para-*H; *d*, 2-CH_3ImCH_3 (labeling as in Fig. 7.25B).

Adapted from Ref. [164].

the relative weight of contact (which causes downfield shifts) and pseudocontact (which causes upfield shifts if the axial ligand field strength is smaller than the equatorial and therefore $\chi_{\parallel} > \chi_{\perp}$) contribution. These experimental data are in good agreement with predictions, the histidine orientation being along the β–δ *meso* axis, and thus forming relatively small angles with methyl groups 1, 5, and 8, and a large angle with methyl group 3 [168]. In sperm whale deoxymyoglobin, methyl 1 and 5 experience upfield pseudocontact shifts whereas methyls 3 and 8 experience downfield pseudocontact shift [169]. This pattern is in agreement with a position of the histidine along the pyrrole II–pyrrole IV axis [170].

By measuring the linewidth at different magnetic fields (from 90 to 360 MHz) for deoxymyoglobin and for deoxyhemoglobin, the Curie contribution to the overall line broadening was separated, and the rotational correlation time was found to be 5×10^{-9} s for deoxymyoglobin and 2.5×10^{-8} s for deoxyhemoglobin [171]. Unlike the linewidth, longitudinal relaxation times are only negligibly affected by Curie relaxation mechanism, and therefore an independent estimate of the electron relaxation time can be made, provided r is known and the Solomon equation [Eq. (4.20)] holds. Such a value for deoxyhemoglobin was found to be 7×10^{-13} s [172].

Fig. 7.59 shows the ^{1}H-NMR spectrum of iron(II) bleomycin [173]. The high resolution and the relatively narrow line widths observed in the spectrum are as expected for high-spin iron(II) complexes. Paramagnetically shifted resonances out to 230 ppm have been observed, and

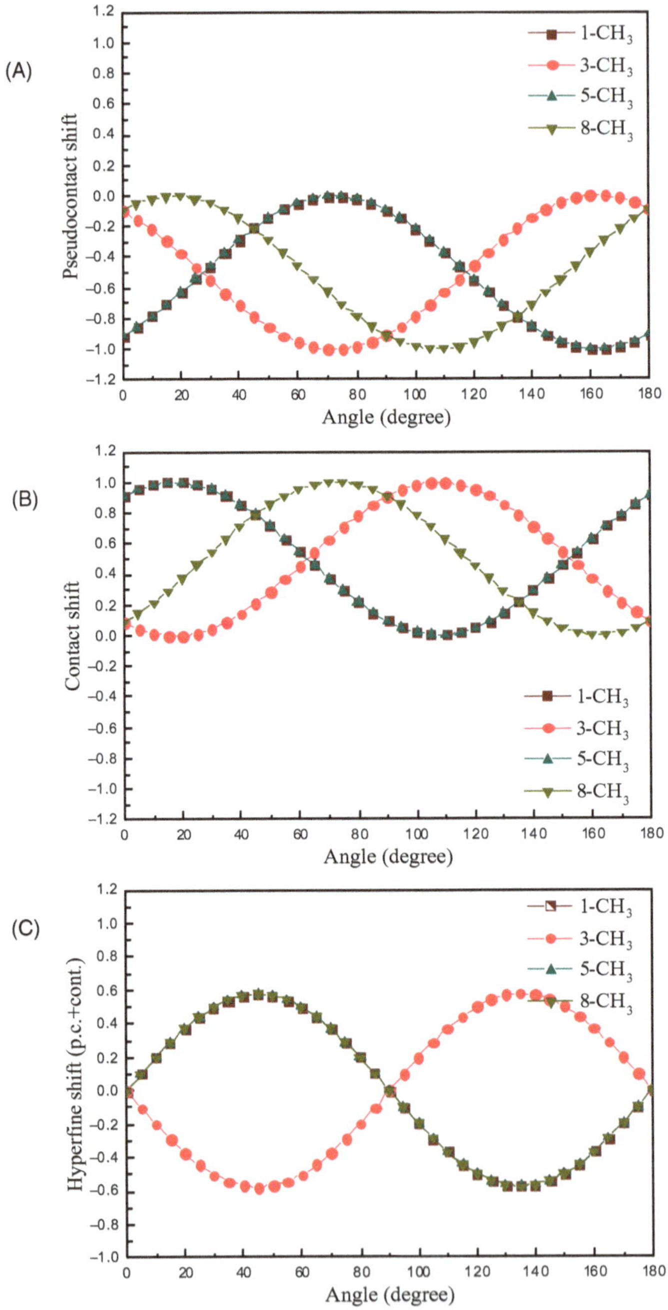

FIGURE 7.58 (A) Dependence of the heme methyl pseudocontact shifts as a function of the angle α between the pyrrole II–pyrrole IV molecular axis and the projection of the His plane on the heme. (B) Dependence of the contact shift as a function of the same α angle. (C) Dependence of the sum of the pseudocontact plus contact shifts on the α angle. The sum in the example is obtained by giving the same weight to both contributions.

Adapted from Ref. [168].

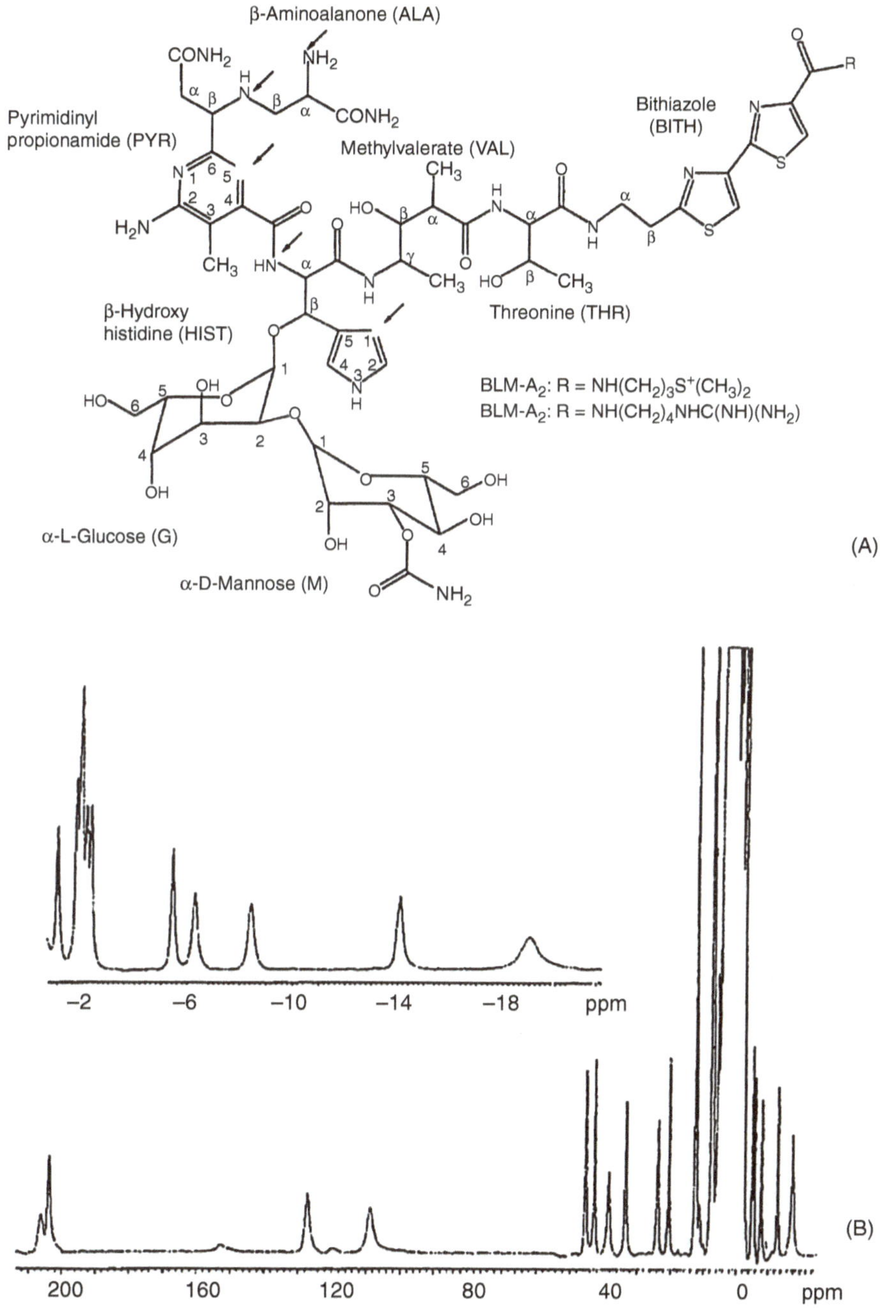

FIGURE 7.59 (A) Schematic structure of bleomycin. The *arrows* indicate the ligands to the metal center. (B) 300 MHz ^{1}H NMR spectrum of iron(II)–bleomycin at 298 K.

Adapted from Ref. [173].

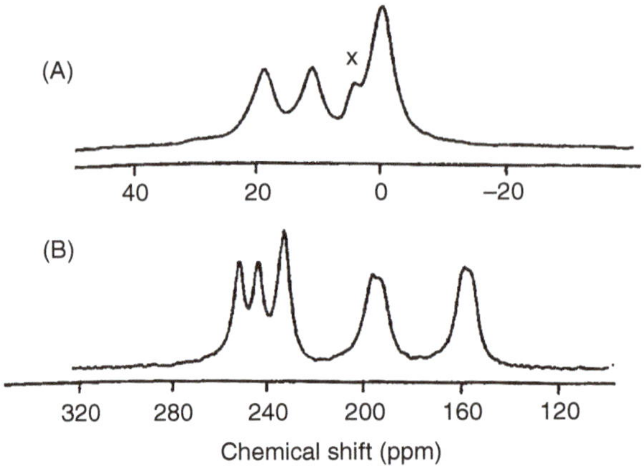

FIGURE 7.60 ^{2}H NMR spectra of reduced [^{2}H$^\alpha$]Cys-labeled rubredoxin (A) and reduced [^{2}H$^{\beta2,\beta3}$]Cys-labeled rubredoxin (B) at 308 K, pH 6.0.

Adapted from Ref. [112].

two-dimensional NMR experiments, together with the measurement of the T_1 values, allow for the estimate of the metal–proton distances and for the identification of the ligands to the metal center (arrows in Fig. 7.59).

In reduced rubredoxin, the eight H$^\beta$ of the four coordinated cysteines are in the 280–150 ppm range, whereas the four H$^\alpha$ (Fig. 7.60) are little shifted (19 ppm, 16 ppm, and two around 0 ppm). The isotropic shifts of the H$^\beta$ nuclei are larger than those of the H$^\alpha$ nuclei. This is consistent with a predominant contact mechanism with σ spin delocalization. Because only one iron ion is present in rubredoxin, the only mechanism that can account for the upfield isotropic shifts is a significant pseudocontact contribution to the hyperfine shift.

Low spin iron(II) is diamagnetic in octahedral symmetry and paramagnetic in tetrahedral symmetry. Some iron(II) compounds undergo a spin transition over a small temperature interval from a diamagnetic system at low temperature to a paramagnetic system at higher temperature (spin-crossover transition). Due to the short T_{1e} in the paramagnetic state of iron(II), it is possible to detect the proton NMR in the paramagnetic state as well as in the diamagnetic state. The width of the paramagnetic spectrum increases with decreasing temperature, following the Curie law, until the spin-crossover transition, where a strong decrease in linewidth occurs and the system becomes diamagnetic [174–176].

7.3.7 d^7 (Co^{2+}, Ni^{3+})

7.3.7.1 Cobalt(II), low spin

Cobalt(II) is a d^7 ion which can be high spin or low spin. Generally, it is low spin in planar or some square pyramidal compounds. The effective electron relaxation times in the low spin state are long enough (10^{-9}–10^{-10} s) so that EPR spectra can be recorded at room temperature [60] and the proton NMR lines are broad. This is due to the high energy of the first excited state (in planar and square pyramidal symmetries), caused by the large difference in energy between the d_{xy}, d_{xz}, d_{yz}

Conf.	L	S	Free ion conf.	Ligand field: Octahedral (O_h)		Tetrahedral (T_d)		Square planar (D_{4h})		Square pyramid. (C_{4v})		Trigonal bipyram. (D_{3h})	
d^7	3	1/2	^2F	↑ – ↓↑ ↓↑ ↓↑	^2E	↓↑ ↑– ↓↑ ↓↑	^2T	– ↑ ↓↑ ↓↑ ↓↑	^2A	– ↑ ↓↑ ↓↑ ↓↑	^2A	– ↓↑ ↑ ↓↑ ↓↑	^2E

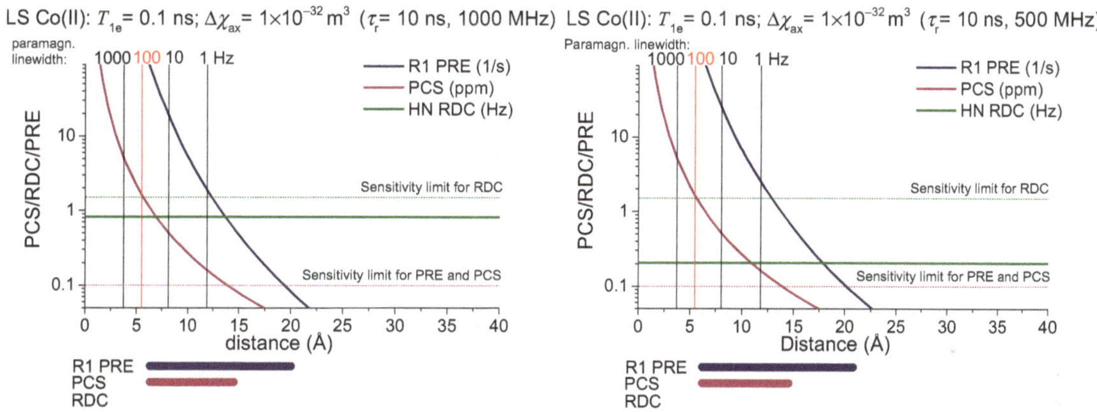

FIGURE 7.61 Population of the electronic d orbitals in different idealized symmetries in low spin cobalt(II) complexes and ^1H R_1 paramagnetic relaxation enhancement (in s^{-1}, *blue line*), maximum pseudocontact shift (in ppm, *pink line*) and maximum HN residual dipolar coupling (in Hz, *green line*) as a function of the distance from the metal ion, for typical values of electron relaxation time and axial magnetic susceptibility anisotropy, for a molecule with τ_r of 10 ns, at 1000 MHz (left panel) or 500 MHz (right panel). The plots also shows the distances with ^1H paramagnetic linewidth of 1, 10, 100, and 1000 Hz. The bars under the plots indicate the range of distances with ^1H R_1 paramagnetic relaxation enhancement, pseudocontact shift, and HN residual dipolar coupling values above the experimental sensitivity limits. The maximum pseudocontact shift is obtained for a nucleus on the *z*-axis of the magnetic susceptibility tensor; the maximum residual dipolar coupling is obtained for an amide vector parallel to the *z*-axis of the magnetic susceptibility tensor.

levels and the $d_{x^2-y^2}$, d_{z^2} levels (Fig. 7.61), and the absence of ZFS. The maximum pseudocontact shifts and amide proton paramagnetic residual dipolar couplings and the ^1H paramagnetic relaxation enhancements and linewidths are shown in Fig. 7.61 for an electron relaxation time of 10^{-10} s and $\Delta\chi_{ax} = 10^{-32}$ m^3.

In small compounds the electron–nucleus correlation time is given by the rotational correlation time (which can be as small as 10^{-11}–10^{-10} s) and some proton NMR spectra on systems like cobalt(II) porphyrins have been extensively studied by NMR spectroscopy. The ^1H NMR spectrum of cobalt(II)

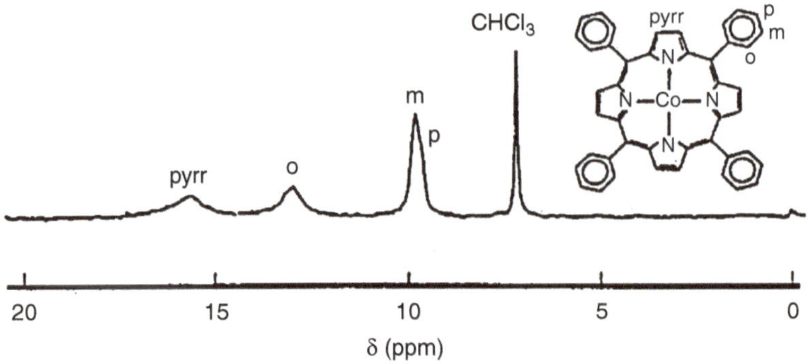

FIGURE 7.62 90 MHz ^{1}H NMR spectrum of tetraphenylporphynato cobalt(II) in $CDCl_3$ at 298 K. Chemical shifts are in ppm from TMS. The signal assignment is also shown.

Adapted from Ref. [178].

tetraphenylporphyrin, which is reported in Fig. 7.62 [177,178], is consistent with a mainly pseudocontact nature of the hyperfine shifts. This finding is expected in view of the small tendency of the unpaired electron in the d_{z^2} orbital to delocalize into the ligand *xy* plane. ^{13}C NMR spectra [177,178], however, show that there is also some spin density on the carbon nuclei of the porphyrin ring. Nitrogen-containing ligands coordinate in the axial position and give rise to five-coordinated adducts [177,178]: the quality of the spectra for such compounds decreases because the electronic relaxation times become longer. Indeed, in D_{4h} or C_{4v} symmetries, the first excited state contains two electrons in the d_{z^2} orbital (the energy of the $d_{x^2-y^2}$ orbital being much larger than that of d_{z^2}), which is destabilized by the fifth ligand (Fig. 7.61). Therefore, the first excited state becomes higher in energy upon base binding to cobalt(II) porphyrins and the electronic relaxation times become longer.

7.3.7.2 Cobalt(II), high spin

High spin cobalt(II), a $S = 3/2$ ion, is more interesting from the NMR point of view as the electron relaxation times are significantly shorter than in the low spin case. The electronic configurations are reported in Fig. 7.63. High spin cobalt(II) is commonly encountered in four- (tetrahedral), five- (either square pyramidal or trigonal bipyramidal), and six-coordinated (octahedral) complexes. The electronic ground state in octahedral geometry is triply degenerate ($^4T_{1g}$), in square pyramidal geometry is doubly degenerate (^4E), and in trigonal bipyramidal and tetrahedral geometries is nondegenerate ($^4A_2$) (Fig. 7.64). When the ion is in low symmetry situations, for instance when it is in a protein environment, the low symmetry removes orbital degeneracy, so that in pseudooctahedral and pseudosquarepyramidal geometries two or one excited states, respectively, are always close to the ground state. Under these conditions, efficient Orbach relaxation mechanisms are operative and electron relaxation is very fast. In six-coordinate complexes, the electron relaxation time is thus around 10^{-12} s. In five-coordinate complexes of C_{4v} symmetry the situation is similar, T_{1e} being slightly longer because the low-lying excited states are relatively farther. In tetrahedral (T_d) and trigonal bipyramidal (D_{3h}) symmetries the separation in energy between the orbitally nondegenerate ground state and the excited states is relatively large, and the ZFS is small. The electron relaxation time is generally one order of magnitude longer, the Orbach-type mechanisms being probably less efficient. Modulation of ZFS may also represent a possible relaxation mechanism. The magnitude of the splitting depends on the closeness of the

Conf.	L	S	Free ion conf.	Ligand field: Octahedral (O_h)		Tetrahedral (T_d)		Square planar (D_{4h})		Square pyramid. (C_{4v})		Trigonal bipyram. (D_{3h})	
d^7	3	3/2	^{4}F	↑ ↑ ↓↑ ↓↑ ↑	^{4}T	↑ ↑ ↑ ↓↑ ↓↑	^{4}A	↑ ↑ ↑ ↓↑ ↓↑	^{4}A	↑ ↑ ↓↑ ↑ ↓↑	^{4}E	↑ ↑ ↑ ↓↑ ↓↑	^{4}A

Ligand field: O_h:

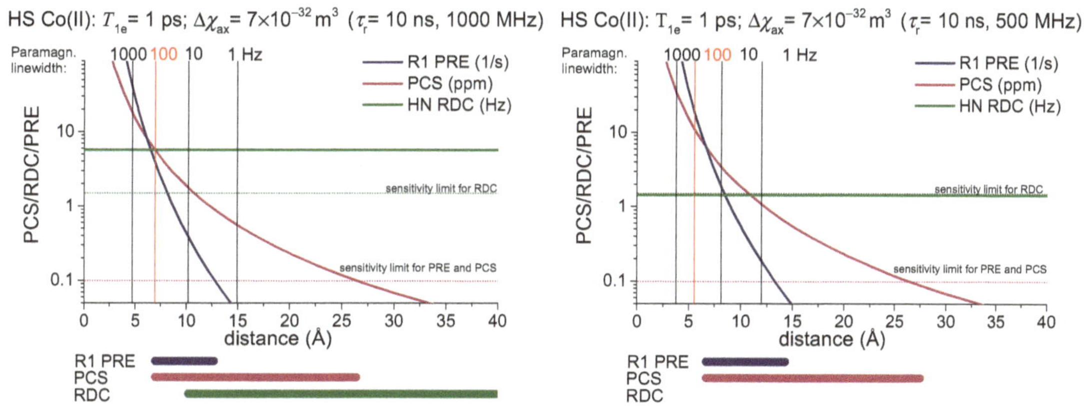

Ligand field: T_d:

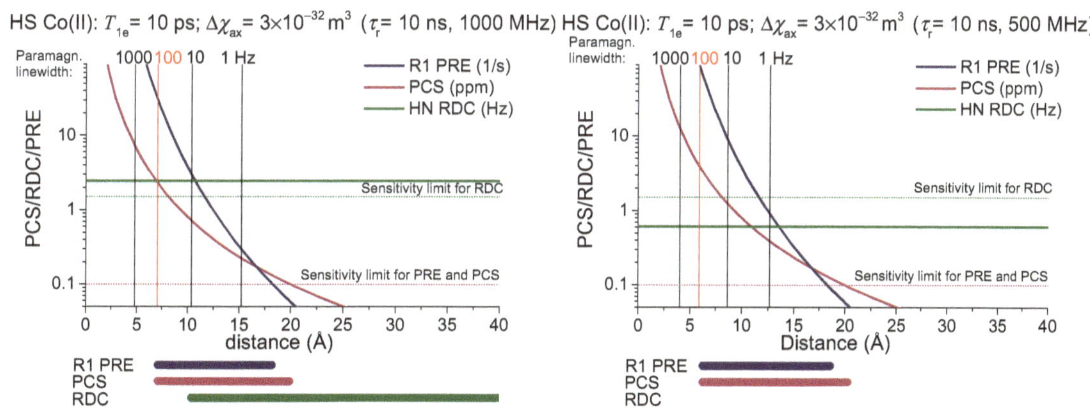

FIGURE 7.63 Population of the electronic d orbitals in different idealized symmetries in high spin cobalt(II) complexes and ^{1}H R_1 paramagnetic relaxation enhancement (in s^{-1}, *blue line*), maximum pseudocontact shift (in ppm, *pink line*) and maximum HN residual dipolar coupling (in Hz, *green line*) as a function of the distance from the metal ion, for typical values of electron relaxation time and axial magnetic susceptibility anisotropy, for a molecule with τ_r of 10 ns, at 1000 MHz (left panel) or 500 MHz (right panel). The plots also shows the distances with ^{1}H paramagnetic linewidth of 1, 10, 100, and 1000 Hz. The bars under the plots indicate the range of distances with ^{1}H R_1 paramagnetic relaxation enhancement, pseudocontact shift, and HN residual dipolar coupling values above the experimental sensitivity limits. The maximum pseudocontact shift is obtained for a nucleus on the z-axis of the magnetic susceptibility tensor; the maximum residual dipolar coupling is obtained for an amide vector parallel to the z-axis of the magnetic susceptibility tensor.

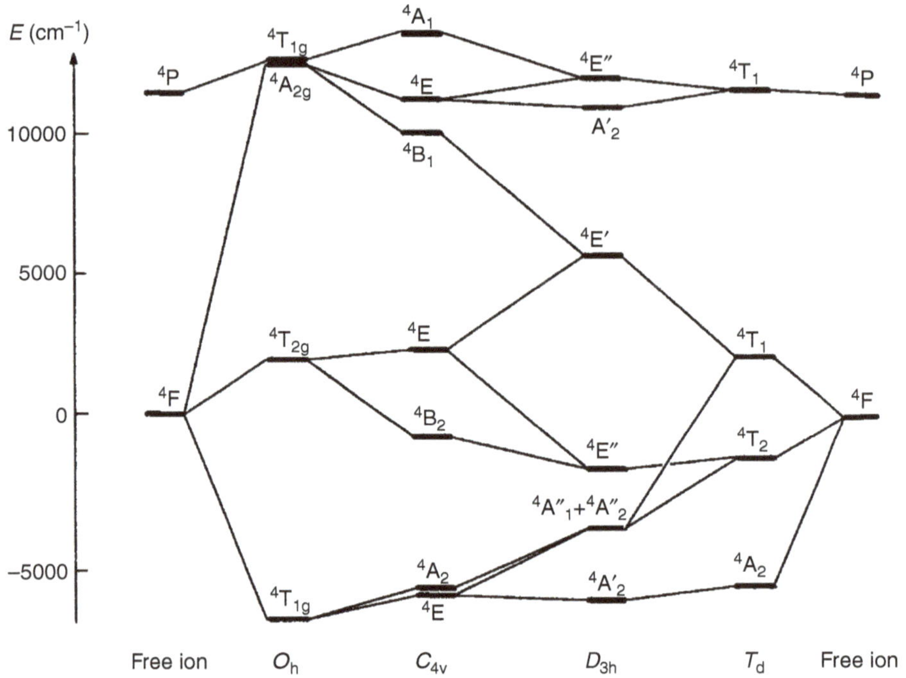

FIGURE 7.64 Splitting of the ^{4}F and ^{4}P free ion terms of high spin cobalt(II) in octahedral (O_h), square pyramidal (C_{4v}), trigonal bipyramidal (D_{3h}), and tetrahedral (T_d) geometries.

Adapted from Refs. [15,179].

low-lying excited states: the farther they are, the smaller is the ZFS. In all cases, no field dependence of the electron relaxation time is observed up to 100 MHz. Fig. 7.63 shows the maximum pseudocontact shifts and amide proton paramagnetic residual dipolar couplings and the ^{1}H paramagnetic relaxation enhancements and linewidths for the typical electron relaxation time and $\Delta\chi_{ax}$ values observed in pseudooctahedral and pseudotetrahedral symmetry, equal to 10^{-12} s and 7×10^{-32} m^3 and to 10^{-11} s and 3×10^{-32} m^3, respectively. The large pseudocontact shift and the moderate paramagnetic relaxation enhancement exerted by high spin pseudooctahedral cobalt(II) is nicely highlighted by the example of the cobalt complex shown in Fig. 7.65, where contact shifts are negligible [180].

The water proton NMRD of $Co(H_2O)_6^{2+}$ is reported in Fig. 7.66. The pseudooctahedral cobalt(II) complex provides almost field-independent water proton R_1 values in the 0.01–60 MHz region [14]. By assuming the validity of the Solomon equation [Eq. (4.17)], both the $\omega_S\tau_c = 1$ and $\omega_I\tau_c = 1$ dispersions can be placed at fields higher than 60 MHz, and therefore an upper limit for τ_c of the order of 10^{-12} s can be set. Since the rotational correlation time, τ_r, is likely to be very similar to that of $Fe(H_2O)_6^{2+}$, $Cu(H_2O)_6^{2+}$, and $Mn(H_2O)_6^{2+}$—about 3×10^{-11} s—a shorter correlation time must dominate the interaction; thus it must be the effective electronic relaxation time, T_{1e} (Table 7.3). Such a low T_{1e} value is consistent with the low water proton R_1 values. If the number of interacting water molecules is the same as for the other aqua ions, T_{1e} at 298 K can be estimated to be about 3×10^{-12} s from the Solomon equation, or up to 6×10^{-12} s from the equation including the effects of probable static ZFS [14]. When measurements are performed in highly viscous ethyleneglycol the observed rates are similar to those obtained in water. This indicates

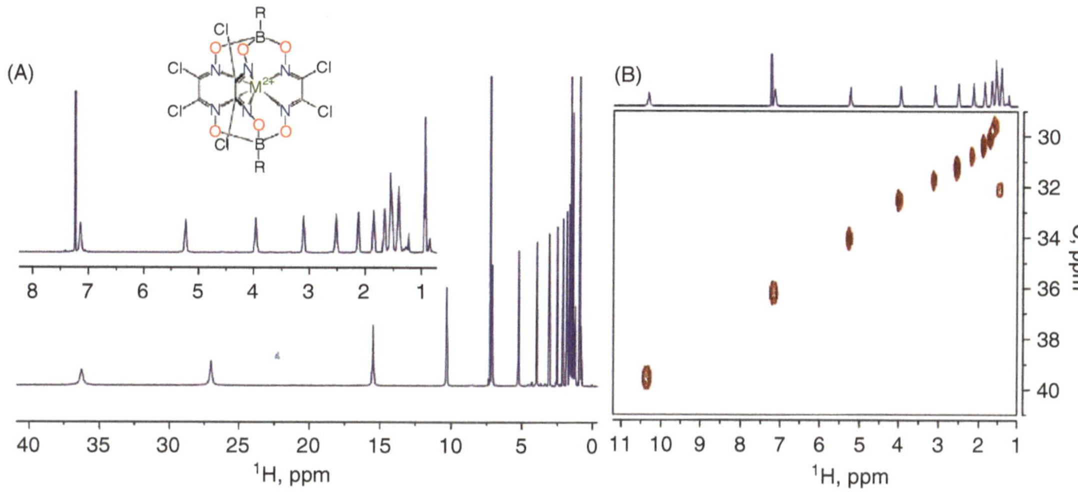

FIGURE 7.65 ^{1}H (A) and fragment of $^1H - ^{13}C$ HSQC (B) NMR spectra (600 MHz, 298 K, 5 mM solution in $CDCl_3$) of the Co^{2+} complex shown in the inset ($M^{2+} = Co^{2+}$, R = n-C_{16}-H_{33}). The sharp signal at 7.25 ppm corresponds to the residual signal of the solvent. The sequence of the pseudocontact shifted ligand proton signals nicely follows the prediction based on the increasing average distance of each CH_2 group in the R moiety.

Reproduced with permission from Ref. [180].

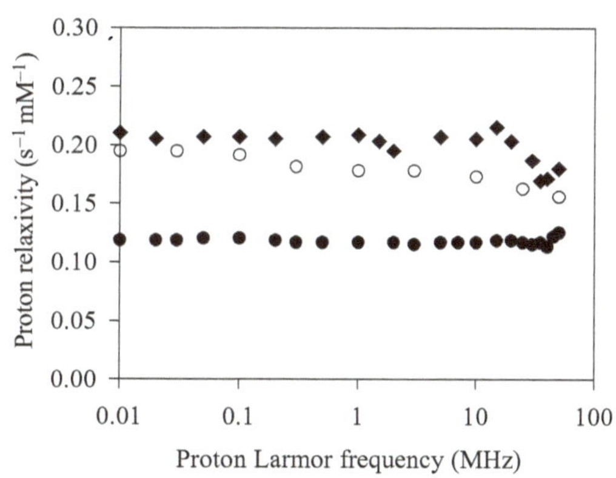

FIGURE 7.66 Solvent ^{1}H NMRD profiles for water solution of $Co(ClO_4)_2 \cdot 6H_2O$ at 298 K (○) as compared to those of ethyleneglycol solutions at 264 K (♦) and 298 K (•).

Adapted from Ref. [14].

that T_{1e} is also similar and that nuclear relaxation is rotation-independent [14]. The resulting picture is fully consistent with very efficient electron relaxation due to Orbach mechanisms (Table 7.3).

With such short T_{1e} values, in 5 or 6 coordinated cobalt(II) proteins, the metal ion provides negligible paramagnetic effects with respect to the diamagnetic contribution. Therefore, NMRD measurements are uninformative. When the protein contains tetracoordinated cobalt(II), then NMRD becomes again relevant. Water ^{1}H R_1 measurements of the high pH species of cobalt(II)-substituted carbonic anhydrase

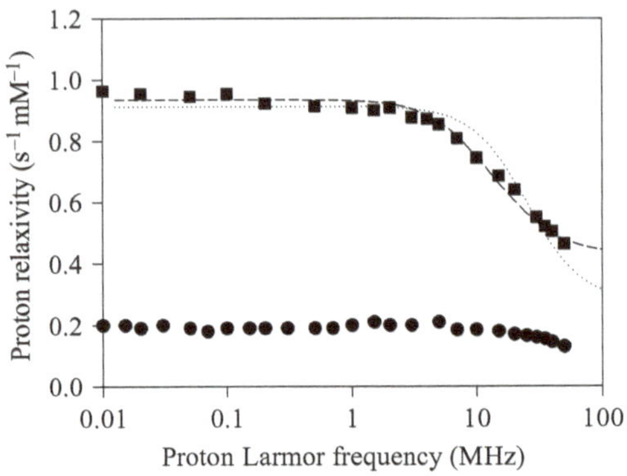

FIGURE 7.67 Water ^{1}H NMRD profiles for cobalt(II) human carbonic anhydrase I at pH 9.9 and 298 K (■) and for solutions of the nitrate adduct of cobalt(II) bovine carbonic anhydrase II at pH 6.0 and 298 K (•). The *dashed line* shows the best fit profile of the former data calculated by including the effect of ZFS, whereas the *dotted line* shows the best fit profile calculated without the effect of ZFS.

Adapted from Refs. [179,181,182].

reveal the hydroxide ion [181] coordinated to the metal together with three histidines in pseudotetrahedral symmetry. The NMRD profile (Fig. 7.67) shows a $\omega_S\tau_c = 1$ dispersion centered around 10 MHz, which qualitatively sets the τ_c value around 10^{-11} s. As the rotational correlation time of the molecule is much longer, this value is a measure of the effective electronic relaxation time. Compared to the hexaaqua complex, the nuclear-relaxing capability of this low symmetry four-coordinated cobalt(II) protein complex is thus relatively large.

A more quantitative analysis of the data would require consideration of the possible effects of zero field splitting, which is known to be sizeable in cobalt(II) complexes (D ranging from a few wavenumbers in tetracoordinated, 10–20 cm^{-1} in five-coordinated and up to 50 cm^{-1} in six-coordinated complexes). Such effects have been taken into account to explain the smoother shape of the dispersion curve with respect to what is predicted by the Solomon equation, and to obtain more accurate values of the best fit parameters (Section 4.7.1) [182].

If a ligand is added in the fifth coordination site of the above system and water is maintained in the coordination sphere [183], the water ^{1}H NMRD profile decreases because the effective electron relaxation time decreases by at least an order of magnitude [14,184]. This is shown by the profile of the nitrate derivative (Fig. 7.67).

Summarizing, data indicate that the six-coordinated $Co(H_2O)_6^{2+}$ chromophore has a T_{1e} of the order of 10^{-12} s, the tetrahedral CoN_3O chromophore has T_{1e} values of the order of 10^{-11} s, and five-coordinated chromophores have a T_{1e} of about $3–5 \times 10^{-12}$ s. These values of T_{1e} follow the order of availability of low-lying excited states: the closer the excited states, the shorter T_{1e}. Therefore, six-and five-coordinated high spin cobalt(II) chromophores are expected to display relatively narrow proton NMR signals, whereas tetrahedral complexes provide broader lines [179,185–188].

In the protein systems, cobalt has been used as a probe for other metal ions like the paramagnetic copper(II) and the diamagnetic zinc(II). For instance, cobalt(II) has been successfully used as spectroscopic probe replacing the copper ion in many blue copper proteins (Section 7.3.1.2) [57,189–194]. The ^{1}H NMR spectrum of cobalt(II)-azurin shows several well-resolved signals sizably downfield and upfield shifted with relatively short T_1 values and belonging to residues directly coordinated to the cobalt ion (Fig. 7.68). Numerous other shifted signals are observed closer to the diamagnetic region of the spectrum, due to residues near cobalt(II) but not bound to it. The two $^{\beta}CH_2$ protons of the coordinated cysteine are found far downfield, as generally found for metal-coordinated cysteines, i.e. in the 200–300 ppm region [189,193,195,196]. As always observed for histidines coordinated to high spin cobalt(II) [189,190], the ring signals are always downfield, between 30 and 80 ppm [188], the sharper being the N–H and the proton in *meta*-like position with respect to the coordinating nitrogen.

The hyperfine shifts of groups bound to the donor atom are largely dominated by the contact interaction, even if pseudocontact shift contributions are sizable and any quantitative use of the shifts

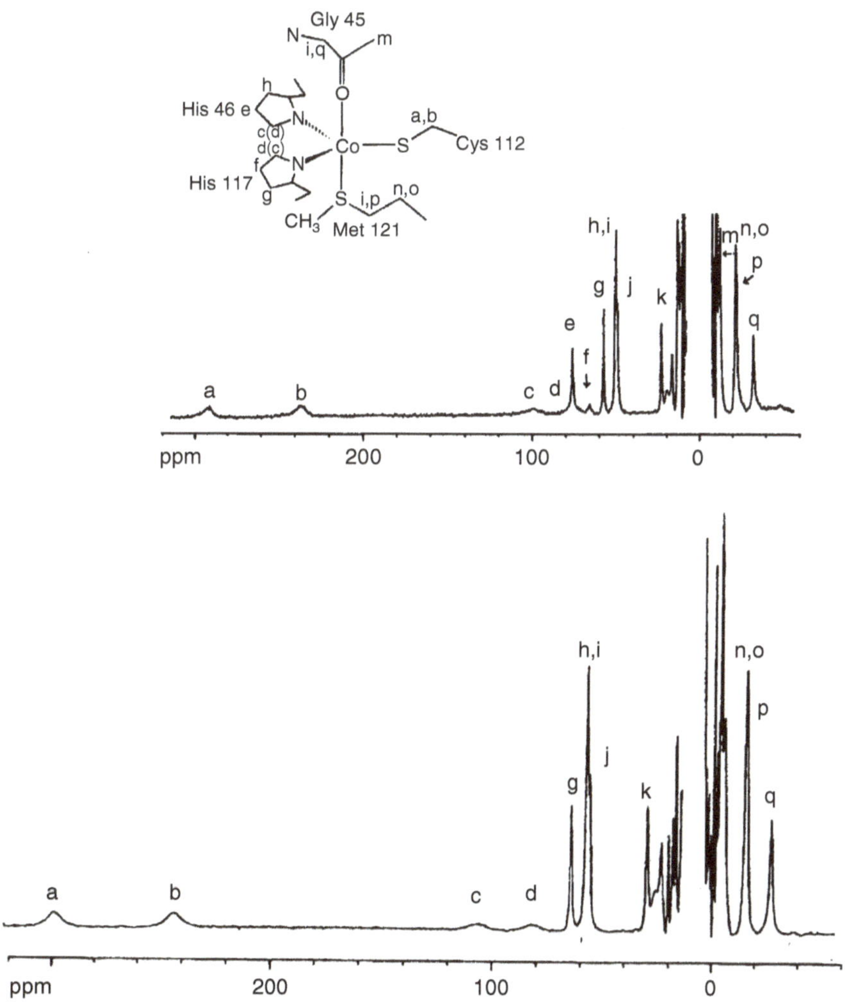

FIGURE 7.68 1D ^{1}H NMR Spectrum (200 MHz, 298 K) of Cobalt(II)–Azurin in H_2O. A schematic drawing of the metal site in *Pseudomonas aeruginosa* native azurin is shown in the upper left corner.

Adapted from Ref. [193].

should rely on the separated contributions. Longitudinal nuclear relaxation times can be used, and have been used in the case of cobalt substitute stellacyanin, to determine metal–proton distances [197]. The contribution of Curie relaxation, estimated from the field dependence of the linewidths, can be used both for assignment and to determine structural restraints [197].

In the case of cobalt substituted Zn-fingers [198], the differences between the chemical shifts for corresponding resonances in the cobalt(II) and zinc(II) complexes allow for the determination of the orientation and anisotropy of the magnetic susceptibility tensor [199]. Similar studies are available for pseudotetrahedral cobalt(II) in the zinc site of superoxide dismutase [200] and five coordinated carbonic anhydrase derivatives [201].

7.3.7.3 Nickel(III), low spin

Nickel(III) is also a d^7 ion and, when stabilized in the form of decamethylnickelocenium ion (Fig. 7.69), it has an orbitally-degenerate $^2E_{1g}$ ground state [65]. As a result, electronic relaxation is fast and NMR lines are remarkably sharp, as shown in Fig. 7.69 [64]. Because of the considerable contribution of

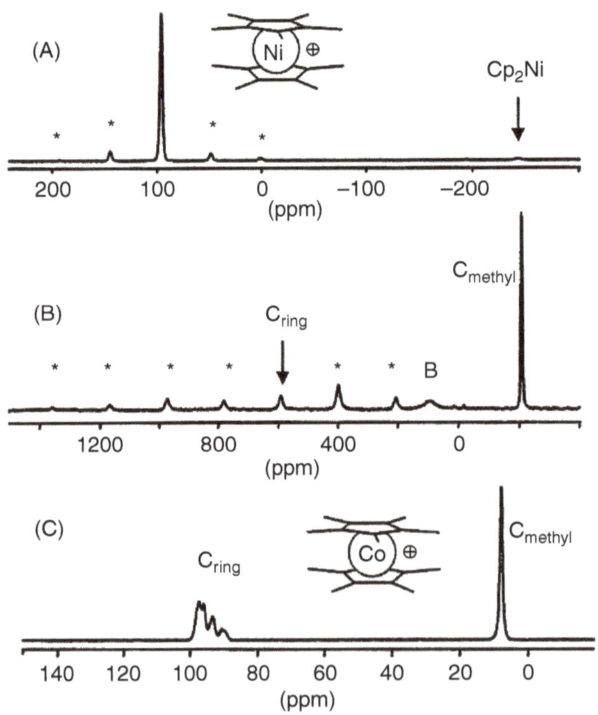

FIGURE 7.69 ^1H (A) and ^{13}C (B) MAS NMR spectra of $[(C_5Me_5)_2Ni]^+PF_6^-$ (Me = methyl, cation shown in the inset) at 300 MHz proton Larmor frequency. The sharpness of the peaks can be appreciated in comparison with the corresponding diamagnetic cobalt(III) compound (C). Spinning sidebands are marked by asterisks, B is the background signal of the probehead and Cp_2Ni is used as internal temperature standard.

Reproduced with permission from [64].

the cyclopentadienyl (Cp) orbitals to the molecular orbitals where the unpaired electrons sit, there is a direct delocalization of the spin density in the Cp rings, with a resulting positive spin density for decamethylnickelocenium. Direct delocalization yields a positive shift of the Cp resonances. This is opposite to what happens in the case of decamethylchromocenium (Fig. 7.24).

REFERENCES

[1] Mulliken RS. J. Chem. Phys. 1955;23:1997.
[2] Mulliken RS. J. Chem. Phys. 1956;24:1118.
[3] Bencini A, Gatteschi D. Electron Paramagnetic Resonance of Exchange-Coupled Systems. Berlin: Springer-Verlag; 1990.
[4] Glättli H, Odehnal M, Ezratty J, Malinovski A, Abragam A. Phys. Lett. A 1969;29:250.
[5] Bertini I, Luchinat C, Brown RD III, Koenig SH. J. Am. Chem. Soc. 1989;111:3532.
[6] Powell DH, Helm L, Merbach AE. J. Chem. Phys. 1991;95:9258.
[7] Freed JH, Kooser RG. J. Chem. Phys. 1968;49:4715.
[8] Kooser RG, Volland WV, Freed JH. J. Chem. Phys. 1969;50:5243.
[9] Bertini I, Felli IC, Luchinat C, Parigi G, Pierattelli R. Chem. Bio. Chem. 2007;8:1422.
[10] Arnesano F, Banci L, Bertini I, Felli IC, Luchinat C, Thompsett AR. J. Am. Chem. Soc. 2003;125:7200.
[11] Hausser R, Noack F. Z. Phys. 1964;182:93.
[12] Luz Z, Shulman RG. J. Chem. Phys. 1965;43:3750.
[13] Pasquarello A, Petri I, Salmon PS, Parisel O, Car R, Toth E, Powell DH, Fischer HE, Helm L, Merbach AE. Science 2001;291:856.
[14] Banci L, Bertini I, Luchinat C. Inorg. Chim. Acta 1985;100:173.
[15] Bertini I, Luchinat C. NMR of Paramagnetic Molecules in Biological Systems. Menlo Park, CA: Benjamin/Cummings; 1986.
[16] Bertini I, Lanini G, Luchinat C, Mancini M, Spina G. J. Magn. Reson. 1985;63:56.
[17] Liebermann RA, Sands RH, Fee JA. J. Biol. Chem. 1982;257:336.
[18] Gaber BP, Brown RD III, Koenig SH, Fee JA. Biochim. Biophys. Acta 1972;271:1.
[19] Bertini I, Briganti F, Luchinat C, Mancini M, Spina G. J. Magn. Reson. 1985;63:41.
[20] Bertini I, Canti G, Luchinat C, Borghi E. J. Inorg. Biochem. 1983;18:221.
[21] Bertini I, Luchinat C. In: Karlin KD, Zubieta J, editors. Biological and Inorganic Copper Chemistry. New York, USA: Adenine Press; 1986. p. 23.
[22] Bertini I, Canti G, Luchinat C. Inorg. Chim. Acta 1981;56:1.
[23] Banci L, Bertini I, Luchinat C, Monnanni R, Scozzafava A. Inorg. Chem. 1988;27:107.
[24] Bertini I, Banci L, Brown RD III, Koenig SH, Luchinat C. Inorg. Chem. 1988;27:951.
[25] Bertini I, Briganti F, Koenig SH, Luchinat C. Biochemistry 1985;24:6287.
[26] Bertini I, Luchinat C, Scozzafava A, Maldotti A, Traverso O. Inorg. Chim. Acta 1983;78:19.
[27] Barker R, Boden N, Cayley G, Charlton SC, Henson R, Holmes MC, Kelly ID, Knowles PF. Biochem. J. 1979;177:289.
[28] Bertini I, Luchinat C, Mincione G, Parigi G, Gassner GT, Ballou DP. J. Biol. Inorg. Chem. 1996;1:468.
[29] Kroes SJ, Salgado J, Parigi G, Luchinat C, Canters GW. JBIC 1996;1:551.
[30] Murthy NN, Karlin KD, Bertini I, Luchinat C. J. Am. Chem. Soc. 1997;119:2156.
[31] Lee H, Polenova T, Beer RH, McDermott AE. J. Am. Chem. Soc. 1999;121:6884.
[32] Szalontai G, Csonka R, Speier G, Kaizer J, Sabolovic J. Inorg. Chem. 2015;54:4663.
[33] Walder BJ, Dey KK, Davis MC, Baltisberger JH, Grandinetti PJ. J. Chem. Phys. 2015;142:014201.

[34] Bertini I, Bren KL, Clemente A, Fee JA, Gray HB, Luchinat C, Malmström BG, Richards JH, Sanders D, Slutter CE. J. Am. Chem. Soc. 1996;46:11658.
[35] Dennison C, Berg A, Canters GW. Biochemistry 1997;36:3262.
[36] Solomon EI. Inorg. Chem. 2006;45:8012.
[37] Holm RH, Kennepohl P, Solomon EI. Chem. Rev. 1996;96:2239.
[38] Koenig SH, Brown RD. Ann. NY Acad. Sci. 1973;222:752.
[39] Andersson I, Maret W, Zeppezauer M, Brown RD III, Koenig SH. Biochemistry 1981;20:3424.
[40] Bertini I, Fernández CO, Karlsson BG, Leckner J, Luchinat C, Malmström BG, Nersissian AM, Pierattelli R, Shipp E, Valentine JS, Vila AJ. J. Am. Chem. Soc. 2000;122:3701.
[41] Kalverda AP, Salgado J, Dennison C, Canters GW. Biochemistry 1996;35:3085.
[42] Salgado J, Kroes SJ, Berg A, Moratal JM, Canters GW. J. Biol. Chem. 1998;273:177.
[43] Bertini I, Ciurli S, Dikiy A, Gasanov R, Luchinat C, Martini G, Safarov N. J. Am. Chem. Soc. 1999;121:2037.
[44] Penfield KW, Gewirth AA, Solomon EI. J. Am. Chem. Soc. 1985;107:4519.
[45] Bertini I, Luchinat C, Xia Z. J. Magn. Reson. 1992;99:235.
[46] La Mar GN, Walker FA. J. Am. Chem. Soc. 1973;95:6950.
[47] Wilson RC, Myers RJ. J. Chem. Phys. 1976;64:2208.
[48] Bertini I, Luchinat C, Xia Z. Inorg. Chem. 1992;31:3152.
[49] Lindner U. Ann. Phys. (Leipzig) 1965;16:319.
[50] Bertini I, Luchinat C, Mancini M, Spina G. In: Gatteschi D, Kahn O, Willett RD, editors. Magneto-Structural Correlations in Exchange-Coupled Systems. Dordrecht: Reidel Publishing Company; 1985. p. 421.
[51] Kowalewski J, Nordenskiöld L, Benetis N, Westlund PO. Progr. Nucl. Magn. Res. Spectrosc. 1985;17:141.
[52] Benetis N, Kowalewski J, Nordenskiöld L, Wennerström H, Westlund PO. Mol. Phys. 1983;48:329.
[53] Kowalewski J, Larsson T, Westlund PO. J. Magn. Reson. 1987;74:56.
[54] Kowalewski J, Egorov D, Kruk A, Laaksonen A, Aski N, Parigi G, Westlund PO. J. Magn. Reson. 2008;195:103.
[55] Nilsson T, Parigi G, Kowalewski J. J. Phys. Chem. 2002;106:4476.
[56] Bertini I, Borghi E, Luchinat C, Monnanni R. Inorg. Chim. Acta 1982;67:99.
[57] Salgado J, Jimenez HR, Moratal Mascarell JM, Kroes SJ, Warmerdam GCM, Canters GW. Biochemistry 1996;35:1810.
[58] Blaszak JA, Ulrich EL, Markley JL, McMillin DR. Biochemistry 1982;21:6253.
[59] Solari E, De Angelis S, Floriani C, Chiesi-Villa A, Rizzoli C. Inorg. Chem. 1992;31:96.
[60] Bencini A, Gatteschi D. In: Melson GA, Figgis BN, editors. Transition Metal Chemistry. New York and Basel: Marcel Dekker, INC; 1982.
[61] Corzilius B, Michaelis VK, Penzel S, Ravera E, Smith A, Luchinat C, Griffin RG. J. Am. Chem. Soc. 2014;136:11716.
[62] Bertini I, Fragai M, Luchinat C, Parigi G. Inorg. Chem. 2001;40:4030.
[63] Bleuzen A, Foglia F, Furet E, Helm L, Merbach A, Weber J. J. Am. Chem. Soc. 1996;118:12777.
[64] Heise H, Köhler FH, Herker M, Hiller W. J. Am. Chem. Soc. 2002;124:10823.
[65] Kollmar C, Kahn O. J. Chem. Phys. 1992;96:2988.
[66] Im S-C, Liu G, Luchinat C, Sykes AG, Bertini I. Eur. J. Biochem. 1998;258:465.
[67] Bocarsly JR, Barton JK. Inorg. Chem. 1989;28:4189.
[68] Bocarsly JR, Barton JK. Inorg. Chem. 1992;31:2827.
[69] Assfalg M, Bertini I, Bruschi M, Michel C, Turano P. Proc. Natl. Acad. Sci. USA 2002;99:9750.
[70] Wheeler WD, Kaizaki S, Legg JI. Inorg. Chem. 1982;21:3248.
[71] Brackett GC, Richards PL, Caughey WS. J. Chem. Phys. 1971;54:4383.
[72] Tsurumaki H, Watanabe Y, Morishima I. J. Am. Chem. Soc. 1993;115:11784.
[73] Luz Z, Fiat D. J. Chem. Phys. 1967;46:469.
[74] Sharp RR, Lohr L, Miller J. Progr. Nucl. Magn. Res. Spectrosc. 2001;38:115.

[75] Abernathy SM, Miller JC, Lohr LL, Sharp RR. J. Chem. Phys. 1998;109:4035.
[76] Koenig SH, Brown RD III, Spiller M. Magn. Reson. Med. 1987;4:252.
[77] La Mar GN, Walker FA. In: Dolphin D, editor. The Porphyrins. New York: Academic Press; 1979. p. 61.
[78] Janson TR, Bouchet LJ, Katz JJ. Inorg. Chem. 1973;12:940.
[79] La Mar GN, Walker FA. J. Am. Chem. Soc. 1975;97:5103.
[80] Groves JT. Proc. Natl. Acad. Sci. USA 2003;100:3569.
[81] McDonald AR, Que L Jr. Coord. Chem. Rev. 2013;257:414.
[82] Rohde JU, In JH, Lim MH, Brennessel WW, Bukowski MR, Stubna A, Münck E, Nam W, Que L Jr. Science 2003;299:1037.
[83] Borgogno A, Rastrelli F, Bagno A. Chem. Eur. J. 2015;21:12960.
[84] Lee Y-M, Kotani H, Suenobu T, Nam W, Fukuzumi S. J. Am. Chem. Soc. 2008;130:434.
[85] Abragam A, Bleaney B. Electron Paramagnetic Resonance of Transition Metal Ions. Oxford: Clarendon Press; 1970.
[86] La Mar GN, Walker FA. J. Am. Chem. Soc. 1973;95:1782.
[87] Bertini I, Galas O, Luchinat C, Messori L, Parigi G. J. Phys. Chem. 1995;99:14217.
[88] Bertini I, Capozzi F, Luchinat C, Xia Z. J. Phys. Chem. 1993;97:1134.
[89] Levanon H, Charbinsky S, Luz Z. J. Chem. Phys. 1970;53:3056.
[90] Levanon H, Stein G, Luz Z. J. Chem. Phys. 1970;53:876.
[91] Koenig SH, Brown RD IIIIn: Sigel H, editor. Metal Ions in Biological Systems, vol. 21. New York: Marcel Dekker, Inc; 1987.
[92] Koenig SH, Brown RD III, Lindstrom TR. Biophys. J. 1981;34:397.
[93] Hershberg RD, Chance B. Biochemistry 1975;14:3885.
[94] Hwang LP, Freed JH. J. Chem. Phys. 1975;63:4017.
[95] Bertini I, Kowalewski J, Luchinat C, Nilsson T, Parigi G. J. Chem. Phys. 1999;111:5795.
[96] Bertini I, Luchinat C, Nerinovski K, Parigi G, Cross M, Xiao Z, Wedd AG. Biophys. J. 2003;84:545.
[97] Zewert TE, Gray HB, Bertini I. J. Am. Chem. Soc. 1994;116:1169.
[98] Heistand RH, Lauffer RB, Fikrig E, Que L Jr. J. Am. Chem. Soc. 1982;104:2789.
[99] La Mar GN, Eaton GR, Holm RH, Walker FA. J. Am. Chem. Soc. 1973;95:63.
[100] Maltempo MM, Moss TH, Cusanovich MA. Biochim. Biophys. Acta 1974;342:290.
[101] Boersma AD, Goff HM. Inorg. Chem. 1982;21:581.
[102] Walker FA, La Mar GN. Ann. NY Acad. Sci. 1973;206:328.
[103] Budd DL, La Mar GN, Langry KC, Smith KM, Nayyir-Mazhir R. J. Am. Chem. Soc. 1979;101:6091.
[104] Mispelter J, Momenteau M, Lhoste J-M. J. Chem. Soc. Dalton Trans. 1981;8:1729.
[105] Pande U, La Mar GN, Lecomte JTJ, Ascoli F, Brunori M, Smith KM, Pandey RK, Parish DW, Thanabal V. Biochemistry 1986;25:5638.
[106] La Mar GN, Jackson JT, Dugad LB, Cusanovich MA, Bartsch RG. J. Biol. Chem. 1990;265:16173.
[107] Lauffer RB, Que L Jr. J. Am. Chem. Soc. 1982;104:7324.
[108] Que L Jr, Lauffer RB, Lynch JB, Murch BP, Pyrz JW. J. Am. Chem. Soc. 1987;109:5381.
[109] Bertini I, Gori G, Luchinat C, Vila AJ. Biochemistry 1993;32:776.
[110] La Mar GN, Jackson JT, Bartsch RG. J. Am. Chem. Soc. 1981;103:4405.
[111] La Mar GN, Budd DL, Smith KM, Langry KC. J. Am. Chem. Soc. 1980;102:1822.
[112] Xia B, Westler WM, Cheng H, Meyer J, Moulis J-M, Markley JL. J. Am. Chem. Soc. 1995;117:5347.
[113] McGarvey BR. Coord. Chem. Rev. 1998;170:75.
[114] Mayer A, Ogawa S, Shulman RG, Yamane T, Cavaleiro JAS, Rocha Gonsalves AMd, Kenner GW, Smith KM. J. Mol. Biol. 1974;86:749.
[115] Bertini I, Capozzi F, Luchinat C, Turano P. J. Magn. Reson. 1991;95:244.
[116] Horrocks WD Jr, Greenberg ES. Biochim. Biophys. Acta 1973;322:38.
[117] Trehwella J, Wright PE. Biochim. Biophys. Acta 1980;625:202.

[118] Emerson SD, La Mar GN. Biochemistry 1990;29:1556.
[119] Turner DL, Williams RJP. Eur. J. Biochem. 1993;211:555.
[120] Shokhirev NV, Walker FA. J. Phys. Chem. 1995;99:17795.
[121] Burns PD, La Mar GN. J. Magn. Reson. 1982;46:61.
[122] Wüthrich K, Keller RM. Biochim. Biophys. Acta 1978;533:195.
[123] Kurland RJ, McGarvey BR. J. Magn. Reson. 1970;2:286.
[124] Bertini I, Luchinat C, Parigi G. Eur. J. Inorg. Chem. 2000;12:2473.
[125] Banci L, Bertini I, Gray HB, Luchinat C, Reddig T, Rosato A, Turano P. Biochemistry 1997;36:9867.
[126] Lin Q, Simonis U, Tipton AR, Norvell CJ, Walker FA. Inorg. Chem. 1992;31:4216.
[127] Simonis U, Dallas JL, Walker FA. Inorg. Chem. 1992;31:5349.
[128] Bertini I, Luchinat C, Parigi G, Walker FA. J. Biol. Inorg. Chem. 1999;4:515.
[129] Shokhirev NV, Walker FA. J. Am. Chem. Soc. 1998;120:981.
[130] Banci L, Bertini I, Ferroni F, Rosato A. Eur. J. Biochem. 1997;249:270.
[131] Arnesano F, Banci L, Bertini I, Felli IC. Biochemistry 1998;37:173.
[132] Arnesano F, Banci L, Bertini I, Felli IC, Koulougliotis D. Eur. J. Biochem. 1999;260:347.
[133] Louro RO, Correia IJ, Brennan L, Coutinho IB, Xavier AV, Turner DL. J. Am. Chem. Soc. 1998;120:13240.
[134] Banci L, Bertini I, Bruschi M, Sompornpisut P, Turano P. Proc. Natl. Acad. Sci. USA 1996;93:14396.
[135] Safo MK, Walker FA, Raitsimiring AM, Walters WP, Dolata DP, Debrunner PG, Scheidt WR. J. Am. Chem. Soc. 1994;116:7760.
[136] Nakamura M, Ikeue T, Fujii H, Yoshimura T. J. Am. Chem. Soc. 1997;119:6284.
[137] Walker FA, Nasri H, Turowska-Tyrk I, Mohanrao K, Watson CT, Shokhirev NV, Debrunner PG, Scheidt WR. J. Am. Chem. Soc. 1996;118:12109.
[138] Wolowiec S, Latos-Grazynski L, Toronto D, Marchon J-C. Inorg. Chem. 1998;37:724.
[139] Piccioli M, Turano P. Coord. Chem. Rev. 2015;284:313.
[140] Brancaccio D, Gallo A, Mikolajczyk M, Zovo K, Palumaa P, Novellino E, Piccioli M, Ciofi-Baffoni S, Banci L. J. Am. Chem. Soc. 2014;136:16240.
[141] Bertrand P, Theodule FX, Gayda J-P, Mispelter J, Momenteau M. Chem. Phys. Lett. 1983;102:442.
[142] Reed CA, Guiset F. J. Am. Chem. Soc. 1996;118:3281.
[143] de Ropp JS, Mandal P, Brauer SL, La Mar GN. J. Am. Chem. Soc. 1997;119:4732.
[144] La Mar GN, de Ropp JS, Smith KM, Langry KC. J. Biol. Chem. 1980;255:6646.
[145] Thanabal V, de Ropp JS, La Mar GN. J. Am. Chem. Soc. 1987;109:265.
[146] Banci L, Bertini I, Turano P, Ferrer JC, Mauk AG. Inorg. Chem. 1991;30:4510.
[147] Caillet-Saguy C, Delepierre M, Lecroisey A, Bertini I, Piccioli M, Turano P. J. Am. Chem. Soc. 2006;128:150.
[148] Caillet-Saguy C, Turano P, Piccioli M, Lukat-Rodgers G, Czjzek M, Guigliarelli B, Izadi-Pruneyre N, Rodgers K, Delepierre M, Lecroisey A. J. Biol. Chem. 2008;283:5960.
[149] Caillet-Saguy C, Piccioli M, Turano P, Lukat-Rodgers G, Wolff N, Rodgers K, Izadi-Pruneyre N, Delepierre M, Lecroisey A. J. Biol. Chem. 2012;287:26932.
[150] Caillet-Saguy C, Piccioli M, Turano P, Izadi-Pruneyre N, Delepierre M, Bertini I, Lecroisey A. J. Am. Chem. Soc. 2009;131:1736.
[151] Kuznets G, Vigonsky E, Weissman Z, Lalli D, Kauffman M, Turano P, Lewinson O, Kornitzer D. PLoS Pathogens 2014;10:e1004407.
[152] Lalli D, Turano P. Acc. Chem. Res. 2013;46:2676.
[153] Luchinat C, Parigi G, Ravera E. J. Biomol. NMR 2014;58:239.
[154] Bertini I, Briganti F, Luchinat C, Xia Z. J. Magn. Reson. 1993;101:198.
[155] Koenig SH, Baglin C, Brown RD III, Brewer CF. Magn. Reson. Med. 1984;1:496.
[156] Koenig SH, Brown RD III. Progr. Nucl. Mag. Reson. Spectrosc. 1990;22:487.
[157] Bertini I, Luchinat C, Parigi G. 1H NMR profiles of paramagnetic complexes and metalloproteins. Adv. Inorg. Chem. 2005;57:105–172.

[158] Meirovitch E, Luz Z, Kalb AJ. J. Am. Chem. Soc. 1974;96:7538.
[159] Banci L, Bertini I, Briganti F, Luchinat C. J. Magn. Reson. 1986;66:58.
[160] Aime S, Anelli PL, Botta M, Brocchetta M, Canton S, Fedeli F, Gianolio E, Terreno E. J. Biol. Inorg. Chem. 2002;7:58.
[161] Kruk D, Kowalewski J. J. Biol. Inorg. Chem. 2003;8:512.
[162] Bhattacharyya L, Brewer CF, Brown RD III, Koenig SH. Biochemistry 1985;24:4985.
[163] La Mar GN, Budd DL, Goff HM. Biochem. Biophys. Res. Commun. 1977;77:104.
[164] Goff H, La Mar GN, Reed CA. J. Am. Chem. Soc. 1977;99:3641.
[165] Yu LP, La Mar GN, Rajarathnam K. J. Am. Chem. Soc. 1990;112:9527.
[166] Satterlee JD, La Mar GN. J. Am. Chem. Soc. 1976;98:2804.
[167] Banci L, Bertini I, Marconi S, Pierattelli R. Eur. J. Biochem. 1993;215:431.
[168] Bertini I, Dikiy A, Luchinat C, Macinai R, Viezzoli MS. Inorg. Chem. 1998;37:4814.
[169] Bougault CM, Dou Y, Ikeda-Saito M, Langry KC, Smith KM, La Mar GN. J. Am. Chem. Soc. 1998;120:2113.
[170] Yang F, Phillips GN Jr. J. Mol. Biol. 1996;256:762.
[171] Johnson ME, Fung LWM, Ho C. J. Am. Chem. Soc. 1977;99:1245.
[172] Hochmann J, Kellerhals H. J. Magn. Reson. 1980;38:23.
[173] Lehmann TE, Ming L-J, Rosen ME, Que L Jr. Biochemistry 1997;36:2807.
[174] Rao PS, Ganguli P, McGarvey BR. Inorg. Chem. 1981;20:3682.
[175] Ozarowski A, Shunzhong Y, McGarvey BR, Mislankar A, Drake JE. Inorg. Chem. 1991;30:3167.
[176] Breuning E, Ruben M, Lehn J-M, Renz F, Garcia Y, Ksenofontov V, Gütlich P, Wegelius E, Rissanen K. Angew. Chem. Int. Ed. 2000;39:2504.
[177] Shirazi A, Goff HM. Inorg. Chem. 1982;21:3420.
[178] Fulton GP, La Mar GN. J. Am. Chem. Soc. 1976;98:2119.
[179] Bertini I, Luchinat C. Adv. Inorg. Biochem. 1985;6:71.
[180] Novikov VV, Pavlov AA, Belov AS, Vologzhanina AV, Savitsky A, Voloshin YZ. J. Phys. Chem. Lett. 2014;5:3799.
[181] Koenig SH, Brown RD III, Bertini I, Luchinat C. Biophys. J. 1983;41:179.
[182] Bertini I, Luchinat C, Mancini M, Spina G. J. Magn. Reson. 1984;59:213.
[183] Bertini I, Canti G, Luchinat C, Scozzafava A. J. Am. Chem. Soc. 1978;100:4873.
[184] Bertini I, Canti G, Luchinat C. Inorg. Chim. Acta 1981;56:99.
[185] Benelli C, Bertini I, Gatteschi D. J. Chem. Soc. Dalton Trans. 1972;661.
[186] Bertini I, Canti G, Luchinat C, Messori L. Inorg. Chem. 1982;21:3426.
[187] Bertini I, Gatteschi D, Wilson LJ. Inorg. Chim. Acta 1970;629.
[188] Bertini I, Canti G, Luchinat C, Mani F. J. Am. Chem. Soc. 1981;103:7784.
[189] Bertini I, Gerber M, Lanini G, Luchinat C, Maret W, Rawer S, Zeppezauer M. J. Am. Chem. Soc. 1984;106:1826.
[190] Bertini I, Luchinat C, Messori L, Scozzafava A. Eur. J. Biochem. 1984;141:375.
[191] Tennent DL, McMillin DR. J. Am. Chem. Soc. 1979;101:2307.
[192] Fernandez CO, Sannazzaro AI, Vila AJ. Biochemistry 1997;36:10566.
[193] Piccioli M, Luchinat C, Mizoguchi TJ, Ramirez BE, Gray HB, Richards JH. Inorg. Chem. 1995;34:737.
[194] Moratal JM, Salgado J, Donaire A, Jimenez HR, Castells J. Inorg. Chem. 1993;32:3587.
[195] Bertini I, Luchinat C, Messori L, Vasak M. J. Am. Chem. Soc. 1989;111:7300.
[196] Moura I, Teixeira M, LeGall J, Moura JJG. J. Inorg. Biochem. 1991;44:127.
[197] Dahlin S, Reinhammar B, Ångstrom J. Biochemistry 1989;28:7224.
[198] Krizek BA, Amann BT, Kilfoil VJ, Merkle DL, Berg JM. J. Am. Chem. Soc. 1991;113:4518.
[199] Harper LV, Amann BT, Vinson VK, Berg JM. J. Am. Chem. Soc. 1993;115:2577.
[200] Bertini I, Luchinat C, Piccioli M, Vicens Oliver M, Viezzoli MS. Eur. Biophys. J. 1991;20:269.
[201] Banci L, Dugad LB, La Mar GN, Keating KA, Luchinat C, Pierattelli R. Biophys. J. 1992;63:530.

CHAPTER

LANTHANOIDS AND ACTINOIDS: SHIFT AND RELAXATION

8

Lanthanoids and actinoids (also called lanthanides and actinides, although the ending "ide" normally indicates a negative ion, and should thus be avoided) constitute a set of potentially powerful probes with fairly similar chemical and structural properties. The different lanthanoid ions have however different characteristics in terms of magnetic susceptibility anisotropy and electronic relaxation, so that the effects they have on the NMR spectra can be largely different. In NMR spectroscopy, paramagnetic lanthanoids are often used as chemical shift and/or line broadening agents to obtain structural and dynamics information.

8.1 ELECTRONIC PROPERTIES

In most compounds, lanthanoids have +3 oxidation state. Ln^{3+} ions have 4f electrons shielded by 5s and 5p electrons, and thus essentially not available for covalent interactions with ligands. Therefore, metal ligand interactions in Ln^{3+} complexes are largely electrostatic, with a geometry mainly determined by steric factors. This determines a large similarity in chemical behavior and complex structure among all the lanthanoid ions; some differences only arise from the progressive decrease in ionic radii on passing from La^{3+} to Lu^{3+}.

The electronic properties of lanthanoids are peculiar, in that spin–orbit interactions are very large, larger in fact than ligand field effects. Therefore, a somewhat different formalism is used.

Spin–orbit interactions strongly couple the orbital L and spin S angular momenta. The resulting free ion terms are characterized by J values which are vector combinations of L and S. The ground state is that with largest S, largest L with equally large S, smallest J for f^1 to f^6 lanthanoids, and largest J for f^8 to f^{13} lanthanoids (Table 8.1). For example, Ce^{3+} has a ground state with $J = 5/2$, since one electron in an f orbital gives rise to a free ion term ^2F with $S = 1/2$ and $L = 3$. Two terms arise through spin–orbit coupling: one with $J = 3 - 1/2 = 5/2$ and the other with $J = 3 + 1/2 = 7/2$. The former term is the ground state, as shown in Table 8.1. The J values of the ground state vary from 5/2 to 8, including $J = 0$ for Eu^{3+} and Sm^{2+}. In general, it can be assumed that the ground term is the only one of interest; however, this is not true for Sm^{3+}, which has an excited J manifold at about 1000 cm^{-1}, and Eu^{3+} and Sm^{2+}, whose ground terms have $J = 0$ and would be diamagnetic if excited terms were not taken into consideration.

In actinoids, the J terms are closer in energy and even more heavily admixed by spin–orbit coupling.

Ligand field effects split the J manifold in a way that is not easily predicted without specific calculations [3]. However, the overall splitting is such that many of the levels are appreciably populated at room temperature. In general, this causes a large magnetic susceptibility anisotropy and therefore large pseudocontact shifts. An elegant procedure that takes J manifold splitting into account in a general way to explain the pseudocontact shifts has been provided by Bleaney [2].

NMR of Paramagnetic Molecules. http://dx.doi.org/10.1016/B978-0-444-63436-8.00009-0

Table 8.1 Some Electronic Properties of Lanthanoid Ions

Ion	Configuration	$^{2S+1}L_J$ of Ground State (Multiplicity in Parentheses)	g_J [a]	$\langle S_z \rangle_J$ [b]	$\langle S_z \rangle_J$ [c]	δ^{pcs} (ppm) [d]
Ce^{3+}	$4f^1$	$^2F_{5/2}$ (6)	6/7	1.07	0.98	+1.6
Pr^{3+}	$4f^2$	$^3H_4$ (9)	4/5	3.20	2.97	+2.7
Nd^{3+}	$4f^3$	$^4I_{9/2}$ (10)	8/11	4.91	4.49	+1.05
Pm^{3+}	$4f^4$	$^5I_4$ (9)	3/5	4.80	4.01	−0.6
Sm^{3+}	$4f^5$	$^6H_{5/2}$ (6)	2/7	1.79	−0.06	+0.17 [e]
Eu^{3+}(Sm^{2+})	$4f^6$	$^7F_0$ (1)	–	–	−10.68	~ −1.0 [e]
Gd^{3+}(Eu^{2+})	$4f^7$	$^8S_{7/2}$ (8)	2	−31.5	−31.50	~0
Tb^{3+}	$4f^8$	$^7F_6$ (13)	3/2	−31.5	−31.82	+20.7
Dy^{3+}	$4f^9$	$^6H_{15/2}$ (16)	4/3	−28.30	−28.54	+23.8
Ho^{3+}	$4f^{10}$	$^5I_8$ (17)	5/4	−22.50	−22.63	+9.4
Er^{3+}	$4f^{11}$	$^4I_{15/2}$ (16)	6/5	−15.30	−15.37	−7.7
Tm^{3+}	$4f^{12}$	$^3H_6$ (13)	7/6	−8.17	−8.21	−12.7
Yb^{3+}	$4f^{13}$	$^2F_{7/2}$ (8)	8/7	−2.57	−2.59	−5.2

[a] *Calculated from Eq. (8.1). The equation does not hold for* f^6 *ions.*
[b] *Calculated according to Eq. (8.4), in* $\frac{\mu_B B_0}{3kT}$ *units.*
[c] *Calculated by inclusion of the excited states [1].*
[d] *Pseudocontact shift predicted from Eq. (8.3) for* $r = 300$ *pm,* $3\cos^2\theta - 1 = 1$, $T = 300$ *K, and* D_z *values for each lanthanoid estimated from Ref. [2].*
[e] *Including contributions from excited J manifolds [2]. For* f^6 *ions, the contribution of the ground state manifold is zero.*

8.2 THE PSEUDOCONTACT CONTRIBUTION TO THE HYPERFINE SHIFTS

Ligand field effects remove the spherical symmetry around the metal ion and cause magnetic susceptibility anisotropy. Pseudocontact shifts are then obtained, as provided by Eq. (2.46). The susceptibility can be viewed as arising from two contributions: one is given by a Zeeman term of the type $g\mu_B B_0 S_z$, and the other is given by the splitting of the magnetic manifold, which provides anisotropy. As the orbital and spin angular momenta are strongly coupled, the g factor in lanthanoid free ions is a function of both L and S and of their vector combination J; it is given by

$$g_J = 1 + \frac{J(J+1) - L(L+1) + S(S+1)}{2J(J+1)} \tag{8.1}$$

and the Zeeman term then becomes $g_J \mu_B B_0 J_z$. Note that when $L = 0$, $g_J = 2$, and when $S = 0$, $g_J = 1$, as expected from general theory. In analogy with Eq. (2.52), the Zeeman term provides a contribution to pseudocontact shift that depends on T^{-1}. In the present approximation of isotropic g_J, this term vanishes. If the ligand field interactions removing the degeneracy of the J manifold at zero magnetic field, D, are much larger than the Zeeman energy and smaller than kT, they provide a contribution that depends on T^{-2} [Eqs. (2.56)–(2.58)]. The general expression is [2]

$$\delta^{pcs} = -\frac{\mu_0}{4\pi}\frac{g_J^2\mu_B^2 J(J+1)(2J-1)(2J+3)}{60(kT)^2} \times \frac{D_z(3\cos^2\theta - 1) + (D_x - D_y)\sin^2\theta\cos 2\varphi}{r^3} \tag{8.2}$$

where r, θ, and φ are the spherical coordinates of the nucleus with respect to the principal directions of the $\boldsymbol{D}$ tensor (Section 1.4), and g_J is given by Eq. (8.1) (Table 8.1). Eq. (8.2) differs from Eqs. (2.56)–(2.58) in that only the term in T^{-2} has remained.

For axial symmetry, Eq. (8.2) becomes

$$\delta^{pcs} = -\frac{\mu_0}{4\pi}\frac{g_J^2\mu_B^2 J(J+1)(2J-1)(2J+3)}{60(kT)^2}\frac{D_z(3\cos^2\theta - 1)}{r^3}. \tag{8.3}$$

The parameter D_z accounts for crystal field effects, assuming that the splitting of the $2J + 1$ levels at zero field does not exceed kT in energy. The shifts at 300 K for a nucleus at 300 pm from the metal and $3\cos^2\theta - 1 = 1$, calculated following Bleaney's treatment [2] and using a plausible value for the ligand field parameter to calculate D_z for each lanthanoid, are reported in the last column of Table 8.1 and in Fig. 8.1. For Sm^{3+}, the shift value has been obtained by also considering the first excited $J = 7/2$ manifold. For Eu^{3+}, the ground state $J = 0$ does not contribute to the shift; instead, the excited levels with $J = 1$ and $J = 2$, which lie at about 400 and 1200 cm^{-1} above it, provide the pseudocontact shift, which has a rather complex temperature dependence.

For f^7 ions, the ligand field splitting is zero under first-order conditions, whereas higher order effects account for the splitting of the $J = 7/2$ level and for a small pseudocontact shift.

Of course, the pseudocontact shifts can be analyzed, as commonly done, using Eqs. (2.46) and (2.47), valid in general for paramagnetic ions. They can thus be determined from the magnetic susceptibility anisotropy tensor and the spherical coordinates of the nucleus in the frame of this tensor, with origin on the paramagnetic ion.

The accuracy of the present point–dipole approximation with respect to a model with delocalized unpaired electrons (Section 2.2.2.2) has also been evaluated for f^n systems [4]. The deviations have been found to be substantially smaller than those estimated for 3d metal ions.

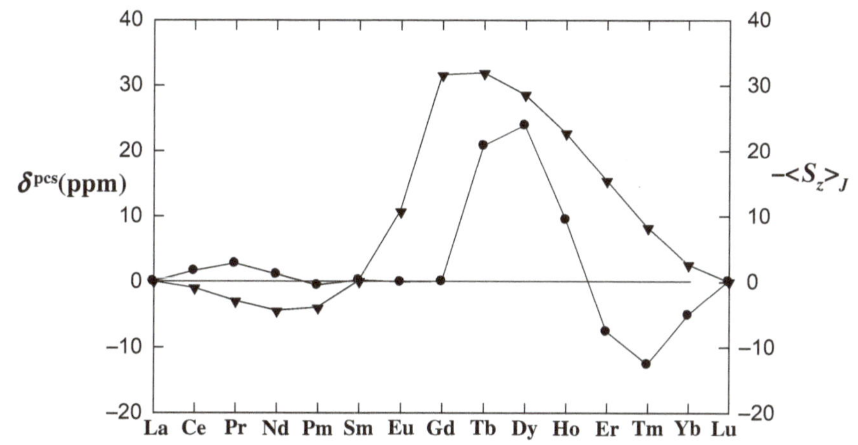

FIGURE 8.1 Calculated patterns of: dipolar shifts (•) [2] and $-\langle S_z\rangle$ (▼) [1] (in $\frac{\mu_B B_0}{3kT}$ units, proportional to the contact shifts) induced by lanthanoid ions.

8.3 THE CONTACT CONTRIBUTION TO THE HYPERFINE SHIFTS

Although the contact contribution can be sizable and sometimes dominant for nuclei directly coordinated to the metal [1], it is often negligible for a nuclei to a few bonds apart from the metal ion, unless efficient spin delocalization mechanisms are operative. The contact shift can be evaluated [Eq. (2.6)] through the expectation value $\langle S_z \rangle_J$ on the ground J manifold, the Zeeman splitting of which is less than kT [5]:

$$\langle S_z \rangle_J = -g_J(g_J - 1)J(J+1)\frac{\mu_B B_0}{3kT}. \tag{8.4}$$

In this approximation, g_J is isotropic. We need to consider $\langle S_z \rangle_J$ and not $\langle J_z \rangle$ because only the spin is delocalized onto the molecule [1]. The contact contribution is therefore

$$\delta^{\text{con}} = \frac{A}{\hbar}\frac{g_J(g_J - 1)\mu_B J(J+1)}{3\gamma_I kT} \tag{8.5}$$

or

$$\delta^{\text{con}} = \frac{1}{\mu_0}\frac{A}{\hbar}\frac{g_J - 1}{\gamma_I \mu_B g_J}\chi \tag{8.6}$$

where χ is the magnetic susceptibility without ligand field splitting [Eq. (1.34)]. Although this simple treatment does not take into account the ligand field splitting of the J manifold, it provides a useful criterion for predicting the general trend of contact shifts along the series. Note that this method does not apply to Sm^{3+}, Eu^{3+}, and Sm^{2+} salts for the reason already outlined. The values of $\langle S_z \rangle_J$, with and without inclusion of the excited states, are reported in Table 8.1 and in Fig. 8.1. It is noteworthy that $\langle S_z \rangle_J$ in lanthanoids can be either negative or positive owing to the large orbital contribution to the determination of the energy levels. Since orbital contributions are larger than spin contributions, $\langle S_z \rangle_J$ is different in sign with respect to the spin-only value when the spin orbit coupling constant is negative, as it occurs in the first half of the series.

It is also often assumed that A is constant along a series of lanthanoid complexes with the same ligand. The mechanism providing spin density delocalization from the metal to the ligands is due to a weak covalent bonding involving the 6*s* metal orbital, which, in turn, can transfer unpaired spin density on nearby nuclei through spin polarization from 4f orbitals [5].

8.4 ANALYSIS OF THE HYPERFINE SHIFTS

8.4.1 SEPARATION OF PSEUDOCONTACT AND CONTACT CONTRIBUTIONS

The following method for separation of pseudocontact and contact shifts, proposed by Dobson et al. [6], is valid for axial systems assuming that: (1) there is a nucleus in such a position in the molecule that the contact contribution is zero because it is far from the paramagnetic center; and (2) the contact hyperfine coupling is constant (±10%) for Nd–Tm.

For nuclei i and j:

$$\frac{\delta_i^{\text{pcs}}}{\delta_j^{\text{pcs}}} = R_{ij}. \tag{8.7}$$

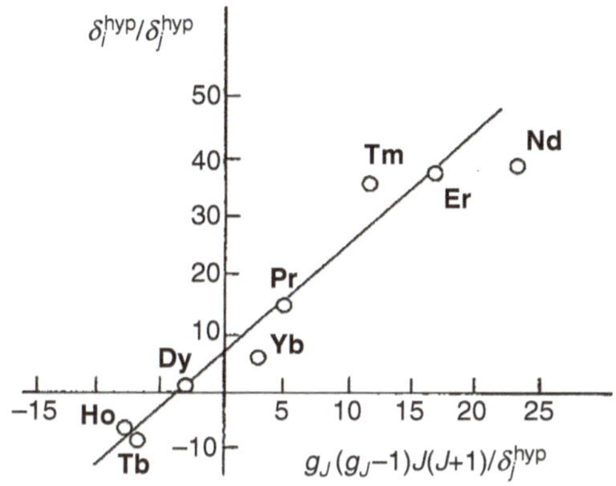

FIGURE 8.2 Observed ratios between the hyperfine shifts of the ^{31}P (δ_i^{hyp}) and ^{1}H (δ_j^{hyp}) resonances of cytidine monophosphate interacting with lanthanoid ions, as a function of $g_J(g_J-1)J(J+1)/\delta_i^{\text{hyp}}$ for each lanthanoid.

Adapted from Ref. [6].

If $\delta_j^{\text{con}} = 0$, then the ratio of the total hyperfine shifts δ^{hyp} for nuclei i and j is

$$\frac{\delta_i^{\text{hyp}}}{\delta_j^{\text{hyp}}} = R_{ij} + \frac{\delta_i^{\text{con}}}{\delta_j^{\text{hyp}}} = R_{ij} + \frac{A}{\hbar}\frac{g_J(g_J-1)\mu_B J(J+1)}{3\gamma_N kT\delta_j^{\text{hyp}}}. \tag{8.8}$$

By plotting $\delta_i^{\text{hyp}}/\delta_j^{\text{hyp}}$ versus $g_J(g_J-1)\,J(J+1)/\delta_j^{\text{hyp}}$ for all the lanthanoids, the intercept gives R_{ij} and hence the ratio between pairs of pseudocontact shifts (Fig. 8.2). Within this approximation, the requirement for axial symmetry was found to hold for many systems, even with low symmetry, possibly as a result of ligand rearrangement which might be fast on the NMR time scale.

Separation of pseudocontact and contact shifts in axial systems can also be achieved, again by using a variety of lanthanoids, without assuming that for a given signal the contact shift is zero, as long as the expected patterns of both pseudocontact and contact shifts (Fig. 8.1) hold. Here, the hyperfine shift can be expressed as the sum of contact and pseudocontact shifts [Eqs. (8.3) and (8.5)]:

$$\delta_{ij}^{\text{hyp}} = \delta_{ij}^{\text{con}} + \delta_{ij}^{\text{pcs}} = K\,A_i\langle S_z\rangle_j + G_i\,D_j \tag{8.9}$$

where the subscript i refers to the observed nucleus and j to the lanthanoid employed [cf. Eq. (2.66)]. $\langle S_z\rangle_j$ and D_j depend only on the lanthanoid, whereas A_i and G_i are assumed to depend only on the nucleus and not on the lanthanoid. G_i is given by $(3\cos^2\theta_i - 1)/r_i^3$ of Eq. (8.3). The unknowns for a single shift are A_i and G_i. If n signals are observed, the unknowns are $2n$; that is, A_i and G_i for the n nuclei. For a second lanthanoid derivative, A_i and G_i are the same, and the experimental data are now $2n$, that is as many as the unknowns. By using a variety of lanthanoids, it is possible, in general, to set up and solve the appropriate number of simultaneous equations and thus evaluate the terms A_i and G_i for each nucleus [7–9]. This procedure is again valid as long as A_i and G_i depend only on the nucleus, and $\langle S_z\rangle_j$ and D_j depend only on the lanthanoid in a predictable way and not on the nucleus under observation. The accuracy of the results depends on how precisely the patterns of Fig. 8.1 are capable of reproducing the contact and pseudocontact shifts within the lanthanoid series. As we have seen, precise estimates of $\langle S_z\rangle$ in the equation for δ^{con} [Eqs. (8.4) and (8.5)] require evaluation of the splitting of the J manifold for the particular complex examined; the equation for δ^{pcs} [Eq. (8.3)], besides being valid only for axial symmetry, assumes that the splitting of the $2J+1$ levels does not exceed kT. It has been pointed out that such an assumption is not correct [10], although a thorough treatment [11] has shown that Eqs. (8.2) and (8.3) are capable of accounting for pseudocontact shifts at or near room temperature with an accuracy of 10–20%. Alternative procedures have been proposed [12–14] for the cases of changes of crystal field parameters through the lanthanoid series, or of structural changes due to lanthanoid contraction and variable coordination number.

When contact contributions can be neglected, the pseudocontact shifts for ligand protons in series of lanthanoid complexes can be reproduced by using the complete equation for nonaxial cases [Eq. (8.2)] and the geometric factors calculated from the X-ray structure [15–21]. The values for the D_z and $D_x - D_y$ parameters obtained through best fitting procedures are in fair to good agreement with the single-crystal anisotropy data [17].

8.4.2 LIMITATIONS OF BLEANEY'S THEORY

Eq. (8.2) is obtained by assuming that the ligand field splitting is much smaller that kT. This assumption is often not fulfilled [22,23] as the ligand field splitting can be as large as 1000 cm^{-1}. Furthermore, the point dipole approximation can be incorrect, especially because the electron density distribution is not spherical in lanthanoids; actually, Eu, Yb, Tm, and Er have a prolate f electron density distribution, whereas Ce, Pr, Tb, and Dy have an oblate distribution [24]. Finally, the assumption that the ligand field splitting is small compared to the spin–orbit coupling, and thus that J is a good quantum number, may be incorrect. If this is the case, the D_z (and $\Delta\chi_{ax}$) values depend not only on the specific lanthanoid, but also on the coordination environment. This actually occurs in a number of coordination complexes with a relatively large ligand field splitting [22]. Furthermore, the principal directions of the $\boldsymbol{D}$ tensor may be oriented differently for the different lanthanoids. This makes the angles θ, and φ in Eq. (8.2) to change for the different lanthanoid ions even if they are bound to the same ligand.

For example, the main axis in Ln-DOTA(H_2O) (Fig. 10.13) changes orientation on passing from Tb to Yb, by rotating up to 90 degrees, as observed at low temperature (2 K) [25,26]. In fact, it is basically perpendicular to the Ln^{3+}-water oxygen direction for Tb and Dy, and almost parallel for Er and Yb. This effect is likely related to the different matching of the ligand geometry with the electron density distribution (prolate or oblate) of the lanthanoid ions. Therefore, the main direction of the magnetic susceptibility anisotropy tensor would depend on the nonspherical electron density distribution of the lanthanoid, so that it can be predicted by a minimization of the electrostatic energy of the system and the knowledge of the crystal structure [27]. When in the plane perpendicular to the Ln^{3+}-water oxygen direction, the orientation of the principal axis strongly depends on the orientation of the hydrogen atoms of the water molecule in the axial coordination site [25].

8.5 NUCLEAR SPIN RELAXATION

In the case of lanthanoids (except gadolinium(III)) and actinoids, the relaxation equations discussed in Chapter 4 are different. For these systems, in fact, the J quantum number substitutes the S quantum number and g_J substitutes g_e. The equations for the dipolar relaxation thus become:

$$R_{1M} = \frac{2}{15}\left(\frac{\mu_0}{4\pi}\right)^2 \frac{\gamma_I^2 g_J^2 \mu_B^2 J(J+1)}{r^6}\left(\frac{7\tau_c}{1+\omega_S^2\tau_c^2}+\frac{3\tau_c}{1+\omega_I^2\tau_c^2}\right) \quad (8.10)$$

$$R_{2M} = \frac{1}{15}\left(\frac{\mu_0}{4\pi}\right)^2 \frac{\gamma_I^2 g_J^2 \mu_B^2 J(J+1)}{r^6} \times \left(4\tau_c + \frac{13\tau_c}{1+\omega_S^2\tau_c^2} + \frac{3\tau_c}{1+\omega_I^2\tau_c^2}\right); \tag{8.11}$$

where $\tau_c = \left(\tau_e^{-1} + \tau_r^{-1} + \tau_M^{-1}\right)^{-1}$, the equations for contact relaxation become:

$$R_{1M} = \frac{2}{3} J(J+1)\left(\frac{A}{\hbar}\right)^2 \frac{\tau_{\text{con}}}{1+\omega_S^2\tau_{\text{con}}^2} \tag{8.12}$$

$$R_{2M} = \frac{1}{3} J(J+1)\left(\frac{A}{\hbar}\right)^2 \left(\tau_{\text{con}} + \frac{\tau_{\text{con}}}{1+\omega_S^2\tau_{\text{con}}^2}\right); \tag{8.13}$$

where $\tau_{\text{con}} = \left(\tau_e^{-1} + \tau_M^{-1}\right)^{-1}$, and the equations for Curie relaxation become:

$$R_{1M} = \frac{2}{5}\left(\frac{\mu_0}{4\pi}\right)^2 \frac{\omega_I^2 g_J^4 \mu_B^4 (J(J+1))^2}{(3kT)^2 r^6} \frac{3\tau_{\text{Curie}}}{1+\omega_I^2\tau_{\text{Curie}}^2} \tag{8.14}$$

$$R_{2M} = \frac{1}{5}\left(\frac{\mu_0}{4\pi}\right)^2 \frac{\omega_I^2 g_J^4 \mu_B^4 (J(J+1))^2}{(3kT)^2 r^6} \left(4\tau_{\text{Curie}} + \frac{3\tau_{\text{Curie}}}{1+\omega_I^2\tau_{\text{Curie}}^2}\right). \tag{8.15}$$

where $\langle J_z \rangle^2$ has replaced $\langle S_z \rangle^2$ in Eqs. (4.30) and (4.31) and $\tau_{\text{Curie}} = \left(\tau_r^{-1} + \tau_M^{-1}\right)^{-1}$. The term $g_J^4 \mu_B^4 (J(J+1))^2$ in Eqs. (8.14) and (8.15) can be more properly replaced by the experimental μ_{eff}^2 values when available [28].

All lanthanoid ions, with the exception of gadolinium(III), are likely to be relaxed by Orbach-type processes, although some very efficient Raman processes are also invoked on the basis of the temperature dependence [29,30]. For gadolinium(III), which has an ^8S free ion ground term, the modulation of the quadratic ZFS is the dominant relaxation mechanism (Section 8.6) [31–35].

As described in the next sections, Curie contribution is often dominating over the dipolar contribution for all lanthanoids(III) except gadolinium(III). Typical R_{1M} and R_{2M} values in small and large molecules, calculated using Eqs. (8.10) and (8.11) and Eqs. (8.14) and (8.15) have been reported in Table 4.1. The relative importance of the dipolar and Curie contributions at 500 and 1000 MHz is shown in Table 8.2 The expected line broadening for a proton at 5 Å from the metal ion, in a macromolecule with τ_r of 4 ns, at a field of 500 MHz, is shown in Fig. 8.3. In these conditions, the Curie contribution is dominating over the dipolar contribution for all lanthanoids except Gd^{3+}, and it can vary over more than three orders of magnitude.

8.6 GADOLINIUM

Gadolinium(III) is an f^7 ion with the seven electrons being distributed one electron per f orbital. In analogy to Mn^{2+}, which has one electron per d orbital and displays the longest electronic relaxation times among the 3d metal ions, gadolinium(III) also has by far the longest electronic relaxation times of all the lanthanoids. Indeed, it is the only ion in the lanthanoid series that can be investigated through EPR, with $T_{1e} \geq 2 \times 10^{-10}$ s, at room temperature [37,38]. This results in a nuclear relaxing capability close to that of manganese(II). Electron relaxation is due to ZFS modulation. The correlation time for

Table 8.2 Dipolar and Curie Contributions to the Longitudinal and Transverse Relaxation Rates for a Proton at 5 Å From a Lanthanoid(III) Ion, at 500 and 1000 MHz (298 K), in a Fast ($\tau_r = 10^{-10}$ s) or Slow ($\tau_r = 10^{-8}$ s) Rotating Molecule

			Small Molecules ($\tau_r = 10^{-10}$ s)				Large Molecules ($\tau_r = 10^{-8}$ s)		
Ln^{3+}	T_{1e} (ps)	$R^{dip}_{1,2M}$ (s^{-1})	R^{Curie}_{1M} (s^{-1})	R^{Curie}_{2M} (s^{-1})	R^{Curie}_{1M} (s^{-1})	R^{Curie}_{2M} (s^{-1})	R^{Curie}_{1M} (s^{-1})	R^{Curie}_{2M} (s^{-1})	R^{Curie}_{2M} (s^{-1})
		500–1000 MHz	500 MHz	500 MHz	1000 MHz	1000 MHz	500–1000 MHz	500 MHz	1000 MHz
Ce^{3+}	0.1–0.4	3–10	1.4	1.7	4.4	6.2	0.15	100	400
Pr^{3+}	0.06–0.3	4–17	5.5	6.8	17	25	0.6	400	1,600
Nd^{3+}	0.2–0.8	10–30	5.7	7.1	18	26	0.6	420	1,700
Sm^{3+}	0.05–0.2	0.2–0.7	0.02	0.02	0.05	0.08	0.002	1.2	5
Gd^{3+}	10^4–10^6	≈10,000 (500 MHz); $R_{1M} = 10^2 - 10^3$, $R_{2M} = 10^5 - 10^6$ (1000 MHz)	130	160	420	600	15	10,000	40,000
Tb^{3+}	0.2–0.8	70–210	300	370	950	1350	33	22,000	90,000
Dy^{3+}	0.4–1	110–280	430	530	1350	1900	50	32,000	125,000
Ho^{3+}	0.2–0.8	80–250	420	520	1350	1900	50	31,000	125,000
Er^{3+}	0.3–0.8	80–200	280	350	900	1300	30	20,000	83,000
Tm^{3+}	0.5–2	60–230	110	135	350	500	12	8,000	32,000
Yb^{3+}	0.2–0.5	15–35	15	17	45	65	1.5	1,000	4,000

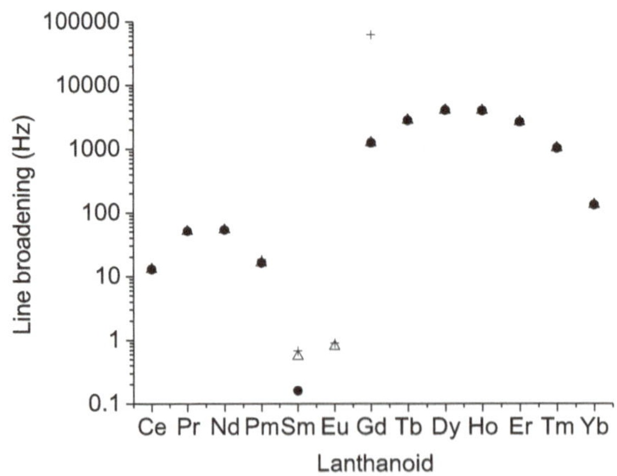

FIGURE 8.3 Expected Curie line broadening for a proton at 5 Å from the metal ion, in a macromolecule with τ_r of 4 ns, at a field of 500 MHz, without (•) and with (Δ) inclusion of the contribution from the first excited *J* manifold. The total line broadening, including the dipolar contribution, is also shown (+).

Adapted from Ref. [36].

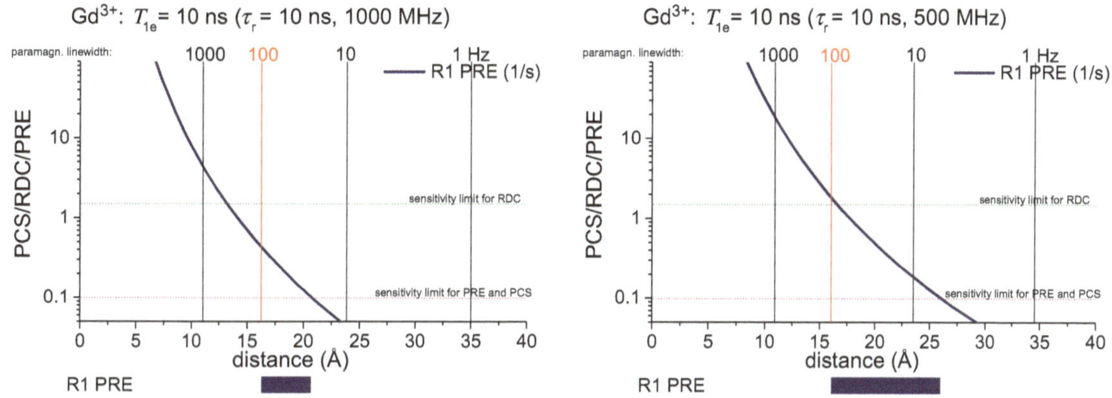

FIGURE 8.4 ^{1}H R_1 paramagnetic relaxation enhancement (in s^{-1}, *blue line*) as a function of the distance from the gadolinium(III) ion, for typical values of electron relaxation time, for a molecule with τ_r of 10 ns, at 1000 MHz (left panel) or 500 MHz (right panel). The plots also shows the distances with ^{1}H paramagnetic linewidth of 1, 10, 100, and 1000 Hz. The bar under the plots indicate the range of distances with ^{1}H R_1 paramagnetic relaxation enhancement values above the experimental sensitivity limits. No pseudocontact shifts and paramagnetic residual dipolar couplings can be observed, due to the basically null magnetic susceptibility anisotropy.

electron relaxation, generally linked to the transient distortions of the Gd^{3+} coordination environment due to random impacts of solvent molecules, could be determined by pseudorotations of the coordination polyhedron, as indicated by molecular dynamics simulations [39]. Fig. 8.4 shows the ^{1}H paramagnetic relaxation enhancement and linewidth at different distances from the metal ion, calculated for an electron relaxation time of 10 ns, at 500 and 1000 MHz. No pseudocontact shifts and paramagnetic residual dipolar couplings can be observed, due to the basically null magnetic susceptibility anisotropy.

Fig. 8.5 shows the NMRD profile of Gd^{3+} water solutions [40]. The observed dispersion corresponds to a τ_c equal to 5×10^{-11} s, which can be ascribed to the reorientation correlation time, but the shape of the dispersion at high fields can be reproduced only by introducing the contribution of a field dependent T_{1e}. The reorientation correlation time and the electron relaxation time must actually be of the same order of magnitude above 10 MHz. The fitting provides $T_{1e(0)} = 1.2 \times 10^{-10}$ s, $\tau_v = 1.6 \times 10^{-11}$ s and 8–9 coordinated water molecules with metal proton distances of 3.1–3.2 Å [41]. Simultaneous fits of ^{17}O, ^{1}H NMRD and EPR measurements have also been performed [42].

The proton NMRD profile of an ethyleneglycol solution containing $GdCl_3$ is also reported in Fig. 8.5 at two different temperatures. The correlation time for electron relaxation, τ_v, is longer than in water. This could indicate that collisions of solvent molecules with the ion are slowed down in viscous solvents. $T_{1e(0)}$, which is related to the magnitude of the instantaneous ZFS induced by collisions, instead, does not change much. Therefore, the decrease of thermal motion as well as the increase in viscosity of the solution are responsible of the increase of both τ_v and τ_r. A correct analysis of the profile should be performed by taking into account the presence of both static and transient ZFS. Data were also analyzed using a spectral density function of the transient ZFS interaction where modulation is allowed with two correlation times (representing the time-scales for collision-induced distortions of the coordination symmetry, and for the nonzero partially averaged ZFS) and an order parameter [43].

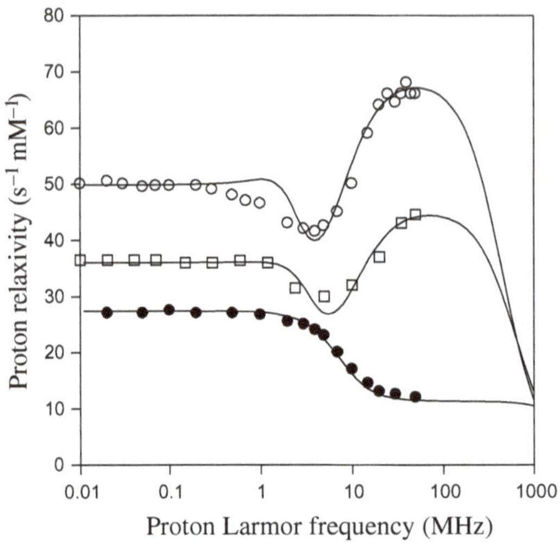

FIGURE 8.5 Solvent ^{1}H NMRD profiles for water solutions of $GdCl_3$ at 298 K (•) and for ethyleneglycol solutions of $GdCl_3$ at 298 K (○) and 312 K (□).

Adapted from Ref. [40].

In macromolecular systems, the field dependent electron relaxation time is usually the correlation time for dipolar relaxation. In these cases, as also found for manganese(II) systems, τ_v seems to be always around 15–25 $\times$ 10^{-12} s at room temperature without any dependence on the reorientation correlation time, i.e. on the molar mass. Common features of the NMRD profiles of gadolinium(III) containing macromolecules are the presence of the ω_S dispersion at about 1 MHz and of a peak centered at about 20–50 MHz. The presence of a peak in the high field region of the NMRD profile actually confirms the field dependence of T_{1e} (Chapter 10). Since gadolinium(III) often exhibits a small static ZFS, the Solomon-Bloembergen-Morgan (SBM) theory (Sections 6.7 and 10.2) is inadequate to fit the experimental profiles, and programs taking into account the effect of both static and transient ZFS have to be employed (Section 6.7). In the analysis of the relaxivity profile of gadolinium(III) complexes, one should also make sure that electron relaxation is within the Redfield limit ($T_{1e} > \tau_v$, Section 4.13), as otherwise good fits can be still obtained but with parameters that may be wrong by orders of magnitude [44,45].

Gadolinium(III) ion is extensively investigated from a relaxometric point of view because its complexes are the best contrast agents for magnetic resonance imaging, as described in Chapter 10. Due to its high relaxing capability, it is also used as NMR line broadening agent (Section 9.2).

8.7 EUROPIUM(II)

The ^{1}H NMRD profiles of europium(II) aqua ions are similar to those of gadolinium(III) aqua ions, with a relaxivity at room temperature of about half [46]. This decrease in relaxivity is ascribed to a shorter reorientation time and to a longer metal ion-protons distance. Electron relaxation times for europium(II) aqua ions are longer than those of gadolinium(III), with a similar transient ZFS (0.056 cm^{-1}), and a much shorter τ_v, of about 1 $\times$ 10^{-12} s at 298 K. This results in an effective electron relaxation time at low fields equal to 7.5×10^{-10} s. The electron relaxation time of europium(II)-DTPA is described by the parameters $\Delta_t = 0.07\ \mathrm{cm}^{-1}$ and $\tau_v = 14 \times 10^{-12}$ s, providing a relatively fast electron relaxation, with $T_{1e(0)} = 3.4 \times 10^{-11}$ s [47].

8.8 OTHER LANTHANOIDS

The f^n configurations of lanthanoids(III) give rise to several free ion terms that upon spin–orbit coupling provide, except for f^7 configurations (Gd^{3+}, Eu^{2+}), several closely spaced energy levels. As the J values for the ground state vary from 5/2 to 8, the multiplicity of the ground state varies from 6 to 17 (Table 8.1), and is further split by crystal field effects. According to the electronic relaxation mechanisms discussed in Section 4.3, the availability of low-lying excited levels accounts for short electronic relaxation times [48]—again with the exception of metal ions with an f^7 configuration—likely due to Orbach-type processes at room temperature. For these fast-relaxing systems, the correlation times for nuclear relaxation are determined by the electronic relaxation times. Values for the latter can thus be estimated from nuclear relaxation measurements.

The NMRD profiles for the aqua ions show small and field independent relaxivity values for proton frequencies up to 50 MHz (Fig. 8.6), thus indicating that τ_c is indeed provided by T_{1e}, as the latter is much smaller than τ_r. The fit of the nuclear relaxation measurements using Eqs. (8.10)–(8.15) provides the values for T_{1e} reported in Fig. 8.7. These very short values confirm an efficient Orbach relaxation

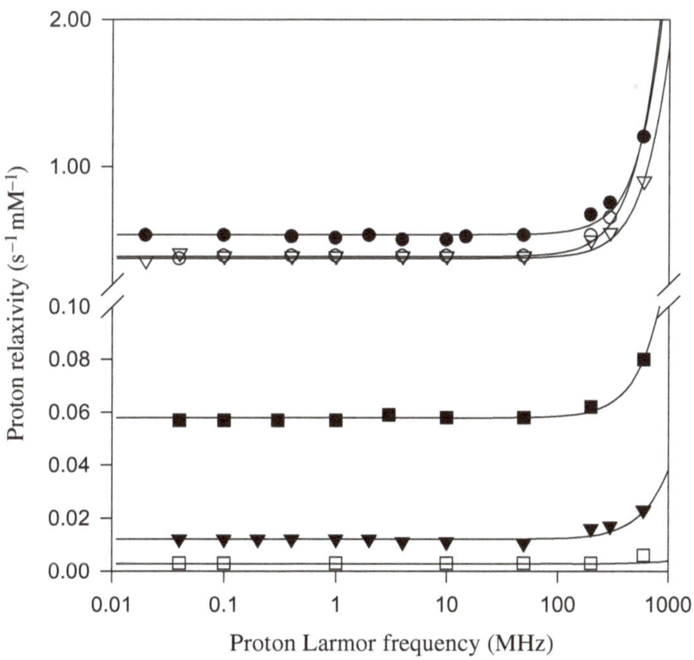

FIGURE 8.6 Water 1H NMRD profiles of some lanthanoid aqua ions at 298 K: (□) Sm, (▼) Pr, (■) Yb, (▽) Er, (○) Ho, (●) Dy.

Adapted from Ref. [49].

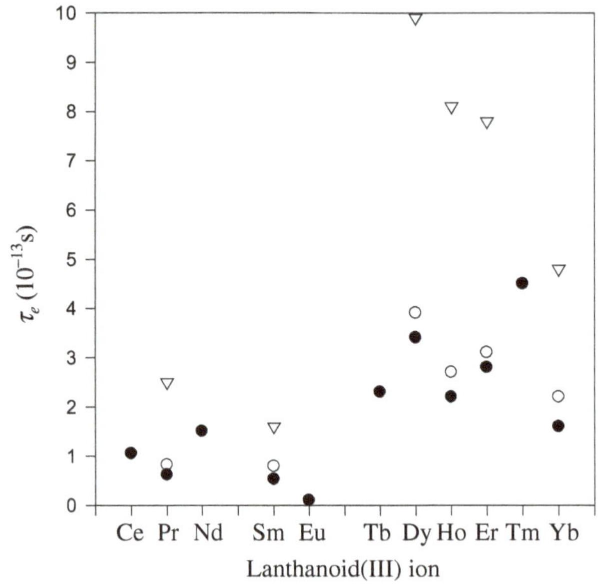

FIGURE 8.7 τ_e values for lanthanoid(III) aqua ions obtained from the fit of the NMRD profiles to Eqs. (8.10) and (8.14) (▼) [49] or estimated from single field measurements (•) [50]. τ_e values estimated by including the effect of ZFS (▽) [49] (using the J and g_J values reported in Table 8.1) are also reported.

mechanism. The high field increase in relaxivity is due to the Curie relaxation, which depends on the square of the external magnetic field [Eq. (8.14)]. The total paramagnetic relaxation rate is thus determined by the sum of the dipolar and Curie contributions. Lanthanoids constitute the only case in which for complexes as small as the aqua ion the field dependence of nuclear *longitudinal* relaxation (Fig. 8.6) (and not only of nuclear *transverse* relaxation) is due to Curie relaxation. Indeed, for small complexes Curie relaxation can only be expected when $T_{1e} \ll \tau_r$, as it occurs for lanthanoids.

From ^{17}O NMR measurements, τ_M was found in the range 10^{-6}–10^{-8} s; it can be thus neglected for the evaluation of the correlation times. This is confirmed by the temperature dependence of the ^{1}H NMRD profiles. Paramagnetic shift measurements of ^{13}C, ^{17}O, and ^{1}H nuclei in some lanthanoid chelating agents point out that the contact interaction between the metal ion and the hydrogen nuclei can be neglected [51].

The very small field dependence of proton relaxivity shown by Pr^{3+}, Yb^{3+}, and Sm^{3+} aqua ions is due to the relatively small effective magnetic moments of such ions. The fit was performed using the proton-lanthanoid distance estimated from the X-ray data, the experimentally available magnetic moments [52], and a coordination number of water molecules equal to nine for the lanthanoid ions in the first half and eight in the second half of the series [53,54]. The best-fit values obtained for the electron relaxation time with qualitative inclusion of the effect of a strong static ZFS are also reported in Fig. 8.7 [49]. Sub-picosecond relaxation times are consistently observed over series of homologous complexes: given that the electronic relaxation time is believed to be proportional to the size of the transient ligand field induced by solvent collision, the similarity of the τ_e values is in agreement with the absence of change in the overall ligand field observed by optical spectroscopy [55].

Short electronic relaxation times result in very little line broadening, besides Curie relaxation, of NMR signals. The hyperfine shifts are mainly pseudocontact in origin and the contact contribution, which arises from covalent bonds, is small owing to the strong radial contraction of the 4f orbitals. The hyperfine coupling constant presumably has a lower value than that of the 3d transition metal ions, for the same type of nuclei in the same geometry. Although some ions have been found to give rise to larger contact contributions than other ions of the series, such contributions are generally neglected for nuclei not directly coordinated to the metal. Separation of the hyperfine shifts can in principle be successfully achieved by utilizing different lanthanoids and assuming that the pseudocontact shifts have axial symmetry (Section 8.4). The combined effects of small line broadening and dominant pseudocontact shifts, which depend on the geometrical coordinates described in Eqs. (8.2) and (8.3), justify the use of these ions and their complexes as shift reagents since the early days of NMR in organic chemistry. As these systems have fast electronic relaxation times, the availability of high magnetic fields enables exploitation of Curie relaxation (Section 4.6) for an independent estimate of the distance of the resonating nucleus from line broadening measurements.

The successful use of lanthanoids as pseudocontact shift reagents is also due to their tendency to interact with molecules containing oxygen and nitrogen as donor atoms. For example, lanthanoid complexes with ligands of the type shown in Fig. 10.13 often interact with organic molecules, possibly of biological interest. Pseudocontact shifts (and relaxation measurements) may then allow mapping of the nuclei in the molecule of interest according to the guidelines given in Chapter 9.

Fig. 8.8 shows the typical values of pseudocontact shifts and paramagnetic amide proton residual dipolar couplings (Chapter 3) expected as a function of the distance from the metal ion, at 500 or 1000 MHz, for a protein with reorientation time of 10 ns. The figure also shows the ^{1}H paramagnetic relaxation enhancement and linewidths calculated for the typical electron relaxation time of the lanthanoids. In the assumption that a safe detection limit of the nuclear peaks in standard NMR spectra corresponds to a paramagnetic linewidth of 100 Hz, relaxation enhancements and pseudocontact shifts can be measured in different ranges of distances from the lanthanoid ion, as shown in the figure, where the sensitivity limit for these observables has been set to 0.1 s^{-1} and 0.1 ppm, respectively. The pseudocontact shift is calculated for a nucleus with $|3\cos^2\theta - 1| = 1$, where θ is the angle between the z-axis of the magnetic susceptibility tensor and the metal-nucleus direction [Eq. (2.46)]; the residual dipolar coupling is calculated for an amide vector with $|3\cos^2\alpha - 1| = 1$, where α is the angle between the z-axis of the magnetic susceptibility tensor and the amide vector [Eq. (3.14)]. Paramagnetic residual dipolar couplings do not depend on the nuclear distance from the lanthanoid. Therefore, they can be observed at all distances, as long as it is possible to resolve the two components of the doublets (Section 12.3.4), if they are larger than the detection sensitivity limit of 1 Hz. The maximum values of residual dipolar couplings are for amide vectors parallel to the z-axis of the magnetic susceptibility anisotropy tensor, i.e. for $\alpha = 0$. This means that the residual dipolar couplings can have values up to twice as much the values shown in the plots.

Fig. 8.9 shows the ratio between pseudocontact shifts and line broadening, which provides an estimate of the resolution of pseudocontact shifted peaks, versus the maximum absolute pseudocontact shifts of a proton at 5, 10, 20, or 30 Å from the different lanthanoids, for a macromolecule with τ_r of 10 ns, at a field of 500 MHz. The maximum pseudocontact shift is obtained for a nucleus on the z-axis of the magnetic susceptibility tensor. The lanthanoids are in green or red if the paramagnetic line broadening is expected smaller or larger than 100 Hz, respectively. Lanthanoids with the smallest magnetic susceptibility anisotropy are those providing the largest resolution for protons at short distances.

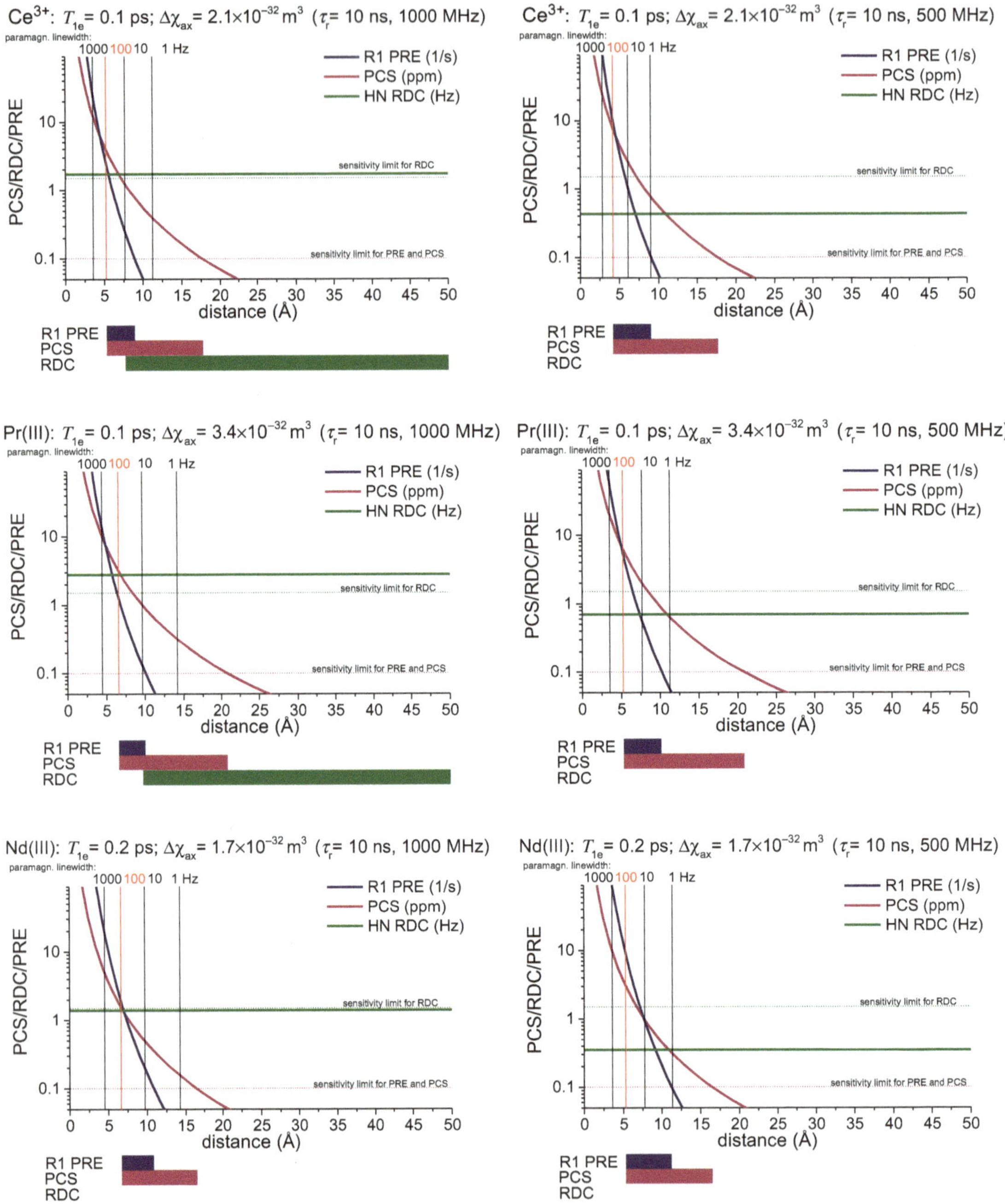

FIGURE 8.8 ^{1}H R_1 paramagnetic relaxation enhancement (in s^{-1}, *blue line*), pseudocontact shift (in ppm, *pink line*) and HN residual dipolar coupling (in Hz, *green line*) as a function of the distance from the lanthanoid ion, for typical values of electron relaxation time and axial magnetic susceptibility anisotropy, for a molecule with τ_r of 10 ns, at 1000 MHz (left panel) or 500 MHz (right panel). The plots also shows the distances with ^{1}H paramagnetic linewidth of 1, 10, 100, and 1000 Hz. The bars under the plots indicate the range of distances with ^{1}H R_1 paramagnetic relaxation enhancement, pseudocontact shift and HN residual dipolar coupling values above the experimental sensitivity limits. The pseudocontact shift is calculated for a nucleus with $3\cos^2\theta - 1 = 1$; the residual dipolar coupling is calculated for an amide vector with $3\cos^2\alpha - 1 = 1$.

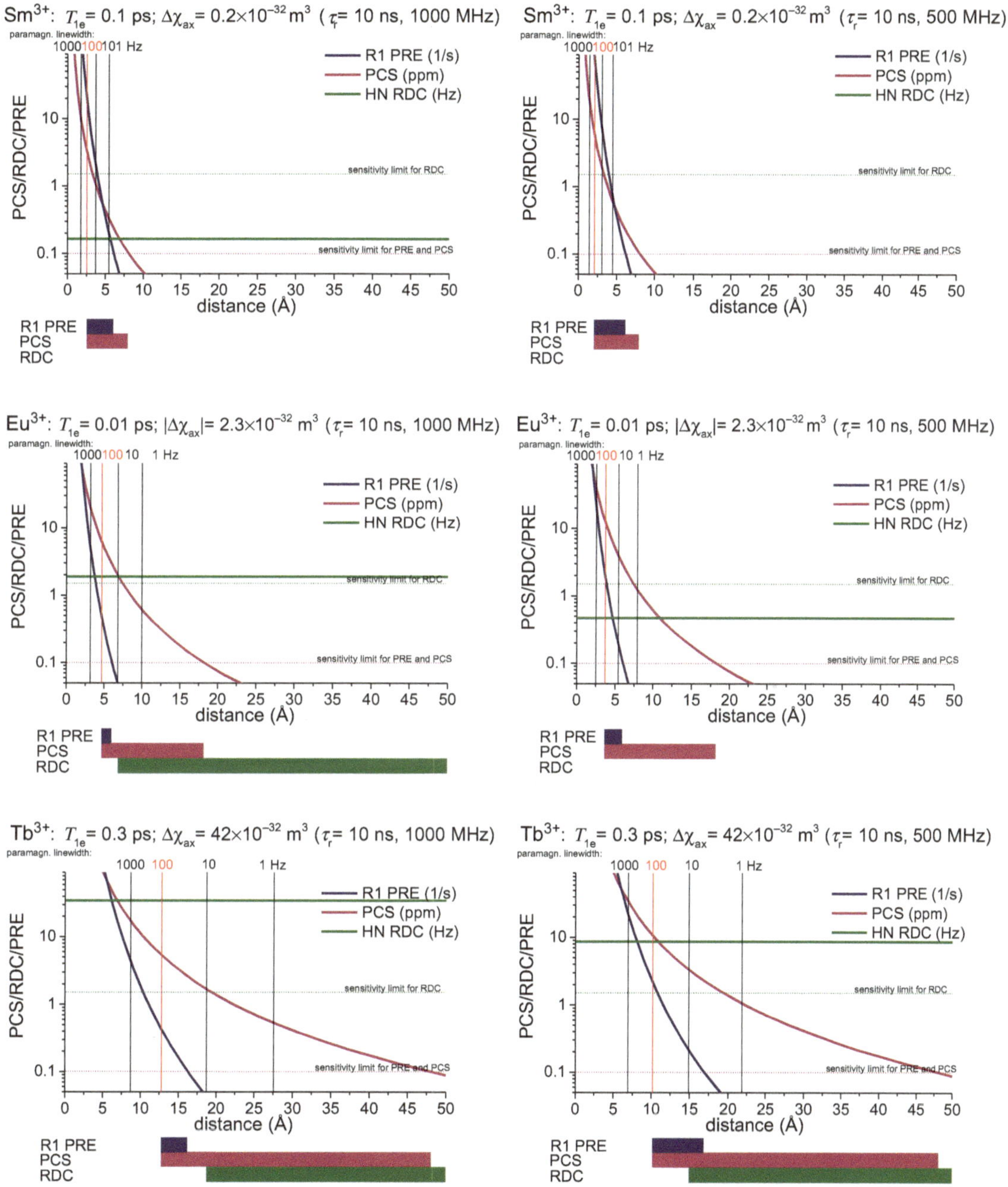

FIGURE 8.8

(*cont.*)

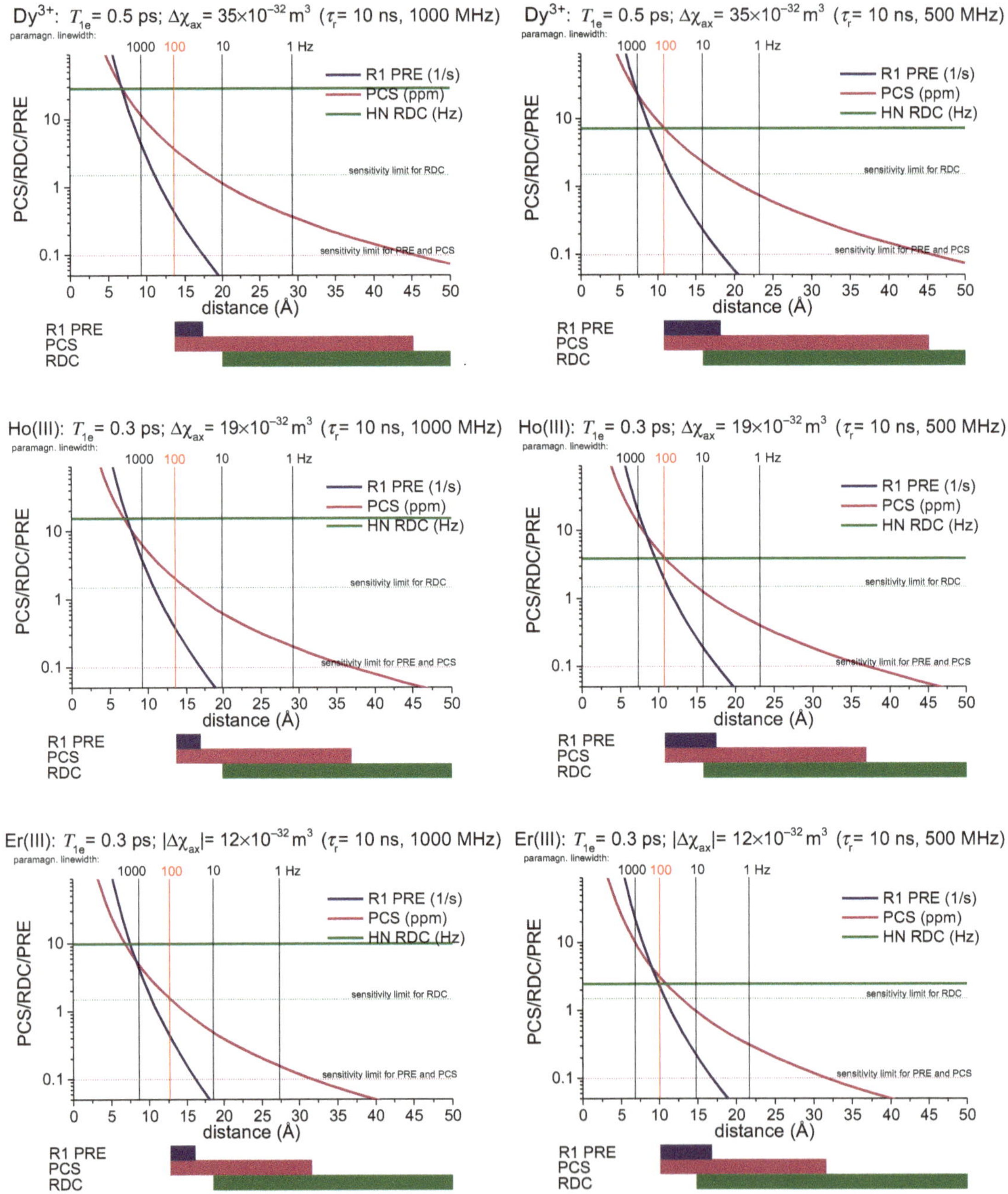

FIGURE 8.8

(cont.)

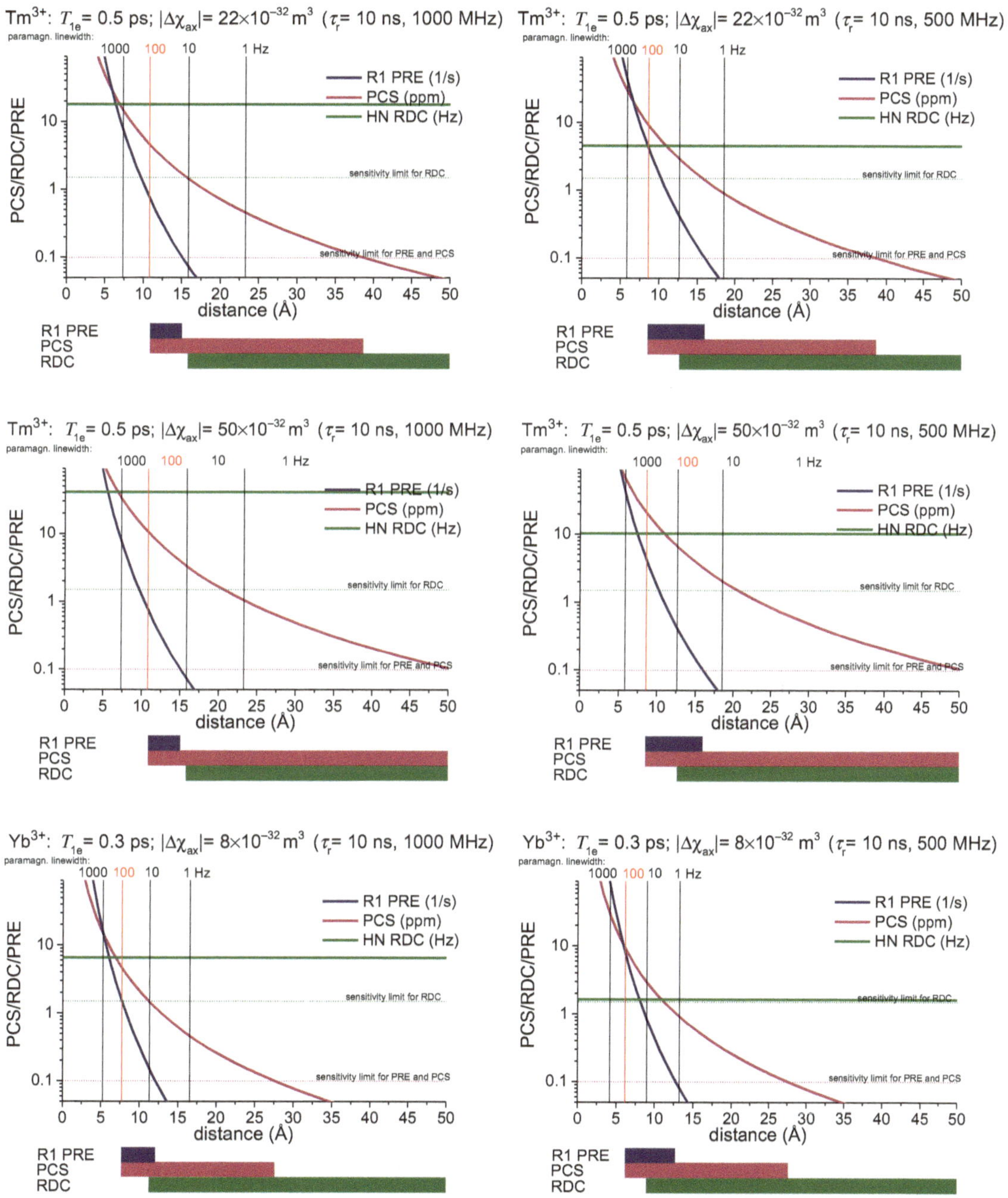

FIGURE 8.8

(cont.)

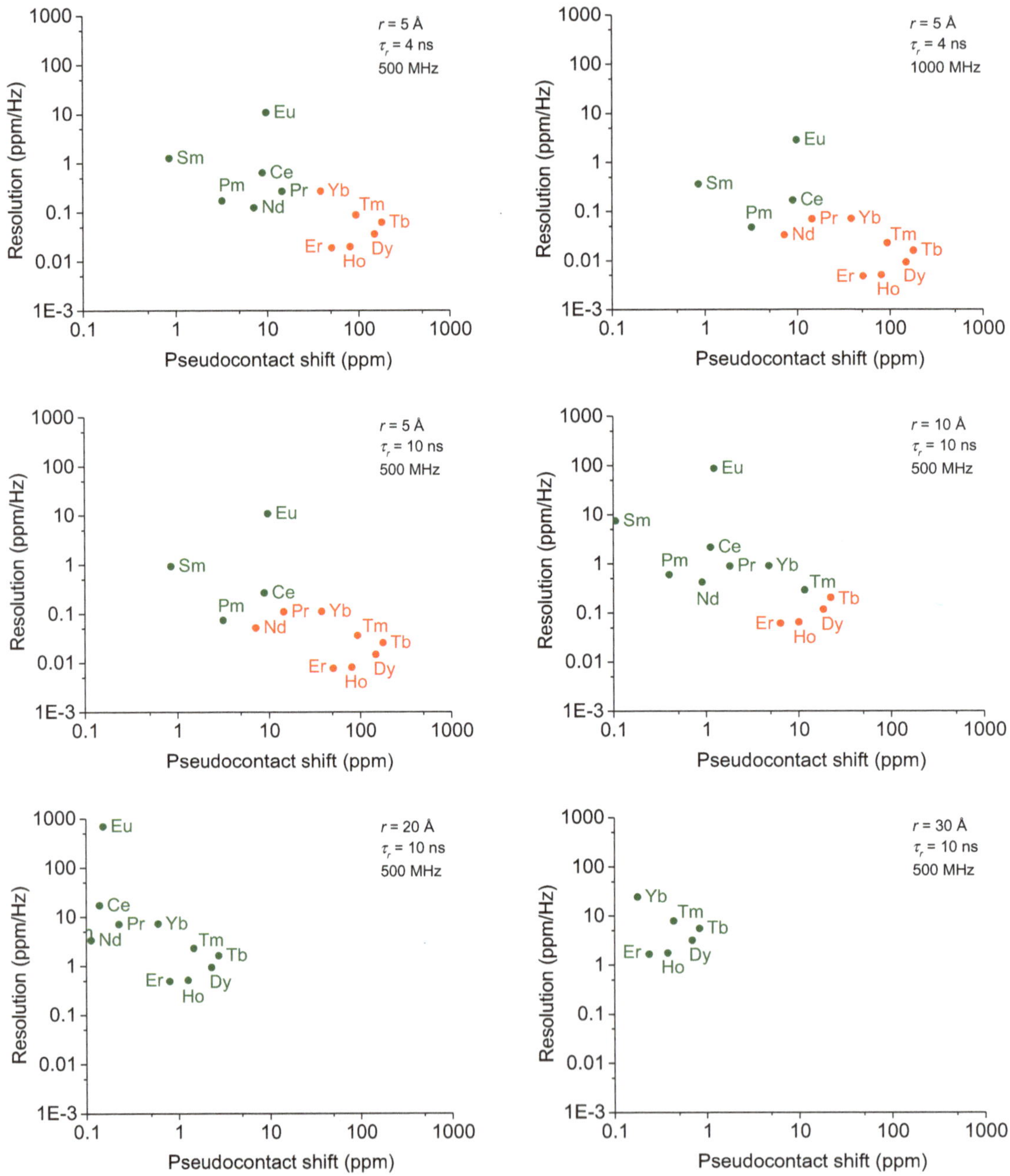

FIGURE 8.9 Resolution, defined as the ratio between pseudocontact shifts and line broadening, versus the maximum absolute pseudocontact shifts for protons at 5, 10, 20, or 30 Å from the different lanthanoids, for a macromolecule with τ_r of 4 or 10 ns, at a field of 500 or 1000 MHz. The lanthanoids with a paramagnetic line broadening smaller or larger than 100 Hz are in green or red, respectively.

They are however useful only to obtain information in a relatively thin spherical shell close to the metal. Lanthanoids with a larger magnetic susceptibility anisotropy should thus be selected to obtain information at larger distances from the metal, when their resolution becomes similarly large. For each shell, the lanthanoids providing spectra with the largest resolution can thus be selected: at 5–15 Å, Ce^{3+} or Eu^{3+} are thus the most convenient; at 15–25 Å, Yb^{3+} or Tm^{3+}; and above 25 Å, Tb^{3+} or Dy^{3+}. Fig. 8.9 also shows the same plots for a macromolecule with τ_r of 4 ns and at a field of 1000 MHz. Clearly, the effect of decreasing the reorientation time is beneficial in terms of resolution and line broadening (because the Curie broadening decreases linearly with the correlation time), whereas the effect of increasing the magnetic field is detrimental (because the Curie broadening increases with the square of the field).

Lanthanoids have been used as substitutes for calcium in calcium binding proteins since the early days in NMR [56,57]. The ionic radius of the Ln^{3+} ions is in fact very similar to that of Ca^{2+}. Remarkable information was obtained on systems such as Yb^{3+}-substituted parvalbumin (which contains a typical calcium binding site called EF-hand [58,59]) from 1D spectroscopy alone [60–65], later complemented by 2D spectroscopy [66]. More recently, pseudocontact shifts and longitudinal proton relaxation times have been used to locate metal ions within a protein structure (Chapter 9). The solution structure of the N-terminal fragment of calmodulin was thus obtained where the two calcium ions were substituted by Ce^{3+}, which has one of the smallest magnetic moments among lanthanoids (about 2.5 μ_B) and a relatively large magnetic anisotropy, yielding a pseudocontact shift/Curie line broadening ratio more favorable than other lanthanoids [67]. The improvement in resolution in the regions close to the paramagnetic centers, determined by the inclusion of the pseudocontact shifts restraints in the structural calculations, was sizable.

The full series of paramagnetic lanthanoid(III) ions (except gadolinium(III) and the radioactive promethium) was substituted to the C-terminal calcium binding site of the EF-hand protein calbindin D_{9k} to obtain the corresponding pseudocontact shifts [68]. The magnetic susceptibility anisotropy tensors for each paramagnetic lanthanoid ion were obtained from the protein structure and the measured pseudocontact shifts (Section 9.3). The obtained tensors show directions of the largest magnetic susceptibility component for Ce, Pr, Nd, Sm, Tb, Dy, and Ho and of the smallest magnetic susceptibility component for Eu, Er, Tm, and Yb all within 15 degree from their average (within 20 degree for Sm), confirming the essential similarity of the coordination environment for all lanthanoids. The observed pattern of axial anisotropies (Table 8.3) is in excellent qualitative agreement with Bleaney's theory [2].

Table 8.3 Magnetic susceptibility anisotropies, $\Delta\chi_{ax}$ and $\Delta\chi_{rh}$, in 10^{-32} m^3, for the eleven lanthanoid derivatives of calbindin D_{9k}, with the *z*-axis of the tensor corresponding to the direction of the largest magnetic susceptibility for Ce, Pr, Nd, Sm, Tb, Dy, and Ho, and of the smallest magnetic susceptibility for Eu, Er, Tm, and Yb

	Ce^{3+}	Pr(III)	Nd(III)	Sm^{3+}	Eu^{3+}	Tb^{3+}	Dy^{3+}	Ho(III)	Er(III)	Tm^{3+}	Yb^{3+}
$\Delta\chi_{ax}$	2.08	3.40	1.74	0.19	−2.34 (2.39)	42.1	34.7	18.5	−11.6 (12.2)	−21.9 (26.0)	−8.26 (8.51)
$\Delta\chi_{rh}$	−0.71	−2.11	−0.46	−0.08	1.63 (−1.52)	−11.2	−20.3	−5.79	8.58 (−7.31)	20.1 (−11.8)	5.84 (−5.34)

The values in parenthesis are calculated for the z-axis corresponding to the direction of the largest absolute magnetic susceptibility, providing the usual convention $|\Delta\chi_{rh}| \le 2|\Delta\chi_{ax}|/3$.

Table 8.3 shows also the $\Delta\chi_{ax}$ and $\Delta\chi_{rh}$ values (in parenthesis) are calculated by defining the *x*-, *y*- and *z*-axes of the principal frame so that the *z* axis corresponds to the direction with the largest absolute anisotropy, providing the usual convention $|\Delta\chi_{rh}| \leq 2|\Delta\chi_{ax}|/3$. Using this convention, the *z*-axes of the magnetic susceptibility anisotropy tensors of Eu, Er, Tm, and Yb are almost perpendicular to the *z*-axis of the tensors of Ce, Pr, Nd, Sm, Tb, Dy, and Ho.

The pseudocontact shifts, together with the paramagnetic residual dipolar couplings, can be very effective to refine protein solution structures. In the case of the lanthanoid derivatives of monolanthanoid-substituted calbindin D_{9k}, since the lanthanoids span a wide range of magnetic anisotropies, the refinement is effective in shells from the metal of ~5–15 Å for Ce^{3+}, ~9–25 Å for Yb^{3+}, ~13–40 Å for Dy^{3+}, as useful pseudocontact shifts and residual dipolar couplings can be observed in these shells [69]. As already discussed, by using different lanthanoids it is possible to enlighten shells at variable distances from the metal itself. The inclusion as structural restraints, together with the diamagnetic restraints, of pseudocontact shifts and paramagnetic residual dipolar couplings from 11 paramagnetic lanthanoids has been shown to provide a sizable improvement in both the precision and the accuracy of the protein conformation (Chapter 9) [70,71]. The pseudocontact shifts and paramagnetic residual dipolar couplings, together with the identification of the secondary structure elements, can be used to determine the backbone solution structure of a metalloprotein like calbindin even in the absence of any NOEs [72–74], or the relative arrangement of a two-domain protein, like calmodulin [75], as well as to obtain information on its conformational heterogeneity [76] (Chapter 9).

Lanthanoid binding tags (Section 9.7) have been developed to be attached to proteins, thus allowing for the measurement of pseudocontact shifts and paramagnetic residual dipolar couplings also for diamagnetic systems. The magnetic susceptibility anisotropy that can be calculated from the experimental data may be smaller than the value expected for the corresponding lanthanoid, if the tag is not rigidly attached, so that motionally averaged data are observed. In these cases, some caution is needed in the analysis of the pseudocontact shifts, as described in Section 9.6. As an example of an efficient lanthanoid binding tag, the Caged Lanthanide NMR Probe 5 (CLaNP-5, Table 9.2) [77], which can bind rigidly to the protein backbone through two cysteine residues [78], provides axial magnetic susceptibility anisotropies of −45, −40, and 9 $\times 10^{-32}$ m^3 for Tb^{3+}, Dy^{3+}, and Yb^{3+}, respectively, and a value of 55 $\times$ 10^{-32} m^3 (i.e., sizably larger than measured for other Tm-substituted proteins) for Tm^{3+} [77,79–81].

REFERENCES

[1] Golding RM, Halton MP. Aust. J. Chem. 1972;25:2577.
[2] Bleaney B. J. Magn. Reson. 1972;8:91.
[3] Luzon J, Sessoli R. Dalton Trans. 2012;41:13556.
[4] Golding RM, Pascual RO, McGarvey BR. J. Magn. Reson. 1982;46:30.
[5] Lewis WB, Jackson JA, Lemons JF, Taube H. J. Chem. Phys. 1962;36:694.
[6] Dobson CM, Williams RJP, Xavier AV. J. Chem. Soc. Dalton Trans. 1973;3:2662.
[7] Reilley CN, Wood BW, Desreux JF. Anal. Chem. 1975;47:2110.
[8] Reuben J, Elgavish GA. J. Magn. Reson. 1980;39:421.
[9] Shelling JG, Bjorson ME, Hodges RS, Taneja AK, Sykes BD. J. Magn. Reson. 1984;57:99.
[10] Horrocks WD Jr. J. Magn. Reson. 1977;26:333.
[11] McGarvey BR. J. Magn. Reson. 1979;33:445.

[12] Di Bari L, Salvadori P. Coord. Chem. Rev. 2005;249:2854.
[13] Di Pietro S, Lo Piano S, Di Bari L. Coord. Chem. Rev. 2011;255:2810.
[14] Berardozzi R, Di Bari L. Inorg. Chem. 2013;52:11514.
[15] Cramer RE, Dubois R, Furuike CK. Inorg. Chem. 1975;14:1005.
[16] Cramer RE, Dubois R, Seff K. J. Am. Chem. Soc. 1974;96:4125.
[17] Cramer RE, Maynard RB. J. Magn. Reson. 1978;31:295.
[18] Stainer MVR, Takats J. J. Am. Chem. Soc. 1983;105:410.
[19] Jenkins BG, Lauffer RB. Inorg. Chem. 1988;27:4730.
[20] Geraldes CFGC, Sherry AD, Kiefer GE. J. Magn. Reson. 1992;97:290.
[21] Lisowski J, Sessler JL, Lynch V, Mody TD. J. Am. Chem. Soc. 1995;117:2273.
[22] Funk AM, Finney K-LNA, Harvey P, Kenwright AM, Neil ER, Rogers NJ, Senanayake PK, Parker D. Chem. Sci. 2015;6:1655.
[23] Mironov VS, Galyametdinov YG, Ceulemans A, Görller-Walrand C, Binnemans K. J. Chem. Phys. 2002; 116:4673.
[24] Rinehart JD, Long JR. Chem. Sci. 2011;2:2078.
[25] Cucinotta G, Perfetti M, Luzon J, Etienne M, Car PE, Caneschi A, Calvez G, Bernot K, Sessoli R. Angew. Chem. Int. Ed. 2012;51:1606.
[26] Boulon ME, Cucinotta G, Luzon J, Degl'Innocenti C, Perfetti M, Bernot K, Calvez G, Caneschi A, Sessoli R. Angew. Chem. Int. Ed. 2013;52:350.
[27] Chilton NF, Collison D, McInnes EJL, Winpenny REP, Soncini A. Nat. Commun. 2013;4:2551.
[28] Vigouroux C, Belorizky E, Fries PH. Eur. Phys. J. D 1999;5:243.
[29] Misra SK, Orhun U. Solid State Commun. 1987;63:867.
[30] Malhotra VM, Buckmaster HA, Dixon JM. J. Phys. C Solid State Phys. 1980;13:3921.
[31] Vishnevskaya GP, Kozyrev BM. J. Struct. Chem. 1966;7:20.
[32] Hudson A, Lewis JWE. Trans. Faraday Soc. 1970;66:1297.
[33] Koenig SH. Magn. Reson. Med. 1991;22:183.
[34] Hernandez G, Tweedle M, Bryant RG. Inorg. Chem. 1990;29:5110.
[35] Aime S, Barbero L, Botta M. Magn. Res. Imaging 1991;9:843.
[36] Geraldes CFGC, Luchinat C. Met. Ions. Biol. Syst. 2003;40:513.
[37] Stephens EM, Grisham CM. Biochemistry 1979;18:4876.
[38] Epstein M, Reuben J, Levitzki A. Biochemistry 1977;16:2449.
[39] Kowall Th, Foglia F, Helm L, Merbach AE. J. Phys. Chem. 1995;99:13078.
[40] Banci L, Bertini I, Luchinat C. Inorg. Chim. Acta 1985;100:173.
[41] Luchinat C, Parigi G, Ravera E. J. Biomol. Nucl. Magn. Reson. 2014;58:239.
[42] Borel A, Yerly F, Helm L, Merbach AE. J. Am. Chem. Soc. 2002;124:2042.
[43] Strandberg E, Westlund PO. J. Magn. Reson. Ser. A 1996;122:179.
[44] Kowalewski J, Luchinat C, Nilsson T, Parigi G. J. Phys. Chem. A 2002;106:7376.
[45] Fries PH, Ferrante G, Belorizky E, Rast S. J. Chem. Phys. 2003;119:8636.
[46] Caravan P, Toth E, Rockenbauer A, Merbach AE. J. Am. Chem. Soc. 1999;121:10403.
[47] Seibig S, Toth E, Merbach AE. J. Am. Chem. Soc. 2000;122:5822.
[48] Fries PH, Belorizky E. J. Chem. Phys. 2012;136:074513.
[49] Bertini I, Capozzi F, Luchinat C, Nicastro G, Xia Z. J. Phys. Chem. 1993;101:198.
[50] Alsaadi BM, Rossotti FJC, Williams RJP. J. Chem. Soc. Dalton Trans. 1980;2147.
[51] Peters JA, Huskens J, Raber DJ. Progr. Nucl. Magn. Reson. Spectrosc. 1996;28:283.
[52] Shannon RD, Prewitt CT. Acta Cryst. B 1969;25:925.
[53] Helm L, Merbach AE. Eur. J. Solid State Inorg. Chem. 1991;28:245.
[54] Kowall T, Foglia F, Helm L, Merbach AE. J. Am. Chem. Soc. 1995;117:3790.
[55] Rogers NJ, Finney K-LNA, Senanayake PK, Parker D. Phys. Chem. Chem. Phys. 2016;18:4370.

[56] Sowadski J, Cornick G, Kretsinger RH. J. Mol. Biol. 1978;124:123.
[57] Horrocks WD Jr, Sudnick DR. J. Am. Chem. Soc. 1979;101:334.
[58] Kretsinger RH, Nockolds CE. J. Biol. Chem. 1973;248:3313.
[59] Moews PC, Kretsinger RH. J. Mol. Biol. 1975;91:201.
[60] Lee L, Sykes BDIn: Darnall DW, Wilkins RG, editors. Advances in Inorganic Biochemistry, Volume 2. New York: Elsevier; 1980. p. 183.
[61] Lee L, Sykes BD. Biochemistry 1980;19:3208.
[62] Lee L, Sykes BD. Biochemistry 1981;20:1156.
[63] Sykes BD, Lee L. Biochemistry 1983;22:4366.
[64] Lee L, Sykes BD. J. Magn. Reson. 1980;41:512.
[65] Lee L, Sykes BD. Biophys. J. 1980;32:193.
[66] Capozzi F, Cremonini MA, Luchinat C, Sola M. Magn. Reson. Chem. 1993;31:S118–27.
[67] Bentrop D, Bertini I, Cremonini MA, Forsén S, Luchinat C, Malmendal A. Biochemistry 1997;36:11605.
[68] Bertini I, Janik MBL, Lee Y-M, Luchinat C, Rosato A. J. Am. Chem. Soc. 2001;123:4181.
[69] Allegrozzi M, Bertini I, Janik MBL, Lee Y-M, Liu G, Luchinat C. J. Am. Chem. Soc. 2000;122:4154.
[70] Bertini I, Donaire A, Jiménez B, Luchinat C, Parigi G, Piccioli M, Poggi L. J. Biomol. Nucl. Magn. Reson. 2001;21:85.
[71] Balayssac S, Bertini I, Luchinat C, Parigi G, Piccioli M. J. Am. Chem. Soc. 2006;128:15042.
[72] Barbieri R, Luchinat C, Parigi G. Chem. Phys. Chem. 2004;21:797.
[73] Hus JC, Marion D, Blackledge M. J. Mol. Biol. 2000;298:927.
[74] Bertini I, Longinetti M, Luchinat C, Parigi G, Sgheri L. J. Biomol. Nucl. Magn. Reson. 2002;22:123.
[75] Bertini I, Kursula P, Luchinat C, Parigi G, Vahokoski J, Willmans M, Yuan J. J. Am. Chem. Soc. 2009;131:5134.
[76] Fragai M, Luchinat C, Parigi G, Ravera E. Coord. Chem. Rev. 2013;257:2652.
[77] Keizers PHJ, Saragliadis A, Hiruma Y, Overhand M, Ubbink M. J. Am. Chem. Soc. 2008;130:14802.
[78] Keizers PH, Desreux JF, Overhand M, Ubbink M. J. Am. Chem. Soc. 2007;129:9292.
[79] Bertini I, Calderone V, Cerofolini L, Fragai M, Geraldes CFGC, Hermann P, Luchinat C, Parigi G, Teixeira JMC. FEBS Lett. 2012;586:557.
[80] Cerofolini L, Fields GB, Fragai M, Geraldes CFGC, Luchinat C, Parigi G, Ravera E, Svergun DI, Teixeira JMC. J. Biol. Chem. 2013;288:30659.
[81] Das Gupta S, Hu X, Keizers PHJ, Liu W-M, Luchinat C, Nagulapalli M, Overhand M, Parigi G, Sgheri L, Ubbink M. J. Biomol. Nucl. Magn. Reson. 2011;51:253.

CHAPTER

PARAMAGNETIC RESTRAINTS FOR STRUCTURE AND DYNAMICS OF BIOMOLECULES

9

Hyperfine shifts, residual dipolar couplings and paramagnetic enhancements of the nuclear relaxation rates, experimentally available when the system contains a paramagnetic site, provide a wealth of information on the structure and mobility of these systems. These paramagnetic restraints are increasingly used in structural biology studies, up to the point that diamagnetic complexes are conveniently made paramagnetic through the use of paramagnetic binding tags. In this chapter, we will discuss the advantages provided by these restraints and present some guidelines for their use.

9.1 WHY USING PARAMAGNETIC RESTRAINTS FOR STRUCTURAL BIOLOGY STUDIES?

Determining the structure of a protein, of a nucleic acid or of a complex system composed of many proteins and/or nucleic acid molecules requires the measurement of a large number of geometrical restraints, which all together, allow the reconstruction of the relative position of all nuclei present in the biological system, i.e. the determination of all nuclear coordinates. In case of systems experiencing some degrees of mobility, the latter can also be detected, and the conformational variability of the system can be characterized. The presence of a paramagnetic center in the investigated system causes hyperfine shifts of the nuclear signals (described in Chapter 2), residual dipolar couplings due to partial self-orientation (described in Chapter 3), paramagnetic enhancement of the nuclear relaxation rates, as well as cross correlation effects (described in Chapter 4). All these effects, being related to relative nuclear positions, can be used to obtain structural information and are thus called paramagnetic restraints.

In nature there are many proteins bearing a paramagnetic metal ion: for these proteins some paramagnetic restraints can be directly obtained (depending on the paramagnetic metal) and used for their structural determination. However, the structural information content provided by the paramagnetic restraints can be so accurate and precious, because not otherwise available, that it is becoming more and more frequent to turn a diamagnetic system into a paramagnetic one, by substituting a naturally coordinated diamagnetic metal with a paramagnetic one, by attaching a paramagnetic probe, or by substituting a paramagnetic metal with unwanted properties with a paramagnetic metal providing the desired restraints. Paramagnetic restraints are particularly precious for protein structure determination when, for any reason, the number of available classical restraints, like NOEs, is limited; or when other types of restraints are less informative, because of short range nature, or inaccurate, like in protein–protein complexes or in complexes between proteins and nucleic acids, or in the presence of internal mobility.

NMR of Paramagnetic Molecules. http://dx.doi.org/10.1016/B978-0-444-63436-8.00007-7

9.2 NUCLEAR RELAXATION PARAMETERS AND STRUCTURAL INFORMATION

9.2.1 MOLECULES CONTAINING PARAMAGNETIC CENTERS

When a nucleus is dipole–dipole coupled to a paramagnetic metal ion that is located several bonds away, there are neither contact nor ligand-centered contributions to relaxation. Measurement of nuclear R_{1M} and R_{2M} can provide in these cases, the distances of the nuclei from the metal ion, if the correlation time of the relaxation process is known. In fact, as shown in Eqs. (4.17), (4.18), (4.30), and (4.31), R_{1M} and R_{2M} are proportional to r^{-6}, where r is the nucleus-metal distance. These equations are here reported as Eq. (9.1) to point out this distance dependence. The proportionality constants depend on both the Solomon and the Curie contributions to relaxation, and thus on the electron relaxation time and on the reorientation time of the molecule:

$$R_{1M} = \frac{k_{\text{Solomon}} + k_{\text{Curie}}}{r^6} \qquad R_{2M} = \frac{k'_{\text{Solomon}} + k'_{\text{Curie}}}{r^6} \tag{9.1}$$

$$k_{\text{Solomon}} = \frac{2}{15}\left(\frac{\mu_0}{4\pi}\right)^2 \gamma_I^2 g_e^2 \mu_B^2 S(S+1)\left[\frac{\tau_c}{1+(\omega_I-\omega_S)^2\tau_c^2} + \frac{3\tau_c}{1+\omega_I^2\tau_c^2} + \frac{6\tau_c}{1+(\omega_I+\omega_S)^2\tau_c^2}\right]$$

$$k_{\text{Curie}} = \frac{2}{5}\left(\frac{\mu_0}{4\pi}\right)^2 \frac{\omega_I^2 g_e^4 \mu_B^4 S^2(S+1)^2}{(3kT)^2} \frac{3\tau_{\text{Curie}}}{1+\omega_I^2\tau_{\text{Curie}}^2}$$

$$k'_{\text{Solomon}} = \frac{1}{15}\left(\frac{\mu_0}{4\pi}\right)^2 \gamma_I^2 g_e^2 \mu_B^2 S(S+1)\left[4\tau_c + \frac{\tau_c}{1+(\omega_I-\omega_S)^2\tau_c^2} + \frac{3\tau_c}{1+\omega_I^2\tau_c^2} + \frac{6\tau_c}{1+(\omega_I+\omega_S)^2\tau_c^2} + \frac{6\tau_c}{1+\omega_S^2\tau_c^2}\right]$$

$$k'_{\text{Curie}} = \frac{1}{5}\left(\frac{\mu_0}{4\pi}\right)^2 \frac{\omega_I^2 g_e^4 \mu_B^4 S^2(S+1)^2}{(3kT)^2}\left[4\tau_{\text{Curie}} + \frac{3\tau_{\text{Curie}}}{1+\omega_I^2\tau_{\text{Curie}}^2}\right]$$

$$\tau_c^{-1} = \tau_e^{-1} + \tau_r^{-1} + \tau_M^{-1}$$

$$\tau_{\text{Curie}}^{-1} = \tau_r^{-1} + \tau_M^{-1}$$

Of course, the same relationships hold if the unpaired electron is located on a radical. In the case of lanthanoids and actinoids the term $g_J^2 J(J+1)$ should replace $g_e^2 S(S+1)$, due to the large spin-orbit interactions which couple the orbital L and spin S angular momenta (Chapter 8).

In general the electron relaxation time, τ_e, and sometimes the reorientation time τ_r, (especially in the presence of fast local mobility, which should also be considered together with the overall molecular tumbling, Section 10.2), is unknown, and thus the proportionality constants cannot be easily calculated. In real cases, it is also difficult to predict whether contact contributions and ligand-centered effects are completely absent for nuclei not far from the paramagnetic metal ion. A protocol is available to use nuclear relaxation rates as structural restraints for the determination of solution structures of paramagnetic metalloproteins (Section 9.5) [1–5].

Instead of obtaining the paramagnetic enhancements R_{1M} and R_{2M} from the difference between the observed nuclear relaxation rates and the relaxation rates of the corresponding nuclei in a diamagnetic analog of the protein, an average diamagnetic value can be estimated, even without a diamagnetic reference sample, from the set of the lowest relaxation rates corresponding to nuclei very far from the paramagnetic ion. For the latter nuclei, in fact, due to the proportionality to r^{-6}, R_{1M}, and R_{2M} approach zero. The error introduced in the distances estimated with this approach is negligible as long as the nuclei are subject to a paramagnetic contribution significantly larger than the diamagnetic contribution to nuclear relaxation, i.e. for nuclei relatively close to the paramagnetic center. On the contrary, the error can be large for nuclei relatively far from the paramagnetic center, but not so far that R_{1M} and R_{2M} are already zero. In these cases, the estimated distances can be used only as upper distance limits, which means that the nuclei can be found anywhere up to such distance from the paramagnetic center. In fact, the possible overestimation of the diamagnetic contribution may determine underestimation of the R_{1M} and R_{2M} values, with consequent estimation of metal-nuclear distances longer than in reality.

A two-time-point measurement provides a more accurate way to determine R_{2M} values [2] by performing measurements on both the paramagnetic sample and a diamagnetic analog (Section 12.1.5): in this case, R_{2M} values larger than expected from Eq. (9.1) can be measured for nuclei very far from the metal, so that lower distance limits can only be safely provided to restrain the position of such nuclei [6] (see also later).

As a consequence of the direct proportionality between NMR signal linewidth and nuclear transverse relaxation rate, the effect of the presence of a paramagnetic metal ion within a molecule is that of increasing all nuclear linewidths according to the relationship

$$\Delta\nu = R_{2M}/\pi. \tag{9.2}$$

Therefore, the line broadening of the NMR signals of the nuclei close to the metal ion can be so severe so as to make the signal undetectable with standard experiments. This is why a "blind" sphere is often present around the paramagnetic metal, in which the NMR lines are too broad to be detected (Fig. 9.1). Due to its dependence on R_{2M}, the size of this blind sphere can be very different depending on the molecular size and on the paramagnetic ion present in the system. In fact, R_{2M} depends on the molecular reorientation time and on the electron relaxation time, which in metalloproteins is often smaller than the reorientation time, and thus dominates the correlation time modulating the dipole-dipole interaction. Paramagnetic metal ions can have electron relaxation times spanning from 10^{-8} s to 10^{-13} s (Chapters 7 and 8). Therefore, in macromolecules, the size of this blind sphere strongly depends on the paramagnetic ion. R_{2M} also depends on the square of γ_I of the detected nucleus, so that different nuclei (^{1}H, ^{13}C, ^{15}N, etc...) have different radii of the blind sphere. In particular, on passing from ^{1}H ($\gamma_H = 2.67 \times 10^8$ rad s^{-1} T^{-1}) to ^{13}C ($\gamma_C = 6.73 \times 10^7$ rad s^{-1} T^{-1}) direct detection, the radius of the blind sphere is reduced by a factor $16^{1/6} \approx 1.6$ (Fig. 9.1). In fact, the line broadening is a factor 16 smaller for ^{13}C nuclei than for ^{1}H nuclei placed at the same distance from the paramagnetic ion, and thus the same line broadening for ^{13}C and ^{1}H nuclei is observed at distances $16^{1/6}$ times larger from the metal ion for ^{13}C than for ^{1}H nuclei. Analogously, the radius of the blind sphere of ^{15}N nuclei ($\gamma_N = 2.71 \times 10^7$ rad s^{-1} T^{-1}) is $97^{1/6} \approx 2.1$ smaller than that of ^{1}H nuclei in the same paramagnetic molecule.

In summary, NMR signals may or may not be observable in paramagnetic systems depending on the distance of the corresponding nuclei from the metal site. When observed, the relative R_{1M} and R_{2M} values, being proportional to the inverse sixth power of the nuclear-metal distance, can be used as structural restraints. However, R_{1M} and R_{2M} are usually measurable only in a spherical shell from the metal where the effect is not too weak to be negligible. The distance from the paramagnetic ion and the width of this shell depend on the type of the detected nuclei, due to their different γ_I. Therefore, measurements performed on both ^{1}H and ^{13}C nuclei may be convenient to obtain complementary distance restraints

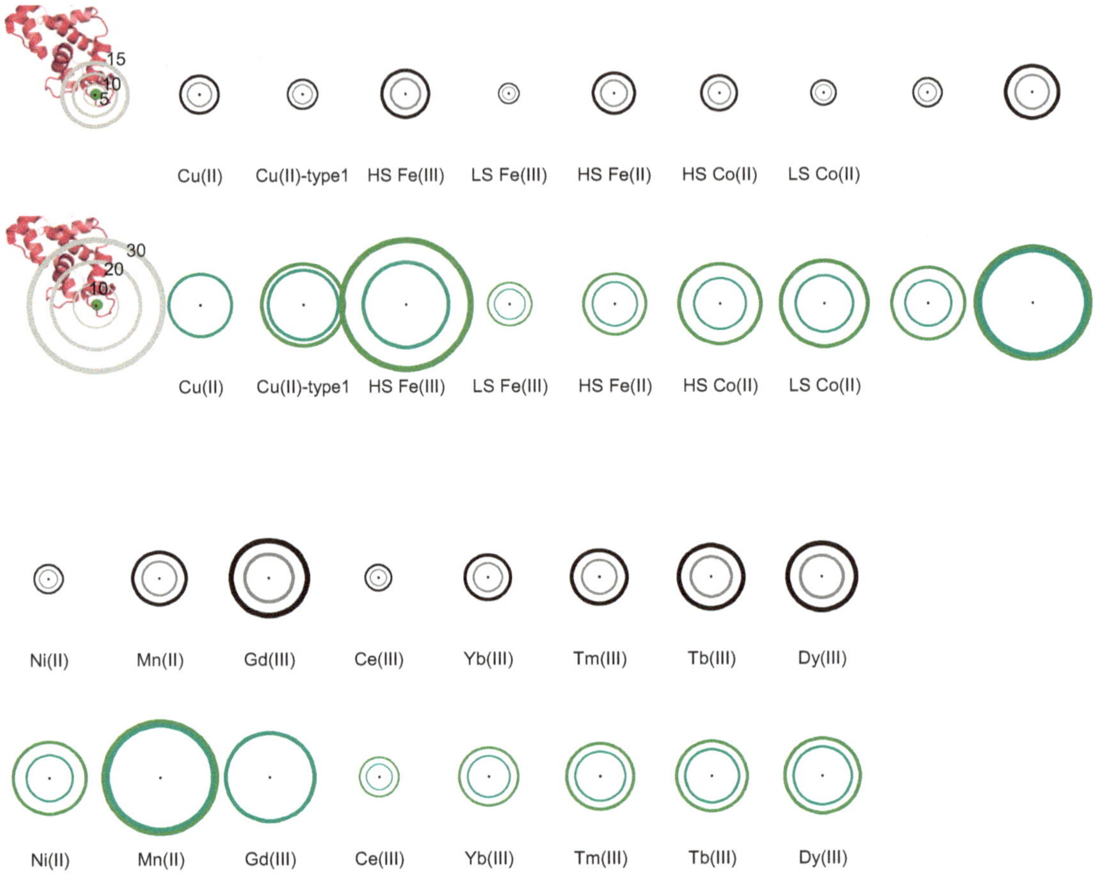

FIGURE 9.1 Blind spheres around the different paramagnetic metal ions for ^{1}H (in *black*) and ^{13}C (in *gray*) nuclei and distances with R_{1M} of 0.1 s^{-1} for ^{1}H (in *green*) and ^{13}C nuclei (in *dark green*). The radius of the blind sphere is fixed at the distance causing a paramagnetic broadening of 100 Hz, for a protein with reorientation time of 30 ns, at 900 MHz. In the first panels the spheres at 5, 10, 15, 20 and 30 Å from the metal are shown.

[2]. The dependence on the sixth power makes R_{1M} and R_{2M} data accurate restraints to determine metal-nuclear distances. If pseudocontact shifts are also available for the same molecule, the two types of restraints can be profitably used together, as both of them depend on the position of the nuclei in a metal-centered frame. Even outside the detectability shell, i.e. if signals are too broad to be detected or if R_{1M} and R_{2M} are so small to be negligible, these data can provide precious structural restraints because they can allow the setting of nuclear upper distance limits or lower distance limits, respectively, from the paramagnetic center (see later). Furthermore, paramagnetic relaxation enhancements can be very useful for the study of protein–protein interactions. In fact, in the presence of a paramagnetic metal in one protein, the observation of paramagnetic NMR line broadening for some residues of the partner protein can permit to determine its orientation in the complex and the residues involved in the interaction surface [7].

Radicals are often used as relaxation agents to obtain structural information. Because their electron relaxation times is about 10^{-7} s, the correlation time for the dipole–dipole relaxation is long, and thus nuclei that are closer than 15 Å from the paramagnetic centers may not be detectable due to the excessive line broadening. R_{2M} data are obtainable on nuclei that are as far as 15–25 Å away from the paramagnetic probe [8]. Therefore, radicals are long-distance relaxation probes. This property has been used, for instance, to obtain structural information in protein–RNA complexes using distance restraints extracted from intermolecular electron–proton dipolar relaxation induced by nitroxide spin-labeled RNA [9]. The presence of extensive signal line broadening and attenuation, for nuclei at distances shorter than 10–15 Å from the radical, provides upper distance limits. Since a number of nitroxide spin labels can be introduced, one at a time, in different protein positions [8,10,11], a network of distance restraints can be obtained and used to recover the protein structure. Caution should however be used in the analysis of the data due to the possible mobility of the attached radicals [12,13]. A possible way to account for the flexibility of the spin labels consists in averaging over a large ensemble of possible conformers [14].

As already seen, in macromolecules, when the reorientation time τ_r is large with respect to the electron relaxation time τ_e, R_{1M} and R_{2M} strongly depend on τ_e. Fig. 9.2 shows R_{1M} and R_{2M} calculated for a proton at distance of 5, 10, 15, or 20 Å from the paramagnetic center, with $S = 1/2$, as a function of the electron relaxation time, τ_e, in a macromolecule with reorientation time τ_r of 10 ns and at a proton Larmor frequency of 900 MHz. The largest R_{1M} is observed for τ_e of the order of 0.2–0.3 ns. The lower detectability limit, which can be fixed at 0.1 s^{-1}, indicates that nonnegligible R_{1M} effects can be observed up to about 20 Å. Fig. 9.3 shows the radius of the ^1H blind sphere as a function of the electron relaxation times, calculated in the same condition as above and assuming a detectability limit of 314 s^{-1}, corresponding to a paramagnetic line width of 100 Hz.

The typical line broadening and R_{1M} values for protons 10 Å away from the various paramagnetic ions are shown in Fig. 9.4, for a macromolecule with τ_r of 10 ns, at 900 MHz of proton Larmor frequency. They result very large for manganese(II), gadolinium(III), and some other lanthanoids

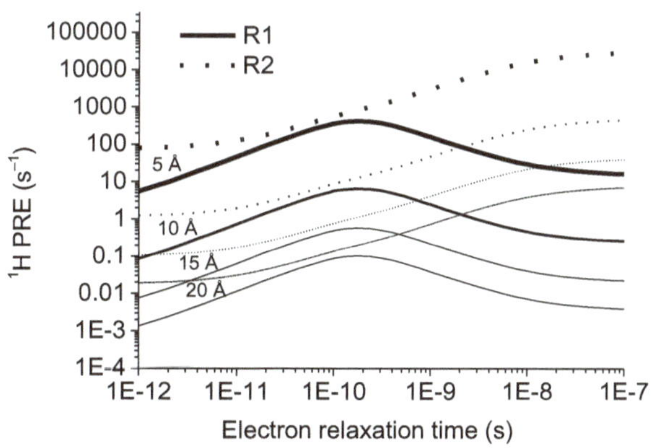

FIGURE 9.2 R_{1M} and R_{2M} for a proton at 5, 10, 15, or 20 Å from the paramagnetic metal ion as a function of the electron relaxation time, τ_e, for a system with $S = 1/2$, $\tau_r = 10$ ns, at 900 MHz proton Larmor frequency. R_{2M} is sizably larger than R_{1M} for low τ_e values because of the non-dispersive Curie contribution, and for high τ_e values because of the non-dispersive term in the Solomon equation.

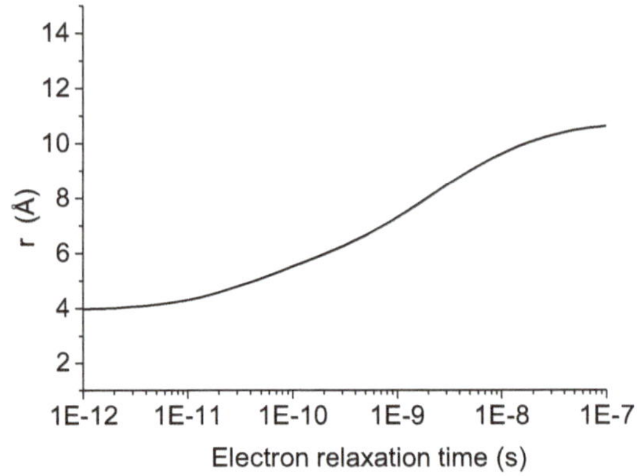

FIGURE 9.3 Radius of the ^{1}H blind sphere as a function of the electron relaxation time, τ_e, assuming a detectability limit of 314 s^{-1}, corresponding to a paramagnetic line width of 100 Hz, for a system with $S = 1/2$, τ_r = 10 ns, at 900 MHz proton Larmor frequency.

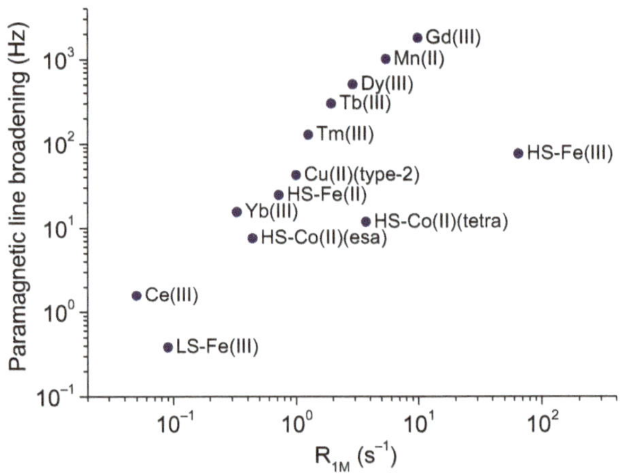

FIGURE 9.4 ^{1}H paramagnetic line broadening and R_{1M} values, for a proton 10 Å away from the different metal ions, calculated using typical values for the electron relaxation time (τ_r = 10 ns, at 900 MHz of proton Larmor frequency).

like terbium(III), thulium(III), and dysprosium(III), and are large also for high spin iron(III), type-2 copper(II), high spin iron(II), ytterbium(III). The radius of the ^{1}H blind sphere is also shown in Fig. 9.5 for the various paramagnetic metal ions, assuming typical values for their electron relaxation times. In agreement with experimental measurements, gadolinium(III) has a bind sphere of about 20 Å,[6] thulium(III) of more than 10 Å [15], copper(II) of almost 10 Å.[2] Due to the reduction of a factor 1.6 in the size of the blind sphere, carbon detection permits to recover all signals of ^{13}C nuclei farther than 9 Å from thulium(III) [15] and all signals in a copper protein except those of the four copper(II) ligands, broadened beyond detection because of contact contributions to relaxation [16]. In cerium(III) metalloproteins, the much smaller blind sphere can allow recovering all signals including those in the

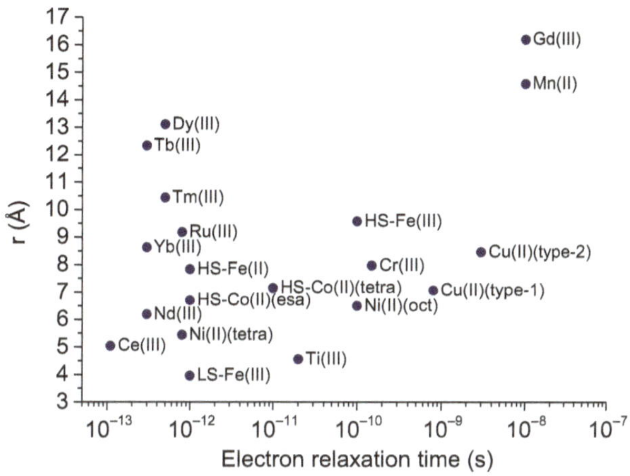

FIGURE 9.5 Radius of the ^{1}H blind sphere for the different metal ions, using typical values for the electron relaxation time, τ_e, and assuming a detectability limit of 314 s^{-1}, corresponding to a paramagnetic line width of 100 Hz, for a system with τ_r = 10 ns, at 900 MHz proton Larmor frequency.

first coordination sphere of the metal ion [17]. A major problem in the solution structure determination of paramagnetic metalloproteins by NMR actually is the characterization of the metal binding site. Paramagnetic relaxation enhancements can be exploited to identify the residues coordinating a paramagnetic metal ion and their coordination geometry with a ^{13}C-^{13}C COSY experiment acquired with a specific combination of acquisition times and recycle delays [18].

R_{1M}-derived distance restraints are used for protein structural calculations since 1996, in the presence of either one or multiple paramagnetic centers. First examples were oxidized high-potential iron-sulfur proteins [19] and ferredoxins [20].

In systems containing a gadolinium(III) ion, R_{2M}-derived distance restraints can be determined. The restraints can be obtained according to the following three classes [6,21]:

1. residues for which the intensity of the NMR signal of the paramagnetic sample is reduced more than fivefold with respect to that of the diamagnetic sample, and residues for which broadening occurs beyond detection, are too close to the paramagnetic center for an accurate distance estimation, and upper distance limits from the metal ion are set to 20 Å;
2. residues for which the intensity of the NMR signal of the paramagnetic sample is reduced less than fivefold with respect to that of the diamagnetic sample, but more than 1.2 fold, are restrained to distances from the paramagnetic center determined from Eq. (9.1), with an error of ± 3–4 Å;
3. residues for which the intensity of the NMR signal of the paramagnetic sample is reduced less than 1.2 fold with respect to that of the diamagnetic sample are too far from the paramagnetic center for an accurate distance estimation, and lower distance limits from the metal ion are set to 30 Å.

Along the same lines, when paramagnetic spin labels were used, residues with intensity reduced to more than a factor 10 can be restrained with upper distance limits of 14 Å and residues with intensity reduced to less than a factor 1.05 can be restrained with lower distance limits of 22 Å [22].

9.2.2 SOLVENT PARAMAGNETIC MOLECULES

Different structural information, still relying on the paramagnetic enhancement of the relaxation rates, can be obtained by using inert and freely soluble paramagnetic complexes as cosolvent in a solution containing the biological system of interest. The paramagnetic enhancements measured for the nuclei of the macromolecule depend on the surface accessibility of the latter by the paramagnetic complexes and on how close to the surface the detected nuclei are (Fig. 9.6). In fact, nuclei located close to the surface experience a strong relaxation enhancement, whereas nuclei located well inside the macromolecule have a very small enhancement. Therefore, these experiments permit to probe surface accessibility and to determine the immersion depth of nuclei from the surface. They are particularly useful to determine the interaction site in protein-protein complexes: the nuclei in the binding region, in fact, experience a much smaller relaxation enhancement when the complex is formed than in the free proteins. Gadolinium(III) complexes (like Gd-DTPA-BMA, diethylenetriamine pentaacetic acid-bis(methylamide)), and stable nitroxide radicals are mostly used as paramagnetic probes [23–32]. They must not bind specifically on the surface of the macromolecules, but rather diffuse freely on the surface. Their concentration can be varied to tune the relaxation enhancements in order to better probe different immersion depths.

If the diffusion is completely free, the interaction between the macromolecular nuclei and the paramagnetic molecule is described by the outer-sphere model described in Chapter 6, and thus the paramagnetic enhancement depends on d^{-3}, where d is the distance of closest approach between the paramagnetic center and the nuclei, approximately given by the distance of the nucleus from the surface plus the radius of the paramagnetic complex. This model holds for flat surfaces and large spherical surfaces. On the other hand, in the presence of deep clefts within the macromolecular surface, which allows only for a single paramagnetic complex to get close to the detected nucleus, the paramagnetic enhancement depends on d^{-6}, as expected from the dipole–dipole coupling between electron and nucleus, and described by the Solomon equation, with d now representing the metal-nuclear distance r [Eq. (9.1)]. In practice, lower distance limits can be derived by imposing a d^{-6} dependence: this ensures that the real distances from the closest paramagnetic centers are equal or larger than the calculated d values [33]. In any case, the nonexchanging nuclei located on the surface of the macromolecule are necessarily those experiencing the maximum paramagnetic enhancement, and thus they can be easily identified.

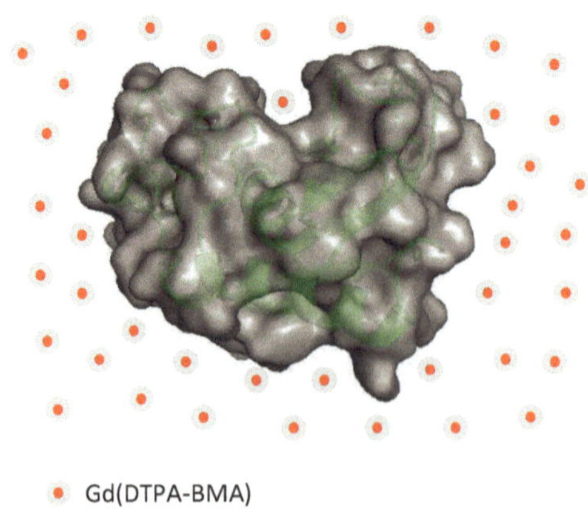

FIGURE 9.6 A molecule is immersed in a solution containing paramagnetic complexes, like Gd(DTPA-BMA). The paramagnetic relaxation enhancements depend on the nuclear distance from the molecular surface.

9.3 HYPERFINE SHIFTS AND STRUCTURAL INFORMATION

In paramagnetic metalloproteins many nuclei can be relatively close to the paramagnetic center as a result of the folding of the protein chain. Such nuclei, if separated from the metal ion by several chemical bonds, can experience only pseudocontact shifts. We are in the lucky situation of having many of these nuclei and therefore many pseudocontact shift values. Pseudocontact shifts can be determined from the difference of the chemical shifts between the paramagnetic protein and a diamagnetic analog, that can be obtained by replacing the paramagnetic ion with a diamagnetic one, i.e. paramagnetic lanthanoids can be replaced with lanthanum(III) or calcium(II), cobalt(II) with zinc(II), etc., or by changing the oxidation state of the paramagnetic metal to a diamagnetic state, i.e. low spin iron(III) can be reduced to low spin iron(II). This determines very minor structural perturbations, which can be determined only a posteriori, usually negligible outside the first coordination sphere of the metal.

If a structure or a structural model is available, the spherical coordinates for each nucleus are known in any arbitrary coordinated system. At this point the unknowns of Eq. (2.46) are $\Delta\chi_{ax}$, $\Delta\chi_{rh}$ and three Euler angles which define the principal axes of the magnetic susceptibility anisotropy tensor with respect to the arbitrary coordinated system. With a minimum of five experimental pseudocontact shifts, the above parameters can be found. This calculation can be performed, for instance, with the program FANTEN available at http://fanten-enmr.cerm.unifi.it:8080/. Besides the structural model of the protein, this analysis requires the assignment of the NMR signals and the chemical shifts of the paramagnetic metalloprotein and of an analogous diamagnetic compound, in order to determine the hyperfine (pseudocontact) shifts from their difference.

In order to avoid considering pseudocontact shifts which are incorrect due to contact contributions to the chemical shifts, all nuclei close in chemical bonds (with less than 5 bonds between nucleus and metal ion, in the absence of π bonds) and/or with direct orbital overlap to the paramagnetic center must be set aside to calculate the anisotropy tensor. The contact contributions of these nuclei can then be determined from the difference between the pseudocontact shifts calculated from the determined anisotropy tensor and the measured hyperfine shifts (see later).

The fact that pseudocontact shifts do not depend on the detected nucleus simplifies the procedure of assignment of the NMR peaks of the paramagnetic complex in heteronuclear experiments, once the peaks of the diamagnetic analog have been assigned. In fact, the signal of nuclei not close to the metal ion have relatively small pseudocontact shifts, and thus they are close to the diamagnetic ones, but slightly shifted along the diagonal in, for example, a ^{1}H-^{15}N HSQC spectrum (Section 12.4). In fact, the shift of the nitrogen nucleus in an amide pair should be very similar to that of the proton nucleus, due to their vicinity with respect to the position of the metal ion. The assignment can thus be easily transferred from the diamagnetic spectrum to the paramagnetic one. From the first estimate of the tensor calculated using the pseudocontact shifts so determined and the available structural model, predictions can be done for the other nuclei (Fig. 9.7). By comparing observed and calculated shifts, a large number of further assignments can be performed, in an iterative fashion. Furthermore, if a structural model is available, the pseudocontact shifts measured from several paramagnetic metals alternatively present in the complex can in principle be used to assign the NMR signals even in the absence of the assignment for the diamagnetic protein [34,35]. In fact, since all pseudocontact shifts must agree with a single magnetic susceptibility anisotropy tensor per metal, a unique correspondence can be determined between the pseudocontact shifts relative to a given nucleus and the coordinates of a nucleus in the structural model.

Only some metal ions, as discussed in Section 2.6, provide $\Delta\chi$ values sufficiently large that pseudocontact shifts can be measured with a good accuracy. Gadolinium(III), manganese(II), and type-2 copper(II), for instance, usually provide negligible pseudocontact shifts; cobalt(II), iron(II), and iron(III)

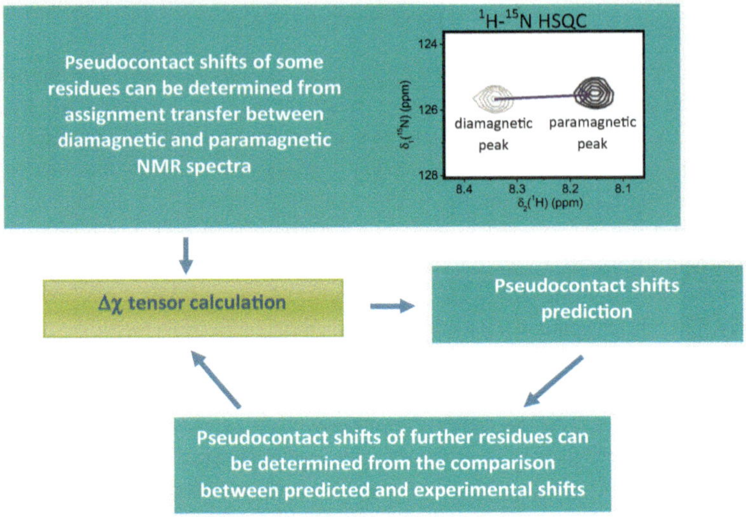

FIGURE 9.7 Scheme for the collection of the pseudocontact shifts in an iterative fashion through comparison between the spectra of paramagnetic and diamagnetic protein analogs.

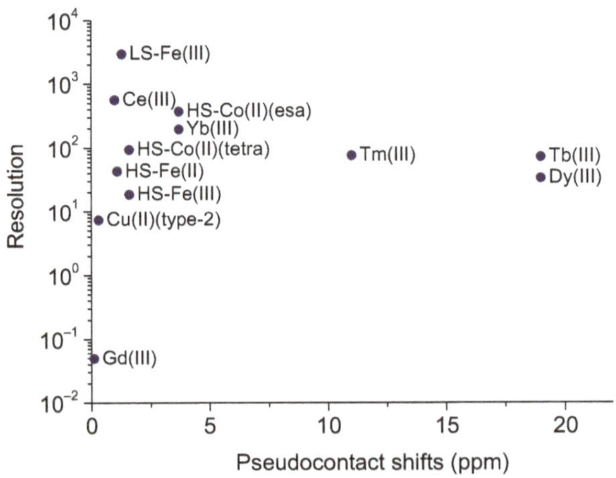

FIGURE 9.8 Resolution versus the typical maximum pseudocontact shift value that can be obtained for a proton 10 Å away from the different paramagnetic metal ions ($\tau_r = 10$ ns, 900 MHz proton Larmor frequency). The resolution is calculated as the frequency shift corresponding to the pseudocontact shift divided by the ^{1}H line broadening, obtained as R_{2M}/π.

provide detectable values, and some lanthanoids, like terbium(III), thulium(III), and dysprosium(III), provide very large pseudocontact shifts. Fig. 9.8 shows the resolution (defined as pseudocontact shift divided by R_{2M}) versus the typical maximum pseudocontact shift value that can be obtained for a proton 10 Å away from the different paramagnetic metal ions. In Table 9.1, typical values of the axial anisotropy of the magnetic susceptibility tensors are reported for the different paramagnetic metal ions.

Table 9.1 Values of the Magnetic Susceptibility per Molecule, as Calculated with Eq. (1.36) at T = 298 K, and Typical Values of the Axial Anisotropy of the Magnetic Susceptibility Tensors for the Different Paramagnetic Metal Ions

	χ (10^{-32} m^3)	$|\Delta\chi_{ax}|$ (10^{-32} m^3)
Ti(III)	2.6	0.4
VO^{2+}	2.6	0.1
Cr(III)	13	< 0.1
Mn(II)	31	< 0.1
Mn(III), high spin	21	0.5
Fe(III), high spin	31	3
Fe(III), low spin	2.6	2.4
Fe(II), high spin	21	2.1
Co(II), high spin, hexacoord.	13	7
Co(II), high spin, tetracoord.	13	3
Co(II), low spin	2.6	1
Ni(II)	7	2
Cu(II)	2.6	0.6
Ce(III)	5.6	2
Pr(III)	11	3.5
Nd(III)	11	1.7
Sm(III)	0.6	0.2
Eu(III)	0[a]	2.4
Gd(III)	55	0.2
Tb(III)	83	40
Dy(III)	99	35
Ho(III)	99	20
Er(III)	80	12
Tm(III)	50	25
Yb(III)	18	8

[a]*Eu(III) is paramagnetic due to the excited states.*

Once the pseudocontact shifts have been measured, they can be used as structural restraints as they depend on the spherical coordinates of the nuclei in the common frame of the magnetic susceptibility anisotropy tensor. Of course, even if the latter tensor is known, both in terms of orientation and anisotropy values, there are infinite positions for a nucleus in agreement with the measured pseudocontact shift value (Fig. 9.9). However, if pseudocontact shifts are measured for the same nucleus after alternative

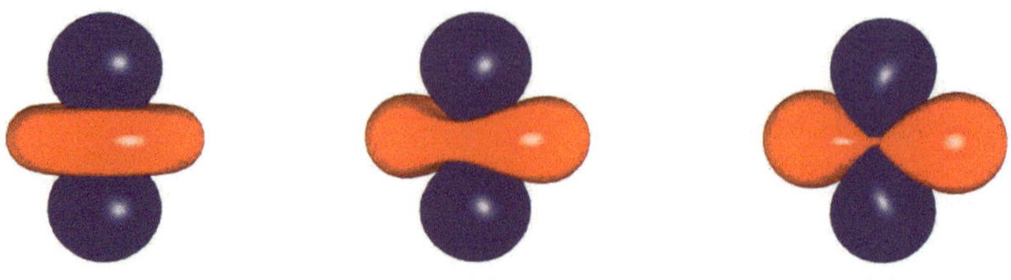

(a) (b) (c)

FIGURE 9.9 Surfaces of constant pseudocontact shift (in absolute value) for (a) an axial susceptibility tensor ($\Delta\chi_{rh}=0$) or a rhombic susceptibility tensor with (b) $\Delta\chi_{rh}=\Delta\chi_{ax}/3$ or (c) $\Delta\chi_{rh}=2\Delta\chi_{ax}/3$. The blue lobes correspond to positive pseudocontact shift values, and the red lobes correspond to negative values.

replacement of different paramagnetic metals (with different, nonproportional, magnetic susceptibility anisotropy tensors), this degeneracy is largely reduced (with two metals) or almost completely removed (with three or more metals, only two positions being in this case in agreement with all data).[1]

Pseudocontact shifts were observed in low spin ferric heme containing proteins [36,37], high spin cobalt(II)-containing proteins [38,39], and paramagnetic lanthanoids systems [40–42] since many decades. They were used for the first time to refine a protein structure in solution starting from a crystallographic structure in 1995 [43], and as restraints into solution structure calculation protocols in the following year [44]. In order to avoid that ligand centered contributions undermine the treatment, an indetermination proportional to the pseudocontact shift values may be assigned to each experimental pseudocontact shift value. In fact, ligand centered contributions may be effective for nuclei close to the metal ion, which experience the largest pseudocontact shifts. As described in Section 2.2.2.2, further contributions to the hyperfine shifts of nuclei close in space to the paramagnetic metal ion, besides the contact shift due to unpaired electron spin density delocalized through chemical bonds and the pseudocontact shift provided by Eq. (9.1), can in fact arise from the ligand-centered pseudocontact shifts and the "through-space" contact shifts resulting when molecular orbitals containing unpaired spin density are overlapped to the orbitals of nuclei, several chemical bonds away from the paramagnetic ion. The separation between contact, "through-space" contact and ligand-centered pseudocontact shifts is very difficult, and may be attempted through density functional calculations.

As already anticipated, once the $\Delta\chi$ values and the directions of the molecular axes are obtained, the metal centered pseudocontact shifts can be calculated even for nuclei that experience also contact and ligand centered pseudocontact shifts. In this way, the contact plus ligand centered pseudocontact shift (the latter is usually small [45]) can be calculated as the difference from the experimental hyperfine shift [46,47].

Contact shifts also contain structural information, as outlined in Section 2.4. However, it is hidden inside the particular mechanism of delocalization of the unpaired electron(s), so that general protocols for the use of contact shifts as structural restraints are not available. Only parametric heuristic equations have been derived for particular cases (Chapter 7). The presence of contact contribution to the hyperfine shift is however in most cases an evidence of through-bond connectivity to the metal ion, and can allow the unambiguous identification of the metal ligands [17].

---

[1]This two-fold degeneracy cannot be removed under any circumstance, as it originates from the quadratic form of the function.

9.4 PARAMAGNETIC RESIDUAL DIPOLAR COUPLINGS AND STRUCTURAL INFORMATION

Residual dipolar couplings are nowadays widely used restraints in structural determination protocols as well as in the analysis of protein mobility, when arising as a consequence of partial orientation induced by external devices present in solution together with the macromolecule under investigation. Paramagnetic residual dipolar couplings (described in Chapter 3) have the same functional form of diamagnetic residual dipolar couplings, and thus contain the same structural information, but have two main advantages with respect to the latter:

1. There is no risk that a perturbation in the protein structure occurs due to possible interactions with the orienting media,
2. The alignment tensor responsible for the paramagnetic residual dipolar couplings is the same as that responsible for pseudocontact shifts, and thus it can be well determined from the latter restraints (Sections 9.3 and 9.5). This holds when paramagnetic residual dipolar couplings are determined from the difference between the ^1J of nuclear pairs in the paramagnetic sample and in a diamagnetic analog, and not from the measurements at two different magnetic fields (Section 3.4).

On the other hand, paramagnetic residual dipolar couplings have the disadvantage that the J-split signal components can be broadened by paramagnetism, and thus measurements can be less precise with respect to diamagnetic complexes, or even impossible in the blind sphere of the paramagnetic metal. However, since R_{2M} decreases with r^{-6}, this effect is negligible far from the paramagnetic ion, i.e. for distances at which the paramagnetic broadening is smaller than 1 Hz (Chapter 7). Furthermore, ^{13}C direct detection experiments can be used to measure residual dipolar couplings with a good precision also for very broad ^{1}H signals [48–50].

Residual dipolar couplings are very useful structural restraints because they depend on the orientation of the internuclear vectors in a common frame defined by the magnetic susceptibility anisotropy tensor. Analogously to what happens for pseudocontact shifts, even if the latter tensor is known, both in terms of orientation and anisotropy values, there are infinite orientations for a nucleus pair in agreement with the measured value (Fig. 9.10). Again, data referred to different tensors corresponding to different

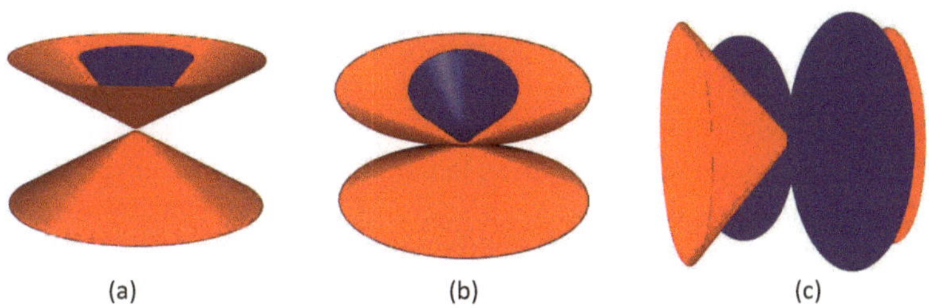

FIGURE 9.10 Surfaces of constant residual dipolar couplings (in absolute value) for (a) an axial susceptibility tensor ($\Delta\chi_{rh}$ =0) or a rhombic susceptibility tensor with (b) $\Delta\chi_{rh}$ =$\Delta\chi_{ax}$/3 or (c) $\Delta\chi_{rh}$ =2$\Delta\chi_{ax}$/3. The different colors of the surfaces correspond to different signs of the residual dipolar couplings. The surfaces indicates the directions of the vectors connecting one nucleus positioned in the origin to the coupled nucleus.

paramagnetic metals permit to largely reduce this degeneracy, up to a single direction (and the opposite one) with three or more tensors.

Analogously to pseudocontact shifts, if the molecular structure or a structural model is available, the α and β angles in Eqs. (3.14) and (3.17) can be determined in any arbitrary coordinate system, and thus $\Delta\chi_{ax}$, $\Delta\chi_{rh}$ and the three Euler angles which define the principal axes of the anisotropy tensor with respect to the arbitrary coordinate system can be found with a minimum of five experimental data. This calculation can be performed with the program FANTEN (http://fanten-enmr.cerm.unifi.it:8080/) to obtain a tensor in best agreement with the residual dipolar couplings alone or with both residual dipolar couplings and pseudocontact shifts simultaneously.

The relative magnitude of the paramagnetic residual dipolar couplings for the different metals parallels that of the pseudocontact shifts, due to the dependence of both data on the same anisotropy values. This means that the larger the maximum value of the pseudocontact shifts of nuclei in a spherical surface at fixed distance from the metal ion, the larger the maximum value observed among all possible orientations of the nuclear pairs (in the absence of mobility, Section 9.6); it does not mean that a large pseudocontact shift of a given nucleus corresponds to a large residual dipolar coupling for a pair containing that nucleus, because these two quantities depend on different structural coordinates.

Detectability of paramagnetic residual dipolar couplings is possible only outside the blind sphere of the metal ion, i.e. for nuclei above a minimum distance from the metal. Differently from pseudocontact shifts, in fact, their detectability is not limited to nuclei inside a spherical shell, because they do not decrease with increasing distances from the metal, but have the same magnitude for the same orientation of the nuclear pair, independently from its distance from the paramagnetic center. Therefore, residual dipolar couplings can be theoretically detected for nuclei above a minimum distance from the metal up to infinity.

The importance of paramagnetic residual dipolar couplings, measured for the protein cyanomet-myoglobin [51], for obtaining structural information, was first outlined in 1995; they were first used for solution structure determination in 1998 for the paramagnetic protein cytochrome b_5 [52].

9.5 PROTOCOLS FOR STRUCTURAL DETERMINATION THROUGH PARAMAGNETIC DATA

Pseudocontact shifts, paramagnetic residual dipolar couplings and paramagnetic relaxation enhancements have been experimentally demonstrated to be consistent within one another and with the diamagnetic restraints (NOEs and dihedral angles determined from ^3J values), and they can thus be collectively used to determine the structure of a protein or to validate or refine an available model. The paramagnetic restraints, although they may in principle be able by themselves to determine the correct fold of the protein if acquired for a large number of nuclei and in the presence of many different paramagnetic metals [53–55], are more profitably used in conjunction with other restraints to improve both the precision and the accuracy of the solution structure of the investigated protein [50,56–58] when ab initio structure calculations are performed. On the other hand, few additional restraints besides the paramagnetic ones can be enough for protein refinement and validation of available models. Finally, in the case of rigid macromolecules or complexes composed of many subunits, when the structures of the latter are known, the paramagnetic restraints can accurately assemble the system, i.e. provide the relative position and orientation of all subunits. The case of systems with internal flexibility is more difficult and will be discussed in Section 9.6.

9.5.1 AB INITIO PROTEIN STRUCTURE DETERMINATION

Protocols for the use of the paramagnetic restraints have been implemented in the two most used programs for the solution structure determination of proteins, Xplor-NIH [5,59] and CYANA [1,50,60] (in the former program inclusion of the paramagnetic data is implemented in the distributed program; in the latter case, the PARAMAGNETIC CYANA version of the program is needed; a patch is downloadable at www.cerm.unifi.it/software-for-download/software-paramagnetic-cyana by owners of a valid CYANA licence). In both programs, the paramagnetic metal and its magnetic susceptibility anisotropy tensor are represented by a pseudoresidue, which is introduced in the protein sequence together with the protein residues, usually at the end of the chain (Fig. 9.11). This pseudoresidue is composed of one pseudoatom, representing the metal ion and the origin of a reference frame, and three further pseudoatoms positioned so that the vectors passing from the first pseudoatom and each of these three pseudoatoms form angles of 90 degrees; these pseudoatoms thus report on the directions of the magnetic susceptibility anisotropy tensor axes. The van der Waals radius of these pseudoatoms is set to zero. In CYANA the pseudoresidue must be separated from the protein residues by a tether of dummy residues with zero van der Waals radius, in order to avoid constraining it close to the last protein residue. As many dummy residues as needed must be chained one after another to span the whole space occupied by the system, and thus allow the pseudoresidue to freely position itself in space (Fig. 9.11).

In case that multiple paramagnetic metals are placed in the system to collect different sets of paramagnetic restraints, an analogous number of these pseudoresidues should be used; if the metals are alternatively placed in the same position, distance restraints can be added to force the pseudoatoms

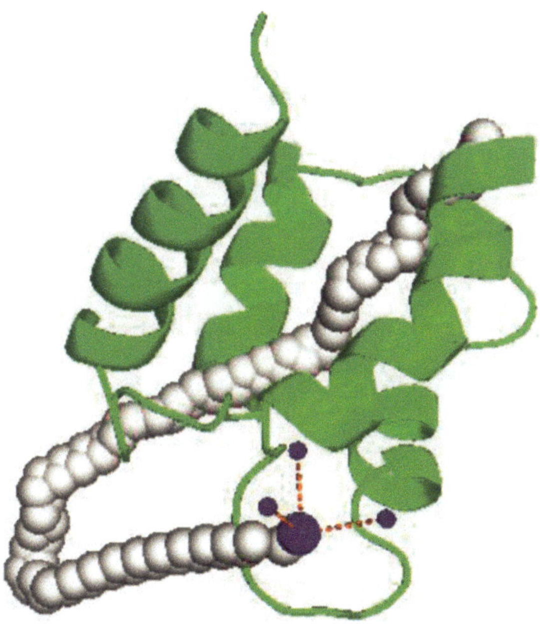

FIGURE 9.11 A pseudoresidue composed of four atoms (in *blue*) representing the position of the metal and the direction of the magnetic susceptibility anisotropy tensor is added to the protein sequence, and connected through dummy residues (in *white*) in order to allow the pseudoatom to be freely placed.

representing the metal ions to stay in the same position. Of course, this is of help to determine a more precise position of the metal ions, and in turn, of all other atoms.

Pseudocontact shifts and residual dipolar couplings are usually provided as restraints as such. From the available model of the protein, the coordinates of the nuclei and of the metal ion as well as the orientation of the magnetic susceptibility tensor corresponding to the metal (determined by the orientation of the pseudoresidue described earlier) are extracted, and the values of pseudocontact shifts and residual dipolar couplings are thus calculated and compared to the experimental ones. Of course the calculations also require the knowledge of the $\Delta\chi_{ax}$ and $\Delta\chi_{rh}$ values. The latter can be estimated initially from literature values determined for the same metal ions, or can be determined through a fit (e.g., with FANTEN) of the pseudocontact shifts/residual dipolar couplings to an available model, for instance determined without the inclusion of the paramagnetic restraints. The program used for structure calculation (Xplor-NIH or CYANA) then moves the coordinates of the protein nuclei in order to minimize the discrepancy between calculated and experimental data. In this way a new model for the protein structure becomes available, and it must be used for a new estimate of the $\Delta\chi_{ax}$ and $\Delta\chi_{rh}$ values. This procedure is repeated until convergence (Fig. 9.12).

If pseudocontact shifts and residual dipolar couplings are measured for the same paramagnetic molecule and upon subtraction of the diamagnetic contribution to chemical shifts and ^1J couplings, respectively, they depend on the same $\Delta\chi_{ax}$ and $\Delta\chi_{rh}$ values (as previously outlined and reported in Section 3.3). This is correct as long as dynamic frequency shifts can be neglected (Sections 4.9 and 12.4). In this case, the $\Delta\chi_{ax}$ and $\Delta\chi_{rh}$ parameters are more conveniently determined from the fit of the pseudocontact shifts to the protein structure and used for protein structure determination also for the residual dipolar couplings restraints. In fact, the pseudocontact shifts can more accurately determine the values of the $\Delta\chi_{ax}$ and $\Delta\chi_{rh}$ parameters as they depend much less than residual dipolar couplings on possible local motions in the protein and on the local structural details (which may not be very accurate). Indeed, if nuclei are placed relatively far from the paramagnetic metal, a small indetermination in their position does not change significantly the metal-nuclear distances and the values of the other structural parameters entering Eq. (2.46), so that the back-calculated $\Delta\chi_{ax}$ and $\Delta\chi_{rh}$ values are not significantly affected by such inaccuracies. On the contrary, a small indetermination in the relative position of coupled nuclei

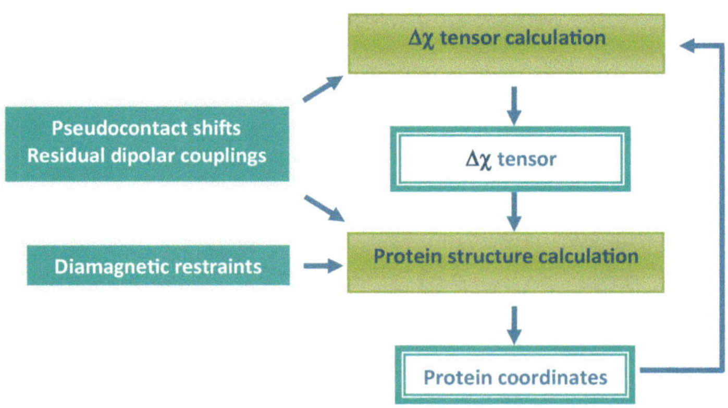

FIGURE 9.12 Protein coordinates are determined by cycling between the programs for anisotropy tensor calculation and protein structure determination, until convergence is reached.

may change sizably the values of the α and β angles in Eq. (3.14), due to the proximity of these atoms, and this has a large effect on the back-calculated $\Delta\chi_{ax}$ and $\Delta\chi_{rh}$ values.

Differently from pseudocontact shifts and residual dipolar couplings, R_{1M} and R_{2M} values are usually not used as such but are converted into distance restraints. This is done because the experimental uncertainty on relaxation rates is roughly proportional to the rates themselves (Section 9.2.1). A direct refinement would tend to overestimate the contribution of large rates, i.e. those of nuclei close to the paramagnetic center, with respect to the contributions of nuclei farther away, because possible discrepancies between calculated and experimental values would be much larger for larger rates. The use of distances is less biased, as they depend on the inverse sixth root of the rates. Conversion of paramagnetic relaxation enhancements in distance restraints is done after evaluation of the constant $k_{\text{Solomon}} + k_{\text{Curie}}$ appearing in Eq. (9.1) from a preliminary structural model. This model is used to estimate the metal-nuclear distances, which are provided to a minimization program to determine the best fit values of $k_{\text{Solomon}} + k_{\text{Curie}}$ against the experimental R_{1M} and R_{2M} data. Such constant is then employed to obtain upper and/or lower distance limits for the metal-nuclear distances from the experimental R_{1M} and R_{2M} data, to be included as restraints in the calculations. The program used for structure calculation then provides a new model for the protein that can be used to determine a more accurate estimate of the $k_{\text{Solomon}} + k_{\text{Curie}}$ constant, and the procedure is repeated iteratively. Alternatively the $k_{\text{Solomon}} + k_{\text{Curie}}$ constant can be evaluated from the estimated correlation times of the system, if available from other sources.

To be noted that when R_{1M} and R_{2M} data are close to zero (within the experimental error) the corresponding nuclei must be relatively far from the metal, so that structural restraints can be set as lower distance limits from the metal; analogously, when pseudocontact shifts are close to zero (within the experimental error) they should also be provided as structural restraints, because still the data contain the information that the corresponding nuclei are far from the metal ion or close to the magic angle.

Restraints are included in structure calculation programs by minimizing a target function defined as:

$$TF = \sum_i w_i[\max(|X_{i,\text{obs}} - X_{i,\text{calc}}| - \text{tol}_i, 0)]^2 \tag{9.3}$$

where the index i runs over the experimental data $X_{i,\text{obs}}$, tol_i indicates the tolerance on the ith restraint, $X_{i,\text{calc}}$ are the back-calculated data, determined from the present structural model, and w_i is the weight (force constant) of the restraint. Specific weights need to be defined whenever restraints of different nature and magnitudes are used together in structural calculations. The choice of the weights is actually critical since it dramatically influences the convergence of the calculations, and should be performed in such a way that all classes of restraints have a comparable contribution to the target function.

As anticipated, the functional form of pseudocontact shifts and residual dipolar couplings, due to the squared dependence on sine and cosine functions, causes the presence of multiple minima and thus difficulty in the convergence of protein structure determination programs. These problems can be alleviated providing multiple sets of data referring to different paramagnetic ions, so that solution degeneracy is strongly decreased (if data referring to three or more anisotropy tensors are available, the number of degenerate solutions is reduced to only two), and practically removed if the restraints imposed by the covalent geometry among the nuclei are considered. The simultaneous use of pseudocontact shifts and residual dipolar couplings is not helpful in removing this degeneracy as both of them depend on the same tensor. Pseudocontact shifts and residual dipolar couplings, containing structural information of different type, but all referred to the same reference frame, are however particularly valuable when used together.

When metal ions with a very low magnetic susceptibility anisotropy are used, like in the case of gadolinium(III), manganese(II), and copper(II) (which on the other hand can provide R_{1M} and R_{2M} restraints), hyperfine shifts are relatively small also at short distances from the metal ion, where signals are usually very broad. In these cases, chemical shifts of detectable signals can be used to obtain restraints on the protein backbone dihedral angles [61,62] as in any diamagnetic protein. In these cases, diamagnetic residual dipolar couplings can be determined as further restraints through partial orientation induced by external orienting agents.

Paramagnetic lanthanoids are used for protein solution structure determination since 1997 [56,63]. A safe procedure for measuring paramagnetic data has been designed by using lanthanoids at the calcium binding sites of proteins with rigid structures [56,64,65]. Several paramagnetic tags have also been developed to be attached to diamagnetic proteins in order to take advantage of the paramagnetic restraints that they determine (Section 9.7). Because the magnetic susceptibility anisotropy of the different lanthanoids spans a wide range of values (Chapter 8), the related paramagnetic restraints can be collected in different shells from the metals. Substitution of different lanthanoids in the same position can thus permit to collect pseudocontact shifts and residual dipolar couplings for different shells, and thus to enlighten the whole protein scaffold, up to more than 40 Å from the ions [64]. By using gadolinium, restraints on the metal–nuclear distances can be obtained from nuclear relaxation.

Paramagnetic cross correlations (Section 4.9), although less commonly used, can also be provided as structural restraints in ab initio structure determination of proteins [66,67]. They can be used as such, being weighted with force constants proportional to the third power of the nucleus-metal distance.

An alternate largely used approach for determining protein structures is through the assembly of short peptide fragments with programs like ROSETTA [68,69]. Structure prediction of large proteins is often prevented by the difficulty of sampling the very large conformational space associated with the search for the global minimum. The inclusion of sparse, long-range experimental restraints has a marked effect in guiding the conformational sampling towards the protein conformation. Pseudocontact shifts can thus be very effective when included as restraints in the ROSETTA fragment assembly method [70], especially if different paramagnetic metals are used and they are located in different positions on the protein surface (GPS-ROSETTA) [71]. Analogously, residuals dipolar couplings [72] and paramagnetic relaxation enhancement [73] can be sufficient, in addition to backbone chemical shift data, possibly complemented by backbone H^N-H^N nuclear Overhauser effects (NOEs), to determine the protein fold.

9.5.2 REFINEMENT OF AVAILABLE PROTEIN MODELS

Paramagnetic data, which can be easily available with a restricted set of NMR experiments, although hardly able by themselves to provide the protein structure, may be conveniently used to validate and refine an existing model. They present clear advantages with respect to other kinds of restraints because their information content is not restricted to local structural details, and they can be measured over a relatively large range of distances from the paramagnetic ion, so that they are considered long-range restraints. They can be more effective than other "diamagnetic" data and more helpful to disclose structural differences with a proposed model, because optimally suited to detect global structural features. Pseudocontact shifts and paramagnetic residual dipolar couplings, in particular, nicely complement each other and can be conveniently used together because they contain structural information of different type, all referred to the same anisotropy tensor. Pseudocontact shifts are particularly effective for the determination of orientation and anisotropy of the magnetic susceptibility tensor and of the nuclear positions with respect to the paramagnetic ion, and residual dipolar couplings for restraining the orientation of the coupled nuclei, i.e. of the substructure to which they belong, once the magnetic susceptibility anisotropy tensor is anchored by pseudocontact shifts (Section 9.5.3). Actually, the direction of the N-^{N}H vectors of amide pairs belonging to α-helices, for instance, reflects the orientation of the helix which they belong to, because N-^{N}H vectors are almost parallel to the axis of the helix itself.

The validation and refinement procedure can be applied to tentative protein structures obtained through modeling on the structure of another protein with some homology in the primary sequence [74]. PARAMAGNETIC CYANA and Xplor-NIH can thus be used to refine the protein model. In the absence of other restraints, the paramagnetic restraints may however drive the protein structure unnecessarily far away from the starting model. Anchoring the protein structure to the starting model can be achieved by including, among the restraints provided for structure calculation, a network of (upper and lower) distance restraints backcalculated from the starting model.

Another technique used for protein refinement through PARAMAGNETIC CYANA is that of including in the protein sequence a "pseudoprotein residue," composed of as many pseudoatoms as the number of atoms of the protein (or a subset), with coordinates equal to the coordinates of the corresponding atoms in the model structure and no van der Waals radius. Analogously to the pseudoresidue defined for including the susceptibility anisotropy tensor, the pseudoprotein residue is linked to the protein sequence through dummy residues, which have the function of allowing it to freely move with respect to the protein residues. Structural calculations can then be performed by imposing upper distance limits between the protein atoms and the corresponding atoms of the pseudoprotein residue, so that they must be very close to one another, together with the paramagnetic restraints and further available experimental restraints [75]. The protein backbone dihedral ϕ and ψ angles, either of the model structure or calculated from the experimentally determined chemical shifts, can also be added as restraints. By doing so, the extent of the disagreement of the experimental restraints with the available model can be checked by anchoring the protein atoms to the coordinates of the model, and the structural changes needed to match the experimental data can be minimized.

The same approach can be used if the initial model is a crystallographic structure of the protein, solved in the solid state. The structural model obtained in the solid state may in fact differ from the protein structure in solution due to packing forces present in the crystal [76,77] and/or absence of conformational heterogeneity due to crystallization of the protein in a single or fewer conformations (Section 9.6) [65,78]. The agreement of residual dipolar couplings and pseudocontact shifts with a structural model [75–77,79–84] can be monitored by restraining the backbone dihedral angles or the nuclear coordinates to remain as close as possible to those of the starting crystal model and at the same time matching the experimental data.

When the structures in solution and in the solid state are very similar, both the NMR restraints and the crystallographic data can be used simultaneously to produce a model consistent with both sets of data. Since 2013, the paramagnetic data and the X-ray data can actually be used together [85] as structural restraints using the macromolecular crystal structure refinement program—REFMAC5 [85–87] available from CCP4 [88]. Agreement with X-ray and NMR data is monitored through the free *R*-factor and the *Q*-factor [89], respectively. The *Q*-factor is calculated from the discrepancies between experimental data and data back-calculated from the available structure, scaled for the size of the experimental data, according to the following equations (for pseudocontact shifts PCS and residual dipolar couplings RDC):

$$Q_{PCS} = \sqrt{\frac{\sum_i (PCS_i^{\text{exp}} - PCS_i^{\text{calc}})^2}{\sum_i (PCS_i^{\text{exp}})^2}}, \qquad Q_{RDC} = \sqrt{\frac{\sum_i (RDC_i^{\text{exp}} - RDC_i^{\text{calc}})^2}{\sum_i (RDC_i^{\text{exp}})^2}}. \tag{9.4}$$

X-ray and NMR data can be combined to produce models that are compatible with both sets of data when: (1) the free *R*-factor is virtually the same of that calculated with X-ray data alone, (2) the NMR restraint violations are minimal (*Q*-factors smaller than 0.2), and (3) deviances from ideal geometry of covalent bonding remain approximately equal to those obtained in the calculations performed without including the NMR data.

Two cases can be encountered:

1. a good fit of all experimental data can be obtained in the refinement procedure, or
2. the paramagnetic data cannot be fitted simultaneously with the X-ray data.

In the first case, the joint refinement against all data is likely to provide a more reliable structural model than those obtained using the two sets separately. In the second case, the violating restraints in the joint refinement provide an indication that real differences may occur.

In this second case, the protein structure is likely to experience a nonnegligible global conformational change on passing from the solid state to solution, for instance involving secondary structural element rearrangements. In these cases, different approaches [76,77] should be applied. These approaches are based on the use of pseudocontact shifts and/or residual dipolar couplings, complemented by backbone dihedral angle restraints, both experimental and model-derived, in order to keep the structure as close as possible to the X-ray model, without restricting the possibility of sizable rearrangements under the driving force of the experimental solution data (Section 9.5.3).

Pseudocontact shifts and residual dipolar couplings have been included as restraints also in the molecular dynamics simulations program AMBER (Assisted Model Building with Energy Refinement) [90–92]. The program is used in NMR structure determination for refining the energetics of the conformers which are found in best agreement with the experimental data.

9.5.3 ASSEMBLY OF RIGID SUBSTRUCTURES

If the system under investigation is composed of subunits (proteins or protein domains) the structure of which is known, and which are rigidly positioned with respect to one another, the paramagnetic restraints can provide their relative position. This requires that at least two paramagnetic metal ions are used to collect these restraints, in order to remove the degeneracy of the solutions. In fact, with a single set of pseudocontact shifts and residual dipolar couplings, the dependence of both restraints on the same squared trigonometric functions of angular parameters causes a fourfold degeneracy with the correct protein chirality in the possible relative positions of two subunits. If a second set of pseudocontact shifts and residual dipolar couplings is generated by using another paramagnetic ion, having a magnetic susceptibility tensor with no principal axis in common with the first metal, another set of four solutions can be found, but there is only a single solution in common between the two sets of data (Fig. 9.13). In this way the correct relative position of the two subunits can be determined [54,93].

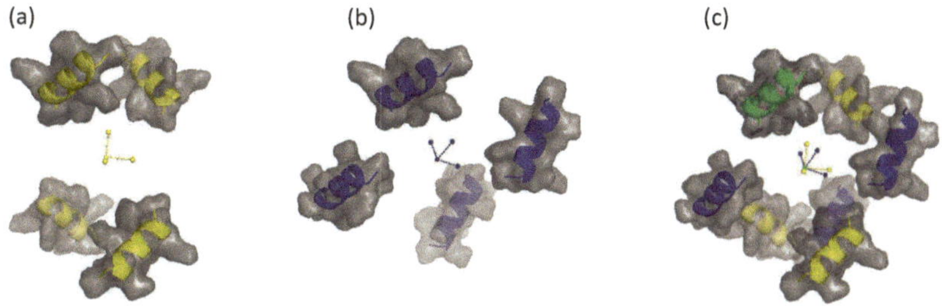

FIGURE 9.13 Pseudocontact shifts and residual dipolar couplings related to a paramagnetic anisotropy tensor determine four possible positions of a protein domain in the frame of that tensor (with opposite sign of two of the three Cartesian coordinates, panel a). Pseudocontact shifts and residual dipolar couplings related another paramagnetic anisotropy tensor (with no axis in common with the first tensor), also determine four possible positions of the same protein domain in the frame of this second tensor (panel b). The correct domain position can be recovered from the single common solution (in green in panel c).

Retrieving the structure of systems composed of multiple subunits can thus be afforded using pseudocontact shifts and residual dipolar couplings according to the following steps (Fig. 9.14):

1. the orientations and anisotropies of the magnetic susceptibility tensors, as well as the positions of the paramagnetic ions, are determined from the pseudocontact shifts of the substructure where the paramagnetic ions are bound;
2. the individual substructures are refined (Section 9.5.2) to be in full agreement with the observed residual dipolar couplings and backbone dihedral angles, if available, as well as with the previously determined anisotropy tensors;
3. the correct relative positions of the subunits are calculated from the multiple sets of pseudocontact shifts and residual dipolar couplings and the previously refined substructures [76]. This procedure is based on the fact that a unique magnetic susceptibility anisotropy tensor per metal is responsible for all pseudocontact shifts/residual dipolar couplings measured in the different subunits.

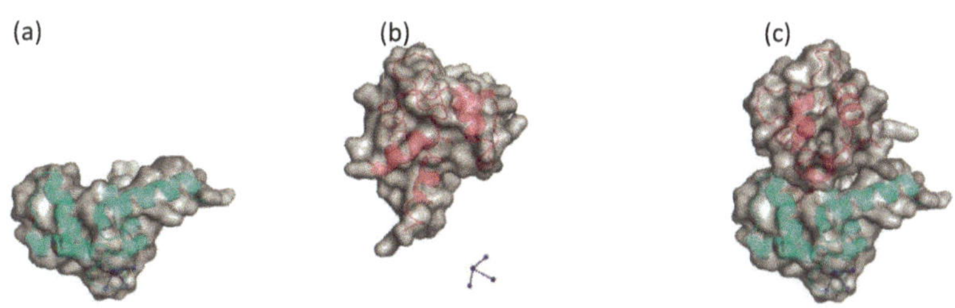

FIGURE 9.14 Once the positions of each protein domain are determined with respect to the same coordinate frame, the protein structure can be reconstructed. In panels a and b the anisotropy tensors are determined from the pseudocontact shifts and residual dipolar couplings related to at least two paramagnetic metals. The tensors are then superimposed in panel c to recover the overall protein structure.

R_{1M}- and R_{2M}-derived distance restraints can be added as further restraints or substituted to either pseudocontact shifts or residual dipolar couplings [94].

This procedure can be used also for determining the structure of a small molecule–protein complex with weak affinity (with micromolar to millimolar dissociation constants). Since the ligand is in fast exchange between the major free state and the minor bound state, the observed pseudocontact shifts, for instance, are weighted averages of the values for the ligand in the free and bound states. Therefore, before being used as structural restraints for determining the ligand position with respect to the protein, they should be converted to the values of the pseudocontact shifts for the bound state from the knowledge of the fraction of bound ligand, as it can be determined from the known concentration of ligand and protein and of the dissociation constant K_D [95]. The latter can be determined from the concentration-dependent ^{1}H and ^{13}C NMR spectra of the ligand. In fact, when the ligand is sufficiently close to the paramagnetic ion to detect sizable paramagnetic effects, increasing ligand concentrations results in narrowing of the peaks and in reduced paramagnetic shifts [96].

9.6 AVERAGE DATA IN THE PRESENCE OF CONFORMATIONAL VARIABILITY

Several proteins and protein complexes exist in different conformations with similar energies [97]. Therefore, their structure cannot be represented by a single conformation but rather by a conformational ensemble. The conformational variability is a key component of protein function [97–103] and should thus be characterized for a better understanding of the mechanism of action.

In the presence of conformational variability, the paramagnetic data are averaged over the values corresponding to the different conformations sampled by the system. The challenge is recovering, from these averaged data, information on the real conformational ensemble.

R_{1M} and R_{2M} are averaged over conformations that interconvert at rates larger than the nuclear relaxation rate. The rate of interconversion over the different conformations must also be larger than the differences in nuclear chemical shifts among the different conformations. Differently, every conformation would produce a separate NMR signal (Sections 6.2 and 6.3). Differences in the nuclear chemical shifts are expected to be mostly due to the different pseudocontact shift contributions, resulting from different nuclear positions in the magnetic susceptibility frame.

If the lifetime of each protein conformation is longer than the correlation time for the relaxation process itself (although much shorter than the nuclear relaxation time), i.e. of the overall tumbling time of the protein and of the electron relaxation time, experimental R_{1M} and R_{2M} are the population-weighted average of the R_{1M} and R_{2M} values corresponding to the different conformations in the ensemble [Eq. (9.1)], i.e.

$$R_{1M} = \langle r^{-6} \rangle (k_{\text{Solomon}} + k_{\text{Curie}})\; R_{2M} = \langle r^{-6} \rangle (k'_{\text{Solomon}} + k'_{\text{Curie}}) \tag{9.5}$$

where $\langle r^{-6} \rangle$ is the average of the r^{-6} values corresponding to the different structures (which is different from $1/\langle r^6 \rangle$ and $\langle r \rangle^{-6}$).

Due to the r^{-6} distance dependence, the contributions to the nuclear R_{1M} and R_{2M} values from conformations where r is small are much larger than those from conformations where r is large. Paramagnetic relaxation enhancement effects can thus be used to monitor the presence of minor conformations by mapping nuclei subject to large paramagnetic broadening, and thus coming close to the paramagnetic center, and those showing no effects, and thus farther from them. Protein conformations can thus be detected even if lowly populated, if in such structures there are nuclei much closer to the paramagnetic center than in conformations of known structure which are highly populated. In this way it is possible to determine whether single conformations must be occupied or cannot be occupied for a given fraction of time [10] and to characterize the lowly sampled protein conformations [104,105]. Spin labelling, or attachment of paramagnetic binding tags, in different protein positions is often required.

On the contrary, if the lifetime of each protein conformation is shorter than, or of the order of, the overall tumbling time of the protein or of the electron relaxation time, this lifetime affects the correlation time modulating the electron-nucleus interaction. This means that experimental R_{1M} and R_{2M} cannot be predicted from the weighted average of the R_{1M} and R_{2M} values corresponding to the individual conformations in the ensemble. In fact, the paramagnetic relaxation enhancements of the various conformations cannot be defined because each of them exists for times that are shorter than the reorientational correlation time.

The effect of the interconversion rate on nuclear relaxation also depends on the degree of spatial restriction of the motion. The Lipari-Szabo model free approach can be used to describe the effects of the different motions on relaxation. According to this model, a squared order parameter, S_s^2, is introduced to quantify the fluctuations in distance and orientation of the vector connecting the paramagnetic center and the observed nucleus [12,106,107]:

$$S_s^2 = \frac{1}{\sum_i w_i r_i^{-6}} \sum_{i,j} \frac{w_i w_j}{r_i^3 r_j^3} \left[\frac{3}{2} \left(\frac{\mathbf{r}_i \cdot \mathbf{r}_j}{r_i r_j} \right)^2 - \frac{1}{2} \right] \tag{9.6}$$

where i and j are indices running on the elements of the conformational ensemble, and w_i are the weights of the conformations. The experimental R_{2M} thus becomes

$$R_{2M} = \langle r^{-6} \rangle [S_s^2 (k'_{\text{Solomon}}(\tau_c) + k'_{\text{Curie}}(\tau_c)) + (1 - S_s^2)(k'_{\text{Solomon}}(\tau_1) + k'_{\text{Curie}}(\tau_1))] \tag{9.7}$$

where $\tau_1^{-1} = \tau_r^{-1} + \tau_e^{-1} + \tau_{\text{inter}}^{-1}$, τ_{inter} being the interconversion time. If the dispersive Lorentzian terms in k'_{Solomon} and k'_{Curie} can be neglected, an effective correlation time, $\tilde{\tau}_c = S_s^2 \tau_c + (1 - S_s^2)\tau_1$, can be defined to take into account both the overall protein reorientation time and the fast motions [12,106–110].

Paramagnetic centers with low magnetic susceptibility anisotropy (which induces very small pseudocontact shifts) and large electron relaxation time (which causes sizable R_{1M} and R_{2M} effects at relatively large distance), like radicals, Mn^{2+} or Gd^{3+}, are used to extract information on the structural variability from the paramagnetic relaxation enhancements.

Experimental residual dipolar couplings are averaged over a structural mobility occurring on time scales faster than, or of the order of, 10^{-2} s (Section 3.1). However, both pseudocontact shifts and residual dipolar couplings average only over conformations with interconversion rate larger than the differences in nuclear chemical shifts among the different conformations (see earlier). In these cases, both the measured pseudocontact shifts and residual dipolar couplings can be calculated as the average over the individual values for each conformation in the ensemble. In the case of lanthanoids with large anisotropy, differences in chemical shifts can be of the order of 10 ppm for nuclei 15 Å far from the metal (placed in another domain), so that, at 500 MHz, averaging occurs if motions occur on time scales faster than $1/(2\pi \times 500 \times 10^6 \times 10^{-5}) = 3 \times 10^{-5}$ s.

If the rate of interconversion is of the order of the differences in chemical shifts among the different structures, conformational exchange causes NMR signal broadening (Sections 6.2 and 6.3). As a result, the modulation of pseudocontact shifts can be related to exchange contributions to the transverse relaxation rates, which can provide information on the amplitude and time scale of the intervening motions through NMR relaxation dispersion experiments [111,112]. For instance, in the Co^{2+}-substituted two-domain protein PA0128, relaxation rate enhancements have been ascribed to a concerted domain motion on a time scale of a few milliseconds [113].

The measurement of paramagnetic residual dipolar couplings provides a formidable effective way to estimate the presence of conformational rearrangement between the substructure containing a paramagnetic ion and another substructure. In fact, in rigid systems, the range of the observed residual dipolar couplings should be the same for all substructures, whereas, in case of motion, the range of the observed residual dipolar couplings values is reduced due to averaging upon the different conformations [114]. All residual dipolar couplings measured for one domain actually quench to zero in the case of overall isotropic reorientation of that domain with respect to the metal-bearing domain.

Since the dependence of residual dipolar couplings on the angular parameters can be written as $P^*\chi P$, where P represents the coupled nuclei vector and χ is the magnetic susceptibility anisotropy tensor [Eq. (3.20)], in the presence of interdomain mobility, paramagnetic residual dipolar couplings can be expressed in the form [115]

$$\Delta\nu^{RDC} \propto < P^*\chi P > \tag{9.8}$$

In the case of relative reorientation of the subunits, the vectors P can be expressed as $P = RP_0$, where R represent the rotation matrices applied to the vector P_0 in a reference conformation. Therefore, we can write the residual dipolar couplings as

$$\Delta\nu^{RDC} \propto < (RP_0)^*\chi(RP_0) > = P_0^* < R^*\chi R > P_0 = P_0^*\tilde{\chi}P_0 \tag{9.9}$$

that is, in the same form as for rigid systems, but with the magnetic susceptibility anisotropy tensor replaced by the "mean tensor" $\tilde{\chi} = < R^*\chi R >$. This "mean tensor" is the average of the magnetic susceptibility anisotropy tensors (differing in their orientation) which are sensed by the nuclei in a subunit which is moving with respect to the subunit containing the paramagnetic metal (Fig. 9.15). In fact, these nuclei see a magnetic susceptibility anisotropy tensor which is moving and sampling all orientations relative to the different conformations in the ensemble. In summary, a mean tensor $\tilde{\chi}$ can always be determined from the experimental residual dipolar couplings, and the ratio of its anisotropy

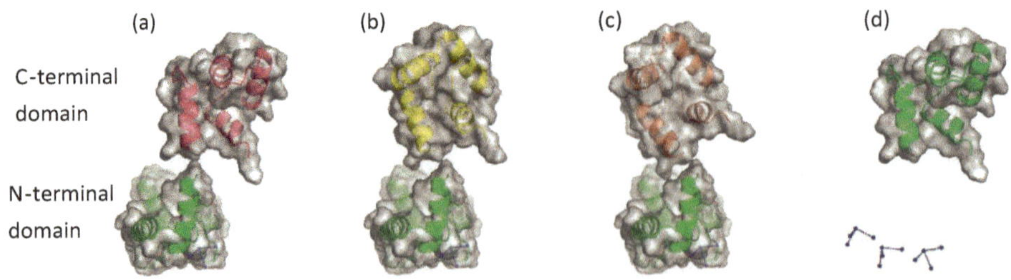

FIGURE 9.15 Due to reorientation of the C-terminal domain with respect to the anisotropy tensor fixed on the N-terminal domain (panels A–C), the nuclei in the C-terminal domain see the anisotropy tensor moving (panel D).

with respect to the anisotropy of the χ tensor represents a generalized order parameter, which decreases (from 1 to 0) as mobility increases [65,93]. On the contrary, motionally averaged pseudocontact shifts cannot be represented by any mean tensor (unless all nuclei have a fixed distance r from the metal ion during the motion, which is highly unlikely).

Recovering the structural ensemble from average data is in mathematical terms an "ill-posed inverse problem," which admits an infinite number of solutions. In fact, possible ensembles of conformations can be easily determined by selecting a number of structures providing averaged pseudocontact shifts and residual dipolar couplings (as well as R_{1M} and R_{2M}, with the warnings detailed earlier) in agreement with the experimental data. However, there is an infinite number of ensembles which may agree equally well with the experimental data [116].

Ensemble averaging approaches can be used to estimate the extent of the conformational heterogeneity and retrieve the conformations more likely sampled by the system, if the ensembles are composed of the smallest number of conformations providing agreement with the experimental data and mobility is not very large [117]. A structural ensemble could thus be determined through the application of the Occam's razor [117]:

1. with a fixed number of structures in the reconstructed ensemble, the conformations and the corresponding weights are chosen for an optimized agreement with the experimental data, providing the lowest possible TF [Eq. (9.3)];
2. the number of structures is then changed, so that a new conformational ensemble is built, with a corresponding lowest possible TF;
3. the most informative ensemble is that determined as the one with the lowest number of structures, providing a TF not significantly larger than that with one additional structure (Fig. 9.16).

This procedure ensures that the selected ensemble is in best agreement with the experimental data and prevents from inclusion of noise structures which are not required for the fit of the data.

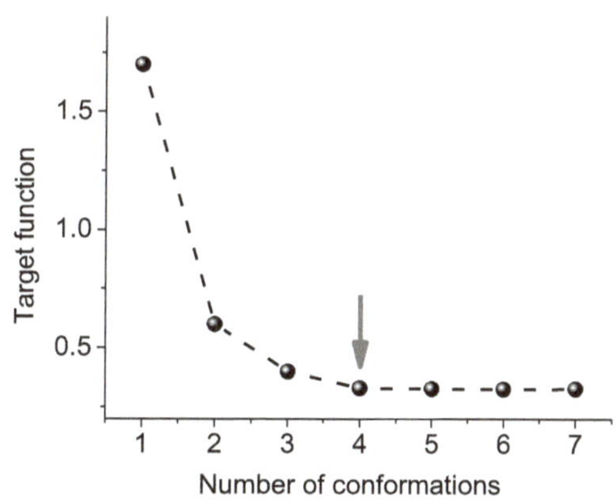

FIGURE 9.16 The agreement between backcalculated and experimental data increases (lower target function) by increasing the number of conformations in the ensemble up to 4 structures. Inclusion of more than 4 conformations is then not justified because further conformations are not required for an improved fit of the data.

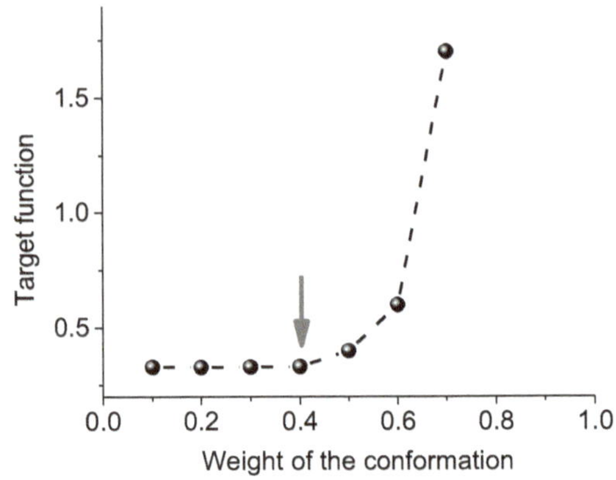

FIGURE 9.17 The maximum occurrence of the analyzed conformation is 0.4 because larger weights causes the target function to increase with respect to the minimum value, whatever the other conformations included in the ensemble are.

To evaluate how much a given conformation can contribute at most to the experimental average, the maximum occurrence (MaxOcc) of conformations can be determined [116,118]. The MaxOcc of a conformation is defined as the largest weight that it can have within any ensemble and be still in agreement with the experimental data. Calculations are performed by searching a conformational ensemble which includes such conformation with a fixed weight and tens of other conformations selected with a minimization program in order to provide averaged data in agreement with the experimental data. These calculations are repeated for increasing weight of the selected conformation, until it becomes impossible to find an ensemble in good agreement with the experimental data (Fig. 9.17). As the other structures in the ensemble only act as "fillers" and have no physical meaning, it is not necessary to restrict their number to a minimum as in the case of application of Occam's razor.

MaxOcc calculations have been performed for flexible two-domain proteins using as experimental restraints pseudocontact shifts and residual dipolar couplings, possibly complemented by R_{2M} or SAXS data [118–120]. The protocol is based on the calculation of the magnetic susceptibility anisotropy tensors from the data acquired for the domain hosting the paramagnetic centers (multiple datasets acquired with different paramagnetic ions are needed), and the knowledge that averaged pseudocontact shifts and residual dipolar couplings measured for the other domain still depend on the same tensors. Paramagnetic lanthanoid ions are commonly used because many of them can be easily replaced in the same position, in order to provide multiple datasets, and have large magnetic susceptibility anisotropy. Using an analogous approach, the MaxOcc can also be calculated for ensembles of structures, as those populating regions defined within the conformational space possibly sampled by the system. Searching for the smallest region providing a MaxOcc of 1, i.e. containing all conformations needed to reproduce the experimental data, permits to determine the minimum size of the conformational space that must be sampled by the system.

A different approach used to calculate structural ensembles is looking for the solution which both explains the data and satisfies the maximum entropy principle, thus ensuring that no additional assumptions on the unknown conformational distribution is made [121–124]. This approach provides the broadest and flattest probability distribution satisfying the experimental restraints; this solution is the one better representing the minimal set of features common to all the infinite solutions satisfying the experimental data. In this case, a very large number of conformations should be allowed in the calculated ensemble.

9.7 TAGGED PROTEINS

Exploiting the advantages provided by the paramagnetic restraints requires the presence of a paramagnetic center in the protein. If it is not naturally present, a paramagnetic center can be attached either by attaching a nitroxide spin label or a paramagnetic binding tag.

Spin labels, like the (1-oxyl-2,2,5,5-tetramethylpyrroline-3-methyl)methanethiosulfonate spin label (MTSSL) or other radicals, can be alternatively attached in several points of the protein. Paramagnetic spin labels attached in several different sites have been used to map the interchange between different structural states [10,14,105,125–127]. In fact, as already mentioned in Section 9.6, by identifying the protein nuclei subject to paramagnetic line broadening, and thus coming close to the spin labels, it is possible to determine which conformations must be occupied or cannot be occupied for a given fraction of time. The flexibility of the spin-labeled group attached to the protein can be taken into account by considering multiple positions of the spin label, and distances should be estimated from the average of the r^{-6} values corresponding to all possible positions [Eq. (9.5)] [13].

Paramagnetic binding tags can be used to obtain pseudocontact shifts and residual dipolar couplings, and can thus provide a larger amount of structural information with respect to spin labels. Lanthanoid tags are particularly important because the whole series of lanthanoid ions (diamagnetic as well as paramagnetic) can be replaced in the same position, so that multiple datasets of paramagnetic data can be easily obtained. Furthermore, some lanthanoids provide the paramagnetic metal ions with the largest magnetic susceptibility anisotropy (Table 9.1), and thus are those providing the largest values for pseudocontact shifts and residual dipolar couplings. A lanthanoid binding tag can be genetically encoded and expressed as fusion tag together with the protein [128–131], or can be a synthetic tag that ligates, for instance, cysteine residues [132,133] (Table 9.2).

The tagging positions on the protein surface must be determined in such a way as to not perturb the investigated system and to provide experimental data with good signal-to-noise ratio for the residues of major interest, so that the latter must be at the proper distance from the tag itself. This is often a challenging task. Furthermore, for a correct analysis of the data, flexibility of the tag should be taken into account. As we have seen in Section 9.6, rigid tags should be used for avoiding reduction in the amplitude of the residual dipolar couplings (which implies a smaller signal-to-noise ratio) and for a proper use of pseudocontact shifts [21,133–136]. In fact, in the presence of extensive tag mobility, it becomes impossible to retrieve the magnetic susceptibility tensor anisotropy from the fit of the observed pseudocontact shifts to the protein structure.

Rigidity can be achieved by two-point attachment of the tag to nonmobile residues on the surface of the protein [137,138]. The attachment is mostly performed through disulfide bridges, after cysteine replacement of the protein residues at the desired binding positions. Alternatively, the tag can be attached in a first position through disulfide bridge and anchored in a second position through the N-terminal fusion [139] or additional coordination of the paramagnetic ion by an amino acid of the protein. Finally, rigidity can also be achieved through one-point attachment of a very bulk metal ion binding cage [136].

Table 9.2 Structures of Lanthanoid Chelating Probes

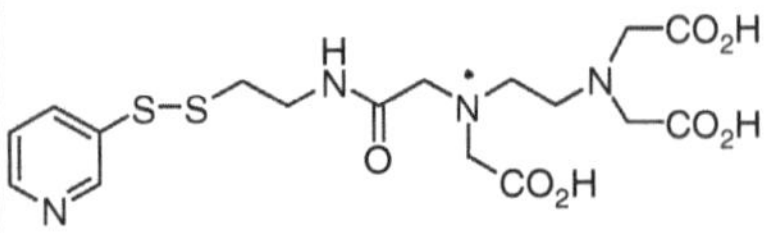

1

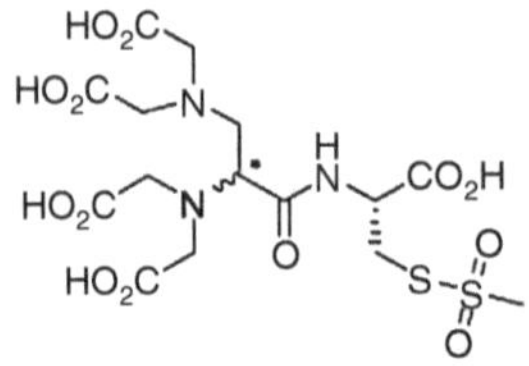

3 (R,R) or (S,R)

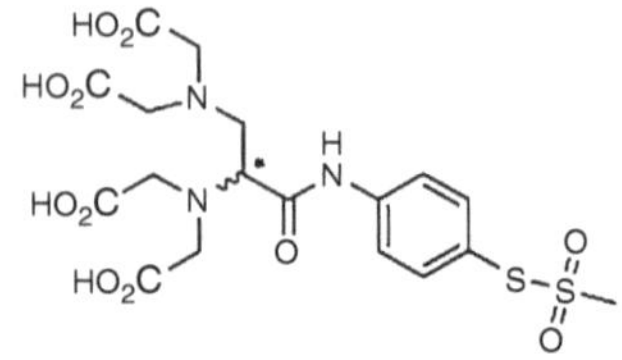

3 (R) or (S)

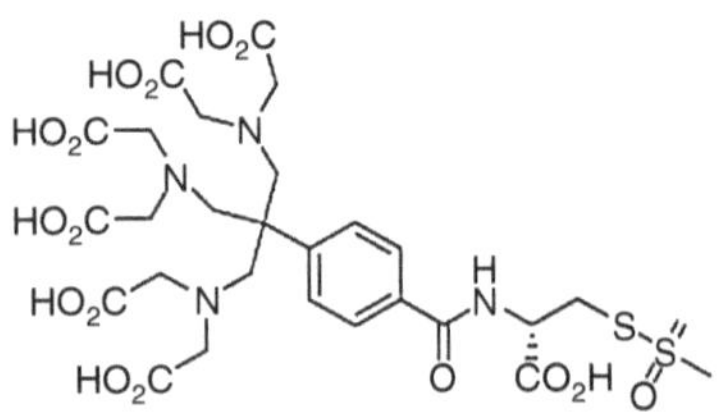

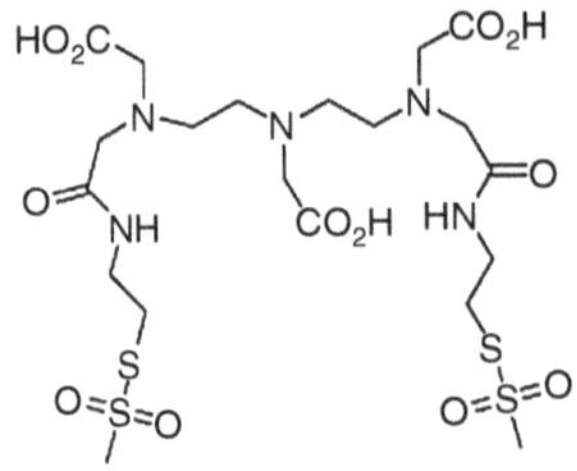

5

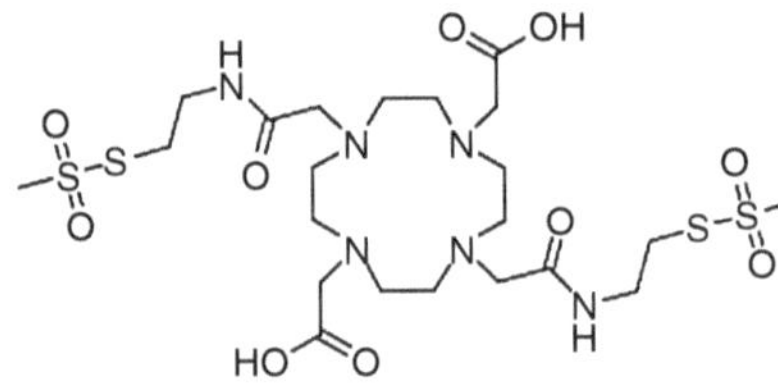

6

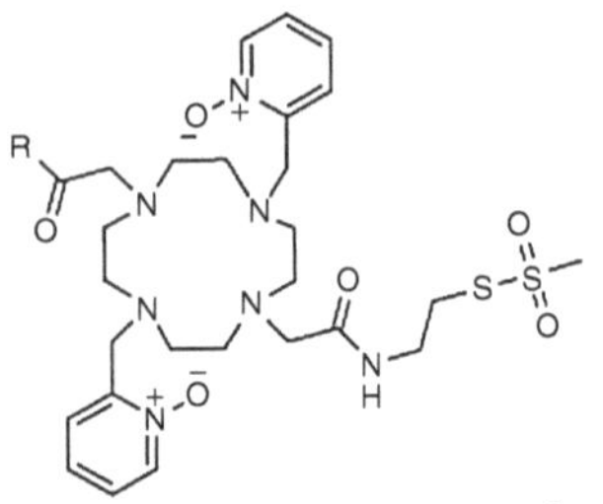

7a R = OH

7b R = 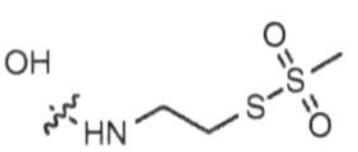

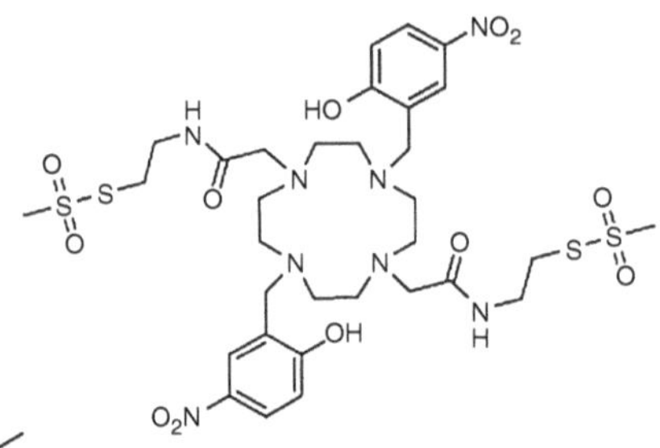

8

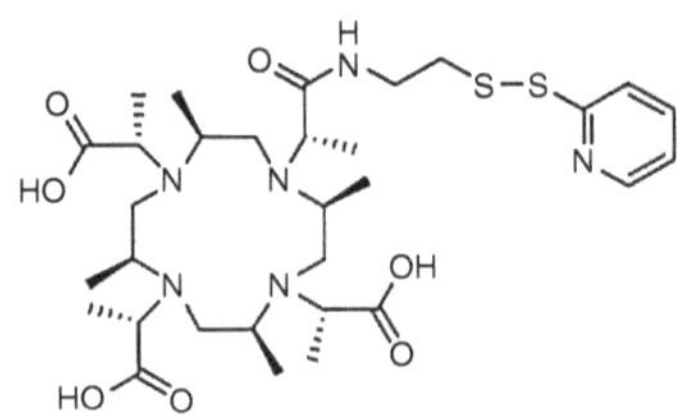

9

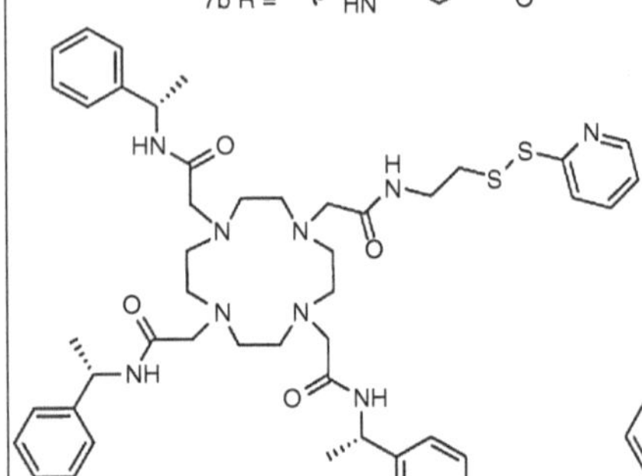

10

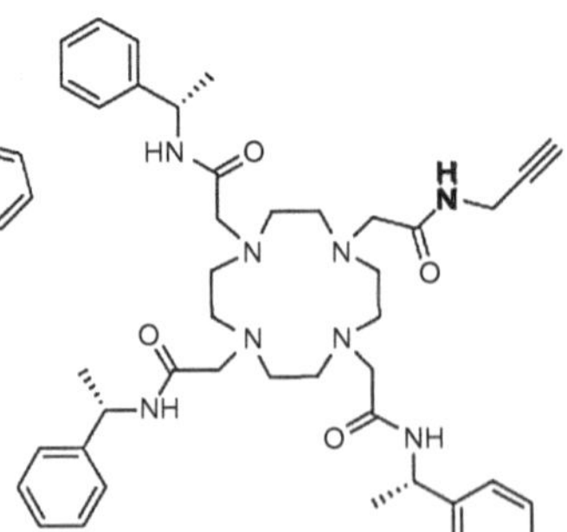

11

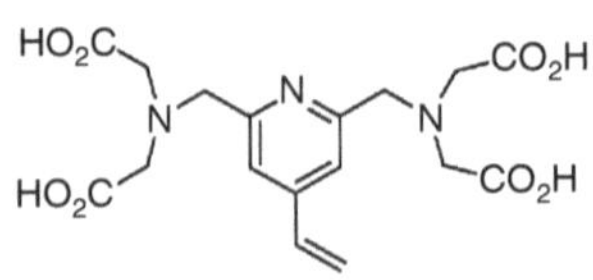

12

(1–3) EDTA-based probes: **1**) *(S-(2-pyridylthio)-cysteaminyl-EDTA,* **2**) *EDTA-CA-MTS,* **3**) *ent-6; (***4**) *TAHA based probe: Cys-Ph-TAHA; (***5**) *DTPA-based probe: CLaNP1; (***6–11**) *DOTA-based probes:* **6**) *CLaNP3,* **7**) *CLaNP5,* **8**) *CLaNP7,* **9**) *DOTAM8,* **10**) *C1,* **11**) *C4; (***12**) *pyridine-core probe: 4VPyMTA.*
Reprinted with permission from Ref. [148].

9.8 STRUCTURAL INFORMATION FROM PSEUDOCONTACT SHIFTS IN THE SOLID STATE

In solution, pseudocontact shifts are determined by the average upon molecular tumbling of the dipole–dipole interactions between nuclear and electron magnetic moments. In the solid state, molecular reorientation being quenched, the averaging can be performed by magic angle spinning (MAS, see Chapter 5). In any case, pseudocontact shifts are described by the same equation [Eq. (2.46)].

In practice, a striking difference between solution and solid state pseudocontact shifts exists: whereas in solution the pseudocontact shift of each nucleus is determined only by the paramagnetic metal belonging to the same molecule, in crystals it is determined by the sum of the contributions from all metal ions close to that nucleus, even those belonging to different molecules (Fig. 9.18). Actually, it may happen, especially if the paramagnetic metal is located close to the periphery of the protein, that for some nuclei intermolecular contributions are even larger than the intramolecular contribution, because these nuclei are closer to the paramagnetic metal of a neighboring molecule than to that one in the same molecule [140–142].

As a result, the pseudocontact shifts in the solid state contain two kinds of information: on the molecular structure, as for the solution case, and on the relative arrangement of the molecules in the crystal. The intramolecular contribution can be directly used as structural restraint for molecular structure determination, analogously to pseudocontact shifts acquired in solution. As expected, in the absence of structural changes, the intramolecular contributions to pseudocontact shifts in the solid state are in excellent agreement with pseudocontact shifts in solution [143]. Conversely, intermolecular contributions can be used to reconstruct the crystal lattice.

The intramolecular contribution can be separated from intermolecular contributions by diluting the paramagnetic ^{13}C,^{15}N-labeled protein molecules into diamagnetic unlabeled protein molecules, so that the intermolecular contributions have vanishingly small intensities [140,144,145]. However, due to the lower concentration of the paramagnetic species, the signal-to-noise ratio will be significantly smaller than for measurements performed without dilution. With a very high dilution, intermolecular contributions are certainly removed, but the signal may not even be detectable. Notably, for not very

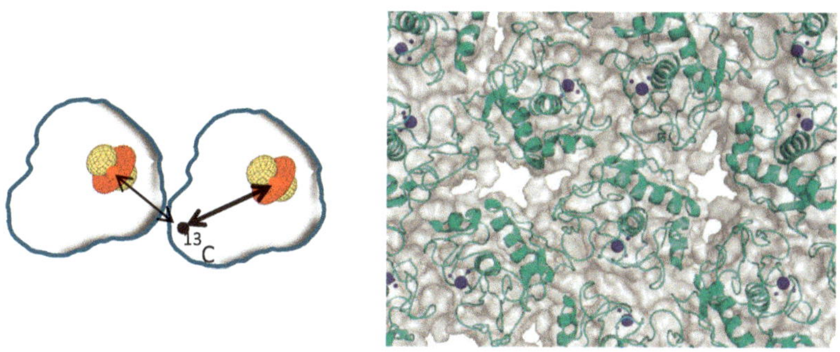

FIGURE 9.18 Pseudocontact shifts in crystals are determined by the sum of the contributions from all paramagnetic metal ions close to the observed nucleus, even those belonging to different molecules. In the figure, the protein molecules are packed in the crystal lattice and the paramagnetic metal ions are represented by spheres surrounded by the pseudoatoms indicating the directions of the anisotropy tensor axes (spheres of smaller radius).

high dilutions, the $^{13}C/^{15}N$ nuclei in paramagnetic proteins will be surrounded by different combinations of paramagnetic and diamagnetic neighbors, leading to the presence of multiple sets of pseudocontact shifts (Fig. 9.19) [146]. The optimal dilution to be used for obtaining a good signal-to-noise ratio for the NMR peak corresponding to intramolecular contribution will thus depend on the number of neighboring molecules around a central molecule and on the magnitude of the intermolecular contributions (which in turn depend on the size of the molecule, the internal position of the paramagnetic ion and the magnetic susceptibility anisotropy).

Pseudocontact shifts are very precious because they permit to recursively solve ambiguities in other kinds of restraints. In fact, pseudocontact shifts are immediately available after comparison of the NMR spectra of the paramagnetic sample and a diamagnetic analog, whereas many distance restraints are initially ambiguous. The first structure of a paramagnetic protein obtained through solid state NMR was that of the catalytic domain of the matrix metalloproteinase 12, where a zinc(II) ion was substituted with a high-spin cobalt(II) ion [147], and it was solved thanks to the availability of intramolecular pseudocontact shifts [144].

As apparent from the simulation shown in Fig. 9.19, both resolution and signal-to-noise ratio are optimal in the absence of dilution, and thus when pseudocontact shifts contain both intramolecular and intermolecular contributions. In this case, they can be used to obtain (1) the structure of the protein and (2) the crystal packing, once the cell parameters are available from powder X-ray diffraction spectra [142]. The following protocol can be employed (if the cell parameters are not available, the protocol can be applied up to step 4: in this case the structure of the protein can still be determined but the crystal packing cannot):

1. an initial structural family is calculated from the pseudocontact shifts and all available unambiguous distance restraints. The calculations are performed by including a given number of pseudoresidues in the protein sequence representing the paramagnetic metal ions and their anisotropy tensor (Section 9.5.1), with anisotropy values fixed to literature values. In this way, both the internal metal and all external metals sensed by the protein nuclei can be considered, and pseudocontact shifts can be back-calculated as the sum of the values corresponding to each paramagnetic metal;
2. the number of metals to be considered is fixed to the lowest value needed to obtain a satisfactory agreement between back-calculated and experimental pseudocontact shifts;
3. the calculated structures are used to iteratively resolve ambiguities in the assignments and thus to determine additional distance restraints;
4. a grid search of the optimal anisotropy values of the magnetic susceptibility tensor is performed, so that they can be fixed to the values providing the best agreement with the experimental pseudocontact shifts;
5. a pseudoresidue is introduced to position freely the crystallographic origin and the reference frame. It is placed in the protein sequence together with the pseudoresidue representing the internal paramagnetic ion. All pseudoresidues representing the external metals are removed. A crystallographic space group among those compatible with the cell parameters is imposed and, using the cell parameter values, all surrounding molecules are positioned in this reference frame, each of them with its own paramagnetic ion and anisotropy tensor. The position of the pseudoresidue representing the crystallographic origin and the reference frame is refined in order to obtain the best agreement of the experimental pseudocontact shifts with the values backcalculated from the sum of all contributions from surrounding molecules;

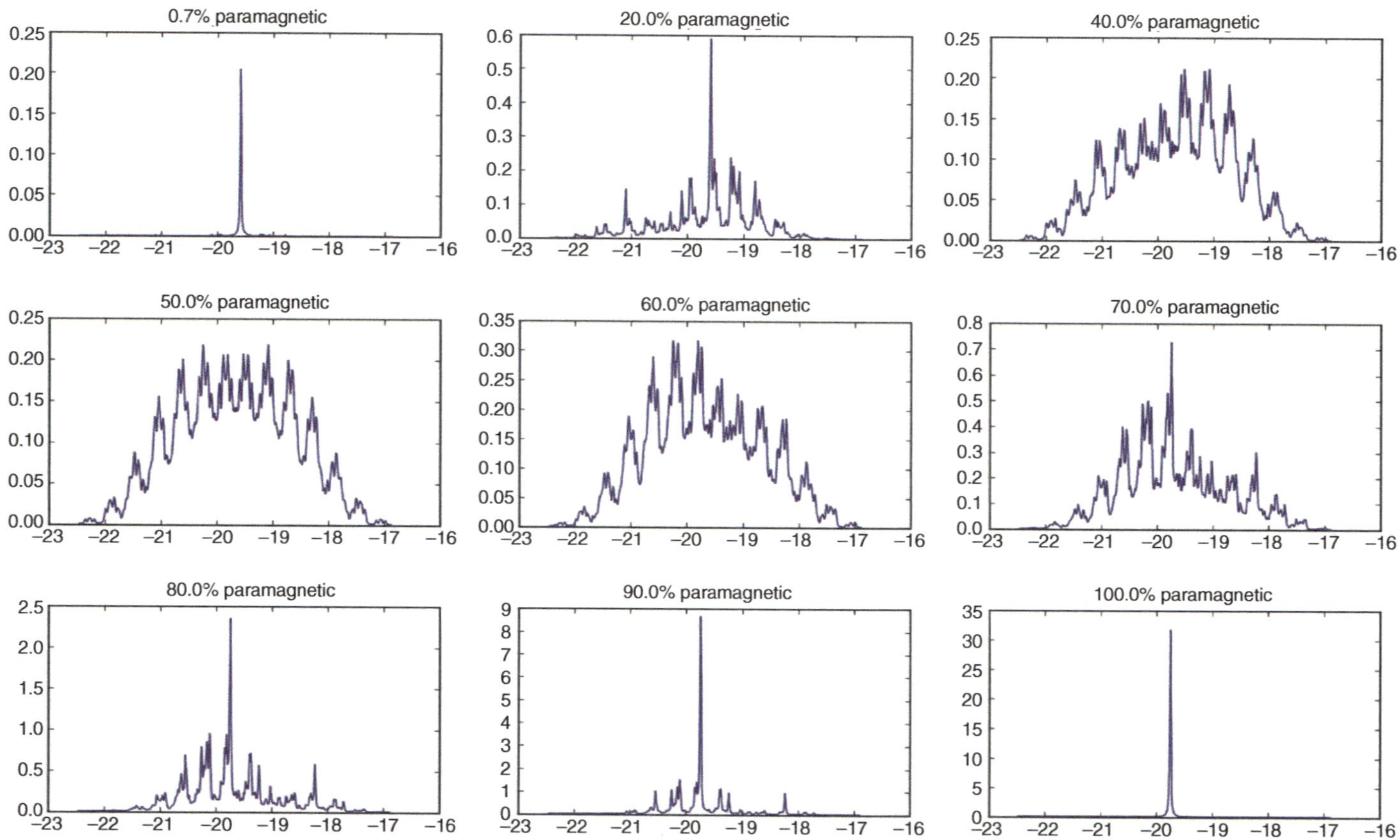

FIGURE 9.19 Simulation showing the appearance of a single pseudocontact-shifted peak of a crystalline labeled paramagnetic protein for decreasing amounts of dilutions into diamagnetic unlabeled proteins, i.e. for increasing amounts of paramagnetic metal occupancy. The *x*-axis shows the chemical shift (in ppm). Note that the intensity scale (*y*-axis) is changed in every frame to allow visualization and that the intensity increases dramatically when 100% occupancy is approached.

Reproduced with permission from Ref. [146].

6. the calculations are repeated for each possible space group, in agreement with the cell parameters, in order to select the one(s) providing the best agreement between experimental and backcalculated pseudocontact shifts and with no compenetrations among the molecules in the crystal. In the case of MMP-12, among the nine different orthorhombic space groups compatible with the unit cell parameters, only four are consistent with the presence of only four molecules per unit cell, as expected from the Matthews coefficient; among these four space groups, two can be excluded from the severe compenetrations among neighboring molecules, and one provided a solution in sizably worse agreement with the experimental data than the correct space group [142].

In summary, a good agreement of all pseudocontact shifts is obtained when the intermolecular contributions are correctly calculated, and thus when the crystallographic frame is correctly positioned at the end of the minimization.

WEB SITES OF MAIN PROGRAMS FOR THE USE OF PARAMAGNETIC RESTRAINTS FOR STRUCTURE AND DYNAMICS

AMBER: ambermd.org; py-enmr.cerm.unifi.it/access/index/amps-nmr

CYANA: www.cyana.org/wiki/index.php/Main_Page - paramagnetic patch: www.cerm.unifi.it/software-for-download/software-paramagnetic-cyana

DipoCoup: http://www.meilerlab.org/index.php/jobs/resources-software

ECHIDNA: http://www.nmr.chem.uu.nl/~christophe/echidna.html

FANTEN: http://fanten-enmr.cerm.unifi.it:8080/

FLEXIBLE-MECCANO: http://www.ibs.fr/science-213/scientific-output/software/flexible-meccano?lang=en

GPS ROSETTA: https://www.rosettacommons.org/

HADDOCK: www.nmr.chem.uu.nl/haddock/; haddock.science.uu.nl/enmr/services/HADDOCK/haddock.php

NUMBAT: www.nmr.chem.uu.nl/~christophe/numbat.html

MaxOcc: py-enmr.cerm.unifi.it/access/index/maxocc

MODULE: http://www.ibs.fr/science-213/scientific-output/software/module/?lang=en

PARASSIGN: protchem.lic.leidenuniv.nl/software/parassign/registration

POSSUM: http://www.nmr.chem.uu.nl/~christophe/possum.html

REDCAT/REDCRAFT: http://ifestos.cse.sc.edu/software.php

REFMAC5: www.ccp4.ac.uk/html/refmac5.html

Xplor-NIH: nmr.cit.nih.gov/xplor-nih; py-enmr.cerm.unifi.it/access/index/xplor-nih

REFERENCES

[1] Bertini I, Luchinat C, Parigi G. Concepts Magn. Reson. 2002;14:259.
[2] Bertini I, Felli IC, Luchinat C, Parigi G, Pierattelli R. ChemBioChem 2007;8:1422.
[3] Bertini I, Luchinat C, Parigi G, Pierattelli R. ChemBioChem 2005;6:1536.
[4] Bertini I, Luchinat C, Rosato A. Progr. Biophys. Mol. Biol. 1996;66:43.
[5] Banci L, Bertini I, Cavallaro G, Giachetti A, Luchinat C, Parigi G. J. Biomol. NMR 2004;28:249.

[6] Vlasie MD, Comuzzi C, van den Nieuwendijk AMCH, Prudêncio M, Overhand M, Ubbink M. Chem. Eur. J. 2007;13:1715.
[7] Almeida RM, Turano P, Moura I, Moura GLC, Pauleta SR. ChemBioChem 2013;14:1858.
[8] Battiste JL, Wagner G. Biochemistry 2000;39:5355.
[9] Ramos A, Varani G. J. Am. Chem. Soc. 1998;120:10992.
[10] Bashir Q, Volkov AN, Ullmann GM, Ubbink M. J. Am. Chem. Soc. 2010;132:241.
[11] Nadaud PS, Helmus JJ, Höfer N, Jaroniec CP. J. Am. Chem. Soc. 2007;129:7502.
[12] Iwahara J, Clore GM. J. Am. Chem. Soc. 2010;132:13346.
[13] Iwahara J, Schwieters CD, Clore GM. J. Am. Chem. Soc. 2004;126:5879.
[14] Volkov AN, Ubbink M, Van Nuland NAJ. J. Biomol. NMR 2010;48:225.
[15] Balayssac S, Jiménez B, Piccioli M. J. Biomol. NMR 2006;34:63.
[16] Bermel W, Bertini I, Felli IC, Kümmerle R, Pierattelli R. J. Am. Chem. Soc. 2003;125:16423.
[17] Bertini I, Lee Y-M, Luchinat C, Piccioli M, Poggi L. ChemBioChem 2001;2:550.
[18] Bertini I, Jiménez B, Piccioli M, Poggi L. J. Am. Chem. Soc. 2005;127:12216.
[19] Bertini I, Couture MMJ, Donaire A, Eltis LD, Felli IC, Luchinat C, Piccioli M, Rosato A. Eur. J. Biochem. 1996;241:440.
[20] Bertini I, Donaire A, Luchinat C, Rosato A. Proteins Struct. Funct. Genet. 1997;29:348.
[21] Vlasie MD, Fernández-Busnadiego R, Prudêncio M, Ubbink M. J. Mol. Biol. 2008;375:1405.
[22] Scanu S, Förster J, Finiguerra MG, Shebestari MH, Huber M, Ubbink M. ChemBioChem 2012;13:1312.
[23] Dick LR, Geraldes CFGC, Sherry AD, Gray CW, Gray DM. Biochemistry 1989;18:7896.
[24] Arumugam S, Hemme CL, Yoshida N, Suzuki K, Nagase H, Berjanskii M, Wu B, Van Doren SR. Biochemistry 1998;37:9650.
[25] Otting G, Liepinsh E, Wüthrich K. J. Am. Chem. Soc. 1991;113:4363.
[26] Petros AM, Mueller L, Kopple KD. Biochemistry 1990;29:10041.
[27] Scarselli M, Bernini A, Segoni C, Molinari H, Esposito G, Lesk AM, Laschi F, Temussi P, Niccolai N. J. Biomol. NMR 1999;15:125.
[28] Yuan T, Ouyang H, Vogel HJ. J. Biol. Chem. 1999;274:8411.
[29] Bernini P, Bertini I, Nepi S, Saccenti E, Schaefer H, Schuetz B, Spraul M, Tenori L. J. Proteome Res. 2009;8:4264.
[30] Madl T, Guettler T, Gorlich D, Sattler M. Angew. Chem. Int. Ed. Engl. 2011;50:3993.
[31] Liepinsh E, Baryshev M, Sharipo A, Ingelman-Sundberg M, Otting G, Mkrtchian S. Structure 2001;9:457.
[32] Respondek M, Madl T, Göbl C, Golser R, Zangger K. J. Am. Chem. Soc. 2007;129:5228.
[33] Madl T, Bermel W, Zangger K. Angew. Chem. Int. Ed. 2009;48:8259.
[34] Skinner SP, Moshev M, Hass MAS, Ubbink M. J. Biomol. NMR 2013;55:379.
[35] Pintacuda G, Keniry MA, Huber T, Park AY, Dixon NE, Otting G. J. Am. Chem. Soc. 2004;126:2963.
[36] Gao YA, Boyd J, Pielak GJ, Williams RJ. Biochemistry 1991;30:1928.
[37] Emerson SD, La Mar GN. Biochemistry 1990;29:1556.
[38] Bertini I, Luchinat C, Piccioli M, Vicens Oliver M, Viezzoli MS. Eur. Biophys. J. 1991;20:269.
[39] Banci L, Dugad LB, La Mar GN, Keating KA, Luchinat C, Pierattelli R. Biophys. J. 1992;63:530.
[40] Campbell ID, Dobson CM, Williams RJP, Xavier AV. Ann. NY Acad. Sci. 1973;222:163.
[41] Campbell ID, Dobson CM, Williams RJP. Proc. Roy. Soc. London A 1975;345:41.
[42] Lee L, Sykes BD. Biochemistry 1983;22:4366.
[43] Gochin M, Roder H. Bull. Magn. Reson. 1995;17:1.
[44] Banci L, Bertini I, Bren KL, Cremonini MA, Gray HB, Luchinat C, Turano P. J. Biol. Inorg. Chem. 1996;1:117.
[45] Banci L, Bertini I, Luchinat C, Pierattelli R, Shokhirev NV, Walker FA. J. Am. Chem. Soc. 1998;120:8472.
[46] Banci L, Bertini I, Luchinat C, Turano P. In: Kadish KM, Smith KM, Guilard R, editors. The Porphyrin Handbook. San Diego, CA: Academic Press; 2000. p. 323.

[47] Arnesano F, Banci L, Bertini I, Felli IC. Biochemistry 1998;37:173.
[48] Ottiger M, Delaglio F, Bax A. J. Magn. Reson. 1998;131:373.
[49] Permi P, Rosevear PR, Annila A. J. Biomol. NMR 2000;17:43.
[50] Balayssac S, Bertini I, Luchinat C, Parigi G, Piccioli M. J. Am. Chem. Soc. 2006;128:15042.
[51] Tolman JR, Flanagan JM, Kennedy MA, Prestegard JH. Proc. Natl. Acad. Sci. USA 1995;92:9279.
[52] Banci L, Bertini I, Huber JG, Luchinat C, Rosato A. J. Am. Chem. Soc. 1998;120:12903.
[53] Hus JC, Marion D, Blackledge M. J. Mol. Biol. 2000;298:927.
[54] Bertini I, Longinetti M, Luchinat C, Parigi G, Sgheri L. J. Biomol. NMR 2002;22:123.
[55] Barbieri R, Luchinat C, Parigi G. ChemPhysChem 2004;21:797.
[56] Bertini I, Donaire A, Jiménez B, Luchinat C, Parigi G, Piccioli M, Poggi L. J. Biomol. NMR 2001;21:85.
[57] Assfalg M, Bertini I, Dolfi A, Turano P, Mauk AG, Rosell FI, Gray HB. J. Am. Chem. Soc. 2003;125:2913.
[58] Bertini I, Luchinat C, Turano P, Battaini G, Casella L. Chem. Eur. J. 2003;9:2316.
[59] Schwieters CD, Kuszewski J, Tjandra N, Clore GM. J. Magn. Reson. 2003;160:65.
[60] Herrmann T, Güntert P, Wüthrich K. J. Mol. Biol. 2002;319:209.
[61] Cornilescu G, Delaglio F, Bax A. J. Biomol. NMR 1999;13:289.
[62] Wishart DS, Case DA. Methods Enzymol. 2001;338:3.
[63] Bentrop D, Bertini I, Cremonini MA, Forsén S, Luchinat C, Malmendal A. Biochemistry 1997;36:11605.
[64] Allegrozzi M, Bertini I, Janik MBL, Lee Y-M, Liu G, Luchinat C. J. Am. Chem. Soc. 2000;122:4154.
[65] Bertini I, Del Bianco C, Gelis I, Katsaros N, Luchinat C, Parigi G, Peana M, Provenzani A, Zoroddu MA. Proc. Natl. Acad. Sci. USA 2004;101:6841.
[66] Bertini I, Cavallaro G, Cosenza M, Kümmerle R, Luchinat C, Piccioli M, Poggi L. J. Biomol. NMR 2002;23:115.
[67] Turano P, Battaini G, Casella L. Chem. Phys. Lett. 2003;373:460.
[68] Simons KT, Kooperberg C, Huang E, Baker D. J. Mol. Biol. 1997;268:209.
[69] Lange OF, Rossi P, Sgourakis NG, Song Y, Lee HW, Aramini JM, Ertekin A, Xiao R, Acton TB, Montelione GT, Baker D. Proc. Natl. Acad. Sci. USA 2012;109:10873.
[70] Schmitz C, Vernon R, Otting G, Baker D, Huber T. J. Mol. Biol. 2012;416:668.
[71] Yagi H, Pilla KB, Maleckis A, Graham B, Huber T, Otting G. Structure 2013;21:883.
[72] Raman S, Lange OF, Rossi P, Tyka M, Wang X, Aramini J, Liu G, Ramelot TA, Eletsky A, Szyperski T, Kennedy MA, Prestegard J, Montelione GT, Baker D. Science 2010;327:1014.
[73] Berardi MJ, Shih WM, Harrison SC, Chou JJ. Nature 2011;476:109.
[74] Bertini I, Faraone-Mennella J, Gray BH, Luchinat C, Parigi G, Winkler JR. J. Biol. Inorg. Chem. 2004;9:224.
[75] Bertini I, Calderone V, Cerofolini L, Fragai M, Geraldes CFGC, Hermann P, Luchinat C, Parigi G, Teixeira JMC. FEBS Lett. 2012;586:557.
[76] Bertini I, Kursula P, Luchinat C, Parigi G, Vahokoski J, Willmans M, Yuan J. J. Am. Chem. Soc. 2009;131:5134.
[77] Chou JJ, Li S, Klee CB, Bax A. Nature Struct. Biol. 2001;8:990.
[78] Bertini I, Fragai M, Luchinat C, Melikian M, Toccafondi M, Lauer JL, Fields GB. J. Am. Chem. Soc. 2012;134:2100.
[79] Chou JJ, Li S, Bax A. J. Biomol. NMR 2000;18:217.
[80] Skrynnikov NR, Goto NK, Yang D, Choy W-Y, Tolman JR, Mueller GA, Kay LE. J. Mol. Biol. 2000;295:1265.
[81] Ulmer TS, Ramirez BE, Delaglio F, Bax A. J. Am. Chem. Soc. 2003;125:9179.
[82] Prestegard JH, Mayer KL, Valafar H, Benison GC. Methods Enzymol. 2005;394:175.
[83] Tian F, Valafar H, Prestegard JH. J. Am. Chem. Soc. 2001;123:11791.
[84] Gottstein D, Kirchner DK, Güntert P. J. Biomol. NMR 2012;52:351.
[85] Rinaldelli M, Ravera E, Calderone V, Parigi G, Murshudov GN, Luchinat C. Acta Cryst. D 2014;D70:958.

[86] Murshudov GN, Vagin AA, Dodson EJ. Acta Crystallogr. D Biol. Crystallogr. 1997;53:240.
[87] Murshudov GN, Skubàk P, Lebedev AA, Pannu NS, Steiner RA, Nicholls RA, Winn MD, Long F, Vagin AA. Acta Cryst. D 2011;67:355.
[88] Winn MD, Ballard CC, Cowtan KD, Dodson EJ, Emsley P, Evans PR, Keegan RM, Krissinel EB, Leslie AGW, McCoy A, McNicholas SJ, Murshudov GN, Pannu NS, Potterton EA, Powell HR, Read RJ, Vagin AA, Wilson KS. Acta Cryst. D 2011;67:235.
[89] Cornilescu G, Marquardt J, Ottiger M, Bax A. J. Am. Chem. Soc. 1998;120:6836.
[90] Case DA, Darden TA, Cheatham TE III, Simmerling CL, Wang J, Duke RE, Luo R, Walker RC, Zhang W, Merz KM, Roberts B, Hayik S, Roitberg A, Seabra G, Swails J, Goetz AW, Kilossvary I, Wong KF, Paesani F, Vanicek J, Wolf RM, Liu J, Wu X, Brozell SR, Steinbrecher T, Gohlke H, Cai Q, Ye X, Hsieh M-J, Cui G, Roe DR, Mathews DH, Seetin MG, Salomon-Ferrer R, Sagui C, Babin V, Luchko T, Gusarov S, Kovalenko A, Kollman PAAMBER 12, 12. San Francisco, CA: University of California; 2012.
[91] Banci L, Bertini I, Gori Savellini G, Romagnoli A, Turano P, Cremonini MA, Luchinat C, Gray HB. Proteins Struct. Funct. Genet. 1997;29:68.
[92] Bertini I, Case DA, Ferella L, Giachetti A, Rosato A. Bioinformatics 2011;27:2384.
[93] Fragai M, Luchinat C, Parigi G, Ravera E. Coord. Chem. Rev. 2013;257:2652.
[94] Simon B, Madl T, Mackereth CD, Nilges M, Sattler M. Angew. Chem. Int. Ed. 2010;49:1967.
[95] Guan J-Y, Keizers PH, Liu W-M, Löhr F, Skinner SP, Heeneman EA, Schwalbe H, Ubbink M, Siegal G. J. Am. Chem. Soc. 2013;135:5859.
[96] John M, Pintacuda G, Park AY, Dixon NE, Otting G. J. Am. Chem. Soc. 2006;128:12910.
[97] Fragai M, Luchinat C, Parigi G. Acc. Chem. Res. 2006;39:909.
[98] Wang LC, Pang YX, Holder T, Brender JR, Kurochkin AV, Zuiderweg ERP. Proc. Natl. Acad. Sci. USA 2001;98:7684.
[99] Eisenmesser EZ, Bosco DA, Akke M, Kern D. Science 2002;295:1520.
[100] Lindorff-Larsen K, Best RB, DePristo MA, Dobson CM, Vendruscolo M. Nature 2005;433:128.
[101] Huang YJ, Montelione GT. Nature 2005;438:36.
[102] Tzeng SR, Kalodimos CG. Nature 2009;462:368.
[103] Li P, Banjade S, Cheng H-C, Kim S, Chen B, Guo L, Llaguno M, Hollingsworth JV, King DS, Banani SF, Russo PS, Jiang Q-X, Nixon BT, Rosen MK. Nature 2012;483:336.
[104] Iwahara J, Clore GM. Nature 2006;440:1227.
[105] Tang C, Schwieters CD, Clore GM. Nature 2007;449:1078.
[106] Lipari G, Szabo A. J. Am. Chem. Soc. 1982;104:4546.
[107] Brüschweiler R, Roux B, Blackledge M, Griesinger C, Karplus M, Ernst RR. J. Am. Chem. Soc. 1992;114:2289.
[108] Clore GM, Szabo A, Bax A, Kay LE, Driscoll PC, Gronenborn AM. J. Am. Chem. Soc. 1990;112:4989.
[109] Ryabov YE, Fushman D. J. Am. Chem. Soc. 2007;129:3315.
[110] Bertini I, Luchinat C, Nagulapalli M, Parigi G, Ravera E. Phys. Chem. Chem. Phys. 2012;14:9149.
[111] Ishima R, Torchia DA. Nature Struct. Biol. 2000;7:740.
[112] Palmer AG III. Chem. Rev. 2004;104:3623.
[113] Wang X, Srisailam S, Ye AA, Lemak A, Arrowsmith C, Prestegard JH, Tian F. J. Biomol. NMR 2007;39:53.
[114] Assfalg M, Bertini I, Turano P, Mauk AG, Winkler JR, Gray BH. Biophys. J. 2003;84:3917.
[115] Longinetti M, Luchinat C, Parigi G, Sgheri L. Inv. Probl. 2006;22:1485.
[116] Bertini I, Ferella L, Luchinat C, Parigi G, Petoukhov MV, Ravera E, Rosato A, Svergun DI. J. Biomol. NMR 2012;53:271.
[117] Clore GM, Schwieters CD. J. Am. Chem. Soc. 2004;126:2923.
[118] Bertini I, Giachetti A, Luchinat C, Parigi G, Petoukhov MV, Pierattelli R, Ravera E, Svergun DI. J. Am. Chem. Soc. 2010;132:13553.
[119] Bertini I, Gupta YK, Luchinat C, Parigi G, Peana M, Sgheri L, Yuan J. J. Am. Chem. Soc. 2007;129:12786.

[120] Cerofolini L, Fields GB, Fragai M, Geraldes CFGC, Luchinat C, Parigi G, Ravera E, Svergun DI, Teixeira JMC. J. Biol. Chem. 2013;288:30659.
[121] Rieping W, Habeck M, Nilges M. Science 2005;309:303.
[122] Fisher CK, Huang A, Stultz CM. J. Am. Chem. Soc. 2010;132:14919.
[123] Vendruscolo M. Curr. Opin. Struct. Biol. 2007;17:15.
[124] Cavalli A, Camilloni C, Vendruscolo M. J. Chem. Phys. 2013;138:094112.
[125] Volkov AN, Worrall JAR, Holtzmann E, Ubbink M. Proc. Natl. Acad. Sci. USA 2006;103:18945.
[126] Tang C, Iwahara J, Clore GM. Nature 2006;444:383.
[127] Fawzi NL, Doucleff M, Suh JY, Clore GM. Proc. Natl. Acad. Sci. USA 2010;107:1379.
[128] Martin LJ, Hähnke MJ, Wöhnert J, Silvaggi NR, Allen KN, Schwalbe H, Imperiali B. J. Am. Chem. Soc. 2007;129:7106.
[129] Veglia G, Opella SJ. J. Am. Chem. Soc. 2000;122:11733.
[130] Barthelmes K, Reynolds AM, Peisach E, Jonker HRA, DeNunzio NJ, Allen KN, Imperiali B, Schwalbe H. J. Am. Chem. Soc. 2011;133:808.
[131] Wöhnert J, Franz KJ, Nitz M, Imperiali B, Schwalbe H. J. Am. Chem. Soc. 2003;125:13338.
[132] Su XC, McAndrew K, Huber T, Otting G. J. Am. Chem. Soc. 2008;130:1681.
[133] Su XC, Huber T, Dixon NE, Otting G. ChemBioChem 2006;7:1599.
[134] Su XC, Man B, Beeren S, Liang H, Simonsen S, Schmitz C, Huber T, Messerle BA, Otting G. J. Am. Chem. Soc. 2008;130:10486.
[135] Zhuang T, Lee HS, Imperiali B, Prestegard JH. Protein Sci. 2008;17:1220.
[136] Häussinger D, Huang J, Grzesiek S. J. Am. Chem. Soc. 2009;131:14761.
[137] Keizers PH, Desreux JF, Overhand M, Ubbink M. J. Am. Chem. Soc. 2007;129:9292.
[138] Hass MAS, Keizers PHJ, Blok A, Hiruma Y, Ubbink M. J. Am. Chem. Soc. 2010;132:9952.
[139] Saio T, Ogura K, Yokochi M, Kobashigawa Y, Inagaki F. J. Biomol. NMR 2009;44:157.
[140] Balayssac S, Bertini I, Bhaumik A, Lelli M, Luchinat C. Proc. Natl. Acad. Sci. USA 2008;105:17284.
[141] Bertini I, Luchinat C, Parigi G. Coord. Chem. Rev. 2011;255:649.
[142] Luchinat C, Parigi G, Ravera E, Rinaldelli M. J. Am. Chem. Soc. 2012;134:5006.
[143] Balayssac S, Bertini I, Lelli M, Luchinat C, Maletta M, Yeo KJ. J. Am. Chem. Soc. 2007;129:2218.
[144] Bertini I, Bhaumik A, De Paepe G, Griffin RG, Lelli M, Lewandowski JR, Luchinat C. J. Am. Chem. Soc. 2010;132:1032.
[145] Li J, Pilla KB, Li Q, Zhang Z, Su X, Huber T, Yang J. J. Am. Chem. Soc. 2013;135:8294.
[146] Bhaumik A, Luchinat C, Parigi G, Ravera E, Rinaldelli M. Cryst. Eng. Comm. 2013;15:8639.
[147] Bertini I, Fragai M, Lee Y-M, Luchinat C, Terni B. Angew. Chem. Int. Ed. 2004;43:2254.
[148] Liu W-M, Overhand M, Ubbink M. Coord. Chem. Rev. 2014;273–274:2.

CHAPTER

RELAXOMETRY AND CONTRAST AGENTS FOR MRI

10

The contrast in a magnetic resonance image is mainly determined by different proton T_1 and T_2 values in different regions of the biological object. The use of contrast agents to further differentiate the relaxation times between healthy and pathological tissues or between different regions, and thus to increase the image contrast, can be highly beneficial and effective depending on the agent distribution in the human body. High sensitivity of the contrast agents can be achieved through optimization of the structural and dynamic parameters of the paramagnetic molecules and exploitation of high-capacity carriers that can be loaded with a huge number of molecules.

10.1 INTRODUCTION

Magnetic resonance imaging (MRI) is a widespread technique used in clinical diagnosis to obtain images of the human body by exciting nuclear spins through radiowaves in the presence of high magnetic fields (usually between 0.2–7 T). The orientation of the various image planes is controlled by gradients applied to the magnetic field. MRI is a widely used imaging technique because it provides images with an excellent resolution without requiring ionizing radiations or radioactive materials. Hydrogen atoms present in any tissues containing water molecules are excited, so that they produce a radiofrequency signal, the amplitude of which depends on the density of the water molecules and on the longitudinal and transverse water proton relaxation rates. This results in an image contrast between different tissues which depends on the differences in such parameters. In order to increase the contrast it can thus be convenient to use paramagnetic agents that can increase the longitudinal and transverse relaxation rates of water protons selectively for specific tissues or parts thereof. Gadolinium(III) complexes are the most used contrast agents, due to the large paramagnetic relaxation enhancements that they induce thanks to the large electron relaxation time (Section 8.6) and electron spin quantum number. These agents are classified as T_1 (positive) agents, together with manganese(II) agents. Effective T_1 contrast agents exhibit a ratio between transverse and longitudinal relaxation rates close to 1. Conversely, iron-oxide particles are classified as T_2 (negative) agents, because they decrease the transverse relaxation time of proton spins much more than the longitudinal relaxation time ($R_2/R_1 \gg 1$). Novel promising contrast agents for MRI are represented by paramagnetic supramolecular scaffolds, which provide synthetic ease, versatility, and good capability of enhancing the contrast. Paramagnetic lanthanoid complexes different from Gd^{3+} are also emerging as contrast agents based on the chemical exchange saturation transfer (CEST) effect.

NMR of Paramagnetic Molecules. http://dx.doi.org/10.1016/B978-0-444-63436-8.00010-7

10.2 RELAXIVITY AND RELAXATION PROFILES

The relaxivity of a paramagnetic complex is defined as the paramagnetic enhancement of the water proton relaxation rates in the presence of 1 mM concentration of the paramagnetic complex, as seen in Section 6.7. The effectiveness of a paramagnetic complex as a T_1 contrast agent can thus be described by its relaxivity, r_1. The enhancement in water proton relaxation rate is thus written as a linear function of the concentration of the paramagnetic metal ion expressed as millimoles per dm^3, i.e. of the concentration of the contrast agent, $[CA]$, in the simplest case that the latter contain only one paramagnetic ion,

$$R_1 = [CA]r_1 + R_{1\mathrm{dia}} \tag{10.1}$$

where $R_{1\mathrm{dia}}$ is the relaxation rate of the solution without the paramagnetic complex.

The paramagnetic relaxation enhancement can thus be determined from the relaxivity of the paramagnetic complex, r_1, in the hypothesis that the parameters from which r_1 depends do not change for the different concentration of the paramagnetic molecules. We report here for the readers' convenience the relevant equations, providing the Solomon-Bloembergen-Morgan (SBM) model as far as the inner-sphere relaxivity is concerned and the Freed model as far as outer-sphere relaxivity is concerned, already introduced in Chapters 4 and 6:

$$r_1 = \frac{[CA]q}{55.6}\frac{1}{T_{1M} + \tau_M} + r_{1\mathrm{os}} \tag{10.2}$$

$$\frac{1}{T_{1M}} = \frac{2}{15}\left(\frac{\mu_0}{4\pi}\right)^2 \frac{\gamma_{\mathrm{H}}^2 g_e^2 \mu_B^2 S(S+1)}{r^6}\left[\frac{7\tau_c}{1+\omega_S^2\tau_c^2} + \frac{3\tau_c}{1+\omega_{\mathrm{H}}^2\tau_c^2}\right] + \frac{2}{3}S(S+1)\left(\frac{A_c}{\hbar}\right)^2 \frac{\tau_{\mathrm{con}}}{1+\omega_S^2\tau_{\mathrm{con}}^2} \tag{10.3}$$

$$\frac{1}{\tau_c} = \frac{1}{\tau_r} + \frac{1}{\tau_M} + \frac{1}{\tau_e} \qquad \frac{1}{\tau_{\mathrm{con}}} = \frac{1}{\tau_M} + \frac{1}{\tau_e} \tag{10.4}$$

$$\frac{1}{\tau_e} = \frac{1}{25}\Delta_t^2\tau_v[4S(S+1)-3]\left(\frac{1}{1+\omega_S^2\tau_v^2} + \frac{4}{1+4\omega_S^2\tau_v^2}\right) \tag{10.5}$$

$$r_{1\mathrm{os}} = \frac{32}{405}\pi\left(\frac{\mu_0}{4\pi}\right)^2 \frac{N_A\gamma_{\mathrm{H}}^2 g_e^2 \mu_B^2 S(S+1)}{d(D_M + D_L)}\left[7J^{tr}(\omega_S) + 3J^{tr}(\omega_H)\right] \tag{10.6}$$

$$J^{tr}(\omega) = \frac{1 + 5z/8 + z^2/8}{1 + z + z^2/2 + z^3/6 + 4z^4/81 + z^5/81 + z^6/648} \tag{10.7}$$

$$z = \sqrt{2\left(\omega\tau_D + \frac{\tau_D}{\tau_e}\right)} \tag{10.8}$$

$$\tau_D = \frac{d^2}{D} \tag{10.9}$$

where $[CA]q/55.6 = P_M$ is the mole fraction of ligand nuclei (i.e. of water protons in our case) coordinated to the paramagnetic metal in the contrast agent. The structural parameters on which the relaxivity depends, as described by the SBM model, are the number (q) of the water molecules coordinated to each paramagnetic ion or bound/coordinated/hydrogen-bonded around the paramagnetic ion, the distance (r) of the corresponding water protons, and the distance of closest approach (d) between the paramagnetic ion and the water protons. The dynamic parameters are the reorientation time of the molecule

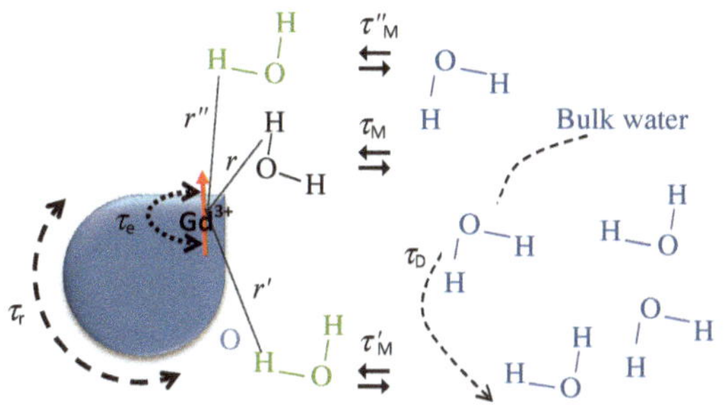

FIGURE 10.1 The relaxivity of bulk water protons can have contributions from first sphere water molecules (in *black*), second sphere water molecules (in *green*) and from the free diffusion of outer-sphere molecules (in *blue*). These contributions depend on proton-paramagnetic metal [Gd^{3+} in this case] distances and the times τ_r, τ_e, τ_M, and τ_D. In this figure, one first sphere water molecule, with metal-proton distance r, and two second sphere water molecules, with distances r' and r'', are shown.

(τ_r), the lifetime of the coordinated water protons (τ_M), and the diffusion coefficients (D) of paramagnetic complexes and water molecules, while the electronic parameters are the transient ZFS (Δ_t) and the correlation time for electron relaxation (τ_v). The contact term [with coupling constant A_c, second term in Eq. (10.3)] is usually negligible due to the absence of sizable unpaired electron delocalization and spin polarization on the water protons, and thus will be considered zero throughout this chapter.

Of course, different water molecules can be coordinated in fixed positions at different distances from the metal ion, with different τ_M values. In this case, Eq. (10.2) should be replaced by Eq. (10.10)

$$r_1 = \sum_i \frac{[CA]q_i}{55.6} \frac{1}{T_{1M,i} + \tau_{M,i}} + r_{1os} \qquad (10.10)$$

where q_i is the number of the water molecules at the same distance r_i from the metal (on which $T_{1M,i}$ depends) and $\tau_{M,i}$ is their lifetime. Inner-sphere water molecules are usually classified as first-sphere or second-sphere water molecules depending on their coordination level with respect to the metal ion. Fig. 10.1 shows a summary of all parameters described earlier.

As discussed in Chapter 4, the correlation times in the Lorentzian dispersion function depend on either T_{1e} or T_{2e}, which may differ at high field, instead of a single τ_e time. T_{2e} actually enters the definition of the correlation time present in the dispersive ω_S term; however, since the corresponding dispersion occurs at relatively low fields, when the difference between T_{1e} and T_{2e} is small, it can be well approximated by T_{1e}, so that a single τ_e can be conveniently used, as reported in Eqs. (10.3)–(10.5). These equations describe the electron relaxation using the pseudorotational model, and their validity is limited to systems in the Redfield limit (Section 4.13), when the Zeeman energy is much larger than the energy related to static ZFS (Section 1.4) and to the hyperfine coupling between nuclear and electron spins (Section 4.7.1), if these effects are present.

The lifetime τ_M enters Eq. (10.2) twice: in the determination of the correlation time τ_c [when τ_r and τ_e are of the same order of longer, Eqs. (10.3) and (10.4)] and in the control of the efficiency of the transmission of the paramagnetic relaxation enhancement from the protons dipole–dipole coupled with the paramagnetic ion to the bulk water protons. For complexes with τ_M as large as to be comparable with T_{1M} (intermediate or slow exchange regime), τ_M represents a limiting factor for relaxivity.

In the SBM model, the electron–nucleus spin system is assumed to be rigidly held within a molecule isotropically rotating in solution. If the molecule cannot be treated as a rigid sphere, its motion is in general anisotropic, and multiple different reorientation correlation times should be considered, depending on the principal values of the rotational diffusion tensor [1]. In many cases, the presence of local reorientation motions should also be considered. A general treatment for the presence of internal motions faster than the reorientation correlation time of the whole molecule is provided by the Lipari-Szabo model-free treatment [2]. Relaxation is calculated as the sum of two terms [3]

$$\frac{1}{T_{1M}}=\frac{2}{15}\left(\frac{\mu_0}{4\pi}\right)^2\frac{\gamma_{\rm H}^2 g_e^2\mu_B^2 S(S+1)}{r^6}\left[S_{LS}^2\left(\frac{7\tau_c}{1+\omega_S^2\tau_c^2}+\frac{3\tau_c}{1+\omega_{\rm H}^2\tau_c^2}\right)+\left(1-S_{LS}^2\right)\left(\frac{7\tau_f}{1+\omega_S^2\tau_f^2}+\frac{3\tau_f}{1+\omega_{\rm H}^2\tau_f^2}\right)\right] \quad (10.11)$$

where S_{LS} is a generalized order parameter, $\tau_f^{-1}=\tau_c^{-1}+\tau_l^{-1}$, and τ_l is the correlation time for the local (fast) metal-proton vector reorientation. In this approach, S_{LS}^2 and τ_l are two new adjustable parameters. The generalized order parameter reports on the degree of spatial restriction of the motion. No local motion means $S_{LS}^2=1$, completely isotropic internal motion means $S_{LS}^2=0$.

Due to the large number of parameters on which the relaxivity depends (Fig. 10.2), the paramagnetic complexes are often analyzed by the acquisition of nuclear magnetic relaxation dispersion (NMRD) profiles (Section 6.7). The field dependence of the relaxivity in fact often permits to determine some of the parameters described above, responsible for the relaxation enhancements. The knowledge of these parameters for a series of different complexes can actually allow for an estimate of the effects that the different structural conformations have over these parameters, making thus possible to design new molecules optimized at the magnetic fields used for medical investigations. On the other hand, the presence of static ZFS may prevent a reliable fit of the low field data (when the Zeeman energy is smaller than the ZFS energy) against the SBM model, and more advanced models should be used (Section 10.4).

Structural parameters	***q***, number(s) of water molecules, and ***r***, distance(s) from the paramagnetic metal ion, for first sphere and second sphere water molecules S_{LS}, order parameter ***d***, distance of closest approach of diffusing water molecules
Dynamic parameters	τ_r, molecular reorientation time τ_l, fast local reorientation time τ_M, lifetime(s) of first sphere and second sphere water molecules D_{diff}, diffusion coefficient
Electronic parameters	Δ_t, transient zero-field splitting (ZFS) τ_v, correlation time for electron relaxation D_{ZFS}, static ZFS θ, angle(s) between the z-axis of the static ZFS and the metal-proton direction(s)

FIGURE 10.2 Summary of the parameters affecting the proton relaxivity of solvent water molecules in solutions containing Gd^{3+} contrast agents.

10.3 THE IDEAL CONTRAST AGENT

The evolution of MRI requires the development of nontoxic contrast agents able to target specific organs or diseased tissues and to determine selectively a large increase in relaxivity. As seen in the previous section, the relaxivity depends on (1) structural parameters, (2) dynamic parameters, (3) electronic parameters, besides on the applied magnetic field. Therefore, optimized agents must be selected in order to match the best conditions resulting from the three types of parameters at the magnetic field of interest. The theory predicts that high relaxivities at the imaging fields (commonly between 0.5 and 3 T) may be observed for systems with relatively long reorientation times and relatively fast water exchange.

Fig. 10.3A shows the relaxivity profile as calculated for one water molecule ($q = 1$) regularly bound to Gd^{3+} ($r = 3.1$ Å) with $\tau_r = 100$ ps, $\tau_M = 10$ ns, $\Delta_t = 0.03$ cm^{-1}, and $\tau_v = 20$ ps, and by increasing τ_r to 1, 10, and 100 ns; Fig. 10.3B–C shows the relaxivity profiles calculated using $\tau_r = 100$ ps and $\tau_r = 100$ ns, respectively, and $\tau_M = 1$, 10, 100, and 1000 ns. These figures clearly indicate the influence of these parameters on the relaxivity profile and show that when the correlation time is dominated by the field dependent electron relaxation time, because τ_r and τ_M are longer, a peak in the high field region appears (Section 4.7.2). The peak is higher when τ_M is much shorter than T_{1M}. Interestingly, the effect of τ_M is not very relevant for fast reorienting molecules (Fig. 10.3B), but it is of paramount importance in slowly tumbling complexes (Fig. 10.3C).

Fig. 10.4 shows the NMRD profile of the complex Gd-EDTA and of the same complex covalently linked to the protein BSA [4]. It is apparent that in the latter system the correlation time for nuclear relaxation is governed by the electron relaxation time, the hump in the high field region being due to its field dependence, and the relaxivity is much higher than in the free complex. Therefore, the strategy adopted since many years for the development of contrast agents with high relaxivity consists in taking advantage of the hump in relaxivity caused by the field dependent electron relaxation by selectively increasing the rotational correlation time when the molecule is in a particular environment, through the formation of adducts with proteins, macromolecules or tissues. The complexes are usually functionalized with groups that increase their ability to bind noncovalently to macromolecules, for instance with the human serum albumin present in the plasma, in order to increase their rotational time, but at the same time to still permit dissociation and excretion of the complex. Importantly, the formed adduct should be rigid, so that the Gd^{3+}-water proton vectors are reoriented by the tumbling of the macromolecule, and not by faster local motions, the presence of which largely decreases the relaxivity peak. Fig. 10.5 shows the effect of decreasing S_{LS}^2 values [Eq. (10.11)] from 1 to 0 on water relaxivity for a Gd-complex with $q = 1$, $\tau_r = 10$ ns, $\tau_M = 10$ ns, $\Delta_t = 0.03$ cm^{-1}, $\tau_v = 20$ ps, and $\tau_l = 100$ ps as a function of the applied magnetic field. The relaxivity peak appears clearly even for small S_{LS}^2 values, but the ratio between the relaxivity at the frequency of the peak and at larger frequencies is much reduced with respect to the $S_{LS}^2 = 1$ case.

As seen in Chapter 6 and shown in Fig. 10.3A–C, relaxivity is also strongly influenced by the exchange time of coordinated water protons with protons of bulk water. The optimal value of τ_M in order to have the maximum value of relaxivity is about 10^{-8} s [5] for fields of 0.5 T. The optimal value for τ_M can be estimated by plotting the relaxivity as a function of τ_M and of τ_e, as shown in Fig. 10.6A–D for magnetic fields of 0.1, 1, 3, and 10 T, with a very large reorientation time (100 ns). The largest relaxivities are obtained for τ_M much shorter than T_{1M}, and τ_c much shorter τ_M. At the magnetic field of 1 T, the maximum relaxivity that can arise in the presence of one water molecule regularly coordinated to Gd^{3+} is about 60 s^{-1} mM^{-1}.

Fig. 10.7 shows the dependence of $1/T_{1M}$ on the correlation time at different magnetic fields: whereas at very low fields an increase in τ_c results in an increased $1/T_{1M}$ value (in the range of values shown in the figure), at intermediate and high fields an increase in τ_c can either determine a decrease or an increase in $1/T_{1M}$. Therefore, the largest relaxivity is achieved for a particular value of τ_c, which decreases

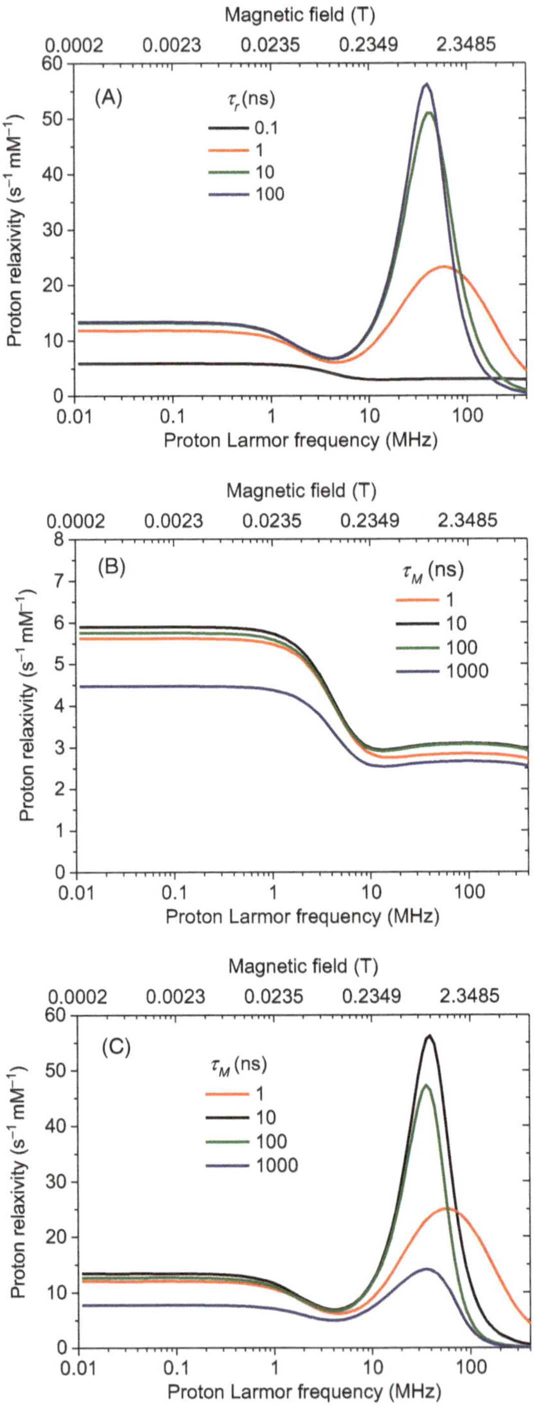

FIGURE 10.3 Relaxivity profiles calculated for one water molecule ($q = 1$) regularly bound to Gd^{3+} ($r = 3.1$ Å) with $\Delta_t = 0.03\ cm^{-1}$, $\tau_v = 20$ ps and (A) $\tau_M = 10$ ns, $\tau_r = 0.1$, 1, 10, and 100 ns (the labels refer to the different τ_r values); (B) $\tau_r = 100$ ps and $\tau_M = 1$, 10, 100, and 1000 ns (the labels refer to the different τ_M values); (C) $\tau_r = 100$ ns and $\tau_M = 1$, 10, 100, and 1000 ns (the labels refer to the different τ_M values).

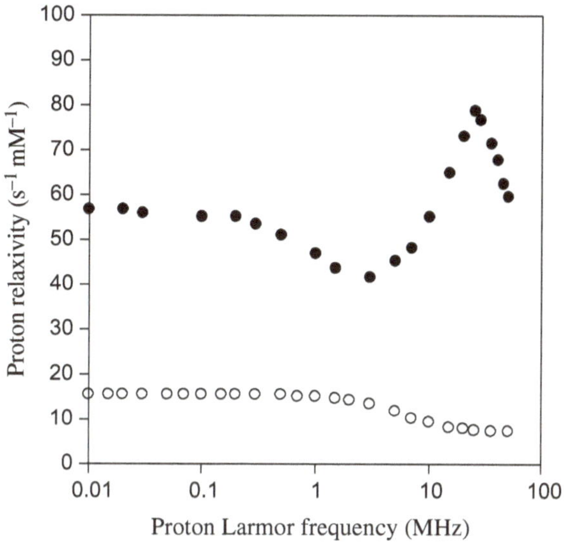

FIGURE 10.4 Water ^1H NMRD profiles for Gd-EDTA (○) and for its covalent adduct with bovine serum albumin (●).

Adapted from Ref. [4].

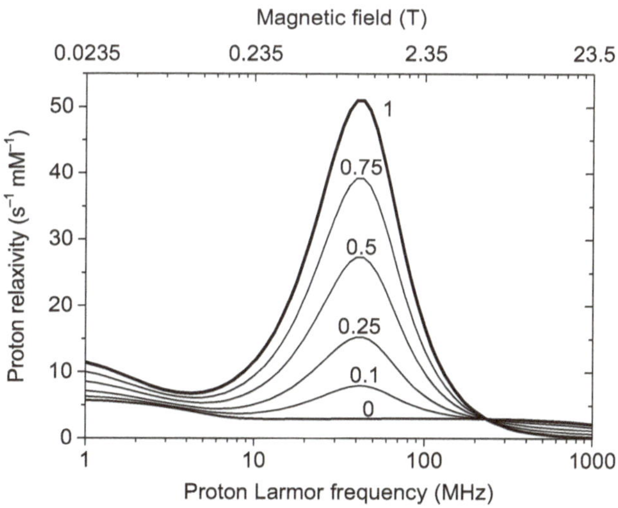

FIGURE 10.5 Relaxivity profiles calculated for one water molecule ($q = 1$) regularly bound to Gd^{3+} ($r = 3.1$ Å) with $\Delta_t = 0.03$ cm^{-1}, $\tau_v = 20$ ps, $\tau_M = 10$ ns, $\tau_r = 10$ ns, $\tau_l = 100$ ps, and different $S_{LS}{}^2$ values (indicated by the labels).

with increasing magnetic field. In particular, the maximum is obtained when the condition $\tau_c = 1/\omega_H$ is met, as can be easily checked through the first derivative of the Lorentzian function. This means that while at 0.1–0.5 T, the optimal τ_c is of the order of ten ns, at 1, 1.5, 3, and 7 T, the optimal τ_c decreases to about 4, 2.5, 1, and 0.5 ns, respectively. Fig. 10.8 show the relaxivity in the presence of one water molecule regularly coordinated to Gd^{3+} as a function of τ_M and of $(1/\tau_e + 1/\tau_r)^{-1}$ (i.e. of τ_e if $\tau_r \gg \tau_e$)

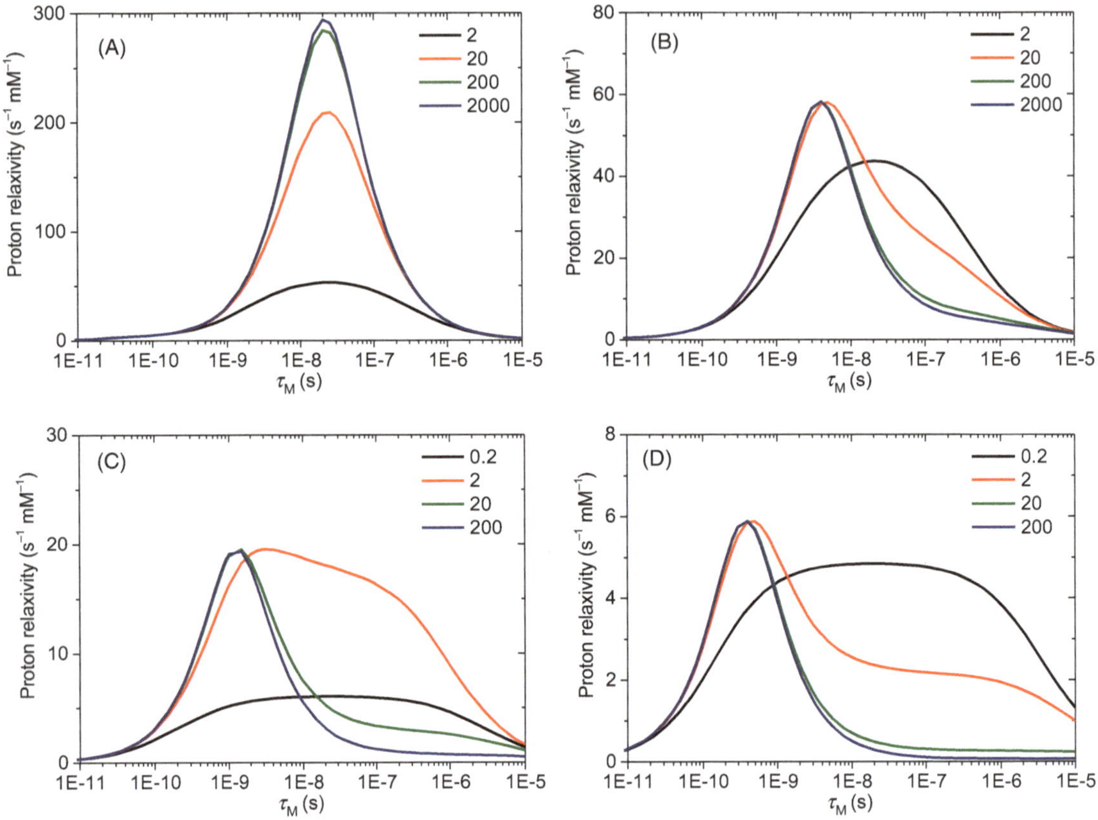

FIGURE 10.6 Proton relaxivity as a function of τ_M at (A) 0.1, (B) 1, (C) 3, and (D) 10 T for τ_r = 100 ns and different τ_e values (q = 1, r = 3.1 Å). The labels refer to the τ_e values in ns.

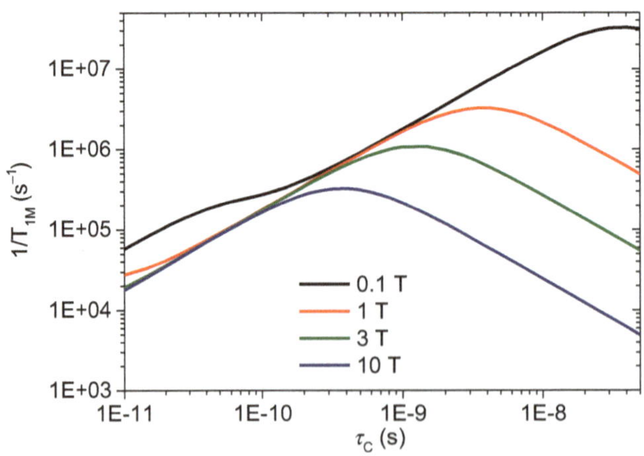

FIGURE 10.7 Paramagnetic enhancement of the relaxation rate for a proton at 3.1 Å from Gd^{3+} as a function of the correlation time at different fields (0.1, 1, 3, and 10 T).

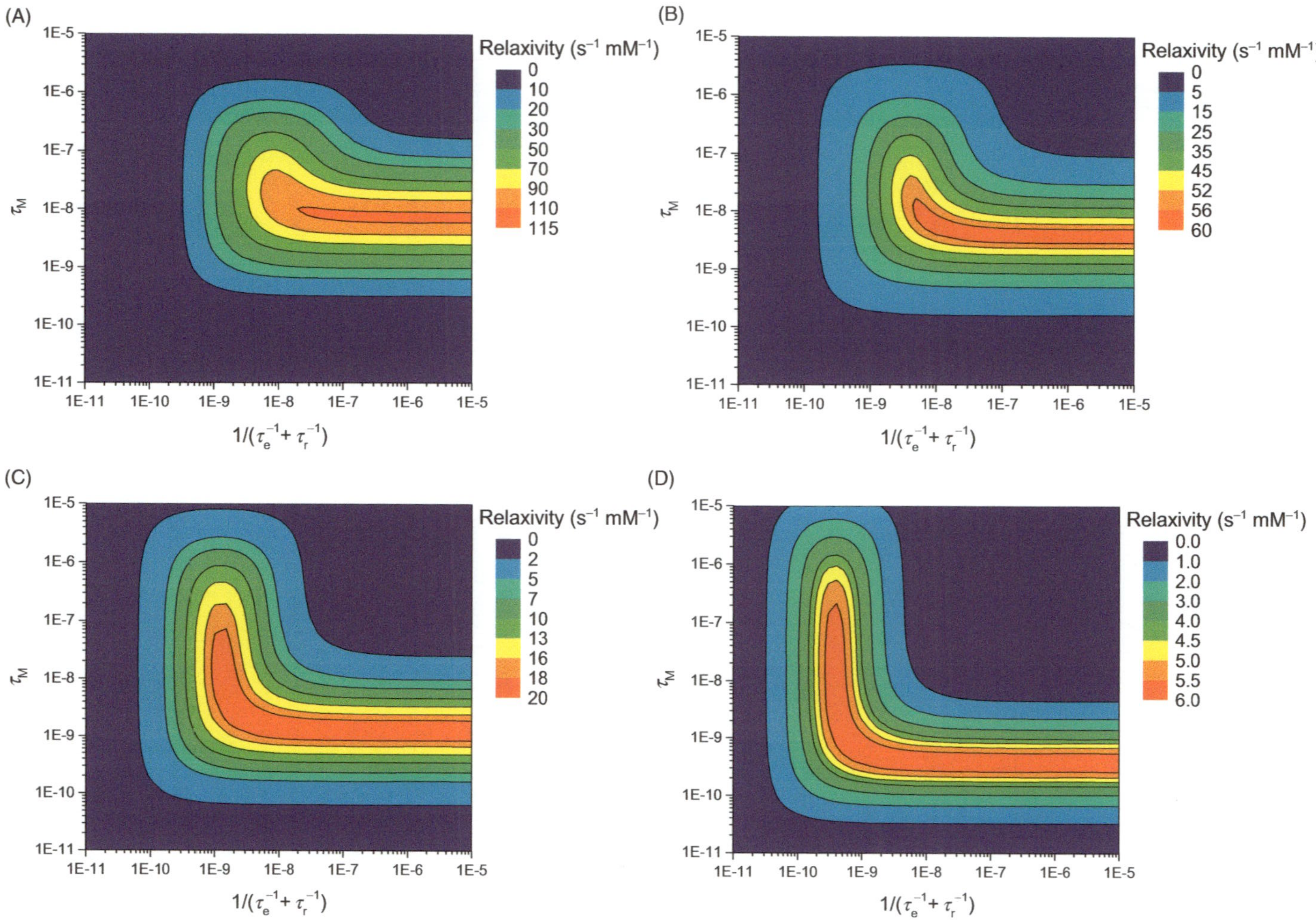

FIGURE 10.8 Proton relaxivity in the presence of one water molecule regularly coordinated to Gd^{3+} as a function of τ_M and of $(1/\tau_e + 1/\tau_r)^{-1}$ at four different fields: (A) 0.5 T, (B) 1 T, (C) 3 T, and (D) 10 T.

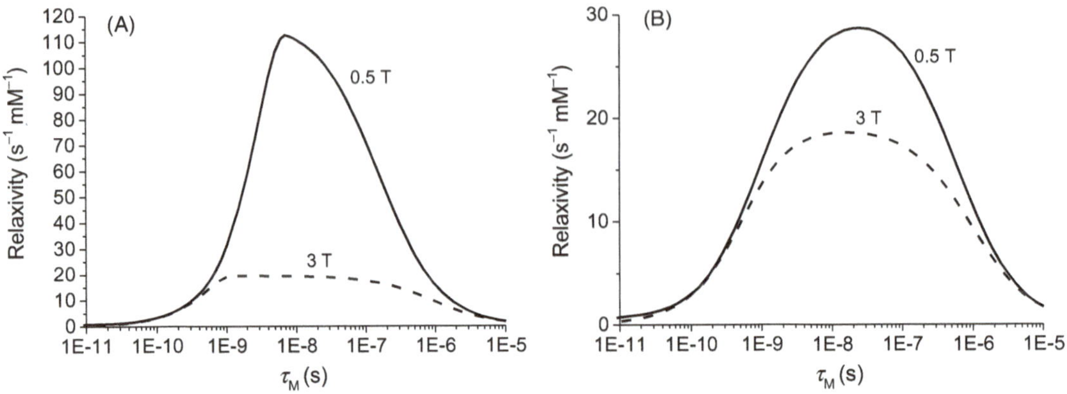

FIGURE 10.9 (A) Proton relaxivity as a function of τ_M at 0.5 and 3 T for optimized τ_r and τ_e values ($q = 1$, $r = 3.1$ Å). The optimized τ_r and τ_e values are determined in order to obtain τ_c as close as possible to $1/\omega_I$. (B) Proton relaxivity as a function of τ_M at 0.5 and 3 T for $\tau_r = 1$ ns.

at four different fields. From these plots, it is clear that how the correlation times affect the relaxation rates in different ways depending on the applied magnetic field. In particular, the range of τ_M values providing maximized relaxivity is much larger at high fields (3–10 T) than at low fields (0.5–1 T), because at high fields similar relaxivities can be obtained for τ_M differing of orders of magnitude if the values of τ_r and τ_e are also properly optimized (Fig. 10.9A) [6]. This broader range of water lifetime providing maximized relaxivity at high fields is reduced if τ_r (or τ_e) are restrained to relatively small values (Fig. 10.9B).

Obviously, the relaxivity increases for a hydration number larger than 1, as q acts as a scaling factor in Eq (10.2). All commercial Gd^{3+}-based contrast agents have $q = 1$ due to the need of avoiding the release of the metal ion under physiological conditions and thus to avoid the toxicity problems related to free Gd^{3+} ion. In fact, a number of coordinated ligands lower than 8 is often (but not always) accompanied by a large decrease of the thermodynamic stability of the complex. Several stable complexes with $q = 2$ have been however developed [7–10]. On the other hand, an increase in the "effective" q can be obtained in the presence of second sphere water molecules (Section 10.4.1), which may provide important contributions to relaxivity, although at larger distances from the metal, if their lifetime is longer than the diffusional correlation time [τ_D, Eq. (10.9)].

The rate of water exchange between coordinated water and the bulk can be estimated by ^{17}O NMR by measuring the transverse relaxation rate of water in the presence of the Gd^{3+} complex as a function of temperature, reminding that, since the oxygen is directly coordinated to the Gd^{3+} ion, R_{2M} is dominated by the contact term [11,12]. Therefore, if the paramagnetic shift $\Delta\omega$ is much smaller than $1/T_{2M}$ (Section 6.3.2), as in the case of Gd^{3+} complexes, the transverse relaxivity (r_2) is provided by

$$r_2 = \frac{[CA]q}{55.6}\frac{1}{T_{2M} + \tau_M} \tag{10.12}$$

$$\frac{1}{T_{2M}} = \frac{1}{3}S(S+1)\left(\frac{A_c}{\hbar}\right)^2 (1/\tau_e + 1/\tau_M)^{-1} \tag{10.13}$$

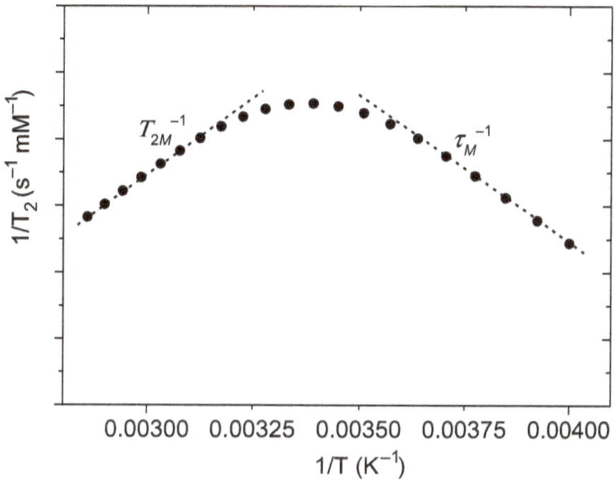

FIGURE 10.10 ^{17}O transverse relaxation rate for water coordinated to Gd^{3+} as a function of temperature.

at fields large enough so that the dispersive term in Eq. (4.28) provides negligible contributions [i.e. for $\omega_S > (1/\tau_e + 1/\tau_M)$]. In the limit of $\tau_M \gg T_{2M}$, r_2 increases with increasing temperature because τ_M decreases [so that the denominator in Eq. (10.12) increases]; on the contrary, in the limit of $\tau_M \ll T_{2M}$, r_2 decreases with increasing temperature because the correlation time decreases (and thus T_{2M} increases). From the temperature dependence of r_2 at a fixed field, τ_M can thus be determined (Fig. 10.10).

It has been shown that τ_M depends on the coordination geometry of the paramagnetic metal and on the steric crowding at the water coordination site and in the second coordination sphere. Therefore, τ_M can be optimized by changing number, hydrophobic nature and steric hindrance of the ligands at the metal site [13,14]. Also the charge of the complex can influence the exchange time and thus the relaxivity, due to the electrostatic nature of the Ln-O coordination bond. As a result, τ_M decreases, i.e. the exchange rate becomes faster, with increasing the negative charge of the complex [12,15].

10.4 NMRD-DERIVED PARAMETERS

10.4.1 COORDINATED PROTONS AND THEIR DISTANCE

The experimental relaxivity profiles are determined by the sum of the contributions from all protons dipole–dipole coupled to the paramagnetic metal [as indicated by the sum in Eq. (10.10)]. These protons can be diffusing around the paramagnetic molecule (outer-sphere relaxation) or localized in well-defined positions with respect to the paramagnetic center (inner-sphere relaxation). In the hypothesis that all "inner-sphere protons" are in fast exchange, i.e. their exchange rate is faster than their relaxation rate, relaxometry measurements can only provide a value for $\sum_i q_i/r_i^6$, where q_i is the number of equivalent water molecules and r_i is their distance. Therefore, from the analysis of the experimental profiles, either a value for the proton(s)–metal ion distance is obtained, in the assumption that the number of protons around the metal ion is known and that all of them are at the same distance, or the hydration number is obtained, if the distance of coordinated water protons is fixed in the small range allowed by coordination chemistry.

In the presence of fast exchanging water molecules/protons (or hydroxide ions/protons) in the first coordination sphere of the paramagnetic metal ion, their contribution to solvent water proton relaxation is usually dominant. However, interactions may be present between the first coordination sphere and further water molecules, for instance through hydrogen-bonding with polar groups of the ligand (Fig. 10.1). If these water molecules, localized in well-defined positions, have lifetimes longer than their diffusional correlation times, a second coordination sphere contribution to relaxation can arise (providing usually a smaller contribution due to the r^{-6} dependence). The picture is even more complicated in the presence of slow-exchanging protons with different exchange time. In case that the first sphere protons are all exchanging with a very low rate, for instance, and second sphere protons are fast exchanging, the latter will provide the largest contribution to relaxation. Furthermore, the assumption of rigid molecular reorientation, underlying Eqs. (10.3)–(10.5), artificially increases the calculated relaxation rates with respect to the correct values that would be calculated if the presence of fast local motions were taken into account. This sometimes results in estimates for proton-metal ion distances which are longer, in order to match the calculated relaxivity values with the experimental ones.

In summary, the relaxation profiles can easily provide information on the presence or absence of protons in the first coordination sphere of a paramagnetic metal ion, but it may be hard to determine from these data the number of coordinated water molecules and their exact distance from the paramagnetic center without further independent information. This is even more so in the presence of static ZFS, when the SBM model cannot account for the low field relaxation data, and the paramagnetic contribution to the relaxation rate of coordinated protons depends on their position with respect to the principal axes of the ZFS tensor (Section 4.7.1).

10.4.2 REORIENTATION, ELECTRON RELAXATION, AND RESIDENCE TIMES

Due to the large number of parameters on which the relaxivity profiles depend, there is the possibility that multiple sets of values for these parameters provide relaxivity profiles in equally good agreement with the experimental data. To remove this degeneracy, it may be convenient to acquire relaxivity profiles at several temperatures, and to fit all of them simultaneously. In this way, we can constrain some parameters (like r, the static ZFS and, possibly, Δ_t) to have the same values for all temperatures, whereas other parameters (like the reorientational correlation time, the exchange correlation time, the correlation time for electron relaxation) must vary with temperature according to the following relationships:

$$\tau_r = A_R \exp(B_R/T) \tag{10.14}$$

$$\tau_v = A_v \exp(B_v/T) \tag{10.15}$$

$$\tau_M = \frac{A_M}{T} \exp(B_M/T) \tag{10.16}$$

where $B_M = \Delta G^{\neq}/R$ [$\Delta G^{\neq}$ is the free energy of activation for the dissociation process, from the Eyring relationship, Eq. (6.19)].

From Eq. (10.16), τ_M decreases significantly when temperature increases. A strong increase in the solvent relaxivity, proportional to $(\tau_M + T_{1M})^{-1}$, when temperature increases is a clear indication of a

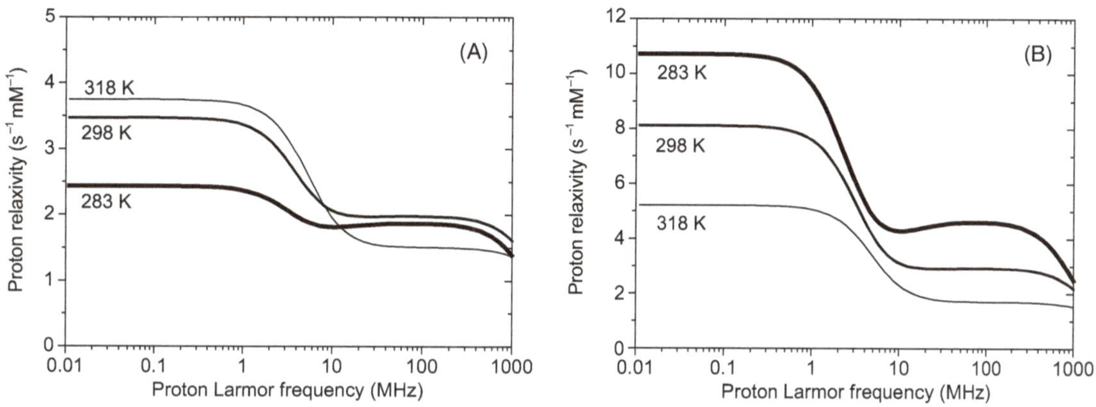

FIGURE 10.11 Relaxivity profiles calculated for one water molecule ($q = 1$) regularly bound to Gd^{3+} ($r = 3.1$ Å) with $\Delta_t = 0.02\ cm^{-1}$, $\tau_v = 16$, 12 or 8.9 ps, $\tau_r = 150$, 94 or 54 ps and (A) $\tau_M = 5.8$, 3.0 or 1.4 μs or (B) $\tau_M = 58$, 30 or 14 ns (283, 298 or 318 K, respectively).

slow or intermediate exchange regime, i.e. of τ_M being of the same order of or larger than T_{1M}. In fact, T_{1M} increases with increasing temperature in all cases where τ_c is dominated by the reorientation time, the proton lifetime or a field independent effective electron relaxation time. T_{1M} may actually decrease at low fields if τ_c is a field dependent effective electron relaxation time, and Δ_t is temperature independent, but it increases in correspondence of the high field relaxivity peak [Eq. (10.5)]. Fig. 10.11 shows the different low field temperature dependence of relaxivity in the slow exchange regime (panel A) or in the fast exchange regime (panel B), calculated by changing the values of τ_M by two orders of magnitude (1–6 μs in panel A, 0.01–0.06 μs in panel B).

Δ_t^2 in the equation describing the field dependence of the electron relaxation as reported in Eq. (10.5) must be provided in s^{-2}. However, Δ_t is often reported in cm^{-1}. To calculate Δ_t in s^{-1} from Δ_t in cm^{-1}, the latter must be multiplied by 2π times the speed of light, c.

Attention should be paid to the fact that that electron relaxation must be within the Redfield limit ($\tau_e > \tau_v$, Section 4.13), as otherwise good fits can be still obtained but with parameters that may be wrong by orders of magnitude [16]. This condition does not need to be fulfilled if the slow motion theory is used (see in subsequent section).

10.4.3 BEST FIT PROGRAMS

From the analysis of the experimental NMRD profiles it is possible to recover some of the parameters determining the relaxivity profile, assuming that the values from the other parameters (or their order of magnitude) are known from other independent measurements or from theoretical predictions. A best fit analysis of the NMRD profiles is thus usually performed against the SBM model. However, as discussed in Section 4.7.1, ZFS effects may cause large deviations from the SBM predictions of the relaxivity at the fields where the Zeeman energy is smaller or of the order of the ZFS energy. A major effect of the introduction of static ZFS is actually that of reducing the amplitude of the $\omega_S \tau_c$

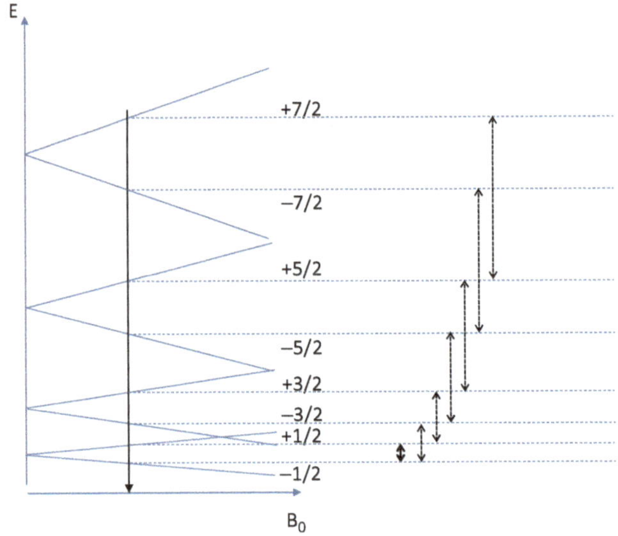

FIGURE 10.12 ZFS removes the degeneracy of the electron levels by splitting the levels with different $|M_S|$ at zero field. The energy values of the electronic spin transitions at a given B_0 are those corresponding to the arrows in the left side of the figure: only the $1/2 \leftrightarrow -1/2$ transition occurs at the electron Zeeman energy.

dispersion because the energy of most electronic spin transitions across ZFS levels is large and can be much larger than the Zeeman energy (Fig. 10.12). In fact, when S is a half-integer, only the $1/2 \leftrightarrow -1/2$ transition occurs at the Zeeman energy, since only the energy of these levels is not affected by the presence of ZFS (in the case of integer S, the $\omega_S \tau_c$ dispersion may be totally quenched). As discussed in Section 4.7.1, the presence of static ZFS may also cause the occurrence of a further dispersion in the relaxivity profile, corresponding to the transition from the dominant ZFS limit to the dominant Zeeman limit.

In the presence of static ZFS (values of 0.02–0.05 cm^{-1} are expected for Gd^{3+} complexes [17]), the so-called modified Florence NMRD model can be used if the reorientation time of the molecule is longer than the electron relaxation time [18–20]. The modified version of the Florence NMRD program takes into account the effect of the ZFS not only in the description of the nuclear relaxation rates with given electron relaxation time values (as the original Florence NMRD program does), but also the effect of the ZFS in the description of the electron relaxation time, with given Δ_t and τ_v values. Other approaches proposed for the analysis of the NMRD profiles in the presence of static ZFS have been developed by Fries and the Grenoble group [21,22] and by Kowalewski and the Stockholm group [20,23,24] (Table 10.1). The latter approach (called slow motion theory) is more general, in the sense that it can be applied basically without any limiting conditions, and does not pass from a description of the electron relaxation time. It was shown that the slow motion theory provides results in good agreement with both the modified Florence NMRD program and the Grenoble group program, in their conditions of applicability [25], thus allowing the use of these simpler and less computationally demanding programs whenever possible.

Table 10.1 Main Programs Available for the Analysis of the NMRD Profiles of Paramagnetic Complexes

Model	Range of Validity
SBM	Absence of static ZFS; in the Redfield limit
Florence NMRD	Slow reorientation ($\tau_r > \tau_e$); in the Redfield limit (ZFS included in nuclear relaxation)
Modified Florence NMRD	Slow reorientation ($\tau_r > \tau_e$); in the Redfield limit (ZFS included in nuclear and electron relaxation)
Grenoble (Fries et al.)	All conditions (ZFS is included; any τ_r; even outside Redfield limit)
Slow motion (Kowalewski et al.)	All conditions (ZFS is included; any τ_r; even outside Redfield limit)

10.5 NMRD PROFILES OF CONTRAST AGENTS

10.5.1 SMALL MOLECULE CONTRAST AGENTS

Manganese(II) ions could be used as such as MRI contrast agent due their high relaxivity (Section 7.3.5.4) and lower toxicity with respect to gadolinium(III) ions. Actually, the only clinically approved manganese contrast agent (MnDPDP), after internalization in hepatocytes, releases free Mn^{2+} ions, which can thus form protein adducts, and allow for the visualization of the healthy liver tissues. Since many biomolecules are able to bind Mn^{2+} efficiently, it is difficult to design stable Mn^{2+} complexes for in vivo applications. On the other hand, accumulation of the Mn^{2+} ions beyond certain levels causes significant neurological risk and a depressive action on heart function. Due to these drawbacks, Mn^{2+} complexes are presently rarely used as contrast agents. A renewed interest in manganese contrast agents is now due to the possibility of confinement of Mn^{2+} ions inside nanocarriers and of their release in selected biological targets.

Gadolinium(III) complexes are the most extensively used and investigated molecules as contrast agents for MRI. In order to avoid toxic effects for the human body, complexes with large thermodynamic and kinetic stability, as Gd-DTPA, Gd-DOTA (Fig. 10.13) and other derivatives, have been designed. Both outer-sphere and inner-sphere contributions are present (Fig. 10.14), with a single water molecule coordinated to the metal. Both Gd-DTPA and Gd-DOTA have similar size, i.e. similar τ_r (60–80 ps), and they have τ_M values (50–500 ns) so that to be negligible for the determination of the water proton relaxivity. However, the relaxivity profiles are somewhat different (about 65% higher for Gd-DOTA than for Gd-DTPA at low fields) due to the different electron relaxation time. From the fit of the experimental NMRD profiles, a much smaller Δ_t is in fact calculated for Gd-DOTA (around 0.02 cm^{-1}; Δ_t^2 around 1.4×10^{19} s^{-2}, Table 10.2) than for Gd-DTPA (0.036 cm^{-1}; $\Delta_t^2 = 4.6 \times 10^{19}$ s^{-2}). This is ascribed to the lower symmetry and larger flexibility in the structure of Gd-DTPA. The spin–orbit interaction is in fact more effective in a less symmetrical structure (Section 7.1.1), and this, combined with a larger flexibility of the metal ligands, also results in larger fluctuations of the static ZFS [26]. At the imaging fields, this difference is not relevant for relaxivity because the electron relaxation time, which increases with field, becomes long compared to τ_r, and almost not affecting the value of τ_c anymore. However, for larger complexes this difference would be very relevant also at high fields.

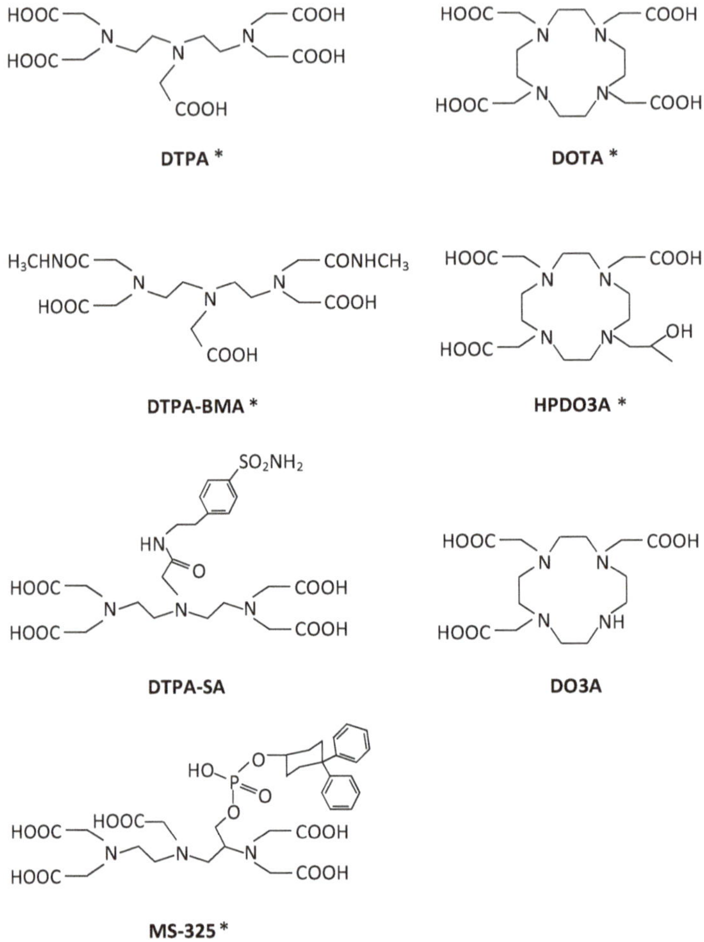

FIGURE 10.13 Some of the Gd^{3+} complexes mentioned in the text. Gd-complexes approved for clinical use are marked with *.

At 0.5 T, the inner sphere relaxivity of Gd-DTPA, Gd-DOTA and similar derivatives is about 2.5–3.5 mM^{-1} s^{-1}, at 25°C, depending mainly on the reorientational correlation time, and thus increasing with the molecular weight of the complex. Several similarly small Gd^{3+} complexes are however known with largely different τ_M parameters, ranging from a few nanoseconds, as in the Gd^{3+} aquaion, to tens of microseconds for the cationic tetraamide derivatives of DOTA [13,30]. The replacement of two carboxylate groups of the DTPA ligand with the two carboxoamide groups present in the DTPA-BMA has the effect of increasing τ_M from about 300 ns to about 2 μs, as the result of the decreased negative charge of the complex from −2 to zero. Therefore τ_M becomes of the same order of T_{1M} and the relaxivity at 0.5 T and 25°C decreases by about 7% with respect to Gd-DTPA [31,32].

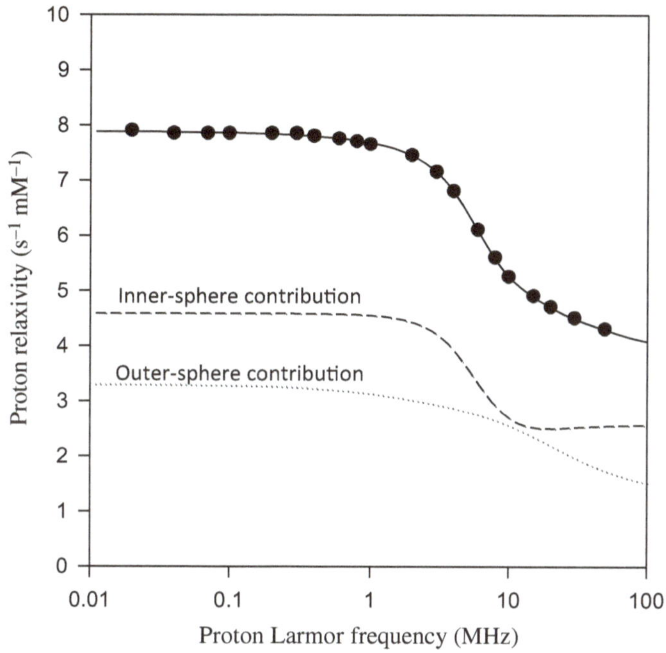

FIGURE 10.14 Water ^{1}H NMRD profiles for Gd-DTPA solutions at 298 K. Dashed line indicates the inner-sphere contribution and dotted line the outer-sphere contribution.

Adapted from Ref. [4].

Table 10.2 Best Fit Parameters of the ^{1}H NMRD Profiles for Gd-DOTA, Gd-DTPA, Gd-DTPA-SA and Gd-DTPA-SA Complexed with Carbonic Anhydrase (CA) at 298 K [4,27–29]

	Gd-DOTA	Gd-DTPA	Gd-DTPA-SA	Gd-DTPA-SA-CA
Δ_t (cm^{-1})	0.014–0.021	0.036	0.042	0.017
τ_v (ps)	11–26	18–25	17	18
τ_r (ps)	70–80	60–75	64	12000
τ_M (ns)	300	300	500	560
ZFS	Not included	Not included	Not included	0.01 cm^{-1}
q	1	1	1	1
r (Å)	3.1	3.1	3.1	3.0
d (Å)	3.6	3.6	3.6	3.6
Diff. coeff. (10^{-9} m^2s^{-1})	2.6	2.6	2.6	2.6

10.5.2 FUNCTIONALIZED CONTRAST AGENTS

As shown in Section 10.3, the highest relaxivities are reached for molecular reorientation times longer than nanoseconds. For this reason large efforts have been and are still devoted to the design of large paramagnetic systems composed of a slowly tumbling macromolecular substrate (that can be properly selected depending on the clinical applications) to which the Gd^{3+} complex can be covalently or noncovalently bound. A strategy to synthesize a selective contrast agent is actually that of functionalizing it in such a way that it is able to specifically react with a biological target. For example, the relaxivity profile of a derivative of Gd-DTPA with a built-in sulfonamide (Gd-DTPA-SA), when free in solution, has only one dispersion at about 10 MHz, which can be easily fit through the outer-sphere and inner-sphere contributions to relaxation (Fig. 10.15). The fit performed with the SBM model (i.e. without including the effect of a static ZFS) provides values for Δ_t (0.042 cm^{-1}), τ_v (17×10^{-12} s), τ_r (64×10^{-12} s), and τ_M (0.5×10^{-6} s) similar to those of Gd-DTPA, with $q = 1$ and $r = 3.1$ Å (Table 10.2). As discussed in Section 10.4, these parameters can change sizably if the fit is performed by considering the effect of static ZFS and the proper theory for fast-rotating systems ($\Delta_t = 0.018$ cm^{-1}, $\tau_v = 17 \times 10^{-12}$ s, $D = 0.02$ cm^{-1}) [33].

However, sulfonamides are known to be specific inhibitors of carbonic anhydrase. The relaxivity profile of Gd-DTPA-SA in the presence of carbonic anhydrase actually shows a largely increased relaxivity, with a dispersion at about 1 MHz and a relaxivity peak in the high field region (Fig. 10.15). This shape indicates that a complex is formed, so that τ_r increases to about 12 ns (as can be calculated from the molecular weight of the protein [34]): in this case, the field dependent electron relaxation time is the correlation time for proton relaxation even at high fields. The profile cannot be well fit with the SBM model, due to the small difference in the relaxivity values before and after the dispersion, and the presence of a small static ZFS must be taken into account [27]. The best fit parameters obtained with the Florence NMRD program are: $D = 0.01$ cm^{-1}, $\Delta_t = 0.017$ cm^{-1}, $\tau_v = 18 \times 10^{-12}$ s, and $\tau_M = 0.56 \times 10^{-6}$ s (Table 10.2). Due to the large τ_r of macromolecular complexes, which exceeds

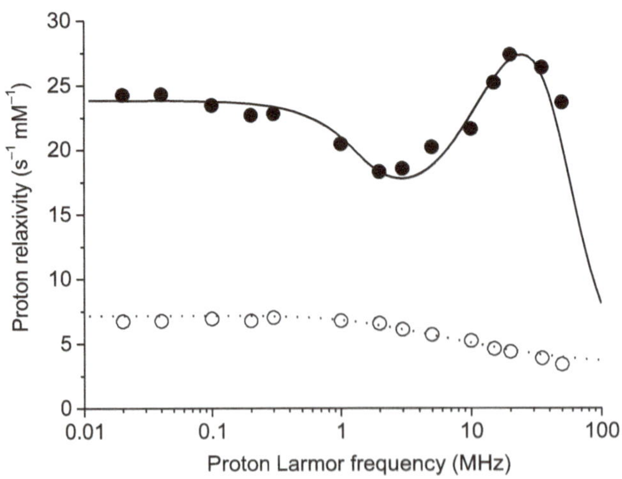

FIGURE 10.15 Water ^{1}H NMRD profiles for solutions of Gd-DTPA-SA at 298 K in the absence (○) and in the presence (●) of 1.36 mM of carbonic anhydrase.

Reproduced with permission from Ref. [27].

the electron relaxation time, the Florence NMRD model can actually be used to correctly analyze the NMRD profile.

Analogously, a DTPA-derivative, MS-325, has been synthesized with the aim of making it bind to serum albumin, and it is thus proposed as a blood pool agent. Blood pool agents are intravascular contrast agents, because due to their large size they are confined in the vascular system for a long time. Albumin binding increases the reorientation time of the complex in such a way that the relaxivity increases to about 35 $s^{-1}mM^{-1}$ at 0.5 T and 35°C, also thanks to the favorable water exchange rate. The NMRD profiles of the complex could be fit to determine the values of the relevant parameters by taking into account the presence of static ZFS and of local mobility (Fig. 10.16) [3].

Another example of functionalized contrast agent is represented by a gadolinium complex holding a galactopyranose group through a glycosidic bond [35]. The presence of this group prevents water molecules from binding. The complex is activated by the enzyme β-galactosidase because the latter cleaves the galactopyranose, thus exposing gadolinium to the solvent. Water can thus coordinate Gd^{3+}, and this enhances the relaxivity by about a factor three, and thus reports on the presence of this enzymatic activity [35] (Fig. 10.17A). The above systems represent examples of contrast agents which are responsive to the presence of an enzyme, the presence of which can thus be monitored by the relaxivity enhancement.

Gd^{3+} agents are also proposed for the detection of specific metal ions. Examples are provided by agents the relaxivity of which increases sizably in the presence of calcium(II) [26] or zinc(II) [36] due to rearrangements in the coordination of Gd^{3+} resulting from metal binding. In the absence of these ions, in fact, Gd^{3+} has no first sphere water molecules, but, if present, these ions selectively bind to the ligand moieties which otherwise are coordinated to Gd^{3+}, thus allowing space for direct coordination of water molecules to the paramagnetic metal (Fig. 10.17B).

Since many pathological events are associated with pH alterations, several agents are proposed for monitoring pH changes. A pH dependent relaxivity is produced by a change in the number of water

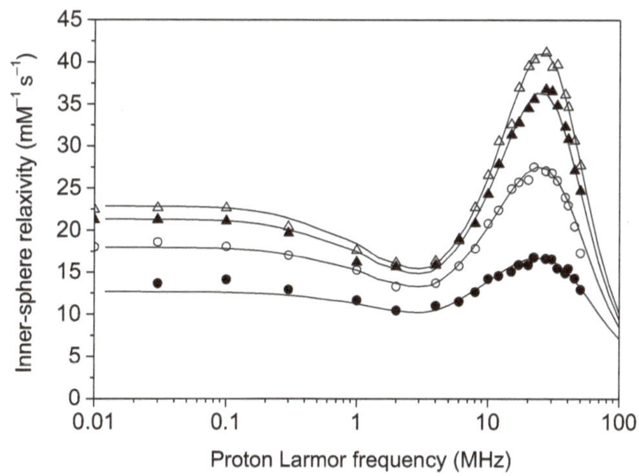

FIGURE 10.16 Inner-sphere relaxivity of MS-325 bound to human serum albumin at 35 (Δ), 25 (▲), 15 (○), and 5 (●)°C.

Reproduced with permission from Ref. [3].

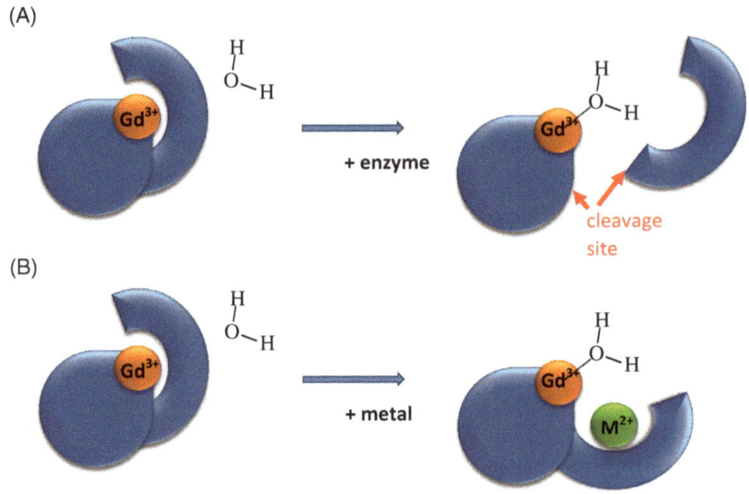

FIGURE 10.17 Examples of contrast agents which are responsive (A) to the presence of an enzyme able to cut the complex, thus exposing gadolinium to the solvent, or (B) to the presence of a metal ion due to ligand rearrangements upon metal binding.

molecules coordinating Gd^{3+} caused by the different coordination ability of the ligands towards Gd^{3+}, depending on their protonated or deprotonated form [37].

Optimization of contrast agents requires that the increase in the molecular reorientation time due to the large size of the complex is not made ineffective by the occurrence of fast local motions (Section 10.3 and Fig. 10.5) and, simultaneously, that the water exchange rate is in a good range. The presence of two water molecules binding each Gd^{3+} ion ($q = 2$) can of course double the relaxivity with respect to complexes with a single water molecule coordinated to the lanthanoid. A complex with $q = 2$ and a rigid core actually shows remarkably higher relaxivity (about 22 $s^{-1}mM^{-1}$ at 1 T and 25°C) with respect to complexes of similar size with larger flexibility [38,39].

10.5.3 HIGH-FIELD CONTRAST AGENTS

Contrast agents optimized for the highest field MRI instrumentation require that the correlation time τ_c is in the subnanoseconds range, rather than several nanoseconds (Section 10.3 and Fig. 10.7). These correlation times can be achieved by using relatively small molecules with a rigid structure, possibly able to bear many paramagnetic metal ions for an improved efficiency. Linking multiple complexes together in fact provides the possibility of a larger increase in the relaxation rate, due to the larger concentration of Gd^{3+} ions. For instance, multiple Gd-HPDO3A-like chelates which have been rigidly connected through click chemistry have been shown to sizably increase high field relaxivity (Fig. 10.18A and Fig. 10.19) [40]. Rigid complexes can also be assembled by transition metal ion coordination of two or three Gd^{3+} complexes with bidentate or tridentate ligands (Fig. 10.18B) [41,42].

A different approach consists in placing the gadolinium ion in the center of the molecule, in order to minimize the effect of the internal motions on the reorientation time of the Gd^{3+}-proton dipole-dipole interaction, which thus depends on the overall molecular weight of the molecule (Fig. 10.18C) [43].

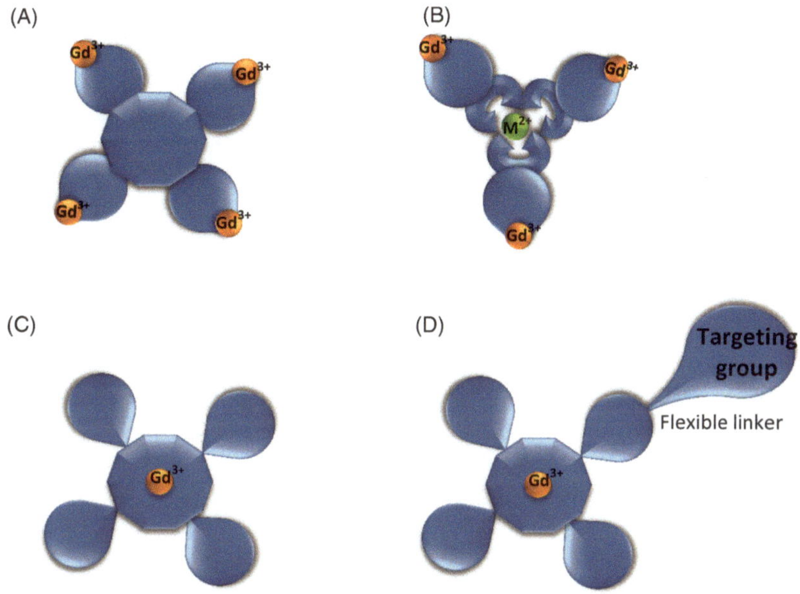

FIGURE 10.18 Strategies to achieve high relaxivity complexes at high fields.

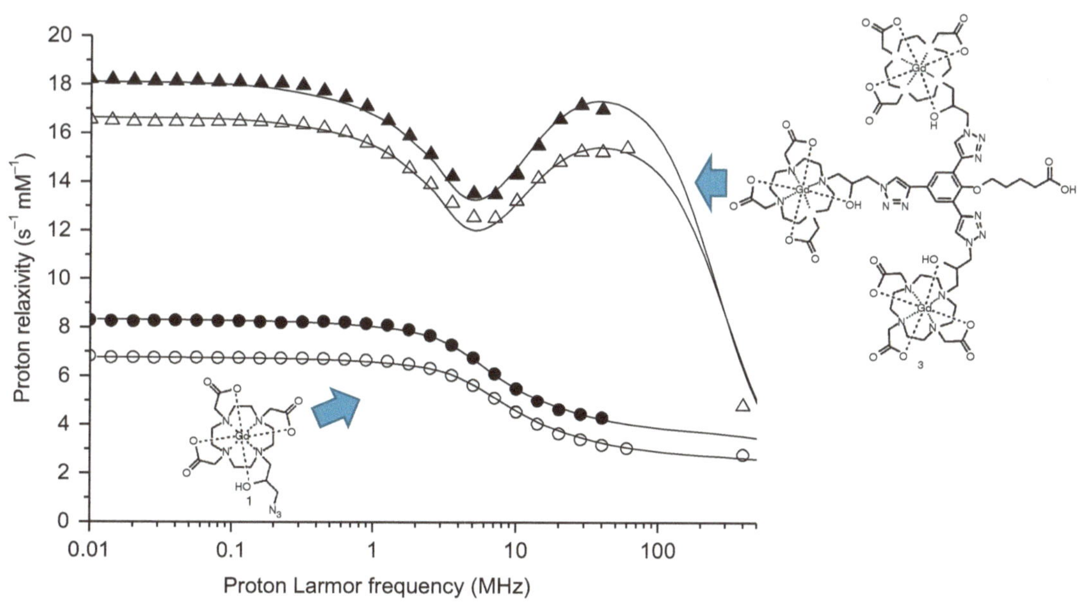

FIGURE 10.19 Water ^{1}H NMRD profiles for complexes 1 (●,○), a HPDO3A derivative, and 3 (▲,Δ) at 298 K (*solid symbols*) and 310 K (*open symbols*).

Reproduced with permission from Ref. [40].

This may results in remarkably high relaxivities [44]. Adjusting the size of the pending groups may actually be used to optimize the effective reorientation time at the field of interest [45]. Agents targeted towards macromolecules should experience internal mobility in the 0.1–1 ns time scale (Fig. 10.7). Therefore, targeting groups, allowing functionalization of the contrast agents, should be linked via flexible groups to partially decouple the motion at the Gd-complex from the slow tumbling of the macromolecule (Fig. 10.18D).

10.6 NANOSYSTEMS

Strategies to increase the sensitivity of MRI and to allow for the monitoring of the presence of a given cellular target pass through the development of high-capacity carriers that can be loaded with a huge number of paramagnetic metal ions. In this way, a sufficiently large amount of paramagnetic metals can be delivered to the desired target to allow its visualization. Suitable nanocarriers can be self-aggregates, dendrimers, micelles, liposomes, nanotubes, proteins, viral capsids (Fig. 10.20). Nanoparticles have different properties than small molecules in biodistribution, confinement effects, surface accessible area, reactivity, and targeting possibilities. They can thus be used to increase the image definition by selectively increasing the nuclear relaxation in selected compartments in tissues, depending on specific biological interests.

Along these lines, Gd-loaded nanoparticles are of potential use as MRI contrast agents, being small enough to pass through the vasculature. In nanovesicular particles, like liposomes or proteins as apoferritin, the Gd complexes can be trapped in the internal cavity without the need of functionalization. Conversely, in dendrimers, silica particles, micelles, or carbon nanotubes, properly designed covalent linkages or (hydrophobic-driven) molecular aggregations are often needed.

The protein apoferritin has a cavity which can be loaded with many Gd complexes: upon the entrapment of Gd-HPDO3A in the inner cavity of the protein, a 20-fold enhancement in relaxivity (ca. 80 $s^{-1}mMGd^{-1}$ at 20 MHz and 25°C) is reached with respect to the relaxivity of water solutions of Gd-HPDO3A [46] (Fig. 10.21A). In liposomes, the paramagnetic molecules can be entrapped in the internal cavity, incorporated within the phospholipid bilayer wall, or both. When the paramagnetic

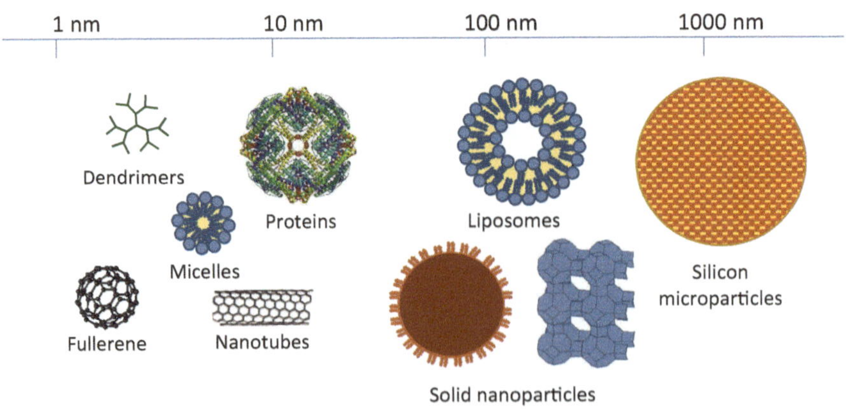

FIGURE 10.20 Nanocarriers used for the development of Gd^{3+}-based contrast agents.

Adapted from Ref. [66].

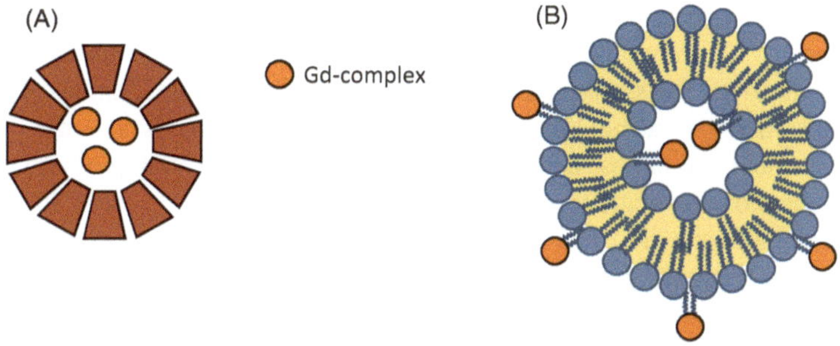

FIGURE 10.21 (A) In nanovesicular particles, Gd complexes can be trapped in the internal cavity. Apoferritin, for instance, has a cavity of about 80 Å of diameter, which can host several paramagnetic molecules. Water can access to the cavity through channels, which are small enough to prevent diffusion of the Gd complexes. (B) In liposomes, the Gd-complexes can be covalently attached to the lipid polar heads, and thus be exposed to the interior or to the exterior of the vesicle. Each vesicle is able to carry a number of paramagnetic molecules as large as 10^4.

moieties are covalently attached to the lipid polar heads, the Gd^{3+} ions can be exposed both to the interior and to the exterior of the vesicle (Fig. 10.21B). These two classes of ions can contribute differently to the overall relaxivity of the bulk water, depending on the water permeability of the liposome bilayer [47]. When paramagnetic Ln^{3+} (different from Gd^{3+}) complexes, like Dy-HPDO3A, are entrapped in the liposome cavity, strong T_2 relaxivity enhancements are observed, arising from the local inhomogeneous magnetic fields created by the nanoparticle [48,49].

Water soluble gadofullerenes (Fig. 10.22) also provide large relaxivity as the result of a number of effects: (1) the large number of protons of the fullerene cage which are dipole–dipole coupled to the Gd^{3+} ions confined into the cage, (2) the long reorientation time of the gadofullerenes, resulting from self-aggregation, and (3) the presence of water molecules in the interstices of the aggregates which are in fast exchange with bulk water molecules [50,51].

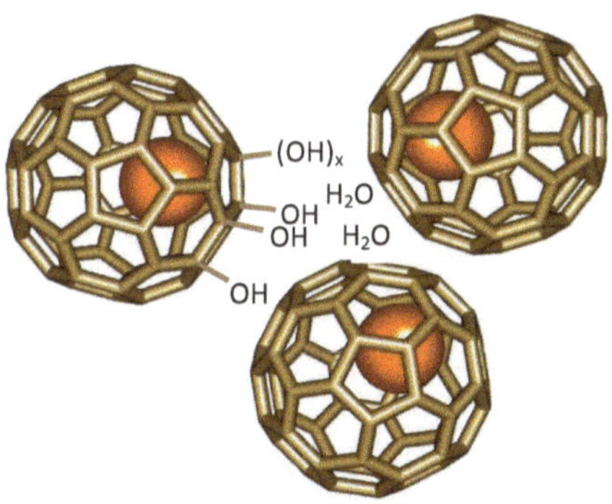

FIGURE 10.22 The high relaxivity of $Gd@C_{60}(OH)_x$ is ascribed to the large number of protons in the fullerene cage and to the presence of water molecules in the interstices of the fullerene aggregates [51]. For clarity, few OH groups have been indicated only in one of the showed fullerene molecules.

In Gd-loaded carbon nanotubes, the relaxivity at MRI fields reaches values as large as 200 $mM^{-1}\,s^{-1}$ also as a result of the large *q* [52]. Peptide amphiphiles, that self-assemble in nanofibers, are used to construct nanostructures with improved relaxivity once conjugated with Gd^{3+} contrast agents. Peptide amphiphiles consist of charged residues (providing solubility), β-sheet-forming amino acids, and hydrophobic tail. In solution, they self-assemble into nanostructures with many different morphologies and physical properties, offering control at subnanometer scale through the selection of amino acids and solution conditions. Peptide amphiphiles assembled with Gd^{3+} chelates exhibit large relaxivities, of the order of 20 $s^{-1}\,mMGd^{-1}$ at 1.5 T [53], depending on the position of the Gd^{3+} chelate, which can sizably affect water lifetime and local mobility.

Gold nanoparticles covered with peptides functionalized with Gd^{3+} complexes represent a more recent approach to improve relaxivity. Gold nanoparticles allow for tunable molecular size and shape, and for the assembly of many paramagnetic complexes on the same molecule, together with reporters and targeting moieties that can allow integrating multimodality, targeting, imaging, and therapy in the same platform [54]. Gold nanoparticles are easily functionalized by thiol derivatives, as cysteine conjugates of the Gd^{3+} ligands (Fig. 10.23A) [54] or thiol-labeled poly deoxy-thymine nucleotides containing conjugation sites for covalent attaching of Gd^{3+} complexes (Fig. 10.23B) [55]. Despite the large reorientation time of these nanocrystals coated with Gd^{3+} chelates, their relaxivity is not as large as expected for a system of this size, due to the occurrence of local fast mobility, likely occurring because of the surface curvature of the particles, which prevents an efficient packing of the Gd^{3+} chelates [54,56,57], and sometimes due to slow exchange of the coordinated water molecules [58]. Nanoparticles with star-shape, instead of spherical, surface provide enhanced relaxivity (about 100 $s^{-1}mM^{-1}$ at 35 MHz and 25°C), also thanks to sizable contributions from second sphere water molecules (Fig. 10.24) [59].

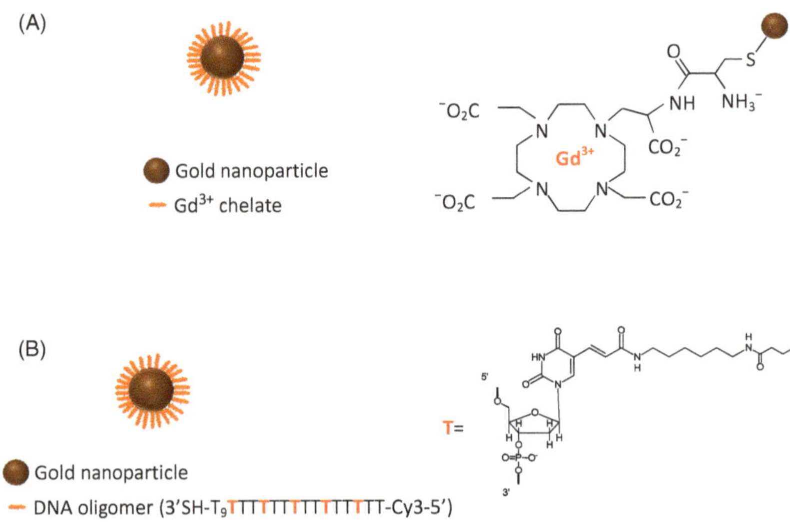

FIGURE 10.23 Gold nanoparticles can be easily functionalized by thiol derivatives. (A) as cysteine conjugates of the Gd^{3+} ligands [54] or (B) thiol-labeled poly deoxy-thymine nucleotides containing conjugation sites with covalently attached Gd^{3+} complexes.

Adapted from Ref. [55].

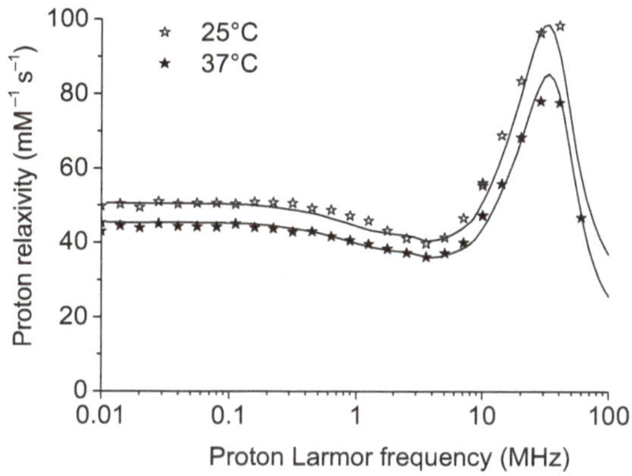

FIGURE 10.24 ^{1}H NMRD profiles for water solutions of Gd^{3+} enriched DNA gold nanostars.

Adapted from Ref. [59].

Larger nanoparticles (of size of the order of 100 nm) are obtained using zeolites (Fig. 10.25). Upon Gd^{3+}-doping, they can be effective for MRI of the gastrointestinal tract [60]. NaY zeolites present a large central cavity, or supercage, with diameter larger than 10 Å [61]. The large relaxivity of Gd^{3+}-loaded zeolite NaY nanoparticles (up to 30 $s^{-1}mM^{-1}$ at 0.5 T and 25°C) is ascribed to the fast exchange of the water molecules coordinated to the immobilized Gd^{3+} ion with the water molecules located inside the zeolite cavities; the latter then diffuse through the zeolite channels and exchange with the

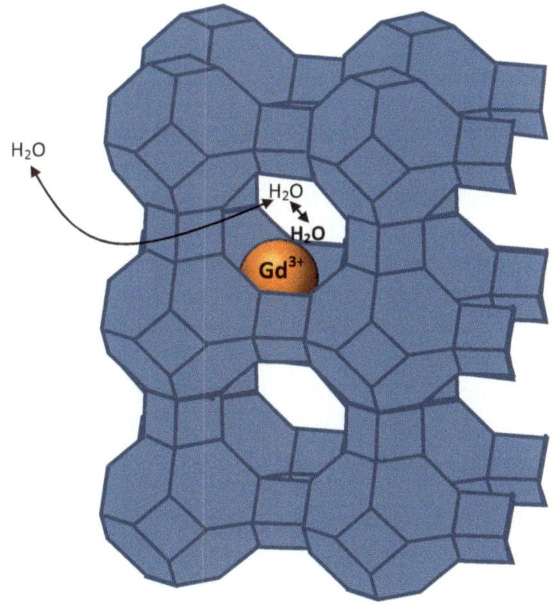

FIGURE 10.25 The water proton relaxivity for suspensions of Gd^{3+}-loaded zeolites is analyzed through a two-step model, where the exchange rate between water molecules coordinated to Gd^{3+} and water molecules inside the zeolite cavities is very different from the exchange rate between water molecules inside and outside the zeolite.

bulk. Since the exchange between Gd^{3+}-bound water and the water inside the zeolite cavity is much faster than the exchange between water inside and outside the zeolite, relaxivity can be described through a two-step model: (1) the interior of the zeolite is considered to contain a concentrated aqueous solution of Gd^{3+}, with 7 water molecules coordinated to the lanthanoid ion and in fast exchange; (2) the residence time of water protons in the interior of the zeolite (of the order of tens of microseconds) is relatively long, but still short enough to allow an efficient propagation of the paramagnetically induced relaxation effect from the interior of the zeolite to the bulk water [61]. The relaxivity would thus be mainly limited by the relatively slow diffusion of water molecules within the zeolite channels.

Confining contrast agents inside nanoporous structures generally represents a way to increase relaxivity. Both Gd^{3+} complexes and carbon nanostructures as gadofullerenes and gadonanotubes for instance show increased relaxivity when confined within the porous structure (10–100 nm of diameter) of silicon microparticles. This can be determined by a reduction in their mobility, by favoring their clustering and mutual interactions, and by a decrease in the mobility of the water molecules which may be possibly get trapped in the interstices of the bundles [62]. The type of Gd^{3+} chelate and the chemical nature of the groups used to functionalize the silica particles can be adapted to modulate the interaction between Gd^{3+} chelates and the surface of the porous structure, which plays a key role in determining the relaxivity because it can affect the exchange rate of the water molecules coordinated to Gd^{3+}, or the extent of fast local mobility [63].

MnO nanoparticles functionalized by conjugation with a tumor-specific antibody are also available [64]. Nanoparticles based on superparamagnetic iron oxide can on the other hand represent effective T_2 contrast agents, as described in Section 10.8.

10.7 TOWARD MOLECULAR IMAGING

Molecular imaging represents the new frontier in the development of contrast agents for MRI [65]. It aims at in vivo characterization and measurement of biological processes at the cellular and molecular level. The detection of altered biochemical processes, which largely anticipate the macroscopic changes that are detected by the current diagnostic modalities, makes early diagnosis of disease possible.

Because of the low sensitivity of MRI, this technique requires the accumulation of a high number of paramagnetic centers at the site of interest (10^7–10^9 Gd-chelates per cell [66]). In fact, the concentration of the specific receptors which are overexpressed in pathological processes rarely exceeds 10^{-9} moles per gram of tissue [67]. With such a very low concentration of the receptor, towards which the contrast agent can be functionalized, the expected relaxation enhancement is negligible unless a large amount of paramagnetic molecules binds to each receptor. To this end different nanoparticles can be used. For example, liposomes can carry a number of paramagnetic molecules as large as 10^4. They can be encapsulated into the interior aqueous volume [68–70] or incorporated by covalent binding to the lipid polar heads (Fig. 10.21) [71–73]. The binding specificity for cell receptors can be assured by conjugating biotinylated antibodies to the surface of liposomes (Fig. 10.26A) [74]. In this way, through avidin–biotin binding, many biotinylated paramagnetic liposomes can bind avidin, which on the other hand can be recognized by a biotinylated antibody to bind a specific receptor. Liposomes can be used also for monitoring the integrity of the vesicles [75] or the vesicle temperature [76] from the relaxivity changes occurring upon vesicle disruption or the change in the permeability of the bilayer, respectively.

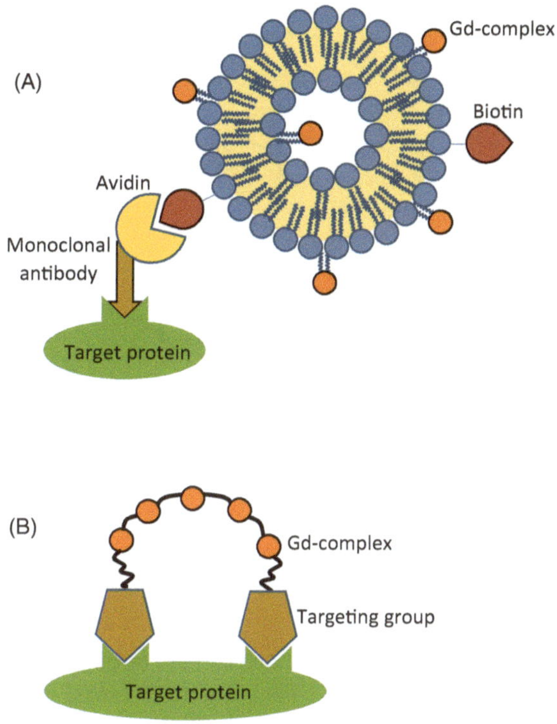

FIGURE 10.26 (A) Scheme of paramagnetic liposomes and of the avidin-biotin targeting approach. The same approach can be used for different types of nanoparticles. (B) Discrete multimers of Gd^{3+} complexes can be conjugated to a couple of targeting vector for binding of a specific receptor with a reduced internal mobility.

(A) Adapted from Ref. [74] (B) Adapted from Ref. [17].

Molecular imaging probes can be obtained by increasing the reorientation time of the Gd^{3+} molecule upon binding to a specific receptor. For instance, a probe for the detection of amyloid plaques associated with Alzheimer's disease is proposed by linking an efficient amyloid marker (the Pittsburgh compound B) to Gd-DO3A: the binding to the amyloid peptides yields increased relaxivities owing to reduced rotational motion [77]. Discrete multimers of Gd^{3+} complexes can also be used once conjugated to two targeting vectors in order to bind a specific receptor and reduce the internal flexibility (Fig. 10.26B) [78].

Superparamagnetic nanosized particles (Section 10.8) represent another class of molecules suitable for molecular imaging because they can be easily conjugated with biological and targeting probes and uptaken by cells [79].

Internalization of Gd^{3+} chelates in cells can occur in a number of ways, depending on the employed carriers (Fig. 10.27). They comprise uptake (1) by specific receptors, as in the case of proteins like apoferritin (Section 10.6), (2) by (receptor mediated) endocytosis, as in the construct between avidin and biotinilated dendrimers bearing Gd chelates, where avidin is internalized thanks to the receptors present on tumor cells [80], or (3) by binding to membrane translocation peptides [81,82], possibly designed to accumulate in tumour cells [83].

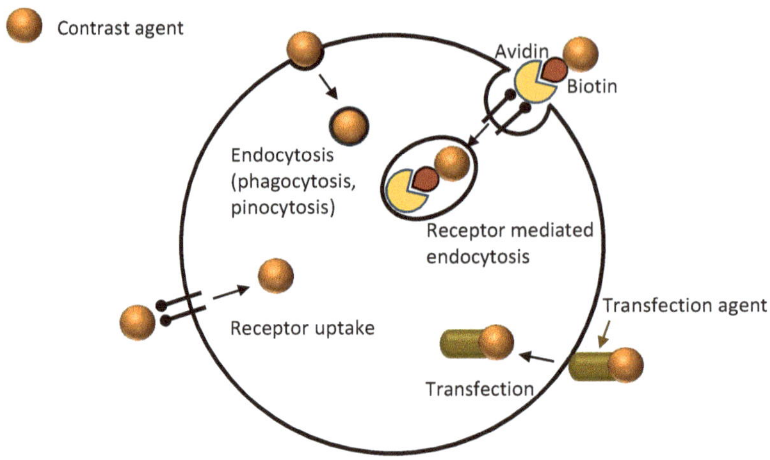

FIGURE 10.27 Schematic overview of the different pathways for cellular contrast agent uptake.

10.8 SUPERPARAMAGNETISM

Superparamagnets are small ferromagnetic crystals (magnetite, Fe_3O_4, for instance), of 10–30 nm diameter. The electron spins are all coupled, so that one single large spin can be considered per particle. Total S' values as high as 10^4 can be reached. As the coupling is ferromagnetic, the ground state is the one with largest S' (Section 11.1). The level immediately above ($S' - 1$) is separated by an energy $S'|J|$ [Eq. (11.2)], which is much larger than kT. Therefore, only the ground state S' is populated. In addition, S' is split by the so-called "anisotropy energy," which is analogous to a negative ZFS. Again, this energy is larger than kT, so that only the most negative M_S state is populated. The resulting magnetic moment aligns along one of the so-called directions of easy magnetization and, in the absence of magnetic field, jumps from one easy direction to another, with a correlation time of the order of few nanoseconds (Néel relaxation process). The mean magnetic moment increases up to saturation for increasing magnetic fields: at low fields it remains locked along the easy directions of magnetization, flipping between them rather than precessing; at high fields, the mean magnetic moment is along the magnetic field direction; at intermediate fields, the direction of the quantization axis depends on the magnitude of the anisotropy energy with respect to the Zeeman energy [84].

Fig. 10.28 shows the typical ^{1}H NMRD profile acquired in water solution containing superparamagnetic particles [84]. A dispersion can be observed when the Zeeman energy equals the anisotropy energy (analogously to what happens for ZFS). A typical feature of superparamagnetism is that the average magnetic moment is proportional to the magnetic field until saturation is reached: this causes an increase of the relaxation rate with increasing field, until the value at saturation, proportional to M_S^2 (Curie relaxation), is reached. The following dispersion is then related to the diffusion time of water molecules diffusing nearby the particles (outer-sphere relaxation, Section 6.6), and is related to the particle size.

In general, superparamagnetic nanoparticles are composed of three parts: (1) the core sphere, responsible for the relaxation enhancement, (2) the water-dispersible shells, which provide compatibility in the biological environment, and (3) the bioactive receptors for targeting purpose [85].

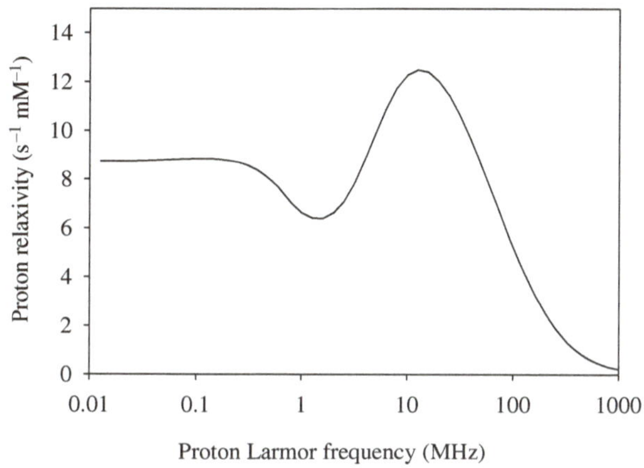

FIGURE 10.28 Water ^{1}H NMRD profile of a superparamagnet.

Superparamagnetic contrast agents determine strong T_2 effects as the result of a dramatic long-range disturbance in magnetic field homogeneity due to the large magnetic susceptibility of the ferromagnetic iron oxide core. Although nanosized iron oxide particles lose their magnetization in the absence of an external magnetic field, when the field is applied, they cause large microscopic field inhomogeneity and thus are very effective in spin dephasing, with consequent shortening of T_2*. The particles can remain in the blood for a time long enough to be used as blood pool agents for magnetic resonance angiography, or accumulate in the liver, spleen, and bone marrow by selective uptake, or in the cells of the reticuloendothelial system [86]. Since their biological distribution depends on their size, they are classified in: (1) micrometer-sized paramagnetic iron oxide (MPIO; several micrometers), (2) superparamagnetic iron oxide (SPIO; hundreds of nanometers), and (3) ultrasmall superparamagnetic iron oxide (USPIO; less than 50 nm) [87].

Superparamagnetic iron(III) oxide (SPIO) agents have an iron oxide core with diameter of 3–5 nm coated by materials like dextran or carboxydextran, so that the diameter of the particle reaches 50–200 nm. The usual clinical targets of SPIO are liver diseases because, due to the negligible uptake by tumour tissues in contrast with the normal liver tissue, a strong contrast between normal and abnormal tissues permits a clear detection of the abnormal tissue. Iron oxide nanoparticles for molecular imaging are usually in the USPIO class, because of the longer circulation time in blood.

10.9 CHEMICAL EXCHANGE SATURATION TRANSFER (CEST)

Contrast agents based on the CEST effect cause a decrease of the signal of water protons, thus creating a negative contrast in the image. These agents are effective not because they alter water relaxivity but because they saturate the water signal, with a result similar to that of a reduction in water density. This is done through the transfer of saturated magnetization from the exchangeable protons of the contrast agent to the water protons. The saturation transfer is induced by a selective radiofrequency irradiation of the exchangeable protons of the contrast agent.

In summary, if a set of exchangeable protons resonate at a frequency different from that of bulk water protons, and their exchange rate with the bulk protons is slow ($k_{ex} < \Delta\omega$), it is possible to selectively saturate the magnetization of this set of protons through irradiation at the proper radiofrequency. Through chemical exchange, the saturation is then transferred to the pool of bulk water protons in the compartments where the contrast agents are present (Fig. 10.29); saturation of the water protons is then responsible of a dramatic decrease in the MRI signal, due to the decreased difference in the population of the +1/2 and −1/2 spin states. On the opposite, the magnetization of bulk water protons is not saturated in the compartments where the contrast agents are not present, so that these compartments retain the whole signal amplitude. The efficiency of CEST contrast agents mainly depends on the number and exchange rate k_{ex} of the saturated protons [66]. Nanoparticles are thus extremely promising as they allow for the storage of thousands of mobile protons, the exchange of which can be tuned depending on their employed system. CEST agents confined within liposomes, for instance, can be highly efficient, depending on the permeability of the liposome membrane, i.e. on the exchange rate of the water protons inside the liposome cavity, which can be modulated by changing the membrane composition.

Paramagnetic lanthanoids (mostly Tm^{3+} and Dy^{3+}) complexes are optimal CEST agents because they are able to largely affect the chemical shifts of the protons in the complex through hyperfine coupling [88]. The pseudocontact shifts depend on the magnitude of the anisotropy of the magnetic susceptibility tensor and on the position of the protons in the frame defined by this tensor (Chapter 2). Therefore, these agents are optimized by selecting lanthanoid complexes with the largest magnetic susceptibility anisotropy (Chapter 8) and with the exchangeable proton(s) on the z-axis of the tensor [in order to maximize the $(3\cos^2\theta - 1)$ term in Eq. (2.46)]. Ln-DOTA and Ln-HPDO3A complexes can be

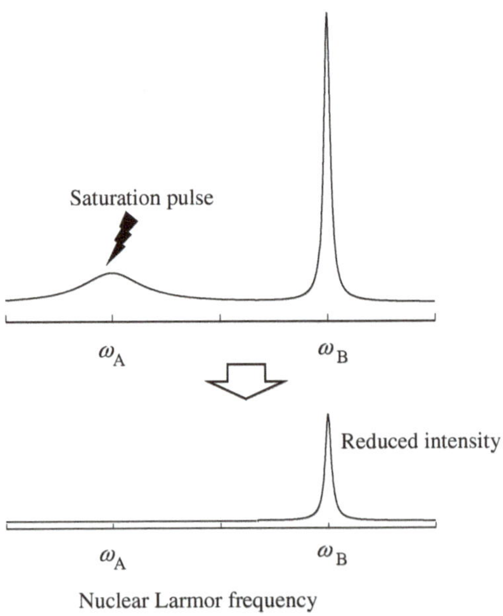

FIGURE 10.29 The CEST effect is due to saturation, through a radiofrequency pulse, of exchange protons at a different frequency (ω_A) from bulk water protons (ω_B), so that chemical exchange results in loss of magnetization of the bulk water protons.

used, where the exchangeable protons are those of the water molecules coordinated to the lanthanoid ions. These complexes can be confined in liposome cavities for improving their performances.

The shift can be sizably increased in nonspherical liposomes, due to the additional contribution arising from bulk magnetic susceptibility (Section 6.8) [89,90]. A further increase in the chemical shift can be obtained by the additional incorporation of CEST complexes within the bilayer: this has the effect of modulating the preferential orientations of the nonspherical liposome in the magnetic field, as a result of the lanthanoid magnetic anisotropy (Chapter 3), with the effect of affecting the shifts [91].

GENERAL REFERENCES

Aime S, Delli Castelli D, Geninatti Crich S, Gianolio E, Terreno E. Pushing the sensitivity envelope of lanthanide-based magnetic resonance imaging (MRI) contrast agents for molecular imaging applications. Acc. Chem. Res. 2009;42:822–31.

Caravan P. Strategies for increasing the sensitivity of gadolinium based MRI contrast agents. Chem. Soc. Rev. 2006;35:512–23.

Caravan P, Ellison JJ, McMurry TJ, Lauffer RB. Gadolinium(III) chelates as MRI contrast agents: structure, dynamics, and applications. Chem. Rev. 1999;99:2293–351.

Gossuin Y, Gillis P, Hocq A, Vuong QL, Roch A. Magnetic resonance relaxation properties of superparamagnetic particles. Wiley Interdiscip. Rev. Nanomed. Nanobiotechnol. 2009;1:299–310.

Laurent S, Forge D, Port M, Roch A, Robic C, Vander Elst L, Muller RN. Magnetic iron oxide nanoparticles: synthesis, stabilization, vectorization, physicochemical characterizations, and biological applications. Chem. Rev. 2008;108:2064–110.

Terreno E, Delli Castelli D, Viale A, Aime S. Challenges for molecular magnetic resonance imaging. Chem. Rev. 2010;110:3019–42.

REFERENCES

[1] Woessner DE. J. Chem. Phys. 1962;37:647.
[2] Lipari G, Szabo A. J. Am. Chem. Soc. 1982;104:4546.
[3] Caravan P, Cloutier N, McDermid S, Ellison JJ, Chasse JM, Lauffer RB, Luchinat C, McMurry TJ, Parigi G, Spiller M. Inorg. Chem. 2007;46:6632.
[4] Koenig SH, Brown RD III. Progr. Nucl. Mag. Reson. Spectrosc. 1990;22:487.
[5] Aime S, Botta M, Fasano M, Terreno E. Chem. Soc. Rev. 1998;27:19.
[6] Caravan P, Farrar CT, Frullano L, Uppal R. Contrast Media Mol. Imaging. 2009;4:89.
[7] Xu J, Franklin SJ, Whisenhunt DW, Raymond KN. J. Am. Chem. Soc. 1995;117:7245.
[8] Hajela S, Botta M, Giraudo S, Xu J, Raymond KN, Aime S. J. Am. Chem. Soc. 2000;122:11228.
[9] Aime S, Calabi L, Cavallotti C, Gianolio E, Giovenzana GB, Losi P, Maiocchi A, Palmisano G, Sisti M. Inorg. Chem. 2004;43:7588.
[10] Raymond KN, Pierre VC. Bioconjug. Chem. 2005;16:3.
[11] Micskei K, Helm L, Brücher E, Merbach AE. Inorg. Chem. 1993;32:3844.
[12] Gonzalez G, Powell DH, Tissieres V, Merbach AE. J. Phys. Chem. 1994;98:53.
[13] Aime S, Botta M, Fasano M, Terreno E. Acc. Chem. Res. 1999;32:941.
[14] Aime S, Barge A, Batsanov AS, Botta M, Delli Castelli D, Fedeli F, Mortillaro A, Parker D, Puschmann H. Chem. Commun. 2002;1120.

[15] Merbach AE. Pure Appl. Chem. 1987;59:161.
[16] Kowalewski J, Luchinat C, Nilsson T, Parigi G. J. Phys. Chem. A 2002;106:7376.
[17] Caravan P. Chem. Soc. Rev. 2006;35:512.
[18] Bertini I, Galas O, Luchinat C, Parigi G. J. Magn. Reson. Ser. A 1995;113:151.
[19] Bertini I, Kowalewski J, Luchinat C, Nilsson T, Parigi G. J. Chem. Phys. 1999;111:5795.
[20] Kowalewski J, Kruk D, Parigi G. Adv. Inorg. Chem. 2005;57:41.
[21] Rast S, Fries PH, Belorizky E, Borel A, Helm L, Merbach AE. J. Chem. Phys. 2001;115:7554.
[22] Fries PH, Belorizky E. J. Chem. Phys. 2007;126:204503.
[23] Larsson T, Westlund PO, Kowalewski J, Koenig SH. J. Chem. Phys. 1994;101:1116.
[24] Nilsson T, Kowalewski J. J. Magn. Reson. 2000;146:345.
[25] Belorizky E, Fries PH, Helm L, Kowalewski J, Kruk D, Sharp RR, Westlund PO. J. Chem. Phys. 2008;128:052315.
[26] Li C, Parigi G, Fragai M, Luchinat C, Meade TJ. Inorg. Chem. 2002;41:4018.
[27] Anelli PL, Bertini I, Fragai M, Lattuada L, Luchinat C, Parigi G. Eur. J. Inorg. Chem. 2000;625–30.
[28] Powell DH, Dhubhghaill OMN, Pubanz D, Helm L, Lebedev YS, Schlaepfer W, Merbach AE. J. Am. Chem. Soc. 1996;118:9333.
[29] Caravan P, Ellison JJ, McMurry TJ, Lauffer RB. Chem. Rev. 1999;99:2293.
[30] Aime S, Barge A, Bruce JI, Botta M, Howard JAK, Moloney JM, Parker D, De Sousa AS, Woods M. J. Am. Chem. Soc. 1999;121:5762.
[31] Aime S, Botta M, Fasano M, Paoletti S, Anelli PL, Uggeri F, Virtuani M. Inorg. Chem. 1994;33:4707.
[32] Pubanz D, Gonzalez G, Powell DH, Merbach AE. Inorg. Chem. 1995;34:4447.
[33] Kruk D, Kowalewski J. J. Magn. Reson. 2003;162:229.
[34] Bertini I, Fragai M, Luchinat C, Parigi G. Magn. Reson. Chem. 2000;38:543.
[35] Urbanczyk-Pearson LM, Femia FJ, Smith J, Parigi G, Luchinat C, Meade TJ. Inorg. Chem. 2008;47:56.
[36] Major JL, Meade TJ, Parigi G, Luchinat C. Proc. Natl. Acad. Sci. USA 2007;104:13881.
[37] Lowe MP, Parker D, Reany O, Aime S, Botta M, Castellano G, Gianolio E, Pagliarin R. J. Am. Chem. Soc. 2001;123:7601.
[38] Ruloff R, von Koten G, Merbach AE. Chem. Commun. 2004;842.
[39] Costa J, Tóth E, Helm L, Merbach AE. Inorg. Chem. 2005;44:4747.
[40] Mastarone DJ, Harrison VSR, Eckermann AL, Parigi G, Luchinat C, Meade TJ. J. Am. Chem. Soc. 2011;133:5329.
[41] Jacques V, Desreux JF. Top. Curr. Chem. 2002;221:123.
[42] Livramento JB, Tóth E, Sour A, Borel A, Merbach AE, Ruloff R. Angew. Chem. Int. Ed. 2005;44:1480.
[43] Fulton DA, Elemento EM, Aime S, Chaabane L, Botta M, Parker D. Chem. Commun. 2006;1064.
[44] Port M, Corot C, Raynal I, Idee JM, Dencausse A, Lancelot E, Meyer D, Bonnemain B, Lautrou J. Invest. Radiol. 2001;36:445.
[45] Fulton DA, O'Halloran M, Parker D, Senanayake K, Botta M, Aime S. Chem. Commun. 2005;474.
[46] Aime S, Frullano L, Geninatti Crich S. Angew. Chem. Int. Ed. Engl. 2002;41:1017.
[47] Laurent S, Vander Elst L, Thirifays C, Muller RN. Langmuir 2008;24:4347.
[48] Delli Castelli D, Gianolio E, Geninatti Crich S, Terreno E, Aime S. Coord. Chem. Rev. 2008;252:2424.
[49] Terreno E, Delli Castelli D, Cabella C, Dastrù W, Sanino A, Stancanello J, Tei L, Aime S. Chem. Biodivers. 2008;5:1901.
[50] Tóth E, Bolskar R, Borel A, Gonzalez G, Helm L, Merbach AE, Sitharaman B, Wilson LJ. J. Am. Chem. Soc. 2005;127:799.
[51] Laus S, Sitharaman B, Tóth E, Bolskar RD, Helm L, Wilson LJ, Merbach AE. J. Phys. Chem. C 2007;111:5633.
[52] Sitharaman B, Wilson LJ. Int. J. Nanomed. 2006;1:291.
[53] Preslar At, Parigi G, McClendon MT, Sefick SS, Moyer TJ, Haney CR, Waters EA, MacRenaris KW, Luchinat C, Stupp SI, Meade TJ. ACS Nano 2014;8:7325.

[54] Ferreira MF, Mousavi B, Ferreira PM, Martins CIO, Helm L, Martins JA, Geraldes CFGC. Dalton Trans. 2012;41:5472.
[55] Song Y, Xu X, MacRenaris KW, Zhang X-Q, Mirkin CA, Meade TJ. Angew. Chem. Int. Ed. 2009;48:9143.
[56] Warsi MF, Adams RW, Duckett SB, Chechik V. Chem. Commun. (Camb.) 2010;46:453.
[57] Warsi MF, Chechik V. Phys. Chem. Chem. Phys. 2011;13:9812.
[58] Moriggi L, Cannizzo C, Dumas E, Mayer CR, Ulianov A, Helm L. J. Am. Chem. Soc. 2009;131:10828.
[59] Rotz MW, Culver KSB, Parigi G, MacRenaris KW, Luchinat C, Odom TW, Meade TJ. ACS Nano 2015;9:3385.
[60] Sur SK, Heinsbergen JF, Bryant RG. J. Magn. Reson. A 1993;103:27.
[61] Platas-Iglesias C, Vander Elst L, Zhou W, Muller RN, Geraldes CFGC, Maschmeyer T, Peters JA. Chemistry 2002;8:5121.
[62] Ananta JS, Godin B, Sethi R, Moriggi L, Liu X, Serda RE, Krishnamurthy R, Muthupillai R, Bolskar RD, Helm L, Ferrari M, Wilson LJ, Decuzzi P. Nat. Nanotech. 2010;5:815.
[63] Carniato F, Tei L, Cossi M, Marchese L, Botta M. Chemistry 2010;16:10727.
[64] Na HB, Lee JH, An K, Park YI, Park M, Lee IS, Nam DH, Kim ST, Kim SH, Kim SW, Lim KH, Kim KS, Kim SO, Hyeon T. Angew. Chem. Int. Ed. Engl. 2007;46:5397.
[65] Weissleder R, Mahmood U. Radiology 2001;219:316.
[66] Aime S, Delli Castelli D, Geninatti Crich S, Gianolio E, Terreno E. Acc. Chem. Res. 2009;42:822.
[67] Spragg DD, Alford DR, Greferath R, Larsen CE, Lee KD, Gurtner GC, Cybulsky MI, Tosi PF, Nicolau C, Gimbrone MA jr. Proc. Natl. Acad. Sci. USA 1997;94:8795.
[68] Unger E, Tilcock C, Ahkong QF, Fritz T. Invest. Radiol. 1990;25:65.
[69] Barsky D, Pütz B, Schulten K, Magin RL. Mag. Reson. Med. 1992;24:1.
[70] Bacic G, Niesman MR, Bennett HF, Magin RL, Swartz HM. Magn. Reson. Med. 1988;6:445.
[71] Fossheim SL, Fahlvik AK, Klaveness J, Muller RN. Mag. Reson. Imaging 2000;17:83.
[72] Tilcock C, Ahkong QF, Koenig SH, Brown RD III, Davis M, Kabalka GW. Magn. Res. Med. 1992;27:44.
[73] Saab-Ismail NH, Simor T, Gaszner B, Lorand T, Szollosy M, Elgavish GA. J. Med. Chem. 1999;42:2852.
[74] Alhaique F, Bertini I, Fragai M, Carafa M, Luchinat C, Parigi G. Inorg. Chim. Acta 2002;331:151.
[75] Viglianti BL, Abraham SA, Michelich CR, Yarmolenko PS, MacFall JR, Bally MB, Dewhirst MW. Magn. Reson. Med. 2004;51:1153.
[76] Terreno E, Sanino A, Carrera C, Delli Castelli D, Giovenzana GB, Lombardi A, Mazzon R, Milone L, Visigalli M, Aime S. J. Inorg. Biochem. 2008;102:1112.
[77] Martins AF, Morfin J-F, Geraldes CFGC, Tóth E. J. Biol. Inorg. Chem. 2014;19:281.
[78] Zhang Z, Greenfield MT, Spiller M, McMurry TJ, Lauffer RB, Caravan P. Angew. Chem. Int. Ed. 2005;44:6766.
[79] Terreno E, Delli Castelli D, Viale A, Aime S. Chem. Rev. 2010;110:3019.
[80] Kobayashi H, Brechbiel MW. Mol. Imag. 2003;2:1.
[81] Bhorade R, Weissleder R, Nakakoshi T, Moore A, Tung C-H. Bioconjug. Chem. 2000;11:301.
[82] Allen MJ, Meade TJ. J. Biol. Inorg. Chem. 2003;8:746.
[83] Heckl S, Pipkorn R, Waldeck W, Spring H, Jenne J, von der Lieth C-W, Corban-Wilhelm H, Debus J, Braun K. Cancer Res. 2003;63:4766.
[84] Roch A, Muller RN, Gillis P. J. Chem. Phys. 1999;110:5403.
[85] Na HB, Song IC, Hyeon T. Adv. Materials 2009;21:2133.
[86] Mendonca Dias MH, Lauterbur PC. Magn. Res. Med. 1986;3:328.
[87] Bulte JWM, Kraitchman DL. NMR Biomed. 2004;17:484.
[88] Zhang S, Merritt M, Woessner DE, Lenkinski RE, Sherry AD. Acc. Chem. Res. 2003;36:783.
[89] Aime S, Delli Castelli D, Lawson D, Terreno E. J. Am. Chem. Soc. 2007;129:2430.
[90] Terreno E, Cabella C, Carrera C, Delli Castelli D, Mazzon R, Rollet S, Stancanello J, Visigalli M, Aime S. Angew. Chem. Int. Ed. 2007;46:966.
[91] Delli Castelli D, Terreno E, Carrera C, Giovenzana GB, Mazzon R, Rollet S, Visigalli M, Aime S. Inorg. Chem. 2008;47:2928.

CHAPTER

MAGNETIC COUPLED SYSTEMS

11

Several polymetallic systems experience magnetic coupling, either ferromagnetic or antiferromagnetic. Such magnetic coupling may affect the hyperfine shift and electron relaxation. Beneficial effects are always expected and observed on nuclear relaxation. A series of examples are discussed.

11.1 THE INDUCED MAGNETIC MOMENT PER METAL ION IN POLYMETALLIC SYSTEMS, THE HYPERFINE CONTACT SHIFT, AND THE NUCLEAR RELAXATION RATES

Compounds containing more than one metal ion with unpaired electrons deserve special attention because NMR is a powerful technique to relate the properties of the whole systems with those of each metal ion. Magnetic coupling occurs when the electronic spin magnetic moments of each metal ion interact with one another. The interaction is again quantized; for example, given S_1 and S_2 as two interacting spins, new spin levels are established, with total S' varying in unitary steps from $S_1 + S_2$ to $|S_1 - S_2|$. A major effect of magnetic spin coupling is the occurrence of new energy levels. Their energy separations vary from almost zero to the order of 10^3 cm^{-1}. At any laboratory temperature, kT may be of the order of the above separation and so the problem of the temperature-dependent population of the levels is introduced. The population P_i of any spin level of energy E_i is given by the Boltzmann law [Eq. (1.25)], which we recall here for convenience

$$P_i = \frac{\exp(-E_i/(kT))}{\sum_i \exp(-E_i/(kT))}.$$

When kT is large with respect to the energy gap, the population of each level is just one over the number of levels or functions. When kT is small with respect to the energy separation, then only the lowest level is occupied. The new energy levels S' are linear combinations of the $|S, M_S>$ functions of each metal ion. The functions and energies can be calculated by the simple Heisenberg Hamiltonian that for a dimer is

$$H = J\, \mathbf{S}_1 \cdot \mathbf{S}_2 \tag{11.1}$$

where J is a constant and $\mathbf{S}_1$ is the spin operator which operates only on the functions of metal 1 [1]. The eigenvalues of Eq. (11.1) and the energy separation between S' and the neighbor $S' - 1$ levels are, respectively,[1]

[1]The eigenvalues of Eq. (11.1) can be derived by squaring the equality $\mathbf{S}' = \mathbf{S}_1 + \mathbf{S}_2$: $\mathbf{S}'^2 = \mathbf{S}_1^2 + \mathbf{S}_2^2 + 2\mathbf{S}_1 \cdot \mathbf{S}_2$ (since $\mathbf{S}_1$ and $\mathbf{S}_2$ commute) and, by rearranging, $\mathbf{S}_1 \cdot \mathbf{S}_2 = 1/2\, [\mathbf{S}'^2 - \mathbf{S}_1^2 - \mathbf{S}_2^2]$. The eigenvalues of the Hamiltonian $J\, \mathbf{S}_1 \cdot \mathbf{S}_2$ are thus given by: $E = J/2[S'(S'+1) - S_1(S_1+1) - S_2(S_2+1)]$ (see Section 1.2). Because $S_1(S_1+1)$ and $S_2(S_2+1)$ are the same for all S' levels, they introduce only an energy offset and can be dropped. The splitting of the levels is thus given by Eq. (11.2).

NMR of Paramagnetic Molecules. http://dx.doi.org/10.1016/B978-0-444-63436-8.00011-9

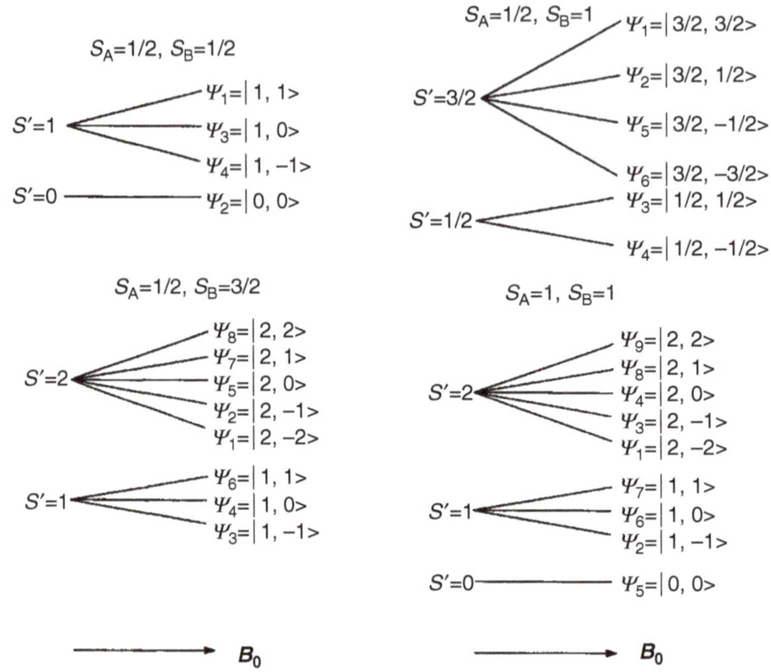

FIGURE 11.1 Electronic spin energy levels for dimetallic systems ½–½, ½–1, ½–³⁄₂, and 1–1.

$$E = \frac{J}{2} S'(S'+1) \quad \text{and} \quad \Delta E = J\, S'. \tag{11.2}$$

When J is positive, the ground state has $S' = |S_1 - S_2|$ and the highest level has $S' = S_1 + S_2$ (Fig. 11.1). We say that the coupling is *antiferromagnetic*. The contrary holds when J is negative and the coupling is *ferromagnetic*.

When more metal ions are coupled, the following Hamiltonian can be used:

$$H = \sum_{j \neq k} J_{jk}\, \boldsymbol{S}_j \cdot \boldsymbol{S}_k \tag{11.3}$$

where all j–k pairwise interactions are taken into consideration. For each ith S', the contribution to the total induced magnetic moment is [in analogy to Eq. (1.33)]:

$$\langle \mu_i \rangle = -\mu_B g_e \langle S'_z \rangle_i P_i \tag{11.4}$$

where P_i, is the population, the total induced magnetic moment being

$$\langle \mu \rangle = -\mu_B g_e \sum_i \langle S'_z \rangle_i P_i. \tag{11.5}$$

The contact hyperfine shift experienced by a nucleus is proportional to the induced magnetic moment of each S' level times the hyperfine coupling constant

$$\delta^{\text{con}} = -\sum_i \frac{A_i}{\hbar \gamma_I B_0} \langle S'_z \rangle_i P_i. \tag{11.6}$$

This equation would be of little help if further considerations were not made. In fact, we do not know the several A_i values. However, the $|S', M_{S'}\rangle$ wavefunctions are linear combinations of the $|S, M_S\rangle$ wavefunctions of each metal ion; therefore we want to show that it is possible to express A_i through the hyperfine constant A_M of the single ions when they are not magnetically coupled. Let us assume that, without loss of generality, a nucleus senses only one metal ion of the cluster, M, i.e. it senses only the induced magnetic moment of that ion. If we are able to project out the $|S, M_S\rangle$ part for that metal from each $|S', M_{S'}\rangle$ wavefunction, then we can use the hyperfine constant A_M that the nucleus would experience if the metal ion were not involved in the magnetic coupling, i.e. if $J = 0$. Indeed the following relationship holds for each S' level:

$$A_i \langle S'_z \rangle_i = A_M \langle S_{Mz} \rangle_i \tag{11.7}$$

where $\langle S_{Mz}\rangle_i$ is the expectation value of the contribution of metal M to the whole wavefunction S'_i and $\langle S'_z\rangle\rangle_i$ is the usual expectation value of S'_z for each i level [2,3]. This relationship is based on the assumption that the establishment of magnetic interactions within a cluster does not change the spin density distribution on the ligands nor on the metal ion. This is apparently the case in the investigated systems. If $|J| \ll kT$, all the levels are equally populated and

$$\sum_i A_i \langle S'_z \rangle_i P_i = A_M \sum_i \langle S_{Mz} \rangle_i P_i = A_M \langle S_{Mz} \rangle \tag{11.8}$$

where $\langle S_{Mz}\rangle$ is the expectation value of S_z for metal M in the absence of magnetic interactions. It follows that the energy of interaction between a nucleus and the induced electronic magnetic moment sensed by it ($\propto A_M\langle S_{Mz}\rangle$) does not change upon establishment of magnetic coupling. The energy of interaction does change if only some of the S_i levels are occupied.

A useful formula to express the contact shift, in the case of a nucleus sensing only one metal ion, is

$$\delta^{\mathrm{con}} = A_M \frac{g_e \mu_B}{\hbar \gamma_I 3kT} \frac{\sum_i C_i S'_i(S'_i+1)(2S'_i+1)\exp(-E_i/(kT))}{\sum_i (2S'_i+1)\exp(-E_i/(kT))} \tag{11.9}$$

where

$$C_i = \langle S_{Mz} \rangle_i / \langle S'_z \rangle_i . \tag{11.10}$$

Eq. (11.9) reduces to Eq. (2.7) in the high temperature limit, as shown in Appendix VIII.

If the nucleus senses another metal ion with a different A_M value, then the total shift will be given by the sum of two terms of the type of Eq. (11.9), each containing its own A_M and C_i values. A_{M1} and A_{M2} can be estimated from the monomeric analogs. The C_{i1} and C_{i2} values can be calculated once and for all for each system with the help of Hamiltonian Eq. (11.3). Their values are reported in Table 11.1 for dimers involving S_1 and S_2 values from ½ to 7⁄2 [4]. These coefficients can be calculated from the following equations:

$$\begin{aligned} C_{i1} &= \left[S'_i(S'_i+1) + S_1(S_1+1) - S_2(S_2+1)\right] / \left[2S'_i(S'_i+1)\right] \\ C_{i2} &= \left[S'_i(S'_i+1) + S_2(S_2+1) - S_1(S_1+1)\right] / \left[2S'_i(S'_i+1)\right]. \end{aligned} \tag{11.11}$$

Table 11.1 Coefficients Relating the Hyperfine Coupling of an S_1 or S_2 Spin System in a Monomer to that in a Coupled System, for each S_i Spin Level [3]

	S_2																				
	1/2			1			3/2			2			5/2			3			7/2		
S_1	S	C_1	C_2	S	C_1	C_2	S	C_1	C_2	S	C_1	C_2	S	C_1	C_2	S	C_1	C_2	S	C_1	C_2
1/2	1	1/2	1/2	3/2	1/3	2/3	2	1/4	3/4	5/2	1/5	4/5	3	1/6	5/6	7/2	1/7	6/7	4	1/8	7/8
	0	–	–	1/2	–1/3	4/3	1	–1/4	5/4	3/2	–1/5	6/5	2	–1/6	7/6	5/2	–1/7	8/7	3	–1/8	9/8
1				2	1/2	1/2	5/2	2/5	3/5	3	1/3	2/3	7/2	2/7	5/7	4	1/4	3/4	9/2	2/9	7/9
				1	1/2	1/2	3/2	4/15	11/15	2	1/6	5/6	5/2	4/35	31/35	3	1/12	11/12	7/3	4/63	59/63
				0	–	–	1/2	–2/3	5/3	1	–1/2	3/2	3/2	–2/5	7/5	2	–1/3	4/3	5/2	–2/7	9/7
3/2							3	1/2	1/2	7/2	3/7	4/7	4	3/8	5/8	9/2	1/3	2/3	5	3/10	7/10
							2	1/2	1/2	5/2	13/35	22/35	3	7/24	17/24	7/2	5/21	16/21	4	1/5	4/5
							1	1/2	1/2	3/2	1/5	4/5	2	1/12	11/12	5/2	1/35	34/35	3	0	1
							0	–	–	1/2	–1	2	1	–3/4	7/4	3/2	–3/5	8/5	2	–1/2	3/2
2										4	1/2	1/2	9/2	4/9	5/9	5	2/5	3/5	11/2	4/11	7/11
										3	1/2	1/2	7/2	26/63	37/63	4	7/20	13/20	9/2	10/33	23/33
										2	1/2	1/2	5/2	12/35	23/35	3	1/4	3/4	7/2	4/21	17/21
										1	1/2	1/2	3/2	2/15	13/15	2	0	1	5/2	–2/35	37/35
										0	–	–	1/2	–4/3	7/3	1	–1	2	3/2	–4/5	9/5
5/2													5	1/2	1/2	11/2	5/11	6/11	6	5/12	7/12
													4	1/2	1/2	9/2	43/99	56/99	5	23/60	37/60
													3	1/2	1/2	7/2	25/63	38/63	4	13/40	27/40
													2	1/2	1/2	5/2	11/35	24/35	3	5/24	19/24
													1	1/2	1/2	3/2	1/15	14/15	2	–1/12	13/12
													0	–	–	1/2	–5/3	8/3	1	–5/4	9/4

3						6	1/2	1/2	13/2	6/13	7/13
						5	1/2	1/2	11/2	64/143	79/143
						4	1/2	1/2	9/2	14/33	19/33
						3	1/2	1/2	7/2	8/21	13/21
						2	1/2	1/2	5/2	2/7	5/7
						1	1/2	1/2	3/2	0	1
						0	–	–	1/2	−2	3
7/2									7	1/2	1/2
									6	1/2	1/2
									5	1/2	1/2
									4	1/2	1/2
									3	1/2	1/2
									2	1/2	1/2
									1	1/2	1/2
									0	–	–

Note that in the ground state of an antiferromagnetic dimer the metal 2 with smaller S_2 always has a negative coefficient, whereas metal 1 has a positive coefficient. This observation is of paramount importance to detect the establishment of antiferromagnetic coupling.

There are no treatments for the pseudocontact term, since the orbital part has never been considered when dealing with magnetic coupled systems. However, Eqs. (11.7) and (11.10) hold for the hyperfine splitting in EPR spectra of both solids and solutions [1]. Therefore the same reasoning is likely to apply to the pseudocontact shift as well. The major complication arises from the point–dipolar nature of the pseudocontact shift treatment, which contrasts with the idea of a polymetallic center.

By analogy with the shift, in dimetallic systems the relaxation rate enhancement of a nucleus sensing one metal ion is given by:

$$R_{1M} \propto E^2 S'(S'+1) f(\omega, \tau_c) \sum_i C_i^2 P_i \tag{11.12}$$

where $E^2 S'(S'+1)$ is a generic notation for the average square of the interaction energy between S' and I, whose modulation with time constant τ_c causes relaxation,[2] and P_i is the Boltzmann partition function.

Eq. (11.12) for the case of dipolar relaxation takes the form:

$$R_{1M} = \frac{2}{15}\left(\frac{\mu_0}{4\pi}\right)^2 \frac{\gamma_I^2 g_e^2 \mu_B^2}{r^6} \times \frac{\sum_i \left[C_i^2 S_i'(S_i'+1)(2S_i'+1)\exp(-E_i/(kT)) \left(\frac{7\tau_{c2}}{1+\omega_S^2\tau_{c2}^2} + \frac{3\tau_{c1}}{1+\omega_I^2\tau_{c1}^2} \right) \right]}{\sum_i \left[(2S_i'+1)\exp(-E_i/(kT)) \right]} \tag{11.13a}$$

where τ_{ci}^{-1} is defined as usual as the sum of rotational, exchange, and electronic relaxation (τ_{c1}^{-1} = longitudinal, τ_{c2}^{-1} = transverse) rates.

Analogously, the equation for contact relaxation takes the form:

$$R_{1M} = \frac{2}{3}\frac{A_M^2}{\hbar^2} \frac{\sum_i \left[C_i^2 S_i'(S_i'+1)(2S_i'+1)\exp(-E_i/(kT)) \left(\frac{\tau_{c2}}{1+\omega_S^2\tau_{c2}^2} \right) \right]}{\sum_i \left[(2S_i'+1)\exp(-E_i/(kT)) \right]} \tag{11.13b}$$

where τ_{ci}^{-1} is now defined as the sum of exchange and electronic relaxation rates.

11.2 ELECTRON RELAXATION AND MAGNETIC COUPLING

A consequence of magnetic coupling can be a change in the electronic relaxation times of the involved metal ions. Magnetic coupling gives rise to new energy levels which can provide new relaxation pathways. New transitions involving spin levels can occur through coupling with the lattice, i.e. solvent collisions, solvent fluctuations, molecular tumbling, phonons when appropriate. This is indeed what

[2] $E^2S(S+1)$ is a generic notation that holds for dipolar, contact or Curie relaxation. In the equation for Curie relaxation there is a $S^2(S+1)^2$ term because a further $S(S+1)$ term is contained in the E^2 term.

happens in polymetallic systems. In this case there is more than one level with the same S' value which allow electronic transitions.

11.2.1 HOMODIMERS

In homodimers, all the levels have different S' values and in principle no further relaxation pathway is occurring. Experiments have been performed aiming at comparing the τ_e value in copper(II) dimetallic systems with respect to monometallic systems [5]. A protein frame has been used in which either only one copper(II) ion is present or two magnetically coupled copper(II) ions are present. In the latter case J is 52 cm^{-1} and all the S' levels are almost equally populated at room temperature. The protein is superoxide dismutase, whose metal ligands are shown in Fig. 11.2. We may have either copper–zinc or copper–copper systems. Nuclear longitudinal relaxation was measured for the water protons since one water molecule interacts with one copper ion. It is shown that in the two cases the copper(II) ion bound to water maintains similar τ_e values. The nuclear relaxation in the dimer is, however, one-half that in the monomer on account of Eq. (11.13a). In the case of two $S = ½$ ions ($C_i = ½$) and $|J| \ll kT$, $P = 3/4$, because only the $S' = 1$ level which is threefold degenerate contributes to paramagnetic relaxation, the $S' = 0$ level being magnetically silent. Since R_{1M} for the monomer depends on $S(S + 1)$ with $S = ½$, and for the dimer it depends on $S'(S' + 1)$ with $S' = 1$ [Eq. (11.13a)], the ratio between the R_{1M} values of the dimer for a nucleus sensing only one metal ion and of the monomer is

$$\frac{C^2S'(S'+1)P}{S(S+1)} = \frac{1/4\times 2\times 3/4}{3/4} = \frac{1}{2}.$$

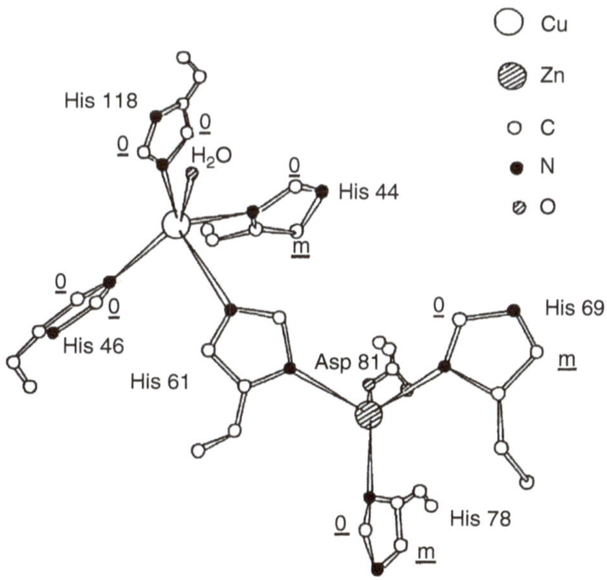

FIGURE 11.2 Schematic structure of the active site of copper–zinc superoxide dismutase. The zinc ion can be substituted, among others, by copper(II), cobalt(II), and nickel(II) ions.

Adapted from Ref. [72].

The constancy of τ_e upon establishing a magnetic coupling in homodimers can be also demonstrated theoretically [6]. It is possible, however, that in the presence of large ZFS of the $S' = 1$ level, new mechanisms of electron relaxation are effective, and τ_e shortens up. That is why sometimes the NMR linewidths of copper(II) dimers are sharper than in monomers by more than a factor 2 [7].

Even for dimetallic centers with each metal ion having $S > ½$, Eq. (11.12) provides a value of R_{1M} which is one-half of that provided by the analogous equation for the monomer. If we repeat the calculation for the hyperfine shifts [Eqs. (2.6) and (11.9)], at variance with nuclear relaxation, we see that the shifts for the coupled and uncoupled systems are the same as long as $|J| \ll kT$. Note that ferromagnetic coupling would have provided the same results because our reasoning was not based on the order of the S' levels.

When $|J|$ is of the order of kT, then the excited S' levels are not fully populated. In the case of antiferromagnetic coupling the magnetic susceptibility decreases and both shifts and nuclear relaxation values decrease. This is a significant advantage when systems with long electron relaxation times are investigated. Typically, copper(II) complexes are not easily investigated by high resolution NMR because τ_e is of the order of 10^{-9} s and τ_r for small complexes is about 10^{-10} s. Therefore, the linewidth at 500 MHz for a proton dipole–dipole coupled at 4 Å is 700 Hz in the absence of contact relaxation. In the case of a copper dimer with a J of 100 cm^{-1}, the shift of a proton sensing only one copper ion according to Eq. (11.9) is 86% of the monomeric case, and the linewidth [which behaves analogously to R_{1M} in Eq. (11.13a)] is 43% of the monomer at room temperature.

Ferromagnetic coupling generally has small $|J|$ values. For the sake of completeness, we consider the case in which there is a ferromagnetic coupling of -100 cm^{-1}. By repeating the calculations, the shifts are 111% of the monomer and the linewidths 55% of the monomer.

11.2.2 HETERODIMERS

When we deal with two different paramagnetic ions, it should be kept in mind that in the absence of magnetic coupling one will have a longer electronic relaxation time and the other will have a shorter one, according to their electronic structure. Such differences are in general very large. The slow relaxing metal ion will be referred to as M_1 and the fast relaxing one as M_2. It is reasonable to believe that, when the absolute value of J (expressed as $|J|/\hbar$) is smaller than τ_{e1}^{-1} (i.e. the electronic relaxation rate of the slow relaxing M_1 ion) no effect on the electronic relaxation of the pair will take place. When $|J|/\hbar > \tau_{e1}^{-1}$ but smaller than τ_{e2}^{-1}, we can approach the system through a first order perturbation treatment. The coupling between S_1 and S_2 is very similar to that between I and S already discussed (Sections 4.4 and 4.5). In both cases we have a scalar coupling which is the contact coupling in the case of the electron–nucleus coupling and the exchange or superexchange coupling in the case of magnetic interactions. Analogous to the Bloembergen Eq. (4.27) for the contact nuclear relaxation we can write the following relationship:

$$\Delta\tau_{e1}^{-1} = \frac{2}{3} S_2(S_2+1)\left(\frac{J}{\hbar}\right)^2 \frac{\tau_{e2}}{1+(\omega_{S1}-\omega_{S2})^2\tau_{e2}^2} \tag{11.14}$$

where $\Delta\tau_{e1}^{-1}$ is the enhancement in the electronic relaxation rate of the slow relaxing ion, ω_{S1} and ω_{S2} are the Larmor frequencies of the M_1 and M_2 ions, and τ_{e2} is the correlation time for the interaction which is given by the electronic relaxation rate of the fast relaxing ion.

When the coupling between S_1 and S_2 is dipolar in origin the equation should again be analogous to the dipolar coupling in the case of nuclear relaxation [Eq. (4.17)]:

$$\Delta\tau_{e1}^{-1} = \frac{2}{15}\left(\frac{\mu_0}{4\pi}\right)^2\left(\frac{1}{\hbar}\right)^2 \frac{g_e^4\mu_B^4 S_2(S_2+1)}{\langle r^3\rangle^2} \times\left(\frac{\tau_{e2}}{1+(\omega_{S1}-\omega_{S2})^2\tau_{e2}^2}+\frac{3\tau_{e2}}{1+\omega_{S1}^2\tau_{e2}^2}+\frac{6\tau_{e2}}{1+(\omega_{S1}+\omega_{S2})^2\tau_{e2}^2}\right) \quad (11.15)$$

where $\langle r^3\rangle$ is the average cube of the interelectronic distance.

It is apparent that the dipolar coupling is a short range coupling and rapidly vanishes with the distance. This does not hold for the scalar coupling. It is also apparent that small $|J|$ values can affect τ_{e1} under favorable conditions. In other words, small $|J|$ values can be revealed through the linewidth of the EPR signal or through NMR relaxation measurements.

Eqs. (11.14) and (11.15) hold as long as $|J|/\hbar < \tau_{e2}^{-1}$. This is a general requirement in relaxation theory called Redfield limit [8]. The coupling energy between the relaxing spin and the lattice (which can be another spin) has to be smaller than the product of $\hbar$ times the reciprocal correlation time. Equivalent to this statement is the following: the relaxation rate of a spin cannot be higher than the reciprocal of the correlation time. In other words, τ_{e1}^{-1} approaches τ_{e2}^{-1} when $|J|$ increases. Outside the Redfield limit, i.e. outside the perturbative limit, each S' level should be regarded as a single spin. The transitions between $M_{S'}$ levels should be evaluated in a manner similar to that discussed for the $S > 1$ multiplets in isolated ions (Section 4.3). If reference is made to the τ_{e1} and τ_{e2} of the single ions in the coupled system they cannot be easily predicted except for their becoming similar and close to the shorter one (Section 11.4).

The absolute value of J, which we have used all over the treatment within the Redfield limit, implies that the theory is the same for ferro- and antiferromagnetic coupling. When $|J| \ll kT$, the sign of J is irrelevant for the nuclear shifts, the electronic relaxation times and nuclear relaxation times. When J is of the order of kT, presumably only electronic relaxation times within the various S' level can be defined, and they are independent of the sign of J, but the nuclear shifts and relaxation rates are different because the various S'_i are differently populated according to the sign of J. Eqs. (11.9) and (11.13a) should be used to predict the particular behavior.

The above treatment is probably valid only in ideal cases. In practical cases spin–orbit coupling causes ZFS, and the S' levels are split at zero magnetic field just like the S levels are. This consideration requires that each S' level will have its own effective electronic relaxation time. Therefore, the above quantitative considerations should be regarded as useful guidelines.

11.2.3 POLYMETALLIC SYSTEMS

In the case of polyionic systems the picture becomes still more complicated. There are no examples of prediction of τ_e values. However, we have empirically learned that τ_e of the entire system is shorter than in the absence of magnetic coupling. High resolution NMR experiments are, in general, successful with magnetically coupled polyionic systems.

11.3 NMR OF DIMETALLIC SYSTEMS

It is convenient to divide this section into systems containing equal or different metal ions. Each of the two cases will be treated according to whether $|J|$ is larger or smaller than kT.

11.3.1 SYSTEMS CONTAINING EQUAL METAL IONS, $|J| < kT$

We take now as an illustrative example the dimers with PMK as ligand. Such a ligand is capable of giving bimetallic complexes of formula $[(PMK)_3M_2]^{4+}$ and with the structure shown in Fig. 11.3 [9]. The ligand provides a pathway for weak magnetic coupling. When $|J| \ll kT$, the shift experienced by a nucleus is simply the sum of the contributions of the two metal ions, i.e. of two terms given by Eq. (11.9) and containing A_{M1} and C_{i1}, and A_{M2} and C_{i2}, respectively. This is indeed the case, as proven by using mixed complexes with the diamagnetic zinc ion [10–12]. The correlation time for the nucleus electron coupling is given by the rotational correlation time in the case of copper(II) complexes ($\tau_r \ll 2 \times 10^{-10}$ s) and by the electronic relaxation time ($\tau_e \approx 6 \times 10^{-13}$ s) in the case of cobalt(II) complexes [13,14]. As explained in the previous section, a decrease is expected in the linewidth, which would be of a factor of two if the proton senses only one paramagnetic ion. Indeed, some decrease in the linewidth is observed. In Table 11.2 the NMR parameters for the CoCo system are reported, together with those of the CoZn, NiZn, CuZn, CoNi, and CoCu systems (Section 11.3.3) [14].

11.3.2 SYSTEMS CONTAINING EQUAL METAL IONS, $|J| \approx kT$

The first systems of this kind investigated in the literature are dicopper(II) dimers [15,16], where J is positive and of the order of hundreds of wavenumbers. The correlation time is τ_r; the nuclear relaxation rates, as well as the shifts, are decreased according to the Boltzmann partition coefficients of Eqs. (11.13a) and (11.9), respectively. A factor larger than two in nuclear relaxation rate between coupled and uncoupled systems should be operative if only one metal ion is sensed by the resonating nucleus. The decrease in magnetic susceptibility due to the population of the lower part of the energy ladder makes these systems more suitable for NMR investigation than the uncoupled ones. The same holds for μ-oxo iron(III) dimers [17].

In Fig. 11.4 the spectra of a copper(II) dimer are shown [16]. The temperature dependence of the shifts is consistent, according to Eq. (11.9), with a J value of about 500 cm^{-1}. In the inset, the ^{2}H NMR spectrum of a selectively deuterated derivative is reported [16], showing the smaller linewidth of the

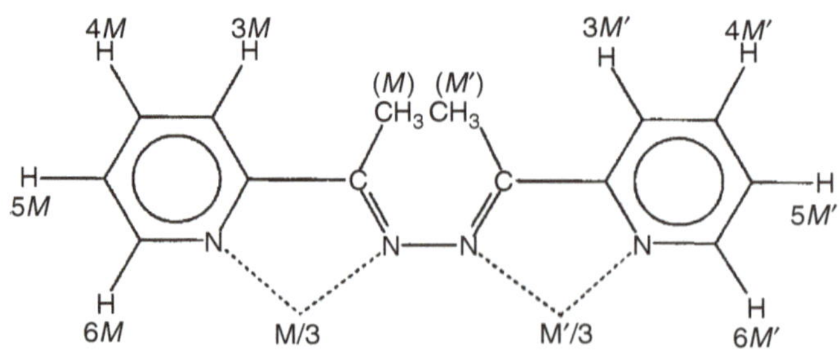

FIGURE 11.3 Schematic structure of $[(PMK)_3M_2]^{4+}$ bimetallic complexes

Adapted from Ref. [8].

Table 11.2 Proton Hyperfine Shifts (ppm) and T_{1M}^{-1} Values (s^{-1}, Estimated Errors in Parentheses) for MM′$(PMK)_3^{4+}$ Complexes [13]

		3-H	4-H	5-H	6-H	$-CH_3$
$CoZn(PMK)_3^{4+}$						
Shifts	(Co)	66.81	2.74	42.31	141.47	44.99
	(Zn)	−9.88	2.41	−10	−0.69[a]	8.08
T_{1M}^{-1}	(Co)	36.1 (0.9)	11.5 (0.3)	29.2 (0.7)	670 (220)	71.9 (1.6)
	(Zn)	6.9 (0.3)	1.41 (0.02)	1.78 (0.06)	3.97 (0.06)[a]	37.3 (0.6)
$NiZn(PMK)_3^{4+}$						
Shifts	(Ni)	62.98	8.54	40.29		39.23
	(Zn)	−9.15	2.76	−8.07	2.11	−10.51
T_{1M}^{-1}	(Ni)	2000 (500)	590 (5)	1630 (110)		5900 (350)
	(Zn)	370 (15)	120 (10)	140 (10)	310 (15)	2600
$CuZn(PMK)_3^{4+}$						
Shifts	(Cu)	39.68	5.68	23.43		17.26
	(Zn)	−4.19	1.31	−3.67	0.87	−3.55
T_{1M}^{-1}	(Cu)		235 (3)	526 (60)		1200 (40)
	(Zn)	118 (10)	43.3 (1)	64 (6)	121 (2)	709 (10)
$Co_2(PMK)_3^{4+}$						
Shifts		56.46	5.08	31.99	141.21	51.88
T_{1M}^{-1}		48.5 (0.5)	17.0 (0.1)	35.8 (0.3)	588 (35)	114 (2)
$CoNi(PMK)_3^{4+}$						
Shifts	(Co)	56.54	4.74	34.47	149.55	33.75
	(Ni)	50.75	11.05	28.91	148.8[b]	44.65
T_{1M}^{-1}	(Co)	52.1 (1.4)	18.6 (0.2)	38.0 (0.6)	645 (17)	180 (0.4)
	(Ni)	112 (3)	35.2 (0.5)	88.5 (1.6)	909 (83)[b]	225 (7)
$CoCu(PMK)_3^{4+}$						
Shifts	(Co)	62.43	3.78	38.61	143.84	41.35
	(Cu)	27.60	8.06	12.21	87.38	24.65
T_{1M}^{-1}	(Co)	40.8 (1.8)	13.5 (0.4)	30.3 (1.2)	529 (28)	94.3 (3.6)
	(Cu)	37.7 (1.8)	8.33 (0.14)	24.2 (0.4)	307 (10)	75.8 (2.3)

[a] *Direct overlap with 5-H (Zn_2).*
[b] *Broad and overlapping with the other 6-H signal.*

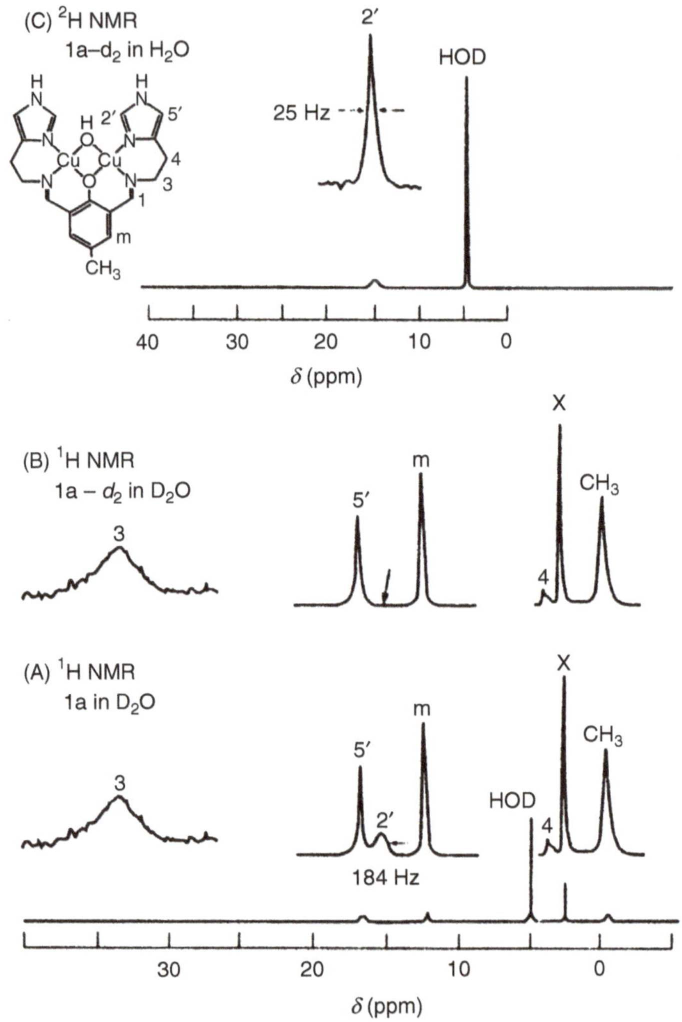

FIGURE 11.4 ^{1}H NMR spectra of a copper(II) dimer (A) and of its selectively deuterated derivative (B) (the arrow pointing to the missing signal). The ^{2}H NMR spectrum of the latter is shown in (C), together with the structure of the compound.

Adapted from Ref. [15].

^{2}H signal. In an increasing number of copper dimers, the NMR lines have been found to be far too sharp to be explained [16,18–23], thus allowing high resolution NMR investigations, being two orders of magnitude sharper than the mononuclear analogs (Fig. 11.5). Since the hyperfine shifts are very large, as expected for the mononuclear analog, and display an essentially normal Curie temperature dependence, the line sharpening must be due to a decrease in the electron relaxation time (from 10^{-9} to 10^{-11} s), and cannot be due to a reduction of paramagnetism caused by a depopulation of the excited states. Hyperfine shifts are in fact very large, spanning from +230 to −14 ppm. This confirm the

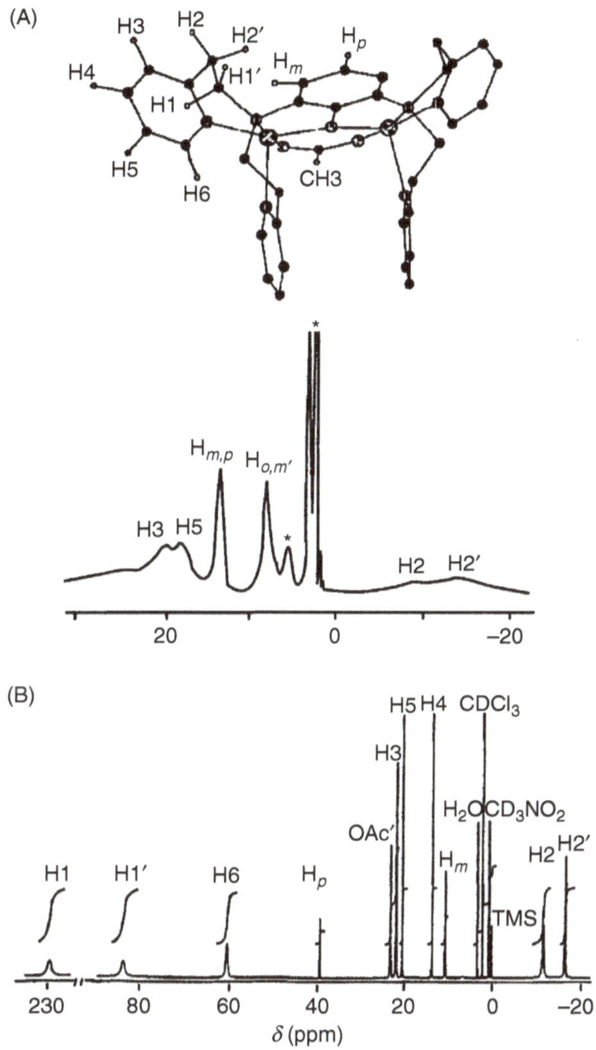

FIGURE 11.5 Crystal structures and ^{1}H NMR spectra of a copper monomer (A) and the corresponding copper dimer (B). The signal assignment is also shown.

Adapted from Ref. [22].

absence of strong antiferromagnetic coupling (which would cause the paramagnetism to be strongly reduced due to depopulation of the excited $S = 1$ state). What causes this decrease is still under investigation. It was suggested that new relaxation pathways can be created upon dimer formation, as a result of modulation of the J value by solvent collision, modulation of the ZFS of the $S' = 1$ state or electron delocalization [23]. Also the so-called Cu_A centers exhibit a similar behavior (Section 7.3.1.1).

When the dimetallic center is embedded in a protein, the correlation time is given by the electronic relaxation time. A studied case is provided by the $[Fe_2S_2]^{2+}$ unit represented in Fig. 11.6 and found

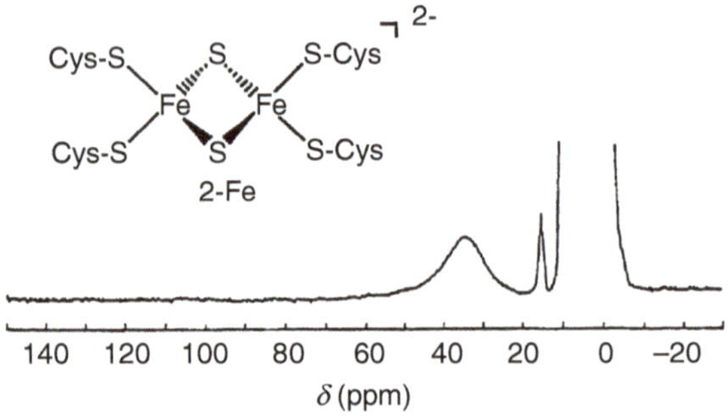

FIGURE 11.6 Schematic structure of the $[Fe_2S_2](SR)_4^{2-}$ unit as a found in oxidized ferredoxins and ^1H NMR spectrum showing the β-CH_2 proton signals at about 40 ppm.

in oxidized ferredoxins. The large correlation time makes the system difficult to investigate by NMR [24,25] despite the sizable decrease in paramagnetism due to a J of about 300 cm^{-1}. The sharpening of the lines of oxidized ferredoxins with respect to the monomeric iron model provided by the oxidized rubredoxin thus arises from the decreased Boltzmann population of the paramagnetic excited states. Indeed, an elegant approach to the assignment of resonances close to the bimetallic cluster has been recently reported [26]. The $[Fe_2S_2]^{2+}$ cluster is expected to induce a sizable hyperfine relaxation on ^1H nuclei within about 10 Å from each of the two iron ions. In line with this, many residues are not observed in conventional double and triple resonance experiments. The application of relaxation-tailored IR-HSQC-AP experiment (Section 12.3.4) can recover several of these residues. This is the case, for instance, for the CIAPIN1 domain of the human protein anamorsin [27,28], a 108 amino acid domain which binds a $[Fe_2S_2]^{2+}$ cluster [27,29], where additional resonances (Fig. 11.7) could be assigned on the basis of the chemical shifts and of the relaxation times, ranging from less than 10 ms up to 150 ms [26]. Using relaxation tailored experiments (Section 12.3.4) [30] paramagnetic relaxation caused by the $[Fe_2S_2]^{2+}$ cluster were used to monitor its transfer from one protein to another, as in the case of hGRX5 and the partner proteins hISCA1 and hISCA2 and to develop a model for the interaction of the two proteins [31].

11.3.3 SYSTEMS CONTAINING DIFFERENT METAL IONS, $|J| < kT$

With the PMK ligand a CoCu derivative has been obtained [14] (Fig. 11.8). From the temperature dependence of the shifts (and magnetic susceptibility measurements in solution), the value of J appears to be positive and much smaller than kT [14]. As expected, the hyperfine shifts are the sum of those of the CuZn and ZnCo systems for each proton (Table 11.2). The NMR lines of the copper domain are now quite sharp, even sharper than those of the cobalt domain (Fig. 11.8). Qualitatively, the data can be accounted for if τ_e of copper is sizably reduced and approaches that of cobalt (and thus the Redfield limit is reached). The magnetic coupling is antiferromagnetic with an $S' = 1$ ground state and an $S' = 2$ excited state. The C_i coefficients are those of Table 11.1, i.e. −1/4 and 1/4 for copper and 5/4 and 3/4 for cobalt. Such coefficients provide a ratio in nuclear R_{1M} of ca. 12 for the protons of the cobalt domain with respect to those of the copper domain, if τ_e were the same [Eq. (11.13a)]. The experimental ratio is about two. The discrepancy can be understood by considering: (1) that the electronic relaxation times for the two metal ions are not identical; (2) that copper(II) tends to give rise to ligand-centered effects

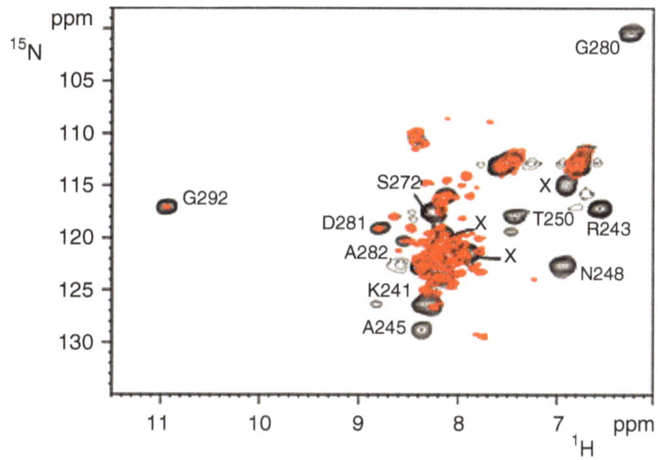

FIGURE 11.7 Overlay of a standard ^{15}N-HSQC (*thin lines*) and IR-^{15}N-HSQC- AP (*thick lines*) experiments acquired on 500 MHz at 298 K on the $[Fe_2S_2]^{2+}$-CIAPIN1 domain. The experimental time for both experiments was ca. 7 h. For the IR-^{15}N-HSQC-AP, acquisition times of 20 and 24 ms were used in t2 and t1 acquisition dimensions, respectively, and recycle and IR delays were 55 and 50 ms, respectively. To rule out effects due to short recycle delays, the reference ^{15}N HSQC has been recorded with a 350 ms overall recycle delay. Peaks are labeled according to their assignment [29]. Unassigned resonances are indicated with X. Signals attributed to D281, A282 and G292 were observed also in the reference experiment but at much lower intensity.

Reproduced with permission from Ref. [30].

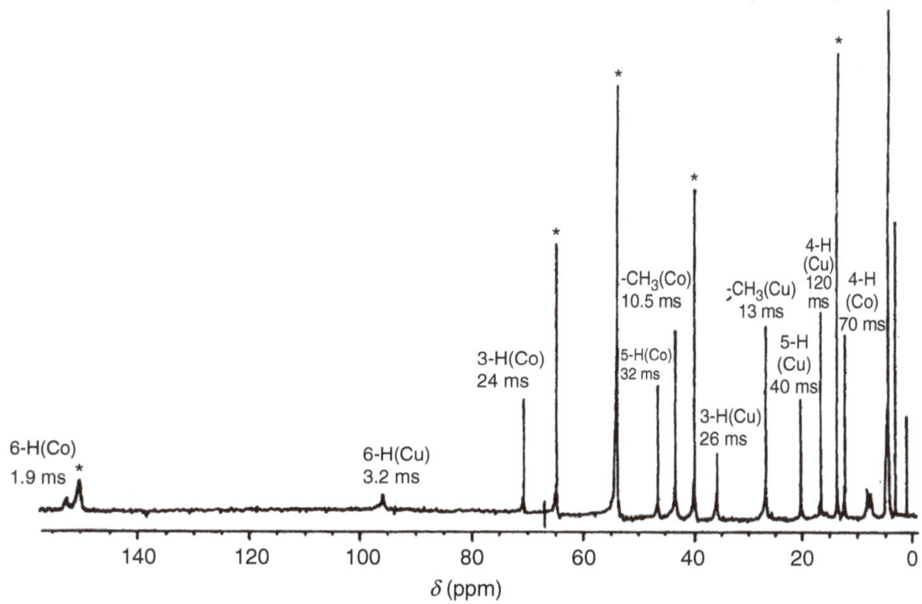

FIGURE 11.8 ^{1}H NMR spectra of the CoCu derivative of PMK (Fig. 11.3).

Adapted from Ref. [13].

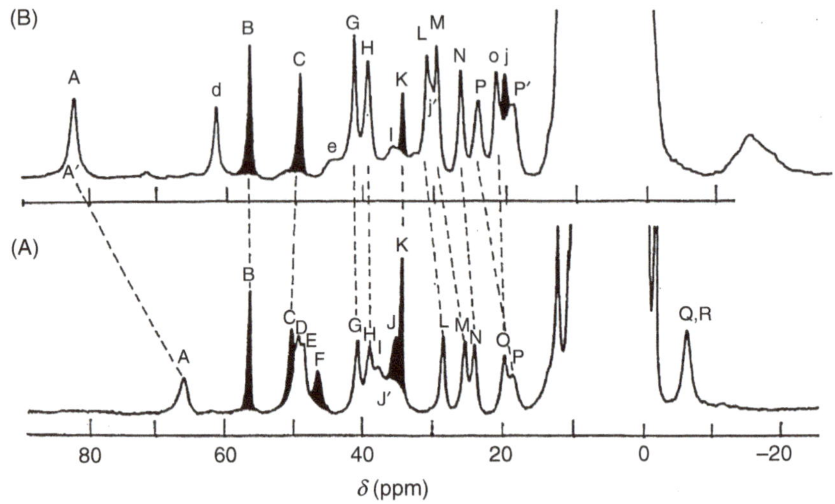

FIGURE 11.9 ^{1}H NMR spectra of Cu_2Co_2 (A) [28] and Cu_2Ni_2 (B) [29] superoxide dismutase. The *dashed lines* relate signals belonging to the various histidine ligands of the copper domain. The *black-shaded signals* disappear when the spectra are recorded in D_2O.

more than cobalt, and therefore to experience higher nuclear relaxation rates than expected on the basis of a pure dipolar mechanism; (3) that the presence of ZFS complicates the prediction strategies.

The above system is similar to that built into the protein superoxide dismutase (Fig. 11.2). The zinc can be replaced by several metal ions. Cobalt(II) and nickel(II) are pertinent here. The cobalt-copper system has $J \approx 17$ cm^{-1} with $S' = 1$ ground state [32]. The electronic relaxation times for the two ions are expected to be either equal or similar. The ^{1}H NMR spectrum is shown in Fig. 11.9A. The assignment has been performed through several steps.

1. The histidine NH protons have been recognized because their signals disappear when the spectrum is recorded in D_2O (Fig. 11.9A) [33].
2. The protons of the copper domain have smaller R_1 than those of the cobalt domain if the metal–proton distances are similar [33].
3. The linewidths have large Curie relaxation contributions which can be factorized out by recording the spectra at different fields. Such contributions and the R_1 values should provide a qualitative order of distances of the protons from the metal ions. However, large spin delocalization effects may provide larger linewidths to protons farther from the metal ion and this has led to a misassignment [34].
4. NOE measurements have allowed the assignment of all histidine protons and of the β-CH_2 of the Asp group [35].
5. NOESY spectra have largely confirmed the previous assignment and have permitted the assignment of the β-CH_2 protons and of some α-CH protons of the histidines [35].

The resulting assignment, which is one of the most successful in paramagnetic metalloproteins, is reported in Table 11.3. The nuclear relaxation rates, which are in a ratio similar to those of the CuCoPMK system, have been interpreted on the basis of large ligand-centered effects on the copper ligands [35].

Table 11.3 Assignment of ^{1}H NMR Signals of Metal Coordinated Residues in Cu_2Co_2SOD (298 K, pH 5.5, Buffer 50 mM Acetate)[a]

Signal	Shift (ppm)	Assignment	Signal (ppm)	Shift	Assignment
A	67.4	His 63 Hδ_2	a′	12.30	His 48 Hβ_2
B	57.0	His 120 HNδ_1	b′	11.21	His 120 Hβ_2
C	50.3	His 46 NHε_2	c′	8.44	His 120 NH
D	49.6	His 71 Hδ_2	d′	6.40	His 48 Hβ_1
E	49.0	His 80 Hδ_2	e′	4.43	His 120 Hα
F	46.7	His 80 HNε_2	f′	3.73	His 48 Hα
G	41.0	His 46 Hδ_2	g′	3.13	His 120 Hβ_1
H	39.5	His 120 Hε_1	h′	1.23	Ala 140 βCH$_3$
I	38.7	Asp 83 Hβ_1	i′	0.56	Val 118 γ_1CH$_3$
J′	37.0	Asp 83 Hβ_2	j′	−0.28	Arg 143 Hγ_1
J	35.6	His 71 HNε_2	k′	−1.51	Val 118 γ_1CH$_3$
K	34.7	His 48 HNδ_1			
L	28.4	His 48 Hδ_2			
M	25.7	His 46 Hε_1			
N	24.3	His 120 Hδ_2			
O	19.8	His 48 Hε_1			
P	18.7	His 46 Hβ_2			
Q	−6.2	His 71 Hβ_2			
R	−6.2	His 46 Hβ_2			

[a]*The tentative assignment of some signals involving residues not directly bound to the metal ions, and falling in the diamagnetic region of Fig. 11.9(A) is also reported [28].*

The value of J is not known in the case of the copper–nickel derivative. However, the same line of discussion should hold for this derivative, whose spectrum is shown in Fig. 11.9B [36]. It is noteworthy that the R_1 values of the protons of the copper domain are smaller in this case than in the cobalt case. This is probably due to a shorter electronic relaxation time of tetrahedral nickel(II) than tetrahedral cobalt(II) (Chapter 7). In this respect it may be interesting to note that in the case of the cobalt–cobalt derivative we have a tetrahedral and a square pyramidal cobalt(II) ion [37,38]. The former has longer electronic relaxation times. Upon establishment of magnetic coupling, the electronic relaxation times of the tetrahedral cobalt(II) ion decrease and tend to reach the values of the square pyramidal cobalt(II) ion. This is evident by comparing the spectra of the protein containing only the tetrahedral cobalt(II) ion with those containing both ions (Fig. 11.10).

Many systems of this kind are available in the literature. Only a small part has been studied by NMR, and sometimes not all the structural and dynamic information has been extracted from the spectra with the support of the available theory.

The results are expected to be the same if the coupling is of ferromagnetic type.

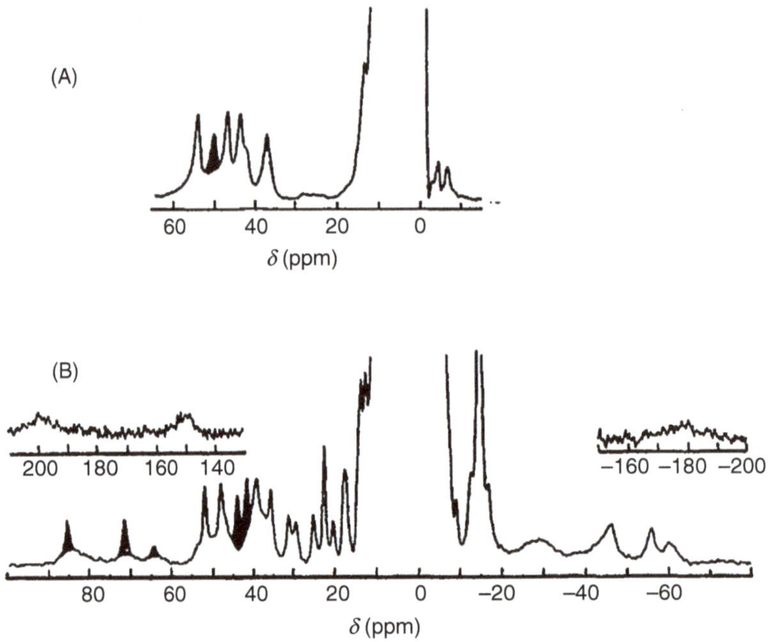

FIGURE 11.10 ^{1}H NMR spectra of E_2Co_2 (A) (E, empty) and Co_2Co_2 superoxide dismutase (B). The *black-shaded signals* disappear when the spectra are recorded in D_2O.

Adapted from Ref. [30,31].

11.3.4 SYSTEMS CONTAINING DIFFERENT METAL IONS, $|J| \approx kT$

If $|J|$ were much larger than kT, only the ground state would be populated. This limit condition is very useful to illustrate the peculiarities of asymmetric dimeric systems. In these systems, one ion will have a spin larger than the other. In the ground state the larger spin will always be aligned along the external magnetic field. Now we can distinguish between ferro- and antiferromagnetic coupling. In the former case the smaller spin is aligned along with the larger spin; and in the latter case the smaller spin is forced to be aligned opposite to the external magnetic field. The second case is the most common. The larger spin provides a "normal" additional magnetic field to the resonating nucleus in a way similar to the case of absence of magnetic coupling. The smaller spin is forced to provide an additional field of opposite sign. From the theory we do expect a negative sign for the coefficient related to the smaller spin [Eqs. (11.10) and (11.11)]. Therefore, if the nuclei coupled to the larger spin experience downfield shift, the nuclei coupled to the smaller spin will experience upfield shifts (Fig. 11.11). In the limit of $|J| \ll kT$ the shifts of the two kinds of nuclei are the same as if there were no magnetic coupling at all. In the intermediate cases the shifts of the nuclei coupled to the smaller spin can be anywhere from downfield to upfield. However, if they are downfield, they will increase with increasing temperature, as the system will tend towards $|J| \ll kT$. The shifts of the nuclei coupled to the larger spin will decrease with increasing temperature similarly to a normal paramagnet [Eqs. (2.6) and (2.7)]. The expected temperature dependence is illustrated in Fig. 11.12, as calculated from Eq. (11.9).

From the point of view of relaxation we expect similar values of τ_e for the two spins, unless ZFS of the magnetically coupled S' levels introduces unpredictable effects.

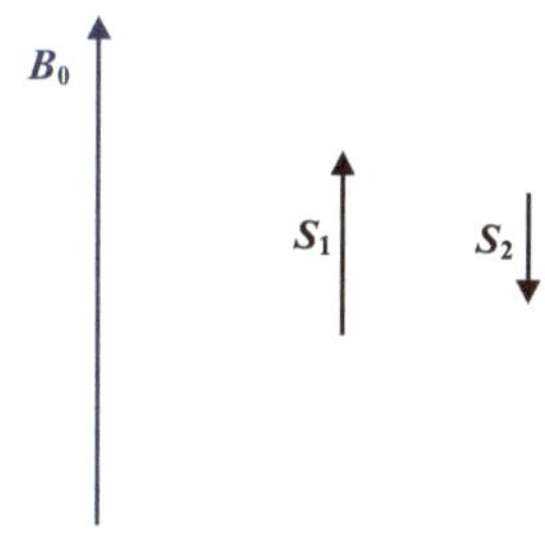

FIGURE 11.11 Pictorial scheme of the magnetic moments corresponding to S_1 and S_2 induced by a magnetic field B_0 in an antiferromagnetically coupled ground state, S_1 being larger than S_2, and $g_e\mu_B B_0 \ll J$.

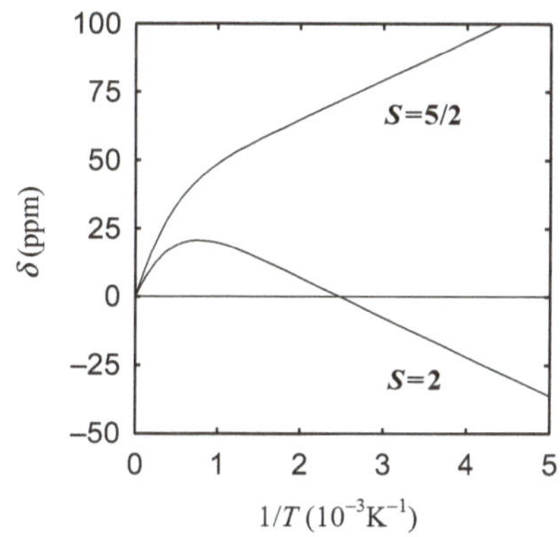

FIGURE 11.12 Predicted temperature dependence of the shifts of nuclei sensing either the $S = 5/2$ or the $S = 2$ spin of an antiferromagnetically coupled pair with $J = 300$ cm^{-1}.

The best case studied is that of reduced $[Fe_2S_2]^+$-containing ferredoxins [39,40]. The systems contain one iron(III) and one iron(II) ions (Fig. 11.13, inset). Both of them are tetrahedral and high spin. The iron(III) has $S = 5/2$ and the iron(II) $S = 2$. The ^{1}H NMR spectrum of one protein is reported in Fig. 11.13 [39]. The four signals far downfield belong to the β-CH_2 protons of the two cysteines bound to iron(III). Such signals have a temperature dependence of Curie type, i.e. the hyperfine shifts increase with decreasing temperature. The sharp signal around 40 ppm is assigned to an α-CH proton of one cysteine of the iron(III) domain. The four sharp signals between 30 and 15 ppm belong to the β-CH_2 protons of the cysteines of the iron(II) domain. The latter four signals have a temperature dependence of antiCurie type, i.e. the hyperfine shifts increase with increasing temperature.

NOE studies have shown the pairwise nature of the β-CH_2 protons of the cysteines of the iron(II) domain [41]; the β-CH_2 proton signals of the iron(III) domain are too broad to observe any dipolar connectivity. Inter-β-CH_2 NOEs have also been observed between two signals of two β-CH_2 of the iron(II) domain, thus providing a key information to assign the cysteines bound to iron(II) to Cys-41 and Cys-46 and, by exclusion, the cysteines bound to iron(III) to Cys-49 and Cys-79. In fact, by analyzing the X-ray structure it appears that Cys-41 and Cys-46 are close enough to one another to give inter-residue NOE. The detection of the iron(III) and iron(II) domains is a piece of information that only NMR could have provided.

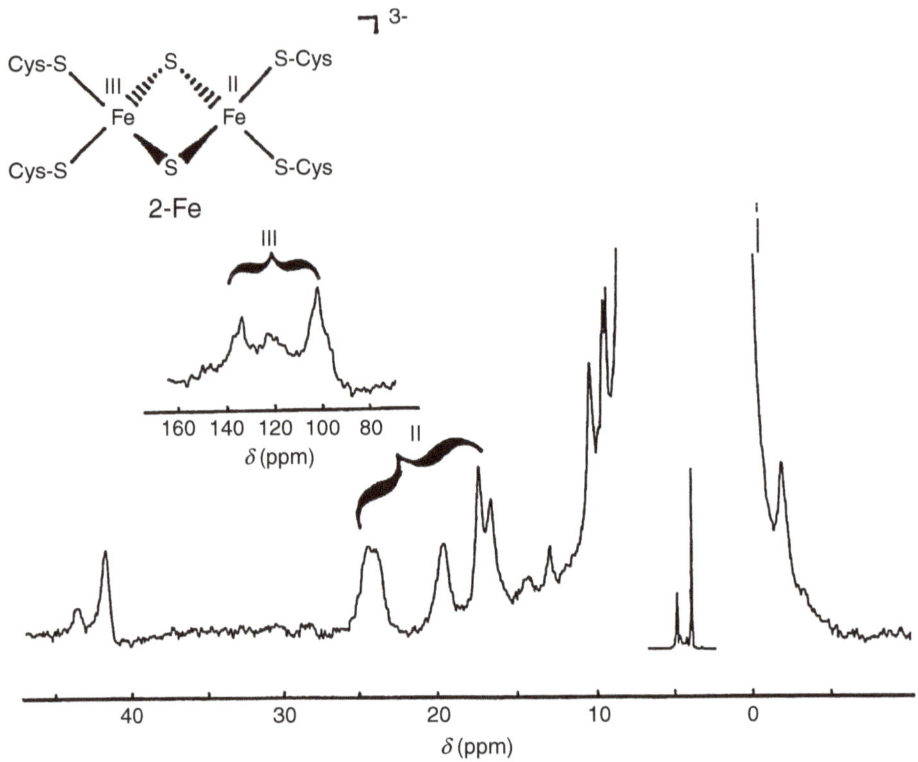

FIGURE 11.13 ^{1}H NMR spectra of reduced Fe_2S_2 ferredoxin containing one iron(II) and one iron(III) ions antiferromagnetically coupled. The roman number labeling II and III refers to the oxidation states of the iron ions to which the cysteines are bound.

Adapted from Ref. [32].

11.4 BEYOND THE REDFIELD LIMIT: $|J|/\hbar > \tau_e^{-1}$

A fast relaxing metal ion in magnetic exchange coupled dimers shortens the electron relaxation time of the slower relaxing metal ion, often to the value of the fast relaxing ion. Electron–electron interactions are thus so strong that the Redfield limit is reached. To describe this situation, linewidths and nuclear relaxation rates can be calculated as a function of the electron relaxation rates of the isolated spins in the hypothesis that the coupling frequency between the spins is large in absolute value compared to the electronic relaxation rates but small compared to the modulation of the electron–lattice interaction:

$$|J|\hbar \geq \tau_{e1}^{-1}, \qquad |J|\hbar \geq \tau_{e2}^{-1}$$
$$|J|\hbar \leq \tau_{v1}^{-1}, \qquad |J|\hbar \leq \tau_{v2}^{-1}.$$

It is further assumed that each electron spin relaxes with its own mechanisms independently of the occurrence of electron–spin coupling. The calculations result in different electronic relaxation times for each level and for each transition in the coupled system (Table 11.4) [6]. The electron relaxation

Table 11.4 Inverse Level Lifetimes and Total Transition Linewidths in Some Magnetic Exchange Coupled Systems[a]

$S_1 = ½, S_2 = ½$

	1	2	3	4				
1	$1/2R_A + 1/2R_B$	$3/4R_A + 3/4R_B$	$3/4R_A + 3/4R_B$	$R_A + R_B$				
2	$3/4R_A + 3/4R_B$	$3/4R_A + 3/4R_B$	$3/4R_A + 3/4R_B$	$3/4R_A + 3/4R_B$				
3	$3/4R_A + 3/4R_B$	$3/4R_A + 3/4R_B$	$3/4R_A + 3/4R_B$	$3/4R_A + 3/4R_B$				
4	$R_A + R_B$	$3/4R_A + 3/4R_B$	$3/4R_A + 3/4R_B$	$1/2R_A + 1/2R_B$				

$S_1 = ½, S_2 = 1$

	1	2	3	4	5	6		
1	$1/2R_A + R_B$	$2/3R_A + 5/3R_B$	$5/6R_A + 4/3R_B$	$2/3R_A + 8/3R_B$	$5/6R_A + 7/3R_B$	$R_A + 3R_B$		
2	$2/3R_A + 5/3R_B$	$13/18R_A + 17/9R_B$	$7/9R_A + 16/9R_B$	$13/18R_A + 20/9R_B$	$7/9R_A + 19/9R_B$	$5/6R_A + 7/3R_B$		
3	$5/6R_A + 4/3R_B$	$7/9R_A + 16/9R_B$	$13/18RA + 14/9R_B$	$7/9R_A + 7/3R_B$	$13/18R_A + 20/9R_B$	$2/3R_A + 8/3R_B$		
4	$2/3R_A + 8/3R_B$	$13/18R_A + 14/9R_B$	$7/9R_A + 7/3R_B$	$13/18R_A + 14/9R_B$	$7/9R_A + 16/9R_B$	$5/6R_A + 4/3R_B$		
5	$5/6R_A + 7/3R_B$	$7/9R_A + 19/9R_B$	$13/18R_A + 20/9R_B$	$7/9R_A + 16/9R_B$	$13/18R_A + 17/9R_B$	$2/3R_A + 5/3R_B$		
6	$R_A + 3R_B$	$5/6R_A + 7/3R_B$	$2/3R_A + 8/3R_B$	$5/6R_A + 4/3R_B$	$2/3R_A + 5/3R_B$	$1/2R_A + R_B$		

$S_1 = ½, S_2 = 3/2$

	1	2	3	4	5	6	7	8
1	$1/2R_A + 3/2R_B$	$5/8R_A + 21/8R_B$	$7/8R_A + 57/32R_B$	$3/4R_A + 15/4R_B$	$3/4R_A + 15/4R_B$	$5/8R_A + 177/32R_B$	$7/8R_A + 39/8R_B$	$R_A + 6R_B$
2	$5/8R_A + 21/8R_B$	$11/16R_A 4–51/16R_B$	$13/16R_A + 87/32R_B$	$9/8R_A + 15/4R_B$	$9/8R_A + 15/4R_B$	$11/16R_A + 147/32R_B$	$13/16R_A + 69/16R_B$	$7/8R_A + 39/8R_B$
3	$7/8R_A + 57/32R_B$	$13/16R_A + 87/32R_B$	$11/16R_A + 2R_B$	$3/4R_A + 117/32R_B$	$3/4R_A + 117/32R_B$	$13/16R_A + 41/16R_B$	$11/16R_A + 147/32R_B$	$5/8R_A + 177/32R_B$
4	$3/4R_A + 15/4R_B$	$9/8R_A + 15/4R_B$	$3/4R_A + 117/32R_B$	$3/4R_A + 15/4R_B$	$3/4R_A + 15/4R_B$	$3/4R_A + 117/32R_B$	$9/8R_A + 15/4R_B$	$3/4R_A + 15/4R_B$
5	$3/4R_A + 15/4R_B$	$9/8R_A + 15/4R_B$	$3/4R_A + 117/32R_B$	$3/4R_A + 15/4R_B$	$3/4R_A + 15/4R_B$	$3/4R_A + U7/32R_B$	$9/8R_A + 15/4R_B$	$3/4R_A + 15/4R_B$
6	$5/8R_A + 177/32R_B$	$11/16R_A + 147/32R_B$	$13/16R_A + 41/16R_B$	$3/4R_A + 117/32R_B$	$3/4R_A + 117/32R_B$	$11/16R_A + 2R_B$	$13/16R_A + 87/32R_B$	$7/8R_A + 57/32R_B$
7	$7/8R_A + 39/8R_B$	$13/16R_A + 69/16R_B$	$11/16R_A + 147/32R_B$	$9/8R_A + 15/4R_B$	$9/8R_A + 15/4R_B$	$13/16R_A + 87/32R_B$	$11/16R_A + 51/16R_B$	$5/8R_A + 21/8R_B$
8	$Ra + 6R_B$	$7/8R_A + 39/8R_B$	$5/8R_A + 177/32R_B$	$3/4R_A + 15/4R_B$	$3/4R_A + 15/4R_B$	$7/8R_A + 57/32R_B$	$5/8R_A + 21/8R_B$	$1/2R_A + 3/2R_B$

(Continued)

Table 11.4 Inverse Level Lifetimes and Total Transition Linewidths in Some Magnetic Exchange Coupled Systems (*cont.*)

$S_1 = 1, S_2 = 1$

	1	2	3	4	5	6	7	8	9
1	$R_A + R_B$	$3/2R_A + 3/2R_B$	$3/2R_A + 3/2R_B$	$15/8R_A + 15/8R_B$	$2R_A + 2R_B$	$15/8R_A + 15/8R_B$	$5/2R_A + 5/2R_B$	$5/2R_A + 5/2R_B$	$3R_A + 3R_B$
2	$3/2R_A + 3/2R_B$	$7/4R_A + 7/4R_B$	$7/4R_A + 7/4R_B$	$15/8R_A + 15/8R_B$	$2R_A + 2R_B$	$15/8R_A + 15/8R_B$	$9/4R_A$ + 9/4RB	$9/4R_A + 9/4R_B$	$5/2R_A + 5/2R_B$
3	$3/2R_A + 3/2R_B$	$7/4R_A + 7/4R_B$	$7/4R_A + 7/4R_B$	$15/8R_A + 15/8R_B$	$2R_A + 2R_B$	$15/8R_A + 15/8R_B$	$9/4R_A + 9/4R_B$	$9/4R_A + 9/4R_B$	$5/2R_A + 5/2R_B$
4	$15/8R_A + 15/8R_B$	$15/8R_A + 15/8R_B$	$15/8R_A + 15/8R_B$	$7/4R_A + 7/4R_B$	$15/8R_A + 15/8R_B$	$7/4R_A + 7/4R_B$	$15/8R_A + 15/8R_B$	$15/8R_A + 15/8R_B$	$15/8R_A + 15/8R_B$
5	$2R_A + 2R_B$	$2R_A + 2R_B$	$2R_A + 2R_B$	$15/8R_A + 15/8R_B$	$2R_A + 2R_B$	$15/8R_A + 15/8R_B$	$2R_A + 2R_B$	$2R_A + 2R_B$	$2R_A + 2R_B$
6	$15/8R_A + 15/8R_B$	$15/8R_A + 15/8R_B$	$15/8R_A + 15/8R_B$	$7/4R_A + 7/4R_B$	$15/8R_A + 15/8R_B$	$7/4R_A + 7/4R_B$	$15/8R_A + 15/8R_B$	$15/8R_A + 15/8R_B$	$15/8R_A + 15/8R_B$
7	$5/2R_A + 5/2R_B$	$9/4R_A + 9/4R_B$	$9/4R_A + 9/4R_B$	$15/8R_A + 15/8R_B$	$2R_A + 2R_B$	$15/8R_A + 15/8R_B$	$7/4R_A + 7/4R_B$	$7/4R_A + 7/4R_B$	$3/2R_A + 3/2R_B$
8	$5/2R_A + 5/2R_B$	$9/4R_A + 9/4R_B$	$9/4R_A + 9/4R_B$	$15/8R_A + 15/8R_B$	$2R_A + 2R_B$	$15/8R_A + 15/8R_B$	$7/4R_A + 7/4R_B$	$7/4R_A + 7/4R_B$	$3/2R_A + 3/2R_B$
9	$3R_A + 3R_B$	$5/2R_A + 5/2R_B$	$5/2R_A + 5/2R_B$	$15/8R_A + 15/8R_B$	$2R_A + 2R_B$	$15/8R_A + 15/8R_B$	$3/2R_A + 3/2R_B$	$3/2R_A + 3/2R_B$	$R_A + R_B$

[a]*The numbers indicate the labeling of the ψ functions as reported in Fig. 11.1.*

rates for the pair are the sum of the rates of the two spins, weighted by coefficients that differ from one transition to another. When one rate is much smaller than the other, the rate of the pair only depends on the rate of the fast relaxing system, being proportional to it through a coefficient. When the fast relaxing system is the one with the larger spin (as for CuCoSOD), the overall rates are larger than for the isolated fast relaxing ion, while they are smaller when the fast relaxing system is the one with the smaller spin.

The obtained electron relaxation times for each level and each transition can then be introduced in equations analogous to Eq. (11.14a), thus providing as many spectral density functions as the number of the levels and transitions in the system. However, it turns out that the spectral densities containing longitudinal electron relaxation times (lifetimes), as well as those containing transverse electron relaxation times (transition linewidths), can be lumped together, for $J \ll kT$ or $J \gg kT$, providing effective average τ_{e1} and τ_{e2} values (Table 11.5), that can be simply introduced in Eq. (11.13a) as such [6]. For instance, for CuCoSOD, the average τ_{e1}^{-1} is 9/16 R_{Cu} +179/96 R_{Co}, where R_{Cu} and R_{Co} are the electron relaxation rate of the uncoupled ions copper(II) and cobalt(II), respectively, i.e. about twice R_{Co}, as R_{Cu} and R_{Co} for the isolated ions are 5×10^8 and 10^{11} s^{-1}, respectively. The experimental τ_{e1}^{-1} value ranges between 1 and 2×10^{11} s^{-1}, in good agreement with predictions.

When the above treatment is applied to homodimers, it shows that, if the electron relaxation rates are the same, they cannot strongly influence each other (Table 11.5) in the absence of other electron relaxation mechanisms. As $S_1 = S_2$, all C_i are equal to 1/2 and therefore a decrease of a factor 2 is expected in nuclear relaxation for $J \ll kT$. On the other hand, for $J \gg kT$ and antiferromagnetic coupling, nuclear relaxation will decrease substantially, as the ground state is diamagnetic ($S' = 0$).

11.5 POLYMETALLIC SYSTEMS

In polymetallic systems, the larger the number of coupled ions, the larger the spreading of the S' levels. As a consequence, even relatively small $|J|$ values give rise to large separations of the S' levels and therefore to depopulation of the highest levels. The general theory is still the same. Analytical solutions, as in dimeric systems, are seldom possible. Often, numerical solutions are possible.

The general Hamiltonian is given by Eq. (11.3), where the sum is over all the j–k pairs. The treatment is in fact based on bimetallic coupling. The S' levels are obtained and the hyperfine coupling for each S' level is given by

$$A_{ij} = A_j \frac{\langle S_{jz} \rangle_i}{\langle S'_z \rangle_i} = A_j C_{ij} \tag{11.16}$$

where A_{ij} is the hyperfine coupling between the ith S' level and a nucleus sensing the jth metal. A_j is the hyperfine coupling between the nucleus and the spin on the jth metal (S_j) in the absence of magnetic coupling. $\langle S_{jz} \rangle_i$ is the contribution of metal j (S_j) to the expectation value of S'_z of the ith level. This equation is quite analogous to Eq. (11.7).

Table 11.5 Effective Electron Relaxation Rates for *J*-Coupled Systems Experienced by Nuclei Sensing the S_1 (N-S_1) or the S_2 (N-S_2) Spins in the HT (High Temperature), AF (Antiferromagnetic, $|J| \gg kT$ and $J > 0$) and F (Ferromagnetic, $|J| \gg kT$ and $J < 0$) limits

N-S_1			N-S_2		
HT	**AF**	**F**	**HT**	**AF**	**F**
$S_1 = ½\ S_2 = ½$					
τ_{s1}^{-1} $1/2R_A + 1/2R_B$ τ_{s2}^{-1} $3/4R_A + 3/4R_B$	Diamagnetic state	$1/2R_A + 1/2R_B$ $3/4R_A + 3/4R_B$	$1/2R_A + 1/2R_B$ $3/4R_A + 3/4R_B$	Diamagnetic state	$1/2R_A + 1/2R_B$ $3/4R_A + 3/4R_B$
$S_1 = ½\ S_2 = 1$					
τ_{s1}^{-1} $107/198R_A + 112/99R_B$ τ_{s2}^{-1} $71/99R_A + 187/99R_B$	$13/18R_A + 14/9R_B$ $7/9R_A + 7/3R_B$	$47/90R_A + 49/45R_B$ $32/45R_A + 83/45R_B$	$73/126R_A + 77/63R_B$ $46/63R_A + 125/63R_B$	$13/18Ra + 14/9R_B$ $7/9R_A + 7/3R_B$	$47/90R_A + 49/45R_B$ $32/45R_A + 83/45R_B$
$S_1 = ½\ S_2 = 3/2$					
τ_{s1}^{-1} $9/16R_A + 179/96R_B$ τ_{s2}^{-1} $43/48R_A + 215/64R_B$	$11/16R_A + 2R_B$ $3/4R_A + 117/32R_B$	$43/80R_A + 147/80R_B$ $37/40R_A + 33/10R_B$	$331/560R_A + 2123/1120R_B$ $69/80R_A + 7677/2240R_B$	$11/16R_A + 2R_B$ $3/4R_A + 117/32R_B$	$43/80R_A + 147/80R_B$ $37/40R_A + 33/10R_B$
$S_1 = 1,\ S_2 = 1$					
τ_{s1}^{-1} $5/4R_A + 5/4R_B$ τ_{s2}^{-1} $7/4R_A + 7/4R_B$	Diamagnetic state	$23/20R_A + 23/20R_B$ $69/40R_A + 69/40R_B$	$5/4R_A + 5/4R_B$ $7/4R_A + 7/4R_B$	Diamagnetic state	$23/20R_A + 23/20R_B$ $69/40R_A + 69/40R_B$

Under certain symmetry conditions, Eq. (11.3) can be rearranged in order to allow for analytical solutions.[3] In trimetallic systems, when $J_{12} \neq J_{13} = J_{23}$, Eq. (11.3) becomes [1]

$$H = J(\mathbf{S}_1 \cdot \mathbf{S}_2 + \mathbf{S}_1 \cdot \mathbf{S}_3 + \mathbf{S}_2 \cdot \mathbf{S}_3) + \Delta J_{12}\mathbf{S}_1 \cdot \mathbf{S}_2 \tag{11.17}$$

where ΔJ_{12} indicates the difference between J_{12} and the other two J values. The energies are

$$E(S'_{12}, S') = \frac{1}{2}\left[J\,S'(S'+1) + \Delta J_{12}S'_{12}(S'_{12}+1)\right] \tag{11.18}$$

where S'_{12} is the subspin value coming out from the coupling of spins 1 and 2.

The coefficients C_{ij} in Eq. (11.16) are then given by [42]

$$\begin{aligned}
C_{i1} &= \left[S'(S'+1) + S'_{12}(S'_{12}+1) - S_3(S_3+1)\right]/\left[2S'(S'+1)\right] \\
&\quad \times\left[S'_{12}(S'_{12}+1) + S_1(S_1+1) - S_2(S_2+1)\right]/\left[2S'_{12}(S'_{12}+1)\right] \\
C_{i2} &= \left[S'(S'+1) + S'_{12}(S'_{12}+1) - S_3(S_3+1)\right]/[2S'(S'+1)] \\
&\quad \times\left[S'_{12}(S'_{12}+1) + S_2(S_2+1) - S_1(S_1+1)\right]/\left[2S'_{12}(S'_{12}+1)\right] \\
C_{i3} &= \left[S'(S'+1) + S_3(S_3+1) - S'_{12}(S'_{12}+1)\right]/\left[2S'(S'+1)\right].
\end{aligned} \tag{11.19}$$

For tetraionic systems, analytical solutions are possible in the two following cases [43–45]:

(a) $J_{12} \neq J_{34} \neq J_{13} = J_{14} = J_{23} = J_{24}$

$$\begin{aligned}
H = {} & J(\mathbf{S}_1 \cdot \mathbf{S}_2 + \mathbf{S}_1 \cdot \mathbf{S}_3 + \mathbf{S}_1 \cdot \mathbf{S}_4 + \mathbf{S}_2 \cdot \mathbf{S}_3 + \mathbf{S}_2 \cdot \mathbf{S}_4 + \mathbf{S}_3 \cdot \mathbf{S}_4) \\
& + \Delta J_{12}(\mathbf{S}_1 \cdot \mathbf{S}_2) + \Delta J_{34}(\mathbf{S}_3\mathbf{S}_4)
\end{aligned} \tag{11.20}$$

$$\begin{aligned}
E(S'_{12}, S'_{34}, S') = {} & \frac{1}{2}[JS'(S'+1) + \Delta J_{12}S'_{12}(S'_{12}+1) \\
& + \Delta J_{34}S'_{34}(S'_{34}+1)]
\end{aligned} \tag{11.21}$$

$$\begin{aligned}
C_{i1} &= \left[S'(S'+1) + S'_{12}(S'_{12}+1) - S'_{34}(S'_{34}+1)\right]/[2S'(S'+1)] \\
&\quad \times\left[S'_{12}(S'_{12}+1) + S_1(S_1+1) - S_2(S_2+1)\right]/\left[2S'_{12}(S'_{12}+1)\right] \\
C_{i2} &= \left[S'(S'+1) + S'_{12}(S'_{12}+1) - S'_{34}(S'_{34}+1)\right]/[2S'(S'+1)] \\
&\quad \times\left[S'_{12}(S'_{12}+1) + S_2(S_2+1) - S_1(S_1+1)\right]/\left[2S'_{12}(S'_{12}+1)\right] \\
C_{i3} &= \left[S'(S'+1) + S'_{34}(S'_{34}+1) - S'_{12}(S'_{12}+1)\right]/[2S'(S'+1)] \\
&\quad \times\left[S'_{34}(S'_{34}+1) + S_3(S_3+1) - S_4(S_4+1)\right]/\left[2S'_{34}(S'_{34}+1)\right] \\
C_{i4} &= \left[S'(S'+1) + S'_{34}(S'_{34}+1) - S'_{12}(S'_{12}+1)\right]/[2S'(S'+1)] \\
&\quad \times\left[S'_{34}(S'_{34}+1) + S_4(S_4+1) - S_3(S_3+1)\right]/\left[2S'_{34}(S'_{34}+1)\right].
\end{aligned} \tag{11.22}$$

---

[3] In case all J values are equal, the eigenvalues of the Hamiltonian $H = J(\mathbf{S}_1 \cdot \mathbf{S}_2 + \mathbf{S}_1 \cdot \mathbf{S}_3 + \mathbf{S}_2 \cdot \mathbf{S}_3)$ can be obtained by trivial extension of the dimer case. As $\mathbf{S}' = \mathbf{S}_1 + \mathbf{S}_2 + \mathbf{S}_3$, we can square the total $\mathbf{S}'$ value to obtain $\mathbf{S}'^2 = \mathbf{S}_1^2 + \mathbf{S}_2^2 + \mathbf{S}_3^2 + 2\mathbf{S}_1 \cdot \mathbf{S}_2 + 2\mathbf{S}_1 \cdot \mathbf{S}_3 + 2\mathbf{S}_2\,\mathbf{S}_3$, and write $H = (J/2)(\mathbf{S}'^2 - \mathbf{S}_1^2 - \mathbf{S}_2^2 - \mathbf{S}_3^2)$, whose eigenvalues are $E = (J/2)[S'(S'+1) - S_1(S_1+1) - S_2(S_2+1) - S_3(S_3+1)]$. Since $S_1(S_1+1)$, $S_2(S_2+1)$ and $S_3(S_3+1)$ are the same for all levels, they introduce only an energy offset and can be dropped. The splitting of the levels is just given by $E = (J/2)S'(S'+1)$. Analytical solutions can still be obtained when one J value (e.g. J_{12}) differs from the other two. The first term of Hamiltonian (11.17), is just as in the above case. Then a subspin S'_{12} can be defined such that $\mathbf{S}'_{12} = \mathbf{S}_1 + \mathbf{S}_2$ and, from its square, $\mathbf{S}_1 \cdot \mathbf{S}_2 = 1/2\,(\mathbf{S}_{12}'^2 - \mathbf{S}_1^2 - \mathbf{S}_2^2)$. Then, from the previous definitions, $H = (J/2)(\mathbf{S}'^2 - \mathbf{S}_1^2 - \mathbf{S}_2^2 - \mathbf{S}_3^2) + (\Delta J_{12}/2)(\mathbf{S}_{12}^2 - \mathbf{S}_1^2 - \mathbf{S}_2^2)$. Dropping again the constant terms, the eigenvalues are: $E = (1/2)[JS'(S'+1) + \Delta J_{12}S'_{12}(S'_{12}+1)]$ (Eq. (11.18)).

Here S'_{12}, S'_{34} and S' are good quantum numbers for the relative subspins.

(b)

$$J_{12} \neq J_{13} = J_{23} \neq J_{14} = J_{24} = J_{34}$$

$$\begin{aligned} H = {} & J(\boldsymbol{S_1}\cdot\boldsymbol{S_2} + \boldsymbol{S_1}\cdot\boldsymbol{S_3} + \boldsymbol{S_1}\cdot\boldsymbol{S_4} + \boldsymbol{S_2}\cdot\boldsymbol{S_3} + \boldsymbol{S_2}\cdot\boldsymbol{S_4} + \boldsymbol{S_3}\cdot\boldsymbol{S_4}) \\ & + \Delta J_{12}(\boldsymbol{S_1}\cdot\boldsymbol{S_2}) + \Delta J_{123}(\boldsymbol{S'_{12}}\cdot\boldsymbol{S_3}) \end{aligned} \tag{11.23}$$

$$\begin{aligned} E(S'_{12}, S'_{123}, S') = {} & \frac{1}{2}[JS'(S'+1) + \Delta J_{12}S'_{12}(S'_{12}+1) \\ & + \Delta J_{123}S'_{123}(S'_{123}+1)] \end{aligned} \tag{11.24}$$

$$\begin{aligned} C_{i1} = {} & [S'(S'+1) + S'_{123}(S'_{123}+1) - S_4(S_4+1)]/[2S'(S'+1)] \\ & \times[S'_{123}(S'_{123}+1) + S'_{12}(S'_{12}+1) - S_3(S_3+1)]/[2S'_{123}(S'_{123}+1)] \\ & \times[S'_{12}(S'_{12}+1) + S_1(S_1+1) - S_2(S_2+1)]/[2S'_{12}(S'_{12}+1)] \\ C_{i2} = {} & [S'(S'+1) + S'_{123}(S'_{123}+1) - S_4(S_4+1)]/[2S'(S'+1)] \\ & \times[S'_{123}(S'_{123}+1) + S'_{12}(S'_{12}+1) - S_3(S_3+1)]/[2S'_{123}(S'_{123}+1)] \\ & \times[S'_{12}(S'_{12}+1) + S_2(S_2+1) - S_1(S_1+1)]/[2S'_{12}(S'_{12}+1)] \\ C_{i3} = {} & 1 - C_{i1} - C_{i2} - C_{i4} \\ C_{i4} = {} & [S'(S'+1) + S_4(S_4+1) - S'_{123}(S'_{123}+1)]/[2S'(S'+1)]. \end{aligned} \tag{11.25}$$

Here ΔJ_{12} indicates the difference between J_{12} and J, ΔJ_{123} the difference between J_{12} (= J_{23}) and J, and S'_{123} the subspin resulting from the coupling of S'_{12} and S_3.

All of these cases have been analyzed in NMR studies [5,46].

When the same metals differ in the oxidation states, i.e. Fe^{3+} and Fe^{2+}, the treatment may be even more complicated [43–49]. We will deal with these systems in subsequent sections.

11.5.1 THE $[(RS)_3Fe_3S_4]^{2-}$ CASE

In this case there are three $S = 5/2$ Fe^{3+} ions antiferromagnetically coupled [42]. One J value (e.g., J_{12}) is larger than the other two (Fig. 11.14). It follows that S'_{12} must be smaller than S_3; the total spin is $S' = 1/2$, and S'_{12} from Mössbauer results to be equal to two [42], i.e. S_1 and S_2 are not completely antiferromagnetically coupled.

The spectrum of a protein containing such a cluster is reported in Fig. 11.15 [50]. There are four signals downfield which are assigned to one β-CH_2 pair of the cysteine bound to iron 3 and to one β-CH_2 proton for each of the two cysteines bound to irons 1 and 2. The geminal protons of the latter are found in the diamagnetic region. The β-CH_2 pair of the cysteine coordinated to iron 3 ($S = 5/2$) experiences a Curie temperature dependence, whereas the other four β-CH_2 protons are antiCurie, in agreement with their belonging to the cysteines coordinated to the $S'_{12} = 2$ pair which is smaller than $S_3 = 5/2$. Other proteins containing Fe_3S_4 clusters have similar NMR properties [51].

11.5.2 THE $[(RS)_{11}Co_4]^{3-}$ CASE

The protein metallothionein (MT) has a relative molecular mass M_r of 6000 and is rich in thiolate groups. It binds up to seven zinc(II), cadmium(II), and cobalt(II) ions. The seven ions are organized in

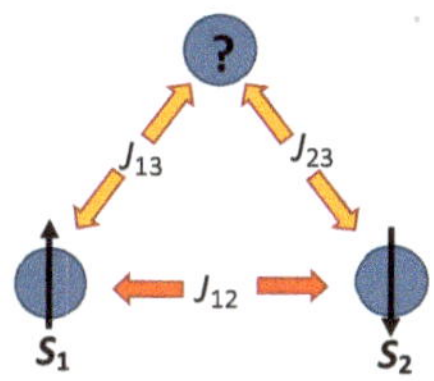

FIGURE 11.14 Magnetic coupling scheme in a Fe_3 system. If $J_{12} > J_{13}, J_{23}, S_1$ and S_2 are antiferromagnetically coupled, and S_3 cannot be antiferromagnetically coupled to both S_1 and S_2.

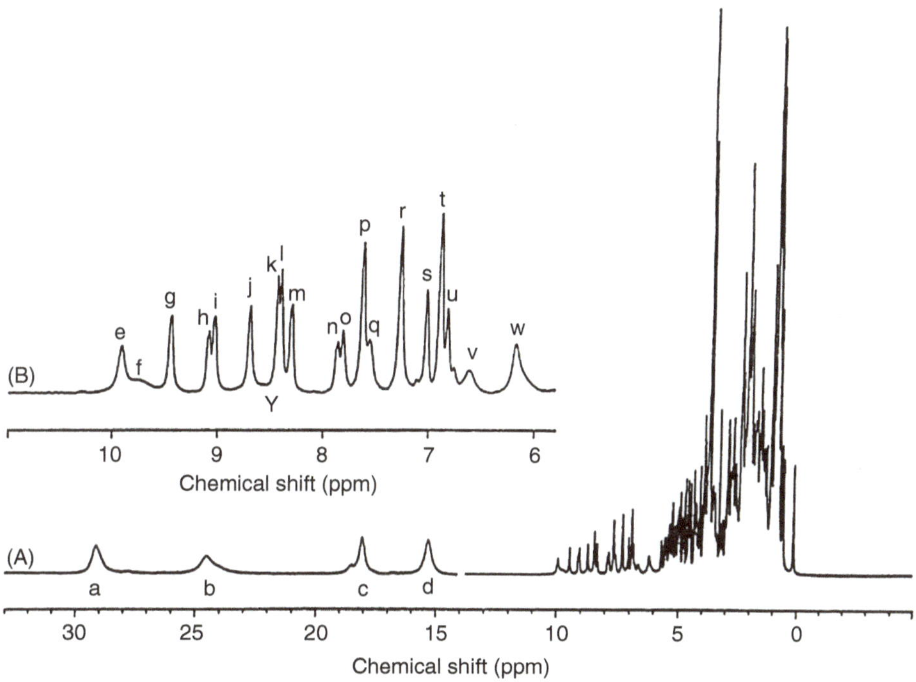

FIGURE 11.15 ^{1}H NMR spectrum of oxidized *D. gigas* ferredoxin II. The two most downfield-shifted signals belong to a cysteine β-CH_2 pair experiencing Curie-type temperature dependence. The details of the aromatic spectral region are shown in B.

Reproduced with permission from Ref. [50].

two clusters. The so-called M_4S_{11} cluster is shown in Fig. 11.16. As suggested from X-ray and NMR measurements, metals 1 and 4 are coordinated by two terminal and two bridging cysteines, whereas metals 2 and 3 are coordinated by one terminal and three bridging cysteines [52–55]. As a consequence, $J_{14} = 0$. If the ions are antiferromagnetically coupled, metals 2 and 3 are overall more antiferromagnetically coupled than metals 1 and 4. This is going to explain some features of the NMR spectra [56,57]. Indeed, it has been proposed that the spectrum of Co_7MT shown in Fig. 11.17 is due only to the Co_4S_{11} cluster. This conclusion is reached on the basis of titration of the apoprotein with cobalt(II) and comparison with the Cd_4Co_3MT species. The spectrum is fascinating, in that signals are spread over

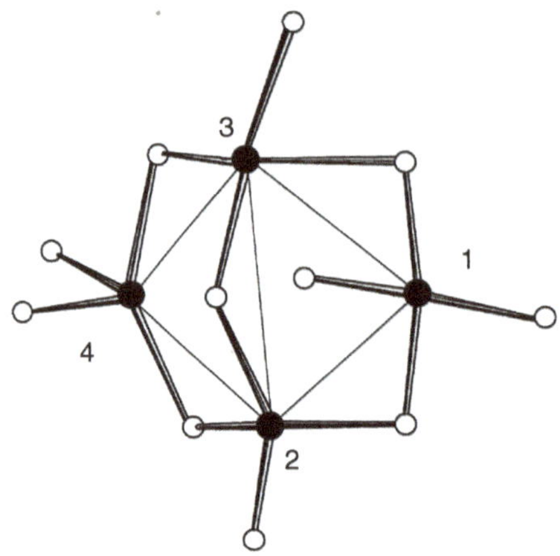

FIGURE 11.16 Schematic structure of the M_4S_{11} cluster in metallothioneins (M = Zn, Cd, Co) [45–48]. There is no bridging ligand between metal 1 and metal 4.

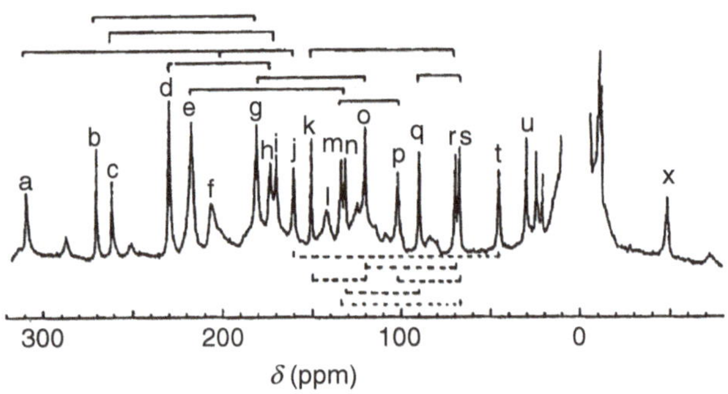

FIGURE 11.17 ^{1}H NMR spectrum of Co_7MT showing the signals of the Co_4S_{11} cluster. The pairwise assignment of the β-CH_2 protons, obtained through NOE and NOESY is also shown.

Adapted from Ref. [49–51].

400 ppm. The pairwise nature of nine out of 11 cysteine β-GH_2s has been demonstrated through NOE and NOESY [58] (Sections 12.3.1). The connected signals are shown in Fig. 11.17.

The temperature dependence of the shifts shows that most of the signals follow a Curie-like behavior, whereas six signals (*k*, *m*, *p*, *q*, *r*, *s*) exhibit a more or less pronounced antiCurie behavior. The ground state of the system is $S' = 0$, since four equal ions are antiferromagnetically coupled. With a J of about 50 cm^{-1}, several excited levels are occupied in the investigated range of temperature. If all the J values were equal we would expect an antiCurie behavior of all the signals. If one J value (namely J_{14}) is zero, we expect, from calculations performed using either Hamiltonian Eqs. (11.20) or (11.23), the four terminal cysteines bound to metals 1 and 4 to exhibit normal Curie behavior; the two terminal cysteines bound to metals 2 and 3, and the cysteine bridging metals 2 and 3 to exhibit antiCurie behavior; and the four cysteines bridging metals 1 and 2, 1 and 3, 2 and 4, and 3 and 4 to experience contrasting effects, and thus to have a less pronounced temperature dependence (Fig. 11.18). As already mentioned, there are indeed six signals exhibiting antiCurie behavior [56,57].

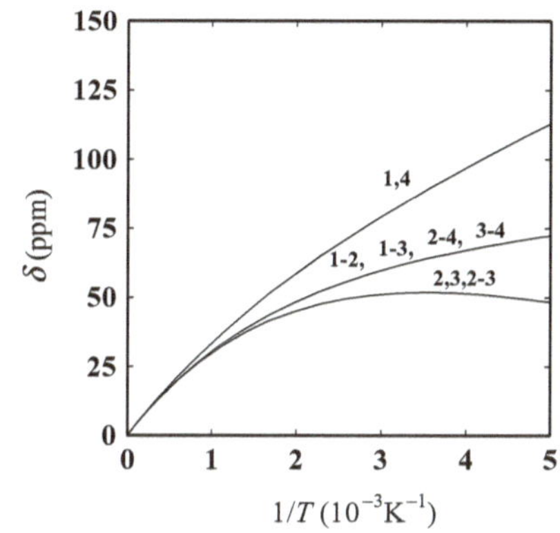

FIGURE 11.18 Calculated hyperfine shifts for cysteine β-CH_2 protons of the Co_4S_{11} cluster from a magnetic coupling model with $J_{14} = 0$ and all other J values equal to 50 cm^{-1}. Numbers refer to the metals to which the cysteines are coordinated (bridging cysteines being identified by two numbers).

11.5.3 THE $[(RS)_4Fe_4S_4]^{X-}$ (X = 1,2,3) CASE

The $[(RS)_4Fe_4S_4]^{2-}$ case is encountered as inorganic compounds with other thiolate ligands and in proteins. The four iron atoms and the four sulfur atoms occupy the vertices of a cube (Fig. 11.19). The tetracoordination of iron is reached through coordination to a further sulfur atom of a cysteine provided by the protein or of an organic thiolate in model complexes. Formally the cluster is constituted by two iron(III) and two iron(II) ions; actually Mössbauer spectroscopy showed that all iron ions are equivalent [59] and that the oxidation number is 2.5. The ground state is $S' = 0$, but magnetically active states are populated at room temperature. The experimental temperature dependence of the shifts is of antiCurie type. NMR has shown that the organic thiolate ligand exchanges fast on the time scale of the difference in shift between the various positions. Only one signal for the CH_2 protons of ethyl thiolate is observed [60,61]. In the case of proteins the rigidity of the ligand makes the two protons distinguishable. As discussed in Section 2.4, the shifts depend on the dihedral angles between the Fe–S–C and the S–C–H planes. NOE, NOESY, and COSY spectra are available which show the connectivities between the β-CH_2 protons, and between the latter and the α-CH, as well as with various amino acidic residues [62,63]. The solution structures of several proteins containing these systems are available.

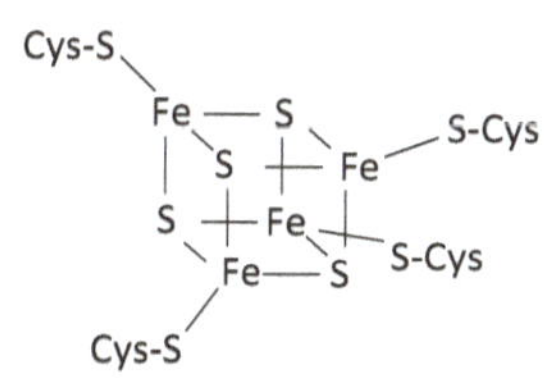

FIGURE 11.19 Cubane-type Fe_4S_4 cluster. The total charge can be +3, +2, or +1 depending on the oxidation states of the iron ions. The cluster shown in the figure is coordinated to four protein cysteines, giving total charge of –1, –2, or –3, respectively.

The $[(RS)_4Fe_4S_4]^-$ cluster formally contains three iron(III) and one iron(II). It is present in a class of proteins called high potential iron-sulfur proteins (HiPIP). It has also been prepared through oxidation of $[(RS)_4Fe_4S_4]^{2-}$ model compounds [64]. Both in the model compound at low temperatures and in proteins there is electron delocalization on one mixed valence pair [65–69]. Therefore, the polymetallic center is constituted by two iron ions at the oxidation state +2.5 and two iron ions at the oxidation state +3. Hamiltonian Eq. (11.20), or a more complicated one [47,49,50], can be used to describe the electronic structure. Indeed, a delocalization operator is sometimes needed in the Hamiltonian [47,49,50]. Consistently with magnetic Mössbauer data the S'_{34} subspin involving the mixed valence pair is 9/2, whereas the S'_{12} subspin involving the iron(III) ions is 4. Mössbauer and EPR data do not exclude $7/2$ and 3, respectively, for the two pairs [64]; in any case, the subspin value of the mixed valence pair is larger than that of the iron(III) pair. This information is vital for the interpretation of the NMR spectra. The β-CH_2 of the cysteines of the mixed valence domain go downfield because they sense the larger spin, whereas the β-CH_2 protons of the cysteines of the iron(III) domain are upfield. Note the similarity with the reduced Fe_2S_2 case (Section 11.3.4). The total spin of the ground state is $S' = 1/2$, and levels with larger S' values are available at room temperature. The T_1 values of the hyperfine shifted signals are accordingly quite short (4–25 ms). The spectrum of the oxidized HiPIP II from *E. halophila* is shown in Fig. 11.20A. The β-CH_2 protons are pairwise assigned through NOEs. The picture is absolutely consistent with what is discussed above [70]. The temperature dependence is consistent with expectations (Fig. 11.21). It should be noted now that electron delocalization may require a further operator in Hamiltonian Eq. (11.21). Its introduction may be relevant for the interpretation of fine aspects of oxidized

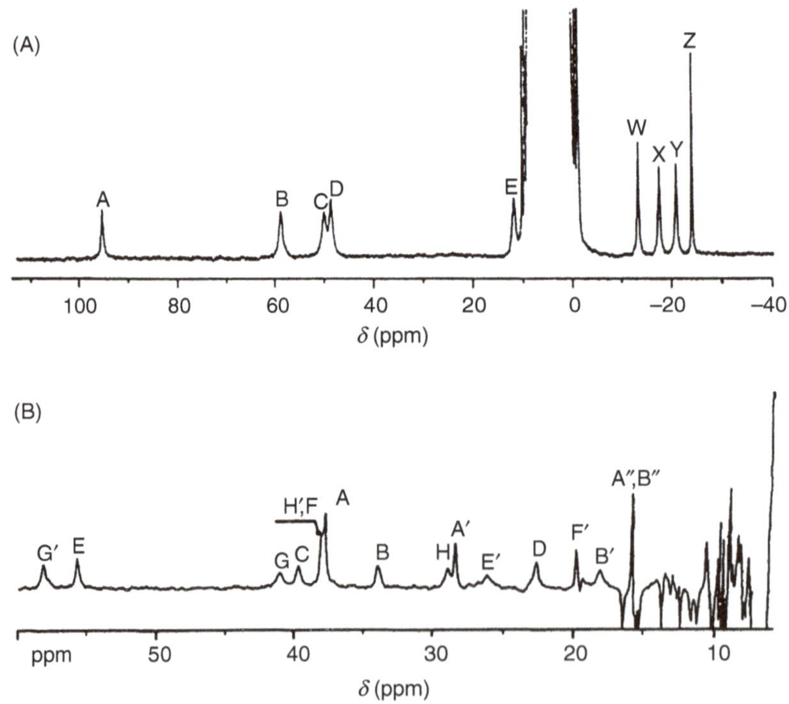

FIGURE 11.20 ^{1}H NMR spectra of (A) oxidized HiPIP II from *E. halophila* [63] and (B) reduced ferredoxin from *C. acidi urici*.

Adapted from Ref. [56].

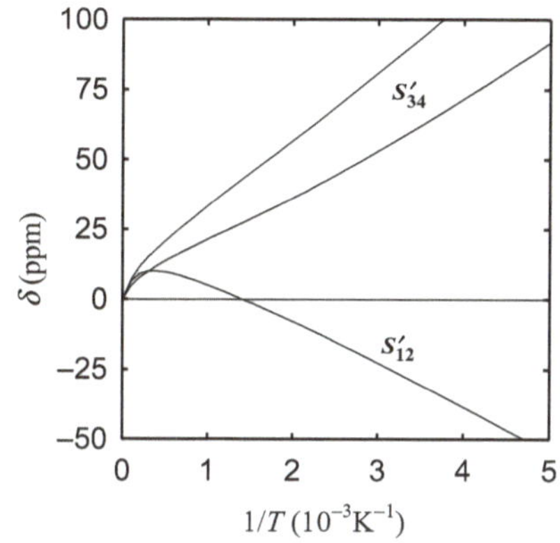

FIGURE 11.21 Calculated temperature dependence of the hyperfine shifts of the cysteine β-CH_2 protons of HiPIP II from *E. halophila* [63]. S'_{12} refers to the protons sensing the ferric pair and S'_{34} to those sensing the mixed valence pair.

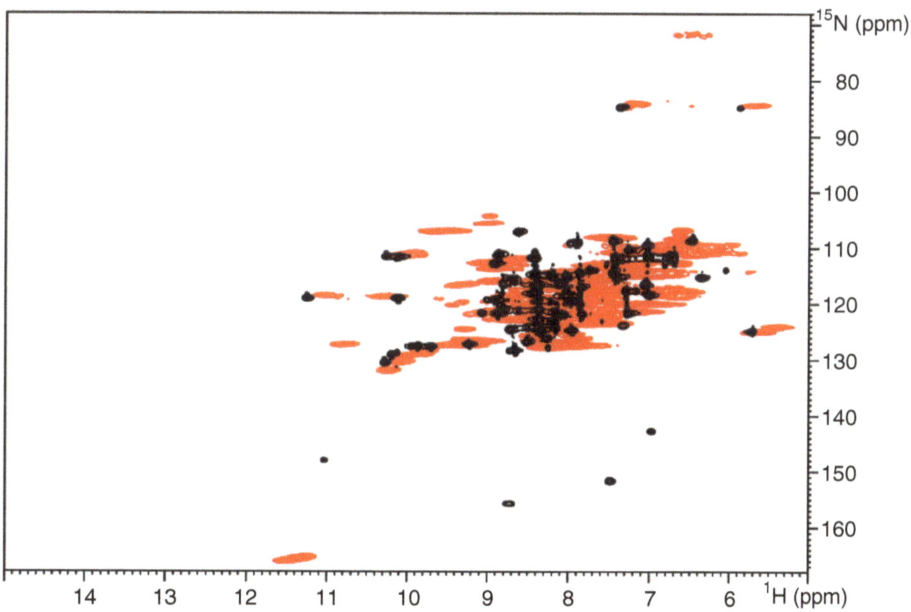

FIGURE 11.22 Overlay of solution state and solid state ^{1}H-^{15}N-HSQC spectra of HiPIP I from *E. halophila.*

Reproduced with permission from Ref. [73].

HiPIPs, but we deliberately avoid this further treatment [46]. The next step in the knowledge of the protein structure is then that of identifying the cysteines bound to the two kinds of iron ions.

The paramagnetic relaxation induced by the presence of the Fe_4S_4 is rather mild, as exemplified by the case of the oxidized HiPIP I from *E. halophila*. In such case no signal of the ^{1}H-^{15}N HSQC spectrum is broadened beyond detection (Fig. 11.22) [71,72]. For 8 resonances, those that display a larger hyperfine shift because hydrogen-bonded to the Fe_4S_4 coordination sphere, a larger signal can

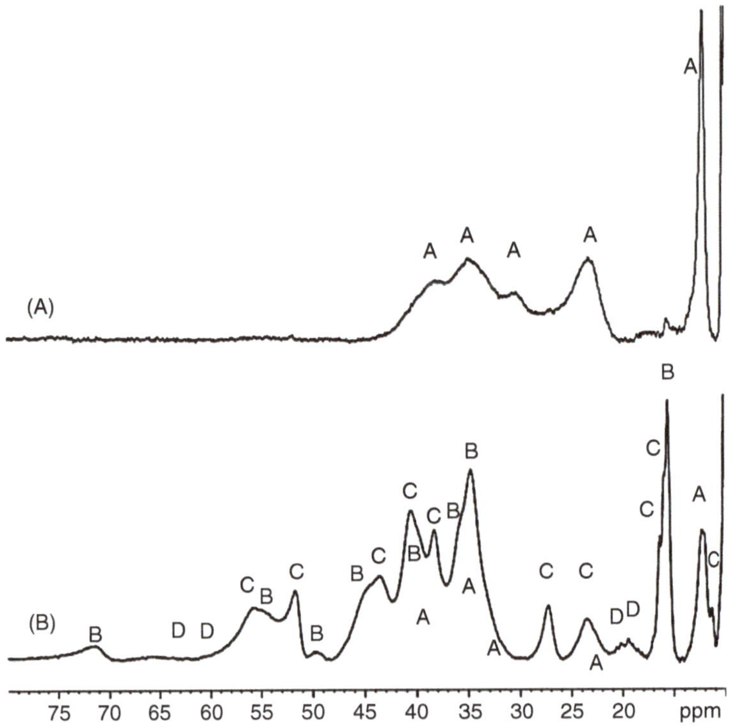

FIGURE 11.23 Paramagnetic 1D 1H NMR spectra of "as purified" $[Fe_2S_2]^{2+}$ hISCA2 (A) and of chemically reconstituted hISCA2 (B) in 50 mM phosphate buffer pH 7.0 at 600 MHz and 283 K. Signals labeled A arise from Hβ and Hα from cysteine ligands coordinated to a $[Fe_2S_2]^{2+}$ cluster. Signals labeled B, C, and D arise from Hβ and Hα from cysteine ligands coordinated to a $[Fe_4S_4]^{+}$ cluster in different species that have been identified according to relative intensities of signals B − D observed in different reconstituted protein samples.

Reproduced with permission from Ref. [82].

be obtained through relaxation-tailored experiments. Intriguingly, this protein yields resolved ^1H-^{15}N HSQC spectra also in the solid state, the full protonation notwithstanding (Fig. 11.22) [73]. This can be explained in the following terms: the hyperfine shift contributes to the breaking of the dipolar network of the protons, making it easier to spin out the interactions [74–76].

The $[(RS)_4Fe_4S_4]^{3-}$ center formally contains three iron(II) and one iron(III) ions. The Mössbauer data indicate that in most cases we are in the presence of two irons at the oxidation state +2.5 and two irons at the oxidation state +2 [77–79]. The system is present in ferredoxins and has been synthesized as model compounds. The ground state in proteins seems to be $S' = ½$. The ^1H NMR spectrum of the two reduced clusters present in *C. acidi urici* ferredoxin is shown in Fig. 11.19B. Note the differences in chemical shifts with respect to Fig. 11.19A; such differences are attributed to different degrees of electron delocalization. In reduced Fe_4S_4 ferredoxins the shifts are all downfield, half of them with Curie and half with antiCurie temperature dependence, just like in the case of the reduced Fe_2S_2 ferredoxins discussed in Section 11.3.4. The successful strategy for the investigation of these systems

[62,63] is similar to that of the oxidized HiPIPs. In model compounds both ½ and $\frac{3}{2}$ ground states have been found [80]. When S is substituted with Se a $\frac{7}{2}$ ground state has been claimed [81]. Recently, an intriguing behavior of a Fe_4S_4 cluster has been reported for hISCA2: hISCA2 can bind either [2Fe-2S] or [4Fe-4S] cluster in a dimeric state and ^1H paramagnetic NMR data show that the level of the dimeric $[4Fe\text{-}4S]^+$ hISCA2 decreases while that of the dimeric $[2Fe\text{-}2S]^{2+}$ hISCA2 species increases, suggesting the occurrence of a [4Fe-4S] → [2Fe-2S] cluster conversion process upon oxidation. Such behavior was successfully monitored by means of NMR: (1) the "as purified" protein shows 5 downfield peaks, assigned to the Hα and Hβ of the cysteins that coordinate a [2Fe-2S] cluster (some of the Hα signals are buried in the diamagnetic envelope); in addition to those, (2) the reconstituted protein shows a number of downfield peaks (Fig. 11.23), compatible with a $[(RS)_4Fe_4S_4]^{3-}$ [82].

11.5.4 THE $[(RS)_3LFe_3NiS_4]^{3-}$ CASE

In this case one iron(II) has been substituted by a nickel(II) ion. As a model complex this is a nice system to investigate [83]. In the model complex a triphenylphosphine is bound to nickel (if L is a neutral ligand the negative charge of the title compound should be decreased by one unit). Interestingly, the shift pattern of the triphenylphosphine moiety is reversed with respect to a nickel(II) complex without magnetic coupling [83]. This can be seen as the result of a spin $S' = \frac{5}{2}$, resulting from the Fe_3S_4 moiety, antiferromagnetically coupled to $S = 1$ of nickel(II). Alternatively, the $S' = \frac{9}{2}$ of the iron(III)–iron(II) mixed valence pair is antiferromagnetically coupled to the iron(II)–nickel(II) pair [84].

This system has been chosen just to pinpoint the variety of systems and of spin coupling schemes available in heteropolymetallic systems.

REFERENCES

[1] Bencini A, Gatteschi D. Electron Paramagnetic Resonance of Exchange-Coupled Systems. Berlin Springer-Verlag; 1990.
[2] Dunham WR, Palmer G, Sands RH, Bearden AJ. Biochim. Biophys. Acta 1971;253:373.
[3] Banci L, Bertini I, Luchinat C. Struct. Bonding 1990;72:113.
[4] Luchinat C, Ciurli S. Biol. Magn. Reson. 1993;12:357.
[5] Bertini I, Banci L, Brown RD III, Koenig SH, Luchinat C. Inorg. Chem. 1988;27:951.
[6] Bertini I, Galas O, Luchinat C, Parigi G, Spina G. J. Magn. Reson. 1998;130:33.
[7] Clementi V, Luchinat C. Acc. Chem. Res. 1998;31:351.
[8] Redfield AG. IBM J. Res. Develop. 1957;1:19.
[9] Stratton WJ. Inorg. Chem. 1970;9:517.
[10] Boyd PDW, Gerlock M, Sheldrick GM. J. Chem. Soc. Dalton Trans. 1974;2:1097.
[11] Gatteschi D, Dei A, Piergentili E. Inorg. Chem. 1979;18:89.
[12] Benelli C, Dei A, Gatteschi D. Inorg. Chem. 1982;108:3298.
[13] Owens C, Drago RS, Bertini I, Luchinat C, Banci L. J. Am. Chem. Soc. 1986;108:3298.
[14] Bertini I, Luchinat C, Owens C, Drago RS. J. Am. Chem. Soc. 1987;109:5208.
[15] Byers W, Williams RJP. J. Chem. Soc. 1973;11:134.
[16] Maekawa M, Kitagawa S, Munakata M, Masuda H. Inorg. Chem. 1989;28:1904.
[17] La Mar GN, Eaton GR, Holm RH, Walker FA. J. Am. Chem. Soc. 1973;95:63.
[18] Holz RC, Brink JM. Inorg. Chem. 1994;33:4609.
[19] Mandal PK, Manoharan PT. Inorg. Chem. 1995;34:270.

[20] Lubben L, Hage R, Meetsma A, Byma K, Feringa BL. Inorg. Chem. 1995;34:2217.
[21] Satcher JH, Balch AL. Inorg. Chem. 1995;34:3371.
[22] Brink JM, Rose RA, Holz RC. Inorg. Chem. 1996;35:2878.
[23] Murthy NN, Karlin KD, Bertini I, Luchinat C. J. Am. Chem. Soc. 1997;119:2156.
[24] Poe M, Phillips WD, Glickson JD, McDonald CC, San Pietro A. Proc. Natl. Acad. Sci. USA 1971;68:68.
[25] Saalmen I, Palmer G. Arch. Biochem. Biophys. 1972;150:767.
[26] Ciofi-Baffoni S, Gallo A, Muzzioli R, Piccioli M. J. Biomol. NMR 2014;58:123.
[27] Banci L, Ciofi-Baffoni S, Mikolajczyk M, Winkelmann J, Bill E, Pandelia ME. J. Biol. Inorg. Chem. 2013;18:883.
[28] Banci L, Bertini I, Ciofi-Baffoni S, Boscaro F, Chatzi A, Mikolajczyk M, Tokatlidis K, Winkelmann J. Chem. Biol. 2011;18:794.
[29] Banci L, Bertini I, Calderone V, Ciofi-Baffoni S, Giachetti A, Jaiswal D, Mikolajczyk M, Piccioli M, Winkelmann J. Proc. Natl. Acad. Sci. USA 2013;110:7136.
[30] Ciofi-Baffoni S, Gallo A, Muzzioli R, Piccioli M. J. Biomol. NMR 2014;58:123.
[31] Banci L, Brancaccio D, Ciofi-Baffoni S, Del Conte R, Gadepalli R, Mikolajczyk M, Neri S, Piccioli M, Winkelmann J. Proc. Natl. Acad. Sci. USA 2014;111:6203.
[32] Morgenstern-Badarau I, Cocco D, Desideri A, Rotilio G, Jordanov J, Dupre' N. J. Am. Chem. Soc. 1986;108:300.
[33] Bertini I, Lanini G, Luchinat C, Messori L, Monnanni R, Scozzafava A. J. Am. Chem. Soc. 1985;107:4391.
[34] Banci L, Bertini I, Luchinat C, Scozzafava A. J. Am. Chem. Soc. 1987;109:2328.
[35] Bertini I, Luchinat C, Piccioli M. Progr. Nucl. Magn. Reson. Spectrosc. 1994;26:91.
[36] Bertini I, Luchinat C, Ming L-J, Piccioli M, Sola M, Valentine JS. Inorg. Chem. 1992;31:4433.
[37] Banci L, Bertini I, Luchinat C, Monnanni R, Scozzafava A, Salvato B. Gazz. Chim. Ital. 1986;116:51.
[38] Banci L, Bertini I, Luchinat C, Monnanni R, Scozzafava A. Inorg. Chem. 1987;26:153.
[39] Bertini I, Lanini G, Luchinat C. Inorg. Chem. 1984;23:2729.
[40] Skjeldal L, Markley JL, Coghlan VM, Vickery LE. Biochemistry 1991;30:9078.
[41] Dugad LB, La Mar GN, Banci L, Bertini I. Biochemistry 1990;29:2263.
[42] Kent TA, Huynh BH, Munk E. Proc. Natl. Acad. Sci. USA 1980;77:6574.
[43] Noodleman L. Inorg. Chem. 1988;27:3677.
[44] Noodleman L. Inorg. Chem. 1991;30:256.
[45] Noodleman L. Inorg. Chem. 1991;30:246.
[46] Bertini I, Ciurli S, Luchinat C. Struct. Bonding 1995;83:1.
[47] Bominaar EL, Borshch SA, Girerd J-J. J. Am. Chem. Soc. 1994;116:5362.
[48] Bertini I, Briganti F, Luchinat C. Inorg. Chim. Acta 1990;175:9.
[49] Blondin G, Girerd J-J. Chem. Rev. 1990;90:1359.
[50] Macedo AL, Moura I, Moura JJG, LeGall J, Huynh BH. Inorg. Chem. 1993;32:1101.
[51] Busse SC, La Mar GN, Yu LP, Howard JB, Smith ET, Zhou ZH, Adams MWW. Biochemistry 1992;31:11952.
[52] Arseniev A, Schultze P, Wörgötter E, Braun W, Wagner G, Vasak M, Kägi JH, Wüthrich K. J. Mol. Biol. 1988;201:637.
[53] Schultze P, Worgotter E, Braun W, Wagner G, Vasak M, Kagi JH, Wüthrich K. J. Mol. Biol. 1988;203:251.
[54] Messerle BA, Bos M, Schaffer A, Vasak M, Kagi JH, Wüthrich K. J. Mol. Biol. 1990;214:781.
[55] Robbins AH, McRee DE, Williamson M, Collett SA, Xuong NH, Furey WF, Wang BC, Stout CD. J. Mol. Biol. 1991;221:1269.
[56] Bertini I, Luchinat C, Messori L, Vasak M. J. Am. Chem. Soc. 1989;111:7296.
[57] Bertini I, Luchinat C, Messori L, Vasak M. J. Am. Chem. Soc. 1989;111:7300.
[58] Bertini I, Luchinat C, Messori L, Vasak M. Eur. J. Biochem. 1993;211:235.
[59] Cammack, R., Sykes, A.G., Advances in Inorganic Chemistry. Iron Sulfur Proteins, Academic Press, San Diego, 1992.

[60] Hagen KS, Watson AD, Holm RH. J. Am. Chem. Soc. 1983;105:3905.
[61] Holm RH, Phillips WD, Averill BA, Mayerle JJ, Herskovitz T. J. Am. Chem. Soc. 1974;96:2109.
[62] Bertini I, Briganti F, Luchinat C, Messori L, Monnanni R, Scozzafava A, Vallini G. Eur. J. Biochem. 1992;204:831.
[63] Bertini I, Capozzi F, Luchinat C, Piccioli M, Vila AJ. J. Am. Chem. Soc. 1994;116:651.
[64] Mouesca J-M, Rius GJ, Lamotte B. J. Am. Chem. Soc. 1993;115:4714.
[65] Middleton P, Dickson DPE, Johnson CE, Rush JD. Eur. J. Biochem. 1980;104:289.
[66] Dickson DPE, Johnson CE, Cammack R, Evans MCW, Hall DO, Rao KK. Biochem. J. 1974;139:105.
[67] Bertini I, Campos AP, Luchinat C, Teixeira M. J. Inorg. Biochem. 1993;52:227.
[68] Papaefthymiou V, Millar MM, Münck E. Inorg. Chem. 1986;25:3010.
[69] Gloux J, Gloux P, Lamotte B, Mouesca J-M, Rius GJ. J. Am. Chem. Soc. 1994;116:1953.
[70] Banci L, Bertini I, Capozzi F, Carloni P, Ciurli S, Luchinat C, Piccioli M. J. Am. Chem. Soc. 1993;115:3431.
[71] Bertini I, Capozzi F, Eltis LD, Felli IC, Luchinat C, Piccioli M. Inorg. Chem. 1995;34:2516.
[72] Bertini I, Donaire A, Felli IC, Luchinat C, Rosato A. Inorg. Chem. 1997;36:4798.
[73] Ravera E, Schubeis T, Martelli T, Fragai M, Parigi G, Luchinat C. J. Magn. Reson. 2015;253:60.
[74] Bertini I, Emsley L, Lelli M, Luchinat C, Mao J, Pintacuda G. J. Am. Chem. Soc. 2010;132:5558.
[75] Wickramasinghe NP, Shaibat MA, Ishii Y. J. Phys. Chem. B 2007;111:9693.
[76] Kervern G, D'Aléo A, Toupet L, Maury O, Emsley L, Pintacuda G. Angew. Chem. Int. Ed. Engl. 2009;48:3082.
[77] Middleton P, Dickson DPE, Johnson CE, Rush JD. Eur. J. Biochem. 1978;88:135.
[78] Mullinger RN, Cammack R, Rao KK, Hall DO, Dickson DPE, Johnson CE, Rush JD, Simonopulos A. Biochem. J. 1975;151:75.
[79] Dickson DPE, Johnson CE, Middleton P, Rush JD, Cammack R, Hall DO, Mullinger RN, Rao KK. J. Phys. Colloq. 1976;37:171.
[80] Holm RH, Ciurli S, Weigel JA. In: Lippard SJ, editor. Progress in Inorganic Chemistry: Bioinorganic Chemistry, Vol. 38. New York: John Wiley & Sons, Inc; 1990.
[81] Meyer J, Moulis J-M, Gaillard J, Lutz M. Adv. Inorg. Chem. 1992;38:73.
[82] Brancaccio D, Gallo A, Mikolajczyk M, Zovo K, Palumaa P, Novellino E, Piccioli M, Ciofi-Baffoni S, Banci L. J. Am. Chem. Soc. 2014;136:16240.
[83] Ciurli S, Ross PK, Scott MJ, Yu SB, Holm RH. J. Am. Chem. Soc. 1992;114:5415.
[84] Bertini I, Luchinat C. Chem. Tracts 1992;4:269.

CHAPTER

HINTS ON EXPERIMENTAL TECHNIQUES

12

This chapter provides some guidelines on how to acquire simple NMR spectra of paramagnetic molecules both in the liquid and in the solid state. Beginners often loose signals or have problems with the baseline, or may observe artifacts, and then the whole approach may have serious problems. On the other hand, when performed properly, NMR experiments on paramagnetic molecules can reveal a wealth of structural and dynamic information as well as the features of the electronic properties of the system.

12.1 HOW TO RECORD 1D NMR SPECTRA OF PARAMAGNETIC MOLECULES

This chapter describes experiments that are aimed mainly at detection of signals showing large paramagnetic effects: however, it is important to stress that especially in biomolecules, there are several other resonances, which may not experience sizeable relaxation, nor extreme shifts, and may be observed using slight modifications on the experiments that are commonly performed on diamagnetic systems.

As it has been extensively discussed in the previous chapters, resonances of spins close to the paramagnetic center are sizably shifted, broadened or both. In general, an NMR approach to a paramagnetic system needs to rely upon some previous knowledge of the electronic structure (Chapters 7, 8, 11), which allows for a rough prediction of shift and relaxation, which in turn allows for preselection of: (1) experiment type and dimensionality, (2) optimal field, (3) spectral window, and (4) length of delays and pulses. Another general guideline is the following, whereas the NMR world is advancing toward more and more complex experiments, with dimensionality up to 7D [1,2]; paramagnetic experiments are usually restrained in dimensionality. High dimensionality requires: (1) a high number of evolution delays and (2) high number of pulses. They might thus be not amenable for systems that (1) have fast relaxation or (2) have resonances so spread so as not to be excited by finite length pulses. In this chapter, we will look in details at the experimental issues connected to these properties.

All over the book we have underlined that paramagnetic molecules have short nuclear relaxation times. This has to be faced when recording the spectra, and proper measures should be taken to minimize, or even put to use, the consequences of the fast nuclear relaxation. For example, a broader line has smaller height than a sharp line, if the number of nuclei is the same and if the spectral width is the same, i.e., the integral is the same. This, in practice, is going to affect the signal to noise (*S/N*) ratio. The decrease in *S/N* can be partially avoided by a proper choice of the acquisition time (or of t_1^{max} and t_2^{max} in 2D spectra). The information content of a FID decays with T_2 of the signal of interest, and therefore acquisition times much longer than T_2 should not be used, unless sharp and ill-resolved signals are also looked for in the spectrum (in this case it is better to perform two experiments: one for the detection of sharp signals and one for broad signals). The reduction in t_1^{max} does not necessarily decrease the number of sampled data points, as often the large spectral widths needed for paramagnetic compounds require a more frequent data point sampling (shortening of the dwell time). Modern instruments use

NMR of Paramagnetic Molecules. http://dx.doi.org/10.1016/B978-0-444-63436-8.00012-0

oversampling techniques, which allow data points to be always sampled at the highest possible speed, thereby optimizing the *S/N* for any chosen spectral width. However, oversampling is effective only for narrow spectral widths.

When the spectral width is of hundreds of parts per million, i.e., more than 10^5 Hz on high field instruments, very short excitation pulses are needed. Of course, high power is needed to reach the r.f. energy corresponding to a 90° pulse in a short time. To best exploit the short relaxation times, it is often convenient to use a full 90° excitation pulse and to recycle fast, because magnetization equilibrium is reached quickly, at the same time suppressing the slower-relaxing signals. With suitable power supplies and purpose-built probes, short ^{1}H 90° pulses can be achieved (as short as 2 μs at 800 MHz) [3]. For solids applications, instrumentation is usually more forgiving in terms of power limits, and 90° pulses shorter than 0.5 μs are possible. When very large spectral widths are needed, pulses of less than 90° can always be used.

In the specific case of solids that have either large quadrupole moments, or are strongly paramagnetic, or both, lines can be as large as a few tens of MHz, and cannot be excited by rectangular pulses. In these cases frequency-stepped acquisition (i.e., the acquisition of several spectra with offset varying across the whole spectral range) can be applied [4]. This method has been used extensively for NMR of paramagnetic materials and quadrupolar nuclei and is also referred to as "spin-echo mapping" [5–12] and variable offset cumulative spectroscopy [13–15].

A large spectral width also requires an adequate ADC. For example, an ADC with a minimum dwell time of 2.5 μs covers a spectral width of 400 kHz. Modern ADCs have dwell times of 1 μs or less, so that very large spectral widths can be acquired with a dynamic range of 16 bit or more. We recall that the dynamic range is the maximal number of powers of two that can be used to digitalize the intensity of a signal. The value of the dynamic range is not important in the presence of signals all of similar intensities. It becomes important when a weak signal is in the presence of a strong signal. This is because when the receiver gain is reduced to accommodate the strong signal within the dynamic range of the ADC, the weak signal and even the noise may have an intensity smaller than the digital threshold of the ADC itself. This problem may sometimes be serious in protein ^{1}H NMR at high field, and in general in the presence of non-deuterated solvents. For these reasons more and more efficient signal suppression techniques have been developed (see later). In the latest generation instruments, however, it has been shown that under nonextreme signal suppression conditions the receiver gain is already high enough that a further increase—that could be made possible by a further reduction of the strongest signals—amplifies the noise to the same extent as the signal [16]. This is even more apparent when large spectral windows are used.

Associated with large spectral widths there are serious baseline problems. These arise from different sources, some of which are unavoidable because they are connected with the physics of the experiment. An ideal experiment would require: (1) a pulse of infinitely short length; (2) no dead time between the end of the pulse and the start of the acquisition; (3) immediate linear response of the receiver at the start of the acquisition; (4) a filter of perfectly rectangular shape; (5) uniform response of the probehead throughout the whole spectral window. In practice, any physical receiver requires a finite time to reach linear response conditions after it is turned on. This time is of the order of microseconds and can be as long as some tens of microseconds. Therefore, if the receiver is turned on immediately after the end of the pulse, and if the dwell time is short because of large spectral width (e.g., 2.5 μs) the first several data points may have an altered intensity. After Fourier transform, this altered intensity is translated into a baseline distortion. To reduce this problem, a dead time (DE) of the order of the time required by the receiver to achieve linearity is introduced before starting the acquisition. This dead time may also be useful to avoid acoustic ringing from the probe, which may be a serious problem at low Larmor frequencies (typically when observing low γ nuclei, or even protons at low fields). Introducing a dead time

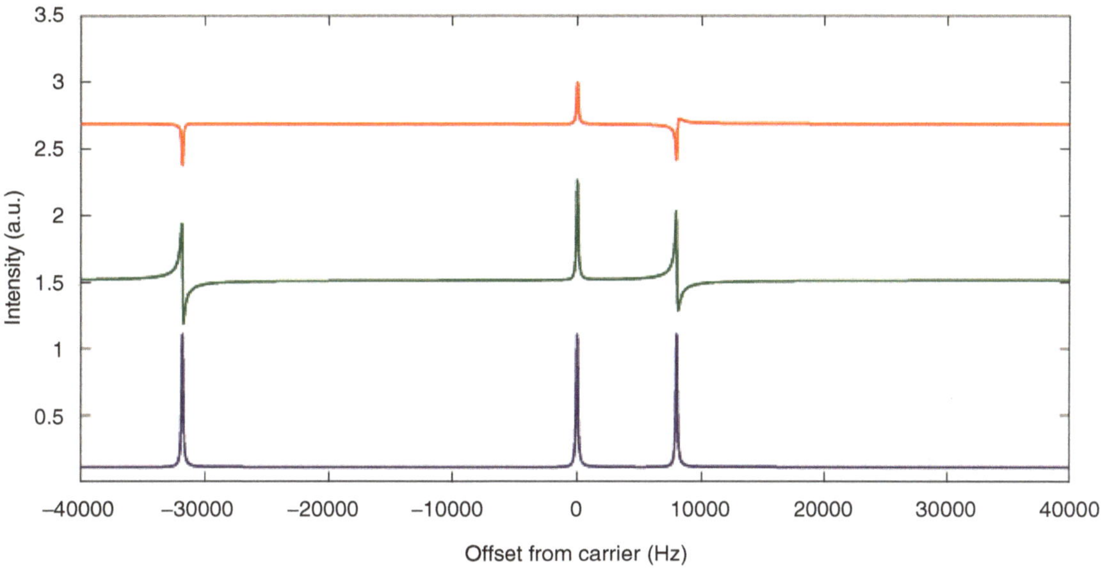

FIGURE 12.1 The effect on Fourier transformation of a finite delay between the last pulse of the sequence and the beginning of the acquisition.

is equivalent to loosing the first few data points in the FID. In turn, this causes a first order dephasing of the signals, i.e., a dephasing that increases with increasing offset from the carrier (Fig. 12.1). First order phase correction again introduces a baseline distortion. Reconstruction of the first points of the FID by linear prediction techniques may be useful in this respect. Of course, if the signals of interest have $T_2 \gg 2\times\mathrm{DE}$, a spin-echo can be introduced in the acquisition, with the echo delay set equal to or larger than the DE. In the case of solids, the stringent limit of the MAS period must be fulfilled, thus the echo delay should be set equal to multiples of the rotor period [17]. In any case, the refocussing efficiency of the echo pulse must be taken care of (see in subsequent paragraph).

The finite duration of the pulse in a way causes the same kind of problems, because nuclei resonating far from the carrier frequency appreciably start precessing during the pulse itself and therefore are dephased. The quality of the probe (Q-factor, reflected power, etc.) also influences the quality of the baseline. Finally, the analog filters used to decrease the noise by filtering out the high frequency noise coming from outside the spectral window are not rectangular, and introduce a baseline distortion. This last problem is alleviated by the latest generation instruments performing digital filtering, although filtering is still limited in terms of spectral width. As an extreme limit, one has to consider that when nuclear relaxation is of the order of the dead time, precious information on such nuclei is lost before acquisition. This can be the case of some paramagnetic solids, but when this occurs the lines are so broad that we are not in the realm of high resolution NMR any longer.

We are now going to discuss some pulse sequences particularly suitable for solution ^{1}H NMR of paramagnetic molecules when the signals of protons experiencing hyperfine coupling are present together with an overwhelming number of protons which are virtually diamagnetic. This is the case of paramagnetic metalloproteins and of water solutions of both paramagnetic macromolecules and small molecules displaying a strong solvent signal. These sequences can be classified according to three criteria. The first is based on presaturation of one signal or of the signals within a chosen window. After presaturation the excitation pulse is applied. Presaturation can also be achieved off-resonance without actually offsetting the carrier frequency. The second approach is based on pulse sequences that do not excite the solvent frequency or a narrow window of frequencies. These sequences are most useful when exchangeable protons may be saturated upon saturation of the solvent. The third approach is that of using pulse sequences that take advantage of the different T_1 values in order to suppress signals with long T_1 values. In the absence of chemical exchange phenomena, the choice of the strategy is not philosophical but is simply based on suitability to specific samples, availability of specific spectrometers, and to some extent, on personal taste. The translation of these sequences to solid state NMR, which is allowed for the latest technological developments of fast MAS (> 40 kHz) is also discussed alongsides.

Of course, all the earlier mentioned methodologies have been developed on ^{1}H-based experiments, because of the higher magnetogyric ratio of the nucleus. However, the same considerations hold true also for all other NMR-sensitive nuclei. The use of ^{13}C or ^{15}N detection is of particular relevance for biomolecular applications: the magnetogyric ratio is small enough to make the nucleus less sensitive to relaxation mechanisms (Chapter 4, see for instance Eq. 4.16), but still large enough so as to warrant a sizeable sensitivity [18,19].

12.1.1 PRESATURATION SEQUENCES

The selective saturation of one signal can be achieved with a soft pulse at a given frequency. Selectivity can be further improved by shaping the pulse, for instance as Gaussian [20], sinc [21] (Fig. 12.2A) or Hermitian [22] functions. Very successful are the compositions of three (G3, Fig. 12.2B) or four (G4) Gaussians (Gaussian cascades) in a single pulse [23], which are further improved by a more appropriate generalized rotation representation (Q3 and Q5) [24]. Shaped pulses are routinely achieved by constructing the desired shape with a relatively large number of short rectangular pulses of variable intensity and phases. A suitable modification of the DANTE [25] sequence allows one to set the carrier frequency at any position with respect to the signal to be presaturated. Such a goal is obtained [26] by subdividing the long presaturation pulse having the low power required to saturate the signal, into a large number of short pulses of the same power with durations corresponding to 1/4 of the reciprocal of the difference in frequency between the carrier and the signal of interest. The phase of the pulses is incremented by 90° each time, for instance with the scheme x, y, $-x$, and $-y$ (Fig. 12.2C). In this way, after the first pulse the z magnetization of the signal is slightly tilted toward the xy plane and its xy projection has simultaneously precessed by 90°. Therefore, the effect of the second pulse is additive, producing a further tilt toward the xy plane, and so on (Fig. 12.3). Other signals with different carrier frequency end up being periodically out of phase with the pulse phase cycle, and the overall effect is zero. The order in which the phase of the pulses is incremented determines the sign of the offset. For instance, an offset of -2000 Hz is selected by using a time of $1/(4 \times 2000) = 0.125$ ms and the pulse phases x, $-y$, $-x$, and y.

For solids, one can expect that presaturation can work rather well if: (1) the signals are substantially shifted from the diamagnetic bulk, lowering the strength of the dipolar network and/or (2) the powder patterns of the signals of interest do not overlap with the signals that are presaturated. Otherwise, presaturation of one resonance may lead to a substantial reduction of the overall signal due to: (1) spin diffusion of the presaturation and (2) passing of some spins in the powder pattern through frequencies which are resonant with the presaturation pulse.

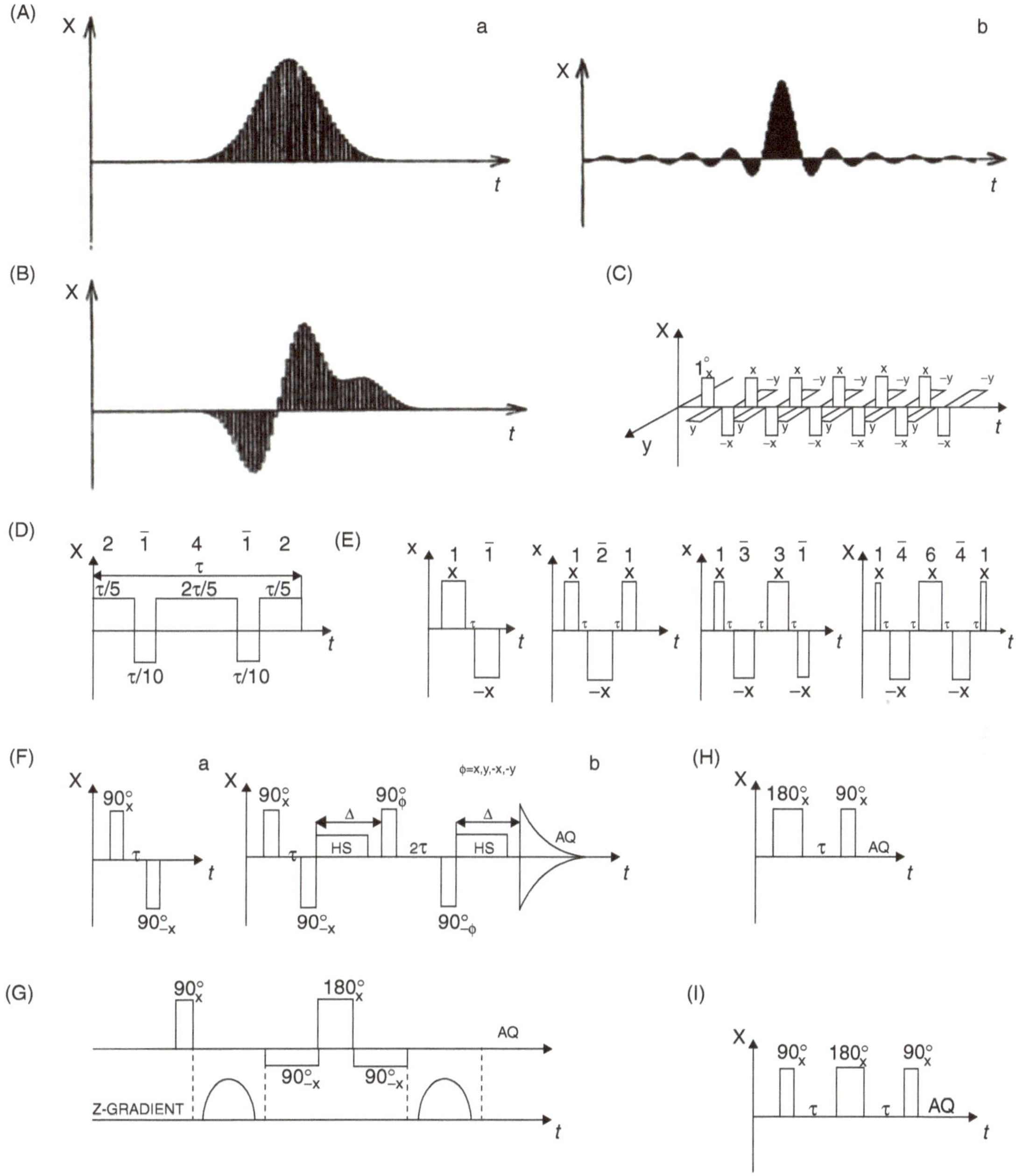

FIGURE 12.2 (A) Gaussian (a) and sinc (b) excitation profiles. (B) Composite (G3) Gaussian pulse. (C) Train of soft pulses modified after the DANTE sequence to achieve selective off-resonance excitation. (D) Redfield $2\bar{1}4\bar{1}2$ sequence. (E) Binomial $1\bar{1}$,$1\bar{2}1$,$1\bar{3}3\bar{1}$,$1\bar{4}6\bar{4}1$ sequences. (F) JR (a) and compensated JR (or $1\bar{1}1\bar{1}$) (b) sequences. (G) Watergate sequence. (H) Weft (Superweft) sequence. (I) Modeft sequence.

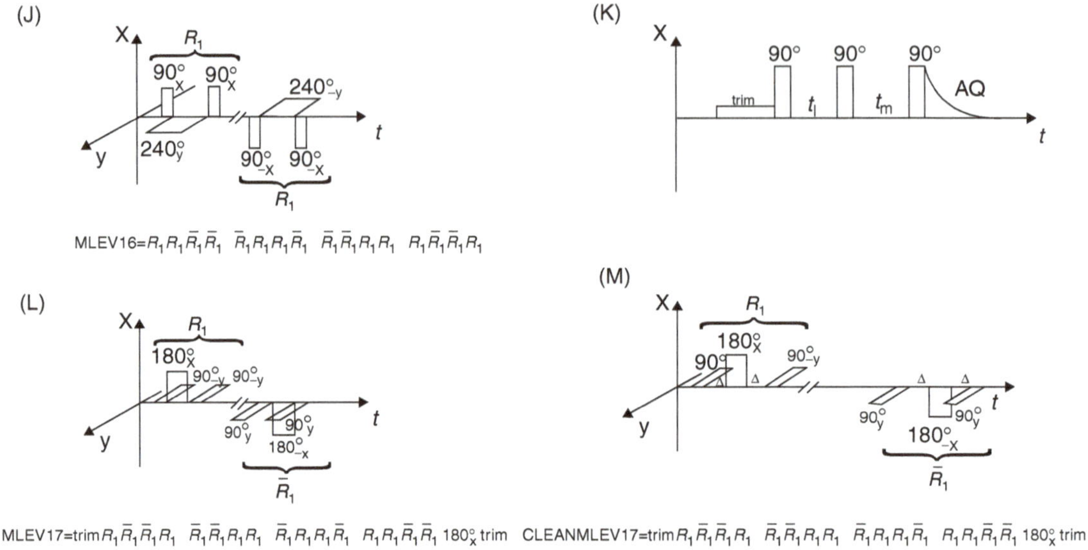

FIGURE 12.2 (*cont.*) (J) MLEV16 sequence. (K) NOESY sequence with trim pulse. (L) MLEV17 sequence with trim pulses. (M) Clean-TOCSY sequence.

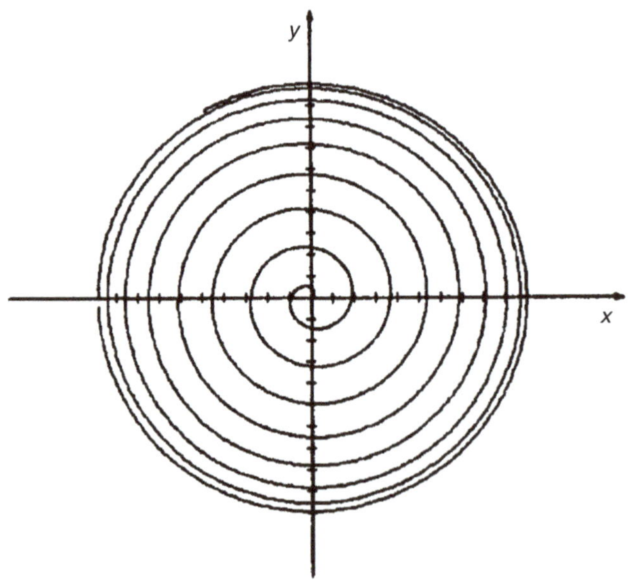

FIGURE 12.3 Saturation effect of the modified DANTE sequence on a signal off-resonance with respect to the carrier frequency. The trajectory of the in-plane component of the magnetization is shown. The *z* magnetization is tilted by the first small angle pulse toward the *x*-axis, and starts precessing toward the *y*-axis. The duration of the pulse corresponds to the duration of a 90° precession, so that at the end of the first pulse the projection lies on the *y*-axis. The phase of the following pulses is rotated in phase with the precession of the signal, in such a way as to "follow" the spin magnetization in its spiral movement toward the *xy* plane.

12.1.2 SELECTIVE NONEXCITATION

It is possible to shape an excitation profile with weak pulses in such a way as to have zero excitation at a certain frequency. This was first due to Redfield [27]. The carrier must be placed in the region of the signals to be excited. The $2\bar{1}4\bar{1}2$ Redfield pulse sequence reported in Fig. 12.2D is a suitable example [27]. The total composite pulse duration must be set equal to the reciprocal of the offset of the signal to be nonexcited. The carrier can also be located on the unwanted signal, and one of the many binomial pulse sequences (Fig. 12.2E) can be used. Binomial pulses are made of strong rather than weak pulses, interleaved by delays [28,29]. The general concept of the effect of strong pulses can be illustrated by the simple "Jump and Return" (JR) sequence (Fig. 12.2F), constituted by a 90° pulse followed by a time τ followed by a –90° pulse [30]. If τ is sizably shorter than T_1, the magnetization of the signal at the carrier is first brought along the y-axis and then back to the z-axis. Signals whose offsets equal $\pm 1/2\tau$ move from the y-axis to the $\pm x$-axis after the time τ; therefore, the second −90° pulse leaves them unaltered (Fig. 12.4). The whole excitation profile is centered at frequencies $\pm 1/2\tau$ and has a width of $1/\tau$. Binomial sequences work similarly, but are constructed in such a way that the total pulse length sums up to a 90° pulse [28,29]. In our experience, the $1\bar{3}3\bar{1}$ sequence has given the best results. A general drawback of all these pulse sequences is that they produce strong baseline distortions in the attempt to phase all signals when they are spread over a large spectral width. A pulse sequence analogous to the JR sequence, which minimizes baseline distortions is the $1\bar{1}1\bar{1}$ sequence [31] (Fig. 12.2F).

A very popular method for selective nonexcitation employs field gradients along the z-axis. After tilting the magnetization of all signals in the xy plane by a nonselective 90° pulse, a field gradient along z applied for a suitable time defocuses all the xy magnetization. If the time of application of the gradient is short enough that no appreciable T_2 relaxation, J modulation, and molecular diffusion take

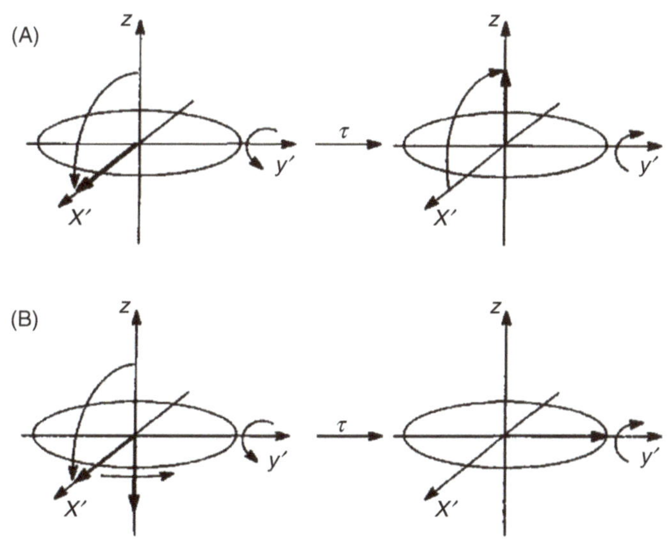

FIGURE 12.4 Effect of the JR sequence on a signal on resonance (A) and off-resonance by $1/2\tau$ Hz (B).

place, the defocusing is reversible, and the *xy* magnetization can be fully recovered by the application of another gradient of opposite sign. Equivalently, the second gradient can be of the same sign of the first, provided that a 180° pulse is applied in between the two. In the so-called Watergate sequence [32], selective nonexcitation is achieved by tailoring the 180° pulse in such a way that all frequencies but the unwanted frequency are irradiated. The simplest scheme employed a $90°_{-x}$(sel)$180°_{x}$(nonsel)$90°_{-x}$(sel) pulse sequence (Fig. 12.2G). The selectivity increases with the length of the 90° pulses.

The approach discussed in this section is preferable to any approach based on solvent suppression when solvent nuclei are in quasi-slow exchange with signals of interest. Saturation of the solvent nuclei transfers magnetization to the observable signals, decreasing their intensity, whereas nonexcitation does not cause any interference.

In solids, signals that have broad powder patterns are in substantial overlap, even if at high MAS they may *appear* as well resolved. Selective excitation of single resonances is possible through long, selective pulses centered at spinning sidebands, rather than at the central band. This works as long as the signals that are to be cancelled out have only a limited subset of crystallite orientations at the frequency that is being irradiated, otherwise it is hardly feasible to selectively excite one signal without touching a substantial fraction of another [33].

12.1.3 SELECTIVE SUPPRESSION OF SIGNALS WITH LONG T_1

Whereas the two strategies of solvent suppression techniques described earlier do not show any particular advantage or disadvantage when applied to paramagnetic systems with respect to diamagnetic systems, the strategies based on differential relaxation properties have obvious intrinsic advantages in paramagnetic systems because of the larger spreading of nuclear relaxation times involved. The simplest approach to take advantage of the fast relaxation times of the signals of interest is that of acquiring the spectrum after a simple 90° pulse with a recycle time short with respect to the T_1 values of the signals to be suppressed, but long with respect to the T_1 values of the paramagnetic signals. Since the latter have also short T_2, short acquisition times are required. The consequence is that the slow relaxing signals are largely saturated, whereas the fast relaxing signals almost fully recover their initial intensity after each cycle. Weft [34] and Superweft [35] pulse sequences (Fig. 12.2H) are based on the 180–τ–90 acquisition pulse sequence. When $\tau = T_1 \ln 2$ the signal with that T_1 has an intensity of zero. If the recycle time is longer than $5T_1$ (i.e., the system has fully recovered at each cycle) zero intensity of the slowly-relaxing signal is maintained all over the experiment. This is the Weft pulse sequence [34]. During the time τ, the fast relaxing signals have essentially recovered their magnetization and are detected. In the Superweft signal, the recycle time is chosen to be short with respect to the T_1 values of the signals to be suppressed, but long with respect to the T_1 values of the signals of interest. It has been shown that a τ value exists that after a few cycles, i.e., at steady state conditions, effectively zeroes the intensities of the signals with long T_1 [35]. An example of application of Superweft is provided in Fig. 12.5A, where the spectrum of a protein containing a heme with low spin iron(III) in D_2O is reported [36]. This is obtained with a Superweft sequence with $\tau = 250$ ms and recycle time = 250 ms. The spectrum shows a number of well-resolved signals outside the diamagnetic region. When the signals of the axially coordinated histidine are looked for, advantage is taken of the fact that they have T_1 values of the order of a few milliseconds, whereas the other hyperfine shifted signals have T_1 values of about 100 ms. Therefore, a Superweft with $\tau = 20$ ms and recycle time of 33 ms provides the spectrum of Fig. 12.5B where only the two nonexchangeable histidine ring protons are evident [36].

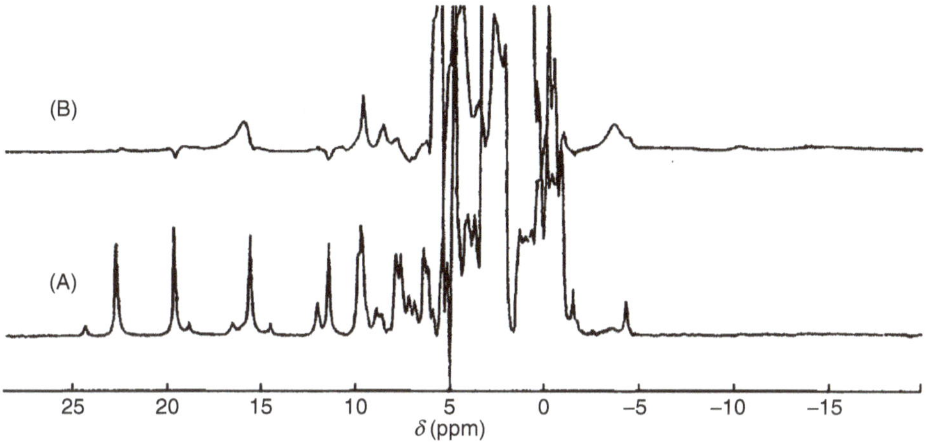

FIGURE 12.5 200 MHz ^{1}H NMR spectra in D_2O of the cyanide adduct of a cytochrome *c* mutant lacking the axial methionine (Ala80cyt *c*) recorded using the Superweft sequence with $\tau = 250$ ms and recycle time of 250 ms (A) and $\tau = 20$ ms and recycle time of 33 ms (B). In the latter spectrum the two signals of the axially coordinated histidine are apparent at 16.1 and –3.4 ppm.

Reproduced with permission from Ref. [36].

Weft or Superweft sequences are also widely used in 2D spectroscopies (see later); in general, it is sufficient to place the 180°–τ part of the sequence in front of any 2D sequence to achieve the desired signal suppression and, in general, represent the best choice for selecting fast relaxing spins.

The Modeft pulse sequence [37] drives the slow relaxing signals to equilibrium when the acquisition pulse is delivered. Modeft means modified driven equilibrium Fourier transform [38]. The pulse sequence is 90 τ 180 τ 90 acquisition (Fig. 12.2I). If τ is short with respect to T_1, the protons are driven to equilibrium. Signals with T_1 short with respect to τ recover after each pulse and the last 90° pulse acts as a normal excitation pulse. In Fig. 12.6 ^{1}H Modeft spectra of anion adducts of a cobalt(II) ($S = 3/2$) protein of relative molecular mass 30,000 are shown [39]. The Modeft sequence is worse than the Superweft sequence in suppressing a single signal like the solvent signal, but performs comparably, and slightly better in some cases, when elimination of more than one slowly relaxing signal is needed. Typically, one might want to record a spectrum of a metalloprotein by the use of the Superweft or the Modeft sequences when the few fast relaxing signals lying underneath the diamagnetic envelope of the many slowly relaxing signals are to be emphasized.

Other strategies to achieve this task are provided by the broadband (BB) saturation sequences (like MLEV16 (Fig. 12.2J) [40], WALTZ 16 [41], GARP [42], DIPSI [43], etc.). The power of the BB saturation pulse can be adjusted in such a way as to effectively saturate the slowly relaxing signals and only marginally the fast relaxing ones. A 90° pulse immediately following the BB saturation sequence then gives a spectrum where the relative intensity of the fast relaxing signals is enhanced. Alternatively, the power of the BB saturation pulse can be high enough to saturate all signals. In this case, the following 90° detection pulse is applied after a time τ, adjusted in such a way that fast relaxing signals are appreciably recovered while slow relaxing signals are not.

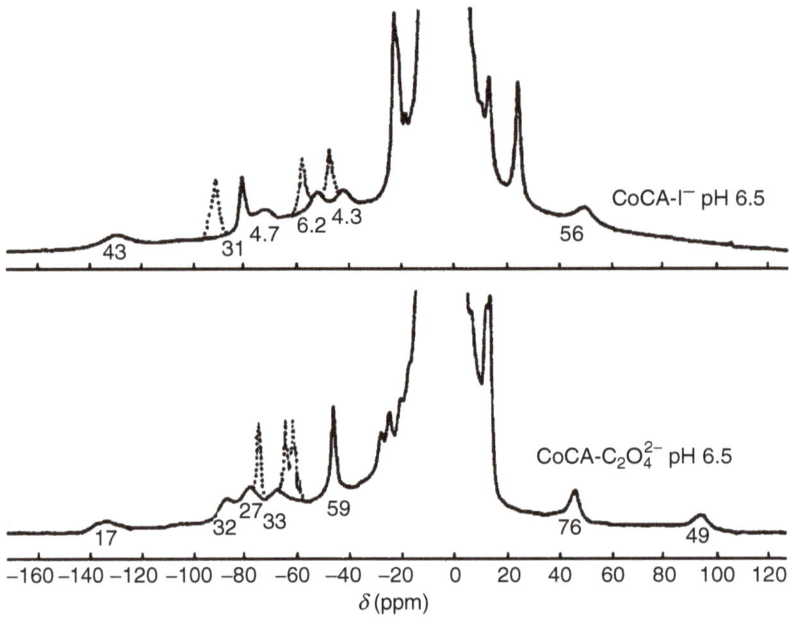

FIGURE 12.6 60 MHz ^{1}H NMR Modeft spectra of cobalt-substituted carbonic anhydrase (M_r 30,000) adducts with iodide and oxalate. The T_1 values for some signals are indicated. The dashed signals disappear in D_2O.

Reproduced with permission from [39].

While the Weft, Superweft, and Modeft sequences are based on differences in T_1, BB saturation techniques are based on the difference in saturability of the signals, which in turn depends on $(T_1T_2)^{1/2}$. Therefore, in cases where $T_2 \ll T_1$ for the paramagnetically affected signals (as can be the case in the presence of strong Curie relaxation), BB techniques may actually give better discrimination. Spin-lock techniques, based on differences in $T_{1\rho}$ ($\cong T_2$), should perform best, although a single spin echo may suffice in accomplishing filtering.

In the case of solids, such approach may fall short in efficiently separating the components: in a crystal lattice, quite often all the protons do have the same relaxation time, given that spin diffusion dominates. Approaches based on T_1 editing have been successfully applied to complex mixtures, containing domains which have substantially different relaxation times [44].

12.1.4 CHOICE OF MAGNETIC FIELD

As the major drawback in the observation of paramagnetic signals is ultimately the width of the lines, it is worth recalling again the various mechanisms causing line broadening. These mechanisms have been illustrated in Sections 4.4–4.6, and their relative importance has been discussed in Section 4.8. The relevant equations describing the three contributions to R_{2M} are Eqs. (4.18) (dipolar), (4.28) (contact), (4.31) (Curie relaxation), plus the equations that describe the effect of exchange between paramagnetic and diamagnetic species Eq. (6.11) (Chemical exchange). The first two contributions to line broadening decrease with magnetic field (Figs. 4.9 and 4.11), while the third increases with magnetic field (Fig. 4.12). This observation suggests that the experimentalist may be able to use the magnetic field as a parameter to optimize signal detection. This attitude has always been peculiar to paramagnetic NMR, whereas it is now becoming popular also in the field of diamagnetic proteins, thanks to the field dependence of the cross-correlation exploited in TROSY-type experiments (see in subsequent paragraph). It is obvious that, in the absence of adverse effects, the higher the field, the better. In fact, if the linewidth (in hertz) is field-independent, the resolution increases linearly with the field: in other words, the linewidth in parts per million *decreases* with increasing field. However, Curie relaxation increases with the square of magnetic field. Therefore, when Curie relaxation is dominant, an increase in field will actually cause a decrease in resolution. In practice, Curie relaxation is dominant when its correlation time τ_r is much larger than τ_e, as in this case, the latter is the dominant correlation time for dipolar and contact mechanisms. Furthermore, the importance of Curie relaxation increases with the electron spin quantum number S (or J for lanthanoids), because dipolar and contact relaxation mechanisms depend on $S(S+1)$ [or $J(J+1)$], whereas Curie relaxation depends on the square of these quantities. Fig. 12.7 illustrates the field dependence of the linewidth (in parts per million) for a middle-sized macromolecular system ($\tau_r = 10^{-8}$ s) containing a fast relaxing paramagnetic metal ion ($\tau_e = 10^{-12}$ s) with S (or J) equal to $\frac{1}{2}, \frac{3}{2}, \frac{5}{2}$, and $\frac{7}{2}$. It appears that the optimal magnetic field decreases from about 400 MHz to 200 MHz, 120 MHz, and 90 MHz. Note that these conditions are relatively common, and far from being extreme. In fact, τ_r values may be easily longer than 10^{-8} s (a τ_r of 10^{-8} s corresponds to a relative molecular mass M_r of 30,000); τ_e values may be shorter, and J values may be larger (e.g., in lanthanoids, where τ_e values close to 10^{-13} s have been reported and J values can be as large as 8).

The case of solids deserves some further considerations, based on what it was said in the earlier paragraphs. As we have seen, Curie-spin interaction and CSA are no more contributing to relaxation,

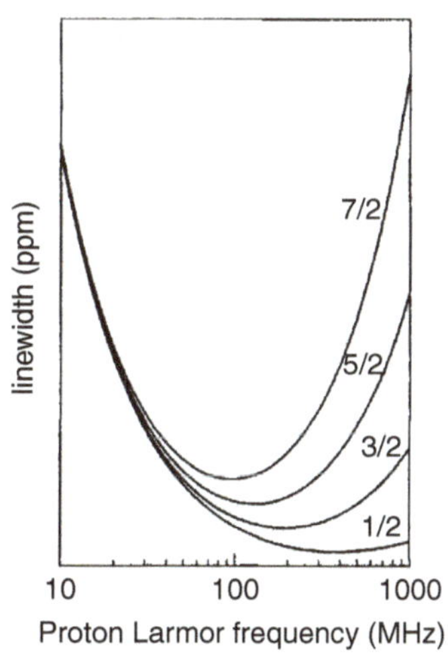

FIGURE 12.7 Calculated field dependence of the linewidth (ppm) for a proton interacting with a paramagnetic metal ion in a macromolecule with $\tau_r = 10^{-8}$ s, $\tau_e = 10^{-12}$ s, and $S = \frac{1}{2}, \frac{3}{2}, \frac{5}{2}$ and $\frac{7}{2}$.

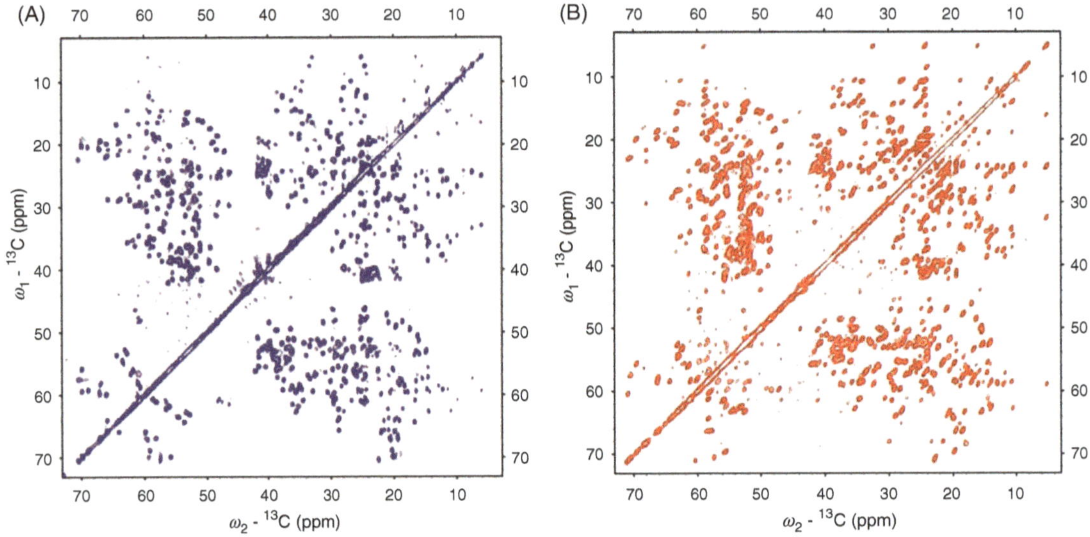

FIGURE 12.8 ^{13}C-^{13}C correlation spectrum obtained for zinc(II) (A, *blue*) and cobalt(II)-substituted (B, *red*) catalytic domain of matrix metalloproteinase 12 obtained as microcrystalline solid.

Reproduced with permission from Refs. [51,52].

but they manifest themselves by creating a powder pattern. Thus one can expect that going to higher field will not be detrimental from the linewidth standpoint. Indeed, the quality of solid state NMR spectra of large systems such as biomolecules, is not directly affected by the size of the system itself, except for the obvious complication arising from an increased number of peaks. This has, for instance, proven beneficial for the study of diamagnetic systems such as ferritin [45], proteasome [46], or ribosome [47], and for some paramagnetic species such as Co^{2+}-loaded MMP12 (Fig. 12.8) [48–50].

However one has to bear in mind that interactions such as Curie-spin and the chemical shielding have frequency spreads that increase (almost) linearly with increased field strength. This raises two fundamental issues for the experimentalist: (1) higher spinning rates will be needed to remove the spinning sidebands at higher fields, and (2) higher power pulses will be needed in order to excite/invert/refocus the powder manifold [53].

For (1), the experimental approach could follow different routes. Intuitively, one way is to reduce the magnetic field, thus reducing the breadth of the powder pattern. This comes at the price of reduced resolution when the signal widths start to get close to their natural linewidths. Another way is to increase the spinning rate. This is surely beneficial in terms of resolution, and usually does not compromise the sensitivity of the experiment. Commercial instruments work routinely up to 60 kHz, and even higher [54]. Another problem is that the efficiency of inversion/refocusing pulses tends to be reduced for high spinning rates, since the spins experience rapid changes of their resonance frequency and adiabaticity of the pulses is lost [55]. A workaround is the use of sideband-specific inversion, which requires long pulses, and several experiments with frequency-stepping are necessary [4,55]. The last approach is to resort to magic angle turning, where spinning sidebands are artificially removed along the indirect dimension of the experiment (Fig. 12.9) [56]. This experiment needs to be perfectly rotor synchronized, thus there is an experimental limitation to the field and spinning rates at which this methodology can work.

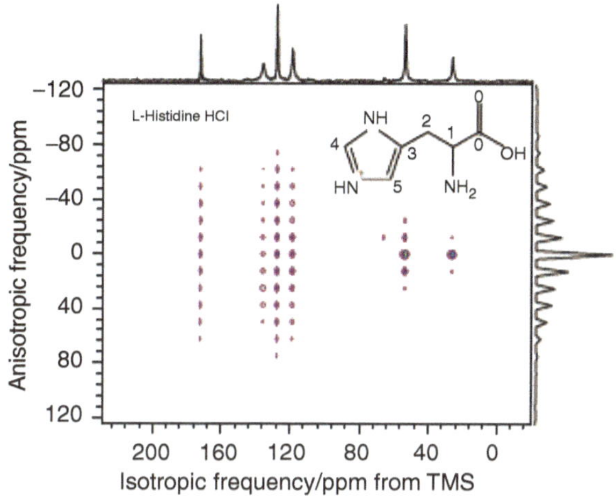

FIGURE 12.9 Two-dimensional MAT spectrum of L-Histidine along with one dimensional projections.

Reproduced with permission from Ref. [56].

12.1.5 EXPERIMENTAL ACCESSIBILITY OF NUCLEAR RELAXATION PARAMETERS

As anticipated in Section 4.2, a nuclear longitudinal relaxation rate R_1 can be defined only when relaxation is an exponential process. This is at variance with nuclear transverse relaxation, which is always exponential and always defined by the transverse relaxation rate R_2. As far as longitudinal relaxation is concerned, when the return to the equilibrium value $M_z(\infty)$ of longitudinal nuclear magnetization after a 180° pulse is exponential, we can write (Section 1.7.4)

$$M_z(t) = M_z(\infty) = 2M_z(\infty)\exp(-R_1 t) \tag{1.53}$$

If we consider that relaxation is further enhanced by interaction with a paramagnetic center, Eq. (1.53) becomes

$$\begin{aligned} M_z(t) &= M_z(\infty) = 2M_z(\infty)\exp(-R_1 t)\exp(-R_{1M} t) \\ &= M_z(\infty) - 2M_z(\infty)\exp[-(R_1 + R_{1M})t] \end{aligned} \tag{12.1}$$

The meaning of Eq. (12.1) is that any further decay function is multiplied by the original decay function if they are independent (also Section 4.2). If both are exponential, then the whole process remains exponential and the rate constant is $R_{1tot} = R_1 + R_{1M}$. As already mentioned in Chapters 1 and 4, longitudinal nuclear relaxation is often a nonexponential process. In our definition of R_1 (Section 1.7.3), it was assumed that all nuclei were independent of one another. In diamagnetic systems nuclei are often coupled to one another, influencing each other in such a way that one cannot be considered a lattice with infinite heat capacity for the other, and R_1 cannot be defined. Paramagnetic relaxation, however, is an exponential process within the broad range of validity of the equations given in this chapter. This is due to the fact that electrons have a much larger magnetic moment and relax so much faster than nuclei that they behave effectively as a lattice with infinite heat capacity. So, Eq. (12.1) takes the more general form

$$M_z(t) = M_z(\infty) - 2M_z(\infty)f(t)\exp(-R_{1M} t) \tag{12.2}$$

where $f(t)$ is a generic nonexponential decay accounting for diamagnetic interactions. In such a case, even if paramagnetic relaxation is exponential, R_{1M} cannot be easily extracted from the analysis of the overall decay. If paramagnetic relaxation is dominant, exponentiality is effectively imposed to the overall nuclear relaxation. If not, the detailed dependence of longitudinal nuclear magnetization as a function of time must be analyzed [57].

Experimental techniques are available to measure magnetization recovery under different conditions. These are, for instance, the selective and nonselective variants of the inversion recovery experiment described in Section 1.7.3.

In a nonselective inversion recovery experiment, the 180° pulse inverts the magnetization of the nuclear spin under consideration as well as that of all other nuclear spins coupled to it. All spins will return to equilibrium and, simultaneously, will influence each other by, for example, dipolar coupling. In a selective experiment, only the nuclear spin under consideration is inverted by the pulse. During its return to equilibrium, it will also be influenced by dipolar coupling with the other, initially unperturbed spins. In both cases, diamagnetic relaxation mechanisms will superimpose to the exponential recovery due to the coupling with the paramagnetic center, and deviations from exponentiality will occur. These deviations will be, however, less severe for a nonselective experiment. Indeed, mutual influence between two nuclear spins I and J occurs through W_0 and W_2 terms (Chapter 4) analogous to the mutual influence between nuclear and electron spins I and S illustrated in Section 4.4 and Fig. 4.7. This influence will be maximal when $2\omega_I\tau_c \gg 1$, i.e., in the slow motion limit, where the W_2 terms become negligible. If we refer to the latter conditions, magnetization recovery of the inverted I spin in a selective experiment occurs by partially decreasing M_z of the neighbor spin J through magnetization transfer (Fig. 12.10). The presence of the neighbor thus "helps" the inverted spin to relax faster (curves B and D, see also Section 12.3.4). This transfer of magnetization is maximal at $t = 0$, and becomes less effective with time (hence nonexponentiality arises). On the contrary, in a nonselective experiment (curves C and E), the transfer of magnetization will be zero at $t = 0$, because both spins I and J are inverted. With time, transfer occurs from the slower relaxing to the faster relaxing spin: the slower will then relax slightly faster and the faster will relax slightly slower (again nonexponentiality occurs). If the two spins accidentally relax with the same rate, no net magnetization transfer occurs and recovery is again exponential. In any case, it is apparent that R_{1M} values can always be better evaluated from nonselective rather than from selective experiments because of this partial compensation effect, and particularly from the initial points of the decay when magnetization transfer is negligible. The stronger the I–J coupling, the worse the selective experiments perform with respect to nonselective ones.

As far as T_1 is concerned, we should therefore remember that longitudinal relaxation can be multi-exponential. Therefore, different information can be obtained, depending on whether a given signal is selectively excited or it is excited together with other signals. We usually refer to T_1 as an experimental T_1 value obtained by assuming that the magnetization recovery is exponential. This is very often a better approximation in paramagnetic systems than in diamagnetic ones, because the unpaired electron acts as a sink for nuclear spin energy, thereby making the whole magnetization recovery closer to a single exponential (Section 1.7.3). When a single signal is excited, we refer to selective T_1, which is thus distinguishable from a nonselective T_1 obtained when more signals are simultaneously excited.

Nonselective T_1 can be measured with the usual inversion recovery 180–τ–90 pulse sequences [58]. It should be noted that, if the frequency range covered is large, the 180 or even the 90 pulse may not be the same all over the spectral region of interest. It may then be convenient to select ranges of excitation. Sometimes it may be convenient to measure T_1 with the Modeft pulse sequence 90–τ–180–τ–90

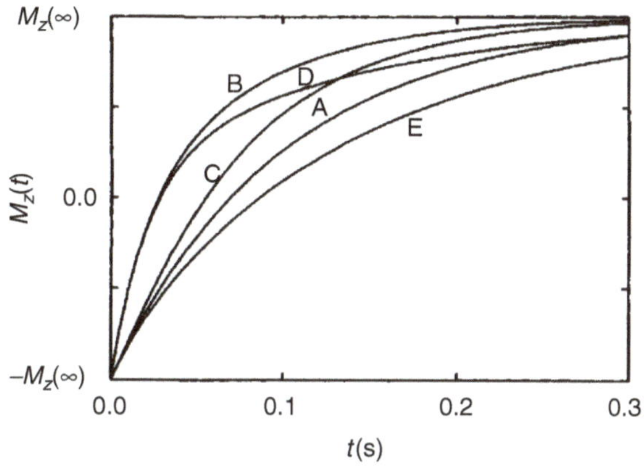

FIGURE 12.10 Exponential recovery (A) of $M_z(t)$ of a nuclear spin I dipole coupled to a paramagnetic metal ion. When I is also coupled to another nuclear spin J, the latter also coupled to the metal ion, nonexponentiality occurs. If J relaxes slower than I, curves B and C are obtained for a selective and a nonselective experiment respectively. If J relaxes slower than I, curves D (selective) and E (nonselective) are obtained. If J relaxes at the same rate as I, a selective experiment gives an intermediate behavior between curves B and D (not shown), while a nonselective experiment gives pure exponential recovery (A). It is apparent that in all cases non-selective experiments perform better than selective experiments, as they are less sensitive to the nonexponentiality introduced by I–J coupling. Conditions: $R_{1M}^{I} = 10\,\mathrm{s}^{-1}$; $R_{1M}^{J} = 20\,\mathrm{s}^{-1}$ (B,C), 5 s^{-1} (D,E) and 10 s^{-1} (A). The I–J cross-relaxation rate σ_{IJ} is –20 s^{-1}.

(Section 1.7.3): by making τ shorter and shorter we eventually have zero intensity also of the fast relaxing signals. The signal intensity as a function of τ is given by [37]

$$M_z(t) = M_z(\infty)[1 - 2\exp(-R_1 t) + (1 - a)\exp(-2R_1 t)] \quad (12.3)$$

where $R_1 = T_1^{-1}$ and a is a parameter that compensates for misadjustments of the first 90° pulse.

Selective T_1 values are generally measured with the 180–τ–90 pulse sequence using a soft 180° pulse. When the nuclear relaxation times are short, it may become impossible to invert the magnetization and at the same time maintain the required selectivity with the soft pulse. When such difficulties arise, a good compromise is to use a soft pulse that can at least saturate the signal. Then, the sequence becomes equivalent to a 90–τ–90 pulse sequence.

For the purpose of biomolecular applications, the number of resonances is usually so large that multidimensional experiments may be required. Of course, ND experiments can be modified to allow for the measurements of T_1 for instance by introducing a 180°–τ module in front of any 2D sequence, and recording several 2D spectra with different τ values [59–62]. In this way, the initial magnetization value of each nucleus at the beginning of the 2D experiment will be a function of τ, and so will be the intensities of the corresponding cross peaks. In heterocorrelated experiments, if the 180°–τ module is used at the proton frequency, then the T_1 of the protons is obtained [61]. To obtain the T_1 of the heteronucleus, whose evolution is usually in the indirect dimension, the τ value is inserted within an HSQC-type sequence (see later). A refocused INEPT provides single quantum (SQ) coherence of nitrogen spins. After t_1 evolution, a 90° pulse rotates nitrogen magnetization on the z-axis. The latter relaxes during the τ value and the resulting magnetization is then converted to proton SQ coherence through an inverse refocused INEPT [63,64].

Fast relaxing nuclei are characterized by a sizable linewidth. The measurement of the linewidth represents the easiest and most straightforward way of measuring T_2^{-1} from the relationship

$$T_2^{-1} = \pi\Delta\nu \tag{12.4}$$

where $\Delta\nu$ is the linewidth at half peak height [Eq. (1.54)]. Such procedure is absolutely adequate in paramagnetic systems when the lines are relatively large, since the relative contribution of field inhomogeneity is in any case small. Of course, when the linewidth becomes narrow, as in a diamagnetic system, the usual Carr–Purcell–Meiboom–Gill (CPMG) technique [65] should be used. CPMG modules can also be used in homonuclear and heteronuclear 2D experiments [66,67].

One time-saving strategy has been proposed by Iwahara and Clore [68], by modification of the abovementioned implementations. Instead of acquiring a full decay curve, one can record only one or two time points. The acquisition of a single time point just requires to compare the intensities of each peak between the paramagnetic and the diamagnetic sample, and the total time during which the ^{1}H magnetization is in the transverse plane [τ in Eq. (12.5a)] will be the total time over which the paramagnetic relaxation enhancement will affect the signal intensity. In strict analogy, acquisition of two time points is performed increasing the delay in the first INEPT step of an HSQC experiment. Then the intensity is compared through the following formulas.

Single time-point[1]		Two time-points	
$\frac{I_p(\tau)}{I_d(\tau)} = \frac{R_{2\text{dia}}}{R_{2\text{dia}} + R_{2M}} \exp(-R_{2M}\tau)$	(12.5a)	$R_{2M} = \frac{1}{t_b + t_a} \ln \frac{I_d(t_b) I_p(t_a)}{I_p(t_b) I_d(t_a)}$	(12.5b)

This approach relies on the reasonable assumption that the T_1 of the nitrogen, which is the other relaxation time for the developing SQ coherence, is negligibly affected by paramagnetic relaxation. Another assumption is that J does not change between the paramagnetic and the diamagnetic sample, i.e., when residual dipolar couplings are not present.

The two time-points approach is more convenient with respect to the single time point measurement for a number of reasons:

1. for the single time-point, both the paramagnetic and the diamagnetic samples should have the same integral intensity (i.e., the same concentration);
2. extracting R_{2M} from the single time-point measurement implies the use of a numerical inversion of Eq. (12.5a) (through the Lambert W function);
3. Eq. (12.5a) implies a Lorentzian lineshape.

However, if the diamagnetic contamination exceeds 3%, further attention is required in the choice of the second time-point, and it might be even necessary to acquire a full decay curve as described earlier.

In the case of T_1 measurements, we have mentioned that the cross relaxation provides multiexponential magnetization recovery (Section 1.7.3). A far less known analogy may occur in the linewidths, when two protons are dipole–dipole coupled and cross correlation occurs between Curie relaxation and

[1]The preexponential factor is needed when comparing intensities, as the intensity of a Lorentzian is related to its area $I_d = \frac{2A}{\pi R_{2\text{dia}}}$; $I_p = \frac{2A}{\pi(R_{2\text{dia}} + R_{2M})}$.

proton–proton dipolar relaxation. In this case, we are in the presence of two overlapping signal components with different linewidths, i.e., of biexponentiality in T_2 [69]. Pulse sequences are available to remove the effects of cross correlation [70]. Such effects are common in paramagnetic metalloproteins where Curie relaxation is usually relevant (in principle, such cross correlation effects can be operative also in the case of T_1, although only to the extent that Curie relaxation on T_1 is effective).

Direct detection of ^{13}C nuclei offers a valuable alternative to the measurement of relaxation rates of C′N multiple-quantum coherences, which may contribute to a more complete description of backbone dynamics in proteins as a complementary tool to conventional techniques that focus on the relaxation of isolated ^{15}N and ^{13}C nuclei. Protonless experiments avoid losses of information caused by fast relaxation or exchange broadening of ^{1}H signals and may therefore offer better sensitivity, particularly in the presence of paramagnetic ions [71].

12.2 MEASUREMENTS OF NUCLEAR OVERHAUSER EFFECT (NOE)

In Chapter 4, we have seen that two dipole–dipole coupled spins relax each other with a law that can be derived from the solution of two coupled differential equations (Eq. 4.55). However, such equation for $\langle I_z(t)\rangle$ can be approximated to a single exponential, not only when $|\sigma_{J(I)}| \ll |\rho_I - \rho_J|$, as is obvious, but also when, on the contrary, $|\sigma_{I(J)}| \gg |\rho_I - \rho_J|$. In these cases, fitting the recovery to an exponential gives a rate constant close to ρ_I [72]. In contrast, in the nonselective experiment, the recovery is closer to $\rho_I + \sigma_{I(J)}$. In order to estimate ρ_I from T_1-type measurements, the best way is that of performing a selective experiment and to measure the variation of $\langle I_z\rangle$ during an initial short time. Under these circumstances the system maximizes the two contributions to the measured ρ_I (i.e., $\rho_{I(J)}$ and $\rho_{I(\text{other})}$), R^I_{1M} remaining independent of the duration of the measurement. However, in order to extract structural information through, for example, Eq. (3.19), it is more appropriate to perform nonselective experiments. This is particularly true in the slow motion limit, where $\sigma_{I(J)} \rightarrow -\rho_{I(J)}$. Under these conditions, and neglecting $\rho_{I(\text{other})}, \rho_I + \sigma_{I(J)} \rightarrow R^I_{1M}$ [Eq. (4.52)]. This can be intuitively understood because cross relaxation in a nonselective experiment contributes less to signal recovery, especially at the beginning of the experiment, and the latter is dominated by paramagnetic effects (R^I_{1M}). In any case, when ρ_I in paramagnetic systems is very large (e.g., 100–1000 s^{-1}), it means that R^I_{1M} dominates the relaxation processes and the results of selective and nonselective experiments are close to one another.

Cross relaxation is the main cause of deviation from exponentiality, both for diamagnetic and paramagnetic nonselective relaxation recovery. However, a nonselective inversion recovery can be pragmatically interpreted on the basis of a single exponential. The problem arises of whether the extracted R^I_{1M} can be used, according to the Solomon equation, to extract metal–proton distances. Though this problem has been stressed in a negative sense [73], we optimistically advise [57,74]: (1) to check exponentiality and (2) to extract distance parameters, with the only caution regarding geminal protons (having large cross relaxation) whose distances from the metal, however, cannot differ much.

Steady state NOEs in paramagnetic small complexes under the conditions of fast rotation ($M_r < 800$–2500 depending on the magnetic field) are small and sometimes may be below detection. In Table 12.1 the steady state NOE intensities are reported for various τ_r values (and the corresponding relative molecular mass values estimated from the Stokes–Einstein relation [Eq. (4.8)] and for various R^I_{1M} values. For each pair of τ_r and R^I_{1M} values, the NOE intensities are calculated for distances of 1.8 and 3.0 Å and for magnetic fields of 200 MHz and 900 MHz. From inspection of the table, it appears that NOE measurements on small complexes may be successful for high concentrations at low magnetic

Table 12.1 Calculated Steady State NOE's (%) for Various Values of x_r and R^I_{1M} [a]

τ_r (s) [M_r]	r_{IJ} (Å)	R^I_{1M} (s^{-1}) 1 ν_0 (MHz) 200	1 900	2 ν_0 (MHz) 200	2 900	4 ν_0 (MHz) 200	4 900	8 ν_0 (MHz) 200	8 900	16 ν_0 (MHz) 200	16 900	32 ν_0 (MHz) 200	32 900	64 ν_0 (MHz) 200	64 900	128 ν_0 (MHz) 200	128 900	256 ν_0 (MHz) 200	256 900
10^{-11}	1.8	7.171	7.078	3.862	3.810	2.009	1.981	1.025	1.010	0.518	0.510	0.260	0.257	0.130	0.129	0.065	0.064	0.033	0.032
[25]	3.0	0.388	0.382	0.195	0.192	0.097	0.096	0.049	0.048	0.024	0.024	0.012	0.012	0.006	0.006	0.003	0.003	0.002	0.002
$10^{-10.5}$	1.8	17.221	15.509	10.406	9.245	5.808	5.114	3.084	2.700	1.591	1.389	0.808	0.705	0.407	0.355	0.205	0.178	0.102	0.089
[80]	3.0	1.197	1.045	0.606	0.528	0.305	0.265	0.153	0.133	0.077	0.067	0.038	0.033	0.019	0.017	0.010	0.008	0.005	0.004
10^{-10}	1.8	30.088	14.955	21.697	9.669	13.929	5.665	8.117	3.098	4.424	1.625	2.317	0.833	1.186	0.422	0.600	0.212	0.302	0.107
[250]	3.0	3.380	1.229	1.750	0.626	0.891	0.316	0.450	0.159	0.226	0.080	0.113	0.040	0.057	0.020	0.028	0.010	0.014	0.005
$10^{-9.5}$	1.8	31.860	–15.959	26.017	–10.409	19.035	-6.139	12.387	-3.373	7.293	–1.774	4.001	–0.911	2.103	–0.461	1.079	–0.232	0.547	–0.117
[800]	3.0	5.700	–1.342	3.063	–0.684	1.591	–0.346	0.811	–0.174	0.410	–0.087	0.206	–0.044	0.103	–0.022	0.052	–0.011	0.026	–0.005
10^{-9}	1.8	-6.365	-57.209	-5.256	-42.124	-3.897	-27.580	-2.569	-16.314	-1.528	-8.979	-0.844	-4.728	-0.445	-2.428	-0.229	-1.231	-0.116	-0.620
[2500]	3.0	-1.198	-6.878	-0.647	-3.577	-0.337	-1.825	-0.172	-0.922	-0.087	-0.463	-0.044	-0.232	-0.022	-0.116	-0.011	-0.058	-0.005	-0.029
$10^{-8.5}$	1.8	-68.209	-83.237	-59.727	-71.890	-47.832	-56.490	-34.206	-39.546	-21.791	-24.718	-12.626	-14.125	-6.858	-7.606	-3.583	-3.955	-1.833	-2.018
[8000]	3.0	-17.482	-19.701	-9.821	-10.941	-5.234	-5.791	-2.706	-2.983	-1.376	-1.514	-0.694	-0.763	-0.349	-0.383	-0.175	-0.192	-0.087	-0.096
10^{-8}	1.8	-92.235	-94.261	-87.379	-89.239	-79.055	-80.645	-66.403	-67.620	-50.302	-51.111	-33.875	-34.342	-20.491	-20.736	-11.447	-11.569	-6.080	-6.140
[25000]	3.0	-43.190	-43.842	-27.726	-28.084	-16.156	-16.339	-8.806	-8.897	-4.611	-4.656	-2.361	-2.384	-1.195	-1.206	-0.601	-0.607	-0.302	-0.304
$10^{-7.5}$	1.8	-97.918	-98.136	-96.139	-96.352	-92.770	-92.970	-86.693	-86.873	-76.650	-76.799	-62.232	-62.341	-45.221	-45.289	-29.236	-29.275	-17.128	-17.147
[80000]	3.0	-71.060	-71.192	-55.183	-55.274	-38.140	-38.194	-23.576	-23.605	-13.368	-13.382	-7.164	-7.171	-3.715	-3.719	-1.893	-1.895	-0.956	-0.956
10^{-7}	1.8	-99.383	-99.406	-98.797	-98.819	-97.646	-97.667	-95.421	-95.442	-91.263	-91.282	-83.947	-83.964	-72.347	-72.360	-56.682	-56.692	-39.554	-39.559
[250000]	3.0	-88.639	-88.658	-79.612	-79.628	-66.140	-66.152	-49.415	-49.423	-32.818	-32.823	-19.631	-19.633	-10.884	-10.885	-5.755	-5.756	-2.963	-2.963

[a] *For each τ_r and R^I_{1M} value two NOEs are calculated for interproton distances of 1.8 and 3.0 Å and for nuclear Larmor frequencies ν_0 of 200 and 900 MHz.*

fields, or at large magnetic fields when rotation is slowed down by using viscous solvents or mixtures of solvents. On the contrary, macromolecules seem perfectly suited for this type of experiment since they fall in the slow motion regime.

In Table 12.2 some significant examples of the NOEs observed in paramagnetic metalloproteins are reported [75]; as expected, as the T_1 values become shorter, the NOE drops to small values, which can be difficult to "extract" from the noise. However, the NOE increases as the molecular size increases. In Fig. 12.11, the NOEs for a protein of M_r 10,000 containing the cluster $Fe_4S_4^{3+}$ are shown when the hyperfine shifted signals due to coordinated β cysteine protons and lying outside the –20–20 ppm spectral window are saturated [76]. The NOE difference spectra show many connectivities with other signals which belong to nearby groups, as well as strong connectivities with the respective geminal protons (not shown).

Table 12.2 Some Examples of NOE Detected on Paramagnetically Shifted Signals in Metalloproteins [75]

Group[a]	H–H distance (Å)	NOE[b] (%)	T_1 (ms)	M_r	τ_r (ns)
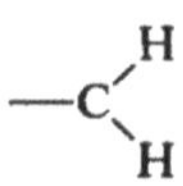 (Asp, Glu, Cys, His)	1.8	6	8.4	9,000	4
		5	3.6	11,000	5
		7	5.6	11,000	5
		43	120	16,000	9
		7.7	2.4	32,000	14
		60	100	36,000	21
		50	80	42,000	22
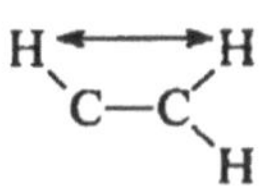 (Cys)	2.2–2.4	4.5	34.0	9,000	4
		1.8	5.6	11,000	5
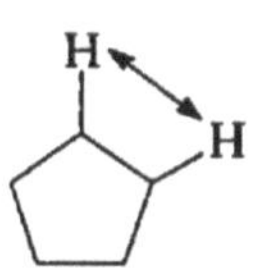 (His)	2.4	6	5.4	30,000	14
		0.9	1.8	32,000	14
		2	4.2	32,000	14
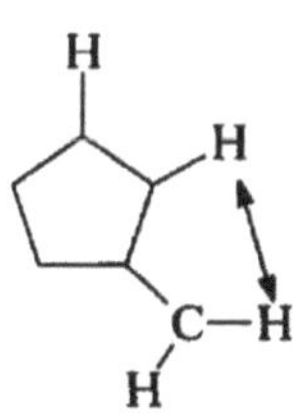 (His)	~3.3	0.6	3.5	32,000	14
		7–10	–	42,000	21

[a]*The groups bearing the two protons belong to metal ion ligands.*
[b]*The NOEs in the present systems are all negative; they are reported here without the minus sign.*

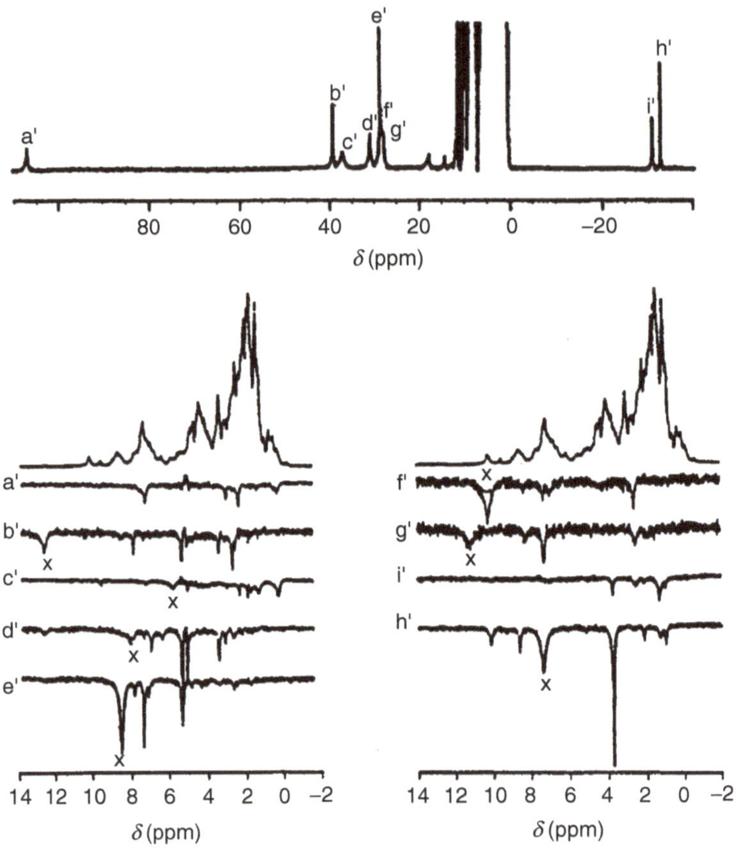

FIGURE 12.11 NOE difference spectra obtained upon saturation of the hyperfine shifted signals corresponding to the β-CH_2 protons of the Fe_4S_4-coordinated cysteines in oxidized HiPIP from *C. vinosum*. Signals marked by × arise from saturation transfer to a small amount of reduced species.

Reproduced with permission from Ref. [72].

The difference spectra in Fig. 12.11 also show signals (marked with ×) arising from saturation transfer to another species in chemical exchange. Indeed, the experiment for detecting steady state NOE is absolutely identical to that used to detect magnetization transfer in the presence of chemical exchange (Section 4.3.4). Therefore, it is not possible to distinguish, from the experimental point of view, real NOE from chemical exchange effects in the limit of slow motion. Actually, the values of NOE are always a fraction of the signal intensity, which in paramagnetic molecules is often small (of the order of 1%), whereas saturation transfer can change the intensity of the signal really up to 100% even in paramagnetic molecules (Section 6.3.4).

In diamagnetic compounds we face the problem that after saturation of signal J the Zeeman population of nuclei I may be largely affected, and this may cause further magnetization transfer on other nuclei of the same nuclear species to which nucleus I is coupled. Such coupling indeed accounts for the $\rho_{I(\text{other})}$ term in Eq. (4.52). It follows that a secondary NOE can be observed on signal K not because J is coupled with K but because I is coupled with K. This phenomenon is called *spin diffusion*. It is not relevant in the fast motion limit, because the NOEs are usually small and, therefore, small is the secondary magnetization transfer to nearby nuclei; furthermore, and more importantly, a positive sign of σ means that a decrease in $\langle J_z \rangle (M_z^J)$ causes an *increase* in $\langle I_z \rangle (M_z^I)$ which, in turn, causes a *decrease* in $\langle K_z \rangle (M_z^K)$, and so on. This alternation in sign eventually results in substantial cancellation of the long range effects, because a nucleus very far from the saturated nucleus will experience both positive and negative contributions depending on whether the number of intermediate spins on the various cross relaxation pathways is even or odd.

In contrast, sizable spin diffusion may be operative in the slow motion regime. Of course, this phenomenon may terribly complicate the analysis of NOEs, until they become useless. After saturating one signal, many or even all signals of the molecule may be affected in a cascade fashion. However, in paramagnetic macromolecular systems, the major source of relaxation for signal *I* can be the coupling with the unpaired electrons; so it returns to equilibrium prior to appreciably transferring polarization to nucleus *K*. Therefore, spin diffusion effects are limited. The counterpart of this simplification is that the NOE is dramatically reduced by the short ρ_I values.

An example is available in the literature where the spin diffusion effects are evaluated in a strongly paramagnetic system [77]. It is the Cu_2Co_2SOD discussed in Section 11.3.3, Fig. 11.19. In Table 12.3, the NOEs at 200 MHz on all the proton ligands are reported when they are saturated one after the other. The T_1 values measured at the same field are also reported [78,79]. The interproton distances are known from the X-ray structure and the value of the reorientational correlation time estimated by the

Table 12.3 Nonselective T_1 and NOE Values in Bovine Cu_2Co_2SOD at 200 MHz [79]

Saturated Signal	Observed Signal	T_1 (obs) (ms)	NOE (%)	*r* (calc) (Å)	*r* (X-ray) (Å)
A Hδ2 His 63	L Hδ2 His 48	4.3	0.9 ± 0.1	2.7 ± 0.2	2.7
A Hδ2 His 63	K Hδ1 His 48	8.0	0.6 ± 0.2	3.2 ± 0.3	3.8
A Hδ2 His 63	R Hβ2 His 46	2.4	0.2 ± 0.1	3.3 ± 0.3	3.4
B Hδ1 His 120	H H∈1 His 120	1.8	1.0 ± 0.2	2.3 ± 0.2	2.4
C Hε2 His 46	G Hδ2 His 46	3.5	1.4 ± 0.2	2.4 ± 0.2	2.5
C Hε2 His 46	M Hε1 His 46	2.7	0.9 ± 0.1	2.5 ± 0.2	2.5
G Hδ2 His 46	C Hε2 His 46	4.2	1.7 ± 0.4	2.4 ± 0.2	2.5
M Hε1 His 46	C Hε2 His 46	4.2	2.2 ± 0.2	2.4 ± 0.2	2.5
R Hβ2 His 46	G Hδ2 His 46	3.5	0.6 ± 0.2	2.8 ± 0.3	3.3
Q Hβ1 His 71	D Hδ2 His 71	3.8	1.4 ± 0.2	2.4 ± 0.2	2.8
L Hδ2 His 48	γ_1 CH_3 Val 118	70.0	33.0 ± 6	2.8 ± 0.3	2.8
R Hβ2 His 46	P Hβ1 His 46	1.6	5.4 ± 0.8	1.7 ± 0.2	1.6
P Hβ1 His 46	R Hβ2 His 46	2.4	9.0 ± 1.0	1.7 ± 0.2	1.6
L Hδ2 His 48	P Hβ1 His 46	1.6	1.3 ± 0.3	2.2 ± 0.2	2.3
P Hβl His 46	L Hδ2 His 48	4.3	5.5 ± 1.0	2.0 ± 0.2	2.3
L Hδ2 His 48	R Hβ2 His 46	2.4	0.5 ± 0.1	2.6 ± 0.3	3.2
R Hβ2 His 46	L Hδ2 His 48	4.3	2.3 ± 0.2	2.3 ± 0.2	3.2

The distances calculated from the NOE values using the two-spin approximation agree with available X-ray distances.

Stokes–Einstein equation [Eq. (3.8)]. The NOEs calculated through the two-spin approximation are in satisfactory agreement with the experimental NOEs. If then a matrix is considered where all the ligand protons plus other protons from nearby groups are allowed to cross relax, the newly calculated NOEs are still close to the experimental values [77]. Spin diffusion is quenched by the fast relaxing nature of the system. The ρ_I values are well above 100 s^{-1}. Steady state NOE is then a safe measurement of proton–proton distances. When, however, the ρ_I values are small, spin diffusion may occur even in paramagnetic systems; tricks are developed to detect spin diffusion and to minimize it, for instance by simply reducing the irradiation time.

The 1D NOE technique is still of practical interest when applied to paramagnetic molecules. Here the signals to be irradiated can be well separated (hyperfine shifted), nuclear relaxation is more affected by the coupling with the unpaired electrons than by the coupling with other protons and, if the molecule is large, large effects can be measured without the drawback of having all the signals in a narrow spectral region. If one is at the extreme limit of detectability of dipolar connectivities (as is often the case in fast relaxing and relatively diluted samples), 1D NOE is the most suitable technique.

As discussed in Sections 1.1 and 1.3, it is difficult to saturate in a selective way a fast relaxing signal. Selectivity here is an absolute requirement because the wings of the excitation profile of a soft pulse may well partially saturate other signals, with the consequence of appearance of artifacts which vary from case to case but may lead to gross misinterpretations. The use of suitably shaped pulses designed to be more selective than a rectangular pulse is advisable. Even with a perfectly selective irradiation, other signals can be partly saturated because their tails may extend down to where the irradiation is applied. To further remove the residual effects of nonperfect selectivity and of tail excitation, the most common procedure is that of alternating experiments with the saturation pulse applied on-resonance and off-resonance, and subtracting the latter from the former. Even so, off-resonance effects on signals very close to the irradiated one cannot be readily discriminated from true NOEs that in macromolecules

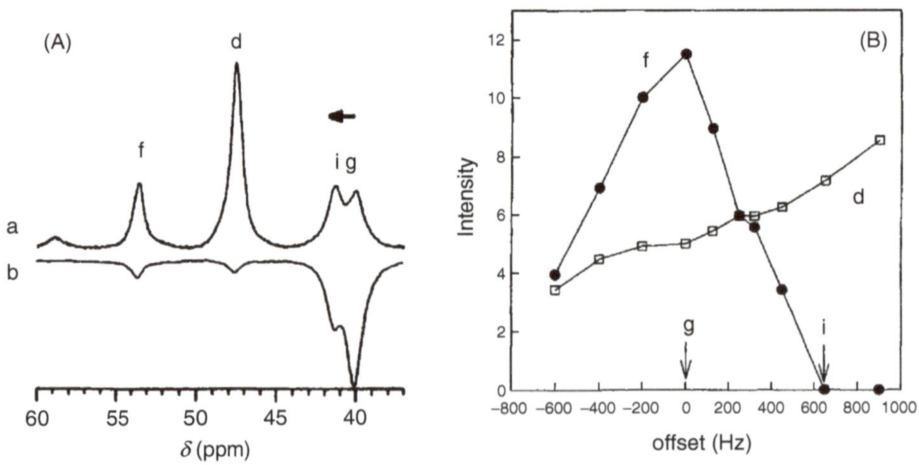

FIGURE 12.12 1D NOE difference experiments on met-aquo myoglobin. (A) Reference spectrum (a) and difference spectrum (b) observed upon saturation of peak *g*. (B) Intensity of negative *f* and *d* signals in (b) as a function of the irradiation frequency. Signal *f* is maximal when the irradiation frequency is on *g*, signal *d* shows a steady increase with increasing frequency.

Reproduced with permission from [80].

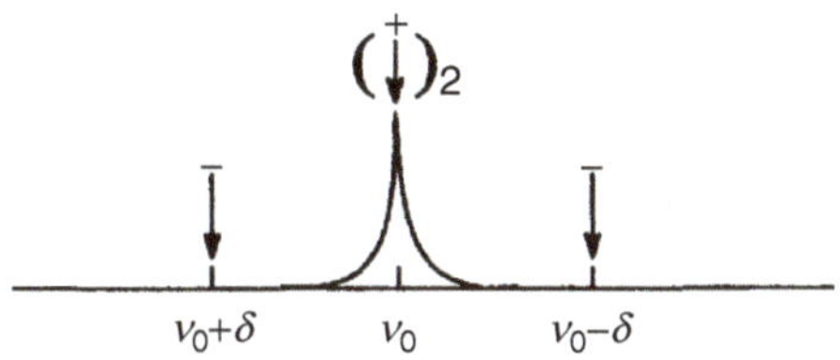

FIGURE 12.13 Irradiation scheme used in NOE experiments to minimize off-resonance effects. The r.f. is alternately placed at ν_0, $\nu_0 + \delta$, $\nu_0 - \delta$. The spectra obtained upon irradiation at $\nu_0 \pm \delta$ are subtracted from those obtained upon irradiation at ν_0.

are negative. Fig. 12.12A provides a clear example [80]. Upon irradiation of signal *g*, signal *f* gives NOE while signal *d* is affected by off-resonance effects. This could be demonstrated by performing an irradiation "profile" (Fig. 12.12B): the effect of *f* is maximal when the irradiation frequency is on *g*, whereas the effect on *d* steadily increases with increasing irradiation frequency [80].

In our experience, the best experimental scheme is to choose two off-resonance positions symmetrical with respect to the irradiated signal and typically offset by one to two times the linewidth of the irradiated signal (Fig. 12.13). It often happens that there is at least one signal too close to the signal to be irradiated to allow for the setting of the two off-resonance frequencies in the optimal position. In fact, if another signal is close to one of the two off-resonance frequencies, in the difference spectrum this signal appears as strongly positive. If this signal gives NOEs with the same signal(s) which experience NOE from the irradiated signal, the two NOEs are opposite in sign and the unwanted one may reduce or cancel the one which is looked for. A general strategy to eliminate this effect is to place the off-resonance offset beyond the unwanted signal. If the offset is far enough, the latter signal is now negative (for macromolecules), since the off-resonance from the central irradiation position is dominant. It is easy to show that an off-resonance position always exists for which the intensity of the unwanted signal is rigorously zero in the difference spectrum. The other offset is, of course, always symmetrical with respect to the irradiated signal. The optimal position can be determined empirically with a small number of scans before starting the real experiment. In Fig. 12.14, the spectrum of the oxidized high potential iron-sulfur protein from *E. halophila* iso I is reported [81], together with the NOE difference spectra obtained upon saturation of signals *y* and *z*, using the above irradiation scheme. Signals *y* and *z* happen to be geminal β-CH_2 protons of the same cysteine residue and to have rather long T_1 values. Therefore, despite the optimization of the off-resonance frequencies, a sizable real NOE is present from one signal to another in both cases. To further eliminate the secondary NOEs from the geminal signal, double difference NOE spectra were obtained by subtracting from one NOE difference spectrum the other one, appropriately scaled.

Approaches of this kind are also particularly important when the NOE to be observed is on a signal close to the irradiated one. In the case of the two β-CH_2 protons of an Asp residue bound to a high spin cobalt(II) in the protein superoxide dismutase, the two signals *D* and *E* are very broad (Fig. 12.15A). The NOE between the two signals was observed by irradiating signal *D* and by recording the difference spectrum with one off-resonance position symmetrical with respect to the position of signal *E* [78]. In principle, in this way any intensity detected at the *E* position is due to a real NOE. In practice, a strong distortion of the baseline at the *E* position is present because of the tails of signal *C*, experiencing off-resonance from *D*. Another experiment is performed by irradiating this third signal with the off-resonance position still symmetrical with respect to *E*, and by further subtracting

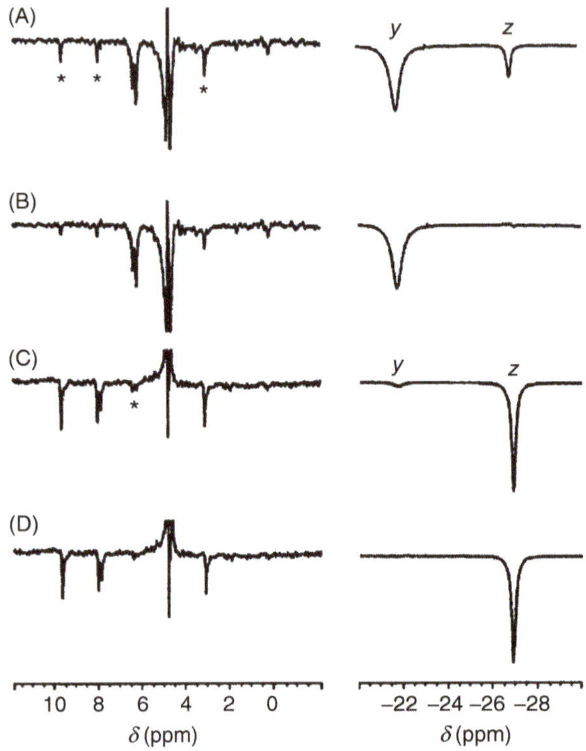

FIGURE 12.14 600 MHz 1D NOE difference spectra of the oxidized form of the high potential Fe_4S_4 protein from *E. halophila* iso I. Spectra (A) and (C) are obtained by irradiating each of the two geminal β-CH_2 protons (*y* and *z*) of the cluster-ligated Cys-39. Spectrum (B) is obtained by subtracting from spectrum (A) spectrum (C) multiplied by a factor such that signal *z* is canceled; spectrum (D) is obtained by subtracting from spectrum (C) spectrum (A) multiplied by a factor such that signal *y* is canceled. Note the reduction of intensity of the signals marked with asterisks.

Reproduced with permission from Ref. [81].

the two NOE difference spectra one from the other. The double difference NOE spectrum is shown in Fig. 12.15B.

It is often necessary to obtain a profile of NOE intensities versus the irradiation time, called NOE buildup. A series of truncated NOEs are thus needed. The only caveat with respect to buildup experiments in a diamagnetic case is that there is a lower limit in the irradiation time because even a truncated NOE needs saturation of the irradiated signals. Since "instantaneous" saturations are not possible, but short saturation times are needed to measure the first points of the buildup, there is the risk that at the beginning of the buildup, the irradiated signal is not 100% saturated [80]. In practice, buildup measurements are feasible when the irradiated signal relaxes faster than the responding signals, but hardly when the reverse is true. Even in the most favorable case, the intensities of the very first points may not be accurate; as a consequence, it is always a risk to use buildup experiments to discriminate between primary and secondary NOEs. However, in paramagnetic systems, secondary NOEs are usually small and, in general, recognizable from other indirect evidence.

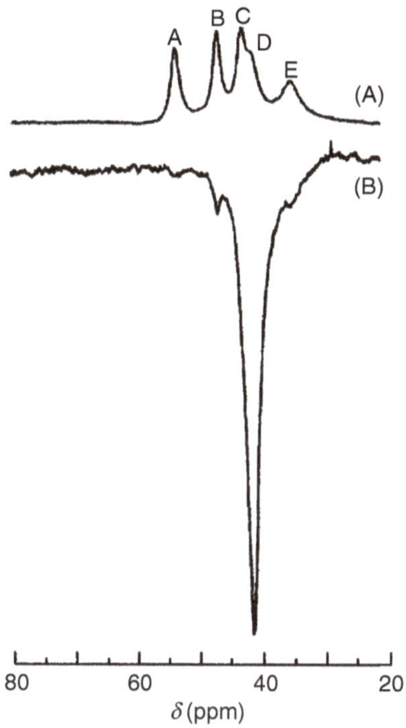

FIGURE 12.15 (A) 200 MHz ^{1}H NMR spectrum of copper-depleted, cobalt-substituted Cu,Zn superoxide dismutase; (B) difference between the difference NOE spectra obtained by irradiating signals *D* and *C*, respectively. The spectra demonstrate that the NOE on *E* arises from *D* and not from C. The off-resonance position was placed symmetrical with respect to *E* in both cases.

Reproduced with permission from [78].

12.3 ND SPECTROSCOPY

Spectroscopy with higher dimensionality is used to obtain some kind of correlation between two or more nuclear spins, for instance through scalar or dipolar connectivities, to improve resolution in crowded regions of spectra or to achieve information about the interactions that one nucleus senses (i.e., scalar couplings, anisotropic interactions in solids). A N-dimensional experiment requires the following steps: preparation, [evolution, mixing]×(N−1) and acquisition Fig. 12.16. The preparation period contains at least one pulse which alters the equilibrium population of at least one nucleus and generates some magnetization in the *xy* plane. The evolution period allows the magnetization to evolve in the *xy* plane according to the Larmor frequency of the nuclei which have experienced the preparation pulse(s), either free or under the effect of additional RF pulses. The evolution period is incremented in successive experiments and constitutes one of the N time domains (t_1,t_{N-1}) on which the Fourier transform is applied at the end of the ND experiment. As in 1D experiments, the resolution in each indirect dimension, i, depends on the maximum value(s) of the indirect evolution, t_i, in that dimension. Historically, this has imposed limits in the overall resolution, as all the increments in the indirect dimension

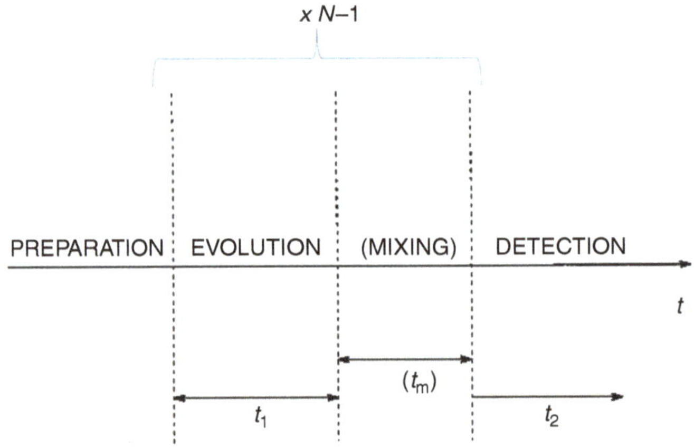

FIGURE 12.16 General scheme of the time periods involved in ND experiments.

were sampled. By the use of specific reconstruction algorithms, however, one can avoid acquiring all the increments (nonuniform sampling, NUS), restraining the total number of increments to be equal, as a rule of thumb, to the number of expected resonances in the spectrum [82,83].

After each evolution, a discontinuity is provided by either a pulse or the start of a spin-lock sequence (see in further sections), then the mixing follows. The mixing period t_m which may be lacking or may coincide with a pulse rather than a time period and which usually (but not always) has a constant duration throughout the experiment, allows for the transfer of magnetization or coherence (see in further sections) between sets of spins which are connected by either dipolar or scalar coupling or by chemical exchange. At the end of the last mixing period, the detection pulse(s) provides the *xy* magnetization which is detected during the detection period as free induction decay. The time axis of the latter constitutes the last time domain (t_n) on which the Fourier transform is applied, and corresponds to the direct dimension.

The requirements for achieving enough signal at the end of the sequence requires that (1) evolved populations/coherences can survive the N-1 evolution periods and (2) mixing is sufficiently efficient. Since each additional dimension thus imposes an additional sensitivity limitation, it is rather clear why NMR of paramagnetic molecules is often restrained to 2/3Ds. The parameters to obtain 2D spectra and beyond are nowadays well optimized for paramagnetic molecules, and useful information is obtained as long as the conditions dictated by the correlation time for the electron–nucleus interaction are not too severe. Sometimes care has to be taken to avoid that the fast return to thermal equilibrium of nuclei wipes out the effects of the internuclear interactions that are sought through ND spectroscopy.

2D experiments are devised in the assumption that the various times involved in the cycle of Fig. 12.16 (with the exception of t_m when present) are small with respect to the nuclear relaxation times. When the latter are short for any reason, for example, in the case of paramagnetic molecules because of the presence of unpaired electrons, the system of spins may have reached the equilibrium, or almost reached the equilibrium, before the detection pulse. Under these circumstances no memory is left for the state of the spins during the preceding steps. As a consequence, cross peaks may be decreased in intensity until below detectability. It is necessary, therefore, to match all the time intervals with the nuclear relaxation times, in order to detect the maximum possible cross peak intensities. The ideal case is that t_1 and t_2 are as short as possible, compatibly with spectral resolution, and in any case not much longer than T_2. Indeed, in a 2D experiment, coherences decay with time constant T_2 during t_1 and t_2, as it happens during acquisition in 1D experiments. As far as the mixing time is concerned, it is again related to nuclear relaxation times in a fashion which will be described experiment by experiment (this consideration does not apply to COSY experiments, for which there is not a formal mixing time). Some gain can be obtained by a more rapid cycling because (1) the spin systems reach equilibrium quite fast, and (2) the overall time required by Fig. 12.16 is much shorter. The first 2D experiment on a system with T_1 in the range between 50 and 100 ms was an EXSY experiment (see in further sections) which appeared in 1984 [84]. In 1985 the first NOESY appeared on a pseudotetrahedral nickel(II) complex [85]; in 1988 the first COSY appeared on a lanthanide complex [86,87]; in 1990 the first TOCSY experiment appeared on a five-coordinated nickel(II) complex [88]. Since then, the reports have been numerous, aiming also at the detection of cross peaks between signals with shorter and shorter T_1 values.

It is rather intuitive that a finite time is required to achieve selective saturation or inversion of a signal by a soft pulse, during which time polarization starts to be exchanged, causing nonlinearity of the response. It should be stressed that this is not the case in all common 2D experiments based on nonselective pulses, which have durations of the order of microseconds instead of milliseconds, as required for selectivity. Selectivity in 2D experiments is intrinsic because of the double frequency labeling along f_1 and f_2.

We are now going to describe some 2D and 3D experiments performed on paramagnetic compounds (Fig. 12.17). We should stress that relatively few laboratories are engaged in making ND spectroscopy suitable for the investigation of this type of compounds. Therefore, whereas the literature is richer and richer in suitable sequences for the various cases in diamagnetic systems, fewer applications are devoted to paramagnetic systems. The following examples will give an overview of the basic achievements which have represented a breakthrough over the years.

12.3.1 SOLUTION NMR—EXSY AND NOESY

The 2D exchange spectroscopy (EXSY) is just an extension of the 1D saturation transfer experiment (Section 4.3.4). The simplest pulse sequence used to obtain EXSY spectra is the NOESY sequence reported in Fig. 12.17A. The first 90° pulse, constituting the preparation period of Fig. 12.7, tilts the z-magnetization along, e.g., the x axis in the rotating frame x' (Fig. 12.18); the variable time interval t_1 follows, during which the spins are labeled according to their frequency. This time must be short with respect to T_2, otherwise not enough inplane magnetization remains to play with. Another 90° pulse is then applied which sends along the z–axis, the component of the magnetization that is along x' at the end of the evolution period. At this point, during the fixed mixing time t_m, the nuclei of a given chemical species (e.g., A) transfer magnetization to the corresponding nuclei of the other species in chemical equilibrium (e.g., B), to an extent that is modulated by the difference in their z-components. After the time t_m, a 90° detection pulse is applied, which brings back to the xy plane the magnetization of the nuclei of the species in chemical equilibrium. The time t_m ideally should be much longer than the exchange time τ_M and much shorter than the T_1 values of the signals. If the exchange time is longer than T_1, then the best compromise is to set $t_m \cong T_1$. The maximal information content in EXSY experiments is in the first t_1 and t_2 points. Therefore, in this case we multiply the FID by a weighting function of cos- or $\cos^2$-type in order to give more "weight" to the first points.

We recall here that the phases of the pulses in the sequence of Fig. 12.17A s, as well as of all other pulse sequences shown in Fig. 12.17 and described later in this chapter, must be properly cycled to

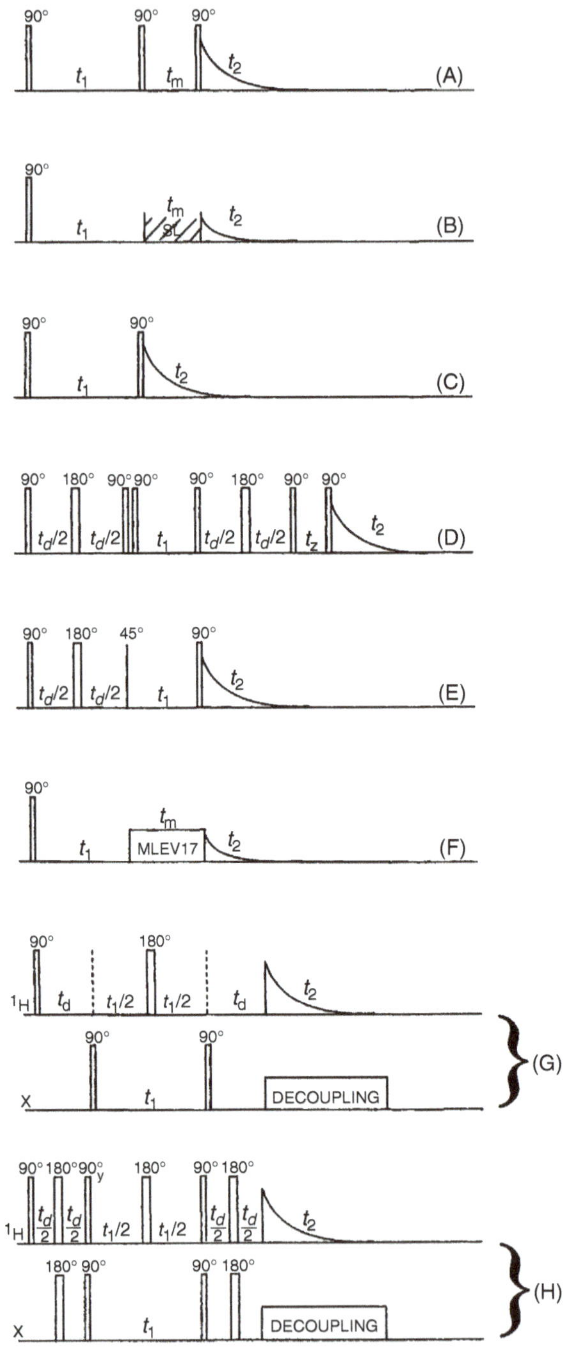

FIGURE 12.17 Some of the most common 2D pulse sequences that can be employed using a proper choice of parameters to record 2D spectra of paramagnetic molecules. (A) NOESY, (B) ROESY, (C) COSY, (D) ISECR COSY, (E) zero-quantum (double quantum) COSY, (F) TOCSY, (G) HMQC, (H) HSQC. Sequences (A), (B) and (F) are also used to obtain EXSY spectra. SL indicates a soft spin-lock sequence, while MLEV17 indicates a train of spin-locking hard pulses that optimizes the development of J_{IJ} coupling. In the reverse heteronuclear experiment (G) the upper and lower levels refer to ^{1}H and heteronucleus, respectively. The phase cycles are not indicated. For clarity of discussion, all initial pulses can be thought to be applied along the y' axis, in such a way that the coherence after the first 90° pulse is always along x'.

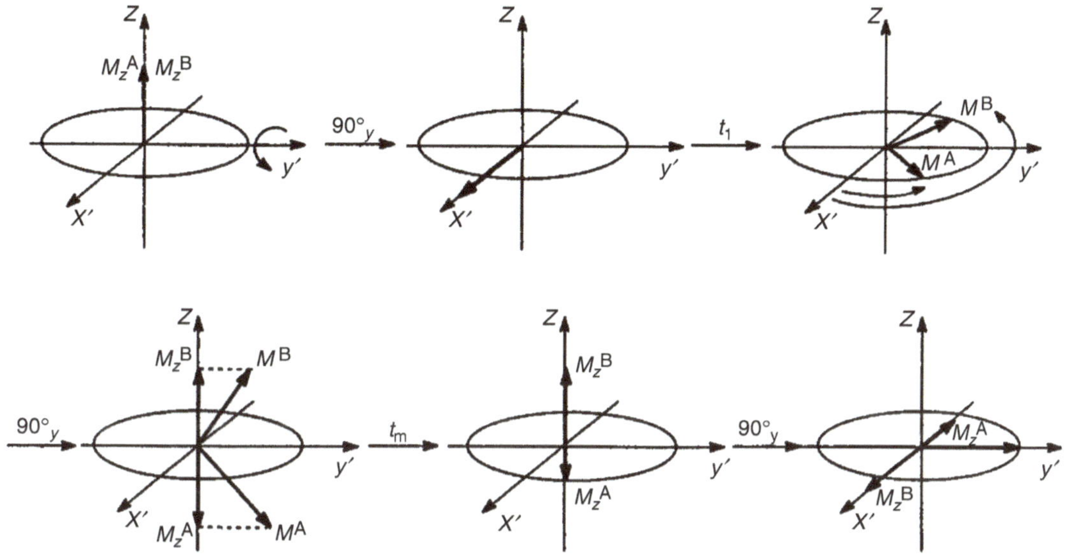

FIGURE 12.18 Vector representation of an EXSY (or NOESY) experiment. The first 90° pulse along y' rotates the equilibrium magnetization of the A and B spins from the z-axis to the x' axis. During t_1, the two transverse coherences precess in the $x'y'$ plane at their characteristic frequencies, and their x' (and y') components result periodically in and out of phase. The second 90° pulse tilts the two x' components along the z-axis. During t_m, magnetization transfer occur, to an extent that depends on how out of phase the two vectors were just before the second 90° pulse (i.e., depending on the value of t_1). The third pulse tilts the two z components, whose intensities have been altered during t_m in a t_1-dependent manner, along x'. Detection can then occur during t_2. In the presence of magnetization transfer, the intensity of (for example) the A signal detected during t_2 is thus modulated not only by its characteristic frequency but also by the frequency of the B signal. A cross peak is thus generated.

achieve selection of the desired connectivities and suppression of artifacts and of other connectivities due to different types of interactions. The criteria to choose the appropriate phase cycling do not depend on the presence of a paramagnetic center in the molecule, and the reader should refer to the many publications on multidimensional NMR for details.

The first EXSY experiment on a paramagnetic system [84] described the chemical exchange between partially reduced species of a cytochrome c containing four hemes (Fig. 12.19). The electron exchange time could be estimated from the relative volumes of the cross peaks. The latter can be determined through numerical integration from the experimental signal intensities over a suitable area of the spectrum. Since the cross peak volume increases with the exchange rate, k_{-1} [$= \tau_M^{-1}$, as defined in Eq. (6.25)], and decreases with $(\rho^A + \rho^B)/2$ (A and B being the two species in chemical exchange), τ_M can in principle be determined from a single 2D experiment:

$$
\begin{aligned}
a_{\text{cross}}(t_m) &= \frac{M_{0J}k_{-1}}{2D}\exp[-(\rho - D)t_m][1-\exp(-2Dt_m)] \\
&= \frac{M_{0J}k_{-1}}{2D}\{\exp[-(\rho - D)t_m] - \exp[-(\rho + D)t_m]\}
\end{aligned}
\tag{12.6}
$$

where M_0 is the intensity of the diagonal peak at zero t_m and

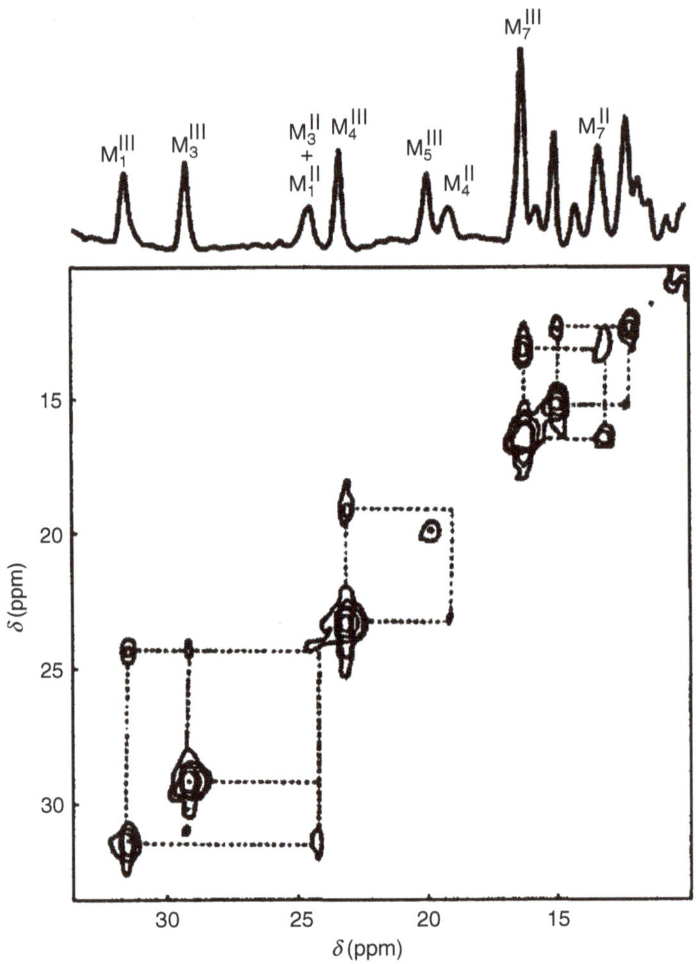

FIGURE 12.19 The first EXSY experiment on a paramagnetic system. The 300 MHz spectrum, taken with mixing time of 50 ms, shows species in chemical exchange belonging to two different redox states of a cytochrome c_3, a protein containing four low spin hemes. The signals marked M_1–M_7 represent various heme methyl groups. EXSY cross peaks are observed between M_i of two species containing two (II) or three (III) oxidized hemes, respectively.

Reproduced with permission from Ref. [84].

$$D=\frac{1}{2}\left[\left(L^A-L^B\right)^2+4k_{-1}k_1\right]^{1/2} \qquad \rho=\frac{\left(L^A+L^B\right)}{2}$$
$$L^A=-\rho^A-k_1 \qquad L^B=-\rho^B-k_{-1}$$

However, if the aim is only that of measuring τ_M, it is more straightforward to perform a single steady state 1D experiment, as explained in Section 6.3.4. An EXSY experiment is shown in Fig. 12.20, relative to the complex praseodymium diethylenetriaminepentaacetate ($(Pr(DTPA)^{2-})$) [86,87]. The

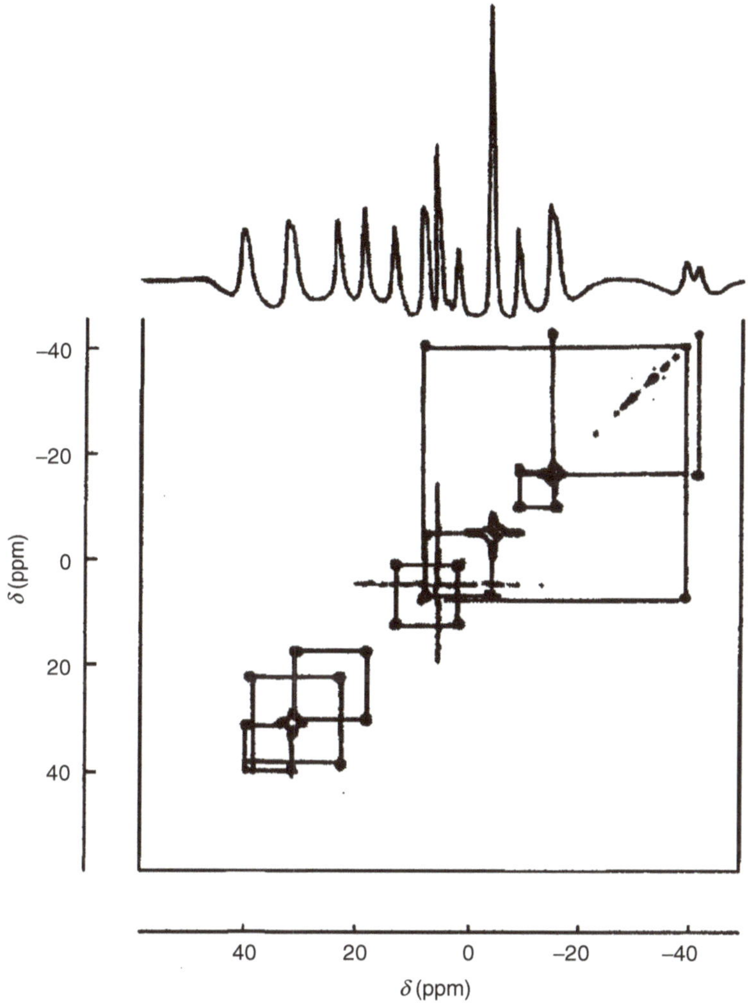

FIGURE 12.20 500 MHz EXSY spectrum of $Pr(DTPA)^{2-}$ recorded with the pulse sequence of Fig. 12.16A.The spectrum demonstrates the presence of an equilibrium between two conformational isomers.

Reproduced with permission from Ref. [86].

complex undergoes chemical exchange between two conformational isomers. The T_1 values of the signals are around 30 ms.

EXSY experiments have the advantage of displaying cross peaks between many signals belonging to two or more complex species. In Fig. 12.21, the EXSY spectrum of a ferredoxin containing two Fe_4S_4 clusters is shown [89]. Here, the cross peaks connect the fully reduced species containing two $Fe_4S_4^+$ clusters with the intermediate species containing two "$Fe_4S_4^{1.5+}$" species and the oxidized species containing two $Fe_4S_4^+$ species. The intermediate species actually contains one oxidized and one reduced cluster, which, however, are in very fast exchange so that, as far as NMR is concerned, only

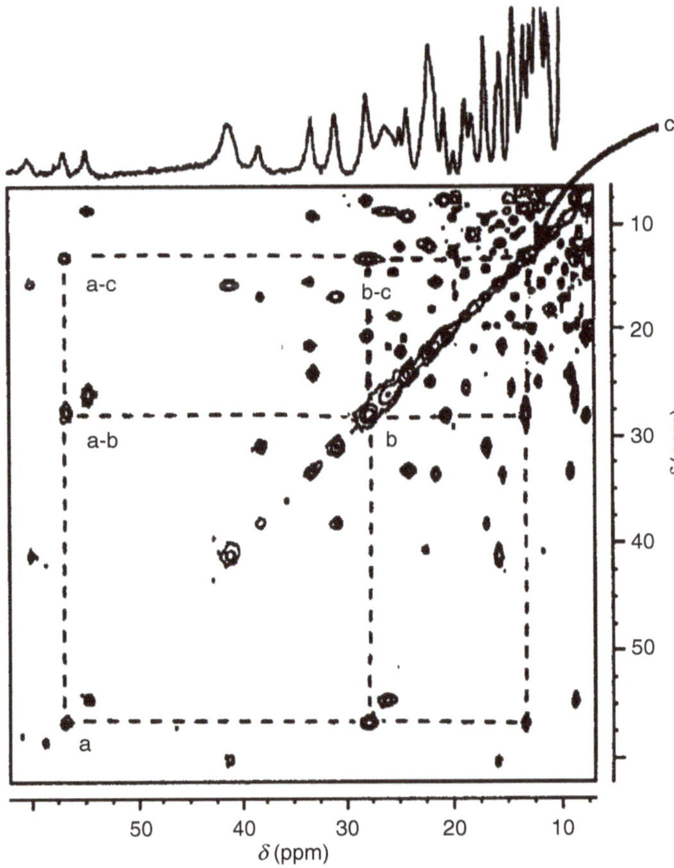

FIGURE 12.21 600 MHz EXSY spectrum of the intermediate reduction product of the $2Fe_4S_4$ ferredoxin from *C. pasteurianum*. Most signals of the intermediate species display cross peaks from both the fully reduced and the fully oxidized species. The sequence used is that in Fig. 12.17A. t_m = 5 ms. A pattern belonging to a single signal exchanging among the three species ((a) fully reduced; (b) half reduced; (c) fully oxidized) is highlighted as an example.

one intermediate species exists. It is noteworthy that we can detect as many as 20 signals of the intermediate species showing two cross peaks connecting them with those of the other two species. One of these patterns is exemplified in Fig. 12.21. Signal (b), arising from the $Fe_4S_4^{1.5+}$ species, is connected to both signal (a), belonging to the fully reduced species, and signal (c), belonging to the fully oxidized species. Signals (a) and (c) are also connected by a "two-step" exchange cross peak, (a)–(c), whose nature is conceptually related to the spin diffusion mechanism described in Section 12.2.

EXSY cross peaks are also obtained in TOCSY experiments (see later) because scalar interactions in the rotating frame are not separable from exchange interactions [90]. An EXSY experiment, performed using a TOCSY sequence (see in further sections) is reported in Fig. 12.22 relative to the

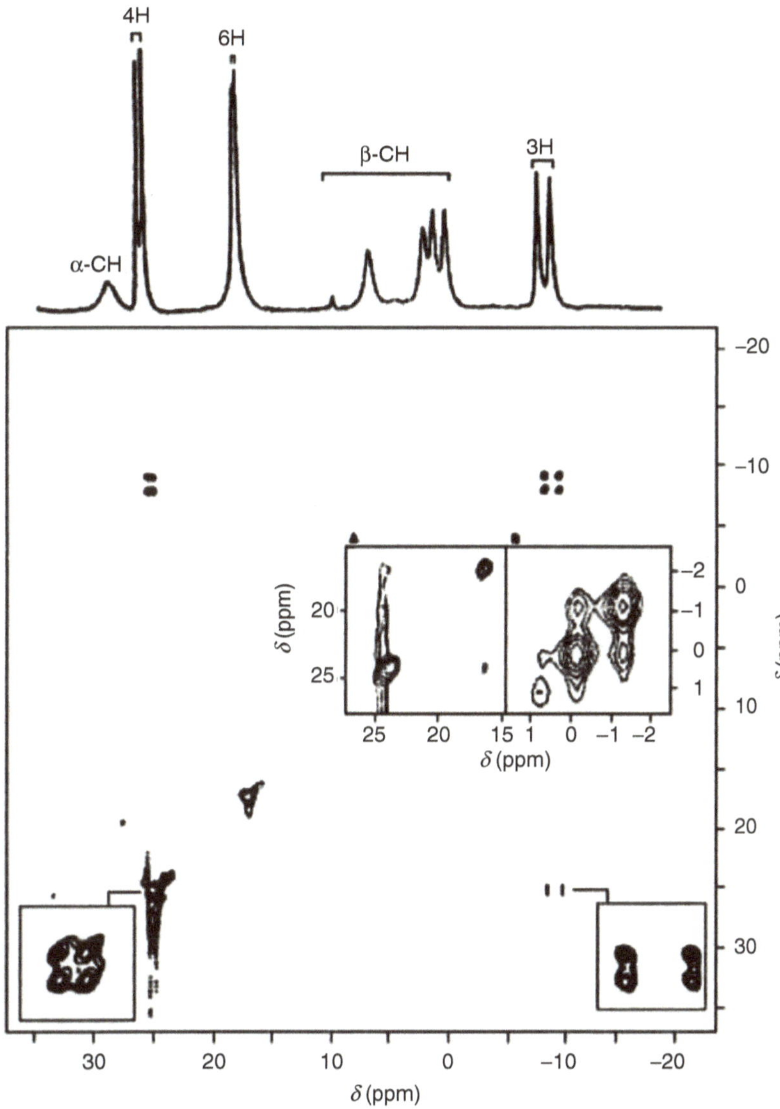

FIGURE 12.22 TOCSY spectrum of the complex 5Cl–NiSAL–MeDPT (scheme in Fig. 12.23) showing exchange cross peaks between the two 4H (T_1 = 5 ms) and between the two 3H (T_1 = 37 ms) salicylaldiminate ring protons. The spin-lock time was 20 ms.

Reproduced with permission from Ref. [88].

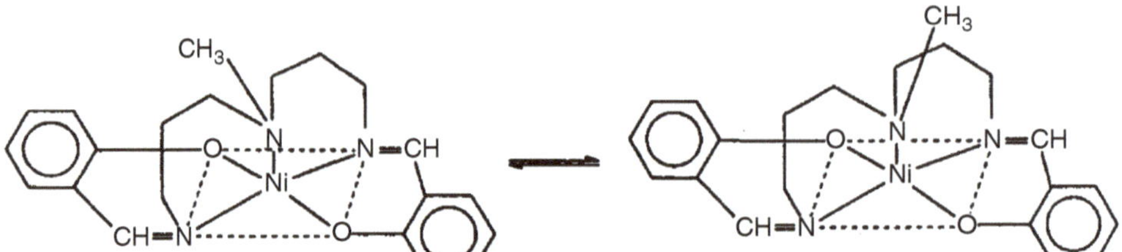

FIGURE 12.23 Conformational equilibrium displayed by 5Cl–Ni–SAL–MeDPT [91]. The two halves of the molecule are non-equivalent owing to the CH_3 substituent on the apical nitrogen. The equilibrium involves two equivalent conformations differing in the chirality of the apical nitrogen.

complex 5Cl–Ni–SAL–MeDPT [88]. This complex, as shown in Fig. 12.23, displays a chemical equilibrium in which the two salicylaldiminate moieties exchange their nonequivalent positions [91]. It is interesting to learn that such complex interconversion occurs with times of the order of the spin-lock time (20 ms) or shorter.

As EXSY cross peak intensities can be as strong as the diagonal peak intensities in favorable cases, EXSY experiments can be performed with relative ease also on nuclei other than protons. EXSY experiments are also sometimes performed with ROESY sequences (see in further sections).

NOESY (Nuclear Overhauser Effect Spectroscopy) relies upon cross-relaxation, and it is among the most widely used experiments for (macro)molecular structure determination. Experiments reveal dipolar interactions between nuclei. Success of the NOESY experiment is thus easily achieved for nuclei with large magnetic moments like protons, or for heteronuclei when the dipole–dipole interaction is very strong.

The basic pulse sequence for the NOESY experiment is just that illustrated in Fig. 12.2K. In the mixing period t_m magnetization transfer occurs between sets of nuclei I and J which are dipole–coupled. After any given t_1 value and the second 90° pulse, the magnetization of the two sets of spins will be along z with different components. The situation is the same as that illustrated for the EXSY experiment of Fig. 12.18, if one substitutes the labels A and B for the two chemically exchanging sites with I and J for the two sets of nuclei. Of note, a positive NOE corresponds to negative NOESY cross peaks and vice versa. This is not surprising if it is remembered that when the NOE is negative, it has the same sign of the irradiated peak in the difference spectra. When τ_r is small, the NOESY cross peaks are small and negative (i.e., of opposite sign with respect to the diagonal peaks). With increasing τ_r they increase in absolute value and then decrease again until the zero value of the cross peak intensity is obtained. Then, the absolute intensities of positive cross peaks increase. This explains why NOESY experiments are best performed on macromolecules, unless very large concentrations can be reached. The best t_m is of the order of T_1. In the case in which the cross peak occurs between signals with sizably different T_1 values t_m can be optimized according to the following equation:

$$t_m^{opt} = \frac{1}{2D} \ln\left(\frac{\rho - D}{\rho + D}\right) \qquad (12.7)$$

where the symbols have the same meaning as in Eq. (12.6). A table is also provided for the reader's convenience (Table 12.4), which reports the best mixing time values for each combination of T_1^I and T_1^J $\left(\rho_I^{-1} \text{ and } \rho_J^{-1}\right)$ values for the pair [92]. Note that when the two T_1 values are very different, t_m^{opt} is

Table 12.4 Optimal Homonuclear NOESY Mixing Times t_m^{opt} for Various T_1 (ρ^{-1}) Values (ms) of the Two Spins I and J, Calculated Using Eq. (12.7)

	T_1^J							
T_1^I	**128**	**64**	**32**	**16**	**8**	**4**	**2**	**1**
128	128 (4.72)	89.0 (3.20)	59.2 (2.02)	38.0 (1.19)	23.7 (0.67)	14.3 (0.36)	8.5 (0.19)	4.9 (0.10)
64	89.0 (3.20)	64.0 (2.36)	44.4 (1.60)	29.6 (1.01)	19.0 (0.59)	11.8 (0.33)	7.2 (0.18)	4.2 (0.09)
32	59.2 (2.02)	44.4 (1.60)	32.0 (1.18)	22.2 (0.80)	14.8 (0.50)	9.5 (0.30)	5.9 (0.17)	3.6 (0.09)
16	38.0 (1.19)	29.6 (1.01)	22.2 (0.80)	16.0 (0.59)	11.1 (0.40)	7.4 (0.25)	4.8 (0.15)	3.0 (0.08)
8	23.7 (0.67)	19.0 (0.59)	14.8 (0.50)	11.1 (0.40)	8.0 (0.29)	5.5 (0.20)	3.7 (0.13)	2.4 (0.07)
4	14.3 (0.36)	11.8 (0.33)	9.5 (0.30)	7.4 (0.25)	5.5 (0.20)	4.0 (0.15)	2.8 (0.10)	1.8 (0.06)
2	8.5 (0.19)	7.2 (0.18)	5.9 (0.17)	4.8 (0.15)	3.7 (0.13)	2.8 (0.10)	2.0 (0.07)	1.4 (0.05)
1	4.9 (0.10)	4.2 (0.09)	3.6 (0.09)	3.0 (0.08)	2.4 (0.07)	1.8 (0.06)	1.4 (0.05)	1.0 (0.04)

Cross peaks intensities (%) for a $\sigma_{I(J)}$ value of $-1\ s^{-1}$ are shown in parentheses.

more than four times longer than the shorter T_1 value. As in EXSY experiments, the maximal information content is in the first t_1 and t_2 points. Again, the most used weighting functions are cos- or $\cos^2$-type.

NOESY experiments may sometimes be a hard task for small molecules, because in the fast motion limit cross relaxation is small and so is the NOE. In Table 12.5 the calculated ^1H–^1H maximal NOESY intensities are reported for a range of τ_r values and R_{1M} values (taken equal for the two spins). From the τ_r values the corresponding relative molecular masses are calculated through the Stokes–Einstein equation [Eq. (4.8)] for $\eta = 1$. The mixing time is always equal to ρ_I ($= \rho_{I(J)} + R_{1M}$) to maximize the effect. The NOEs are reported for two magnetic fields and for two proton–proton distances of 1.8 and 3.0 Å. The purpose of this table is to show under which circumstance NOESY spectra of paramagnetic molecules can be attempted. The crossing point between fast and slow motion regimes occurs at $M_r \approx 500$ at 1000 MHz, $M_r \approx 800$ at 600 MHz, and at $M_r \approx 2500$ at 200 MHz. The maximal intensities in the fast motion regime are always small in absolute value and therefore the observability of cross peaks is predictable but critical. The researcher dealing with low molar mass compounds may either use low fields or increase τ_r by increasing solvent viscosity and use the highest possible field. Rotating frame NOE (ROE) can also assist in this sense. A ROE experiment measures the transfer of transverse

Table 12.5 Calculated Maximal Intensities of NOESY Cross Peaks (%)[a] for Various Values of τ_r and R^I_{1M} (= R^I_{1M})

τ_r (s) $[M_r]$	r_{IJ} (Å)	R^I_{1M} (s^{-1}): 1, ν_0 (MHz)		2, ν_0 (MHz)		4, ν_0 (MHz)		8, ν_0 (MHz)		16, ν_0 (MHz)		32, ν_0 (MHz)		64, ν_0 (MHz)		128, ν_0 (MHz)		256, ν_0 (MHz)	
		200	600	200	600	200	600	200	600	200	600	200	600	200	600	200	600	200	600
10^{-11}	1.8	−2.640	−2.626	−1.421	−1.413	−0.739	−0.735	−0.377	−0.375	−0.190	−0.189	−0.096	−0.095	−0.048	−0.048	−0.024	−0.024	−0.012	−0.012
[25]	3.0	−0.143	−0.142	−0.072	−0.071	−0.036	−0.036	−0.018	−0.018	−0.009	−0.009	−0.004	−0.004	−0.002	−0.002	−0.001	−0.001	−0.001	−0.001
$10^{-10.5}$	1.8	−6.366	−6.089	−3.835	−3.647	−2.138	−2.025	−1.135	−1.072	−0.585	−0.552	−0.297	−0.280	−0.150	−0.141	−0.075	−0.071	−0.038	−0.036
[80]	3.0	−0.440	−0.416	−0.223	−0.210	−0.112	−0.106	−0.056	−0.053	−0.028	−0.027	−0.014	−0.013	−0.007	−0.007	−0.004	−0.003	−0.002	−0.002
10^{-10}	1.8	−11.236	−8.248	−8.045	−5.587	−5.141	−3.400	−2.989	−1.909	−1.628	−1.017	−0.852	−0.526	−0.436	−0.267	−0.221	−0.135	−0.111	−0.068
[250]	3.0	−1.244	−0.772	−0.644	−0.396	−0.328	−0.200	−0.165	−0.101	−0.083	−0.051	−0.042	−0.025	−0.021	−0.013	−0.010	−0.006	−0.005	−0.003
$10^{-9.5}$	1.8	−11.920	0.890	−9.679	0.616	−7.045	0.381	−4.568	0.216	−2.685	0.116	−1.472	0.060	−0.774	0.031	−0.397	0.015	−0.201	0.008
[800]	3.0	−2.098	0.088	−1.127	0.045	−0.585	0.023	−0.298	0.012	−0.151	0.006	−0.076	0.003	−0.038	0.001	−0.019	0.001	−0.010	0.000
10^{-9}	1.8	2.343	19.647	1.934	14.379	1.434	9.406	0.945	5.581	0.562	3.084	0.310	1.629	0.164	0.838	0.084	0.425	0.043	0.214
[2500]	3.0	0.441	2.366	0.238	1.233	0.124	0.630	0.063	0.319	0.032	0.160	0.016	0.080	0.008	0.040	0.004	0.020	0.002	0.010
$10^{-8.5}$	1.8	27.084	33.735	23.302	28.378	18.275	21.644	12.831	14.784	8.080	9.111	4.657	5.176	2.525	2.782	1.319	1.446	0.674	0.738
[8000]	3.0	6.464	7.238	3.619	4.005	1.926	2.118	0.996	1.091	0.506	0.554	0.255	0.279	0.128	0.140	0.064	0.070	0.032	0.035
10^{-8}	1.8	38.951	39.972	36.394	37.299	32.208	32.938	26.264	26.779	19.296	19.610	12.702	12.872	7.591	7.677	4.220	4.262	2.238	2.259
[25000]	3.0	16.387	16.633	10.331	10.459	5.969	6.033	3.244	3.275	1.697	1.712	0.869	0.876	0.440	0.443	0.221	0.223	0.111	0.112
$10^{-7.5}$	1.8	42.060	42.175	41.073	41.183	39.238	39.339	36.040	36.127	31.041	31.109	24.401	24.446	17.209	17.235	10.909	10.923	6.332	6.339
[80,000]	3.0	28.398	28.455	21.347	21.383	14.373	14.393	8.754	8.764	4.932	4.937	2.638	2.640	1.367	1.368	0.696	0.697	0.352	0.352
10^{-7}	1.8	42.884	42.896	42.554	42.565	41.909	41.920	40.678	40.689	38.432	38.442	34.639	34.647	28.998	29.004	21.987	21.991	14.933	14.935
[250000]	3.0	37.050	37.059	32.481	32.488	26.145	26.150	18.928	18.931	12.291	12.293	7.268	7.269	4.012	4.012	2.118	2.119	1.090	1.090

For each τ_r and R^I_{1M} value two NOE's are calculated for interproton distances of 1.8 and 3.0 Å and for nuclear Larmor frequencies ν_0 of 200 and 600 MHz

[a]*The largest theoretical cross peak intensity for a NOESY experiment is 50%, at variance with a transient NOE experiment (Appendix VII). To predict the maximal values for transient NOE experiments the values of this table should be multiplied by two and the sign should be reversed. Cf. also Table 12.1 for steady state NOE experiments.*

magnetization because of cross-relaxation within a pair of spins when spin-locked [93]. The spin-locking field is usually applied off-resonance so as to avoid TOCSY-like transfers (see in further sections). The process is similar to the transient NOE (Appendix VII), but its extent depends on the rotating frame relaxation rate (ρ_ρ) and the rotating frame cross-relaxation rate (σ_ρ), defined as:

$$\rho_{\rho I(J)} = \left(\frac{\mu_0}{4\pi}\right)^2 \frac{\hbar^2\gamma_I^2\gamma_J^2 J(J+1)}{30r_{IJ}^6}\left[\tau_c + \frac{9\tau_c}{1+(\omega_{1I}+\omega_{1J})^2\tau_c^2} + \frac{18\tau_c}{1+\omega_I^2\tau_c^2} + \frac{12\tau_c}{1+(\omega_I+\omega_J)^2\tau_c^2}\right]. \tag{12.8a}$$

$$\sigma_{\rho I(J)} = \left(\frac{\mu_0}{4\pi}\right)^2 \frac{\hbar^2\gamma_I^2\gamma_J^2 J(J+1)}{30r_{IJ}^6}\left[-\tau_c + \frac{9\tau_c}{1+(\omega_{1I}+\omega_{1J})^2\tau_c^2} + \frac{12\tau_c}{1+(\omega_I+\omega_J)^2\tau_c^2}\right]. \tag{12.8b}$$

ROE has some disadvantages with respect to NOE experiments, because of the need of spin-locking signals over a broad spectral width, and because of its lesser efficiency; however it does not have any zero-crossing (because $\omega_I\tau_c \ll 1$, $\sigma_\rho > 0$) and this may constitute a precious help for the measurement of dipolar interactions in small paramagnetic complexes. As an example of a system in the fast motion regime, a portion of the ROESY spectrum of a small lanthanide diporphyrin complex is shown in Fig. 12.24, together with portions of the NOESY, TOCSY, and HMQC spectra [94]. The homonuclear correlation subspectra refer to phenyl rings attached in the meso position to one of the two coordinated porphyrins (Fig. 12.24, inset). The flip rate of the rings is slow on the chemical shift scale but fast on the relaxation time scale. Therefore, separate signals are observed for the exo and endo and ortho and meta ring protons. Strong and positive EXSY cross peaks are apparent in ROESY and NOESY spectra (as well as in the TOCSY spectrum, Section 12.3.3), whereas all dipole–dipole cross peaks are negative in the ROESY and NOESY spectra. In the latter, only one dipole–dipole cross peak is observed, on account of the rotational correlation time of the system being close to the null point for the NOE interaction [94].

As discussed in Section 12.2, spin diffusion is not a problem in small complexes. In macromolecules, the problems are negligible when the nuclear R_{1M} values are much larger than cross relaxation. Such problems become increasingly significant when the R_{1M} values get smaller. Under these circumstances it may be useful to perform the NOESY experiments at different mixing times. At short mixing times (short relative to T_1) only primary NOEs are detected whose intensities are proportional to σ. At increasing mixing time, the build up of these signals can be followed and new cross peaks may appear. In macromolecular systems the best conditions for measuring NOESY cross peaks are met, because τ_r is long (Table 12.5). In Fig. 12.25, the NOESY spectrum of an Fe_4S_4 protein of M_r 6000 is shown where cross peaks are apparent between signals with T_1 values of a few milliseconds [95]. With mixing time of the order of few milliseconds NOESY cross peaks are observed between signals with short T_1, whereas NOESY cross peaks are observed between signals with longer T_1 values if longer mixing times are used. The observation of cross peaks between fast relaxing and slow relaxing signals can be accomplished with mixing times appropriately chosen according to Table 12.4. The difference in T_2, which often parallels the difference in T_1, provides elliptical cross peaks (as shown in Fig. 12.26A). The maximal value of t_1 $\left(t_1^{\max}\right)$ is best chosen to be equal to or less than one of the two T_2 values, and the maximal value of t_2 $\left(t_2^{\max}\right)$ to be equal to or less than that of the other. Under these circumstances, one of the two cross peaks on one side of the diagonal is enhanced whereas the other is sacrificed (Fig. 12.26B,C). The difficulty, which is a limit of the technique, is that it is often not possible to see in the same 2D map cross peaks arising from signals with strongly different T_1 and T_2 values. If reference

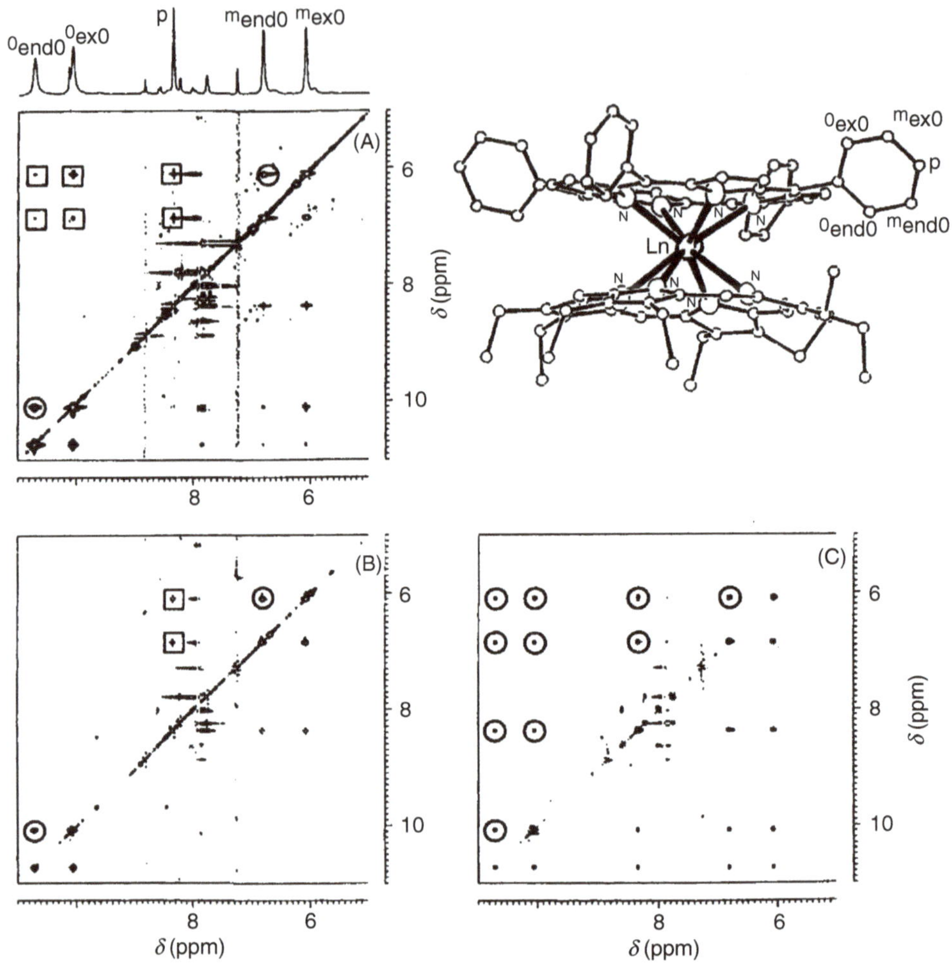

FIGURE 12.24 ROESY (A), NOESY (B) and TOCSY (C) cross peaks observed in the phenyl ring 2D pattern of the complex shown in the inset (Ln = Yb^{3+}). Negative cross peaks are highlighted by squares, and positive cross peaks by circles.

Adapted from Ref. [94].

is made to the system of Fig. 12.25 we see that only some cross peaks are evident between β-CH_2 protons of cysteines and between β-CH_2 and β-CH protons. If some other cross peaks are observed between β-CH_2 and other more diamagnetic protons, then cross peaks between the latter and diamagnetic protons are absolutely missing. When we increase the mixing time in order to detect the latter part, the cross peaks with the β-CH_2 are sizably reduced in intensity and will be eventually lost. The procedure used up to now is that of firmly recognizing signals farther from the paramagnetic center to which the β-CH protons are dipolarly connected (mostly through 1D NOE), and then try to relate them to known dipolar patterns observed in the diamagnetic region of NOESY spectra recorded with longer t_m, t_1^{max} and t_2^{max}. Please recall that we are implying that the spectra of the "diamagnetic" signals in the protein can be easily observed by routine experiments. When setting a NOESY experiment on a

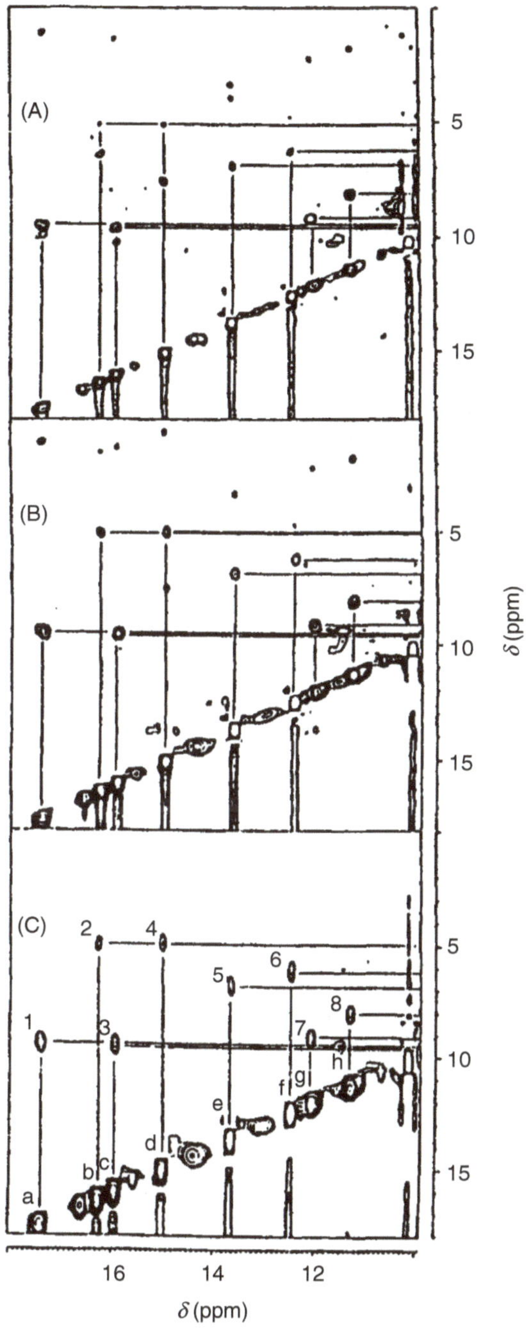

FIGURE 12.25 NOESY spectra of the oxidized form of 2(Fe_4S_4) ferredoxin from *C. pasteurianum* taken using mixing times of 20 (A), 10 (B) and 5 (C) ms. Signals (a)–(h) belong to β-CH protons of cluster-coordinated cysteines and cross peaks 1–8 connect them to their geminal partners, which lie in the diamagnetic region of the spectrum. The other cross peaks arise from α-CH protons of the cysteines and from other nearby residues. The cross peaks between the fastest relaxing signals (β-CH protons with their geminal partners) are maximal at the shortest mixing time, whereas those between the former and more slowly relaxing signals increase at longer mixing times.

Reproduced with permission from [95].

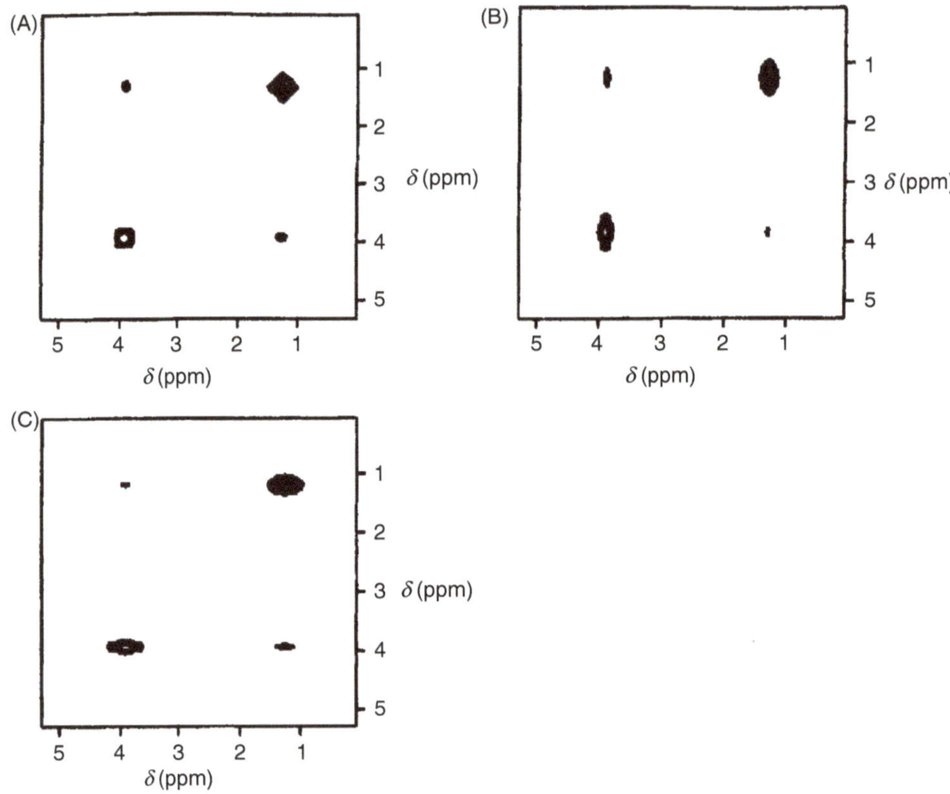

FIGURE 12.26 Computer-simulated NOESY cross peaks between signals with different T_2 values. Conditions: $\rho_1^I + \rho_1^J = 40\,\mathrm{s}^{-1}$; $\rho_2^I = 50\,\mathrm{s}^{-1}$; $\rho_2^J = 200\,\mathrm{s}^{-1}$; $\sigma_{IJ} = 5\ \mathrm{s}^{-1}$. (A) $t_1^{max} = t_2^{max} = 0.01\,\mathrm{s}$; (B) $t_1^{max} = T_2^I$, $t_2^{max} = T_2^J$; (C) $t_1^{max} = T_2^J$, $t_2^{max} = T_2^I$. Note that the upper left cross peak has maximal intensity in case (B), the lower right cross peak has maximal intensity in case (C), whereas in case A the two intensities are equal and intermediate between the larger and the smaller of cases (B) and (C). For simplicity, $t_m = T_1^I = T_1^J$ has been taken. A $\cos^2$ weighting function has been applied.

paramagnetic molecule the mixing time t_m should be of the order of the T_1 of the signals between which cross peaks are looked for. When the T_1 values are very different, and this is common in paramagnetic molecules, especially if there are protons close to the paramagnetic center and protons far from it, the best mixing time should be calculated case by case from Eq. (12.7). For the reader's convenience, Table 12.4 is provided, with the relative T_1 values spanning two orders of magnitude. Note that, when the T_1 values are very different, the best mixing time, which is always intermediate between the two T_1 values, can be as large as five times the shorter T_1.

A comment is due to the comparison of the intensity of a steady state NOE with that of a transient NOE in the case of signals with different T_1 values. This is because the intensity of a NOESY cross peak is related to the intensity of a transient NOE (Appendix VII) and because in paramagnetic systems we may be at the lower limit of cross peak detectability. Therefore, a steady state NOE obtained by saturating the fast relaxing signal is always advisable to extend the detectability of its dipolar connectivities

Table 12.6 Comparison Between Steady State (SS) and Maximal Transient (TR) NOE Intensities (%) for Various T_1 (ρ^{-1}) Values (ms) of the Two Spins *I* and *J*, Calculated Using Eqs. (4.52) and (7.26) (the negative signs are omitted)

		T_1^J							
T_1^I		**128**	**64**	**32**	**16**	**8**	**4**	**2**	**1**
128	SS	12.80	6.40	3.20	1.60	0.80	0.40	0.20	0.10
	TR	9.44	6.40	4.03	2.38	1.33	0.72	0.37	0.19
64	SS	12.80	6.40	3.20	1.60	0.80	0.40	0.20	0.10
	TR	6.41	4.71	3.20	2.02	1.19	0.67	0.36	0.19
32	SS	12.80	6.40	3.20	1.60	0.80	0.40	0.20	0.10
	TR	4.03	3.20	2.35	1.60	1.01	0.59	0.33	0.18
16	SS	12.80	6.40	3.20	1.60	0.80	0.40	0.20	0.10
	TR	2.38	2.02	1.60	1.18	0.80	0.50	0.30	0.17
8	SS	12.80	6.40	3.20	1.60	0.80	0.40	0.20	0.10
	TR	1.33	1.19	1.01	0.80	0.59	0.40	0.25	0.15
4	SS	12.80	6.40	3.20	1.60	0.80	0.40	0.20	0.10
	TR	0.72	0.67	0.59	0.50	0.40	0.29	0.20	0.13
2	SS	12.80	6.40	3.20	1.60	0.80	0.40	0.20	0.10
	TR	0.37	0.36	0.33	0.30	0.25	0.20	0.15	0.10
1	SS	12.80	6.40	3.20	1.60	0.80	0.40	0.20	0.10
	TR	0.19	0.19	0.18	0.17	0.15	0.13	0.10	0.07

The calculations are performed for a $\sigma_{I(J)}$ value of -1 s^{-1}. Steady state NOE is always superior to transient NOE, except when T_1^I of the irradiated signal is more than twice the T_1^J of the responding signal (upper right part of the table).

as much as possible. Table 12.6 shows the advantage of steady state versus transient NOE for a range of ratios of T_1 values. It appears that steady state NOE is always superior to transient NOE [compare Eqs. (7.26) and (4.52)], except when the T_1 of the irradiated signal is longer than that of the responding signal by more than a factor of two. In paramagnetic metalloproteins, the faster relaxing signals are often outside the diamagnetic envelope and are separated one from the other, so that they can be irradiated quite easily. The slower relaxing signals are often in the diamagnetic envelope. It is therefore a good practice to measure the steady state NOE by irradiating each of them prior to performing 2D experiments. Of course, NOE experiments are advantageous over NOESY experiments if one is looking for dipolar connectivities from only a few specific signals.

Sometimes the shifts of a few signals are so large that it is inconvenient to include them into the spectral window because this will either decrease the resolution or, if resolution is maintained by increasing the data points in the two dimensions, cause cumbersome spectral manipulations. Then, it may be convenient to use a smaller spectral width and allow the excluded signals to appear as folded

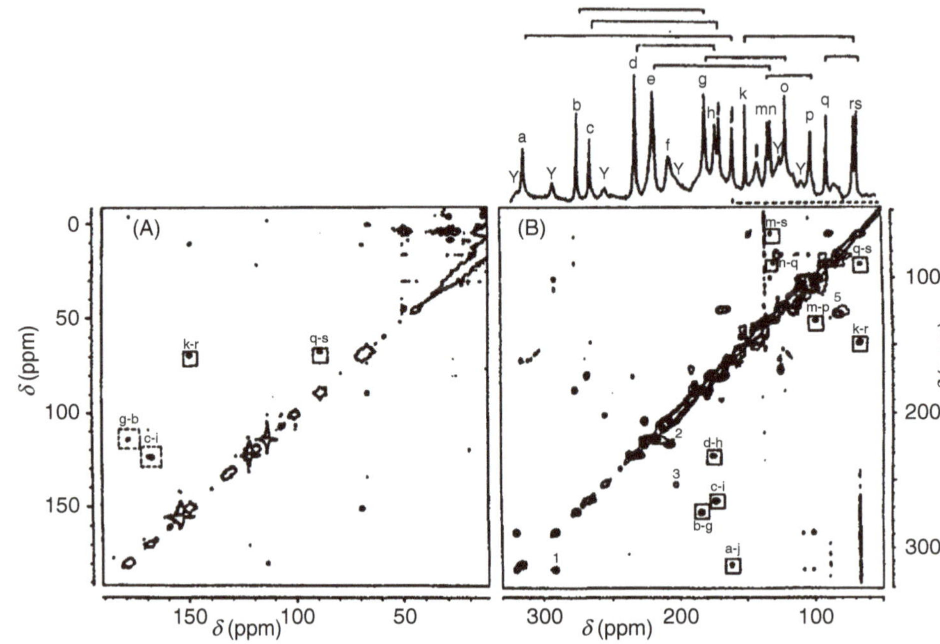

FIGURE 12.27 600 MHz NOESY spectra of the four-cobalt cluster of cobalt-substituted metallothionein. The spectra have been recorded with t_m = 7 ms using a specially built probe featuring a 90° pulse of 3.7 μs. (A) Spectrum obtained with a 125 kHz 16 bit ADC by folding the spectrum in both dimensions; cross peaks g–b and c–i are folded. (B) Spectrum recorded using a fast ADC over a 250 kHz spectral width. The nine observed connectivities between geminal protons of cobalt-coordinated cysteine β-CH_2 are summarized in the inset. Cross peaks marked with numbers are EXSY cross peaks to a minor species labeled Y in the inset.

Reproduced with permission from Ref. [97].

images, or "ghost" signals. This can be easily accomplished in the f_1 dimension by increasing the time increment in t_1 between one experiment and the next [96]. It can also be accomplished in the f_2 dimension by likewise increasing the dwell time, provided the filter width is chosen large enough to include all the signals. Ghost signals give "ghost" cross peaks. An example of this is illustrated in Fig. 12.27A, showing the 600 MHz NOESY spectrum of the four-cobalt cluster of Co_7MT, with MT indicating the protein metallothionein ($M_r \approx 6000$) [97].

Sometimes the signals are outside the maximal spectral width allowed by the ADC of the instrument, situation which is usually coped with recovering a folded spectrum is a necessity. In older instruments, an ADC of the latest generation (Section 12.1) may not available, in this case another possibility is that of using an ADC with smaller dynamic range but larger spectral window. Fig. 12.27B shows the 600 MHz NOESY spectrum of the above protein complex [97] covering a spectral width of 250 kHz, recorded with a 12 bit ADC. At least nine out of the 11 expected cross peaks between geminal protons of cysteine β-CH_2 (see inset of Fig. 12.27) are observed. Some of the signals giving rise to cross peaks have T_1 values as short as 1 ms.

Of course, these considerations regarding the folded spectra and decreased dynamic range of the ADC hold for any kind of 2D spectroscopy. We have mentioned them here because such problems happen to have been afforded in the NMR literature on paramagnetic compounds in connection with NOESY spectra.

In summary, when performing a NOESY spectrum of a paramagnetic molecule containing both fast and slow relaxing nuclei, it is convenient: (1) to record a NOESY spectrum with a mixing time which matches the short relaxation times in order to detect connectivities between fast relaxing signals; (2) to record a NOESY spectrum with a mixing time which matches the relaxation times of the slow relaxing nuclei in order to detect connectivities between slow relaxing signals; (3) to record a NOESY with a mixing time taken from Table 12.4 depending on the T_1 of the fast and slow relaxing nuclei in order to detect connectivities between fast and slow relaxing signals; and/or (4) to measure the steady state NOEs by saturating the fast relaxing signals in order to be sure to detect connectivities with slow relaxing nuclei as much as possible.

A further parameter in measuring NOESY spectra is the recycle time. In small molecules it should be ideally five times the longest T_1 of the signals for which connectivities are looked for, to let the nuclei of interest recover their magnetization. In macromolecules such time is better set approximately equal to the longest T_1 among the signals of interest: the loss in cross peak intensity is overcome by the gain in number of cycles within the same experimental time. Of course, this intrinsically means increased sensitivity for observation of connectivity between the fast relaxing nuclei in a paramagnetic system. When recording a NOESY spectrum with a short recycle time, the transverse magnetization of the slow relaxing signals has no time to disappear before the next scan; this gives rise to various artifacts in the 2D spectrum. For this reason, a continuous wave (CW) pulse of the duration of some milliseconds, called trim pulse, can be used to destroy transverse coherence before the following experiment. A convenient pulse sequence is shown in Fig. 12.2K. The intensity of the pulse should be set according to the spectral region over which the trim is needed. Typically, trim pulses of the order of 1 ms are used for paramagnetic systems.

In a NOE–NOESY experiment (Section 12.3.4), the connectivities between a fast relaxing signal and a set of slow relaxing signals are obtained simultaneously with those within the above set of slow relaxing signals. Therefore the NOESY parameters should be set according to the slow relaxing signals.

NOESY experiments aimed at identifying dipolar connectivities between bound water molecules and protein protons have been developed in the late 1990s [98]. Among them, the ePHOGSY experiment [99] has been successfully applied on paramagnetic metalloproteins [100,101].

When the relaxation is too severe to allow for direct observation of protons, a case that can easily occur when the molar mass is large and in the case of slow relaxing paramagnetic centers, ^{13}C-^{13}C NOESY experiments can be performed as well. These experiments are particularly suitable for the observation of amino acid side chains in high-molar mass systems [102]. For instance, a ^{13}C-based strategy exploiting the paramagnetic broadening due to the presence of iron(III) species was used to identify the binding sites of ferric species in ferritin (M_r ~500,000) and to trace their pathway to the central cavity [45].

12.3.2 SOLUTION NMR—COSY

This experiment has been extensively used in the past, but is still a valuable help, especially when dealing with strongly paramagnetic signals, as it is the simplest possible 2D sequence, and easily shows otherwise inaccessible correlations (sometimes even too many... see later). If two signals A and B are coupled by scalar interactions, a cross peak is expected. Under these conditions each signal of the

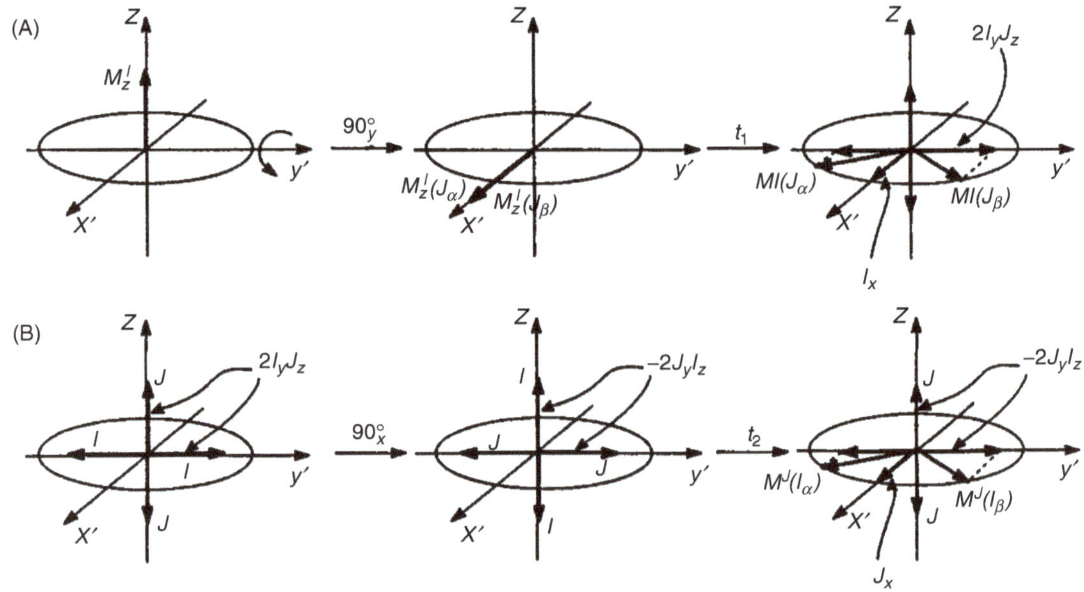

FIGURE 12.28 Vector representation of a COSY experiment. The first 90° pulse along y' rotates the equilibrium magnetization of the I and J spins from the z axis to the x' axis. During t_1, the transverse doublet components of, for example, signal I (assumed to process at the rotating frame frequency) separate. Their x' projections are in phase, while their y' projections are in antiphase. Antiphase coherence is associated with the two opposite components of the J spin (hence the label $2I_yJ_z$ for the resulting four-vector pattern). The second 90° pulse along x' interchanges the I and J spins, producing a $-2J_xI_z$ antiphase product operator. During t_2 the latter generates a detectable IP coherence of the J spin, whose initial intensity had been modulated by the I spin Larmor frequency during t_1, thus originating a cross peak.

two-spin system I–J is split into two components, whose separation in frequency is determined by the scalar coupling constant J_{IJ}. The simplest COSY pulse sequence is 90—τ—90 acquisition. There is an initial 90° pulse followed by the evolution time t_1, during which antiphase coherence of the scalar-coupled spins builds up. The concept of antiphase coherence is recalled in Fig. 12.28. The coherence of the two components of, for example, the I doublet, rotates in the xy plane at different frequency. In a frame rotating at the frequency of the center of the doublet, the two components will rotate in opposite directions with frequency $\pm J_{IJ}/2$. The two inplane components arise from I spins whose scalar coupled partners have opposite M_z components (α and β). The projections of the two I components on the rotating frame axis on which the magnetization was tilted by the first 90° pulse (e.g., the x'-axis) oscillate in phase (Fig. 12.28A) while those on the y'-axis oscillate in antiphase. The inphase (IP) and antiphase projections are called single quantum IP and single quantum antiphase coherences. In the product operator language they are indicated as I_x and $2I_yJ_z$ (Appendix IX). The two are cyclically interconverted into one another during the time t_1.

When the second 90° pulse is applied, the antiphase magnetization of the I and J spins is interchanged (Fig. 12.28B). During t_2, the new antiphase coherence, $-2J_yI_z$, is again cyclically converted into a single quantum coherence J_x, which is a detectable magnetization of the J spin. The mixing

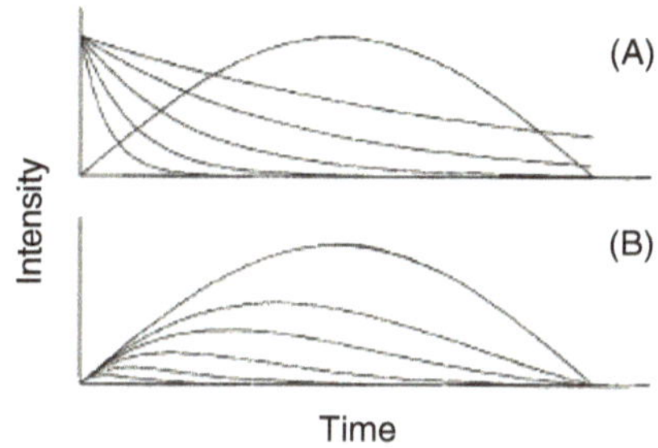

FIGURE 12.29 (A) A $\sin(\pi Jt)$ function and various exponential decay functions. (B) Composite functions obtained by the product of sin and exponential functions, describing the evolution of M_I, M_J and characterized by a progressive left shift of the maxima. The positions of the maxima are given by Eq. (12.10).

Reproduced with permission from Ref. [70].

formally occurs with the second 90°, and we have seen that the appropriate coherences must develop during both t_1 and t_2. If it were not for transverse relaxation, the antiphase coherence of the I spin would develop during t_1 proportionally to $\sin(\pi J_{IJ}t_1)$, and the maximal antiphase intensity would be obtained for $t_1 = 1/2J_{IJ}$. This sinusoidal dependence is, however, damped by a term $\exp(-\rho_2^I t_1)$ (Fig. 12.29). An analogous behavior is shown by the build up of the single quantum coherence J_x during t_2. Overall, the intensity of the cross peak builds up and decays according to the following relationship:

$$I(t) = \sin(\pi J_{IJ}t_1)\exp\left(-\rho_2^I t_1\right)\sin(\pi J_{IJ}t_2)\exp\left(-\rho_2^J t_2\right) \quad (12.9)$$

To a good approximation, ρ_2^I is the transverse relaxation rate of spin I and ρ_2^J is the transverse relaxation rate of spin J. The two cross peaks are distinguished according to which of the two T_2^I or T_2^J constants appears in each dimension. From Eq. (12.9), it appears that the maximum information is not contained in the first data points. In fact, considering that Eq. (12.9) represents the evolution of the signal, the derivative of I (t) with respect to t_1 (or t_2) provides the value of t_1 (or t_2) with maximum intensity of the signal:

$$t_{1(2)} = \frac{1}{\pi J_{IJ}}\operatorname{atan}\left(\frac{\pi J_{IJ}}{\rho_2^{J(I)}}\right) \quad (12.10)$$

For $T_2 \gg 1/(2J_{IJ})$, it is well known that the maximal intensity is obtained when the center of the t_1 and t_2 intervals is at $1/(2J_{IJ})$, i.e., for $t_1^{\max} = t_2^{\max} = 1/J_{IJ}$. For $T_2 \ll 1/(2J_{IJ})$, maximal intensity is obtained for $t_1^{\max} = 2T_2^I$ and $t_2^{\max} = 2T_2^J$. Of course, at these short time values, the sine build up of the interaction is small. Weighting functions of sin, $\sin^2$, or matched-filter-type are used to maximize the signal-to-noise ratio, of course also processing a different number of indirect points to enhance the S/N of individual crosspeaks. In any case, since the T_2 values in paramagnetic compounds are small, the cross peak intensity is drastically reduced.

The coherence transfer provides cross peaks which are antiphase for the various J_{IJ}-split components. The antiphase nature of the cross peaks then leads to partial or total cancellation of the cross peaks themselves, especially if they are phased in the absorption mode. This behavior can be simulated (Fig. 12.30) using appropriate treatments of the time evolution of the spin system, for instance using

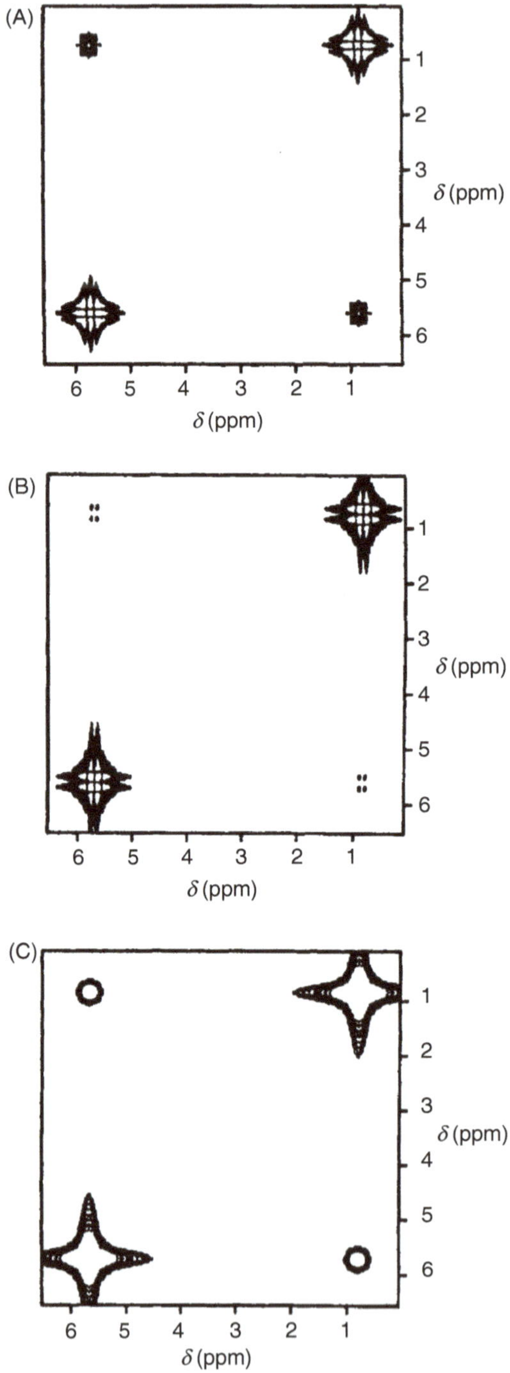

FIGURE 12.30 Calculated shapes of cross peaks in COSY spectra characterized by $T_2 < 1/2J_{IJ}$ (conditions: $J_{IJ} = 8$ Hz; $T_2 = 5$ ms; $t_1^{max} = T_2^{max} = 10$ ms). (A) Phase-sensitive spectra, phased in dispersion mode; (B) phase-sensitive spectra, phased in absorption mode; (C) magnitude mode spectra. Sin^2 weighting functions are used.

the density matrix formalism [103,104]. In the case in which T_2^{-1} is larger than J, one would think that the antiphase character of the cross peaks in the TPPI mode cancels partially or totally the cross peaks. Indeed, much of the intensity of the signal is lost. It has been pointed out that, if the cross peaks are phased in dispersion mode rather than in absorption mode, the loss of intensity is drastically reduced [105]. Accurate phasing may be difficult, however, when the cross peak intensity is small, as happens in paramagnetic compounds. Acquisition and Fourier transform in the so-called magnitude mode gives results that are of comparable quality but are absolutely insensitive to phasing. Indeed, in magnitude mode the intensity is given by

$$I = (A^2 + D^2)^{1/2} \tag{12.11}$$

where A and D stand for the pure absorption (or real) and pure dispersion (or imaginary) components of the signal respectively. In Fig. 12.31A part of the magnitude COSY spectrum of the pseudotetrahedral complex NiSAL–i-prop (scheme in the inset) is reported. COSY cross peaks are evident between adjacent protons in the aromatic ring. Although the lines in this case are not very broad, it appears that the phase-sensitive spectrum, routinely phased in absorption mode, has a higher resolution than the magnitude mode spectrum but is of lower quality (Fig. 12.31A,B). A better quality is obtained by phasing the spectrum in dispersion mode (Fig. 12.31C). A good quality spectrum is also obtained with the IP cross peaks COSY (ISECR COSY) sequence shown in Fig. 12.17D [106], which provides good resolution and IP cross peaks (Fig. 12.31D).

The occurrence of coherence transfer phenomena may give rise to COSY cross peaks even in the absence of scalar coupling [69]. When reference was made to the dipolar coupling scheme between two protons of Fig. 4.18, it was implicitly assumed that the two single quantum longitudinal transition probabilities of, for instance, nucleus I, w_1^I, were equal, i.e., they were independent of the spin state of nucleus J. Similar assumptions are made when dealing with transverse transition probabilities. When the coupling is dipolar in origin, the two transitions are degenerate. Indeed, a signal belonging to a nucleus dipole−coupled to another nucleus actually results from the superposition of two coincident components with equal T_1 and T_2, although one never thinks of it. There are a number of cases of dipole–coupled pairs, however, where the two components, still coincident due to the absence of scalar couplings, have different T_1 and T_2. This phenomenon originates from cross correlation between the dipolar coupling operative within the pair and other interactions experienced by the two nuclei and modulated by the same motion responsible for the modulation of the dipolar coupling [107]. In principle, a third nuclear spin K dipole-coupled to the other two spins I and J is sufficient to differentiate the linewidths. In practice, the effect is small and hardly detectable. A more efficient mechanism is the cross correlation with the chemical shift anisotropy relaxation mechanism.

In paramagnetic systems, a major nuclear relaxation mechanism is Curie relaxation (Section 4.6). In paramagnetic macromolecules Curie spin relaxation is often the dominant contribution to T_2. Cross correlation between proton–proton dipolar coupling and Curie relaxation may cause the two degenerate signal components of each nucleus to have markedly different linewidths [69]. Under these conditions, the common COSY experiment of Fig. 12.17C yields strong cross peaks between signals coupled by dipolar *but not by scalar* interactions. The origin of the phenomenon can be better understood if one recalls that the components of COSY cross peaks are in antiphase (for instance, Fig. 12.31B,C). It is obvious that, if the two components coincide, total cancellation occurs. However, if the two components

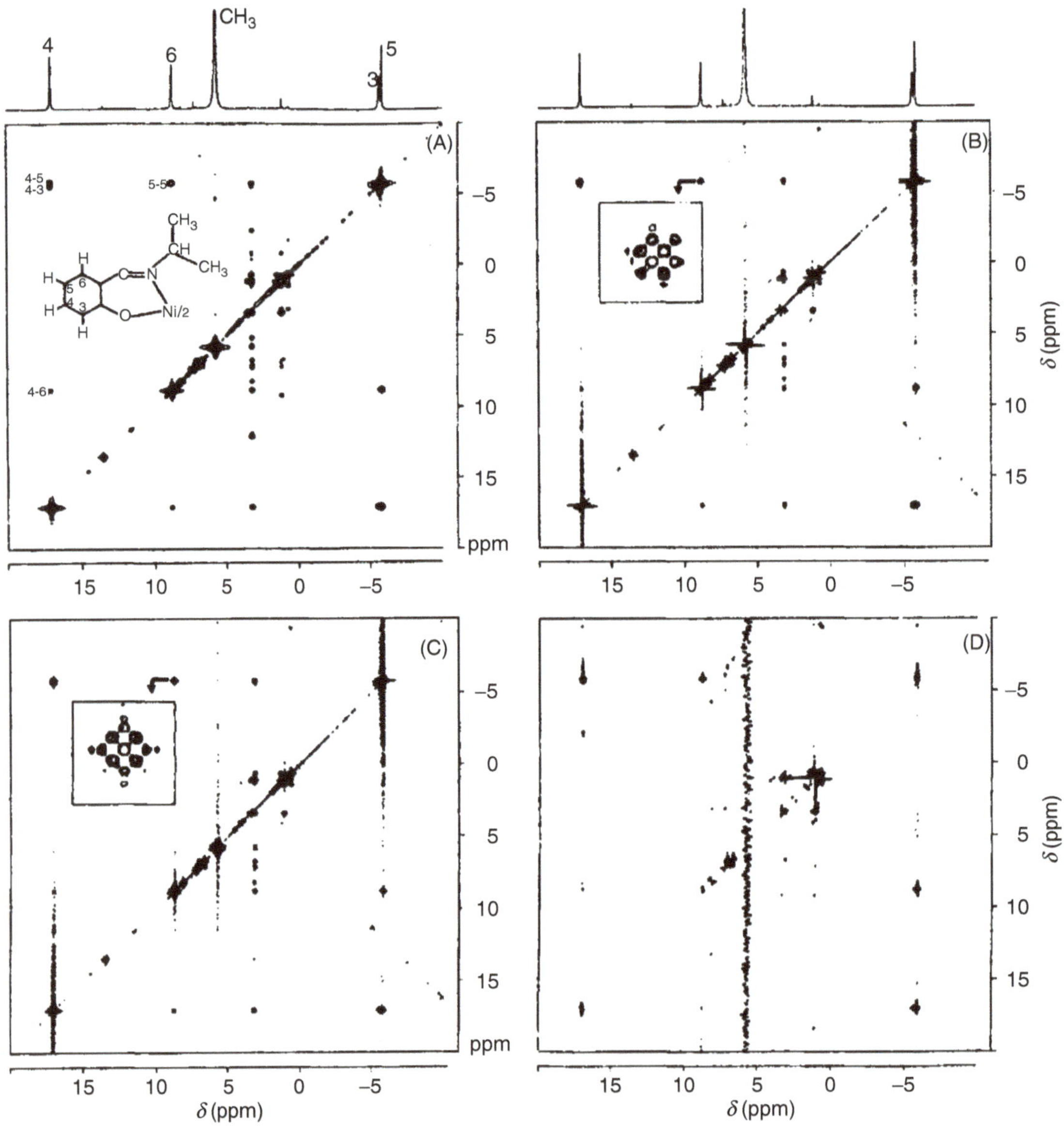

FIGURE 12.31 Aromatic ring part of the COSY spectra of the complex shown in the inset. (A) Magnitude mode spectrum; (B) phase-sensitive spectrum, absorption mode; (C) phase-sensitive spectrum, dispersion mode; (D) ISECR spectrum (sequence in Fig. 12.17D). Sin^2 weighting functions have been used for spectra (A)–(C) and a cos^2 weighting function for spectrum (D). Peaks in (A) and (D) are in phase and positive. The positive components of the 6–5 peak in (B) and (C) are shown in the enlargements.

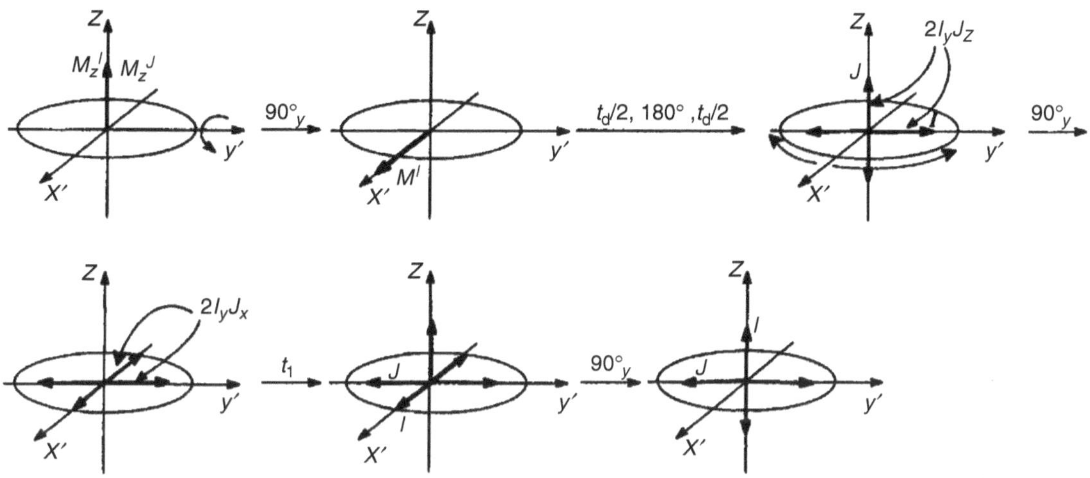

FIGURE 12.32 Vector representation of a DQ (ZQ) experiment. The first 90° pulse along y' rotates the equilibrium magnetization of the I (and J) spins from the z axis to the x' axis. After a time $t_d = 1/2J_{IJ}$ (interrupted by a refocusing 180° pulse), the antiphase coherence $2I_yJ_z$ (see Appendix IX) is at its maximum. Another pulse along y' (shown as a 90° pulse for clarity, although 45° pulses are more commonly used) then transforms the antiphase coherence into a DQ (ZQ) coherence (the $2I_yJ_x$ component is shown). During t_1 the DQ (ZQ) evolves until a further 90° pulse along y' transforms the $-2I_xJ_y$ component (shown at its maximum for clarity) into a $2J_yI_z$ antiphase coherence. During t_2 the latter generates a detectable IP coherence of the J spin, whose initial intensity had been modulated by the DQ (ZQ) Larmor precession frequency during t_1, thus originating a cross peak.

have different linewidths, they do not cancel and may give rise to very strong cross peaks, as illustrated in Fig. 12.32.

The amount of cross correlation between the dipolar coupling of nuclei A and X and Curie relaxation of the A spin is [108]:

$$\Delta\nu = \frac{2}{15\pi}\left(\frac{\mu_0}{4\pi}\right)^2 \frac{B_0\gamma_A^2\gamma_X\mu_B^2 g_e^2\hbar S(S+1)}{r_r^3 r_{AX}^3 kT} \times\left(4\tau_c + \frac{3\tau_c}{1+\omega_0^2\tau_c^2}\right)\frac{3\cos^2\theta_{SAX}-1}{2} \tag{4.37}$$

where r_{AX} and r_r are the AX internuclear distance and the distance of the A nucleus from the unpaired electron, θ_{SAX} is the angle between the A-electron and the AX vectors, S is the electron spin quantum number, and τ_c is the correlation time for molecular tumbling. For lanthanoids S and g_e must be replaced with J and g_J, respectively (Chapter 8). So, cross correlation can contain structural information. Sequences are available to measure the different transverse relaxation rates of the two components [109]. A sensitivity enhanced TROSY pulse sequence was modified to include a transverse ^{1}H relaxation period prior to detection. Relaxation rate constants are then calculated from the measured intensity ratios of the individual doublet lines. The dipolar coupling with the time-averaged magnetic moment of the electron can be seen as a contribution to the CSA (Section 4.6). Therefore, the effect described above cannot be distinguished from the general CSA–dipole–dipole cross correlation. Such cross correlation also provides a contribution to the signal splitting which adds to that of the ^1J coupling of, for example, the NH group. As long as this is kept in mind, COSY cross peaks between very broad lines can be looked for, and interpreted as dipolar correlation cross peaks.

In several kinds of correlation spectroscopy, it is customary to also exploit the evolution of either zero quantum (ZQ) or double quantum (DQ) coherences. As shown in Fig. 4.18, they correspond to the transition $- + \rightarrow + -$ and $- - \rightarrow + +$, respectively. The typical pulse sequence is described in Fig. 12.17E, where after the first 90° pulse we wait for a time $t_d/2$, then we send a 180° pulse which has the only role of refocusing the spins. After the same time $t_d/2$ we send a 45° pulse in order to dislocate the antiphase coherence out of the yz plane in such a way as to create ZQ and DQ coherences (Fig. 12.17E). In the product operator formalism, DQ and ZQ coherences are proportional to $I_xJ_y \pm I_yJ_x$, respectively. I_yJ_x can be visualized as shown in Fig. 12.31. The time t_d should be optimized to either $1/2J_{IJ}$ or T_2. The DQ (ZQ) coherence is let evolve for a time t_1. At this point we can apply a 90° pulse to transform back into an antiphase coherence either the DQ or the ZQ coherence depending on the phase cycling (Fig. 12.33).

Since there is a waste of time in the phase cycling (and some extra loss of signal due to the overall length of the sequence), there is no advantage in using these or other COSY techniques in the place of the simple COSY experiment described in Fig. 12.17C. However, the relaxation of the ZQ transitions may be little affected by the presence of unpaired electrons and under certain conditions may also have some advantages as far as resolution is concerned [110].

12.3.3 SOLUTION NMR—TOCSY

TOCSY is a rotating frame experiment designed to detect scalar connectivities over a large range of J_{IJ} values, especially useful for small (ca. 1 Hz) J_{IJ} values. The most common pulse sequence is shown in Fig. 12.17F [111]. The spin lock is achieved by applying a train of relatively high power pulses (the MLEV17 sequence being one of the most used sequences [112]) in such a way as to continuously refocus the chemical shift evolution of the various signals in the xy plane. Analogously to ROESY experiments, the magnetization during the spin-lock (mixing) time disappears with $T_{1\rho}$ (i.e., essentially T_2, Section 4.4). It follows that coherence transfer in the xy plane, which is built up with a $\sin(\pi J_{IJ}t)$ function, also decreases with time constant $\rho_{1\rho} = (\rho^I_{1\rho} + \rho^J_{1\rho})/2$:

$$I = \sin(\pi J_{IJ}t_m)\exp(-\rho_{1\rho}t_m). \tag{12.12}$$

A TOCSY experiment on a paramagnetic molecule is reported in Fig. 12.22 for the 5Cl–Ni–SAL–MeDPT complex [88]. Cross peaks between signals with linewidths of the order of 100 Hz were easily detected. In particular, the couplings of each aromatic proton with its neighbors are evident. TOCSY cross peaks between signals with similarly broad lines can also be detected in proteins (Fig. 12.34) [113].

In TOCSY experiments, the problem of overheating the sample is more serious than in ROESY experiments because of the large irradiation energy required by the spin-lock pulse, and in extreme cases the probe coil may not stand the pulse power for the required time. Each individual component of the pulse train must have enough power to irradiate the whole spectral window of interest. Other spin

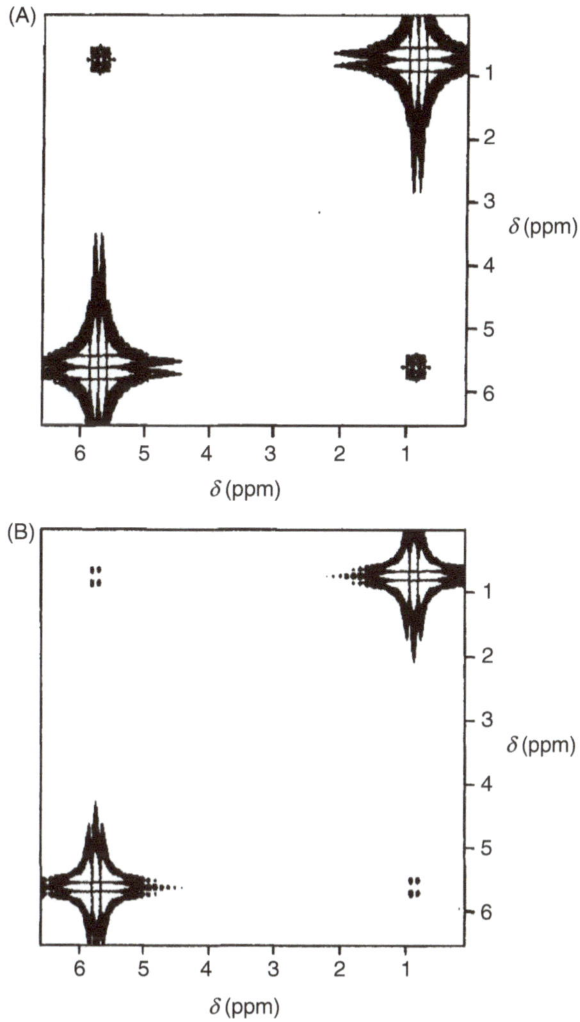

FIGURE 12.33 Simulated COSY cross peaks originating from a pair of signals dipole-coupled in the presence of cross relaxation with Curie relaxation and in the absence of scalar coupling. The two degenerate components are in antiphase, but they do not cancel out due to their different linewidths. Note that absorption mode phasing (A) and dispersion mode phasing (B) show opposite patterns with respect to the case of scalar coupling (Fig. 12.29A,B).

lock sequences constructed from WALTZ-type [42] or DIPSY-type [43] sequences may allow some reduction of the power needed, the spectral window being the same.

Another improvement with respect to these problems is represented by the so-called Clean-TOCSY experiment [114]. Clean-TOCSY was originally devised to avoid the build up of ROE during a TOCSY experiment. In small molecules ROE cross peaks in a TOCSY spectrum may be mistaken for scalar

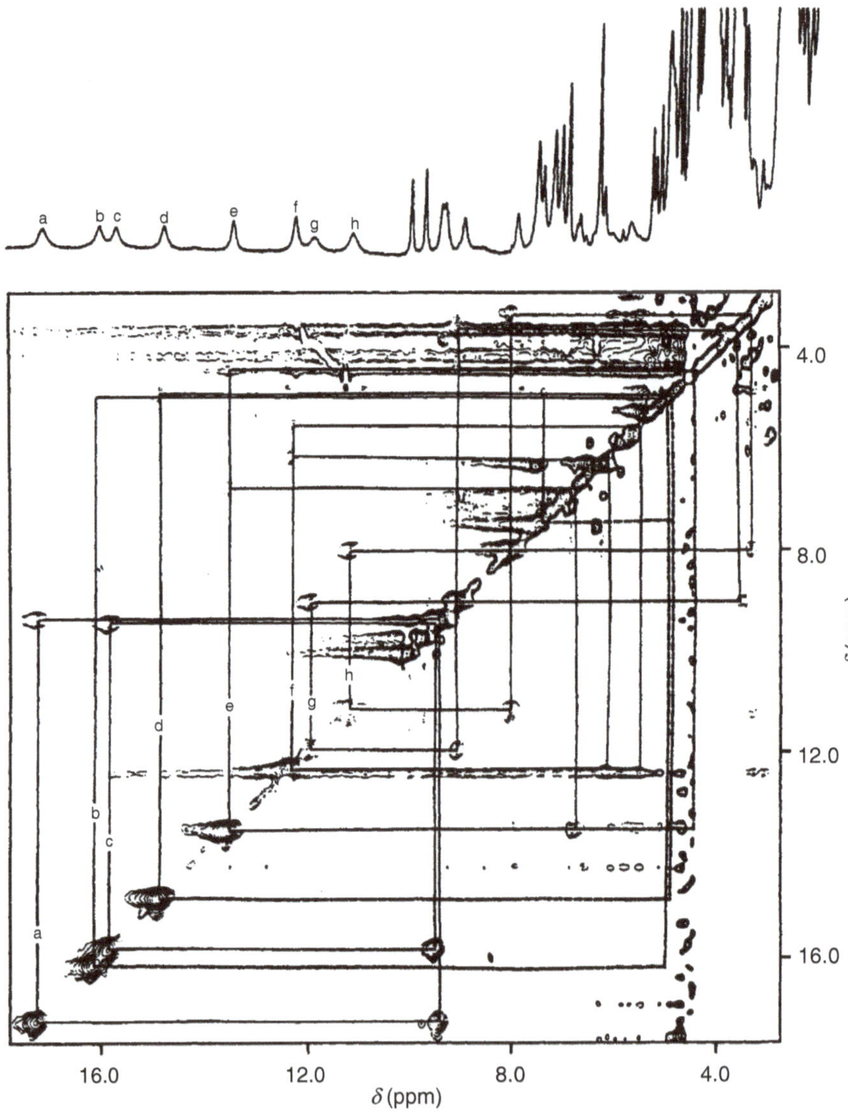

FIGURE 12.34 TOCSY spectrum of the oxidized form of ferredoxin from *C. pasteurianum*, showing all the cysteine ligand β-CH_2–β-CH_2 and some β-CH_2–α-CH connections.

Reproduced with permission from Ref. [113].

TOCSY cross peaks, while in large molecules, where ROESY and TOCSY have opposite signs, the presence of ROESY may accidentally cancel TOCSY cross peaks. In Clean-TOCSY, short delays are interleaved between the pulses of the MLEV17 sequence (Fig. 12.2M) in such a way that the ROE effect is lost. This has also the side-effect that, for the same overall spin lock time t_m, less energy (power × time) is delivered to the sample. Therefore, the length of each pulse can be, for instance, halved, and its power doubled without delivering to the sample more energy than in the original TOCSY experiment. In principle, even full high power pulses can be used, provided that enough delay is interleaved between them.

12.3.4 SOLUTION NMR—HETERONUCLEAR CORRELATION EXPERIMENTS

Heteronuclear correlation experiments are by and large the most important class of experiments when dealing with biological solids, be them paramagnetic or diamagnetic. In heteronuclear correlation experiments, the frequency labeling along the indirect dimension(s) is accomplished with frequencies of nuclei different from the detected nucleus. Polarization/magnetization is exchanged though coupling, which may be scalar or dipolar, or cross correlation. A crucial aspect related to the successful use of heteronuclear experiments in paramagnetic systems is the efficiency of polarization transfer between proton and the attached heteronucleus. Polarization transfer allows to obtain transverse magnetization of the heteronucleus with a larger sensitivity, close to that of proton. In fact, the sensitivity gain due to polarization transfer with respect to the direct excitation of the heteronucleus depends on γ_H/γ_X, where γ_H is the proton magnetogyric ratio and γ_X is the magnetogyric ratio of spin X. One (and perhaps the most popular) way to achieve polarization transfer in solution is to excite the H spin and then let the two-spins system HX system under the J coupling. This can be done, for instance by the insensitive nuclei enhancement by polarization transfer (INEPT) experiment. The ^{1}H spins are excited, and scalar coupling start building an antiphase coherence. A pair of synchronized 180° pulses refocuses chemical shift evolution but allows for evolution of the scalar coupling. Finally, a pair of synchronized 90° pulses converts the coherence from $2I_xS_z$ to $2I_zS_x$. By and large the most commonly used sequence based on this heteronuclear polarization transfer is the HSQC and variations thereof. In the absence of relaxation the maximum of polarization transfer occurs for $t_d = 1/(2J_{HX})$.

In one of the most common heterocorrelation pulse sequences for reverse detection, called heteronuclear multiple quantum coherence (HMQC) (Fig. 12.17G) [115,116], ^{1}H–^{13}C MQ (multiple quantum) coherence is generated by first applying a 90° pulse on protons and, after a time t_d chosen equal to $1/2J_{IJ}$, by applying a 90° pulse on carbon (Fig. 12.35). At this point, the MQ coherence is let evolve for a variable time t_1. A refocusing 180° pulse on proton is applied in the middle of the evolution period, i.e., at $t_1/2$. A further 90° pulse on carbon converts the MQ coherence into an antiphase magnetization (Fig. 12.35), which is let evolve for another time t_d to fully develop the observable single quantum coherence on proton, which is then detected during the acquisition time t_2.

In paramagnetic compounds, in the absence of contact relaxation, heteronuclei are relaxed much less than protons by dipolar coupling with the unpaired electron(s), all other things being equal, because of the γ_I^2 dependence. By looking at the evolution of coherence during the experiment, we can predict the relaxation of the system during the sequence. When the linewidths are larger than J_{IJ}, the first antiphase on proton decays during t_d essentially with ρ_2^I for proton (the heteronucleus being still along the z-axis). The MQ decays with the sum of $\rho_2^I + \rho_2^J$, whereas during the second t_d the magnetization decays again with ρ_2^I. In cases where $\rho_2^I \gg \rho_2^J$ (for instance, large Curie relaxation effects on protons, Section 4.6), direct detection may be less disfavored than reverse detection, because during the two t_d intervals ρ_2^J instead of ρ_2^I is involved.

An example of an HMQC spectrum is reported in Fig. 12.36 [110]. ^{1}H–^{13}C connectivities can be detected for signals for which $T_2 < 1/2J_{IJ}$ by setting $t_d = T_2$.

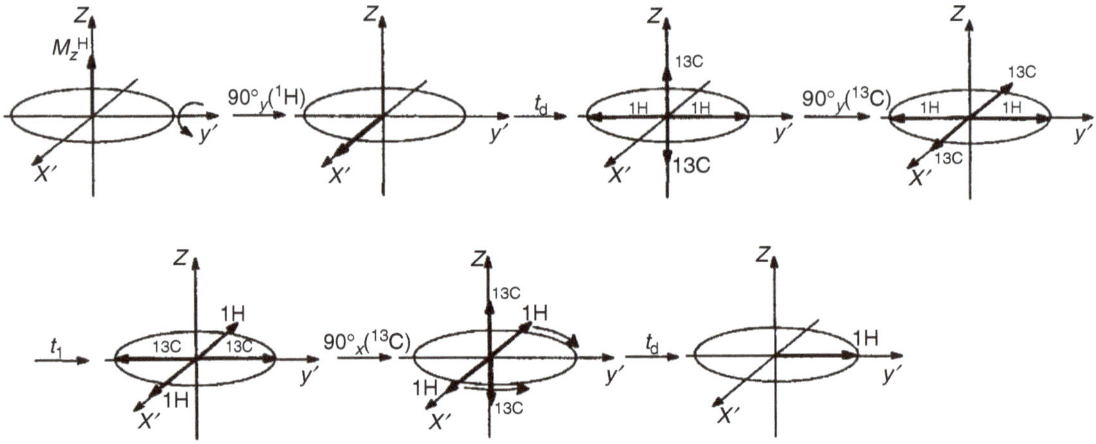

FIGURE 12.35 Vector representation of a ^{1}H–^{13}C HMQC experiment. The first 90° pulse along y' rotates the equilibrium magnetization of the proton spin, I^H, from the z-axis to the x' axis. After a time $t_d = 1/2J_{HX}$, the antiphase coherence $2I_y^H I_z^C$ (Appendix IX) is at its maximum. A 90° pulse on carbon along y' then transforms the antiphase coherence into a MQ (multiple quantum) coherence (the $2I_y^H I_x^C$ component is shown). During t_1 the MQ evolves (with a 180° refocusing pulse on proton in the middle), until a further 90° pulse on carbon along x' transforms the $-2I_x^H I_y^C$ component (shown at its maximum for clarity) into a $2I_x^H I_z^C$ antiphase coherence. After the time t_d, IP coherence of the proton spin develops. The latter is detected during t_2. Its initial intensity is modulated by the carbon Larmor frequency during t_1 (if proton refocusing has been used), thus originating a proton–carbon cross peak.

In principle, HSQC experiments (Fig. 12.17H) should be more effective than HMQC. This is because during the transfer delay t_d the 180° pulses on ^{1}H and ^{13}C eliminate contributions to relaxation arising from field inhomogeneity, diffusion, and cross-correlation effects. Furthermore, the antiphase magnetization evolving during in HSQC experiments roughly decays with the sum of $\rho_2^I + \rho_2^J$ and therefore should relax during t_1 slower than DQ (ZQ). Practically, in our hands we never found substantial difference in the amount of information obtained in HSQC versus HMQC experiments. A possible reason could be that the smaller number of pulses of HMQC makes the latter sequence more robust with respect to experimental missetting. On the other hand it should be stressed that the critical step in this experiment is the polarization transfer from I to J occurring during t_d, which is the same in both experiments.

Usually, HSQC/HMQC spectra are recorded in decoupled mode, i.e., the splitting due to the ^1J coupling between the proton and the heteronucleus, which would give rise to a four-peak pattern in the spectrum, is removed. This is achieved by (1) inserting a 180° pulse on ^{1}H in the middle of the t_1 evolution of the heteronuclear spectrum and (2) broadband decoupling the heteronucleus during proton acquisition in t_2. To measure the extent of the H–X coupling, it is possible to record a spin-state selective experiment, the most common being the so-called IPAP. The experiment is carried out by acquiring two datasets in an interleaved manner [117]. The first experiment is a regular HSQC (Fig. 12.17H) with ^{1}H coupled to X during t_1 evolution. The X chemical shifts are modulated by cos ($J\,t_1$), and the two components of the doublets have the same sign. The spectrum is called IP. In the second experiment, an

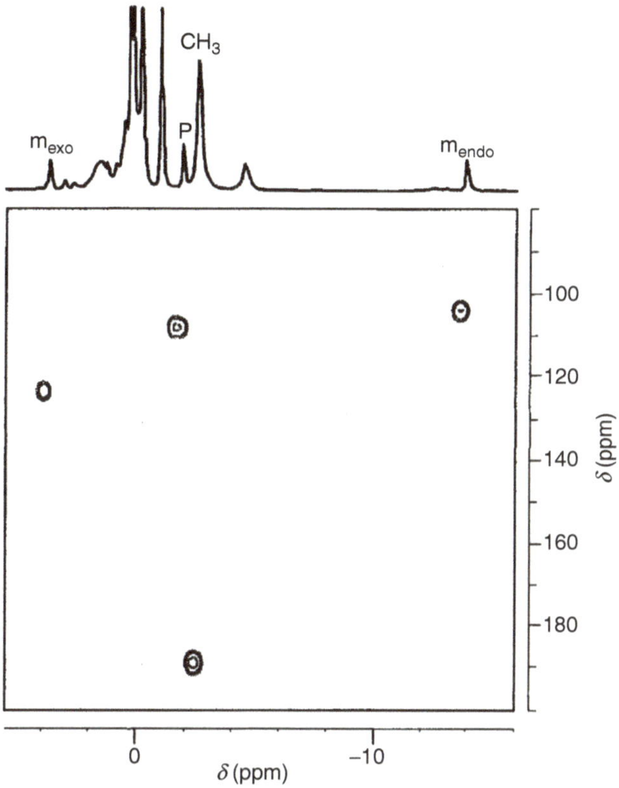

FIGURE 12.36 600 MHz natural abundance ^{1}H–^{13}C reverse HMQC spectrum of the complex shown in Fig. 12.24 (Ln = Dy^{3+}).The cross peaks involve protons with linewidths of the order of, or larger than, J_{HC}.

Reproduced with permission from Ref. [110].

extra INEPT step is added to transfer the X antiphase term to an IP term prior to the X evolution. After the X evolution period, only the term modulated by sin ($J\ t_1$) is detected, and the two components of the doublets have the opposite sign. The spectrum is called antiphase (AP). The 90° purge pulse on ^{1}H prior to the t_1 period is used to transfer the remaining residual $\cos(\pi J\Delta)I_z^{\mathrm{H}}I_x^{\mathrm{X}}$ to unobservable multiple quantum terms. A more detailed description in product operator formalism, such as how to obtain the IP and AP spectra, can be found in the original paper [117]. The addition and subtraction of the IP and AP spectra give two simplified spectra just like regular HSQCs with the X chemical shifts displaced by either $+J/2$ or $-J/2$ (Fig. 12.37). The couplings can be readily measured from the frequency difference of the corresponding peaks in the two spectra, which makes IPAP the most commonly used experiment to measure RDCs, both in paramagnetic and in diamagnetic systems. A caveat has to be added: as the X frequency is additionally split by the coupling to neighboring protons, in the presence of cross-relaxation, the components of these additional multiplets may have different breadths, thus

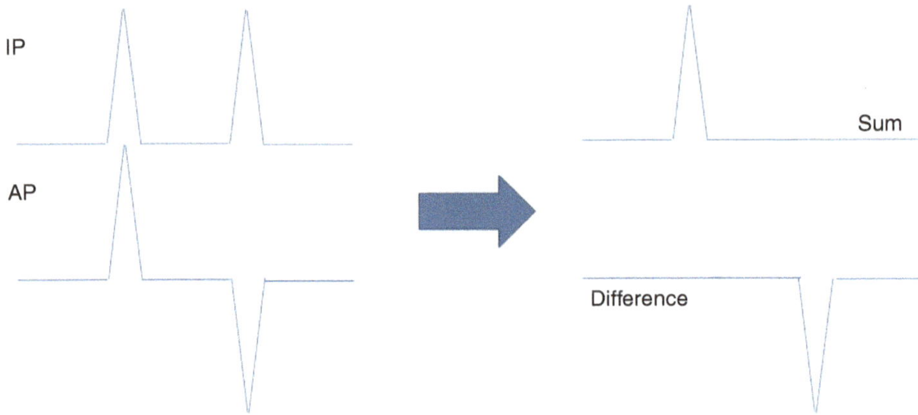

FIGURE 12.37 Schematics of the IPAP sum/subtraction processing.

different intensities, leading to a systematic error in the *J* estimate. To solve this problem, an additional selective decoupling pulse that does not affect the proton for which *J* is measured can be added [118].

When relaxation is fast, the IPAP scheme might be too long to allow for the detection of the signal. When this is the case, a shorter experiment, which is called spin state selection (S^3E) can be used [119,120].

One may also wish to select only one of the four components: this is possible using selective coherence transfer techniques. This latter strategy is used in TROSY experiments [121], and is particularly convenient when the four components of the multiplet have different linewidths, because the sharper one can be selected. In turn, this occurs whenever there are cross-correlation phenomena [122]. One common example of cross-correlation giving rise to different linewidths is in ^{1}H–^{15}N heterocorrelated spectra of proteins. The nuclei of backbone NH groups experience considerable cross-correlation between the ^{1}H–^{15}N dipole–dipole coupling and the chemical shift anisotropy (CSA). As both are predictable, it can be also predicted that in a coupled HSQC spectrum the lower-right peak of the multiplet will be always the sharpest, and TROSY experiments that select this component will give the best results in terms of spectral resolution.

In paramagnetic systems, there can be nonnegligible contributions to the CSA of NH nuclei from their dipolar coupling with the time-averaged magnetic moment of the electron (Section 4.6). As a consequence, it cannot be predicted a priori which will be the sharpest component for each NH peak in a heterocorrelated experiment, and TROSY is less useful, unless four different TROSY spectra are acquired by selecting a different component each time [121].

When paramagnetic contributions strongly enhance R_2 proton relaxation, the efficiency of polarization transfer decreases and the maximum transfer occurs for shorter t_d [123]. This is analogous to what already described for the COSY experiments, and is at the basis of the proton R_2 measurement as described by Eq. (12.5). At increasing R_2 proton relaxation, the loss of efficiency of polarization transfer can be larger than the sensitivity gain due to polarization transfer, as highlighted in Fig. 12.38. This will make the use of PT-based sequences unfavorable with respect to direct excitation of the heteronucleus. Recently, HSQC-based experiments were proposed without the last antiphase—IP conversion step. Although acquiring antiphase magnetization yields reduced intensity with respect to the acquisition of an IP singlet, especially when the two components are so broad so as to overlap significantly,

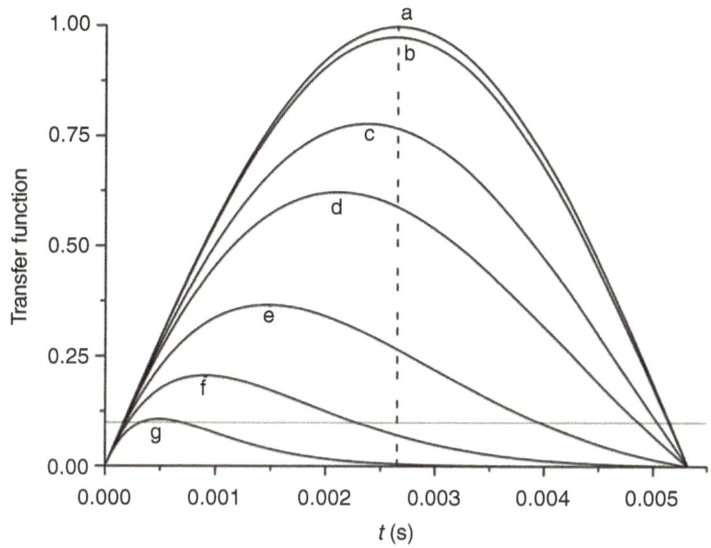

FIGURE 12.38 Efficiency of an INEPT transfer function at different ^{1}H T_2 values: b 100 ms, c 10 ms, d 5 ms, e 2 ms, f 1 ms, g 0.5 ms. Relaxation is neglected in a. A *dashed line* is shown at the 2.65 ms of INEPT step (94 Hz for ^{1}H-^{15}N *J* coupling). A *solid line* is shown in correspondence of 10% transfer efficiency.

Reproduced with permission from Ref. [124].

when the doublet is phased in dispersion mode, it gives rise to a "pseudosinglet" originated by the sum of the two dispersive components of the doublet [124].

Indeed, paramagnetic broadening is proportional to the square of the nuclear magnetogyric ratio γ. Therefore, the inverse detection of heteronucleus (i.e., the transfer of magnetization from the X nucleus to the proton followed by proton detection) may prevent the observation of the X signal because of proton T_2 relaxation. In fact, the relative values of γ for ^{1}H, ^{13}C, and ^{15}N nuclei are 1:1/4:1/10, and thus the relative contribution to overall relaxation arising from the hyperfine interaction is 1:1/16:1/100, respectively. Therefore, to identify ^{13}C or ^{15}N resonances that escaped detection through inverse detection technique, direct detection of ^{13}C or ^{15}N can be used [125].

Not long ago, a new approach based on selective nonexcitation has been introduced to reduce the time burden of NMR experiments, which works also in the case of diamagnetic systems or paramagnetic systems with short electron relaxation times [126,127] (see the concept of non-exponentiality in Section 12.1.5). In this family of experiments (BEST or SOFAST), all the pulses on the H channel are selective, and thus affect the spin state of the signals that one wants to acquire (say, for instance, the amide protons), whereas all other protons in the molecule are still along the *z*-axis. Coupling of the excited spins with all other spins being aligned with the field sizeably shortens the longitudinal relaxation of the excited spins, making it possible to recycle faster.

12.3.5 SOLUTION NMR—3D EXPERIMENTS

As the variety of 3D and higher dimensionality experiments increases with the number of dimensions, it is inappropriate to describe any of them in detail here. However, it may be useful to summarize a few guidelines on how to select the best high-dimensionality experiments for a paramagnetic system. (1) Among different sequences that provide the same—or almost the same—type of information, always select the simplest in terms of number of pulses and the shortest in terms of duration. This suggestion is obviously dictated by the need to minimize loss of magnetization/coherence due to paramagnetic relaxation. (2) In macromolecules, always select the sequences that minimize the duration of inplane evolution of spins, as T_2 is usually shorter than T_1, and can be much shorter when paramagnetic relaxation is dominated by Curie relaxation. (3) If possible, let the longer inplane evolution be that of heteronuclei rather than that of protons, as relaxation depends on the square of the magnetogyric ratio and is therefore 16 times and 100 times less for carbons and nitrogens (respectively) than it is for protons (Section 4.4). (4) Whenever possible, substitute a 3D experiment with a few 2D experiments…

We have seen that one limitation of 2D spectroscopy when dealing with paramagnetic compounds showing a wide range of T_1 (and T_2) values is that the parameters cannot be optimized for the detection of all connectivities in a single experiment. Often, at least two experiments are necessary. An advantage of 3D spectroscopy which is peculiar to paramagnetic systems, and in particular to paramagnetic macromolecules, is that one can cover the whole range of T_1 and T_2 values present in the spectrum by optimizing the spectral parameters in one 2D plane (spectral resolution, maximal evolution time, mixing time, etc.) to the detection of connectivities between fast relaxing signals, and those in another 2D plane to the detection of connectivities with slow relaxing signals. Of course, a clear advantage only appears when fast relaxing signals are many and severely overlapped with slow relaxing signals.

An obvious extension to 3D spectroscopy from 2D spectroscopy is the homonuclear NOESY–NOESY [128]. There are two t_1 variable times and one t_2, which after Fourier transform provide three frequency domains. The 3D NOESY–NOESY spectrum of met-myoglobin cyanide, which contains low spin iron(III) in a heme moiety (Chapter 7), was measured already more than 20 years ago [129]. In Fig. 12.39, a slice of the 3D spectrum is shown at the 12-CH_3 height. On the diagonal it shows all the dipolar connectivities between 12-CH_3 and other protons; off-diagonal there are dipolar connectivities within the latter set of protons, all belonging to the porphyrin ring.

Results similar to those shown in the slice of Fig. 12.39 can be obtained with the so-called NOE–NOESY sequence [130]. Here a hyperfine shifted signal, for example, 12-CH_3 of the above compound, is selectively saturated, and then the NOESY pulse sequence is applied. The NOESY difference spectrum obtained by subtracting a NOESY spectrum without presaturation of the 12-CH_3 signal is shown in Fig. 12.40. Here, some more cross peaks are evident with respect to the 3D NOESY–NOESY experiment because secondary NOEs develop much more when the primary NOEs from the 12-CH_3 signal evolve in a steady state experiment like the NOE–NOESY rather than in a transient-type experiment like the NOESY–NOESY. In Fig. 12.40, dipolar connectivity patterns are apparent among protons belonging to Phe-33 (2) Phe-43 (1, 6, 7, 8), Phe-46 (3), and propionate 13 (9, 10). Some of these assignments could only be performed by exploiting the NOE–NOESY experiment.

Other variants are NOE–COSY and NOE–TOCSY. However, these experiments are much less sensitive than NOE–NOESY.

In principle, all the combinations of homonuclear 2D spectroscopies can be performed to originate a 3D spectrum (COSY–COSY, NOESY–COSY, NOESY–TOCSY, etc.). The considerations made in this chapter for the most basic experiments can be easily extended to their combinations. The general

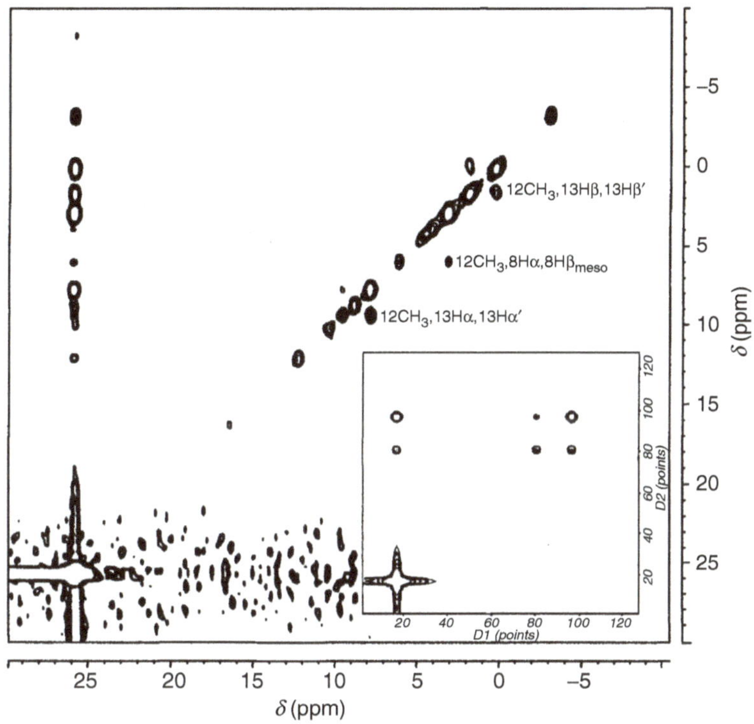

FIGURE 12.39 Cross-section of the 600 MHz 3D NOESY–NOESY spectrum of met-myoglobin cyanide. The slice is taken at the 12-CH_3 height. The inset shows the simulated cross-section involving the 12-CH_3, 13-Hα and 13-Hα′ signals.

Reproduced with permission from Ref. [129].

guideline should always be that the more complex the pulse sequence is, the more the experimental sensitivity will suffer from fast nuclear relaxation.

The most useful advantage of 3D spectroscopy, however, is that heteronuclei can be included along one dimension. If one of the two 2D sequences is a heterocorrelated sequence, then simultaneous heteronucleus–proton connectivities can be observed. In the case of macromolecules, where there are solubility problems, heteronuclear enrichment is probably required, especially when HSQC-TOCSY and HSQC-NOESY combination is no more enough to resolve assignment ambiguities. Of note, heterocorrelated spectroscopies, of the type of those used in diamagnetic systems, have been applied with success, even using natural abundance ^{13}C, on a paramagnetic protein containing the $[Fe_4S_4]^{2+}$ polymetallic center [61,131].

For proteins showing not enough dispersion of resonances in ^{15}N-^{1}H-^{1}H correlation experiments a series of 3D sequences correlating different heteronuclei have been developed, to achieve sequential connectivities along the backbone of the protein. In order to correlate the backbone connectivities,

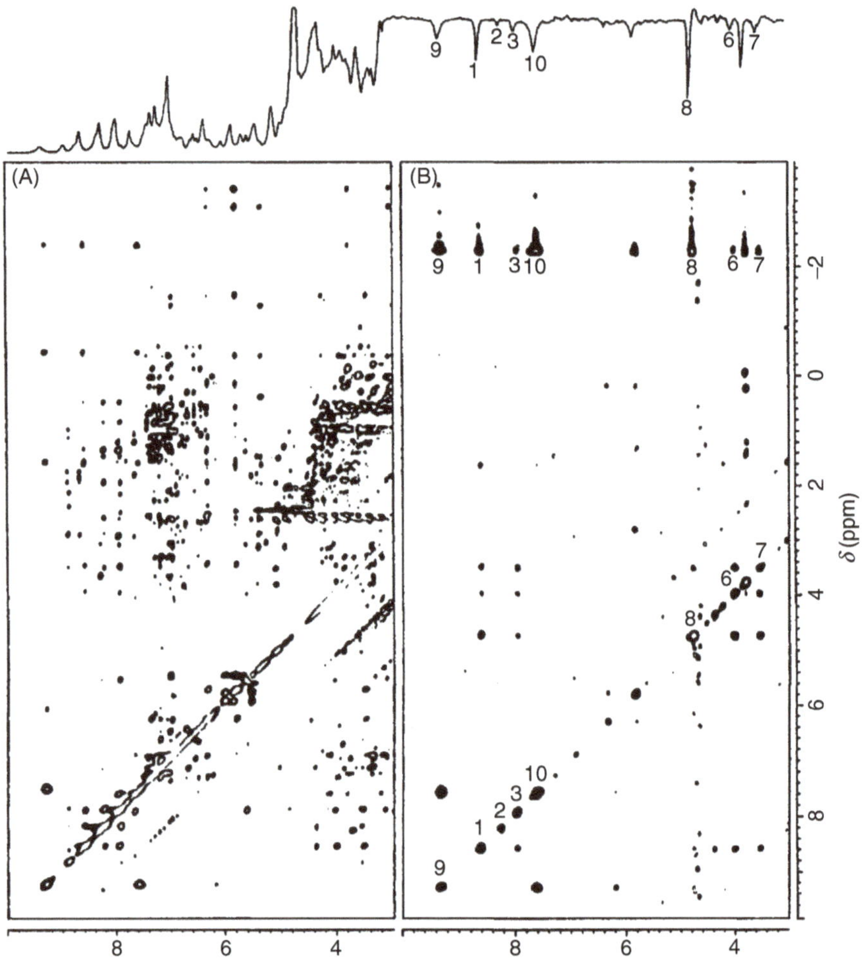

FIGURE 12.40 NOESY spectrum (A) and NOE–NOESY spectrum (B) of met-myoglobin cyanide. The latter spectrum is obtained by pre-irradiation of the 12-CH_3 signal.

Reproduced with permission from Ref. [130].

homo- and heteronuclear INEPT steps are added. These experiments, which are usually referred to as triple-resonance experiments, are assigned in pairs:

1. HNCO and HN(CA)CO yield information about the H_i^N-N_i-CO_i and H_i^N-N_i-CO_{i-1};
2. HNCA and HN(CO)CA yield information about the H_i^N-N_i-$C\alpha_i$ and H_i^N-N_i-$C\alpha_{i-1}$;
3. CBCANH and CBCA(CO)NH, where Cβ and Cα are coevolved, yield access to the sidechain shifts, which also give information about the secondary structure content of the protein.

Protocols for setting up backbone and sidechain assignment experiments are available [132]. These experiments are now routine also for the solid state, where the heteronuclear INEPT steps are replaced by CP steps [133]. The homonuclear transfers can be either performed by spin-locking [134], longitudinal mixing [135](see later) or INEPT [136].

This set of experiments constitutes the main strategy for assigning protein resonances before going further into structural characterization (Chapter 9), even if the protein bears a paramagnetic center. Of course, the considerations about relaxation that we made at the beginning of this chapter still hold true, and the following changes to the experimental conditions should be considered: (1) reduce the increments of the indirect dimensions: this reduces the resolution, but hyperfine shifted signals may be "naturally" resolved; (2) decrease the time allowed for the buildup of the correlated signals: both for CP and INEPT this comes at the price of reduced efficiency of the transfer, but transverse relaxation effects have smaller impact—this is particularly relevant for INEPT through the smaller J couplings (N_i-$C\alpha_i$ is about -11 Hz and $C\alpha$-$C\beta$ is about 35 Hz, which correspond to 45 and 14 ms of total transverse evolution respectively); (3) in the case of solid state, where shorter pulses can be obtained, use shorter pulses. The reduced sensitivity will be partly mitigated by the possibility of faster recycling of the experiment.

12.3.6 SOLID-STATE NMR—HOMONUCLEAR EXPERIMENTS

By and large, the most popular approach to detection of through-space or through-bond proximities in MAS-NMR has been the detection of carbon–carbon correlations. As seen in Section 5.1.2, under magic angle spinning (MAS) a large share of the dipolar interaction is averaged out. It is however possible to revert the effect of MAS by rotor-synchronized π pulses, which reintroduces the dipolar coupling (Section 5.1.4). Under this model there are the RFDR-like experiment [137] (radio-frequency driven recoupling). Because of the symmetry of the pulse sequence, the recoupling efficiency is higher toward the largest coupling, and scales following complex laws [138]. For this reason, RFDR experiments can be regarded as through-bond experiments. The recoupling efficiency is strictly dependent on the bandwidth of refocusing of the π pulses [139]. Of course, in the case of paramagnetic systems where the signals may be spread over a more significant range shorter pulses are needed to maintain the recoupling efficiency.

Another way to mix carbon–carbon magnetization is via spin diffusion [140]. Spin diffusion depends on the overlap of the two signals for which magnetization transfer is sought for. Under MAS, the overlap is strongly reduced or cancelled. However, lines can be broadened if the coupling to ^{1}H is reintroduced, which may happen turning off the decoupling during the mixing time (proton-driven spin diffusion—PDSD) [141], or by active recoupling by rotor-synchronized π pulses or by a rotation-matched continuous wave field (dipolar-aided rotational resonance—DARR) [142,143]. Again, in the case of paramagnetic solids, the overlap of the lines is further reduced because of hyperfine shift, and longer mixing times need to be used to increase the recoupling .While the reduction of the overlap between resonances is detrimental for the success of a DARR experiment, it makes it possible to record ^{1}H-^{1}H correlation experiments [144]. An experiment based on a scheme similar to the NOESY in solution has been applied to probe spatial proximity in Cu^{2+}-cyclam complex (Fig. 12.41A). Such experiments present intense diagonal peaks, which may prevent the observation of cross-peaks between spins which have close resonance frequencies. To solve this potential drawback, another experiment has been recently proposed, based on double-quantum recoupling [145]. The resulting spectrum, acquired on copper(II)(Ala)$_2$, has no diagonal and, in addition favors recoupling between nearest neighbor spins (Fig. 12.41B).

12.3.7 SOLID-STATE NMR—HETERONUCLEAR CORRELATION EXPERIMENTS AND PROTON DETECTION

Heteronuclear correlation in solids encompasses a variety of experiments. Nitrogen–carbon correlation experiments have been at the grounds of biomolecular applications of solid state NMR for nearly two decades, when the first ^{1}H detected experiments have started to appear, enabled by extensive deuteration and fast MAS [146–148]. For paramagnetic systems, ^{1}H-based experiments have been demonstrated in the pioneering studies by Ishii and coworkers [149–151], and have reached maturity substantially before: this happened because the hyperfine interaction sizably reduces the HX and HH couplings, and makes it possible to accomplish decoupling via moderate MAS (> 20 kHz) [152] (Section 5.1.3).

Standard nitrogen-carbon correlation experiments in two or more dimensions are described extensively by Barbet-Massin and Pintacuda [153], together with the protocols for efficiently setting them up on any diamagnetic biomolecular system. When applying such a protocol to a paramagnetic system, the following additional considerations must be made:

1. the presence of a paramagnetic center reduces the transverse relaxation time, yielding a reduced $T_{1\rho}$, which in turn forces the use of shorter contact time in cross-polarization experiments;
2. spin-lock efficiency decreases for signals which are more spread, thus band-selective cross-polarization could be a better choice in these systems [154];

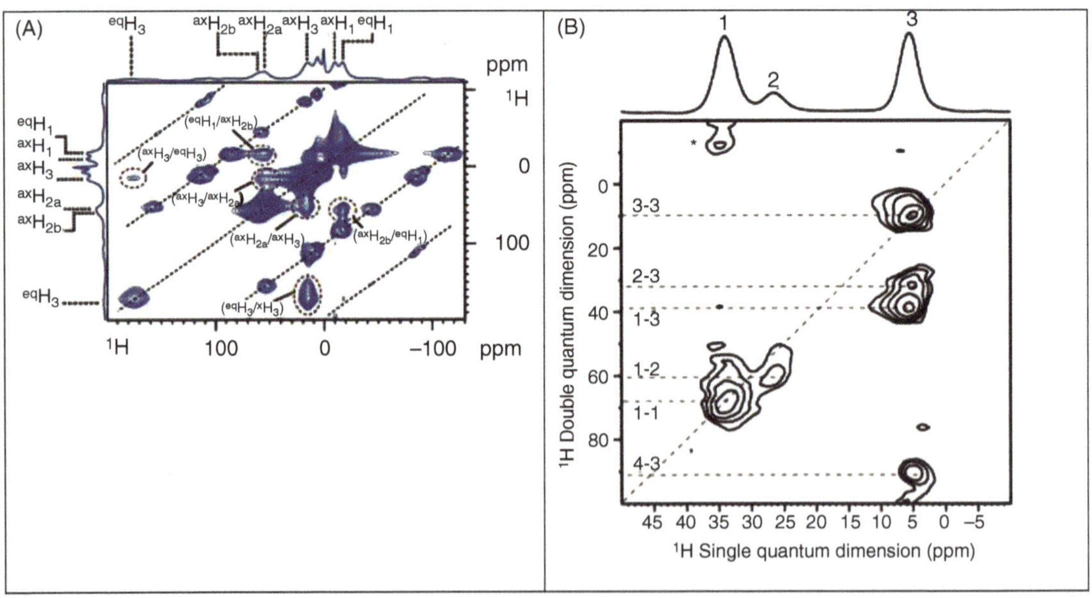

FIGURE 12.41 Solid-state ^{1}H-^{1}H correlation experiments on Cu-Cyclam (A), acquired using a NOESY-like pulse sequence, and Cu-(L-ala)$_2$ (B), acquired using a single-quantum double-quantum correlation, with reduce diagonal peaks intensity.

Reproduced with permission from Refs. [144,145].

3. spin-diffusion efficiency is reduced by the reduction of the strength of the dipolar network (see earlier) thus, as possibly opposed to intuition and to what has been discussed above, when preparing 3D experiments with spin-diffusion based mixing, longer mixing times may be needed.

On the other hand, as we have seen throughout the book, the longitudinal relaxation time gets shorter, and faster recycle is allowed, with a subsequent reduction of the total experimental time [149].

A significant breakthrough is the measurement of (paramagnetic) biomolecular solids has been the introduction of ^{1}H-detected CP-based experiments. These experiments, which are in general performed on highly deuterated systems so as to reduce the strength of ^{1}H–^{1}H coupling are the solid-state analog of the HSQC experiment. As mentioned above, the presence of a paramagnetic center which induces large shifts tends to break the ^{1}H dipolar network, yielding more efficient MAS decoupling. This said, in line with the consideration drawn in Section 12.3.5, the protocols for experimental setup do not significantly change with respect to the diamagnetic experiments presented in [133,155], with the only exceptions of shorter recycle delays and contact pulses.

In line with what we have seen for solution NMR, there are experiments tailored to reveal a few signals which are highly shifted, fast relaxing or both. As anticipated in Chapter 5, these experiments are not based on cross-polarization. Basically, the dipolar interaction is reintroduced through high power, broadband, refocusing pulses. These pulses refocus H and C chemical shift but allow for evolution of H–C dipolar coupling, similarly to what happens during an INEPT transfer (Section 12.3.4). The result is a net polarization transfer.

12.3.8 SOLID-STATE NMR—LINESHAPE ANALYSIS METHODS

As already highlighted in Chapter 5, the lineshape of a static powdered sample reports on all the interactions as sensed by each nucleus. Usually, MAS is used to reduce the spectral complexity, but in samples containing one or few kind of different nuclei, such information has been exploited to recover structural information, especially in the case of paramagnetic systems. Some historical examples regarding ^{1}H and ^{19}F experiments are given in references [156–160].

In particular, a huge amount of investigations have accumulated regarding ^{2}H lineshape analysis experiments. The ^{2}H nucleus has nuclear spin $I = 1$ and a quadrupolar moment of $Q = 2.8 \times 10^{-31}$ m^2, which interacts with the field gradient at the position of the deuteron. As the quadrupole spitting for deuterons in crystals (about 200 kHz) is generally larger than the linewidth (about 1 kHz), the absorption lines of nonequivalent deuterons in crystals can be observed separately. Thus, as shown in Fig. 12.42, each deuteron appears as a doublet spaced by $\frac{3}{8}(\cos^2\theta - 1)e^2Qq_{zz}/h$ for a given orientation θ of the electronic field gradient with respect to B_0. Of course in a paramagnetic species the center of the doublet is shifted with respect to the diamagnetic analog. An interesting example along this lines is the detection of five nonequivalent water molecules in room-temperature spectra of $CuSO_4 \cdot (^2H_2O)_5$ [161]. In this case, five pairs of lines were observed, which are shown to result from fast reorientation about the metal/water bond (10 lines are observed in the low-temperature spectrum). Each quadrupolar doublet was assigned by selective irradiation of the 0 - > 1 transition (Fig. 12.43) [162]. Other remarkable examples are given in Section 7.3.1.1.

When the powder pattern is much larger than the spinning frequency, a situation that is most probably encountered in systems bearing a paramagnetic center with a large magnetic susceptibility (Table 9.1), the lineshape can be recovered by the a fit of the intensities of the spinning sidebands [164]. When MAS frequency is larger than the frequency spread of the anisotropic interaction,

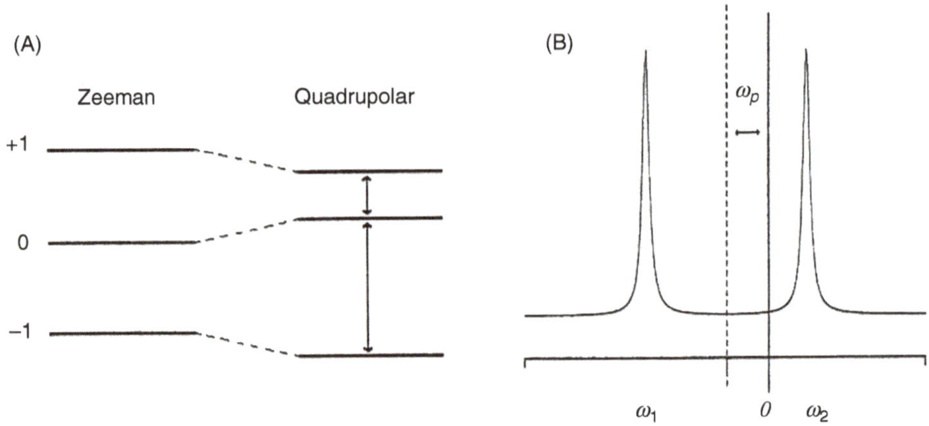

FIGURE 12.42 (A) Schematic representation of the energy levels for a deuteron ($I = 1$) resulting from the Zeeman interaction with the magnetic field and with the quadrupole moment of the nucleus. (B) Spectral splitting pattern for a deuteron in a paramagnetic system. ωp is the displacement from the diamagnetic position by the interaction with the unpaired electron.

Reproduced with permission from Ref. [163].

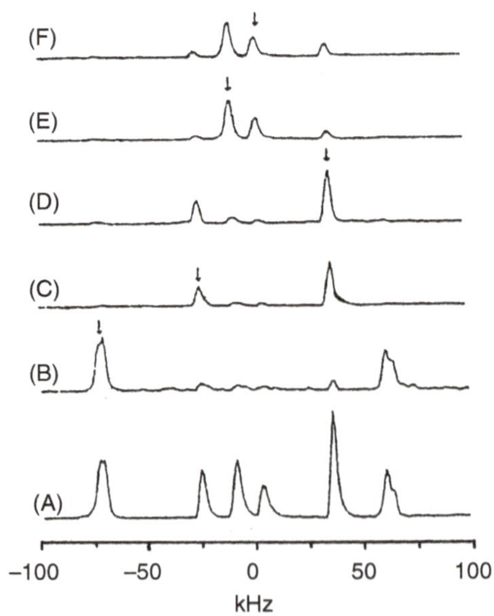

FIGURE 12.43 ^{2}H NMR of a single crystal of $CuSO_4 \cdot (^2H_2O)_5$ (A). Difference spectra (B)–(F) are taken by selectively irradiating at the frequencies marked by arrows.

Reproduced with permission from Ref. [162].

spinning sidebands are erased, but reintroduction of the anisotropic components which are removed as an effect of MAS is also feasible in multidimensional experiments (Section 5.1.4). These are based on reverting the effect of MAS by rotation-synchronized pulses. The relevant theory has been given in detail by Levitt [165].

12.4 COLLECTING PARAMAGNETIC RESTRAINTS IN BIOMOLECULES

Paramagnetic restraints are obtained from the comparison of the spectra of the paramagnetic system and of a diamagnetic analog. The latter is obtained by removing the paramagnetic center, substituting the paramagnetic metal with a diamagnetic one or by changing the oxidation state of the paramagnetic center (L.S. Fe^{2+}/L.S. Fe^{3+}, Cu^{+}/Cu^{2+}, reduction of S-(1-oxyl-2,2,5,5-tetramethyl-2,5-dihydro-1H-pyrrol-3-yl)methyl Methanethiosulfonate (MTSSL), etc.). The election experiment for such kind of measurements is the ^{1}H-^{15}N-HSQC and variations thereof, although for larger biomolecules other experiments might be needed (TROSY [121], CRIPT [166], CRINEPT [167], methyl-TROSY [168]). Since all the measurements are based on spectral comparison, the complete assignment of the spectrum is needed. However, protocols exist to achieve assignment of the paramagnetic spectrum by the prediction of pseudocontact shifts/residual dipolar couplings.

Hyperfine shifts are calculated from the difference between the chemical shifts of the nuclei in the paramagnetic spectrum and in the diamagnetic one. For nuclei that are far removed from the paramagnetic center, the pseudocontact shift is usually modest, and the assignment for these resonances can be easily transferred from one spectrum to another. From this subset of pseudocontact shifts and a structural model, it is possible to predict the magnetic susceptibility anisotropy tensor and in turn to predict other shifts. Another hint comes from the nature of the pseudocontact shift: since it does not depend on the observed nucleus, and since the amide proton is at limited distance from its nitrogen, ^{1}H and ^{15}N have similar shifts, and thus appear to be moving "diagonally." The whole approach can be easily made into an iterative protocol to increase the number of available pseudocontact shifts and refine the available structural model at the same time. Pseudocontact shifts can be measured in the solid state as well: in some cases (yielded by fast MAS, deuteration or both) the same approach as in solution can be followed. Otherwise, NC correlation experiments can be used as well. CC correlation experiments show a sizably larger number of resonances, thus their interpretation can be less straightforward and may require several iterations of structure refinement/pseudocontact shifts prediction.

Residual dipolar couplings are obtained as the difference in the measured ^{1}J (or, less commonly ^{2}J) values at the same field and temperature between the paramagnetic system and the diamagnetic analog. $^{1}J_{\mathrm{NH}}$ residual dipolar couplings are routinely measured via the IPAP approach (Section 12.3.5). Since residual dipolar couplings are the result of nonisotropic reorientation, they do not exist in the solid state.

Cross correlation also impacts the observed J coupling: in general, for paramagnetic proteins, the following contributions to the ^{1}J must be considered:

$$^{1}J_{\mathrm{NH}}(B_0) = {}^{1}J_{\mathrm{NH}}^{SC} + \Delta v_{RDC}^{dia} + \Delta v_{RDC}^{para} + \Delta v_{DFS}^{dia} + \Delta v_{DFS}^{para} \tag{12.13}$$

where $^{1}J_{\mathrm{NH}}^{SC}$ is the scalar coupling, $\Delta v_{RDC}^{dia} + \Delta v_{RDC}^{para}$ is the residual dipolar coupling due to the total magnetic anisotropy of the molecule which includes diamagnetic and paramagnetic contributions, and $\Delta v_{DFS}^{dia} + \Delta v_{DFS}^{para}$ is the dynamic frequency shift due to cross correlation between CSA and dipole–dipole interactions (in diamagnetic systems) and to cross correlation between Curie and dipole–dipole interactions (in paramagnetic systems). Therefore, if the $^1J_{NH}$ of a paramagnetic protein are measured at two different magnetic fields and then subtracted, information on the total residual dipolar coupling is obtained (Sections 3.4 and 4.9), whereas if the $^1J_{NH}$ of a paramagnetic protein are subtracted of the $^1J_{NH}$ of the diamagnetic analog at the same field, the sum of the paramagnetic contribution to the residual dipolar coupling and of the dynamic frequency shift (DFS) due to cross correlation between Curie relaxation and dipolar coupling is obtained, and this could be used to back-calculate the values of the DFS, in case the paramagnetic susceptibility tensor is known from the measurements of pseudocontact shifts. However, often only a few NH pairs are expected to experience appreciable DFS, and attempts to obtain a reliable experimental verification from back-calculations have been unsuccessful. Attempts to detect DFS contributions in small paramagnetic DNA-drug complexes were qualitatively but not quantitatively successful [169].

The last structurally-relevant restraints that are usually acquired in biological macromolecules are the paramagnetic relaxation enhancements. These restraints are acquired measuring the difference in recovery of longitudinal magnetization or decay of transverse magnetization in heteronuclear 2D experiments. The experimental details are given in Section 12.1.5.

12.5 NUCLEAR MAGNETIC RELAXATION DISPERSION (NMRD)

The information content of nuclear longitudinal relaxation measurements in both paramagnetic and diamagnetic systems can be greatly increased by performing such measurements as a function of the magnetic field. For paramagnetic species, the reason is apparent from the functional form of the equations discussed in Chapter 4 and from the relevant experimental data, reported in Chapter 7. The field dependence of a relaxation rate is called relaxation dispersion, and is abbreviated as NMRD. In principle, NMRD would be helpful for any chemical system, but practical limitations, as it will be shown, restrict its use, with a few exceptions, to water protons.

12.5.1 CHANGING THE MAGNETIC FIELD

The basic instrument for performing routine NMRD experiments must contain: (1) a variable field magnet (usually an electromagnet); (2) a frequency synthesizer, and a broadband transmitter, and receiver; and (3) one (or more) broadband tunable probes covering a large frequency range. A variable field external lock or, at least, a reasonably good flux stabilizer for the magnetic field are also desirable although not strictly necessary (see in subsequent section). Once these requirements are met, the experiment can be done by first measuring the water proton T_1 at the regular proton frequency—100 MHz for a 2.35 T magnet, for example,with a usual pulse sequence—such as that used in an inversion-recovery experiment, for example. Then the electromagnet current is reduced to bring the field to a lower value (for instance, 1.88 T, corresponding to 80 MHz resonance frequency), the synthesizer is set to 80 MHz, the transmitter and receiver are retuned accordingly (changing the probehead and the matching/tuning circuitries if necessary), the pulse durations are readjusted, and the T_1 measurement is repeated. Using this procedure, it is possible to go as low as is permitted by the lowest frequency probe (typically 2–4 MHz); but sensitivity problems will ultimately arise. Of course the set of data may be implemented by measurements in the 200 MHz–1 GHz range, each on a different commercial instrument, if available. Except at 100 MHz, the standard lock system is useless, be it an external ^{1}H lock or an internal

^{2}H lock, since both operate at a fixed frequency and once the field has been lowered, the lock signal remains far off-resonance. This problem may be partially overcome by choosing magnetic field values at which an easily observable nucleus resonates at the same frequency at which ^{2}H resonates when the field is set at 2.35 T; in this way the ^{2}H lock channel may be used. Alternatively, by using an additional frequency synthesizer and an especially built tunable external lock probe, the proton signal of an external lock can be followed over the entire field range.

If none of these instrumental systems is used the field will be unlocked, and care should be taken to keep it as stable as possible. Long-term stability is not usually required, as the water signal can be observed with a single pulse, but short-term field variations during the sampling of the FID (typically from 0.1 to 10 s) will be reflected in the signal lineshape. Since a T_1 experiment requires reliable signal intensities to follow the recovery of magnetization on the z-axis, it is better to use the areas of the peaks rather then their height; alternatively, the initial intensity of the FID can be used directly in the diode-detection mode, instead of the phase-detection mode. In the diode-detection mode, several transients can be summed up to average out the noise without the worry of field fluctuations, which would alter the shape of the FID in the phase-detection mode. Since the field homogeneity of the magnet is in general optimized only for the usual high field working conditions, and for the dramatic decrease in sensitivity at low fields, the experiments are usually limited to single and intense resonance lines, which in general are solvent lines.

Another option for measurements at variable fields is offered by the use of shuttle probeheads [170–174]. In this experimental setup, only a single high field NMR instrument is used. The experiment is performed as follows: first the sample is allowed to build up the magnetization at high field, i.e., in the position where the RF coils are in any NMR probehead. Subsequently, the sample is mechanically or pneumatically shuttled at a given position along the z-axis that, due to the stray field, corresponds to a lower magnetic field. Here the magnetization is let decay for a certain (variable) time, then the sample is brought back into high field. In this way the measurement of the magnetization is performed at high field, with enough sensitivity and resolution to resolve biomolecular resonances [175]. To acquire a full decay curve, the sample is shuttled several times and kept at the lower field for several different time values. Different fields will be sampled by bringing the sample closer or farther from the center of the magnet.

12.5.2 FIELD-CYCLING RELAXOMETRY

Besides the complications inherent with the type of measurements just discussed, and their being rather tedious, the main disadvantage of performing NMRD by changing the operating magnetic field is the rather high lower limit of practically accessible fields. This problem has been overcome by using a completely different approach, developed in the 1950s by Redfield (e.g., [176]), pursued later to a high degree of refinement and reliability by Koenig and Brown [177,178] and marketed in a version derived from a design by F. Noack [179]. According to this approach, a solenoid without any iron core to generate a magnetic field is used. In this way the field can be turned on and off, or changed from one value to any other, rapidly; typical rise or fall times for the field are from 1 to 0.1 ms MHz^{-1}. Depending on the design of the solenoid, the highest magnetic field obtainable ranges between 0.5 and 1.2 T (from 20 to 50 MHz for protons).

T_1 is defined as the time constant for the first-order process in which the Boltzmann equilibrium of the macroscopic magnetization along the z-axis is reached from *any* nonequilibrium situation.

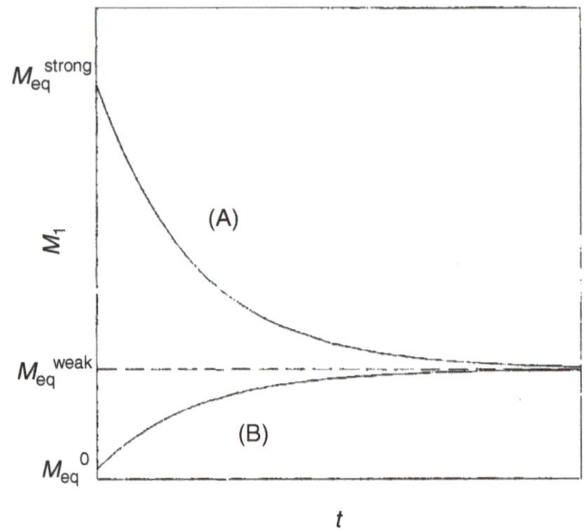

FIGURE 12.44 The time constant for the process of reaching the equilibrium value of the macroscopic magnetization at a certain weak field M_{eq}^{weak} is the same starting from a higher field (A) or from zero field (B).

Thus, the onset of the equilibrium when either a sample is placed into a magnetic field or after its magnetization has been reversed, for instance, follows the same rate law. Now, suppose a sample is taken from a strong magnetic field and rapidly put into a weak one: the magnetization will decay according to Fig. 12.44A. The time constant required to bring it to the equilibrium value at that field is the same as that for a sample that started at zero field (Fig. 12.44B). T_1 will then be the time constant for the field at which the magnetization evolves toward equilibrium: that is, the weak field (or measuring field).

If data points could be taken along the curve in Fig. 12.44A, this would measure the T_1 at the weak field by sampling magnetization values of the order of magnitude of those obtained in the high field magnet. This can, in fact, be accomplished by using the previously described solenoid, and a transmitter and receiver system tuned at a single frequency value (4 MHz in the following example). The procedure is as follows (Fig. 12.45):

1. The magnetic field is turned on and set to a value corresponding to, say, 50 MHz (Fig. 12.45A). It is maintained at this value for a time of the order of 5 $T_{1(50)}$ (where $T_{1(50)}$ is the T_1 at 50 MHz) to allow the magnetization on the z-axis to build up substantially (Fig. 12.45B).
2. The field is then lowered to a value (as low as 0.01 MHz, as in this example) corresponding to the desired Larmor frequency and kept at this value for a time τ (Fig. 12.45A), during which M_z decays toward its new equilibrium value, M_∞ (Fig. 12.45B).
3. The field is then raised to a value corresponding to a fixed observation frequency (4 MHz here, Fig. 12.45A). As soon as the field is set to 4 MHz a 90° or a 90°–t–180° pulse sequence is applied (with t of the order of milliseconds) and the receiver turned on (enlargement in Fig. 12.45B). The height of the echo that appears is proportional to the value of M_z at the time τ if the raise time of the magnetic field from 0.01 to 4 MHz and the time t are negligible compared to $T_{1(4)}$.

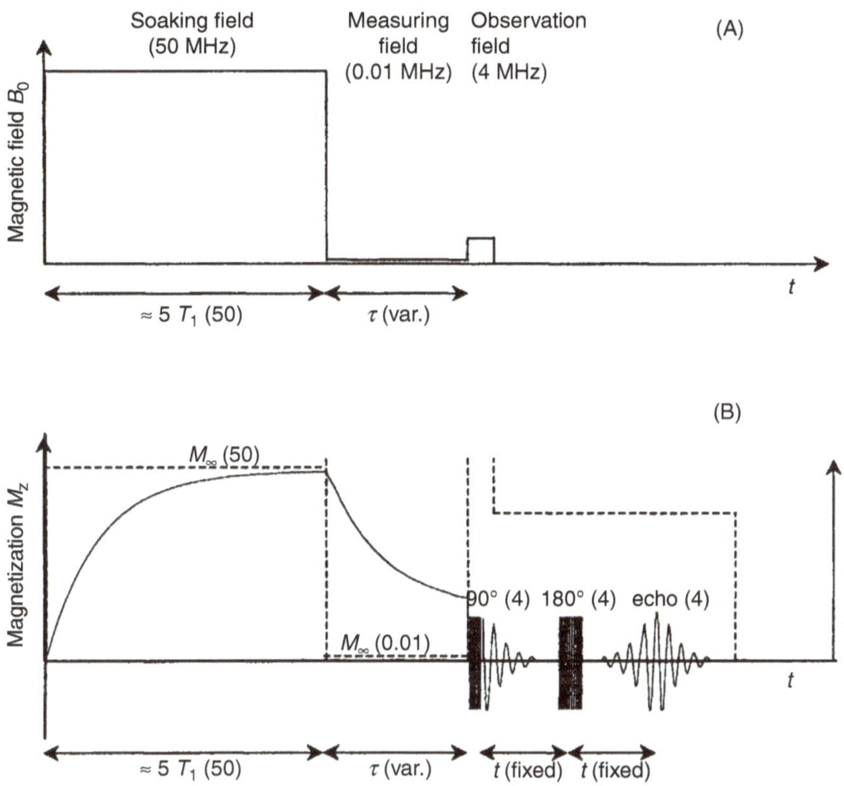

FIGURE 12.45 Pulsed field-water proton T_1 measurements. Sequence for measuring T_1 at fields lower than 22.5 MHz.

4. The field is then turned off to let the magnetization dissipate (and the system cool down), and the whole procedure is repeated using a different value of τ. T_1 can then be calculated for that particular magnetic field from the heights of the echoes, as a function of τ.

In summary, T_1 at 0.01 MHz is measured at 4 MHz by taking advantage of the sensitivity obtainable at 50 MHz. Of course, the same kind of measurement can be performed at any field value, provided it is not too close to 50 MHz, as the observed variation in M would then be too small. It should be remembered that the observation field is always 4 MHz, so only one transmitter and receiver system is needed, and there is no need to retune the coil, either.

When measuring relaxation rates at fields higher than 25 MHz, there is a larger change in magnetization when starting from zero magnetic field and going up as compared to starting from 50 MHz and going down. Therefore, for fields higher than 25 MHz the procedure is simplified as follows (Fig. 12.46): the magnetic field is turned on and set to the value of interest (for instance, 30 MHz, Fig. 12.46A) for a time τ, during which magnetization starts building up according to $T_{1(30)}$ (Fig. 12.46B); the field is then switched to the 4 MHz observation value and steps 3 and 4 are executed. Again, everything is repeated for different values of τ.

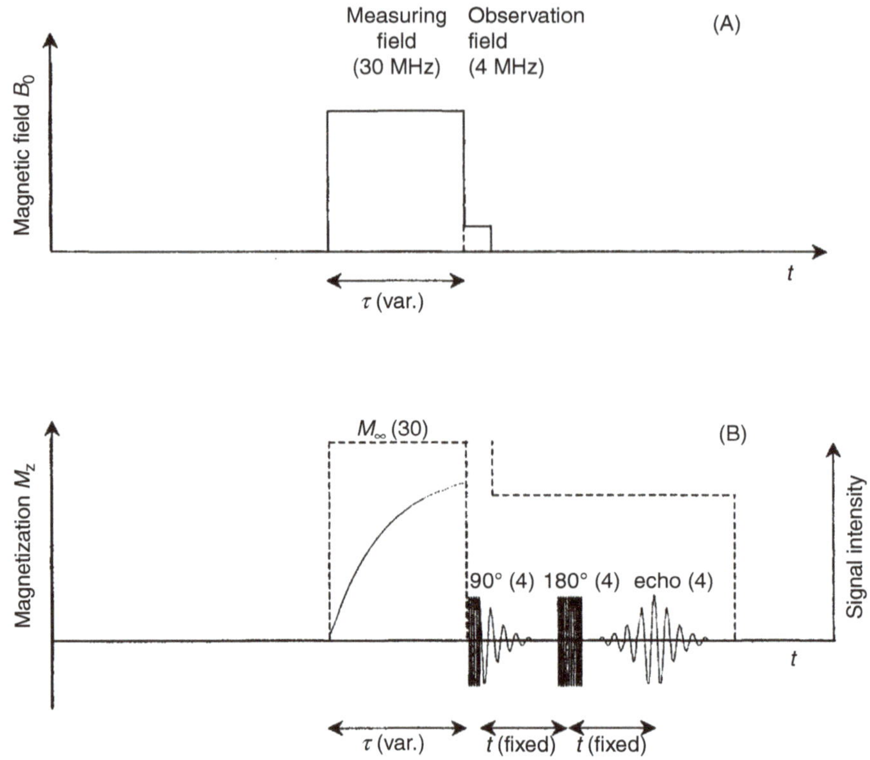

FIGURE 12.46 Pulsed field-water proton T_1 measurements. Sequence for measuring T_1 at fields higher than 22.5 MHz.

REFERENCES

[1] Nowakowski M, Saxena S, Stanek J, Zerko S, Kozminski W. Progr. Nucl. Magn. Reson. Spectrosc. 2015; 90–91:49.
[2] Zerko S, Kozminski W. J. Biomol. NMR 2015;63:283.
[3] Luchinat C, Piccioli M, Pierattelli R, Engelke F, Marquardsen T, Ruin R. J. Magn. Reson. 2001;150:161.
[4] Pell AJ, Clement RJ, Grey CP, Emsley L, Pintacuda G. J. Chem. Phys. 2013;138:114201.
[5] Sananes MT, Tuel A. Solid State Nucl. Magn. Reson. 1996;6:157.
[6] Sananes MT, Tuel A, Volta JC. J. Catal. 1994;145:251.
[7] Sananes MT, Tuel A, Volta JC. J. Catal. 1994;148:395.
[8] Tong YY. J. Magn. Reson. A 1996;119:22.
[9] Canesson L, Tuel A. Chem. Commun. 1997;2:241.
[10] Canesson L, Boudeville Y, Tuel A. J. Am. Chem. Soc. 1997;119:10754.
[11] Mali G, Ristic A, Kaucic V. J. Phys. Chem. B 2005;109:10711.
[12] Kim J, Middlemiss DS, Chernova NA, Zhu BYX, Masquelier C, Grey CP. J. Am. Chem. Soc. 2010;2010:5.
[13] Massiot D, Farnan I, Gautuer N, Trumeau D, Trokiner A, Coutures P. Solid State Nucl. Magn. Reson. 1995;4:241.
[14] Ash JT, Grandinetti PJ. Magn. Reson. Chem. 2006;44:823.

[15] Schurko RW, Wi S, Frydman L. J. Phys. Chem. A 2002;106:51.
[16] Piccioli M. J. Magn. Reson. Ser. B 1996;110:202.
[17] Kervern G, Steuernagel S, Engelke F, Pintacuda G, Emsley L. J. Am. Chem. Soc. 2007;129:14118.
[18] Babini E, Bertini I, Capozzi F, Felli IC, Lelli M, Luchinat C. J. Am. Chem. Soc. 2004;126:10496.
[19] Caillet-Saguy C, Delepierre M, Lecroisey A, Bertini I, Piccioli M, Turano P. J. Am. Chem. Soc. 2006;128:150.
[20] Bauer C, Freeman R, Frenkiel T, Keeler J, Shaka AJ. J. Magn. Reson. 1984;58:442.
[21] Temps AJ Jr, Brewer CF. J. Magn. Reson. 1984;56:355.
[22] Warren WS. J. Chem. Phys. 1984;81:5437.
[23] Emsley L, Bodenhausen G. Chem. Phys. Lett. 1990;165:469.
[24] Emsley L, Bodenhausen G. J. Magn. Reson. 1992;97:135.
[25] Morris GA, Freeman R. J. Magn. Reson. 1978;29:433.
[26] Kay LE, Marion D, Bax A. J. Magn. Reson. 1989;84:72.
[27] Redfield AG, Kunz SD, Ralph EK. J. Magn. Reson. 1975;19:114.
[28] Turner DL. J. Magn. Reson. 1983;54:146.
[29] Hore PJ. J. Magn. Reson. 1983;55:283.
[30] Plateau P, Guéron M. J. Am. Chem. Soc. 1982;104:7310.
[31] Sklenar V, Bax A. J. Magn. Reson. 1987;74:469.
[32] Piotto M, Saudek V, Sklenar V. J. Biomol. NMR 1992;2:661.
[33] Pell AJ, Pintacuda G, Emsley L. J. Chem. Phys. 2011;135:144201.
[34] Patt SL, Sykes BD. J. Chem. Phys. 1972;56:3182.
[35] Inubushi T, Becker ED. J. Magn. Reson. 1983;51:128.
[36] Bren KL, Gray HB, Banci L, Bertini I, Turano P. J. Am. Chem. Soc. 1995;117:8067.
[37] Hochmann J, Kellerhals H. J. Magn. Reson. 1980;38:23.
[38] Becker ED, Ferretti JA, Farrar TC. J. Am. Chem. Soc. 1969;91:7784.
[39] Bertini I, Canti G, Luchinat C, Mani F. J. Am. Chem. Soc. 1981;103:7784.
[40] Levitt MH, Freeman R, Frenkiel T. J. Magn. Reson. 1982;47:328.
[41] Shaka AJ, Keeler J, Frenkiel T, Freeman R. J. Magn. Reson. 1983;52:335.
[42] Shaka AJ, Barker PB, Freeman R. J. Magn. Reson. 1985;64:547.
[43] Shaka AJ, Lee CJ, Pines A. J. Magn. Reson. 1988;77:274.
[44] Nishiyama Y, Frey MH, Mukasa S, Utsumi H. J. Magn. Reson. 2010;202:135.
[45] Turano P, Lalli D, Felli IC, Theil EC, Bertini I. Proc. Natl. Acad. Sci. USA 2010;107:545.
[46] Mainz A, Religa TL, Sprangers R, Linser R, Kay LE, Reif B. Angew. Chem. Int. Ed. 2013;52:8746.
[47] Barbet-Massin E, Huang CT, Daebel V, Hsu ST, Reif B. Angew. Chem. Int. Ed. Engl. 2015;54:4367.
[48] Balayssac S, Bertini I, Bhaumik A, Lelli M, Luchinat C. Proc. Natl. Acad. Sci. USA 2008;105:17284.
[49] Bertini I, Bhaumik A, De Paepe G, Griffin RG, Lelli M, Lewandowski JR, Luchinat C. J. Am. Chem. Soc. 2010;132:1032.
[50] Bertini I, Emsley L, Lelli M, Luchinat C, Mao J, Pintacuda G. J. Am. Chem. Soc. 2010;132:5558.
[51] Balayssac S, Bertini I, Lelli M, Luchinat C, Maletta M, Yeo KJ. J. Am. Chem. Soc. 2007;129:2218.
[52] Balayssac S, Bertini I, Falber K, Fragai M, Jehle S, Lelli M, Luchinat C, Oschkinat H, Yeo KJ. ChemBioChem 2007;8:486.
[53] Pell AJ, Pintacuda G. Prog. Nucl. Magn. Reson. Spectrosc. 2015;84-85:33.
[54] Pandey MK, Zhang R, Hashi K, Ohki S, Nishijima G, Matsumoto S, Noguchi T, Deguchi K, Goto A, Shimizu T, Maeda H, Takahashi M, Yanagisawa Y, Yamazaki T, Iguchi S, Tanaka R, Nemoto T, Miyamoto T, Suematsu H, Saito K, Miki T, Ramamoorthy A, Nishiyama Y. J. Magn. Reson. 2015;261:1.
[55] Pell AJ, Kervern G, Emsley L, Deschamps M, Massiot D, Grandinetti PJ, Pintacuda G. J. Chem. Phys. 2011;134:024117.
[56] Walder BJ, Dey KK, Kaseman DC, Baltisberger JH, Grandinetti PJ. J. Chem. Phys. 2013;138:174203.
[57] Bertini I, Donaire A, Luchinat C, Rosato A. Proteins Struct. Funct. Genet. 1997;29:348.

[58] Vold RL, Waugh JS, Klein MP, Phelps DE. J. Chem. Phys. 1968;48:3831.
[59] Arseniev A, Sobol AG, Bystrov VF. J. Magn. Reson. 1986;70:427.
[60] Bertini I, Donaire A, Felli IC, Luchinat C, Rosato A. Magn. Reson. Chem. 1996;34:948.
[61] Bertini I, Couture MMJ, Donaire A, Eltis LD, Felli IC, Luchinat C, Piccioli M, Rosato A. Eur. J. Biochem. 1996;241:440.
[62] Huber JG, Moulis J-M, Gaillard J. Biochemistry 1996;35:12705.
[63] Kay LE, Torchia DA, Bax A. Biochemistry 1989;28:8972.
[64] Peng JW, Wagner G. J. Magn. Reson. 1992;98:308.
[65] Meiboom S, Gill D. Rev. Sci. Instr. 1960;29:688.
[66] Kay LE, Nicholson LK, Delaglio F, Bax A, Torchia DA. J. Magn. Reson. 1992;97:359.
[67] Korzhnev DM, Tischenko EV, Arseniev AS. J. Biomol. NMR 2000;17:231.
[68] Iwahara J, Tang C, Clore GM. J. Magn. Reson. 2007;184:185.
[69] Bertini I, Luchinat C, Tarchi D. Chem. Phys. Lett. 1993;203:445.
[70] Bertini I, Luchinat C, Piccioli M, Tarchi D. Concepts Magn. Reson. 1994;6:307.
[71] Mori M, Kateb F, Bodenhausen G, Piccioli M, Abergel D. J. Am. Chem. Soc. 2010;132:3594.
[72] Granot J. J. Magn. Reson. 1982;49:257.
[73] La Mar GN, de Ropp JSIn: Berliner LJ, Reuben J, editors. Biological Magnetic Resonance, vol. 12. New York: Plenum Press; 1993. p. 1.
[74] Bertini I, Luchinat C, Rosato A. Progr. Biophys. Mol. Biol. 1996;66:43.
[75] Banci L. In: Berliner LJ, Reuben J, editors. Biological Magnetic Resonance. New York: Plenum; 1993. p. 79.
[76] Bertini I, Capozzi F, Ciurli S, Luchinat C, Messori L, Piccioli M. J. Am. Chem. Soc. 1992;114:3332.
[77] Banci L, Bertini I, Luchinat C, Piccioli M. FEBS Lett. 1990;272:175.
[78] Banci L, Bertini I, Luchinat C, Viezzoli MS. Inorg. Chem. 1990;29:1438.
[79] Bertini I, Luchinat C, Piccioli M. Progr. Nucl. Magn. Reson. Spectrosc. 1994;26:91.
[80] Lecomte JTJ, Unger SW, La Mar GN. J. Magn. Reson. 1991;94:112.
[81] Banci L, Bertini I, Capozzi F, Carloni P, Ciurli S, Luchinat C, Piccioli M. J. Am. Chem. Soc. 1993;115:3431.
[82] Kazimierczuk K, Zawadzka A, Kozminski W, Zhukov I. J. Magn. Reson. 2007;188:344.
[83] Kazimierczuk K, Zawadzka A, Kozminski W. J. Magn. Reson. 2008;192:123.
[84] Santos H, Turner DL, Xavier AV, LeGall J. J. Magn. Reson. 1984;59:177.
[85] Peters W, Fuchs M, Sicius H, Kuchen W. Angew. Chem. 1985;24:231.
[86] Jenkins BG, Lauffer RB. J. Magn. Reson. 1988;80:328.
[87] Jenkins BG, Lauffer RB. Inorg. Chem. 1988;27:4730.
[88] Luchinat C, Steuernagel S, Turano P. Inorg. Chem. 1990;29:4351.
[89] Bertini I, Briganti F, Luchinat C, Messori L, Monnanni R, Scozzafava A, Vallini G. Eur. J. Biochem. 1992;204:831.
[90] Kessler H, Oschkinat H, Griesinger C, Bermel W. J. Magn. Reson. 1986;70:106.
[91] Bertini I, Sacconi L, Speroni GP. Inorg. Chem. 1972;11:1323.
[92] C. Luchinat, M. Piccioli, In: G.N. La Mar (Ed.) NMR of Paramagnetic Macromolecules. NATO ASI Series. Kluwer Academic, Dordrecth, 1995.
[93] Bothner-By AA, Stephens RL, Lee J, Warren CD, Jeanloz RW. J. Am. Chem. Soc. 1984;106:811.
[94] Bertini I, Coutsolelos AG, Dikiy A, Luchinat C, Spyroulias GA, Troganis A. Inorg. Chem. 1996;35:6308.
[95] Bertini I, Briganti F, Luchinat C, Messori L, Monnanni R, Scozzafava A, Vallini G. FEBS Lett. 1991;289:253.
[96] Nagayama K, Kumar A, Wüthrich K, Ernst RR. J. Magn. Reson. 1980;40:321.
[97] Bertini I, Luchinat C, Messori L, Vasak M. Eur. J. Biochem. 1993;211:235.
[98] Otting G. Progr. Nucl. Magn. Reson. Spectrosc. 1997;31:259.
[99] Dalvit C, Hommel U. J. Magn. Reson. Ser. B 1995;109:334.
[100] Bertini I, Dalvit C, Huber JG, Luchinat C, Piccioli M. FEBS Lett. 1997;415:45.

[101] Bertini I, Huber JG, Luchinat C, Piccioli M. J. Magn. Reson. 2000;147:1.
[102] Matzapetakis M, Turano P, Theil EC, Bertini I. J. Biomol. NMR 2007;38:237.
[103] Bax A. Two Dimensional Nuclear Magnetic Resonance in LiquidsDordrecht: Reidel; 1982.
[104] Shriver J. Concepts Magn. Reson. 1992;4:1.
[105] Xavier AV, Turner DL, Santos H. Methods Enzymol. 1993;227:1.
[106] Talluri S, Scheraga HA. J. Magn. Reson. 1990;86:1.
[107] Wimperis S, Bodenhausen G. Mol. Phys. 1989;66:897.
[108] Ghose R, Prestegard JH. J. Magn. Reson. 1997;128:138.
[109] Boisbouvier J, Gans P, Blackledge M, Brutscher B, Marion D. J. Am. Chem. Soc. 1999;121:7700.
[110] Bertini I, Luchinat C, Rosato A. Chem. Phys. Lett. 1996;250:495.
[111] Braunschweiler L, Ernst RR. J. Magn. Reson. 1983;53:521.
[112] Bax A, Davis DG. J. Magn. Reson. 1985;65:355.
[113] Sadek M, Brownlee RTC, Scrofani SDB, Wedd AG. J. Magn. Reson. 1993;101:309.
[114] Griesinger C, Otting G, Wüthrich K, Ernst RR. J. Am. Chem. Soc. 1988;110:7870.
[115] Müller L. J. Am. Chem. Soc. 1979;101:4481.
[116] Bax A, Griffey RH, Hawkins BL. J. Magn. Reson. 1983;55:301.
[117] Ottiger M, Delaglio F, Bax A. J. Magn. Reson. 1998;131:373.
[118] Yao L, Ying J, Bax A. J. Biomol. NMR 2009;43:161.
[119] Meissner A, Duus JO, Sørensen OW. J. Biomol. NMR 1997;10:89.
[120] Lerche MH, Meissner A, Poulsen FM, Sørensen OW. J. Magn. Reson. 1999;140:259.
[121] Pervushin K, Riek R, Wider G, Wüthrich K. J. Am. Chem. Soc. 1998;120:6394.
[122] Brutscher B. Concepts Magn. Reson. 2000;12:207.
[123] Bertini I, Luchinat C, Macinai R, Piccioli M, Scozzafava A, Viezzoli MS. J. Magn. Reson. Ser. B 1994;B104:95.
[124] Ciofi-Baffoni S, Gallo A, Muzzioli R, Piccioli M. J. Biomol. NMR 2014;58:123.
[125] Bertini I, Lee Y-M, Luchinat C, Piccioli M, Poggi L. ChemBioChem 2001;2:550.
[126] Schanda P, Kupce E, Brutscher B. J. Biomol. NMR 2005;33:199.
[127] Lescop E, Schanda P, Brutscher B. J. Magn. Reson. 2007;187:163.
[128] Boelens R, Vuister GW, Koning TMG, Kaptein R. J. Am. Chem. Soc. 1989;111:8525.
[129] Banci L, Bermel W, Luchinat C, Pierattelli R, Tarchi D. Magn. Reson. Chem. 1993;31:S3–7.
[130] Bertini I, Dikiy A, Luchinat C, Piccioli M, Tarchi D. J. Magn. Reson. Ser. B 1994;103:278.
[131] Banci L, Bertini I, Eltis LD, Felli IC, Kastrau DHW, Luchinat C, Piccioli M, Pierattelli R, Smith M. Eur. J. Biochem. 1994;225:715.
[132] Ferella L, Rosato A, Turano P. In: Bertini I, McGreevy KS, Parigi G, editors. NMR of Biomolecules-Towards Mechanistic Systems Biology. Wiley-Blackwell: Weinheim; 2012. p. 53.
[133] Knight MJ, Webber AL, Pell AJ, Guerry P, Barbet-Massin E, Bertini I, Felli IC, Gonnelli L, Pierattelli R, Emsley L, Lesage A, Hermann T, Pintacuda G. Angew. Chem. Int. Ed. 2011;50:11697.
[134] Shi C, Fasshuber HK, Chevelkov V, Xiang S, Habenstein B, Vasa SK, Becker S, Lange A. J. Biomol. NMR 2014;59:15.
[135] Schuetz A, Wasmer C, Habenstein B, Verel R, Greenwald J, Riek R, Böckmann A, Meier BH. ChemBioChem 2010;11:1543.
[136] Barbet-Massin E, Pell AJ, Retel JS, Andreas LB, Jaudzems K, Franks WT, Nieuwkoop AJ, Hiller M, Higman V, Guerry P, Bertarello A, Knight MJ, Felletti M, Le MT, Kotelovica S, Akopjana I, Tars K, Stoppini M, Bellotti V, Bolognesi M, Ricagno S, Chou JJ, Griffin RG, Oschkinat H, Lesage A, Emsley L, Herrmann T, Pintacuda G. J. Am. Chem. Soc. 2014;136:12489.
[137] Bennett AE, Griffin RG, Ok JH, Vega S. J. Chem. Phys. 1992;96:8624.
[138] Bayro MJ, Huber M, Ramachandran R, Davenport TC, Meier BH, Ernst M, Griffin RG. J. Chem. Phys. 2009;130:114506.

[139] Bayro MJ, Ramachandran R, Caporini MA, Eddy MT, Griffin RG. J. Chem. Phys. 2008;128:052321.
[140] Bloembergen N. Physica 1949;15:386.
[141] Takegoshi K, Nakamura S, Terao T. Chem. Phys. Lett. 1999;307:295.
[142] Takegoshi K, Nakamura S, Terao T. Chem. Phys. Lett. 2001;344:631.
[143] Takegoshi K, Nakamura S, Terao T. J. Chem. Phys. 2003;118:2325.
[144] Swamy SKK, Karczmarska A, Makowska-Janusik M, Kassiba A, Dittmer J. ChemPhysChem 2013;14:1864.
[145] Li S, Trebosc J, Lafon O, Zhou L, Shen M, Pourpoint F, Amoureux JP, Deng F. J. Magn. Reson. 2015;251:36.
[146] Reif B, Griffin RG. J. Magn. Reson. 2003;160:78.
[147] Chevelkov V, van Rossum BJ, Castellani F, Rehbein K, Diehl A, Hohwy M, Steuernagel S, Engelke F, Oschkinat H, Reif B. J. Am. Chem. Soc. 2003;125:7788.
[148] Zhou DH, Shah G, Cormos M, Mullen C, Sandoz D, Rienstra CM. J. Am. Chem. Soc. 2007;129:11791.
[149] Wickramasinghe NP, Shaibat M, Ishii Y. J. Am. Chem. Soc. 2005;127:5796.
[150] Wickramasinghe NP, Shaibat MA, Ishii Y. J. Phys. Chem. B 2007;111:9693.
[151] Wickramasinghe NP, Shaibat MA, Jones CR, Casabianca LB, de Dios AC, Harwood JS, Ishii Y. J. Chem. Phys. 2008;128:052210.
[152] Ishii Y, Wickramasinghe NP, Chimon S. J. Am. Chem. Soc. 2003;125:3438.
[153] Barbet-Massin E, Pintacuda G. In: Bertini I, McGreevy KS, Parigi G, editors. NMR of Biomolecules: Towards Mechanistic Systems Biology. Weinheim, Germany: Wiley-VCH Verlag GmbH & Co. KGaA; 2012.
[154] Laage S, Marchetti A, Sein J, Pierattelli R, Sass HJ, Grzesiek S, Lesage A, Pintacuda G, Emsley L. J. Am. Chem. Soc. 2008;130:17216.
[155] Knight MJ, Pell AJ, Bertini I, Felli IC, Gonnelli L, Pierattelli R, Hermann T, Emsley L, Pintacuda G. Proc. Natl. Acad. Sci. USA 2012;109:11095.
[156] Fyfe CA. Solid State NMR for ChemistsGuelph, Ontario, Canada: C.E.C. Press; 1983.
[157] McGarvey BR, Nagy S. Inorg. Chem. 1987;26:4198.
[158] Anderson SR. J. Organomet. Chem. 1974;71:263.
[159] Sandreczki T, Ondercin D, Kreilick RW. J. Magn. Reson. 1979;34:171.
[160] Reuveni A, McGarvey BR. J. Magn. Reson. 1979;34:181.
[161] Soda G, Chiba T. J. Chem. Phys. 1969;50:439.
[162] Wittebort RJ. J. Magn. Reson. 1989;83:626.
[163] Aime S, Bertini I, Luchinat C. Coord. Chem. Rev. 1996;150:221.
[164] Herzfeld J, Berger AE. J. Chem. Phys. 1980;73:6021.
[165] M.H. Levitt, In: D.M. Grant, R.K. Harris (Eds.) Chichester 165, 2002.
[166] Riek R, Wider G, Pervushin K, Wüthrich K. Proc. Natl. Acad. Sci. USA 1999;96:4918.
[167] Riek R, Pervushin K, Wüthrich K. TIBS 2000;25:462.
[168] Ollerenshaw JE, Tugarinov V, Kay LE. Magn. Reson. Chem. 2003;41:843.
[169] Bertini I, Cavallaro G, Cosenza M, Kümmerle R, Luchinat C, Piccioli M, Poggi L. J. Biomol. NMR 2002;23:115.
[170] Redfield AG. Magn. Reson. Chem. 2003;41:753.
[171] Redfield AG. J. Biomol. NMR 2012;52:159.
[172] Victor K, Kavolius V, Bryant RG. J. Magn. Reson. 2004;171:253.
[173] Chou CY, Chu M, Chang CF, Huang TH. J. Magn. Reson. 2012;214:302.
[174] Reese M, Türke M-T, Tkach I, Parigi G, Luchinat C, Marquardsen T, Tavernier A, Höfer P, Engelke F, Griesinger C, Bennati M. J. Am. Chem. Soc. 2009;131:15086.
[175] Charlier C, Khan SN, Marquardsen T, Pelupessy P, Reiss V, Sakellariou D, Bodenhausen G, Engelke F, Ferrage F. J. Am. Chem. Soc. 2013;135:18665.
[176] Anderson AG, Redfield AG. Phys. Rev. 1959;116:583.
[177] Hallenga K, Koenig SH. Biochemistry 1976;15:4255.
[178] Brown RD III, Brewer C, Koenig SH. Biochemistry 1977;16:3883.
[179] Ferrante G, Sykora S. Adv. Inorg. Chem. 2005;57:405.

NMR PROPERTIES OF NUCLEI

I

Isotope	Spin (I)	Natural Abundance (%)	Relative Sensitivity[a]	g_I	γ_I[b]	NMR Frequency (MHz at 2.3488 T)	Q (10^{-28} m^2)
^{1}H	1/2	99.98	1.00	5.5856912	2.6751987×10^{8}	100.000	–
^{2}H	1	1.56×10^{-2}	9.65×10^{-3}	0.8574376	4.106593×10^{7}	15.351	2.875×10^{-3}
^{3}H	1/2	0	1.21	5.957920	2.853473×10^{8}	106.664	–
^{3}He	1/2	1.3×10^{-4}	0.442	–4.255248	-2.037999×10^{8}	76.181	–
^{6}Li	1	7.42	8.50×10^{-3}	0.8220467	3.937092×10^{7}	14.717	-6.44×10^{-4}
^{7}Li	3/2	92.58	0.294	2.170949	1.039750×10^{8}	38.866	-3.66×10^{-2}
^{9}Be	3/2	100	1.39×10^{-2}	–0.7852	-3.7606×10^{7}	14.057	5.3×10^{-2}
^{10}B	3	19.58	1.99×10^{-2}	0.600217	2.87466×10^{7}	10.746	8.472×10^{-3}
^{11}B	3/2	80.42	0.165	1.792425	8.58460×10^{7}	32.090	4.065×10^{-2}
^{13}C	1/2	1.108	1.59×10^{-2}	1.404822	6.72822×10^{7}	25.150	–
^{14}N	1	99.63	1.01×10^{-3}	0.4037607	1.933763×10^{7}	7.228	1.56×10^{-2}
^{15}N	1/2	0.37	1.04×10^{-3}	–0.5663784	-2.712600×10^{7}	10.140	–
^{17}O	5/2	3.7×10^{-2}	2.91×10^{-2}	–0.75752	-3.62803×10^{7}	13.562	-2.578×10^{-2}
^{19}F	1/2	100	0.834	5.257732	2.518127×10^{8}	94.129	–
^{21}Ne	3/2	0.257	2.46×10^{-3}	–0.441197	-2.11306×10^{7}	7.899	1.029×10^{-1}
^{23}Na	3/2	100	9.27×10^{-2}	1.478347	7.080361×10^{7}	26.467	1.01×10^{-1}
^{25}Mg	5/2	10.13	2.68×10^{-3}	–0.34218	-1.63883×10^{7}	6.126	0.22
^{27}Al	5/2	100	0.207	1.456602	6.976216×10^{7}	26.077	0.140
^{29}Si	1/2	4.7	7.86×10^{-3}	–1.11058	-5.3190×10^{7}	19.883	–
^{31}P	1/2	100	6.65×10^{-2}	2.26320	1.08393×10^{8}	40.518	–
^{33}S	3/2	0.76	2.27×10^{-3}	0.429214	2.05567×10^{7}	7.684	-6.4×10^{-2}
^{35}Cl	3/2	75.53	4.72×10^{-3}	0.5479157	2.624176×10^{7}	9.809	-8.249×10^{-2}
^{37}Cl	3/2	24.47	2.72×10^{-3}	0.4560820	2.184349×10^{7}	8.165	-6.493×10^{-2}
^{39}K	3/2	93.1	5.10×10^{-4}	0.2609772	1.249918×10^{7}	4.672	4.9×10^{-2}

(Continued)

NMR of Paramagnetic Molecules. http://dx.doi.org/10.1016/B978-0-444-63436-8.00013-2

Isotope	Spin (I)	Natural Abundance (%)	Relative Sensitivity[a]	g_I	γ_I[b]	NMR Frequency (MHz at 2.3488 T)	Q (10^{-28} m^2)
^{41}K	3/2	6.88	8.43×10^{-5}	0.1432466	6.860621×10^{6}	2.565	6.0×10^{-2}
^{43}Ca	7/2	0.145	6.42×10^{-3}	–0.37636	-1.80253×10^{7}	6.738	–
^{45}Sc	7/2	100	0.302	1.358995	6.508741×10^{7}	24.330	–0.22
^{47}Ti	5/2	7.28	2.10×10^{-3}	–0.31539	-1.51053×10^{7}	5.646	0.29
^{49}Ti	7/2	5.51	3.78×10^{-3}	–0.31548	-1.51094×10^{7}	5.648	0.24
^{50}V	6	0.24	5.58×10^{-2}	0.557908	2.67203×10^{7}	9.988	7×10^{-2}
^{51}V	7/2	99.76	0.384	1.47183	7.0491×10^{7}	26.350	-5.2×10^{-2}
^{53}Cr	3/2	9.55	9.08×10^{-4}	–0.31636	-1.51517×10^{7}	5.664	2.2×10^{-2}
^{55}Mn	5/2	100.00	0.176	1.3813	6.6155×10^{7}	24.729	0.40
^{57}Fe	1/2	2.19	3.42×10^{-5}	0.1812459	8.680549×10^{6}	3.245	–
^{59}Co	7/2	100.00	0.278	1.322	6.332×10^{7}	23.67	0.404
^{61}Ni	3/2	1.19	3.59×10^{-3}	–0.50001	-2.39475×10^{7}	8.952	0.162
^{63}Cu	3/2	69.09	9.34×10^{-2}	1.4822	7.0988×10^{7}	26.536	–0.209
^{65}Cu	3/2	30.91	0.115	1.5878	7.6046×10^{7}	28.426	–0.195
^{67}Zn	5/2	4.11	2.87×10^{-3}	0.350191	1.677198×10^{7}	6.269	0.150
^{69}Ga	3/2	60.4	6.97×10^{-2}	1.34439	6.43881×10^{7}	24.069	0.168
^{71}Ga	3/2	39.60	0.143	1.70818	8.18112×10^{7}	30.581	0.106
^{73}Ge	9/2	7.76	1.41×10^{-3}	–0.1954371	-9.360221×10^{6}	3.499	–0.173
^{75}As	3/2	100.00	2.54×10^{-2}	0.95965	4.5961×10^{7}	17.180	0.29
^{77}Se	1/2	7.58	7.03×10^{-3}	1.07012	5.1252×10^{7}	19.158	–
^{79}Br	3/2	50.54	7.94×10^{-2}	1.404266	6.72556×10^{7}	25.140	0.293
^{81}Br	3/2	49.46	9.95×10^{-2}	1.513707	7.24971×10^{7}	27.100	0.27
^{83}Kr	9/2	11.55	1.90×10^{-3}	–0.215704	-1.033089×10^{7}	3.862	0.270
^{85}Rb	5/2	72.15	1.06×10^{-2}	0.54121	2.59207×10^{7}	9.689	0.274
^{87}Rb	3/2	27.85	0.177	1.83416	8.78449×10^{7}	32.837	0.132
^{87}Sr	9/2	7.02	2.71×10^{-3}	–0.242849	-1.16309×10^{7}	4.348	0.15
^{89}Y	1/2	100.00	1.19×10^{-4}	–0.2748306	-1.316268×10^{7}	4.920	–
^{91}Zr	5/2	11.23	9.49×10^{-3}	–0.52145	-2.49741×10^{7}	9.335	–
^{93}Nb	9/2	100.00	0.488	1.37122	6.5673×10^{7}	24.549	–0.36
^{95}Mo	5/2	15.72	3.27×10^{-3}	–0.3657	-1.7514×10^{7}	6.547	-1.9×10^{-2}
^{97}Mo	5/2	9.46	3.49×10^{-3}	–0.3734	-1.7884×10^{7}	6.785	–0.102
^{99}Ru	5/2	12.72	1.13×10^{-3}	–0.2565	-1.2286×10^{7}	4.592	7.6×10^{-2}

Isotope	Spin (I)	Natural Abundance (%)	Relative Sensitivity[a]	g_I	γ_I[b]	NMR Frequency (MHz at 2.3488 T)	Q (10^{-28} m^2)
^{101}Ru	5/2	17.07	1.59×10^{-3}	−0.2875	-1.3770×10^{7}	5.147	0.44
^{103}Rh	1/2	100.00	3.17×10^{-5}	−0.17680	-8.468×10^{6}	3.165	–
^{105}Pd	5/2	22.23	1.13×10^{-3}	−0.2568	-1.230×10^{7}	4.60	0.8
^{107}Ag	1/2	51.82	6.72×10^{-5}	−0.227140	-1.08786×10^{7}	4.066	–
^{109}Ag	1/2	48.18	1.02×10^{-4}	−0.2613810	-1.251852×10^{7}	4.679	–
^{111}Cd	1/2	12.75	9.66×10^{-3}	−1.1897712	-5.698264×10^{7}	21.300	–
^{113}Cd	1/2	12.26	1.11×10^{-2}	−1.2446010	-5.960865×10^{7}	22.282	–
^{113}In	9/2	4.28	0.351	1.22864	5.8844×10^{7}	21.996	0.846
^{115}In	9/2	95.72	0.353	1.23129	5.8971×10^{7}	2.044	0.861
^{115}Sn	1/2	0.35	3.56×10^{-2}	−1.83766	-8.8012×10^{7}	32.899	–
^{117}Sn	1/2	7.61	4.60×10^{-2}	−2.00208	-9.5887×10^{7}	35.843	–
^{119}Sn	1/2	8.58	5.27×10^{-2}	−2.09456	-1.00316×10^{8}	37.499	–
^{121}Sb	5/2	57.25	0.163	1.34536	6.4434×10^{7}	24.086	−0.20
^{123}Sb	7/2	42.75	4.66×10^{-2}	0.72851	3.4891×10^{7}	13.043	−0.26
^{123}Te	1/2	0.87	1.84×10^{-2}	−1.47358	-7.0575×10^{7}	26.381	–
^{125}Te	1/2	6.99	3.22×10^{-2}	−1.77656	-8.5086×10^{7}	31.806	–
^{127}I	5/2	100.00	9.54×10^{-2}	1.12531	5.38953×10^{7}	20.146	−0.789
^{129}Xe	1/2	26.44	2.16×10^{-2}	−1.555952	-7.45204×10^{7}	27.856	–
^{131}Xe	3/2	21.18	2.82×10^{-3}	0.461241	2.209056×10^{7}	8.258	−0.120
^{133}Cs	7/2	100.00	4.84×10^{-2}	0.7377209	3.533224×10^{7}	13.207	-3×10^{-3}
^{135}Ba	3/2	6.59	5.00×10^{-3}	0.558629	2.675484×10^{7}	10.001	0.18
^{137}Ba	3/2	11.32	7.00×10^{-3}	0.624910	2.992930×10^{7}	11.188	0.28
^{138}La	5	0.089	9.41×10^{-2}	0.74278	3.5575×10^{7}	13.298	0.51
^{139}La	7/2	99.911	6.06×10^{-2}	0.79520	3.8085×10^{7}	14.236	0.22
^{141}Pr	5/2	100.00	0.303	1.6544	7.924×10^{7}	29.62	-5.89×10^{-2}
^{143}Nd	7/2	12.17	3.39×10^{-3}	−0.3043	-1.457×10^{7}	5.45	−0.484
^{145}Nd	7/2	8.30	7.93×10^{-4}	−0.1874	-8.98×10^{6}	3.36	−0.253
^{147}Sm	7/2	14.97	1.52×10^{-3}	−0.23280	-1.1150×10^{7}	4.168	−0.18
^{149}Sm	7/2	13.83	8.52×10^{-4}	−0.19191	-9.192×10^{6}	3.436	5.2×10^{-2}
^{151}Eu	5/2	47.82	0.179	1.3887	6.6509×10^{7}	24.861	1.15
^{153}Eu	5/2	52.18	1.54×10^{-2}	0.6132	2.9368×10^{7}	10.978	2.94
^{155}Gd	3/2	14.73	1.48×10^{-4}	−0.1727	-8.273×10^{6}	3.092	1.59

(Continued)

Isotope	Spin (I)	Natural Abundance (%)	Relative Sensitivity[a]	g_I	γ_I[b]	NMR Frequency (MHz at 2.3488 T)	Q (10^{-28} m^2)
^{157}Gd	3/2	15.68	3.34×10^{-4}	–0.2265	-1.0850×10^{7}	4.056	2.03
^{159}Tb	3/2	100.00	6.94×10^{-2}	1.343	6.431×10^{7}	24.04	1.18
^{161}Dy	5/2	18.88	4.75×10^{-4}	–0.1922	-9.205×10^{6}	3.441	2.44
^{163}Dy	5/2	24.97	1.30×10^{-3}	0.2690	1.2885×10^{7}	4.817	2.57
^{165}Ho	7/2	100.00	0.204	1.1923	5.710×10^{7}	21.35	2.73
^{167}Er	7/2	22.94	5.11×10^{-4}	–0.16186	-7.752×10^{6}	2.898	2.827
^{169}Tm	1/2	100.00	5.70×10^{-4}	–0.4632	-2.218×10^{7}	8.293	–
^{171}Yb	1/2	14.31	5.52×10^{-3}	0.98734	4.7287×10^{7}	17.676	–
^{173}Yb	5/2	16.13	1.35×10^{-3}	–0.271956	-1.30250×10^{7}	4.869	2.8
^{175}Lu	7/2	97.41	3.13×10^{-2}	0.63791	3.0552×10^{7}	11.421	5.68
^{176}Lu	7	2.59	4.05×10^{-2}	0.456	2.18×10^{7}	8.16	8
^{177}Hf	7/2	18.5	1.40×10^{-3}	0.22671	1.0858×10^{7}	4.059	4.5
^{179}Hf	9/2	13.75	5.47×10^{-4}	–0.14242	-6.821×10^{6}	2.550	5.1
^{181}Ta	7/2	99.99	3.74×10^{-2}	0.6771	3.243×10^{7}	12.12	3.9
^{183}W	1/2	14.40	7.50×10^{-5}	0.2355694	1.128231×10^{7}	4.217	–
^{185}Re	5/2	37.07	0.139	1.2748	6.1057×10^{7}	22.823	2.36
^{187}Re	5/2	62.93	0.143	1.2879	6.1681×10^{7}	23.057	2.24
^{187}Os	1/2	1.64	1.24×10^{-5}	0.12930	6.193×10^{6}	2.315	–
^{189}Os	3/2	16.10	2.44×10^{-3}	0.439955	2.107112×10^{7}	7.876	0.91
^{191}Ir	3/2	37.30	2.65×10^{-5}	0.0974	4.665×10^{6}	1.744	0.78
^{193}Ir	3/2	62.70	3.42×10^{-5}	0.1061	5.080×10^{6}	1.899	0.70
^{195}Pt	1/2	33.80	1.04×10^{-2}	1.21898	5.8382×10^{7}	21.823	–
^{197}Au	3/2	100.00	2.76×10^{-5}	0.098772	4.73056×10^{6}	1.768	0.594
^{199}Hg	1/2	16.84	5.94×10^{-3}	1.0117702	4.845750×10^{7}	18.114	–
^{201}Hg	3/2	13.22	1.49×10^{-3}	–0.373483	-1.788753×10^{7}	6.686	0.455
^{203}Tl	1/2	29.50	0.196	3.244514	1.553920×10^{8}	58.086	–
^{205}Tl	1/2	70.50	0.202	3.2764268	1.5692047×10^{8}	58.657	–
^{207}Pb	1/2	22.60	9.06×10^{-3}	1.16438	5.57666×10^{7}	20.846	–
^{209}Bi	9/2	100.00	0.144	0.91347	4.3749×10^{7}	16.354	–0.46
^{235}U	7/2	0.72	1.2×10^{-4}	–0.10	-4.8×10^{6}	1.8	4.55

[a] *NMR sensitivity at constant field and equal number of nuclei, relative to* ^1H *nuclei, given by* $(4/3)(|g_I|/|g_H|)^3 I(I+1)$.
[b] γ_I *is defined by:* $\gamma_I = g_I\mu_N/\hbar$, *where the nuclear magneton,* μ_N *is* 5.050824×10^{-27} $J\,T^{-1}$, *and* $\hbar$ *is* $1.0545877 \times 10^{-34}$ $J\,s\,rad^{-1}$.

DIPOLAR COUPLING BETWEEN TWO SPINS II

The Hamiltonian for the dipolar coupling between two magnetic moments associated with the spins I and J is given by an equation similar to Eq. (1.1)

$$H = -\frac{\mu_0}{4\pi}\frac{\hbar^2\gamma_I\gamma_J}{r^3}[3(\boldsymbol{I}\cdot\boldsymbol{r})(\boldsymbol{J}\cdot\boldsymbol{r}) - \boldsymbol{I}\cdot\boldsymbol{J}] \tag{II.1}$$

where $\boldsymbol{r}$ is a unit vector oriented along the line of length r connecting $\boldsymbol{I}$ and $\boldsymbol{J}$. In the most general case, the $\boldsymbol{I}$ and $\boldsymbol{J}$ vectors are not parallel, and each of them will form different angles with $\boldsymbol{r}$ and with the applied magnetic field, $\boldsymbol{B}_0$. If the Zeeman interaction is the dominant among all interactions involving the spins, the quantization axis of both spins will be along $\boldsymbol{B}_0$ (Fig. II.1). The first spin $\boldsymbol{I}$ creates a dipole $\boldsymbol{\mu}_1$ which precesses around the field $\boldsymbol{B}_0$ and thus has a static component along the field and a rotating component in the plane perpendicular to the field. The static component of $\boldsymbol{\mu}_1$ produces at the site of the dipole $\boldsymbol{\mu}_2$, related to the second spin $\boldsymbol{J}$, a small static field that depends on the relative position of the spins. For large $\boldsymbol{B}_0$, only the component parallel or antiparallel to $\boldsymbol{B}_0$ significantly changes the net static field. From Eq. (II.1), when $\theta_I = \theta_J = \gamma$, it results that the dipolar interaction energy depends on the interspin distance r, on the eigenstates of the I and J spins, and on the angle γ between the interspin vector $\boldsymbol{r}$ and the external magnetic field (along which both I and J are quantized)

$$E \propto 3\cos^2\gamma(t) - 1. \tag{II.2}$$

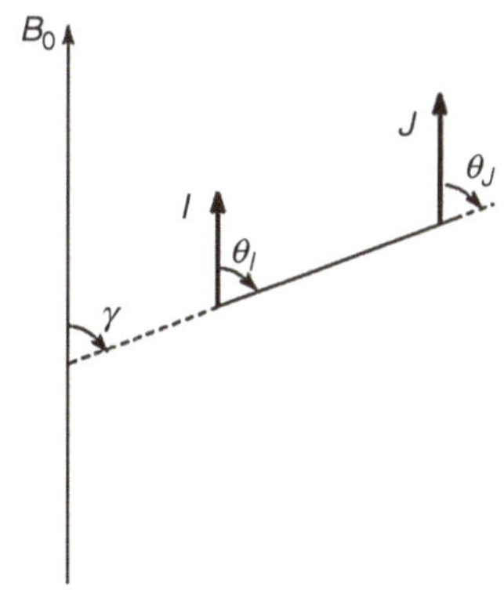

FIG. II.1 Reciprocal orientation of a nuclear spin *I* and an electron spin *J*, in the presence of an external magnetic field

NMR of Paramagnetic Molecules. http://dx.doi.org/10.1016/B978-0-444-63436-8.00014-4

In general, dipolar relaxation occurs whenever these quantities are modulated by time-dependent phenomena. In solids r and γ are fixed for a single I–J pair. In liquids, γ varies upon rotation of the molecule bearing the I–J pair, and r may vary because of chemical exchange.

It is interesting to note that on rapid (with respect to the anisotropy of the interaction energy in frequency units) rotation, the dipolar interaction energy averages to zero. In contrast, the square of this quantity (which determines the extent of the contribution to relaxation) does not average to zero. The average values obtained by integration of the functions results in

$$\frac{1}{2}\int_{-1}^{1}(3\cos^2\gamma - 1)\mathrm{d}(\cos\gamma) = \frac{1}{2}(x^3 - x)\Big|_{-1}^{1} = 0 \tag{II.3}$$

$$\begin{aligned}\frac{1}{2}\int_{-1}^{1}(3\cos^2\gamma - 1)^2\mathrm{d}(\cos\gamma) &= \frac{1}{2}\int_{-1}^{1}(9x^4 + 1 - 6x^2)\,\mathrm{d}x \\ &= \frac{1}{2}\left(\frac{9}{5}x^5 + x - \frac{6}{3}x^3\right)\Bigg|_{-1}^{1} = \frac{4}{5}.\end{aligned} \tag{II.4}$$

DERIVATION OF THE EQUATIONS FOR CONTACT SHIFT AND RELAXATION IN A SIMPLE CASE

III

The spin Hamiltonian for two contact coupled $S = ½$, $I = ½$ spins is given by

$$\begin{aligned} H &= g_e \mu_B B_0 S_z - \hbar\gamma_I B_0 I_z + A\boldsymbol{I}\cdot\boldsymbol{S} \\ &= \tfrac{1}{2}Z_e - \tfrac{1}{2}Z_N + A[S_z I_z + \tfrac{1}{2}(S_+ I_- + S_- I_+)] \end{aligned} \quad \text{(III.1)}$$

where Z_e and Z_N are the electron and nuclear Zeeman energies. From the definition of ladder operators

$$\begin{aligned} I_+ &= I_x + iI_y \quad & I_- &= I_x - iI_y \\ S_+ &= S_x + iS_y \quad & S_- &= S_x - iS_y \end{aligned} \quad \text{(III.2)}$$

we obtain

$$\begin{aligned} I_x &= \frac{1}{2}(I_+ + I_-) \quad & I_y &= \frac{2}{2i}(I_+ - I_-) \\ S_x &= \frac{1}{2}(S_+ + S_-) \quad & S_y &= \frac{2}{2i}(S_+ - S_-). \end{aligned} \quad \text{(III.3)}$$

The energy matrix is

$M_S,\ M_I$	$\lvert \frac{1}{2},\frac{1}{2}\rangle$	$\lvert \frac{1}{2},-\frac{1}{2}\rangle$	$\lvert -\frac{1}{2},\frac{1}{2}\rangle$	$\lvert -\frac{1}{2},-\frac{1}{2}\rangle$
$\langle \frac{1}{2},\frac{1}{2}\rvert$	$\frac{1}{2}Z_e - \frac{1}{2}Z_N + \frac{1}{4}A$	0	0	0
$\langle \frac{1}{2},-\frac{1}{2}\rvert$	0	$\frac{1}{2}Z_e + \frac{1}{2}Z_N - \frac{1}{4}A$	$\frac{1}{2}A$	0
$\langle -\frac{1}{2},\frac{1}{2}\rvert$	0	$\frac{1}{2}A$	$-\frac{1}{2}Z_e - \frac{1}{2}Z_N - \frac{1}{4}A$	0
$\langle -\frac{1}{2},-\frac{1}{2}\rvert$	0	0	0	$-\frac{1}{2}Z_e + \frac{1}{2}Z_N + \frac{1}{4}A$

Diagonalization of the above matrix yields the following energies and eigenfunctions:

$$\begin{aligned} E_1 &= \tfrac{1}{2}Z_e - \tfrac{1}{2}Z_N + \tfrac{1}{4}A & \Psi_1 &= \lvert \tfrac{1}{2},\tfrac{1}{2}\rangle \\ E_2 &= -\tfrac{1}{4}A + \tfrac{1}{2}R & \Psi_2 &= c_1\lvert \tfrac{1}{2},-\tfrac{1}{2}\rangle + c_2\lvert -\tfrac{1}{2},\tfrac{1}{2}\rangle \\ E_3 &= -\tfrac{1}{4}A - \tfrac{1}{2}R & \Psi_3 &= -c_2\lvert \tfrac{1}{2},-\tfrac{1}{2}\rangle + c_1\lvert -\tfrac{1}{2},\tfrac{1}{2}\rangle \\ E_4 &= -\tfrac{1}{2}Z_e + \tfrac{1}{2}Z_N + \tfrac{1}{4}A & \Psi_4 &= \lvert -\tfrac{1}{2},-\tfrac{1}{2}\rangle \end{aligned}$$

NMR of Paramagnetic Molecules. http://dx.doi.org/10.1016/B978-0-444-63436-8.00015-6

where

$$R = (A^2 + (Z_e + Z_N)^2)^{1/2}$$
$$c_1 = \left[\frac{1}{2}\left(1 + \frac{Z_e + Z_N}{R}\right)\right]^{1/2} \quad c_2 = \left[\frac{1}{2}\left(1 - \frac{Z_e + Z_N}{R}\right)\right]^{1/2}.$$

Nuclear transition energies are given by $E_2 - E_1$ and $E_4 - E_3$. By referring for simplicity to the high field limit, that is, when $Z_e \gg A$, we have

$$E_2 - E_1 = Z_N - \tfrac{1}{2}A \quad E_4 - E_3 = Z_N + \tfrac{1}{2}A.$$

The first transition is between states characterized by positive Z_e, the second between states with negative Z_e. Note that the nuclear transitions are separated by A and independent of S. However, the nucleus is only able to experience an average additional field from the magnetic moment of the electron, owing to its fast relaxation between the two electronic states. The average transition energy is thus given by

$$\Delta E_{av} = P_+(Z_N - \tfrac{1}{2}A) + P_-(Z_N + \tfrac{1}{2}A) \tag{III.4}$$

where P_+ and P_- are the Boltzmann population of the electronic levels, as defined in Eq. (1.25). Eq. (III.4) can thus be rewritten as

$$\Delta E_{\text{av}} = \frac{\exp[-(Z_e/2kT)]}{\exp[-(Z_e/2kT)] + \exp(Z_e/2kT)}(Z_N - \tfrac{1}{2}A) + \\ + \frac{\exp(Z_e/2kT)}{\exp[-(Z_e/2kT)] + \exp(Z_e/2kT)}(Z_N + \tfrac{1}{2}A) \tag{III.5}$$

and, in the limit $Z_e \ll 2kT$:

$$\Delta E_{\text{av}} = \tfrac{1}{2}(1 - Z_e/2kT)(Z_N - \tfrac{1}{2}A) + \tfrac{1}{2}(1 + Z_e/2kT)(Z_N + \tfrac{1}{2}A) \\ = Z_N + \tfrac{1}{2}A\frac{Z_e}{2kT} = \hbar\gamma_I B_0 + A\frac{g_e\mu_B B_0}{4kT}. \tag{III.6}$$

The contribution to the nuclear energy due to coupling with the unpaired electron, relative to the nuclear Zeeman energy equals the contact shift in ppm and is thus given by

$$\delta^{\text{con}} = A\frac{g_e\mu_B}{4\hbar\gamma_I kT} \tag{III.7}$$

which can be generalized for $S \neq ½$ to obtain Eq. (2.7):

$$\delta^{\text{con}} = \frac{A}{\hbar}\frac{g_e\mu_B S(S+1)}{3\gamma_I kT} \tag{III.8}$$

which is the simplest equation for the contact shift.

Contact interactions also give rise to relaxation. The perturbing Hamiltonian for contact interaction, H' [last term in Eq. (III.1)], is analogous to the first term of the perturbing Hamiltonian of the dipolar interaction [Appendix V, Eq. (V.10)] except that the part containing the ladder operators is multiplied by 1/2 instead of −1/4. The transition probability w_0 (Fig. 4.7) is provided by (Appendix V)

$$w_0 = \frac{1}{\hbar^2}\left|\int_{-t}^{t}\langle +-|H'(0)|-+\rangle\langle -+|H'(t')|+-\rangle e^{-i(\omega_I-\omega_S)t'}\,dt'\right|. \quad \text{(III.9)}$$

The only part of H' extracting $|+-\rangle$ from $|-+\rangle$ is that containing I_+S_-; at variance with the dipolar case, $w_1^I = w_1^S = 0$ and $w_2 = 0$. Therefore, in this case, since $T_1^{-1} = w_0 + 2w_1^I + w_2$ (Eq. (4.45) and Appendix VII), we have $T_1^{-1} = w_0$.

By integration of Eq. (III.9), we obtain

$$T_1^{-1} = \frac{A^2}{2\hbar^2}\frac{\tau_c}{1+(\omega_I-\omega_S)^2\tau_c^2}. \quad \text{(III.10)}$$

The corresponding equation for T_2^{-1} can be obtained analogously:

$$T_2^{-1} = \frac{A^2}{4\hbar^2}\left(\frac{\tau_c}{1+(\omega_I-\omega_S)^2\tau_c^2} - \tau_c\right). \quad \text{(III.11)}$$

DERIVATION OF THE PSEUDOCONTACT SHIFT EQUATIONS IN DIFFERENT FORMS

IV

We refer to the generic geometric arrangement of the electron and nuclear magnetic moments depicted in Fig. 2.6. The reference frame is defined by the main directions of the magnetic susceptibility anisotropy tensor; α is the angle of the external field direction $\boldsymbol{\kappa}$ with the z-axis; β is the angle from the x-axis of the projection of the external field direction on the xy plane; θ is the angle between the metal–nucleus vector $\boldsymbol{r}$ and the z-axis; φ is the angle from the x-axis of the projection of $\mathbf{r}$ on the xy plane. The dipolar shift results

$$\begin{aligned}\delta^{\mathrm{dip}} = \frac{1}{4\pi r^3}\Big[\chi_{\parallel}\cos^2\alpha\left(3\cos^2\theta-1\right)+\chi_{\perp}\sin^2\alpha\left(3\sin^2\theta\cos^2(\varphi-\beta)-1\right)+\\ +\tfrac{3}{4}(\chi_{\parallel}+\chi_{\perp})\sin 2\alpha\sin 2\theta\cos(\varphi-\beta)\Big].\end{aligned} \tag{2.30}$$

Eq. (2.30) gives the dipolar shift in the solid state. Note that when χ is isotropic it reduces to

$$\begin{aligned}\delta^{\mathrm{dip}} &= \frac{1}{4\pi r^3}\chi[3(\cos\alpha\cos\theta+\sin\alpha\sin\theta\cos(\varphi-\beta))^2-1]\\ &= \frac{1}{4\pi r^3}\chi(3\cos^2\gamma-1)\end{aligned} \tag{IV.1}$$

which gives the dipolar shift in magnetically isotropic solids and which averages zero in solution (Eq. (2.27) and Appendix II). The angle γ is the angle between $\boldsymbol{\kappa}$ and $\boldsymbol{r}$. Note that $\boldsymbol{\kappa}\cdot\boldsymbol{r} = r\cos\gamma = r(\cos\alpha\cos\theta+\sin\alpha\sin\theta\cos(\varphi-\beta))$.

To derive the pseudocontact shift in solution δ^{pcs} we must now take the rotational average of Eq. (2.30). By recalling that

$$\frac{1}{2\pi}\int_0^{2\pi}\cos\Omega\,\mathrm{d}\Omega = 0 \quad \frac{1}{2\pi}\int_0^{2\pi}\cos^2\Omega\,\mathrm{d}\Omega = \frac{1}{2\pi}\frac{\Omega}{2}\bigg|_0^{2\pi} = \frac{1}{2}$$

integration of Eq. (2.30) in $\mathrm{d}\beta$ gives

$$\delta = \frac{1}{4\pi r^3}[\chi_{\parallel}\cos^2\alpha(3\cos^2\theta-1)+\chi_{\perp}\sin^2\alpha(\tfrac{3}{2}\sin^2\theta-1)]. \tag{IV.2}$$

We then need integrating over the solid angle α ($\mathrm{d}\cos\alpha = \sin\alpha\,\mathrm{d}\alpha$), and recalling that

NMR of Paramagnetic Molecules. http://dx.doi.org/10.1016/B978-0-444-63436-8.00016-8

$$\frac{1}{2}\int_0^{\pi}\cos^2\alpha\sin\alpha\,d\alpha=\frac{1}{3}\qquad \frac{1}{2}\int_0^{\pi}(1-\cos^2\alpha)\sin\alpha\,d\alpha=\frac{2}{3}$$

we obtain

$$\begin{aligned}\delta^{pcs}&=\frac{1}{4\pi r^3}\left[\chi_{\parallel}\tfrac{1}{3}(3\cos^2\theta-1)+\chi_{\perp}\tfrac{1}{3}(3-3\cos^2\theta-2)\right]\\&=\frac{1}{12\pi r^3}\left(\chi_{\parallel}-\chi_{\perp}\right)\left(3\cos^2\theta-1\right)\end{aligned}\tag{2.31}$$

which is Eq. (2.31).

In the case of a rhombic paramagnetic susceptibility tensor, the pseudocontact shifts, written again in a reference frame defined by the main directions of the $\boldsymbol{\chi}$ tensor, are provided by

$$\delta^{pcs}=\frac{1}{12\pi r^3}[(3\sin^2\theta\cos^2\varphi-1)\tilde{\chi}_{xx}+(3\sin^2\theta\sin^2\varphi-1)\tilde{\chi}_{yy}+(3\cos^2\theta-1)\tilde{\chi}_{zz}]\tag{2.45}$$

where $\tilde{\chi}_{xx}$, $\tilde{\chi}_{yy}$, and $\tilde{\chi}_{zz}$ are the principal components of the paramagnetic susceptibility tensor.

The term multiplying $\tilde{\chi}_{xx}$ can be rewritten as

$$\begin{aligned}&3\sin^2\theta\cos^2\varphi-1=3\sin^2\theta\cos^2\varphi-\frac{3}{2}\sin^2\theta+\frac{3}{2}\sin^2\theta-1=\frac{3}{2}\sin^2\theta(2\cos^2\varphi-1)+\frac{3}{2}(1-\cos^2\theta)-1\\&=\frac{3}{2}\sin^2\theta\cos2\varphi-\frac{1}{2}(3\cos^2\theta-1)\end{aligned}$$

where the relationship $\cos2\varphi=2\cos^2\varphi-1$ has been used. Analogously, the term multiplying $\tilde{\chi}_{yy}$ can be rewritten as

$$\begin{aligned}&3\sin^2\theta\sin^2\varphi-1=3\sin^2\theta\sin^2\varphi-\frac{3}{2}\sin^2\theta+\frac{3}{2}\sin^2\theta-1=\frac{3}{2}\sin^2\theta(2\sin^2\varphi-1)+\frac{3}{2}(1-\cos^2\theta)-1\\&=\frac{3}{2}\sin^2\theta(-2\cos^2\varphi+1)-\frac{1}{2}(3\cos^2\theta-1)=-\frac{3}{2}\sin^2\theta\cos2\varphi-\frac{1}{2}(3\cos^2\theta-1)\end{aligned}$$

so that

$$\delta^{pcs}=\frac{1}{12\pi r^3}\left[\left(\frac{3}{2}\sin^2\theta\cos2\varphi-\frac{1}{2}(3\cos^2\theta-1)\right)\tilde{\chi}_{xx}+\left(-\frac{3}{2}\sin^2\theta\cos2\varphi-\frac{1}{2}(3\cos^2\theta-1)\right)\tilde{\chi}_{yy}+(3\cos^2\theta-1)\tilde{\chi}_{zz}\right].\tag{IV.3}$$

Therefore

$$\delta^{pcs}=\frac{1}{12\pi r^3}\left[(3\cos^2\theta-1)\left(\tilde{\chi}_{zz}-\frac{\tilde{\chi}_{xx}}{2}-\frac{\tilde{\chi}_{yy}}{2}\right)+\frac{3}{2}\sin^2\theta\cos2\varphi(\tilde{\chi}_{xx}-\tilde{\chi}_{yy})\right]\tag{IV.4}$$

and thus Eq. (2.46) is obtained.

From Eq. (2.35), providing the expression of the pseudocontact shift valid in any reference frame with the metal ion in the origin,

$$\delta^{pcs}=\frac{1}{4\pi r^5}[(3z^2-r^2)\chi_{zz}/3+(3x^2-r^2)\chi_{xx}/3+(3y^2-r^2)\chi_{yy}/3+2xy\chi_{xy}+2xz\chi_{xz}+2yz\chi_{yz}]\tag{2.35}$$

using the relationships

$$3x^2 - r^2 = 3x^2 - \frac{3(x^2 + y^2 + z^2)}{2} + \frac{r^2}{2} = \frac{3}{2}(x^2 - y^2) - \frac{1}{2}(3z^2 - r^2)$$

$$3y^2 - r^2 = 3y^2 - \frac{3(x^2 + y^2 + z^2)}{2} + \frac{r^2}{2} = -\frac{3}{2}(x^2 - y^2) - \frac{1}{2}(3z^2 - r^2)$$

we obtain

$$\delta^{\text{pcs}} = \frac{1}{4\pi r^5}\Big[(3z^2 - r^2)\chi_{zz}/3 + \left(\frac{3}{2}(x^2 - y^2) - \frac{1}{2}(3z^2 - r^2)\right)\chi_{xx}/3 + \left(-\frac{3}{2}(x^2 - y^2) - \frac{1}{2}(3z^2 - r^2)\right)\chi_{yy}/3 + 2xy\chi_{xy} + 2xz\chi_{xz} + 2yz\chi_{yz}\Big] \quad \text{(IV.5)}$$

which can be rewritten as

$$\delta^{\text{pcs}} = \frac{1}{4\pi r^5}[(3z^2 - r^2)(\chi_{zz} - \chi_{xx}/2 - \chi_{yy}/2)/3 + \frac{3}{2}(x^2 - y^2)(\chi_{xx} - \chi_{yy})/3 + 2xy\chi_{xy} + 2xz\chi_{xz} + 2yz\chi_{yz}] \quad \text{(IV.6)}$$

and thus, since the trace of the tensor is independent on the reference frame and

$$\chi_{zz} - \frac{\chi_{xx} + \chi_{yy}}{2} = \frac{3}{2}(\chi_{zz} - \bar{\chi}), \quad \text{(IV.7)}$$

we obtain Eq. (2.37)

$$\delta^{\text{pcs}} = \frac{1}{4\pi r^3}\left[(\chi_{zz} - \bar{\chi})\frac{2z^2 - x^2 - y^2}{2r^2} + (\chi_{xx} - \chi_{yy})\frac{x^2 - y^2}{2r^2} + \chi_{xy}\frac{2xy}{r^2} + \chi_{xz}\frac{2xz}{r^2} + \chi_{yz}\frac{2yz}{r^2}\right] \quad (2.37)$$

which is the equation describing the pseudocontact shift, valid for a generic reference frame with the metal ion in the origin, in Cartesian coordinates.

APPENDIX

RELAXATION BY DIPOLAR INTERACTION BETWEEN TWO SPINS

V

TRANSITION PROBABILITIES ALONG THE Z DIRECTION

A system constituted by two spins, I and S, is considered. The Hamiltonian is given by

$$H = -\hbar\gamma_I \boldsymbol{B_0} \cdot \boldsymbol{I} - \hbar\gamma_S \boldsymbol{B_0} \cdot \boldsymbol{S} + H' = H_0 + H' \tag{V.1}$$

where the first two terms are the Zeeman terms for each spin and are time-independent, and H' is the interaction Hamiltonian, which is time-dependent. For the dipolar interaction, H' is given by the first term of the equation written in note 2 of Section 4.4; the interaction depends, besides other variables, on the magnitude of r and orientation of $\boldsymbol{r}$, the vector connecting the atoms bearing the I and S spins. H' is a time-dependent Hamiltonian, whose modulation induces spin relaxation. In the present case we shall assume, without loss of generality, that the only time-dependent part in H' is the reciprocal orientation of the two spins with respect to the external magnetic field, that is, the direction of $\boldsymbol{r}$, while r is constant. In this way H' can only be modulated by rotation and not, for instance, by chemical exchange which would make r time-dependent. Even in this case the treatment is analogous and the final result identical. If $|m_i\rangle$ and $|m_j\rangle$ are two eigenstates of the unperturbed Hamiltonian, H_0, with energies E_i and E_j, the transition probability per unit time between these two states, w_{ij}, is, in the first order, given by the equation written in note 3 of Section 4.4, which can be rewritten as

$$w_{ij} = \frac{1}{\hbar^2}\left| \int_{-t}^{t} \langle m_j \mid H'(0) \mid m_i \rangle \langle m_i \mid H'(t') \mid m_j \rangle e^{-i\omega_{ij}t'} \, \mathrm{d}t' \right| \tag{V.2}$$

where $\omega_{ij} = (E_i - E_j)/\hbar$. This integral can be calculated if the perturbing Hamiltonian $H'(t')$ fluctuates randomly; when the fluctuations are rapid, w_{ij} is independent of time.

In the coupled system I–S there are four possible eigenstates according to the values of m_{Iz} (first term in the ket) and m_{Sz} (second term in the ket), as shown in Fig. 4.7A. The transition probabilities in this system are w_1^I, w_1^S, w_0 and w_2.

The perturbing Hamiltonian H' is given by

$$H' = k[\boldsymbol{I} \cdot \boldsymbol{S} - 3(\boldsymbol{I} \cdot \boldsymbol{r})(\boldsymbol{S} \cdot \boldsymbol{r})] \tag{V.3}$$

where $k = (\mu_0/4\pi)\hbar\gamma_I\gamma_S/r^3$ will be omitted in the following calculations ($\boldsymbol{r}$ is the unit vector oriented along the line of length r connecting I and S). By developing the vector products we obtain

$$I_xS_x + I_yS_y + I_zS_z - 3(I_xl + I_ym + I_zn)(S_xl + S_ym + S_zn) \tag{V.4}$$

where l, m and n are the direction cosines of $\boldsymbol{r}$. Eq. (V.4) can be rearranged into

NMR of Paramagnetic Molecules. http://dx.doi.org/10.1016/B978-0-444-63436-8.00017-X

$$I_xS_x(1-3l^2)+I_yS_y(1-3m^2)+I_zS_z(1-3n^2)-3(I_xS_y+I_yS_x)lm+ \\ -3(I_xS_z+I_zS_x)ln-3(I_yS_z+I_zS_y)mn. \tag{V.5}$$

We shall now further rearrange Eq. (V.5) so as to distinguish terms which leave the total z component, $M_z = m_{Iz} + m_{Sz}$, unchanged, terms for which $\Delta M_z = \pm 1$, and terms for which $\Delta M_z = \pm 2$. This distinction is important because these terms will contribute to w_0, w_1^I and w_1^S, and w_2, respectively. The direction cosines can be expressed by the polar angles, according to the relationships

$$l = \sin\theta\cos\varphi \\ m = \sin\theta\sin\varphi \\ n = \cos\theta. \tag{V.6}$$

Using Eqs. (III.2), (III.3) and (V.6), Eq. (V.5) can be rearranged to

$$I_zS_z(1-3\cos^2\theta)+\tfrac{1}{4}(I_+S_+ + I_+S_- + I_-S_+ + I_-S_-)(1-3\sin^2\theta\cos^2\varphi)+ \\ -\tfrac{1}{4}(I_+S_+ - I_+S_- - I_-S_+ + I_-S_-)(1-3\sin^2\theta\sin^2\varphi)+ \\ +\tfrac{3}{2}i(I_+S_+ - I_-S_-)\sin^2\theta\sin\varphi\cos\varphi+ \\ -\tfrac{3}{2}(I_+S_z + I_-S_z + I_zS_+ + I_zS_-)\sin\theta\cos\theta\cos\varphi+ \\ +\tfrac{3}{2}i(I_+S_z - I_-S_z + I_zS_+ - I_zS_-)\sin\theta\cos\theta\cos\varphi \tag{V.7}$$

By recalling that $\cos^2\alpha + \sin^2\alpha = 1$, $\cos^2\alpha - \sin^2\alpha = \cos 2\alpha$, $\sin\alpha\cos\alpha = ½\sin 2\alpha$, and $\cos\alpha \pm i\sin\alpha = e^{\pm i\alpha}$, Eq. (V.7) can be further rearranged to

$$I_zS_z(1-\cos^2\theta)-\tfrac{1}{4}(I_+S_- + I_-S_+)(1-3\sin^2\theta)-I_+S_+(\tfrac{3}{4}\sin^2\theta^{-2i\varphi})+ \\ -I_-S_-(\tfrac{3}{4}\sin^2\theta e^{2i\varphi})-(I_+S_z - I_zS_+)(\tfrac{3}{2}\sin\theta\cos\theta\, e^{-i\varphi})+ \\ -(I_-S_z + I_zS_-)(\tfrac{3}{2}\sin\theta\cos\theta\, e^{i\varphi}). \tag{V.8}$$

By defining now

$$F_0(t) = k(1-3\cos^2\theta(t)) \\ F_1(t) = -\tfrac{3}{2}k\sin\theta(t)\cos\theta(t)e^{-i\varphi(t)} \\ F_2(t) = -\tfrac{3}{4}k\sin^2\theta(t)e^{-2i\varphi(t)} \tag{V.9}$$

with $k = (\mu_0/4\pi)\hbar\gamma_I\gamma_S/r^3$, and recalling that the time dependence is indeed contained in θ and φ, the final dipolar Hamiltonian is

$$H' = [I_zS_z - \tfrac{1}{4}(I_+S_- + I_-S_+)]F_0 + [I_+S_z + I_zS_+]F_1 + \\ +[I_-S_z + I_zS_-]F_1^* + I_+S_+F_2 + I_-S_-F_2^*. \tag{V.10}$$

Note that the F_i functions are just the spherical harmonics of order i.

We can now express the transition probabilities. Let us start from

$$w_0 = \frac{1}{\hbar^2}\left|\int_{-t}^{t}\langle +-|H'(0)|-+\rangle\langle -+|H'(t')|+-\rangle e^{-i(\omega_I-\omega_S)t'}\,dt'\right|. \tag{V.11}$$

The only part of H' extracting $|+-\rangle$ from $|-+\rangle$ is that containing I_+S_-, and therefore

$$w_0 = \frac{1}{\hbar^2}\left|\int_{-t}^{t} \tfrac{1}{16}\langle F_0(0)F_0^*(t')\rangle e^{-i(\omega_I-\omega_S)t'}\,dt'\right|. \tag{V.12}$$

Analogously,

$$w_1^I = \frac{1}{\hbar^2}\left|\int_{-t}^{t} \tfrac{1}{4}\langle F_1(0)F_1^*(t')\rangle e^{-i\omega_I t'}\,dt'\right| \tag{V.13}$$

$$w_1^S = \frac{1}{\hbar^2}\left|\int_{-t}^{t} \tfrac{1}{4}\langle F_1(0)F_1^*(t')\rangle e^{-i\omega_S t'}\,dt'\right| \tag{V.14}$$

$$w_2 = \frac{1}{\hbar^2}\left|\int_{-t}^{t} \langle F_2(0)F_2^*(t')\rangle e^{-i(\omega_I+\omega_S)t'}\,dt'\right|. \tag{V.15}$$

We shall recall that the time dependence in the perturbing Hamiltonian is now confined to the F functions, which are only affected by rotation. If we assume that the rotation is random we can write the expectation value of the products $F_i(0)F_i(t)$ as an exponential function (see note 3 of Section 4.4)

$$\langle F_i(0)F_i^*(t)\rangle = \langle |F_i(0)|^2\rangle e^{-t/\tau_c} \tag{V.16}$$

so that we can extract $\langle|F_i|^2\rangle$ from the integrals in Eqs. (V.12)–(V.15).[1] In turn, the $\langle|F_i|^2\rangle$ can be easily computed; we show here the calculations for $\langle|F_0|^2\rangle$:

$$\begin{aligned}\langle |F_0|^2\rangle &= k^2\langle|1-3\cos^2\theta|^2\rangle \\ &= \tfrac{1}{2}k^2\int_{-1}^{1}(1-6\cos^2\theta+9\cos^4\theta)\,d(\cos\theta) = \tfrac{4}{5}k^2.\end{aligned} \tag{V.17}$$

The other averages are obtained in a similar manner:

$$\langle|F_1|^2\rangle = \langle|F_2|^2\rangle = \tfrac{3}{10}k^2. \tag{V.18}$$

Before performing the integration we should distinguish between the two cases of I and S being like or unlike spins. In the case of like spins $T_1^{-1} = 2(w_1 + w_2)$, and we have integrals of the type

$$\int_{-t}^{t}\langle F_\alpha(0)F_\alpha^*(t')\rangle e^{-i\alpha\omega t'}\,dt' \tag{V.19}$$

where $\omega = \omega_I = \omega_S$ and $\alpha = 1$ in the case of w_1^I (or w_1^S) and $\alpha = 2$ in the case of w_2. These integrals must be evaluated for $t \gg \tau_c$. Under these conditions Eqs. (V.14) and (V.15) become

---

[1]This holds when $\tau_c = \tau_r$; otherwise, if $\tau_c < \tau_r$, the average should be performed on the integral value and the F function cannot be extracted from the integral. However, the final results are unchanged if H_0 is isotropic.

$$
\begin{aligned}
w_1 &= \frac{\tau_c}{2\hbar^2}\left\langle |F_1|^2 \right\rangle \frac{1}{1+\omega^2\tau_c^2} \\
w_2 &= \frac{2\tau_c}{\hbar^2}\left\langle |F_2|^2 \right\rangle \frac{1}{1+4\omega^2\tau_c^2}
\end{aligned} \tag{V.20}
$$

and thus

$$
T_1^{-1} = \frac{3}{10}\left(\frac{\mu_0}{4\pi}\right)^2 \frac{\hbar^2\gamma^4}{r^6}\left(\frac{\tau_c}{1+\omega^2\tau_c^2} + \frac{4\tau_c}{1+4\omega^2\tau_c^2}\right). \tag{V.21}
$$

In the case of unlike spins, in general, relaxation does not follow an exponential behavior. However, a special case of primary interest in this book is when S is an electron spin. In this case $\langle S_z(t)\rangle = \langle S_z(0)\rangle$ during the entire relaxation process of I. T_1^{-1} is given by $w_0 + 2w_1^I + w_2$ [Eq. (4.45) and Appendix VII]. In analogy to Eq. (V.20) w_0 is given by

$$
w_0 = \frac{1}{16}\frac{2\tau_c}{\hbar^2}\left\langle |F_0|^2 \right\rangle \frac{1}{1+(\omega_S-\omega_I)^2\tau_c^2}. \tag{V.22}
$$

Thus, by analogy with the like-spins case,

$$
T_1^{-1} = \frac{1}{10}\left(\frac{\mu_0}{4\pi}\right)^2 \frac{\hbar^2\gamma_I^2\gamma_S^2}{r^6}\left(\frac{3\tau_c}{1+\omega_I^2\tau_c^2} + \frac{\tau_c}{1+(\omega_I-\omega_S)^2\tau_c^2} + \frac{6\tau_c}{1+(\omega_I+\omega_S)^2\tau_c^2}\right) \tag{V.23}
$$

which, given that $\omega_S \gg \omega_I$, can be approximated by

$$
T_1^{-1} = \frac{1}{10}\left(\frac{\mu_0}{4\pi}\right)^2 \frac{\hbar^2\gamma_I^2\gamma_S^2}{r^6}\left(\frac{7\tau_c}{1+\omega_S^2\tau_c^2} + \frac{3\tau_c}{1+\omega_I^2\tau_c^2}\right). \tag{V.24}
$$

TRANSITION PROBABILITIES IN THE *XY* PLANE

In calculating longitudinal relaxation, we started from four eigenstates $|--\rangle, |-+\rangle, |+-\rangle, |++\rangle$. For transverse relaxation we are interested in the eigenstates of I_x (or I_y) and S_x (or S_y). We can define the states $|\alpha\rangle$ and $|\beta\rangle$ so that

$$
\begin{aligned}
I_x|\alpha\rangle &= \tfrac{1}{2}|\alpha\rangle \quad S_x|\alpha\rangle = \tfrac{1}{2}|\alpha\rangle \\
I_x|\beta\rangle &= \tfrac{1}{2}|\beta\rangle \quad S_x|\beta\rangle = \tfrac{1}{2}|\beta\rangle.
\end{aligned}
$$

However, here $|\alpha\alpha\rangle$ $|\alpha\beta\rangle$ $|\beta\alpha\rangle$ $|\beta\beta\rangle$ are not eigenstates of the energy. Nevertheless, they are still orthogonal and can always be expanded in eigenvectors of the energy. For example, in the present case of spin 1/2 we have:

$$
\begin{aligned}
|\alpha\rangle &= \tfrac{1}{\sqrt{2}}\left(|+\rangle + |-\rangle\right) \\
|\beta\rangle &= \tfrac{1}{\sqrt{2}}\left(|+\rangle - |-\rangle\right).
\end{aligned}
$$

In general, two orthogonal states $|a\rangle$ and $|b\rangle$ are expanded

$$|a\rangle = \sum_i a_i \, |m_i\rangle \quad |b\rangle = \sum_i b_i |m_i\rangle$$

and, therefore, the transition probability between the states $|a\rangle$ and $|b\rangle$ is

$$u_{ab} = \frac{1}{\hbar^2}\left|\int_{-t}^{t}\sum_{ij}\langle m_j \,|\, H'(0) \,|\, m_i\rangle\langle m_i \,|\, H'(t') \,|\, m_j\rangle a_i b_j \mathrm{e}^{-i\omega_{ij}t'}\mathrm{d}t'\right|.$$

This equation is analogous to Eq. (V.2), and through a procedure analogous to that already reported, the equations for T_2^{-1} in the like-spin case (Eq. (4.13)) and unlike-spin case [Eq. (4.18)] are obtained.

CALCULATION OF $\langle S_Z \rangle$: CURIE'S LAW

VI

The expectation values of physical operators O for which spin states are eigenfunctions are given by

$$\langle O \rangle = \frac{\sum_i \langle i | O | i \rangle \mathrm{e}^{-E_i/(kT)}}{\sum_i \mathrm{e}^{-E_i/(kT)}}. \tag{VI.1}$$

In the case of the operator S_z, Eq. (VI.1) becomes

$$\langle S_z \rangle = \frac{\sum_{S,M_S} \langle S, M_S | S_z | S, M_S \rangle \mathrm{e}^{-E_{S,M_S}/(kT)}}{\sum_{S,M_S} \mathrm{e}^{-E_{S,M_S}/(kT)}}. \tag{VI.2}$$

where the eigenfunctions $|S, M_S\rangle$ are the spin functions of the S multiplet without any splitting at zero magnetic field and E_{S,M_S} are the energies. The latter are non degenerate in the presence of a static magnetic field and equal to $g_e\mu_B B_0 M_S$. Now, for current values of magnetic fields, $E_{S,M_S}/(kT) \ll 1$; therefore, $\mathrm{e}^{E_{S,M_S}/(kT)} \approx 1 - E_{S,M_S}/(kT)$, and Eq. (VI.2) can be written as

$$\langle S_z \rangle = \frac{\sum_{S,M_S} \langle S, M_S | S_z | S, M_S \rangle \quad (1 - E_{S,M_S}/(kT))}{\sum_{S,M_S} (1 - E_{S,M_S}/(kT))}. \tag{VI.3}$$

The matrix elements in the numerator yield the M_S values, that is, numbers ranging from $-S$ to S. The terms in the sum will thus be of the type

$$-S - \frac{g_e\mu_B B_0}{kT} S^2, -(S-1) - \frac{g_e\mu_B B_0}{kT}(S-1)^2, \ldots, S - \frac{g_e\mu_B B_0}{kT} S^2 \tag{VI.4}$$

and it can be seen that the sum of $-S$, $-(S-1)$,..., S is zero. The numerator becomes

$$-\frac{g_e\mu_B B_0}{kT} \sum_{M_S=-S}^{S} M_S^2. \tag{VI.5}$$

In the denominator the terms containing the energy cancel, and we are left with $\sum_{M_S} 1 = 2S+1$. Recalling that $\sum M_S^2 = (2S+1)S(S+1)/3$, Eq. (VI.3) becomes

$$\langle S_z \rangle = -\frac{g_e\mu_B B_0}{3kT} S(S+1) \tag{VI.6}$$

which gives the magnetic field and temperature dependence of $\langle S_z \rangle$, known as Curie's law.

NMR of Paramagnetic Molecules. http://dx.doi.org/10.1016/B978-0-444-63436-8.00018-1

DERIVATION OF THE EQUATIONS RELATED TO NOE VII

A necessary step to derive Eq. (4.44) and other equations in Chapter 4 is the obtainment of a suitable expression for the time dependence of $\langle I_z \rangle$ in a dipole–coupled two spin system. By referring to Fig. 4.18, $\langle I_z \rangle$ and $\langle J_z \rangle$ are given by

$$\begin{aligned} \langle I_z \rangle &= K(P_{++} + P_{+-} - P_{-+} - P_{--}) \\ \langle J_z \rangle &= K(P_{++} + P_{-+} - P_{+-} - P_{--}). \end{aligned} \tag{VII.1}$$

The rate of variation of $\langle I_z \rangle$ with time, $\mathrm{d}\langle I_z(t) \rangle/\mathrm{d}t$, is proportional to the rate of population increase of levels + + and + − minus the rate of population increase of levels − + and − −:

$$\frac{\mathrm{d}\langle I_z(t) \rangle}{\mathrm{d}t} = K\left(\frac{\mathrm{d}P_{++}}{\mathrm{d}t} + \frac{\mathrm{d}P_{+-}}{\mathrm{d}t} - \frac{\mathrm{d}P_{-+}}{\mathrm{d}t} - \frac{\mathrm{d}P_{--}}{\mathrm{d}t}\right). \tag{VII.2}$$

Among the various transitions shown in Fig. 4.18, we can neglect those involving only changes in spin state of J. For instance, the first of the four terms in the right-hand side of Eq. (VII.2) is:

$$\frac{\mathrm{d}P_{++}}{\mathrm{d}t} = -(w_1^I + w_2)P_{++} + w_1^I P_{-+} + w_2 P_{--} + c. \tag{VII.3}$$

The constant c can be evaluated by setting $\mathrm{d}P_{++}/\mathrm{d}t = 0$:

$$c = (w_1^I + w_2)P_{++}(\infty) - w_1^I P_{-+}(\infty) - w_2 P_{--}(\infty) \tag{VII.4}$$

and Eq. (VII.3) becomes

$$\frac{\mathrm{d}P_{++}}{\mathrm{d}t} = -(w_1^I + w_2)\Delta P_{++} + w_1^I \Delta P_{-+} + w_2 \Delta P_{--} \tag{VII.5}$$

where

$$P - P(\infty) = \Delta P. \tag{VII.6}$$

Analogous equations can be written for the other three terms in Eq. (VII.2). By grouping all the terms together, Eq. (VII.2) becomes

$$\begin{aligned} \frac{\mathrm{d}\langle I_z(t) \rangle}{\mathrm{d}t} &= K[-2\Delta P_{++}(w_1^I + w_2) - 2\Delta P_{+-}(w_1^I + w_0) \\ &\quad + 2\Delta P_{-+}(w_1^I + w_0) + 2\Delta P_{--}(w_1^I + w_2)] \end{aligned} \tag{VII.7}$$

NMR of Paramagnetic Molecules. http://dx.doi.org/10.1016/B978-0-444-63436-8.00019-3

which can be rearranged to

$$\begin{aligned}\frac{\mathrm{d}\langle I_z(t)\rangle}{\mathrm{d}t} &= 2K[(\Delta P_{--} - \Delta P_{++})(w_1^I + w_2) \\ &+ (\Delta P_{-+} - \Delta P_{+-})(w_1^I + w_0)].\end{aligned} \tag{VII.8}$$

The factor 2 derives from the fact that a single transition decreases the population of the starting state and at the same time increases the population of the target state, thereby changing the population difference by twice as much.

However, from Eq. (VII.1) we can write

$$\begin{aligned}\langle I_z\rangle + \langle J_z\rangle &= -2K(P_{--} - P_{++}) \\ \langle I_z\rangle - \langle J_z\rangle &= -2K(P_{-+} - P_{+-})\end{aligned} \tag{VII.9}$$

and, using Eq. (VII.6)

$$\begin{aligned}\langle I_z(t)\rangle - \langle I_z(\infty)\rangle + \langle J_z(t)\rangle - \langle J_z(\infty)\rangle &= -2K(\Delta P_{--} - \Delta P_{++}) \\ \langle I_z(t)\rangle - \langle I_z(\infty)\rangle - \langle J_z(t)\rangle + \langle J_z(\infty)\rangle &= -2K(\Delta P_{-+} - \Delta P_{+-}).\end{aligned} \tag{VII. 10}$$

Substituting Eq. (VII.10) into Eq. (VII.8) gives

$$\begin{aligned}\frac{\mathrm{d}\langle I_z(t)\rangle}{\mathrm{d}t} &= -\left[\langle I_z(t)\rangle - \langle I_z(\infty)\rangle + \langle J_z(t)\rangle - \langle J_z(\infty)\rangle\right](w_1^I + w_2) + \\ &-\left[\langle I_z(t)\rangle - \langle I_z(\infty)\rangle - \langle J_z(t)\rangle + \langle J_z(\infty)\rangle\right](w_1^I + w_0)\end{aligned} \tag{VII.11}$$

that finally becomes

$$\begin{aligned}\frac{\mathrm{d}\langle I_z(t)\rangle}{\mathrm{d}t} &= -\left[\langle I_z(t)\rangle - \langle I_z(\infty)\rangle\right](w_0 + 2w_1^I + w_2) + \\ &-\left[\langle J_z(t)\rangle - \langle J_z(\infty)\rangle\right](w_2 - w_0).\end{aligned} \tag{VII.12}$$

Eq. (VII.12) is the starting point to derive not only the equations relevant for the NOE phenomenon (Section 4.14) but also Eq. (4.16) and the following ones (Section 4.4). A somewhat different form of Eq. (VII.12) has been encountered when dealing with transfer of magnetization between two sites in chemical exchange (Section 6.3.4).

According to the definitions given in Eqs. (4.45) and (4.46), Eq. (VII.12) can be rewritten as

$$\frac{\mathrm{d}\langle I_z(t)\rangle}{\mathrm{d}t} = -\left[\langle I_z(t)\rangle - \langle I_z(\infty)\rangle\right]\rho_{I(J)} - \left[\langle J_z(t)\rangle - \langle J_z(\infty)\rangle\right]\sigma_{I(J)}. \tag{VII.13}$$

Eq. (4.44), relevant for steady state NOE, can easily be obtained from Eq. (VII.13) by setting $\mathrm{d}\langle I_z(t)\rangle/\mathrm{d}t = 0$ and $\langle J_z(t)\rangle = 0$:

$$0 = -\left[\langle I_z(t)\rangle - \langle I_z(\infty)\rangle\right]\rho_{I(J)} + \langle J_z(\infty)\rangle\sigma_{I(J)} \tag{VII.14}$$

which rearranges to

$$\langle I_z(t)\rangle = \langle I_z(\infty)\rangle + \left[\sigma_{I(J)}/\rho_{I(J)}\right]\langle J_z(\infty)\rangle. \tag{4.44}$$

For the homonuclear case, generalized by substituting $\rho_{I(J)}$ with the total relaxation rate ρ_I in Eq. (VII.13) and (VII.14), we obtain

$$\frac{\langle I_z(t)\rangle - \langle I_z(\infty)\rangle}{\langle I_z(\infty)\rangle} = \frac{\sigma_{I(J)}}{\rho_I}. \tag{4.52}$$

Up to now steady state NOEs have been considered, that is, when one signal is saturated for a long time with respect to T_1 of the nucleus on which NOE is going to be measured. Let's consider here what happens when the saturation time is short and variable. The resulting NOE is called truncated NOE because not enough time is left for full polarization transfer. The equation for truncated NOE can also be derived from Eq. (VII.13), again generalized by substituting $\rho_{I(J)}$ with the total relaxation rate ρ_I, by setting $\langle J_z(t)\rangle = 0$ (instantaneous saturation). Integration then gives

$$\langle I_z(t)\rangle = A\exp(-\rho_I t) + B. \tag{VII.15}$$

To evaluate the constants A and B we start by deriving again Eq. (VII.15):

$$\frac{d\langle I_z(t)\rangle}{dt} = -A\rho_I \exp(-\rho_I t). \tag{VII.16}$$

From Eq. (VII.15) we can then write

$$A\exp(-\rho_I t) = \langle I_z(t)\rangle - B \tag{VII.17}$$

and, by substituting Eq. (VII.17) into Eq. (VII.16), we obtain

$$\frac{\mathrm{d}\langle I_z(t)\rangle}{\mathrm{d}t} = -\rho_I\langle I_z(t)\rangle + \rho_I B. \tag{VII.18}$$

However, from Eq. (VII.13) under instantaneous saturation of J we can write

$$\frac{\mathrm{d}\langle I_z(t)\rangle}{\mathrm{d}t} = -\left[\langle I_z(t)\rangle - \langle I_z(\infty)\rangle\right]\rho_I + \langle J_z(\infty)\rangle\sigma_{I(J)} \tag{VII. 19}$$

and, by equating the right-hand sides of Eqs. (VII.18) and (VII.19)

$$B = \langle I_z(\infty)\rangle + [\sigma_{I(J)}/\rho_I]\langle J_z(\infty)\rangle. \tag{VII.20}$$

By substituting Eq. (VII.20) back into Eq. (VII.15) we obtain

$$\langle I_z(t)\rangle = A\exp(-\rho_I t) + \langle I_z(\infty)\rangle + [\sigma_{I(J)}/\rho_I]\langle J_z(\infty)\rangle. \tag{VII.21}$$

At $t = 0$, $\langle I_z(0)\rangle = \langle I_z(\infty)\rangle$, and therefore

$$A = -[\sigma_{I(J)}/\rho_I]\langle J_z(\infty)\rangle \tag{VII.22}$$

from which Eq. (VII.21) becomes

$$\langle I_z(t)\rangle = \langle I_z(\infty)\rangle + (\sigma_{I(J)}/\rho_I)\langle J_z(\infty)\rangle[1-\exp(-\rho_I t)] \tag{VII.23}$$

by setting $\langle J_z(\infty)\rangle = \langle I_z(\infty)\rangle$ and rearranging we obtain the equation for truncated NOE:

$$\frac{\langle I_z(t)\rangle - \langle I_z(\infty)\rangle}{\langle I_z(\infty)\rangle} = \eta_{I(J)}(t) = [\sigma_{I(J)}/\rho_I][1-\exp(-\rho_I t)]. \tag{VII.24}$$

It can be easily verified that for t long with respect to ρ_I, the exponential term tends to zero and we are back to the case of steady state NOE [Eq. (4.52)]. For t tending to zero, the NOE tends to zero. For times short with respect to ρ_I, $1 - \exp(-x) \approx 1 - (1 - x) = x$, and

$$\eta_{I(J)}(t) = \sigma_{I(J)}t. \tag{VII.25}$$

For irradiation times of J short with respect to the relaxation time of I the NOE extent is independent of the relaxation time of the nucleus and provides a direct measurement of $\sigma_{I(J)}$. The value of ρ_I can then be obtained from the steady state NOE measurement.

One last observation is related to the obtainment of the intensity of transient NOE, which is related to the acquisition of 2D NOESY experiments (Chapter 12). By transient NOE one refers to the effect occurring when only one spin in a *I-J* system is driven outside equilibrium. In such a case the time evolution of the *z*-magnetization of the perturbed spin, *J*, now also plays a role. The magnitude of the NOE as a function of time is thus the result of two competing effects. One is the return to equilibrium of nucleus *J* which causes a decrease in the absolute value of polarization. In fact, when *J* is at equilibrium, no polarization transfer occurs any more. The larger the difference in spin population with respect to the equilibrium value, the larger is the buildup of NOE. At $t = 0$ the cross relaxation contribution is twice that of a steady state experiment, since in the transient case the polarization is inverted, whereas in the steady state case it is zero. The second competing effect is the buildup of NOE, which is of the type of Eq. (VII.24). The resulting transient NOE can be obtained as a particular solution of Eq. (4.55) for $<I_z(\text{t})>$ as

$$\eta_{I(J)}(t) = -\frac{\sigma_{I(J)}}{D}\exp[-(\rho - D)t][1-\exp(-2Dt)]. \tag{VII.26}$$

where ρ and D are defined as in Eq. (4.56).

APPENDIX

MAGNETICALLY COUPLED DIMERS IN THE HIGH-TEMPERATURE LIMIT VIII

CONTACT SHIFT

We perform here a sample calculation to show that Eq. (11.9)

$$\delta^{\text{con}} = A_1 \frac{g_e \mu_B}{\hbar \gamma_I 3kT} \frac{\sum_i C_{i1} S_i'(S_i'+1)(2S_i'+1)\exp(-E_i/(kT))}{\sum_i (2S_i'+1)\exp(-E_i/(kT))} \quad (11.9)$$

reduces to Eq. (VIII.1)

$$\delta^{\text{con}} = A_1 \frac{1}{\hbar \gamma_I B_0} \langle \hat{S}_{1z} \rangle = -A_1 \frac{g_e \mu_B}{\hbar \gamma_I 3kT} S_1'(S_1'+1) \quad (\text{VIII.1})$$

in the high-temperature limit. If $E_i/(kT) \ll 1$ the exponential terms in the numerator and denominator are all ca. 1 and Eq. (11.9) becomes

$$\delta^{\text{con}} = A_1 \frac{g_e \mu_B}{\hbar \gamma_I 3kT} \frac{\sum_i C_{i1} S_i'(S_i'+1)(2S_i'+1)}{\sum_i (2S_i'+1)}. \quad (\text{VIII.2})$$

Let us take $S_1 = 2$ and $S_2 = 5/2$, as in the case of reduced Fe_2S_2 (Section 11.3.4). From Table 11.1 it is

$S' = 1/2$	$S' = 3/2$	$S' = 5/2$	$S' = 7/2$	$S' = 9/2$
$C_{11} = -4/3$	$C_{12} = 2/15$	$C_{13} = 12/35$	$C_{14} = 26/63$	$C_{15} = 4/9$
$C_{21} = 7/3$	$C_{22} = 13/15$	$C_{23} = 23/35$	$C_{24} = 37/63$	$C_{25} = 5/9$

and the contribution from metal 1 ($S_1 = 2$) is

$$\begin{aligned}
&\frac{-4/3 \cdot 3/4 \cdot 2}{2+4+6+8+10} + \frac{2/15 \cdot 15/4 \cdot 4}{2+4+6+8+10} + \frac{12/35 \cdot 35/4 \cdot 6}{2+4+6+8+10} \\
&\quad + \frac{26/63 \cdot 63/4 \cdot 8}{2+4+6+8+10} + \frac{4/9 \cdot 99/4 \cdot 10}{2+4+6+8+10} \\
&= \frac{1}{30}(-2+2+18+52+110) \\
&= \frac{180}{30} = 6 = S_1(S_1+1).
\end{aligned}$$

As expected, this contribution is the same as that of the isolated ion.

NMR of Paramagnetic Molecules. http://dx.doi.org/10.1016/B978-0-444-63436-8.00020-X

NUCLEAR RELAXATION

In the high temperature limit and assuming a single correlation time, Eqs. (11.13a) and (11.13b) can be simplified to

$$R_{1M_1} = \frac{2}{15}\left(\frac{\mu_0}{4\pi}\right)^2 \frac{\gamma_I^2 g_e^2 \mu_B^2}{r^6} \frac{\sum_i \left[C_{i1}^2 S_i'(S_i'+1)(2S_i'+1)\left(\frac{7\tau_{c2}}{1+\omega_S^2\tau_{c2}^2} + \frac{3\tau_{c1}}{1+\omega_I^2\tau_{c1}^2} \right)\right]}{\sum_i [(2S_i'+1)]} \tag{VIII.3}$$

$$R_{1M_1} = \frac{2}{3}\frac{A_1^2}{\hbar^2} \frac{\sum_i \left[C_{i1}^2 S_i'(S_i'+1)(2S_i'+1)\left(\frac{\tau_{c2}}{1+\omega_S^2\tau_{c2}^2} \right)\right]}{\sum_i [(2S_i'+1)]}. \tag{VIII.4}$$

These equations are related to the equations for dipolar and contact relaxation in the case of isolated metal ions [Eq. (4.20) and Eq. (4.27), respectively] by a coefficient. By comparison of Eq. (4.20) with Eq. (VIII.3) and Eq. (4.27) with Eq. (VIII.4), this coefficient is found to be [1]

$$X_1 = \frac{1}{S_1(S_1+1)} \frac{\sum_i \left[C_{i1}^2 S_i'(S_i'+1)(2S_i'+1) \right]}{\sum_i [(2S_i'+1)]}.$$

Therefore, in the high-temperature limit, the equations for nuclear relaxation are related to those for the isolated systems only by a multiplicative coefficient [1].

REFERENCES

[1] Owens C, Drago RS, Luchinat C, Bertini I, Banci L. J. Am. Chem. Soc. 1986;108:3298.

APPENDIX

REFERENCE TABLES

IX

PHYSICAL AND MATHEMATICAL CONSTANTS

Quantity	Symbol	Value	
		SI	CGS emu
Physical constants			
Permeability of the vacuum	μ_0	$4\pi \times 10^{-7}$ kg m s^{-2} A^{-2}	1
Speed of light in the vacuum	c	2.9979×10^{8} m s^{-1}	2.9979×10^{10} cm s^{-1}
Elementary charge (absolute value of the electron charge)	e	1.6022×10^{-19} A s	1.6022×10^{-20} abcoulombs
Planck constant	h	6.6261×10^{-34} J s	6.6261×10^{-27} erg s
	$\hbar = h/2\pi$	1.0546×10^{-34} J s rad^{-1}	1.0546×10^{-27} erg s rad^{-1}
Avogadro constant	N_A	6.0221×10^{23} mol^{-1}	6.0221×10^{23} mol^{-1}
Electron rest mass	m_e	0.9109×10^{-30} kg	0.9109×10^{-27} g
Proton rest mass	m_p	1.6726×10^{-27} kg	1.6726×10^{-24} g
Free-spin electron g factor[a]	g_e	2.0023	2.0023
Electron Bohr magneton	μ_B	9.2740×10^{-24} J T^{-1}	9.2740×10^{-21} erg G^{-1}
Nuclear Bohr magneton	μ_N	5.0508×10^{-27} J T^{-1}	5.0508×10^{-24} erg G^{-1}
Free-electron magnetic moment	$\mu_e = -\frac{g_e\mu_B}{2}$	-9.2848×10^{-24} J T^{-1}	-9.2848×10^{-21} erg G^{-1}
Proton magnetic moment	$\mu_p = \frac{g_p\mu_N}{2}$	1.4106×10^{-26} J T^{-1}	1.4106×10^{-23} erg G^{-1}
Electron-to-proton magnetic moments ratio	$\lvert\mu_e/\mu_p\rvert$	658.21	658.21
Free-electron magnetogyric ratio[a]	γ_e	-1.7609×10^{11} rad s^{-1} T^{-1}	-1.7609×10^{7} rad s^{-1} G^{-1}
Proton magnetogyric ratio	γ_p	2.6752×10^{8} rad s^{-1} T^{-1}	2.6752×10^{4} rad s^{-1} G^{-1}
Boltzmann constant	k	1.3807×10^{-23} J K^{-1}	1.3807×10^{-16} erg K^{-1}
Hyperfine coupling constant of the hydrogen atom	$(A/(h))_{\mathrm{H}}$ $(A/(\hbar))_{\mathrm{H}}$	1.4204×10^{9} Hz 8.9247×10^{9} rad s^{-1}	1.4204×10^{9} Hz 8.9247×10^{9} rad s^{-1}

(*Continued*)

NMR of Paramagnetic Molecules. http://dx.doi.org/10.1016/B978-0-444-63436-8.00021-1

Quantity	Symbol	Value
Mathematical constants		
	e	2.7183
	π	3.1416

[a]*The electron spin possesses a negative γ and therefore a negative g_e. However, it is common practice to use a positive g_e and to revert the sign of all equations involving g_e [Eq. (1.10)].*

CONVERSION FACTORS

	To Convert From	To	Multiply by
Length	Ångström (Å)	meter (m)	1×10^{-10}
Pressure	atmosphere (atm)	Pascal (Pa) (kg m^{-1} s^{-2})	1.01325×10^{5}
Mass	atomic mass unit (amu)	kilogram (kg)	1.6606×10^{-27}
Energy	calories (cal)	Joule (J)	4.1840
	electron volt (eV)	Joule (J)	1.6022×10^{-19}
	kilowatt-hour (kWh)	Joule (J)	3.6×10^{6}
Viscosity	centipoise	kg m^{-1} s^{-1}	1×10^{-3}
Angles	degree (deg)	radian (rad)	0.017453
Volume	liter (1)	cubic meter (m^3)	1×10^{-3}
Frequency	radian/second (rad s^{-1})	cycle/second (cps) (Hz)	0.15915
	wavenumber (cm^{-1})	energy (J)	1.9865×10^{-23}
		frequency (s^{-1}) (Hz)	2.9979×10^{10}
		frequency (rad s^{-1})	1.8837×10^{11}
	cycle/second (cps) (Hz)	energy (J)	6.6262×10^{-34}
Temperature	Kelvin (K)	energy (J)	1.3807×10^{-23}
		wavenumber (cm^{-1})	0.69467

SOME PHYSICAL QUANTITIES AND THEIR SI UNITS

Electric charge (quantity of electricity)	Q	C	A s
Electric current	i	A	A
Electric potential	V	V	kg m^2 s^{-3} A^{-1} (= J A^{-1} s^{-1})
Energy	E	J	kg m^2 s^{-2}

Frequency	ν	Hz	s^{-1}
Frequency (angular velocity)	ω	rad s^{-1}	rad s^{-1}
Magnetic field strength	H	A m^{-1}	A m^{-1}
Magnetic induction (flux density)	B	T	kg A^{-1} s^{-2}
Magnetic moment	μ	J T^{-1}	A m^2
Magnetic susceptibility	χ	m^3	m^3
Magnetization	M	J T^{-1} m^{-3}	A m^{-1}
Magnetogyric ratio	Γ	rad s^{-1} T^{-1}	rad A s kg^{-1}
Molar magnetic susceptibility	χ_M	m^3 mol^{-1}	m^3 mol^{-1}
Power	P	W	kg m^2 s^{-3} (=J s^{-1})
Viscosity (dynamic)	η	kg m^{-1} s^{-1}	kg m^{-1} s^{-1}

Index

F

G

H

I

J

K

L

U

V

W

X

Y

Z

FSC
www.fsc.org
MIX
Papier | Fördert
gute Waldnutzung
FSC® C083411

Kolibri 360 GmbH
Ferdinand-Jühlke-Straße 7
99095 Erfurt, Deutschland
produktsicherheit@kolibri360.de